T0213427

Lecture Notes in Computer Science   10132

Commenced Publication in 1973
Founding and Former Series Editors:
Gerhard Goos, Juris Hartmanis, and Jan van Leeuwen

More information about this series at http://www.springer.com/series/7409

Editors
Laurent Amsaleg
CNRS–IRISA
Rennes
France

Gylfi Þór Guðmundsson
Reykjavík University
Reykjavik
Iceland

Cathal Gurrin
Dublin City University
Dublin
Ireland

Björn Þór Jónsson
Reykjavik University
Reykjavik
Ireland

Shin'ichi Satoh
National Institute of Informatics
Tokyo
Japan

ISSN 0302-9743   ISSN 1611-3349   (electronic)
Lecture Notes in Computer Science
ISBN 978-3-319-51810-7   ISBN 978-3-319-51811-4   (eBook)
DOI 10.1007/978-3-319-51811-4

Library of Congress Control Number: 2016962021

LNCS Sublibrary: SL3 – Information Systems and Applications, incl. Internet/Web, and HCI

Printed on acid-free paper

This Springer imprint is published by Springer Nature
The registered company is Springer International Publishing AG
The registered company address is: Gewerbestrasse 11, 6330 Cham, Switzerland

Laurent Amsaleg · Gylfi Þór Guðmundsson
Cathal Gurrin · Björn Þór Jónsson
Shin'ichi Satoh (Eds.)

MultiMedia Modeling

23rd International Conference, MMM 2017
Reykjavik, Iceland, January 4–6, 2017
Proceedings, Part I

 Springer

Preface

These proceedings contain the papers presented at MMM 2017, the 23rd International Conference on MultiMedia Modeling, held at Reykjavik University during January 4–6, 2017. MMM is a leading international conference for researchers and industry practitioners for sharing new ideas, original research results, and practical development experiences from all MMM related areas, broadly falling into three categories: multimedia content analysis; multimedia signal processing and communications; and multimedia applications and services.

MMM conferences always include special sessions that focus on addressing new challenges for the multimedia community. The following four special sessions were held at MMM 2017:

- SS1: Social Media Retrieval and Recommendation
- SS2: Modeling Multimedia Behaviors
- SS3: Multimedia Computing for Intelligent Life
- SS4: Multimedia and Multimodal Interaction for Health and Basic Care Applications

MMM 2017 received a total 198 submissions across four categories; 149 full-paper submissions, 34 special session paper submissions, eight demonstration submissions, and seven submissions to the Video Browser Showdown (VBS 2017). Of all submissions, 68% were from Asia, 27% from Europe, 3% from North America, and 1% each from Oceania and Africa.

Of the 149 full papers submitted, 35 were selected for oral presentation and 33 for poster presentation, which equates to a 46% acceptance rate overall. Of the 34 special session papers submitted, 24 were selected for oral presentation and two for poster presentation, which equates to a 76% acceptance rate overall. In addition, five demonstrations were accepted from eight submissions, and all seven submissions to VBS 2017. The overall acceptance percentage across the conference was thus 54%, but 46% for full papers and 23% of full papers for oral presentation.

The submission and review process was coordinated using the ConfTool conference management software. All full-paper submissions were reviewed by at least three members of the Program Committee. All special session papers were reviewed by at least three reviewers from the Program Committee and special committees established for each special session. All demonstration papers were reviewed by at least three reviewers, and VBS papers by two reviewers. We owe a debt of gratitude to all these reviewers for providing their valuable time to MMM 2017.

We would like to thank our invited keynote speakers, Marcel Worring from the University of Amsterdam, The Netherlands, and Noriko Kando from the National Institute of Informatics, Japan, for their stimulating contributions.

We also wish to thank our organizational team: Demonstration Chairs Esra Acar and Frank Hopfgartner; Video Browser Showdown Chairs Klaus Schoeffmann, Werner Bailer, Cathal Gurrin and Jakub Lokoč; Sponsorship Chairs Yantao Zhang and Tao Mei;

Proceedings Chair Gylfi Þór Guðmundsson; and Local Organization Chair Marta Kristín Lárusdóttir.

We would like to thank Reykjavik University for hosting MMM 2017. Finally, special thanks go to our supporting team at Reykjavik University (Arnar Egilsson, Ýr Gunnlaugsdóttir, Þórunn Hilda Jónasdóttir, and Sigrún Heba Ómarsdóttir) and CP Reykjavík (Kristjana Magnúsdóttir, Elísabet Magnúsdóttir and Ingibjörg Hjálmfríðardóttir), as well as to student volunteers, for all their contributions and valuable support.

The accepted research contributions represent the state of the art in multimedia modeling research and cover a very diverse range of topics. A selection of the best papers will be invited to submit extended versions to a special issue of *Multimedia Tools and Applications*. We wish to thank all authors who spent their valuable time and effort to submit their work to MMM 2017. And, finally, we thank all those who made the (sometimes long) trip to Reykjavík to attend MMM 2017 and VBS 2017.

January 2017

Björn Þór Jónsson
Cathal Gurrin
Laurent Amsaleg
Shin'ichi Satoh
Gylfi Þór Guðmundsson

Organization

Organizing Committee

General Chairs

Björn Þór Jónsson	Reykjavik University, Iceland
Cathal Gurrin	Dublin City University, Ireland

Program Chairs

Laurent Amsaleg	CNRS–IRISA, France
Shin'ichi Satoh	NII, Japan

Demonstration Chairs

Frank Hopfgartner	University of Glasgow, UK
Esra Acar	Technische Universität Berlin, Germany

VBS 2017 Chairs

Klaus Schöffmann	Klagenfurt University, Austria
Werner Bailer	Joanneum Research, Austria
Cathal Gurrin	Dublin City University, Ireland
Jakub Lokoč	Charles University in Prague, Czech Republic

Sponsorship Chairs

Yantao Zhang	Snapchat
Tao Mei	Microsoft Research Asia

Proceedings Chair

Gylfi Þ. Guðmundsson	Reykjavik University, Iceland

Local Chair

Marta K. Lárusdóttir	Reykjavik University, Iceland

Local Support

Reykjavik University Event Services, CP Reykjavik

Steering Committee

Phoebe Chen (Chair)	La Trobe University, Australia
Tat-Seng Chua	National University of Singapore, Singapore

Kiyoharu Aizawa	University of Tokyo, Japan
Cathal Gurrin	Dublin City University, Ireland
Benoit Huet	EURECOM, France
R. Manmatha	University of Massachusetts, USA
Noel E. O'Connor	Dublin City University, Ireland
Klaus Schöffmann	Klagenfurt University, Austria
Yang Shiqiang	Tsinghua University, China
Cees G.M. Snoek	University of Amsterdam, The Netherlands
Meng Wang	Hefei University of Technology, China

Special Session Organizers

SS1: Social Media Retrieval and Recommendation

Liqiang Nie	National University of Singapore, Singapore
Yan Yan	University of Trento, Italy
Benoit Huet	EURECOM, France

SS2: Modeling Multimedia Behaviors

Peng Wang	Tsinghua University, China
Frank Hopfgartner	University of Glasgow, UK
Liang Bai	National University of Defense Technology, China

SS3: Multimedia Computing for Intelligent Life

Zhineng Chen	Chinese Academy of Sciences, China
Wei Zhang	Chinese Academy of Sciences, China
Ting Yao	Microsoft Research Asia, China
Kai-Lung Hua	National Taiwan University of Science and Technology, Taiwan, R.O.C.
Wen-Huang Cheng	Academia Sinica, Taiwan, R.O.C.

SS4: Multimedia and Multimodal Interaction for Health and Basic Care Applications

Stefanos Vrochidis	ITI-CERTH, Greece
Leo Wanner	Pompeu Fabra University, Spain
Elisabeth André	University of Augsburg, Germany
Klaus Schöffmann	Klagenfurt University, Austria

Program Committee

Esra Acar	Technische Universität Berlin, Germany
Amin Ahmadi	DCU/Insight Centre for Data Analytics, Ireland
Le An	UNC Chapel Hill, USA
Ognjen Arandjelović	University of St. Andrews, UK
Anant Baijal	SAMSUNG Electronics, South Korea

Werner Bailer	Joanneum Research, Austria
Ilaria Bartolini	University of Bologna, Italy
Jenny Benois-Pineau	University of Bordeaux/LABRI, France
Milan Bjelica	Unoiversity of Belgrade, Serbia
Laszlo Böszörmenyi	Klagenfurt University, Austria
Benjamin Bustos	University of Chile, Chile
K. Selçuk Candan	Arizona State University, USA
Shiyu Chang	UIUC, USA
Savvas Chatzichristofis	Democritus University of Thrace, Greece
Edgar Chávez	CICESE, Mexico
Wen-Huang Cheng	Academia Sinica, Taiwan, R.O.C.
Gene Cheung	National Institute of Informatics, Japan
Wei-Ta Chu	National Chung Cheng University, Taiwan, R.O.C.
Vincent Claveau	IRISA-CNRS, France
Kathy M. Clawson	University of Sunderland, UK
Claudiu Cobarzan	Babes-Bolyai University, Romania
Michel Crucianu	Cnam, France
Peng Cui	Tsinghua University, China
Rossana Damiano	Università di Torino, Italy
Petros Daras	Centre for Research and Technology Hellas, Greece
Wesley De Neve	Ghent University, Belgium
François Destelle	DCU Insight, Ireland
Cem Direkoglu	Middle East Technical University, Northern Cyprus Campus, Turkey
Lingyu Duan	Peking University, China
Jianping Fan	UNC Charlotte, USA
Mylène Farias	University of Brasília, Brazil
Gerald Friedland	ICSI/UC Berkeley, USA
Weijie Fu	Hefei University of Technology, China
Lianli Gao	University of Electronic Science and Technology, China
Yue Gao	Tsinghua University, China
Guillaume Gravier	CNRS, IRISA, and Inria Rennes, France
Ziyu Guan	Northwest University of China, China
Gylfi Þór Guðmundsson	Reykjavik University, Iceland
Silvio Guimarães	PUC Minas, Brazil
Allan Hanbury	TU Wien, Austria
Shijie Hao	Hefei University of Technology, China
Alex Hauptmann	Carnegie Mellon University, USA
Andreas Henrich	University of Bamberg, Germany
Nicolas Hervé	Institut National de l'Audiovisuel, France
Richang Hong	Hefei University of Technology, China
Frank Hopfgartner	University of Glasgow, UK
Michael Houle	National Institute of Informatics, Japan
Jun-Wei Hsieh	National Taiwan Ocean University, Taiwan, R.O.C.
Zhenzhen Hu	Nanyang Technological University, Singapore

Yannick Prié   University of Nantes, France
Jianjun Qian   Nanjing University of Science and Technology, China
Xueming Qian   Xi'an Jiaotong University, China
Georges Quénot   LIG-CNRS, France
Miloš Radovanović   University of Novi Sad, Serbia
Michael Riegler   Simula Research Lab, Norway
Stevan Rudinac   University of Amsterdam, The Netherlands
Mukesh Kumar Saini   Indian Institute of Technology Ropar, India
Jitao Sang   Chinese Academy of Sciences, China
Klaus Schöffmann   Klagenfurt University, Austria
Pascale Sébillot   IRISA/INSA Rennes, France
Jie Shao   University of Electronic Science and Technology,
                             China
Xiaobo Shen   Nanjing University of Science and Technology, China
Koichi Shinoda   Tokyo Institute of Technology, Japan
Mei-Ling Shyu   University of Miami, USA
Alan Smeaton   Dublin City University, Ireland
Lifeng Sun   Tsinghua University, China
Yongqing Sun   NTT Media Intelligence Laboratories, Japan
Sheng Tang   Institute of Computing Technology, Chinese Academy
                             of Sciences, China
Shuhei Tarashima   NTT, Japan
Wei-Guang Teng   National Cheng Kung University, Taiwan, R.O.C.
Georg Thallinger   Joanneum Research, Austria
Qi Tian   University of Texas at San Antonio, USA
Christian Timmerer   Alpen-Adria-Universität Klagenfurt, Austria
Dian Tjondronegoro   Queensland University of Technology, Australia
Shingo Uchihashi   Fuji Xerox Co., Ltd., Japan
Feng Wang   East China Normal University, China
Jinqiao Wang   Chinese Academy of Sciences, China
Shikui Wei   Beijing Jiaotong University, China
Lai Kuan Wong   Multimedia University, Malaysia
Marcel Worring   University of Amsterdam, The Netherlands
Hong Wu   University of Electronic Science and Technology
                             of China, China
Xiao Wu   Southwest Jiaotong University, China
Changsheng Xu   Chinese Academy of Sciences, China
Toshihiko Yamasaki   The University of Tokyo, Japan
Bo Yan   Fudan University, China
Keiji Yanai   University of Electro-Communications, Tokyo, Japan
Kuiyuan Yang   Microsoft Research, China
You Yang   HUST, China
Jun Yu   Hangzhou Dianzi University, China
Maia Zaharieva   Vienna University of Technology, Austria
Matthias Zeppelzauer   University of Applied Sciences St. Poelten, Austria
Zheng-Jun Zha   University of Science and Technology of China, China

Cha Zhang	Microsoft Research, USA
Hanwang Zhang	National University of Singapore, Singapore
Tianzhu Zhang	CASIA, Bangladesh
Cairong Zhao	Tongji University, China
Ye Zhao	Hefei University of Technology, China
Lijuan Zhou	Dublin City University, Ireland
Shiai Zhu	University of Ottawa, Canada
Xiaofeng Zhu	Guangxi Normal University, China
Arthur Zimek	University of Southern Denmark, Denmark
Roger Zimmermann	National University of Singapore, Singapore

Demonstration, Special Session and VBS Reviewers

Shanshan Ai	Beijing Jiaotong University, China
Alberto Messina	RAI CRIT, Italy
François Brémond	Inria, France
Houssem Chatbri	Dublin City University, Ireland
Jingyuan Chen	National University of Singapore, Singapore
Yi Chen	Helsinki Institute for Information Technology, Finland
Yiqiang Chen	Chinese Academy of Sciences, China
Zhineng Chen	Chinese Academy of Sciences, China
Zhiyong Cheng	National University of Singapore, Singapore
Mariana Damova	Mozaika, Romania
Stamatia Dasiopoulou	Pompeu Fabra University, Spain
Monika Dominguez	Pompeu Fabra University, Spain
Ling Du	Tianjin Polytechnic University, China
Jana Eggink	BBC R&D, UK
Bailan Feng	Chinese Academy of Sciences, China
Fuli Feng	Chinese Academy of Sciences, China
Min-Chun Hu	National Cheng Kung University, Taiwan, R.O.C.
Lei Huang	Ocean University of China, China
Marco A. Hudelist	Klagenfurt University, Austria
Bogdan Ionescu	Politehnica University of Bucharest, Romania
Eleni Kamateri	CERTH, Greece
Hyowon Lee	Singapore University of Technology and Design, Singapore
Andreas Leibetseder	Alpen-Adria-Universität Klagenfurt, Austria
Na Li	Dublin City University, Ireland
Xirong Li	Renmin University of China, China
Wu Liu	Beijing University of Posts and Telecommunications, China
Jakub Lokoč	Charles University in Prague, Czech Republic
Mathias Lux	Klagenfurt University, Austria
Georgios Meditskos	CERTH, Greece
Wolfgang Minker	Ulm University, Germany
Bernd Münzer	Klagenfurt University, Austria

Adrian Muscat	University of Malta, Malta
Yingwei Pan	University of Science and Technology of China, China
Zhengyuan Pang	Tsinghua University, China
Stefan Petscharnig	Alpen-Adria Universität Klagenfurt, Austria
Manfred Jürgen Primus	Alpen-Adria-Universität Klagenfurt, Austria
Zhaofan Qiu	University of Science and Technology of China, China
Amon Rapp	University of Toronto, Canada
Fuming Sun	Liaoning University of Technology, China
Xiang Wang	National University of Singapore, Singapore
Hongtao Xie	Chinese Academy of Sciences, China
Yuxiang Xie	National University of Defense Technology, China
Shiqiang Yang	Tsinghua University, China
Yang Yang	University of Electronic Science and Technology of China, China
Changqing Zhang	Tianjin University, China

External Reviewers

Duc Tien Dang Nguyen	Dublin City University, Ireland
Yusuke Matsui	National Institute of Informatics, Japan
Sang Phan	National Institute of Informatics, Japan
Jiang Zhou	Dublin City University, Ireland

Contents – Part I

SS1: Social Media Retrieval and Recommendation

SS2: Modeling Multimedia Behaviors

SS3: Multimedia Computing for Intelligent Life

SS4: Multimedia and Multimodal Interaction for Health and Basic Care Applications

Contents – Part II

Demonstrations

Video Browser Showdown

Full Papers Accepted for Oral Presentation

3D Sound Field Reproduction at Non Central Point for NHK 22.2 System

Song Wang[1,2], Ruimin Hu[1,2(✉)], Shihong Chen[1,2], Xiaochen Wang[1,2],
Yuhong Yang[1,2], Weiping Tu[1,2], and Bo Peng[3]

[1] State Key Laboratory of Software Engineering, School of Computer Science,
Wuhan University, Wuhan 430072, China
wangsongf117@163.com, hrm1964@163.com
[2] National Engineering Research Center for Multimedia Software,
School of Computer Science, Wuhan University, Wuhan 430072, China
[3] Military Economy Academy, Wuhan 430072, China

Abstract. Reducing channel number is convenient for NHK 22.2 system in loudspeaker layout and good for the application of NHK 22.2 in family environment. In 2011, Akio Ando has proposed a down-mixing method which could simplify 22.2 multichannel system to 10.2 and 8.2 multichannel system, but this method only could perfect reproduce 3D sound field at a central listening point. In practice, people may stay at a non central point, Ando's method could not maintain sound physical properties at non central point well. Conventional non central zone sound filed reproduction methods such as pressure matching method and particle velocity matching method have theoretical limitations. This paper propose a general down-mixing method basing on the position of listening point and sound physical properties, it could produce a sweet spot at any non central point in reconstruction field and reduce channel number. In experiments, the proposed method simplifies 22 channel system to 10 channel system, experimental results demonstrate that it performs better than traditional method at non central point sound field reconstruction.

Keywords: Sound field reproduction · Non central point · Multichannel system · Sound physical property

1 Introduction

As the development of 3D movie and 3D television, 3D audio needs to keep pace with the development of 3D video to bring perfect audio-visual experience for people. There are following 3D audio technologies: Ambisonics [1–3], Wave Field Synthesis (WFS) [4–6], Head Related Transfer Function (HRTF) [7–9],

The research was supported by National Nature Science Foundation of China (61231015), National High Technology Research and Development Program of China (863 Program) (2015AA016306), National Nature Science Foundation of China (61471271, 61671335).

L. Amsaleg et al. (Eds.): MMM 2017, Part I, LNCS 10132, pp. 3–14, 2017.
DOI: 10.1007/978-3-319-51811-4_1

Vector Based Amplitude Panning (VBAP) [10,11] and so on. Among them, VBAP uses three loudspeakers to synthesis a virtual sound source by vector view [10,11]. VBAP is of high computational efficiency and accuracy sound image reconstruction. Also, it is easy to place loudspeakers by VBAP. [12–14] has generalized VBAP in its application.

NHK in Japan has proposed the 22.2 system, which has three vertical layers of loudspeakers with 2 low frequency effects (LFEs), namely 9 loudspeakers at the upper layer, 10 loudspeakers at the middle layer, 3 loudspeakers at the bottom layer and 2 LFEs. In 2011, Ando proposed a down-mixing method that describes VBAP from physical view, and it can convert 22-channel signals into 10-, 8-channel signals [15]. The author assumes that all loudspeakers locate on a sphere, the listening point is fixed at the center of this sphere. This method uses three loudspeakers in reproduced system to replaced a loudspeaker in original system. In the replacing process, it maintains sound pressure and the direction of particle velocity at the central listening point. But, three loudspeakers replacing a loudspeaker could only get a local solution, Ando uses the local solution to form the global solution; particle velocity is vector, magnitude of particle velocity at the listening point is not maintained. To solve these problems, in 2015, Wang proposed a global model which maintains sound pressure, the direction and magnitude of particle velocity at the central listening point as much as possible in down-mixing processing [16], and gets a global optimal solution. In Song's method, all loudspeakers are also assumed to locate on a same sphere, the listening point is a fixed point at the center of this sphere.

Reducing channel number is convenient for the application of NHK 22.2 in family environment. Although Ando's and Song's method could reduce the channel number of 22.2 multichannel system, they all only reproduce sound field better at the center. But as we all know, in practical listening environment, people may not always stay at a constant point, they may walk to or stay at a non central position, then Ando's and Song's method may not be fit for these situations.

There are some non central point or non central zone sound field reconstruction methods. The existing pressure matching method [17] matches pressure in a same given zone between original sound field and reproduced sound field. From its principle, it can be used for 3D non central region sound field reconstruction. It is reported that when loudspeakers used for reproduction are non-uniformly put or minimum angular distance between the adjacent loudspeakers and the direction of incoming target sound field is longer, existing pressure matching method shows reproduction failure [17]. To counter these problems, particle velocity matching method [18,19] has been proposed by matching particle velocity in a same given zone between original sound field and reproduced sound field. We call it 'PVMSZ' for short. It also could be used for 3D non central region sound field reconstruction, though only central zone sound field reconstruction tests are made in [18,19]. For 22.2 multichannel system, the sweet spot is at the center, the best listening region is the central region, the listening experience of non central region is worse than central region. But by PVMSZ in non central point or non central region sound field reconstruction, the best situation is that the non central region of original sound field is correctly recovered, which makes listener at a non central point or non central region could not get the same listening experience as he or she stay at a central point or central region.

To resolve above problem, this paper proposes a reproduction method basing on the location of listening point and sound physical property which could make a non central point become a sweet spot, besides simplifying 22.2 multichannel system to less channel system for improving the flexibility of 22.2 multichannel system application in home environment. This paper is outlined as follows: Sect. 2 describes basic definition about sound physical properties, Sect. 3 proposes a listening point position and sound physical property based error minimum method and gets its solution, Sect. 4 presents some objective and subjective experiments, and compares the results between different methods.

2   Basic Definition

Fourier transform of the sound pressure produced by single loudspeaker (a loudspeaker can be seen as a point source) at the receiving point is expressed as:

$$p(\boldsymbol{r},\omega) = G\frac{e^{-ik|\boldsymbol{r}-\boldsymbol{\xi}|}}{|\boldsymbol{r}-\boldsymbol{\xi}|}s(\omega) \tag{1}$$

Particle velocity produced by single loudspeaker at the receiving point is:

$$u(\boldsymbol{r},\omega) = G(1 + \frac{1}{ik|\boldsymbol{r}-\boldsymbol{\xi}|})\frac{e^{-ik|\boldsymbol{r}-\boldsymbol{\xi}|}}{|\boldsymbol{r}-\boldsymbol{\xi}|^2}\begin{pmatrix} r_x - \xi_x \\ r_y - \xi_y \\ r_z - \xi_z \end{pmatrix}s(\omega) \approx G\frac{e^{-ik|\boldsymbol{r}-\boldsymbol{\xi}|}}{|\boldsymbol{r}-\boldsymbol{\xi}|^2}\begin{pmatrix} r_x - \xi_x \\ r_y - \xi_y \\ r_z - \xi_z \end{pmatrix}s(\omega) \tag{2}$$

where G is a proportion coefficient, $\boldsymbol{r} = (r_x, r_y, r_z)$ is the coordinate of the listening point, $\boldsymbol{\xi} = (\xi_x, \xi_y, \xi_z)$ is the coordinate of single loudspeaker, k is the wave number, $k = 2\pi f/c$, f is the frequency of sound, c is the speed of sound in air, $s(\omega)$ is the Fourier transform of the input signal to the loudspeaker.

Particle velocity and sound pressure are linear function of input signal, so the sound pressure or particle velocity of m loudspeakers at the listening point could be obtained by simple superposition.

3   Listening Point Position and Sound Physical Property Based Error Minimum Method and Solution

We first study the case that a virtual sound source (single loudspeaker) is replaced by m loudspeakers. In cartesian coordinates, the virtual sound source and m loudspeakers are on a same sphere, the center of the sphere is $O(0, 0, 0)$ ($\boldsymbol{o} = (0, 0, 0)$), the radius of the sphere is R, the listening point is at any point L with coordinate $\boldsymbol{r} = (r_x, r_y, r_z)$, $|\boldsymbol{r}| \leq R$ in the sphere, coordinate of the virtual sound source is $\boldsymbol{\xi} = (\xi_x, \xi_y, \xi_z)$, coordinates of m loudspeakers are $\boldsymbol{\xi}^{(j)} = (\xi_x^{(j)}, \xi_y^{(j)}, \xi_z^{(j)})$, $j = 1, 2, \cdots, m$.

We suppose that the sound pressure produced by a virtual sound source at the center O is the same as the sound pressure produced by m loudspeakers at the listening point L, we can get:

$$G\sum_{j=1}^{m}\frac{e^{-ik|r-\xi^{(j)}|}}{|r-\xi^{(j)}|}s_j(\omega) = G\frac{e^{-ik|o-\xi|}}{|o-\xi|}s(\omega) \qquad (3)$$

where $s_j(\omega) = w_j s(\omega)$. By Eq. (3), we could get:

$$\begin{pmatrix} \frac{\cos(k|r-\xi^{(1)}|)}{|r-\xi^{(1)}|} \cdots \frac{\cos(k|r-\xi^{(m)}|)}{|r-\xi^{(m)}|} \\ \frac{\sin(k|r-\xi^{(1)}|)}{|r-\xi^{(1)}|} \cdots \frac{\sin(k|r-\xi^{(m)}|)}{|r-\xi^{(m)}|} \end{pmatrix} W = \begin{pmatrix} \frac{\cos(k|o-\xi|)}{|o-\xi|} \\ \frac{\sin(k|o-\xi|)}{|o-\xi|} \end{pmatrix} \qquad (4)$$

where $W = (w_1, w_2, \cdots, w_m)^T$ is the signal distribution weight vector, by which signals of m loudspeakers could be obtained when these m loudspeakers replace a virtual sound source, T is the transposition of the matrix.

Suppose that the particle velocity produced by a virtual sound source at the center O is the same as the particle velocity produced by m loudspeakers at the listening point L, we can get:

$$G\sum_{j=1}^{m}\frac{e^{-ik|r-\xi^{(j)}|}}{|r-\xi^{(j)}|^2}\begin{pmatrix} r_x - \xi_x^{(j)} \\ r_y - \xi_y^{(j)} \\ r_z - \xi_z^{(j)} \end{pmatrix} s_j(\omega) = G\frac{e^{-ik|o-\xi|}}{|o-\xi|^2}\begin{pmatrix} -\xi_x \\ -\xi_y \\ -\xi_z \end{pmatrix} s(\omega) \qquad (5)$$

From Eq. (5), we could get:

$$\tilde{H}_1 W = H_1 \qquad (6)$$

where

$$\tilde{H}_1 = \begin{pmatrix} \frac{e^{-ik|r-\xi^{(1)}|}\left(r_x-\xi_x^{(1)}\right)}{|r-\xi^{(1)}|^2} \cdots \frac{e^{-ik|r-\xi^{(m)}|}\left(r_x-\xi_x^{(m)}\right)}{|r-\xi^{(m)}|^2} \\ \frac{e^{-ik|r-\xi^{(1)}|}\left(r_y-\xi_y^{(1)}\right)}{|r-\xi^{(1)}|^2} \cdots \frac{e^{-ik|r-\xi^{(m)}|}\left(r_y-\xi_y^{(m)}\right)}{|r-\xi^{(m)}|^2} \\ \frac{e^{-ik|r-\xi^{(1)}|}\left(r_z-\xi_z^{(1)}\right)}{|r-\xi^{(1)}|^2} \cdots \frac{e^{-ik|r-\xi^{(m)}|}\left(r_z-\xi_z^{(m)}\right)}{|r-\xi^{(m)}|^2} \end{pmatrix}, \quad H_1 = \begin{pmatrix} \frac{e^{-ik|o-\xi|}(-\xi_x)}{|o-\xi|^2} \\ \frac{e^{-ik|o-\xi|}(-\xi_y)}{|o-\xi|^2} \\ \frac{e^{-ik|o-\xi|}(-\xi_z)}{|o-\xi|^2} \end{pmatrix}$$

From Eq. (6), we could get:

$$\tilde{H}_2 W = H_2 \qquad (7)$$

where

$$\tilde{H}_2 = \begin{pmatrix} \frac{\cos(k|r-\xi^{(1)}|)\left(r_x-\xi_x^{(1)}\right)}{|r-\xi^{(1)}|^2} \cdots \frac{\cos(k|r-\xi^{(m)}|)\left(r_x-\xi_x^{(m)}\right)}{|r-\xi^{(m)}|^2} \\ \frac{\cos(k|r-\xi^{(1)}|)\left(r_y-\xi_y^{(1)}\right)}{|r-\xi^{(1)}|^2} \cdots \frac{\cos(k|r-\xi^{(m)}|)\left(r_y-\xi_y^{(m)}\right)}{|r-\xi^{(m)}|^2} \\ \frac{\cos(k|r-\xi^{(1)}|)\left(r_z-\xi_z^{(1)}\right)}{|r-\xi^{(1)}|^2} \cdots \frac{\cos(k|r-\xi^{(m)}|)\left(r_z-\xi_z^{(m)}\right)}{|r-\xi^{(m)}|^2} \\ \frac{\sin(k|r-\xi^{(1)}|)\left(r_x-\xi_x^{(1)}\right)}{|r-\xi^{(1)}|^2} \cdots \frac{\sin(k|r-\xi^{(m)}|)\left(r_x-\xi_x^{(m)}\right)}{|r-\xi^{(m)}|^2} \\ \frac{\sin(k|r-\xi^{(1)}|)\left(r_y-\xi_y^{(1)}\right)}{|r-\xi^{(1)}|^2} \cdots \frac{\sin(k|r-\xi^{(m)}|)\left(r_y-\xi_y^{(m)}\right)}{|r-\xi^{(m)}|^2} \\ \frac{\sin(k|r-\xi^{(1)}|)\left(r_z-\xi_z^{(1)}\right)}{|r-\xi^{(1)}|^2} \cdots \frac{\sin(k|r-\xi^{(m)}|)\left(r_z-\xi_z^{(m)}\right)}{|r-\xi^{(m)}|^2} \end{pmatrix}, \quad H_2 = \begin{pmatrix} \frac{\cos(k|o-\xi|)(-\xi_x)}{|o-\xi|^2} \\ \frac{\cos(k|o-\xi|)(-\xi_y)}{|o-\xi|^2} \\ \frac{\cos(k|o-\xi|)(-\xi_z)}{|o-\xi|^2} \\ \frac{\sin(k|o-\xi|)(-\xi_x)}{|o-\xi|^2} \\ \frac{\sin(k|o-\xi|)(-\xi_y)}{|o-\xi|^2} \\ \frac{\sin(k|o-\xi|)(-\xi_z)}{|o-\xi|^2} \end{pmatrix}$$

Then we divide the first to fifth rows by the sixth row in Eq. (7), we could get:

$$
\begin{aligned}
\frac{\sum_{j=1}^{m} \frac{\cos\left(k|\boldsymbol{r}-\boldsymbol{\xi}^{(j)}|\right)\left(r_x-\xi_x^{(j)}\right)}{|\boldsymbol{r}-\boldsymbol{\xi}^{(j)}|^2} w_j}{\sum_{j=1}^{m} \frac{\sin\left(k|\boldsymbol{r}-\boldsymbol{\xi}^{(j)}|\right)\left(r_z-\xi_z^{(j)}\right)}{|\boldsymbol{r}-\boldsymbol{\xi}^{(j)}|^2} w_j} &= \frac{\cos(k|\boldsymbol{o}-\boldsymbol{\xi}|)(-\xi_x)}{\sin(k|\boldsymbol{o}-\boldsymbol{\xi}|)(-\xi_z)} \\[4pt]
\frac{\sum_{j=1}^{m} \frac{\cos\left(k|\boldsymbol{r}-\boldsymbol{\xi}^{(j)}|\right)\left(r_y-\xi_y^{(j)}\right)}{|\boldsymbol{r}-\boldsymbol{\xi}^{(j)}|^2} w_j}{\sum_{j=1}^{m} \frac{\sin\left(k|\boldsymbol{r}-\boldsymbol{\xi}^{(j)}|\right)\left(r_z-\xi_z^{(j)}\right)}{|\boldsymbol{r}-\boldsymbol{\xi}^{(j)}|^2} w_j} &= \frac{\cos(k|\boldsymbol{o}-\boldsymbol{\xi}|)(-\xi_y)}{\sin(k|\boldsymbol{o}-\boldsymbol{\xi}|)(-\xi_z)} \\[4pt]
\frac{\sum_{j=1}^{m} \frac{\cos\left(k|\boldsymbol{r}-\boldsymbol{\xi}^{(j)}|\right)\left(r_z-\xi_z^{(j)}\right)}{|\boldsymbol{r}-\boldsymbol{\xi}^{(j)}|^2} w_j}{\sum_{j=1}^{m} \frac{\sin\left(k|\boldsymbol{r}-\boldsymbol{\xi}^{(j)}|\right)\left(r_z-\xi_z^{(j)}\right)}{|\boldsymbol{r}-\boldsymbol{\xi}^{(j)}|^2} w_j} &= \frac{\cos(k|\boldsymbol{o}-\boldsymbol{\xi}|)}{\sin(k|\boldsymbol{o}-\boldsymbol{\xi}|)} \\[4pt]
\frac{\sum_{j=1}^{m} \frac{\sin\left(k|\boldsymbol{r}-\boldsymbol{\xi}^{(j)}|\right)\left(r_x-\xi_x^{(j)}\right)}{|\boldsymbol{r}-\boldsymbol{\xi}^{(j)}|^2} w_j}{\sum_{j=1}^{m} \frac{\sin\left(k|\boldsymbol{r}-\boldsymbol{\xi}^{(j)}|\right)\left(r_z-\xi_z^{(j)}\right)}{|\boldsymbol{r}-\boldsymbol{\xi}^{(j)}|^2} w_j} &= \frac{(-\xi_x)}{(-\xi_z)} \\[4pt]
\frac{\sum_{j=1}^{m} \frac{\sin\left(k|\boldsymbol{r}-\boldsymbol{\xi}^{(j)}|\right)\left(r_y-\xi_y^{(j)}\right)}{|\boldsymbol{r}-\boldsymbol{\xi}^{(j)}|^2} w_j}{\sum_{j=1}^{m} \frac{\sin\left(k|\boldsymbol{r}-\boldsymbol{\xi}^{(j)}|\right)\left(r_z-\xi_z^{(j)}\right)}{|\boldsymbol{r}-\boldsymbol{\xi}^{(j)}|^2} w_j} &= \frac{(-\xi_y)}{(-\xi_z)}
\end{aligned} \tag{8}
$$

Equation (8) could guarantee that direction of particle velocity of virtual sound source at center O and that of m loudspeakers at the listening point L are the same. From Eq. (8), we could get:

$$
\begin{aligned}
&\sum_{j=1}^{m} w_j \left[\cos\left(k|\boldsymbol{r}-\boldsymbol{\xi}^{(j)}|\right)\sin\left(k|\boldsymbol{o}-\boldsymbol{\xi}|\right)\left(r_x-\xi_x^{(j)}\right)(-\xi_z)\right. \\
&\quad \left. -\sin\left(k|\boldsymbol{r}-\boldsymbol{\xi}^{(j)}|\right)\cos\left(k|\boldsymbol{o}-\boldsymbol{\xi}|\right)\left(r_z-\xi_z^{(j)}\right)(-\xi_x)\right]/|\boldsymbol{r}-\boldsymbol{\xi}^{(j)}|^2 = 0 \\
&\sum_{j=1}^{m} w_j \left[\cos\left(k|\boldsymbol{r}-\boldsymbol{\xi}^{(j)}|\right)\sin\left(k|\boldsymbol{o}-\boldsymbol{\xi}|\right)\left(r_y-\xi_y^{(j)}\right)(-\xi_z)\right. \\
&\quad \left. -\sin\left(k|\boldsymbol{r}-\boldsymbol{\xi}^{(j)}|\right)\cos\left(k|\boldsymbol{o}-\boldsymbol{\xi}|\right)\left(r_z-\xi_z^{(j)}\right)(-\xi_y)\right]/|\boldsymbol{r}-\boldsymbol{\xi}^{(j)}|^2 = 0 \\
&\sum_{j=1}^{m} w_j \left[\cos\left(k|\boldsymbol{r}-\boldsymbol{\xi}^{(j)}|\right)\sin\left(k|\boldsymbol{o}-\boldsymbol{\xi}|\right) - \sin\left(k|\boldsymbol{r}-\boldsymbol{\xi}^{(j)}|\right)\cos\left(k|\boldsymbol{o}-\boldsymbol{\xi}|\right)\right] \\
&\quad \cdot \left(r_z-\xi_z^{(j)}\right)/|\boldsymbol{r}-\boldsymbol{\xi}^{(j)}|^2 = 0 \\
&\sum_{j=1}^{m} w_j \left[\left(r_x-\xi_x^{(j)}\right)(-\xi_z) - \left(r_z-\xi_z^{(j)}\right)(-\xi_x)\right] \\
&\quad \cdot \sin\left(k|\boldsymbol{r}-\boldsymbol{\xi}^{(j)}|\right)/|\boldsymbol{r}-\boldsymbol{\xi}^{(j)}|^2 = 0 \\
&\sum_{j=1}^{m} w_j \left[\left(r_y-\xi_y^{(j)}\right)(-\xi_z) - \left(r_z-\xi_z^{(j)}\right)(-\xi_y)\right] \\
&\quad \cdot \sin\left(k|\boldsymbol{r}-\boldsymbol{\xi}^{(j)}|\right)/|\boldsymbol{r}-\boldsymbol{\xi}^{(j)}|^2 = 0
\end{aligned} \tag{9}
$$

Together with Eq. (4), we could get:

$$
AW = B \tag{10}
$$

where

$$A = \begin{pmatrix} t_{11} & t_{12} & \cdots & t_{1m} \\ t_{21} & t_{22} & \cdots & t_{2m} \\ t_{31} & t_{32} & \cdots & t_{3m} \\ t_{41} & t_{42} & \cdots & t_{4m} \\ t_{51} & t_{52} & \cdots & t_{5m} \\ \frac{\cos(k|r-\xi^{(1)}|)}{|r-\xi^{(1)}|} & \frac{\cos(k|r-\xi^{(2)}|)}{|r-\xi^{(2)}|} & \cdots & \frac{\cos(k|r-\xi^{(m)}|)}{|r-\xi^{(m)}|} \\ \frac{\sin(k|r-\xi^{(1)}|)}{|r-\xi^{(1)}|} & \frac{\sin(k|r-\xi^{(2)}|)}{|r-\xi^{(2)}|} & \cdots & \frac{\sin(k|r-\xi^{(m)}|)}{|r-\xi^{(m)}|} \end{pmatrix},$$

$$\begin{cases} t_{1j} = \Big[\cos\left(k|r-\xi^{(j)}|\right)\sin\left(k|o-\xi|\right)\left(r_x - \xi_x^{(j)}\right)(-\xi_z) \\ \qquad - \sin\left(k|r-\xi^{(j)}|\right)\cos\left(k|o-\xi|\right)\left(r_z - \xi_z^{(j)}\right)(-\xi_x)\Big]/|r-\xi^{(j)}|^2 \\ t_{2j} = \Big[\cos\left(k|r-\xi^{(j)}|\right)\sin\left(k|o-\xi|\right)\left(r_y - \xi_y^{(j)}\right)(-\xi_z) \\ \qquad - \sin\left(k|r-\xi^{(j)}|\right)\cos\left(k|o-\xi|\right)\left(r_z - \xi_z^{(j)}\right)(-\xi_y)\Big]/|r-\xi^{(j)}|^2 \\ t_{3j} = \Big[\cos\left(k|r-\xi^{(j)}|\right)\sin\left(k|o-\xi|\right) - \sin\left(k|r-\xi^{(j)}|\right)\cos\left(k|o-\xi|\right)\Big] \\ \qquad \cdot\left(r_z - \xi_z^{(j)}\right)/|r-\xi^{(j)}|^2 \\ t_{4j} = \Big[\left(r_x - \xi_x^{(j)}\right)(-\xi_z) - \left(r_z - \xi_z^{(j)}\right)(-\xi_x)\Big]\sin\left(k|r-\xi^{(j)}|\right)/|r-\xi^{(j)}|^2 \\ t_{5j} = \Big[\left(r_y - \xi_y^{(j)}\right)(-\xi_z) - \left(r_z - \xi_z^{(j)}\right)(-\xi_y)\Big]\sin\left(k|r-\xi^{(j)}|\right)/|r-\xi^{(j)}|^2 \\ \qquad\qquad\qquad\qquad j = 1, 2, \cdots, m \end{cases}$$

$$B = \left(0, 0, 0, 0, 0, \frac{\cos(k|o-\xi|)}{|o-\xi|}, \frac{\sin(k|o-\xi|)}{|o-\xi|}\right)^T$$

The square error of particle velocity magnitude at the listening point L is defined as:

$$\begin{aligned} &C(w_1, w_2, w_3, \cdots, w_m) \\ &= \|u(o, \omega) - \sum_{j=1}^m u_j(r, \omega)\|_2^2 \\ &= \|Gs(\omega)\|_2^2 \Big[\big(\frac{\cos(k|o-\xi|)(-\xi_x)}{|o-\xi|^2} - \sum_{j=1}^m \frac{\cos(k|r-\xi^{(j)}|)(r_x - \xi_x^{(j)})}{|r-\xi^{(j)}|^2}w_j\big)^2 \\ &+ \big(\frac{\cos(k|o-\xi|)(-\xi_y)}{|o-\xi|^2} - \sum_{j=1}^m \frac{\cos(k|r-\xi^{(j)}|)(r_y - \xi_y^{(j)})}{|r-\xi^{(j)}|^2}w_j\big)^2 \\ &+ \big(\frac{\cos(k|o-\xi|)(-\xi_z)}{|o-\xi|^2} - \sum_{j=1}^m \frac{\cos(k|r-\xi^{(j)}|)(r_z - \xi_z^{(j)})}{|r-\xi^{(j)}|^2}w_j\big)^2 \\ &+ \big(\frac{\sin(k|o-\xi|)(-\xi_x)}{|o-\xi|^2} - \sum_{j=1}^m \frac{\sin(k|r-\xi^{(j)}|)(r_x - \xi_x^{(j)})}{|r-\xi^{(j)}|^2}w_j\big)^2 \\ &+ \big(\frac{\sin(k|o-\xi|)(-\xi_y)}{|o-\xi|^2} - \sum_{j=1}^m \frac{\sin(k|r-\xi^{(j)}|)(r_y - \xi_y^{(j)})}{|r-\xi^{(j)}|^2}w_j\big)^2 \\ &+ \big(\frac{\sin(k|o-\xi|)(-\xi_z)}{|o-\xi|^2} - \sum_{j=1}^m \frac{\sin(k|r-\xi^{(j)}|)(r_z - \xi_z^{(j)})}{|r-\xi^{(j)}|^2}w_j\big)^2\Big] \end{aligned} \qquad (11)$$

where $u(o, \omega)$ is particle velocity of the virtual sound source at the center O, $u_j(r, \omega)(j = 1, 2, \cdots, m)$ is particle velocity of the jth loudspeaker at the listening point L. For a given signal, $\|Gs(\omega)\|_2^2$ is a constant. So to make the error square of particle velocity magnitude C minimum, we should make the formula in bracket of Eq. (11) minimum. Then the solution of replacing a virtual sound source by m loudspeakers is equivalent to solving the following question:

$$\min_{W} \tfrac{1}{2}\|\tilde{H}_2 W - H_2\|_2^2$$
$$s.t. \quad AW = B \tag{12}$$

Equation (12) could guarantee sound pressure and the direction of particle velocity at the listening point L is the same as that at center O, and make the square error of particle velocity magnitude between the listening point L and center O as little as possible. Equation (12) is a least squares problems with equality constraints and it could be worked out by existing mature algorithms such as Guass-Newton Algorithm [20].

Then, by above method, n-channel system in original system is simplified to m-channel $(m < n)$ in reproduced system step by step, the final distribution coefficients of m loudspeakers in reproduced system are got by superposition of each loudspeaker's distribution coefficients in every replacement process respectively. We call the proposed method Listening Point Position and Sound Physical Property Based Error Minimum (LPPSPP) method, which is a general method.

4 Experiments

The performance of PVMSZ and LPPSPP method are measured by objective and subjective tests in non central point sound field reconstruction. NHK 22.2 multichannel system arrangement without LFE channels is simplified to 10-channel system as shown in Fig. 1, which is proposed by Ando [15]. The radius of the spheres is $2\,\mathrm{m}$, the center of the sphere is at $O(0,0,0)$, the non central listening point is supposed at $L(0.5, 0.5, 0)$. Sound speed c is $340\,\mathrm{m/s}$, the radius of human head is about $0.085\,\mathrm{m}$.

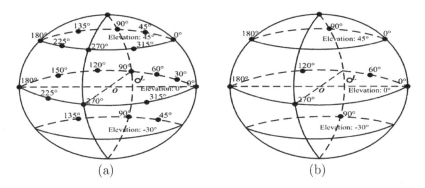

Fig. 1. Loudspeakers arrangement. • denotes loudspeakers, ○ denotes listening point. (a): 22 loudspeakers; (b): 10 loudspeakers.

4.1 Objective Experiments

The original signal is 1000 Hz single frequency signal in objective experiments.

The particle velocity magnitude error at non central listening point is:

$$Pvme\left(\boldsymbol{r},\omega\right) = \frac{\left|\left|\boldsymbol{u}_{ori}\left(\boldsymbol{o},\omega\right)\right| - \left|\boldsymbol{u}_{rep}\left(\boldsymbol{r},\omega\right)\right|\right|}{\left|\boldsymbol{u}_{ori}\left(\boldsymbol{o},\omega\right)\right|} \times 100\% \tag{13}$$

where $\boldsymbol{u}_{ori}$ is particle velocity of original system, $\boldsymbol{u}_{rep}$ is particle velocity of reproduced system. For 1000 Hz single frequency signal, compared with the original 22-channel at O, the particle velocity magnitude error at L are 0.00% by LPPSPP method, 10.53% by PVMSZ method. More comparisons are shown in Fig. 2. From it we can see that the particle velocity magnitude error by LPPSPP method is much lower than that by PVMSZ.

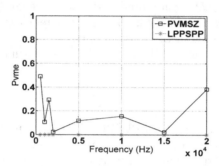

Fig. 2. Particle velocity magnitude error comparison at non central listening point L.

The particle velocity direction has been obtained by normalization:

$$\boldsymbol{u}_{oI}\left(\boldsymbol{o},\omega\right) = \frac{\boldsymbol{u}_{ori}\left(\boldsymbol{o},\omega\right)}{\left|\boldsymbol{u}_{ori}\left(\boldsymbol{o},\omega\right)\right|}, \qquad \boldsymbol{u}_{rI}\left(\boldsymbol{r},\omega\right) = \frac{\boldsymbol{u}_{rep}\left(\boldsymbol{r},\omega\right)}{\left|\boldsymbol{u}_{rep}\left(\boldsymbol{r},\omega\right)\right|} \tag{14}$$

Then the particle velocity direction error is defined as:

$$Pvde\left(\boldsymbol{r},\omega\right) = \frac{cos^{-1}\left(\boldsymbol{u}_{oI}\left(\boldsymbol{o},\omega\right)\cdot\boldsymbol{u}_{rI}\left(\boldsymbol{r},\omega\right)\right)}{\pi} \times 100\% \tag{15}$$

The particle velocity direction error are compared in Fig. 3. The radius of the red circle is 0.085 m. It shows that the particle velocity direction error by LPPSPP method is less than PVMSZ method around the non central point L.

Reproduction sound field of 22 channel system with 10 channel system for 1000 Hz on a sphere with radius 0.085 m are compared in Fig. 4, which indicates that the reproduced field by LPPSPP method is much closer to original field than PVMSZ method.

The relative mean square error (RMSE) of the reproduction is used as the sound pressure error metric:

$$\varepsilon(k\alpha) = \frac{\iiint\limits_{V} |S_r(\boldsymbol{r}+\boldsymbol{s},\omega) - S_d(\boldsymbol{s},\omega)|^2 dV}{\iiint\limits_{V} |S_d(\boldsymbol{s},\omega)|^2 dV} \tag{16}$$

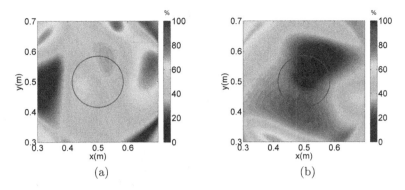

Fig. 3. Particle velocity direction error for 1000 Hz at plain z = 0. (a): PVMSZ method, (b): LPPSPP method. (Color figure online)

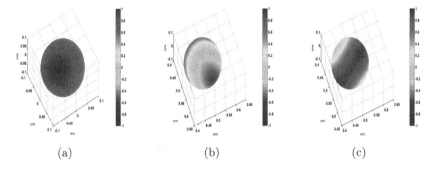

Fig. 4. Original sound field and reproduced sound field comparison. (a): original field, the center of sphere is O, (b): PVMSZ method, (c): LPPSPP method, the center of sphere is L for (b) and (c).

where the integration region V is a spherical ball of radius α and center O, $s = (s_x, s_y, s_z)$ is a point in V, r is the coordinate of any listening point L, $S_d(s, \omega)$ and $S_r(r + s, \omega)$ are the original and reproduced sound fields respectively.

RMSE of the reproduction field is displayed in Fig. 5. When α is 0.085 m, which is the size of human head radius, the RMSE of PVMSZ and LPPSPP method are respectively 97.30%, 3.92%. So at a region with human head radius, the RMSE of LPPSPP method is 93.38% lower than that of PVMSZ method. But when α grows bigger, the RMSE of proposed method increase much faster.

To better analysis the influence of non central region's location on RMSE by PVMSZ and LPPSPP, mean RMSE is adopted. It is the average of many non central region's RMSE, when the centers of these non central regions are a constant distance from the central point O. Suppose these non central region are all 3D spherical region with human head radius 0.085 m. The mean RMSE comparison is shown in Fig. 6. The distances variation range from centers of these non central regions to central point O is 0 m to 2 m, the mean RMSE variation range by PVMSZ is from 41.62% to 153.07%, the mean RMSE variation range by

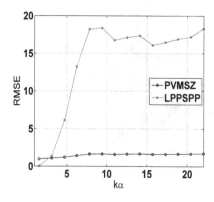

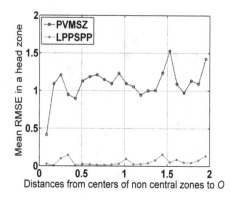

Fig. 5. The relative mean square error comparison.

Fig. 6. The mean relative mean square error comparison in a human head region.

LPPSPP is from 0.54% to 15.39%. The mean RMSE by LPPSPP is much more lower than that by PVMSZ and the change trend of mean RMSE by LPPSPP is more gentle than the change trend of mean RMSE by PVMSZ. It means that the influence of non central region's location on RMSE by LPPSPP is much limited relative to that by PVMSZ.

From Figs. 2 to 6, we can see that the sound field reproduced by LPPSPP method performs better than PVMSZ method in a 3D spherical region with human head radius. The reason is that PVMSZ method matches particle velocity between the same non central region of original sound field and reproduced sound field, and non central region sound field in original system is not the best. LPPSPP method aims at making physical properties of sound at the non central listening point to approximate that at central listening point as much as possible.

4.2 Subjective Experiments

Comparison Mean Opinion Score (CMOS) is used to test PVMSZ method and LPPSPP method, the test material consists of Ref/A/B, in which Ref is the original sound signal, A is signal generated by LPPSPP method, B is signal generated by PVMSZ method. Ref is played back by 22 channel system, A and B are played back by 10 channel system as shown in Fig. 1. We compare the sound image of A and B which is closer to Ref. The score has 7 levels, which are listed in Table 1. 10 listeners performed the listening test, including 5 males and 5 females. All of them actively work in the audio field. The center of listeners' heads is at origin O in testing when they listen Ref, the center of listeners' heads is at non central point L in testing when they listen A and B. The test results consist of an average score and a 95% confidence interval.

A white noise is used as original test sequence whose sample rate is 48 kHz, bit depth is 16 bits, intensity is 12 dB, and length is 10 s. The test results is given in Fig. 7. We can see that it is statistically comparable to PVMSZ method in a 95% confidence interval sense. The average scores of our method are higher

Table 1. Levels comparison standard

Comparison of the stimuli	Score
Sound image of A is much closer to Ref than B	+3
Sound image of A is closer to Ref than B	+2
Sound image of A is slightly closer to Ref than B	+1
Sound image of A to Ref is the same as B	0
Sound image of A is slightly further to Ref than B	−1
Sound image of A is further to Ref than B	−2
Sound image of A is much further to Ref than B	−3

than traditional method, the average value of CMOS is 1.9. The result means that location accuracy of proposed method is better than traditional method and that is in accordance with objective test results.

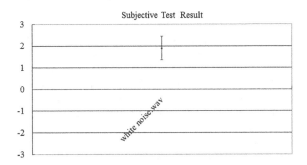

Fig. 7. CMOS score.

5   Conclusions

To apply 22.2 multichannel system in family better, this paper proposes a listening point position and sound physical property based error minimum method (LPPSPP method). This method can make sound pressure and the direction of particle velocity at non central listening point in reproduced system the same as that at central listening point in original system, and minimize the square error of particle velocity magnitude between non central listening point in reproduced system and central listening point in original system. Compared with PVMSZ method, objective and subjective experimental results show that the proposed method has lower sound pressure and particle velocity error in a human head region. The future work is to improve reproduction performance of the proposed method better.

References

1. Gerson, M.A.: Periphony: with-height sound reproduction. J. Audio Eng. Soc. **21**(1), 2–10 (1973)
2. Zhang, W., Abhayapala, T.D.: Three dimensional sound field reproduction using multiple circular loudspeaker arrays: functional analysis guided approach. IEEE/ACM Trans. Audio Speech Lang. Process. **22**(7), 1184–1194 (2014)
3. Grimm, G., et al.: Evaluation of spatial audio reproduction schemes for application in hearing aid research. Acta Acust. United Acust. **101**(13), 842–854 (2015)
4. Berkhout, A.J.: A holographic approach to acoustic control. J. Audio Eng. Soc. **36**(12), 977–995 (1988)
5. Berkhout, A.J.: Acoustic control by wave field synthesis. J. Acoust. Soc. Am. **93**(5), 2764–2778 (1993)
6. Lee, J.-M., et al.: Wave field synthesis of a virtual source located in proximity to a loudspeaker array. J. Acoust. Soc. Am. **134**(3), 2106–2117 (2013)
7. Blauert, J.: Spatial Hearing. MIT Press, Cambridge (1983)
8. Yin, F., et al.: Review on 3D audio technology. J. Commun. **32**(2), 130–138 (2011)
9. Rasumow, E., et al.: Smoothing individual head-related transfer functions in the frequency and spatial domains. J. Acoust. Soc. Am. **135**(4), 2012–2025 (2014)
10. Pulkki, V.: Virtual sound source positioning using vector base amplitude panning. J. Audio Eng. Soc. **45**(6), 456–466 (1997)
11. Pulkki, V.: Localization of amplitude-panned virtual sources II: two-and three-dimensional panning. J. Audio Eng. Soc. **49**(9), 753–767 (2001)
12. Wang, S., et al.: Sound intensity and particle velocity based three-dimensional panning methods by five loudspeakers. In: Proceedings of 2013 IEEE International Conference on Multimedia and Expo, San Jose, CA, United States (2013)
13. Wang, S., et al.: Three-dimensional panning by four loudspeakers and its solution. In: Proceedings of 2014 IEEE International Conference on Multimedia and Expo Workshops, Chengdu, China (2014)
14. Wang, S., Hu, R., Chen, S., Wang, X., Yang, Y., Tu, W.: 3D panning based sound field enhancement method for Ambisonics. In: Ho, Y.-S., Sang, J., Ro, Y.M., Kim, J., Wu, F. (eds.) PCM 2015. LNCS, vol. 9314, pp. 135–145. Springer, Heidelberg (2015). doi:10.1007/978-3-319-24075-6_14
15. Ando, A.: Conversion of multichannel sound signal maintaining physical properties of sound in reproduced sound field. IEEE Trans. Audio Speech Lang. Process. **19**(6), 1467–1475 (2011)
16. Wang, S., et al.: A down-mixing method for 22.2 multichannel system reproduction. In: Proceedings of 2015 IEEE International Conference on Acoustics, Speech and Signal Processing, South Brisbane, Austrilian (2015)
17. Seo, J., et al.: 21-channel surround system based on physical reconstruction of a three dimensional target sound field. In: Proceedings of 128th International Convention of the Audio Engineering Society, London (2010)
18. Shin, M., et al.: Control of velocity for sound field reproduction. In: Proceedings of 52nd International Conference of Audio Engineering Society, Guildford, United Kingdom (2013)
19. Shin, M., et al.: Velocity controlled sound field reproduction by non-uniformly spaced loudspeakers. J. Sound Vib. **370**, 444–464 (2016)
20. Nocedal, J., Wright, S.J.: Numerical Optimization. Spring, Berlin (2000)

A Comparison of Approaches for Automated Text Extraction from Scholarly Figures

Falk Böschen[1]([⊠]) and Ansgar Scherp[1,2]([⊠])

[1] Kiel University, Kiel, Germany
{fboe,asc}@informatik.uni-kiel.de
[2] ZBW - Leibniz Information Centre for Economics, Kiel, Germany
a.scherp@zbw.eu

Abstract. So far, there has not been a comparative evaluation of different approaches for text extraction from scholarly figures. In order to fill this gap, we have defined a generic pipeline for text extraction that abstracts from the existing approaches as documented in the literature. In this paper, we use this generic pipeline to systematically evaluate and compare 32 configurations for text extraction over four datasets of scholarly figures of different origin and characteristics. In total, our experiments have been run over more than 400 manually labeled figures. The experimental results show that the approach BS-4OS results in the best F-measure of 0.67 for the Text Location Detection and the best average Levenshtein Distance of 4.71 between the recognized text and the gold standard on all four datasets using the Ocropy OCR engine.

Keywords: Scholarly figures · Text extraction · Comparison

1 Introduction

Scholarly figures are data visualizations in scientific papers such as bar charts, line charts, and scatter plots [7]. Many researchers use a semi-supervised text extraction approach [6,18]. However, semi-supervised approaches do not scale with the amount of scientific literature published today. Thus, unsupervised methods are needed to address the task of text extraction from scholarly figures. This task is challenging due to the heterogeneity in the appearances of the scholarly figures such as varying colors, font sizes, and text orientations. Nevertheless, extracting text from scholarly figures provides additional information that is not contained in the body text [4]. To the best of our knowledge, no comparison of the different approaches for text extraction from scholarly figures has been conducted so far.

Based on the related work, we have defined a generic pipeline of six sequential steps that abstracts from the various works on text extraction from scholarly figures. We have re-implemented and systematically evaluated the most relevant approaches for text extraction from scholarly figures as described in the literature. In total, 32 configurations of the generic pipeline have been investigated.

© Springer International Publishing AG 2017
L. Amsaleg et al. (Eds.): MMM 2017, Part I, LNCS 10132, pp. 15–27, 2017.
DOI: 10.1007/978-3-319-51811-4_2

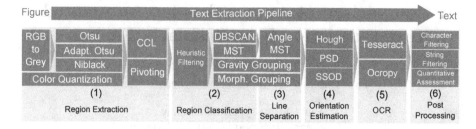

Fig. 1. Generic pipeline for text extraction from figures abstracted from the literature

Figure 1 shows the pipeline and the investigated methods for each step. We assess each pipeline configuration with regard to the accuracy of the text location detection via precision, recall, and F1-measure. In addition, we evaluate the text recognition quality using Levenshtein distance based on the evaluation methodology of the Born-Digital Image Track of the ICDAR Robust Reading Competition. In summary, the contributions of the paper are: (i) A systematic comparison of in total 32 configurations of a generic pipeline for text extraction from scholarly figures. Each configuration consists of a combination of six to nine methods from a total of 21 different methods that we have implemented and evaluated. (ii) We make available four manually labeled datasets of scholarly figures that allow reproducing and extending our results[1]. (iii) Furthermore, we make available the implementation of our generic pipeline which allows to use it on other datasets.

2   Related Work

An early work on text extraction from scholarly figures is by Huang et al. [9]. They use connected component labeling and a series of filters to extract regions from the figure that represents text. Subsequently, text lines are found by using a derivation of Newton's formula for "Gravity" from classical physics and OCR is applied. Sas and Zolnierek [17] propose a three-stage approach for text extraction from figures. Their approach binarizes the figure, applies connected component labeling, and filters the extracted regions using pre-defined thresholds. Tesseract[2] is used for OCR and after a rotation of 90° the process is repeated. Finally, we have developed a pipeline called TX for unsupervised text extraction from scholarly figures [2,3]. The pipeline combines an adaptive binarization method with connected component labeling and DBSCAN clustering to find text. A minimum spanning tree algorithm is used to estimate text lines followed by a Hough transformation for calculating the orientation, followed by OCR with Tesseract. A recent approach for semi-automatic text extraction from cartographic maps is proposed by Chiang et al. [6]. Cartographic maps use text

---

[1] http://www.kd.informatik.uni-kiel.de/en/research/software/text-extraction.
[2] https://github.com/tesseract-ocr/.

elements for city and street names, regions, and landmarks. In contrast to other approaches, Chiang et al. apply a color quantization algorithm and separate the text in the map from the rest using a semi-automatic extraction that requires a positive and a negative example for each text color. Text lines are detected using morphological operators and recognized using the commercial OCR engine AbbyyFineReader[3]. A text detection algorithm for biomedical images was proposed by Xu and Krauthammer [19] as part of the Yale Image Finder. Their pivoting algorithm uses vertical and horizontal histogram projection analysis to recursively split the image while classifying each region into text or non-text. Lu et al. [14] developed a retrieval engine for scholarly figures in chemistry. Their system works only on 2D plots and uses connected component labeling and fuzzy rules. Another approach for text extraction from figures by Jayant et al. [10] uses classic connected component labeling, support vector machines, minimum spanning trees, Adobe Photoshop (for preprocessing) as well as commercial OCR engines. They extract figures from books and their approach makes the assumptions that these figures have the same style throughout a book.

3   Pipeline Structure

Based on the discussion of the related work, we derived a generic pipeline for text extraction from scholarly figures as shown in Fig. 1. The pipeline consists of six steps and can be implemented through different methods, which are described below. This allows to create different configurations of the pipeline and to conduct a fair comparison of these configurations.

The first step of the pipeline takes a scholarly figure (color raster image) as input. The figure is converted into a binary image using either Color Quantization [5], by reducing the number of colors in an image and taking each resulting color channel as a separate binary image, or by converting the image to greyscale using the formula $Y = 0.2126R + 0.7152G + 0.0722B$ and subsequently applying a binarization method. For binarization, we use Otsu's Method [15], which finds the binarization threshold by maximizing the intra-class variance, Niblack's Method [13] which is often used for document image binarization, and Adaptive Otsu Binarization [3], which hierarchically applies Otsu's Method to adapt to local inhomogenities. The output of the first step is a set of regions, where each region is a set of connected pixels. They can be extracted using classic Connected Component Labeling (CCL) [16], which iterates over the pixel of an image and connects adjacent foreground pixel into regions. Another option is the Pivoting Histogram Projection method [19], which iteratively splits the binary image by analyzing the horizontal and vertical projection profiles. The second step takes these regions as input and computes a feature vector for each region, consisting of coordinates of the center of mass, dimension, and area occupation, to classify them into text or graphics. Heuristic Filtering [17] can be applied, prior to more complex algorithms, to preprocess the set of regions and remove outliers. The

---

[3] http://www.abbyy.com/ocr-sdk/.

classification of the remaining regions can be achieved using clustering methods like DBSCAN [3], since text should be more dense in the feature space, or Minimum Spanning Tree (MST) clustering [10]. Other approaches are Grouping Rules based on Newtons Gravity Formula [9] from classical physics or the Morphological Method [6], which uses morphological operators to merge regions on pixel level. Subsequently, the generated sets of regions that are classified as text are fed into the third pipeline step to determine individual lines of text if necessary. For this step, we have only found one method in the literature, the Angle-Based MST clustering [3]. It computes a MST on the centers of mass of the regions and removes those edges that are not inside a predefined range of 60° around the main orientation. The fourth step of the pipeline computes the orientation for each text line using one of the following methods: The Hough Transformation [3] can be used on the centers of mass of a text line's regions to transform them into Hough space, where the maximal value determines the orientation. A different option is to minimize the Perpendicular Squared Distance of the bounding box of a text line to identify its orientation [10]. The third option is the Single String Orientation Detection algorithm [6] which determines the text line orientation using morphological operators. In the fifth step, existing OCR engines are used to recognize the horizontal text lines. We have evaluated the Tesseract OCR engine and Ocropy[4], since both are freely available, frequently updated, and allow to reproduce our results without limitations. We used the English language models that are provided by the OCR engines and we deactivated any kind of layout analysis. The recognized text is post-processed in the sixth and last step of the pipeline. Here, we apply either Special Character Filtering that removes all special characters from the text, since they often appear when text was incorrectly recognized, Special Character Filtering per String [17] that removes complete text lines, if they contain too many special characters, or Quantitative OCR Assessment [10]. The latter analyzes the difference between the number of characters (regions) that went into the OCR process and the number of recognized characters in order to decide whether to discard a text line.

4   Pipeline Configurations

From the methods defined in the previous section, one can create various pipeline configurations. Some methods are restricted in how they can be combined as illustrated in Fig. 1.

Configurations Based on the Discussion of the State-of-the-art in Sect. 2: Each of the seven configurations is identified by (x), an acronym created from the contributing author(s). The first configuration (SZ13) is inspired by the work of Sas and Zolnierek [17]. It uses Otsu's method for binarization, followed by CCL. Subsequently, it applies heuristic filtering similar to the original approach. The decision tree used by Sas and Zolnierek is replaced by the line generation

---

[4] https://github.com/tmbdev/ocropy.

approach based on MST. Since the original work by Sas and Zolnierek does not include a method for orientation estimation, we do not use any replacement in step 4. Tesseract is used as OCR engine, since it was also used in the original paper. In the post processing step, all strings are removed that contain too many special characters.

The second configuration (Hu05) is based on the work of Huang et al. [9]. After region extraction using Otsu binarization and CCL, the Heuristic Filter method is applied, and the regions are grouped using the Gravity method. Finally, the grouped regions are processed with Tesseract.

Based on the work of Jayant et al. [10], the configuration (Ja07) starts with Otsu's method and CCL. Subsequently, it clusters the regions using a MST and approximates the orientation by minimizing the perpendicular squared distance. Text recognition is achieved by applying Tesseract.

Different from the previous configurations, the fourth configuration (CK15) – inspired by Chiang and Knoblock [6] – uses Color Quantization to generate multiple binary images, followed by a CCL. Subsequently, it applies heuristic filtering and Morphological Clustering on the regions. This step differs from the original paper, where the relevant color levels were manually selected. Thus, we assess all extracted binary images. The orientation of each cluster is estimated using the SSOD method, followed by Tesseract OCR, and quantitative post-processing.

Similar to the previous pipeline configuration, the fifth configuration (Fr15), inspired by Fraz et al. [8], starts with Color Quantization and CCL. The original approach uses a supervised SVM to form words, which we replaced with unsupervised methods from our methods set. The extracted regions are filtered and DBSCAN is applied, followed by a MST clustering into text lines. The orientation of each text line is calculated using Hough method and the text is recognized using Tesseract.

All configurations so far use CCL to extract regions. The sixth configuration (XK10), motivated by Xu and Krauthammer [19], uses the pivoting algorithm after binarization with adaptive Otsu. The regions are filtered using heuristics and grouped into lines using DBSCAN and MST. This differs from the original work, which only applied heuristic filtering to remove the graphic regions. The reason behind this is that the authors only aimed at finding text regions and not to recognize the text. Thus, we filled the rest of the pipeline steps with suitable methods. The orientation of each line is estimated via Hough and OCR is conducted with Tesseract.

Finally, configuration (BS15) resembles our own work [3]. It uses adaptive Otsu for binarization and CCL for region extraction. Heuristic Filtering is applied on the regions and DBSCAN groups them into text elements. Text lines are generated using the angle-based MST approach and the orientation of each line is estimated via Hough transformation, before applying Tesseract's OCR.

Influence of Individual Methods: In order to evaluate the influence of the individual methods, we chose the pipeline configuration (BS15) as basis for systematical modification, since our evaluation showed that it produces the best results, as reported in Sect. 6. The systematic modifications are organized along the six steps of the generic pipeline in Fig. 1. Each of the systematic configurations has an identifier (BS-XYZ) based on the original configuration, where X is a number that refers to the associated pipeline step and YZ uniquely identifies the method. The systematically modified configurations are described below. *Modifications of Step (1):* The binarization and region extraction is evaluated with the following configurations: (BS-1NC) differs from (BS15) by using Niblack instead of adaptive Otsu for binarization. Configuration (BS-1OC) uses the third option for binarization, Otsu's method. Color quantization is combined with the pivoting region extraction in (BS-1QP). *Modification over Steps (2) and (3):* The next step is the region classification and generation of text lines. Configuration (BS-2nF) differs from the base configuration by not applying the optional heuristic filtering method. Configuration (BS-2CG) uses the Gravity Grouping instead of DBSCAN and MST. Configuration (BS-2CM) applies MST to cluster regions and create text lines. Morphological text line generation is used in configuration (BS-23M). *Modifications of Step (4):* The following two configurations assess the methods for estimating the orientation of a text line: Configuration (BS-4OP) uses the Perpendicular Squared Distance method and configuration (BS-4OS) uses the Single String Orientation Detection method to estimate the orientation. *Modifications of Step (5):* For all configurations, both OCR engines are used to generate the results. The identifier of a configuration is extended to (BS-XYZ-T) or (BS-XYZ-O), when referencing the configurations that use Tesseract or Ocropy, respectively. Furthermore, we assess the direct impact of the OCR engine on the recognition results with configuration (BS15-O), which only differs with respect to the OCR method from the base configuration by using the Ocropy OCR engine instead of Tesseract. *Modifications of Step (6):* The last step of the pipeline is the post-processing. We use three configurations to evaluate the different post-processing methods: Configuration (BS-6PC) uses the Special Character Filter method for post-processing. Configuration (BS-6PS) uses the String Filter method for post-processing. Configuration (BS-6PQ) uses the Quantitative Assessment method for post-processing.

5   Evaluation

Datasets: We have used four datasets of varying origin and characteristics with in total 441 figures in our evaluation. We have created the **EconBiz** dataset, a corpus of 121 scholarly figures from the economics domain. We obtained these figures from a corpus of 288,000 open access publications from EconBiz[5] by extracting all images, filtering them by size and other constraints, and randomly selecting the subset of 121 figures. The dataset resembles a wide variety of scholarly figures from bar charts to maps. The figures were manually labeled to create the necessary gold standard information. We manually labeled the **DeGruyter** dataset as well, which comprises scholarly figures from books provided by DeGruyter[6]

---

[5] https://www.econbiz.de/.
[6] http://www.degruyter.com/.

under a creative commons license[7]. We selected ten books, mostly from the chemistry domain, which contain figures with English text and selected 120 figures randomly from these books. The gold standard for these figures was created using the same tool which has been used for the creation of the EconBiz dataset. The Chart Image Dataset[8] consists of two subsets. The CHIME-R dataset comprises 115 real images that were collected on the Internet or scanned from paper. It has mostly bar charts and few pie charts and line charts. The gold standard was created by Yang [20]. The CHIME-S dataset consists of 85 synthetically generated images. This set mainly contains line charts and pie charts and few bar charts. The gold standard was created by Jiuzhou [11].

We have also looked at ImageNet, TREC, ImageClef and ICDAR datasets. But none of them can be used to evaluate the specific challenges of scholarly figures. They either do not have the necessary ground truth information about the contained text or the dataset does not consist of scholarly figures. But we adopted the evaluation scheme of the Born-Digital Images track of the ICDAR Robust Reading Competition (RRC) [12], which is described below.

Procedure: We have selected three measures to evaluate the pipeline configurations and compare their results. Our gold standard consists of text elements which represent single lines of text taken from a scholarly figure. Each text line consists of one or multiple words which are separated by blank space. Each word may consist of any combination of characters and numbers. Every text line is defined by a specific position, size, and orientation. Each pipeline configuration generates a set of text line elements as well. These text lines need to be matched to the gold standard. Since we do not have pixel information per character, we match the extraction results with the gold standard by using the bounding boxes. This is based on the first evaluation task of the ICDAR RRC and evaluates the text localization on text line level. We iterate over all text lines in the gold standard and take all matches that are above the so-called intersection threshold. Our matching procedure calculates the intersection area between all pairs of the pipeline output and gold standard text lines. If the intersection comprises at least ten percent of the combined area of both text elements, than it is considered a match. This reduces the error introduced through elements which are an incorrect match and only have a small overlap with the gold standard. But it still allows to handle text lines that are broken into multiple parts. We look at each gold standard element and take all elements from the pipeline as matches that are above the intersection threshold. Thus, a gold standard element can have multiple matching elements and an element from the pipeline can be assigned to multiple elements from the gold standard if it fulfills the matching constraint for each match. We have defined three measures to assess these matches. The first two measures analyze the text localization. The third measure compares the recognized text, similar to the word recognition task of the ICDAR RRC, although we compare text lines and not individual words. First, we evaluate how accurate the configurations are at the Text Location Detection.

---

[7] http://www.degruyter.com/dg/page/open-access-policy.

[8] https://www.comp.nus.edu.sg/~tancl/ChartImageDataset.htm.

If at least one match is found for an element from the gold standard set, it counts as a true positive, regardless of what text was recognized. If no match was found, it is considered as false negative. A false positive is an element from the pipeline output which has no match. From these values, we compute precision, recall, and F1-measure. This measure is a binary evaluation and assesses only whether a match to an element exists or not. In addition, we report the Element Ratio (ER) which is the number of elements recognized by the pipeline divided by the number of elements in the gold standard and the Matched Element Ratio (MER) which is the number of matched items from the pipeline divided by the number of elements of the gold standard. These ratios give an idea whether gold standard elements get matched by multiple elements and whether the configuration tends to find more elements or less elements than it actually should find.

Second, we investigate the matching in more detail by assessing the Text Element Coverage. For each gold standard text element, we take the pixel of the bounding boxes and compute their overlap to calculate precision, recall, and F1-measure over all of its matches. The true positives in this case are the overlapping pixel and the false positives are those pixel from the text elements from the pipeline which are not overlapping. The false negatives are the pixels of the gold standard element which were not covered by a text element from the pipeline. The values are averaged over all gold standard text elements in a figure.

Third, we assess the Text Recognition Quality by computing the Levenshtein distance between the extracted text and the gold standard. We calculate the distance for each match and report the average for the whole figure. Since multiple text elements from the pipeline can be matched to a gold standard text line, we have to combine their text into one string. We combine the elements using their position information. Besides a (local) Levenshtein Distance per match, we also compute a global Levenshtein distance over all extracted text. This means that for each figure, we combine all characters from the text elements of the gold standard and add them to one string. Likewise, we create a string from the text elements extracted by the pipeline. The characters in both strings are sorted alphabetically and we compute the Levenshtein Distance between these strings. This approximates the overall number of operations needed to match the strings without considering position information. Since the global Levenshtein Distance depends on the number of characters inside a figure, we normalize it to an operations per character (OPC) score, which is computed by dividing the global Levenshtein Distance by the number of characters in the gold standard. This makes the results comparable across scholarly figures with different amounts of characters.

6 Results

We have executed all configurations listed in Sect. 4 over the datasets described in Sect. 5. For reasons of simplicity, we are only reporting the average values for Text Location Detection, Text Element Coverage, and Text Recognition Quality over all datasets. The detailed results per dataset can be found in our Technical Report [1]. We compute the average Precision/Recall/F1-measure over the

Table 1. Average Precision (Pr), Recall (Re), and F1 values for Text Location Detection and Text Element Coverage, Element Ratio (ER), and Matched Element Ratio (MER) over all datasets for configurations from the literature

Config.	Text Location Detection					Text Element Coverage		
	Pr	Re	F1 (SD)	ER	MER	Pr	Re	F1 (SD)
SZ13	0.63	0.47	0.54 (0.23)	0.80	0.59	0.52	**0.59**	0.47 (0.21)
Hu05	0.61	0.43	0.48 (0.28)	0.77	0.57	**0.79**	0.54	**0.57 (0.20)**
Ja07	0.59	0.45	0.49 (0.28)	0.83	0.51	0.41	0.32	0.32 (0.21)
BS15	0.66	**0.55**	**0.58 (0.25)**	1.04	0.69	0.60	0.49	0.50 (0.24)
CK15	0.52	0.50	0.53 (0.23)	1.37	0.60	0.53	0.41	0.42 (0.21)
Fr15	0.55	0.51	0.54 (0.25)	1.44	**0.72**	0.65	0.54	0.54 (0.23)
XK10	**0.73**	0.35	0.45 (0.26)	0.43	0.39	0.33	0.34	0.30 (0.22)

elements of each figure. We report the average Precision/Recall/F1-measure in terms of mean and standard deviation over all figures. The local Levenshtein distance is reported as the average of the mean values per figure and the average standard deviation. The global Levenshtein distance is defined by the mean and standard deviation over all figures and the average of the normalized OPC score.

First, we report the results of the configurations from the literature. Subsequently, we present the results for the systematically modified configurations. The Text Location Detection and Text Element Coverage results for the configurations from the literature computed over all datasets are reported in Table 1. The best result, based on the F1-measure, is achieved by configuration (BS15) with a F1-measure of 0.58. The coverage assessment in Table 1 shows the best precision of 0.79 for (Hu05), the best recall of 0.59 for (SZ13), and the best F1-measure of 0.57 for (Hu05). The text recognition quality is presented in Table 2. We obtain the best results with (BS15) with 0.67 operations per character (OPC), an average global Levenshtein of 108.81, and an average local

Table 2. Average local Levenshtein (L), global Levenshtein (G), and Operations Per Character (OPC) over all datasets for the configurations from the literature using Tesseract

Config.	$AVG_L(SD)$	$AVG_G(SD)$	OPC
SZ13	6.67 (4.82)	122.28 (141.03)	0.70
Hu05	6.65 (5.41)	126.35 (138.95)	0.71
Ja07	7.92 (5.56)	150.25 (140.59)	1.13
BS15	6.23 (4.93)	**108.81 (108.53)**	**0.67**
CK15	**6.07 (5.08)**	120.12 (125.87)	0.71
Fr15	6.72 (6.02)	135.64 (201.31)	0.85
XK10	7.06 (5.41)	125.45 (134.88)	0.74

Table 3. Systematically modified configurations: Average Precision (Pr), Recall (Re), and F1 values for Text Location Detection and Text Element Coverage, Element Ratio (ER), and Matched Element Ratio (MER) over all datasets

Config.	Text Location Detection					Text Element Coverage		
	Pr	*Re*	*F1 (SD)*	*ER*	*MER*	*Pr*	*Re*	*F1 (SD)*
BS15	0.66	0.55	0.58 (0.25)	1.04	0.69	0.60	0.49	0.50 (0.24)
BS-1NC	0.64	0.52	0.57 (0.25)	0.96	0.64	0.59	0.44	0.47 (0.24)
BS-1OC	0.67	0.40	0.49 (0.26)	0.74	0.53	0.46	0.40	0.38 (0.26)
BS-1QP	0.61	0.44	0.48 (0.25)	0.96	0.75	0.41	0.57	0.42 (0.23)
BS-2nF	0.60	0.46	0.51 (0.23)	0.86	0.52	0.59	0.54	0.50 (0.21)
BS-2CG	0.62	0.50	0.55 (0.27)	0.90	0.64	0.76	0.54	0.57 (0.20)
BS-2CM	0.61	0.54	0.59 (0.25)	1.19	0.74	0.57	0.47	0.47 (0.24)
BS-23M	0.67	0.55	0.62 (0.23)	1.08	0.65	0.60	0.47	0.48 (0.22)
BS-4OP	0.62	0.53	0.57 (0.24)	**1.01**	0.66	0.49	0.40	0.41 (0.20)
BS-4OS	0.67	**0.63**	**0.67 (0.22)**	1.27	**0.88**	**0.77**	**0.63**	**0.65 (0.17)**
BS-6PC	**0.69**	0.54	0.59 (0.25)	0.97	0.70	0.59	0.49	0.49 (0.24)
BS-6PS	0.67	0.55	0.60 (0.25)	**1.01**	0.69	0.59	0.49	0.49 (0.24)
BS-6PQ	0.66	0.38	0.48 (0.25)	0.60	0.43	0.39	0.29	0.31 (0.21)

Table 4. Average local Levenshtein (L), global Levenshtein (G), and Operations Per Character (OPC) over all datasets for the systematic configurations

Config.	Tesseract			Ocropy		
	$AVG_L(SD)$	$AVG_G(SD)$	OPC	$AVG_L(SD)$	$AVG_G(SD)$	OPC
BS15	6.23 (4.93)	108.81 (108.53)	0.67	5.47 (4.98)	108.55 (106.64)	0.64
BS-1NC	6.27 (4.95)	117.58 (124.23)	0.69	5.70 (5.09)	117.46 (128.73)	0.66
BS-1OC	6.55 (5.06)	131.58 (142.74)	0.75	6.16 (5.21)	131.39 (143.16)	0.73
BS-1QP	8.31 (6.14)	154.54 (168.10)	1.09	7.06 (5.62)	136.40 (132.05)	0.82
BS-2nF	6.55 (4.94)	111.30 (105.13)	0.75	6.29 (5.50)	120.71 (109.18)	0.76
BS-2CG	6.68 (5.65)	108.86 (102.93)	0.66	6.22 (5.75)	130.21 (127.87)	0.69
BS-2CM	6.30 (5.29)	115.43 (113.79)	0.69	5.85 (5.34)	110.74 (107.23)	0.67
BS-23M	6.15 (5.12)	104.61 (105.97)	0.63	5.52 (5.10)	106.71 (104.05)	0.64
BS-4OP	8.30 (5.59)	147.91 (129.55)	1.04	7.23 (5.60)	135.21 (122.48)	0.85
BS-4OS	**5.47 (4.39)**	**96.29 (99.44)**	**0.58**	**4.71 (4.66)**	**95.49 (94.80)**	**0.53**
BS-6PC	5.96 (4.88)	105.50 (107.16)	0.61	5.46 (5.00)	109.07 (104.57)	0.63
BS-6PS	6.20 (4.90)	108.06 (109.38)	0.64	5.45 (4.96)	106.38 (103.29)	0.63
BS-6PQ	6.07 (5.03)	120.78 (122.44)	0.67	5.79 (4.97)	126.92 (124.06)	0.71

Levenshtein of 6.23. The best local Levenshtein of 6.07 is achieved by configuration (CK15). For the systematically modified configurations, Table 3 shows the Text Location Detection results and the Text Element Coverage. Table 4 shows the Text Recognition Quality. The best location detection F1-measure of 0.67 is achieved by (BS-4OS), which is also supported by the coverage assessment with the highest F1-measure of 0.65. Configuration (BS-4OS-O) also produces the best text recognition results with an average local Levenshtein of 4.71 and an OPC of 0.53. In addition, configuration (BS-4OS-O) shows the best results of 95.49 for the average global Levenshtein Distance. Comparing the different, systematically modified configurations per step of the pipeline shows that the only major improvement is achieved by (BS-4OS). Please note, a performance analysis of the different configurations can be found in our Technical Report [1].

7 Discussion

Comparing the different configurations from the literature shows that the best performing configuration is (BS15). A possible reason is that our pipeline does not make many assumptions about the figures, e.g. figure type, font, or color. Thus performing better on the heterogeneous datasets. In the following, we will discuss the results for the individual pipeline steps based on the results from the systematically modified configurations. Comparing the configurations for the first pipeline step leads to the conclusion that the adaptive binarization works best, because it can adapt to local variations of the appearance in a figure. Otsu's method is too simple and Niblack's method is more suited for document images which have fewer color variations. The lower results for the pivoting algorithm can be explained with the larger regions and the possibility that a region can be a mixture of text and graphic elements due to the only horizontal and vertical subdivision. Looking at step 2 and 3 of the pipeline, only the morphological clustering shows slightly better results than the DBSCAN-MST combination, most likely due to its processing on pixel level. The overall best results, when also considering the systematic configurations, are achieved by (BS-4OS). This can be explained by the fact that the orientation estimation via Hough works on the centers of mass of character regions, which is an aggregated region representation, while the SSOD in (BS-4OS) computes the orientation on the original pixels. Thus, it avoids a possible error, which could be induced by the pixel aggregation. When comparing the OCR engines from step 5, Ocropy generally produces better results than Tesseract. Ocropy seems to be more conservative, having built in much more restrictions about what input to accept and when to execute the OCR. Furthermore, each OCR engine comes with its own English language model and we did not evaluate their influence. The methods for post-processing do not improve the results. One reason might be the simplicity of methods. Thus, some more advanced techniques may be developed in the future. Overall, there are many more options for the different pipeline steps, e.g., other binarization methods, different clustering algorithms, or post-processing methods that could be used. However, we made a selection of relevant approaches and methods to limit the combinatorial complexity. On the other hand, as stated in the introduction, we provide the datasets and the implementation of the generic pipeline that was used in our experiment to the public. This allows for integrating and comparing new methods as well as the reproduction of our results.

Acknowledgement. This research was co-financed by the EU H2020 project MOV-ING (http://www.moving-project.eu/) under contract no 693092.

References

1. Böschen, F., Scherp, A.: A systematic comparison of different approaches for unsupervised extraction of text from scholarly figures [extended report]. Technical report 1607, Christian-Albrechts-Universität zu Kiel (2016). http://www.uni-kiel.de/journals/receive/jportal_jparticle_00000290
2. Böschen, F., Scherp, A.: Formalization and preliminary evaluation of a pipeline for text extraction from infographics. In: Bergmann, R., Görg, S., Müller, G. (eds.) LWA 2015 Workshop: KDML, pp. 20–31. CEUR (2015)
3. Böschen, F., Scherp, A.: Multi-oriented text extraction from information graphics. In: DocEng, pp. 35–38. ACM (2015)
4. Carberry, S., Elzer, S., Demir, S.: Information graphics: an untapped resource for digital libraries. In: SIGIR, pp. 581–588. ACM (2006)
5. Chiang, Y., Knoblock, C.A.: A general approach for extracting road vector data from raster maps. IJDAR **16**(1), 55–81 (2013)
6. Chiang, Y., Knoblock, C.A.: Recognizing text in raster maps. GeoInformatica **19**(1), 1–27 (2015)
7. Choudhury, S.R., Giles, C.L.: An architecture for information extraction from figures in digital libraries. In: WWW, pp. 667–672 (2015)
8. Fraz, M., Sarfraz, M.S., Edirisinghe, E.A.: Exploiting colour information for better scene text detection and recognition. IJDAR **18**(2), 153–167 (2015)
9. Huang, W., Tan, C.L., Leow, W.K.: Associating text and graphics for scientific chart understanding. In: ICDAR, pp. 580–584. IEEE Computer Society (2005)
10. Jayant, C., Renzelmann, M., Wen, D., Krisnandi, S., Ladner, R.E., Comden, D.: Automated tactile graphics translation: in the field. In: ASSETS, pp. 75–82 (2007)
11. Jiuzhou, Z.: Creation of synthetic chart image database with ground truth. Honors year project report, National University of Singapore (2006). https://www.comp.nus.edu.sg/~tancl/ChartImageDatabase/Report_Zhaojiuzhou.pdf
12. Karatzas, D., Gomez-Bigorda, L., Nicolaou, A., Ghosh, S.K., Bagdanov, A.D., Iwamura, M., Matas, J., Neumann, L., Chandrasekhar, V.R., Lu, S., Shafait, F., Uchida, S., Valveny, E.: ICDAR 2015 competition on robust reading. In: ICDAR, 23–26 August 2015, pp. 1156–1160. IEEE Computer Society (2015)
13. Khurshid, K., Siddiqi, I., Faure, C., Vincent, N.: Comparison of Niblack inspired binarization methods for ancient documents. In: Document Recognition and Retrieval (DRR), pp. 1–10. SPIE (2009)
14. Lu, X., Kataria, S., Brouwer, W.J., Wang, J.Z., Mitra, P., Giles, C.L.: Automated analysis of images in documents for intelligent document search. IJDAR **12**(2), 65–81 (2009)
15. Otsu, N.: A threshold selection method from gray-level histograms. TSMC **9**(1), 62–66 (1979)

16. Samet, H., Tamminen, M.: Efficient component labeling of images of arbitrary dimension represented by linear bintrees. IEEE TPAMI **10**(4), 579–586 (1988)
17. Sas, J., Zolnierek, A.: Three-stage method of text region extraction from diagram raster images. In: Burduk, R., Jackowski, K., Kurzynski, M., Wozniak, M., Zolnierek, A. (eds.) Proceedings of the 8th International Conference on Computer Recognition Systems CORES 2013, vol. 226, pp. 527–538. Springer, Heidelberg (2013)
18. Savva, M., Kong, N., Chhajta, A., Fei-Fei, L., Agrawala, M., Heer, J.: ReVision: automated classification, analysis and redesign of chart images. In: UIST, pp. 393–402. ACM (2011)
19. Xu, S., Krauthammer, M.: A new pivoting and iterative text detection algorithm for biomedical images. J. Biomed. Inform. **43**, 924–931 (2010)
20. Yang, L., Huang, W., Tan, C.L.: Semi-automatic ground truth generation for chart image recognition. In: Bunke, H., Spitz, A.L. (eds.) DAS 2006. LNCS, vol. 3872, pp. 324–335. Springer, Heidelberg (2006). doi:10.1007/11669487_29

A Convolutional Neural Network Approach for Post-Processing in HEVC Intra Coding

Yuanying Dai, Dong Liu$^{(\boxtimes)}$, and Feng Wu

CAS Key Laboratory of Technology in Geo-Spatial Information Processing and
Application System, University of Science and Technology of China, Hefei, China
daiyy@mail.ustc.edu.cn, {dongeliu,fengwu}@ustc.edu.cn

Abstract. Lossy image and video compression algorithms yield visually
annoying artifacts including blocking, blurring, and ringing, especially at
low bit-rates. To reduce these artifacts, post-processing techniques have
been extensively studied. Recently, inspired by the great success of convolutional neural network (CNN) in computer vision, some researches were
performed on adopting CNN in post-processing, mostly for JPEG compressed images. In this paper, we present a CNN-based post-processing
algorithm for High Efficiency Video Coding (HEVC), the state-of-the-
art video coding standard. We redesign a Variable-filter-size Residue-
learning CNN (VRCNN) to improve the performance and to accelerate
network training. Experimental results show that using our VRCNN as
post-processing leads to on average 4.6% bit-rate reduction compared to
HEVC baseline. The VRCNN outperforms previously studied networks
in achieving higher bit-rate reduction, lower memory cost, and multiplied
computational speedup.

Keywords: Artifact reduction · Convolutional neural network (CNN) ·
High Efficiency Video Coding (HEVC) · Intra coding · Post-processing

1 Introduction

Lossy image and video compression algorithms, such as JPEG [16] and High
Efficiency Video Coding (HEVC) [13], by nature cause distortion and yield artifacts especially at low bit-rates. For example, due to block-based coding, there
are visible discontinuities at block boundaries in compressed images, which are
known as blocking artifacts; due to loss of high-frequency components, compressed images often become blurred than the original. Other artifacts include
ringing, color bias, and so on. These compression artifacts may severely decrease
the perceptual quality of reconstructed image or video, and thus how to reduce or
remove these artifacts is an important problem and has been extensively studied
in the literature.

In HEVC, the state-of-the-art video coding standard, there are two post-
processing techniques for artifact reduction, namely deblocking [11] and sample adaptive offset (SAO) [4]. The differences between deblocking and SAO are

© Springer International Publishing AG 2017
L. Amsaleg et al. (Eds.): MMM 2017, Part I, LNCS 10132, pp. 28–39, 2017.
DOI: 10.1007/978-3-319-51811-4_3

twofold. First, deblocking is specifically designed to reduce blocking artifacts, but SAO is designed for general compression artifacts. Second, deblocking does not require any additional bit, but SAO requires to transmit some additional bits for signaling the offset values. Both techniques contribute to the improvement of the visual quality of reconstructed video, and also help to improve the objective quality and equivalently achieve bit-rate saving.

Recently, convolutional neural network (CNN) achieved great success in high-level computer vision tasks such as image classification [9] and object detection [5]. Inspired by the success, it was also proposed to utilize CNN for low-level computer vision tasks such as super-resolution [3,8] and edge detection [18].

More recently, Dong et al. proposed an artifact reduction CNN (AR-CNN) [2] approach for reducing artifacts in JPEG compressed images. The AR-CNN is built upon their previously designed super-resolution CNN (SRCNN) [3], and reported to achieve more than 1 dB improvement over JPEG images. Wang et al. [17] investigated another network structure for JPEG artifact reduction. Furthermore, Park and Kim [12] proposed to utilize the SRCNN network to replace the deblocking or SAO in HEVC, and reported achieving bit-rate reduction. However, the results in [12] were achieved by training a network with several frames of a video sequence and then testing the network with the same sequence, which cannot reveal the generalizability of the trained network.

In this paper, we present a redesigned CNN for artifact reduction in HEVC intra coding. We propose to integrate variable filter size into the designed CNN to improve its performance. We also utilize the recently proposed residue learning technique [6] to accelerate the training of CNN. Moreover, we trained the network with a collection of natural images and tested the network with the standard video sequences, so as to demonstrate the generalizability of the network. Our proposed Variable-filter-size Residue-learning CNN (VRCNN) can be adopted as post-processing to replace deblocking and SAO, as it reduces general compression artifacts and requires no additional bit. Experimental results show that VRCNN achieves on average 4.6% bit-rate reduction compared to deblocking and SAO in HEVC baseline. The VRCNN also outperforms previously studied networks in achieving higher bit-rate reduction, lower memory cost, and multiplied computational speedup.

The remainder of this paper is organized as follows. Section 2 presents the details of the designed VRCNN. Section 3 discusses the details of training and using VRCNN. Section 4 gives out the experimental results, followed by conclusions in Sect. 5.

2 Our Designed CNN

Currently, there are several existing networks for artifact reduction: AR-CNN [2], D^3 [17], and SRCNN [12]. Note that AR-CNN is built upon SRCNN, and the SRCNN was originally designed for super-resolution [3]. The D^3 network was specifically designed for JPEG as it utilized the JPEG built-in 8×8 discrete cosine transform (DCT), and thus not suitable for HEVC which adopts variable block size transform. In the following, we first discuss on AR-CNN, and then presents our redesigned VRCNN.

2.1 AR-CNN

AR-CNN is a 4-layer fully convolutional neural network. It has no pooling or full-connection layer, so the output can be of the same size as the input given proper boundary condition (the boundary condition of convolutions will be discussed later). Denote the input by Y, the output of layer $i \in \{1, 2, 3, 4\}$ by $F_i(Y)$, and the final output by $F(Y) = F_4(Y)$, then the network can be represented as:

$$F_1(Y) = g(W_1 * Y + B_1) \tag{1}$$
$$F_i(Y) = g(W_i * F_{i-1}(Y) + B_i), i \in \{2, 3\} \tag{2}$$
$$F(Y) = W_4 * F_3(Y) + B_4 \tag{3}$$

where W_i and B_i are the weights and biases parameters of layer i, $*$ stands for convolution, and $g()$ is a non-linear mapping function. In recent CNNs, the rectified linear unit (ReLU) [10] is often adopted as the non-linear mapping, i.e. $g(x) = \max(0, x)$.

Table 1. The configuration of AR-CNN [2]

Layer	Layer 1	Layer 2	Layer 3	Layer 4
Filter size	9×9	7×7	1×1	5×5
# filters	64	32	16	1
# parameters	5184	100352	512	400
Total parameters	106448			

The four layers in AR-CNN are claimed to perform four steps of artifact reduction: feature extraction, feature enhancement, mapping, and reconstruction (as discussed in [2]). Accordingly, the configuration of AR-CNN is summarized in Table 1. Note that the amount of (convolutional) parameters in each layer is calculated as (number of filters in the last layer) × (number of filters in this layer) × (filter size).

2.2 VRCNN

Since AR-CNN is designed for JPEG, but our aim is to perform artifact reduction for HEVC, we redesign the CNN structure and name it VRCNN. The structure of VRCNN is shown in Fig. 1 and its configuration is given in Table 2. It is also a 4-layer fully convolutional neural network, like AR-CNN. We now discuss the differences between VRCNN and AR-CNN.

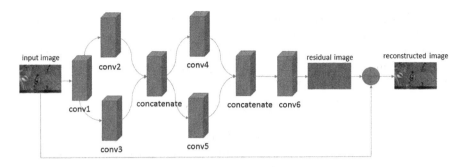

Fig. 1. The structure of VRCNN, a 4-layer fully convolutional neural network.

Table 2. The configuration of VRCNN

Layer	Layer 1	Layer 2		Layer 3		Layer 4
Conv. module	conv1	conv2	conv3	conv4	conv5	conv6
Filter size	5×5	5×5	3×3	3×3	1×1	3×3
# filters	64	16	32	16	32	1
# parameters	1600	25600	18432	6912	1536	432
Total parameters	54512					

The rationale cause of compression artifacts in JPEG and HEVC is the quantization of transformed coefficients. The transform is block wise, thus the quantization error of one coefficient affects only the pixels in the same block. As JPEG adopts fixed 8×8 DCT, but HEVC adopts variable block size transform[1], which shall be taken into account to reduce the quantization error. Therefore, we propose to adopt variable filter size in the second layer, because this layer is designed to make the "noisy" features "cleaner" [2]. Specifically, we replace the second layer of AR-CNN (fixed 7×7 filters) with the combination of 5×5 and 3×3 filters. The outputs of different-sized filters are concatenated to be fed into the next layer. Similarly, we also adopt variable filter size in the third layer that performs "restoration" of features [2]. The fixed 1×1 filters in AR-CNN are replaced by combination of 3×3 and 1×1 filters. Note that the first and the last layers of VRCNN do not use variable filter size, because these two layers perform feature extraction and final reconstruction, respectively [2], which are not affected by variable block size transform of HEVC.

The variable filter size technique, i.e. combination of filters of different sizes in one layer of CNN, has been proposed earlier in CNNs for image classification, e.g. the well-known GoogleNet [15], where different-sized filters are to provide multi-scale information of the input image. In our VRCNN, variable filter size is proposed to suit for HEVC variable block size transform, and thus used only in

---

[1] HEVC adopts 4×4, 8×8, 16×16, up to 32×32 DCT, and allows the choice of discrete sine transform (DST) at 4×4.

selected layers. To the best of our knowledge, VRCNN is the first network that uses variable filter size for artifact reduction.

In addition, we propose to integrate the recently developed residue learning technique [6] into VRCNN. That is, the output of the last layer is added back to the input, and the final output is:

$$F(\boldsymbol{Y}) = W_4 * F_3(\boldsymbol{Y}) + B_4 + \boldsymbol{Y} \tag{4}$$

In other words, the CNN is designed to learn the residue between output and input rather than directly learning the output. In the case of artifact reduction, the input (before filtering) and the output (after filtering) shall be similar to the other to a large extent, therefore, learning the difference between them can be easier and more robust. Our empirical study indeed confirms that residue learning converges much faster. Note that residue learning is also a common strategy in super-resolution with or without CNN [8,14].

Last but not the least, to integrate CNN into the in-loop post-processing of HEVC, it is very important to control the network complexity. For that purpose, our designed VRCNN is greatly simplified than AR-CNN. Comparing Tables 1 and 2, VRCNN uses more filters, but at smaller sizes. As a result, the amount of parameters is greatly reduced in VRCNN. We notice that recent work on super-resolution also uses smaller filters but much more (20) layers [8], while VRCNN has 4 layers like AR-CNN.

3   Training and Using VRCNN

We propose to adopt VRCNN for post-processing in HEVC to replace the original deblocking and SAO. In order to make a fair comparison with the original deblocking and SAO, we train the VRCNN on a collection of natural images, and test it on the HEVC standard test sequences. The training and testing images have no overlap so as to demonstrate the generalizability of the trained network.

3.1   Training

An original image $\boldsymbol{X}_n$, where $n \in \{1, \ldots, N\}$ indexes each image, is compressed with HEVC intra coding, while turning off deblocking and SAO, and the compressed image is regarded as the input to VRCNN, i.e. $\boldsymbol{Y}_n$. The objective of training is to minimize the following loss function:

$$L(\Theta) = \frac{1}{N} \sum_{n=1}^{N} \|F(\boldsymbol{Y}_n|\Theta) - \boldsymbol{X}_n\|^2 \tag{5}$$

where Θ is the whole parameter set of VRCNN, including $W_i, B_i, i \in \{1, 2, 3, 4\}$. This loss is minimized using stochastic gradient descent with the standard back-propagation.

In order to accelerate the training, we also adopt the adjustable gradient clipping technique proposed in [8]. That is, the learning rate α is set large, but the actual gradient update is restricted to be in the range of $[-\tau/\alpha, \tau/\alpha]$ where τ is a constant (set to 0.01 in our experiments). The key idea beneath this technique is to clip the gradient when α is large, so as to avoid exploding. As training goes on, the learning rate α becomes smaller and then the range is too large to be actually used.

3.2 Using VRCNN

We integrate a trained VRCNN into HEVC intra coding. The deblocking and SAO are turned off, and the compressed intra frame is directly fed into the trained VRCNN, producing the final reconstructed frame. Unlike SAO, the VRCNN needs no additional bit, but still can reduce general compression artifacts as demonstrated by experimental results. Therefore, in all-intra coding setting, VRCNN can be made in-loop or out-of-loop. One remaining issue is the boundary condition for convolutions. In this work, we follow the practice in [8], i.e. padding zeros before each convolutional module so that the output is of the same size as the input. Zero-padding seems quite simple but works well in experiments.

4 Experimental Results

4.1 Implementation

We use the software Caffe [7] for training VRCNN as well as comparative networks on a NVIDIA Tesla K40C graphical processing unit (GPU). A collection of 400 natural images, i.e. the same set as that in [2], are used for training. Each original image is compressed by HEVC intra coding (deblocking and SAO turned off) at four different quantization parameters (QPs): 22, 27, 32, and 37. For each QP, a separate network is trained out. Only the luminance channel (i.e. Y out of YUV) is considered for training. Due to the limited memory of the GPU, we do not use the entire image as a sample. Instead, the original image X_n and compressed image Y_n are both divided into 35×35 sub-images without overlap. The corresponding pair of sub-images is regarded as a sample, so we have in total 46,784 training samples. Note that different from [2], we use zero padding before each convolution so that the output is of the same size as the input, and therefore the loss is computed over the entire sub-image.

During network training, the weights are initialized using the method in [6]. Training samples are randomly shuffled and the mini-batch size is 64. The momentum parameter is set to 0.9, and weight decay is 0.0001. The base learning rate is set to decay exponentially from 0.1 to 0.0001, changing every 40 epochs. Thus, in total the training takes 160 epochs and uses around 1.5 h on our GPU. The bias learning rate is set to 0.01, 0.01, and 0.1, for QP 27, 32, and 37, respectively. For QP 22, the network is not trained from scratch but rather fine-tuned from the network of QP 27. For this fine tuning, the base learning rate is 0.001, bias learning rate is 0.0001, and training finishes after 40 epochs.

For comparison, we also trained other two networks, AR-CNN [2] and VDSR [8], using the same training images. The AR-CNN is designed for JPEG artifact reduction, so we cannot reuse their trained network for HEVC, but we re-train the network from scratch using the source code provided by the authors. The training proceeds in 2,500,000 iterations and takes almost 5 days. The VDSR is proposed for super-resolution and claimed to outperform SRCNN (the basis of AR-CNN), so we also include it for comparison. We also re-train the network from scratch and manually tune the training hyper-parameters. The training takes about 6 h to finish 38,480 iterations. It can be observed that the training of our VRCNN and VDSR is significantly faster than that of AR-CNN, because both VRCNN and VDSR adopt residue learning. And the training of our VRCNN is also faster than VDSR since our network is much more simple.

After training, we integrate the network into HEVC reference software HM[2] and test on the HEVC standard test sequences. Five classes, 20 sequences are used for test. Class F is not used as it is screen content. For each sequence, only the first frame is used for test. Four QPs are tested: 22, 27, 32, 37, and for each QP the corresponding network is used. For AR-CNN and VDSR, we also train a separate network for each QP. Note that the training is performed on the luminance channel (Y) but the trained network is also used for chrominance channels (U and V).

Compared to [12], which uses the same sequences for training and test, our experiments can better reveal the generalizability of the trained network and make fair comparison with the original deblocking and SAO. Moreover, the results are not complete for all HEVC standard test sequences [12]. Therefore, we do not include its results in the following.

4.2 Comparison with HEVC Baseline

We first compare our VRCNN used as post-processing against the original deblocking and SAO. To evaluate the coding efficiency, we use the BD-rate measure [1] on luminance and chrominance channels independently. The results are summarized in Table 3. It can be observed that the VRCNN achieves significant bit-rate reduction on all the test sequences. For the luminance (Y), as high as 7.6% BD-rate is achieved on the RaceHorses sequence, and on average 4.6% BD-rate is achieved on all the sequences. For the chrominance (U and V), the BD-rate is more significant for several sequences, reaching as high as 11.5% on the RaceHorses sequence. Note that the network is trained only on the luminance channel, this result shows that the network can be readily used for the chrominance channels, too.

We also compare the visual quality of reconstructed images as shown in Fig. 2. A portion of each image is enlarged as inset in the bottom-right corner of each image. It can be observed that the image before post-processing Fig. 2(b) contains obvious blocking and ringing artifacts. The processed image

---

[2] HM   version   16.0,   https://hevc.hhi.fraunhofer.de/svn/svn_HEVCSoftware/tags/ HM-16.0/.

Table 3. The BD-rate results of our VRCNN compared to HEVC baseline

Class	Sequence	BD-rate		
		Y (%)	U (%)	V (%)
Class A	Traffic	−5.6	−3.5	−4.1
	PeopleOnStreet	−5.4	−5.9	−5.7
	Nebuta	−0.9	−4.9	−4.1
	SteamLocomotive	−1.9	−0.5	−0.3
Class B	Kimono	−2.5	−1.5	−1.4
	ParkScene	−4.4	−3.3	−2.5
	Cactus	−4.6	−3.9	−6.3
	BasketballDrive	−2.5	−3.7	−5.3
	BQTerrace	−2.6	−3.3	−3.0
Class C	BasketballDrill	−6.9	−5.8	−6.8
	BQMall	−5.1	−5.3	−5.3
	PartyScene	−3.6	−4.4	−4.4
	RaceHorses	−4.2	−6.7	−11.0
Class D	BasketballPass	−5.3	−4.4	−6.5
	BQSquare	−3.8	−4.2	−6.4
	BlowingBubbles	−4.9	−8.4	−7.9
	RaceHorses	−7.6	−8.5	−11.5
Class E	FourPeople	−7.0	−5.3	−5.2
	Johnny	−5.9	−5.0	−5.5
	KristenAndSara	−6.7	−6.1	−6.2
Class Summary	Class A	−3.5	−3.7	−3.6
	Class B	−3.3	−3.2	−3.7
	Class C	−5.0	−5.5	−6.9
	Class D	−5.4	−6.4	−8.1
	Class E	−6.5	−5.5	−5.6
Overall	**All**	**−4.6**	**−4.7**	**−5.5**

by HEVC baseline Fig. 2(c) greatly reduces blocking, but ringing is still visible. The processed image by VRCNN Fig. 2(f) suppresses all kinds of artifacts and produces better visual quality than Fig. 2(c).

4.3   Comparison with Other Networks

We also compare our VRCNN with AR-CNN and VDSR to demonstrate the advantage of our redesigned network structure. First, the coding efficiency of each network is evaluated using the BD-rate measure. The results are summarized in Table 4. It can be observed that AR-CNN performs slightly worse than

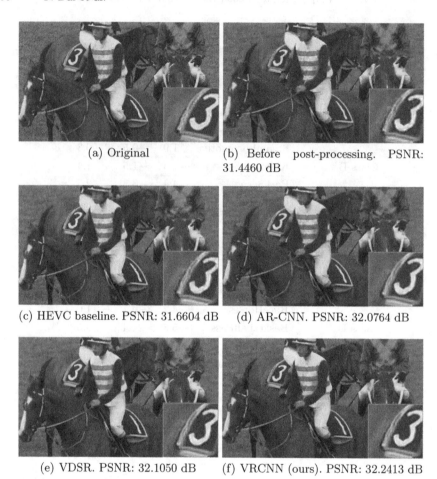

(a) Original

(b) Before post-processing. PSNR: 31.4460 dB

(c) HEVC baseline. PSNR: 31.6604 dB

(d) AR-CNN. PSNR: 32.0764 dB

(e) VDSR. PSNR: 32.1050 dB

(f) VRCNN (ours). PSNR: 32.2413 dB

Fig. 2. The first frame of RaceHorses, compressed at QP 37, and post-processed by HEVC baseline as well as different CNNs.

the original deblocking and SAO in HEVC baseline, but VDSR also demonstrates significant gain. Note that VDSR is proposed for super-resolution and claimed to outperform SRCNN, while AR-CNN is built upon SRCNN, this result is reasonable because VDSR is much deeper (20 layers) than AR-CNN (4 layers). However, our proposed VRCNN, being also 4-layer, outperforms AR-CNN significantly, and also outperforms VDSR slightly, in terms of BD-rate. Since our VRCNN features variable filter size and residue learning compared to AR-CNN, this result demonstrates that carefully designed shallow network may still be competitive with deep network for artifact reduction.

The reconstructed images using AR-CNN and VDSR are also shown in Fig. 2 for comparison. The image obtained by AR-CNN contains slight blocking artifacts, but the image obtained by VDSR and our VRCNN have eliminated most compression artifacts. The visual quality comparison is consistent with the objective BD-rate measure.

Table 4. The BD-rate results of AR-CNN and VDSR compared to HEVC baseline

Network		BD-rate		
		Y (%)	U (%)	V (%)
AR-CNN	Class A	0.9	2.1	2.1
	Class B	1.0	3.3	4.5
	Class C	−0.6	2.6	4.0
	Class D	−0.8	1.9	2.0
	Class E	0.4	5.5	6.1
	Overall	**0.2**	**3.0**	**3.7**
VDSR	Class A	−2.8	−3.2	−3.1
	Class B	−2.7	−2.7	−3.3
	Class C	−4.1	−4.8	−5.7
	Class D	−4.4	−5.6	−7.3
	Class E	−5.7	−5.7	−6.1
	Overall	**−3.8**	**−4.3**	**−4.9**

We also compare the computational complexity of different networks. This comparison was performed on a personal computer with Intel core i7-4790K central processing unit (CPU) at 4 GHz and NVIDIA GeForce GTX 750Ti GPU with 2 GB memory. Due to the limited memory of GPU, we cannot process large images using Caffe's GPU mode on this computer. Thus we used the sequence *Suzie* at resolution 176×144, the first 10 frames are used for test under all-intra setting, and the decoding time results are summarized in Table 5. Note that Caffe can work in CPU or GPU mode, both modes are tested. The reported decoding time includes both CPU computation and GPU computation if have. Since most computations of decoding are performed by CPU, post-processing, if using Caffe's GPU mode, is the last step, thus the transmission time between CPU and GPU is not negligible. Overall, it can be observed that our VRCNN is more than 2× faster than VDSR, since the latter is much deeper. Moreover, though VRCNN and AR-CNN are both 4-layer, VRCNN is slightly slower because in the second and third layers there are filters of different sizes, causing some troubles

Table 5. The results of decoding time (seconds per frame) of AR-CNN, VDSR and VRCNN

Network	Mode	
	CPU	GPU
AR-CNN	0.72	0.33
VDSR	2.15	1.27
VRCNN (ours)	0.98	0.45

Table 6. The sizes of trained networks (number of bytes required to store) of AR-CNN, VDSR and VRCNN

Network	Size
AR-CNN	417 KB
VDSR	2600 KB
VRCNN (ours)	214 KB

for parallel computing. The decoding time using CNNs does not meet real-time requirement on current main-stream personal computers, which calls for further efforts on optimizing the computational architecture.

Last but not the least, since the trained CNN is used for post-processing, especially at the decoder side, its memory cost is an important issue. We also compare the sizes of trained networks of AR-CNN, VDSR, and our VRCNN. The results are given in Table 6. Obviously, our VRCNN requires the lowest memory cost on storing the network because it is much shallower than VDSR and also has much less parameters than AR-CNN (shown in Tables 1 and 2).

5   Conclusion

In this paper, we have presented a convolutional neural network for post-processing in HEVC intra coding. The proposed network VRCNN outperforms the previously studied AR-CNN or VDSR in achieving higher bit-rate reduction, lower memory cost, and multiplied computational speedup. Compared to the HEVC baseline, VRCNN achieves on average 4.6% BD-rate (in luminance) on the HEVC standard test sequences. Our future work is planned in two directions. First, we will extend VRCNN for HEVC inter coding, i.e. processing P and B frames. Second, we will investigate how to further simplify the network while maintaining its coding efficiency.

Acknowledgment. This work was supported by the National Program on Key Basic Research Projects (973 Program) under Grant 2015CB351803, by the Natural Science Foundation of China (NSFC) under Grant 61331017, Grant 61390512, and Grant 61425026, and by the Fundamental Research Funds for the Central Universities under Grant WK2100060011 and Grant WK3490000001.

References

1. Bjontegaard, G.: Calcuation of average PSNR differences between RD-curves. VCEG-M33 (2001)
2. Dong, C., Deng, Y., Loy, C.C., Tang, X.: Compression artifacts reduction by a deep convolutional network. In: ICCV, pp. 576–584 (2015)
3. Dong, C., Loy, C.C., He, K., Tang, X.: Learning a deep convolutional network for image super-resolution. In: Fleet, D., Pajdla, T., Schiele, B., Tuytelaars, T. (eds.) ECCV 2014. LNCS, vol. 8692, pp. 184–199. Springer, Heidelberg (2014). doi:10. 1007/978-3-319-10593-2_13

4. Fu, C.M., Alshina, E., Alshin, A., Huang, Y.W., Chen, C.Y., Tsai, C.Y., Hsu, C.W., Lei, S.M., Park, J.H., Han, W.J.: Sample adaptive offset in the HEVC standard. IEEE Trans. Circ. Syst. Video Technol. **22**(12), 1755–1764 (2012)
5. Girshick, R., Donahue, J., Darrell, T., Malik, J.: Rich feature hierarchies for accurate object detection and semantic segmentation. In: CVPR, pp. 580–587 (2014)
6. He, K., Zhang, X., Ren, S., Sun, J.: Deep residual learning for image recognition. In: CVPR, pp. 770–778 (2016)
7. Jia, Y., Shelhamer, E., Donahue, J., Karayev, S., Long, J., Girshick, R., Guadarrama, S., Darrell, T.: Caffe: convolutional architecture for fast feature embedding. In: ACM Multimedia, pp. 675–678. ACM (2014)
8. Kim, J., Lee, J.K., Lee, K.M.: Accurate image super-resolution using very deep convolutional networks. In: CVPR, pp. 1646–1654 (2016)
9. Krizhevsky, A., Sutskever, I., Hinton, G.E.: Imagenet classification with deep convolutional neural networks. In: NIPS, pp. 1097–1105 (2012)
10. Nair, V., Hinton, G.E.: Rectified linear units improve restricted Boltzmann machines. In: International Conference on Machine Learning (ICML), pp. 807–814 (2010)
11. Norkin, A., Bjontegaard, G., Fuldseth, A., Narroschke, M., Ikeda, M., Andersson, K., Zhou, M., Van der Auwera, G.: HEVC deblocking filter. IEEE Trans. Circ. Syst. Video Technol. **22**(12), 1746–1754 (2012)
12. Park, W.S., Kim, M.: CNN-based in-loop filtering for coding efficiency improvement. In: 2016 IEEE 12th Image, Video, and Multidimensional Signal Processing Workshop (IVMSP), pp. 1–5. IEEE (2016)
13. Sullivan, G.J., Ohm, J.R., Han, W.J., Wiegand, T.: Overview of the high efficiency video coding (HEVC) standard. IEEE Trans. Circ. Syst. Video Technol. **22**(12), 1649–1668 (2012)
14. Sun, J., Zheng, N.N., Tao, H., Shum, H.Y.: Image hallucination with primal sketch priors. In: CVPR, vol. 2, pp. 729–736. IEEE (2003)
15. Szegedy, C., Liu, W., Jia, Y., Sermanet, P., Reed, S., Anguelov, D., Erhan, D., Vanhoucke, V., Rabinovich, A.: Going deeper with convolutions. In: CVPR, pp. 1–9 (2015)
16. Wallace, G.K.: The JPEG still picture compression standard. IEEE Trans. Consum. Electr. **38**(1), xviii–xxxiv (1992)
17. Wang, Z., Chang, S., Liu, D., Ling, Q., Huang, T.S.: D3: Deep dual-domain based fast restoration of JPEG-compressed images. In: CVPR, pp. 2764–2772 (2016)
18. Xie, S., Tu, Z.: Holistically-nested edge detection. In: ICCV, pp. 1395–1403 (2015)

A Framework of Privacy-Preserving Image Recognition for Image-Based Information Services

Kojiro Fujii[✉], Kazuaki Nakamura, Naoko Nitta, and Noboru Babaguchi

Graduate School of Engineering, Osaka University, Suita, Japan
{fujii,k-nakamura,naoko,babaguchi}@nanase.comm.eng.osaka-u.ac.jp

Abstract. Nowadays mobile devices such as smartphones are widely used all over the world. Moreover, the performance of image recognition has dramatically increased by deep learning technologies. From these backgrounds, we think that the following scenario of information services could be realized in the near future: users take a photo and send it to a server, who recognizes the location in the photo and returns the users some information about the recognized location. However, this kind of client-server-based image recognition can cause a privacy issue because image recognition results are sometimes privacy sensitive. To tackle the privacy issue, in this paper, we propose a novel framework for privacy-preserving image recognition in which the server cannot uniquely identify the recognition result but users can do so. An overview of the proposed framework is as follows: First users extract a visual feature from their taken photo and transform it so that the server cannot uniquely identify the recognition result. Then users send the transformed feature to the server, who returns a candidate set of recognition results to the users. Finally, the users compare the candidates and the original visual feature for obtaining the final result. Our experimental results demonstrate the effectiveness of the proposed framework.

Keywords: Image recognition · Privacy protection · Feature transformation · Information services

1 Introduction

Image recognition including object recognition and scene recognition have been one of the hottest topics in the area of computer vision in the past decades. Recently, the performance of image recognition has increased dramatically with the development of deep learning technologies [1]. Moreover, nowadays mobile devices such as smartphones are widely used all over the world and their computational capacity is still growing. From these backgrounds, image recognition-based information services working on mobile devices are investigated and several prototypes are developed. One example is a tourist assistance system proposed by Zeng et al. [2], in which users can get guide information by taking a photo

© Springer International Publishing AG 2017
L. Amsaleg et al. (Eds.): MMM 2017, Part I, LNCS 10132, pp. 40–52, 2017.
DOI: 10.1007/978-3-319-51811-4_4

of a landmark, street, building, and so on and sending it to a cloud server that hosts image recognition services. This kind of client-server-based information services are advantageous in that they can provide the latest information only by updating the server's information database and recognition criteria. However, this framework can cause a privacy issue because image recognition results are sometimes privacy sensitive. In this paper, we aim to tackle the privacy issue in client-server-based image recognition. To clarify our focus more specifically, we first introduce our assumed scenario.

Assumed Scenario. Similar with the scenario of Zeng et al. [2], we focus on photo-based information services based on the client-server architecture. As a *field* for the services, we assume a certain public space in which only a limited number of *spots* are included, where a service provider knows how many and what kind of *spots* exist in the space. A typical example of such a space is a shopping mall that has various kinds of stores. In this example, each store in the shopping mall is a *spot*, and the shopping mall itself is a *field*. Other examples include a theme park consisting of a group of entertainment attractions and a city that has a lot of places for sightseeing (e.g. Kyoto city).

In the above *field*, the service provider creates a server system consisting of a database and an image recognizer. In the database, information for each *spot* such as a product list, bargain products, congestion level, and customer evaluations (e.g. tweets for the *spot*) are stored and updated in real-time. To get the information, users take a photo of a *spot*, extract a visual feature from the photo, and send it to the server using their own smartphone. When receiving the visual feature from the users, the server identifies the *spot* in the photo using the image recognizer and returns the corresponding information in the database to the users. Figure 1 shows the overview of this scenario.

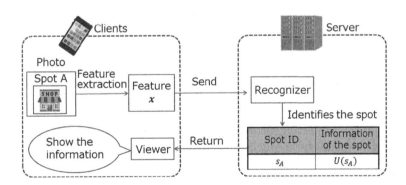

Fig. 1. Image recognition-based information service assumed in this paper

Privacy Issue. Basically, users have to be in or in front of a *spot* when taking its photo. This means the server can get to know the users' current location in terms of *spot* ID at the *spot*-identification stage. Moreover, when the users send a visual feature of the photo to the server, some identifiers of the users' smartphone such as IP address are also sent automatically, which can be used for making a correspondence between current and past results of *spot*-identification. This means the server can get to know the location history of the users.

Because the location history is a kind of users' privacy information that reflects their interests and preference, it should be protected so that the server cannot get to know. This requires a privacy-preserving recognition framework in which the server cannot uniquely identify the recognition result but users can do so. We aim to establish such a framework without any restrictions on recognition algorithms.

The contribution of this paper is summarized as follows: First, this paper raises a novel problem, i.e., privacy protection of recognition results in client-server-based image recognition. Second, this paper provides a general framework for the problem that is independent of recognition algorithms; our proposed framework is only based on transformation of visual features. Third, this paper proposes a concrete method of the feature transformation, which will be a basis of future works in this novel research area.

The remainder of this paper is organized as follows: In Sect. 2, we briefly review some previous works related to privacy protection from the aspect of visual information processing. Next, we describe our proposed framework for privacy-preserving image recognition in detail in Sect. 3 and experimentally evaluate its performance in Sect. 4. Finally we conclude this paper with several future works in Sect. 5.

2   Related Works

There are various kinds of privacy sensitive information in today's society such as medical records held by hospitals, transaction histories held by banks, personal profiles held by social networking service or cloud service providers, and so on. Multimedia contents such as video including human face or voice and 3-dimensional models of human body are also privacy sensitive information. Since different types of methods are generally required for protecting different kinds of data, privacy protection is related to a wide range of information technologies. In this section, we limit the range to visual information processing and briefly review several related works.

Methods for protecting privacy information in visual contents, especially images and video, has been widely studied in the past decade. One typical process is to abstract privacy sensitive regions such as human faces and entire bodies by blocking out, silhouetting, pixelization, complete removal, and so on [3]. Chinomi et al. [4] propose a system called PriSurv, which adaptively applies such operations to surveillance video data based on the relationship between people in the video and a viewer. Mitsugami et al. [5] also focus on surveillance video and propose to replace people in the video with rod-like symbols for protecting their privacy. Similar abstraction techniques are also used for dealing with the privacy issue of Google Street View [6,7]. More recently, Zhang et al. [8] propose an anonymous camera consisting of an infrared camera, a RGB camera, and a liquid crystal on silicon (LCoS) device, which can abstract face regions in video at the capturing phase by optical masking techniques.

Most of the above studies aim to protect the privacy of people appearing in visual contents captured by a camera. In contrast, some other studies focus on the privacy of owners of visual contents. For instance, in a general procedure of content-based image retrieval, in which users send an image as a query to a server and the server returns a set of images that have the similar content with the query, the query itself should not be disclosed to the server because it reflects the users' personality such as interests and preference. This can be achieved by cryptographic techniques; that is, the users first encrypt a query image and send it to the server which calculates the similarity between the query and each image in the database in the encrypted domain. To this end, Lu et al. [9] propose to use order preserving encryption (OPE) E_{ope}, which ensures $E_{ope}(x) < E_{ope}(y)$ if two plaintexts x and y satisfie $x < y$, and the Jaccard similarity. Thanks to OPE, the Jaccard similarity computed in the encrypt domain can reflect the similarity in the plaintext domain. Instead of OPE, Zhang et al. [10] employs homomorphic encryption (HE) E_{he}. Since their HE satisfies additive and multiplicative homomorphism, i.e., $E_{he}(x + y) = E_{he}(x) + E_{he}(y)$ and $E_{he}(xy) = E_{he}(x)E_{he}(y)$, for any plaintext pair (x, y), the encrypted version of Euclidian distance between a query and an gallery image can be calculated without decryption. The encrypted distances calculated on the server side are then returned to the users and decrypted on the user side for obtaining the final result. Chu et al. [11], who focus on the task of video retrieval, also use HE. They regard a video as a set of shots and calculate the similarity between two videos in the encrypted domain using a bipartite graph that represents the relationships between the shots. HE is also employed for the face recognition task in order to protect privacy information contained in face images [12–14], but these methods has a disadvantage that only Euclidian distance-based recognizers such as k-NN and Eigenface can be used in their frameworks.

There also are several methods for privacy-preserving image retrieval that are not based on cryptographic techniques. In the method of Weng et al. [15], a query image is first transformed to a hash code on the user side. Next, a part of bits in the hash code are removed and remaining bits are sent to a server. The server compares the sent bits with the hash code of each gallery image in a database, and returns the user a set of images containing the same bits with those sent by the user. Finally, the user screens the results from the server using the original hash code of the query for removing mismatched images. Fanti et al. [16] also propose a similar framework with that of Weng et al. In their methods, the server cannot get the complete information about a query even when the query is not encrypted, because only a part of the query is sent to the server. Moreover, the server cannot get to know which gallery images are truly related to the original query. This is also a desirable property for protecting users' privacy.

There are only a few studies focusing on privacy protection in the context of general image recognition for generally improving the performance of image recognition. Liu et al. [17] propose to use images that are dispersed in a network in a privacy-preserving manner. In their framework, each data holder trains an image recognizer only using their own data, and external users use each data holder's recognizer as a weak classifier, whose recognition results are integrated into the final result. This framework aims to protect the privacy of data holders; they do not focus on the privacy of external users who want to get a recognition result in contrast to our study.

3   Privacy-Preserving Image Recognition

In this section, we describe the proposed framework for privacy-preserving image recognition in detail, which is inspired by the image retrieval method of Weng et al. [15]. Note that we use the term "clients" instead of "users" in the remainder for a contrast to "server".

3.1   Notation

Before describing the proposed method in detail, we first summarize the notation used in this paper briefly.

Let $S = \{s_i | i = 1, 2, \cdots, N\}$ be a set of *spots* in a *field*, where s_i is the *spot* ID of i-th *spot* and N is the number of the *spots*. We assume that, for each *spot* s_i, a set of its typical visual features $T(s_i) = \{t_j(s_i) \in \mathbb{R}^n | j = 1, 2, \cdots, M\}$ is publicly available (available for both a server and clients), where n is the dimensionality of the feature vectors and M is the number of available typical features per *spot*. Let p be a photo of some *spot* that clients take with their smartphone, and let $x \in \mathbb{R}^n$ be a visual feature extracted from p.

3.2   Overview of the Proposed Framework

In the scenario described in Sect. 1, users' current location is unavoidably leaked to a server if the server can uniquely identify the *spot* in photo p. Therefore, we propose to transform feature x into $x' \in \mathbb{R}^n$ before sending it to the server in order to degrade the *spot*-identification performance on the server side. The overview of the proposed framework is as follows (see also Fig. 2):

(1) Clients extract visual feature x from the taken photo p. Note that x is effective enough for identifying the *spot* in p by the server's image recognizer.
(2) The extracted feature x is transformed into x' on the clients' smartphone, which is then sent to the server. With the transformation, the effectiveness of the original feature x is degraded so that the server cannot uniquely identify the *spot* in p from x'. Note that the server cannot get to know the original feature x because the transformation is done on the client side.
(3) Because x' is less effective, the server does not uniquely identify the *spot* in p but choose a set of its candidates. The server returns the users the set of candidate *spots* $\hat{S} \subset S$ as well as the *spot* information of each candidate $s \in \hat{S}$ in the database simultaneously.

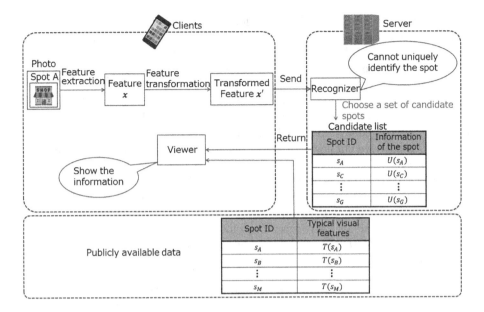

Fig. 2. Overview of the proposed framework

(4) For each candidate $s \in \hat{S}$ returned from the server, the clients compare the original feature x with $t_j(s)(j = 1, \cdots, M)$ and decide the final recognition result uniquely. Since this process is also done on the client side, the server cannot get to know the final result. This means the server gets to know only several candidates of the users' current location.

The above framework does not restrict the recognition algorithm; many kinds of algorithms including SVM, neural networks, boosting, and naïve Bayes can be used in both the server and the client sides unlike the previous works [12–14] that only allow Eigenface-based face recognition. The visual feature transformation from x into x' plays a key role in the proposed framework. We describe how to design the transformation in detail in the next section.

3.3 Visual Feature Transformation for Privacy Protection

In the image retrieval method of Weng et al. [15], they remove a part of bits from a retrieval query for preserving the users' privacy. This is equivalent with feature selection in the context of pattern recognition; that is, a part of dimensions are removed from feature vector x and the symbol '$*$' which means "do not care" is padded to the removed dimensions, which is used as x'. However, this method is not suitable to our scenario because the server can get to know which dimensions were removed on the client side, and therefore can re-train a new recognizer that is specialized for x' using $\{T(s_i)|i = 1, \cdots, N\}$. The new recognizer increases the *spot*-identification performance on the server side, which is not desirable from the aspect of privacy protection. Therefore we should employ a transformation method that makes the server unable to judge whether visual features sent from the clients are original version or transformed version. To this end, we focus on a linear subspace of the original feature space of x. Let L be a $n \times m$ matrix for projecting x onto a certain m-dimensional subspace, where $m < n$. Note that L satisfies $L^T L = I_m$, where I_m is the m-dimensional unit matrix. Using L, the projection of x on the subspace can be represented as $y = LL^T x \in \mathbb{R}^n$. It cannot be judged without L whether y is a projection of some other vector x or not. Hence, we employ LL^T as an operator of the transformation and use $LL^T x$ as x'.

Now the problem boils down to how to design the matrix L. To degrade the *spot*-identification performance on the server side, a set of projected features $T'(s_i) = \{t' = LL^T t | t \in T(s_i)\}$ for *spot* s_i should not be separable from $T'(s_l)$ for several other *spots* s_l. However, if too many *spots* are not separable from s_i, the performance of *spot*-identification and its computational cost on the client side become unacceptable. Based on this consideration, we design L as follows:

(1) Divide a set of *spots* S into K disjoint subsets so that each subset has at least two *spots*, which we referred to as $C_1, C_2, \cdots, C_K$ in the remainder.
(2) Find L that maximizes $\text{tr}(L^T \Sigma_b(S)L)$ and $\text{tr}(L^T \Sigma_w(C_k)L)(k = 1, \cdots, K)$ as well as minimizes $\text{tr}(L^T \Sigma_w(S)L)$ and $\text{tr}(L^T \Sigma_b(C_k)L)(k = 1, \cdots, K)$, where $\Sigma_w(S)$ and $\Sigma_b(S)$ are the within- and between-class scatter matrices for all *spots* in S and $\Sigma_w(C_k)$ and $\Sigma_b(C_k)$ are the within- and between-class scatter matrices for the *spots* in k-th subset C_k. Note that $\text{tr}(\Psi)$ denotes the trace of a square matrix Ψ.

Simultaneously minimizing between-class variance and maximizing within-class variance for each subset C_k in the step (2), two *spots* s_i and s_l are expected to be hardly separable if both of them belong to the same subset, i.e., $s_i, s_l \in C_k$. At the same time, simultaneously maximizing between-class variance and minimizing within-class variance for S, two *spots* s_i and s_l are expected to be easily separable if they are not in the same subset.

For convenience of formulation, we attempt to maximize $\text{tr}(L^T(\epsilon I_n + \Sigma_w(S))^{-1}L)$ and $\text{tr}(L^T(\epsilon I_n + \Sigma_b(C_k))^{-1}L)$ instead of minimizing $\text{tr}(L^T \Sigma_w(S)L)$ and $\text{tr}(L^T \Sigma_b(C_k)L)$, by which the problem of designing feature transformation boils down to finding the $n \times m$ matrix L that maximizes

$$\text{tr}\left(L^T\left[\Sigma_b(S) + \alpha(\epsilon I_n + \Sigma_w(S))^{-1} + \sum_{k=1}^{K}\left\{\beta \Sigma_w(C_k) + \gamma(\epsilon I_n + \Sigma_b(C_k))^{-1}\right\}\right]L\right)$$

under the constraint of $L^T L = I_m$, where positive constants α, β, and γ are the weight parameters for each term. ϵI_n is a regularization parameter for $\Sigma_w(S)$ and $\Sigma_b(C_k)$ so that they have the inverse matrix. This maximization problem can be solved by finding m-largest eigenvalues of positive semi-definite matrix

$$\Phi = \Sigma_b(S) + \alpha(\epsilon I_n + \Sigma_w(S))^{-1} + \sum_{k=1}^{K} \left\{ \beta \Sigma_w(C_k) + \gamma \left(\epsilon I_n + \Sigma_b(C_k)\right)^{-1} \right\} \quad (1)$$

and their eigenvectors $u_1, \cdots, u_m$. Using the eigenvectors, the optimal $\hat{L}$ is obtained as

$$\hat{L} = (u_1 \cdots u_m)$$

Note that the server cannot get to know the $\hat{L}$ because the clients freely determine a partition for S, i.e., $C_1, \cdots, C_k$, in the step (1) and the partition result is not sent to the server in our proposed framework. Moreover, if needed, the clients can also change the partition for S anytime they want. This means $\hat{L}$ used by each client is not always same, which could improve the robustness of the proposed framework to statistical attacks by the server.

4 Experiments for Performance Evaluation

To evaluate the performance of the feature transformation proposed in the previous section, we conducted an experiment.

4.1 Experimental Setting

Since it is difficult to conduct experiments in the real environment such as actual shopping malls or theme parks due to issues of personal rights (e.g. people's portrait rights), we picked up 10 famous places in the world and virtually regarded them as *spots*. The picked up places were as follows: *Colosseo, Arc de Triomphe, Ginkaku-ji temple, Kaminarimon, Kinkaku-ji temple, Notre Dame de Paris, Palais Garnier, Parthenon, Leaning Tower of Pisa,* and *Taj Mahal.* For each of these places, we gathered its photos from Flickr and extract their visual features. More specifically, we applied AlexNet CNN model pre-trained on ImageNet to the gathered photos using Caffe [18] and extracted 4096-dimensinal feature vectors by choosing the output of the "fc7" layer, which were then compressed to 256-dimensional vectors by principal component analysis (PCA). The number of the extracted visual features was 120 per place (or *spot*), 100 of which were used as $T(s_i)$ and remaining 20 features were used as x for testing the performance.

Performance of the proposed method of visual feature transformation was evaluated with the following two criteria: *spot*-identification accuracy on the server side (A_s) and that on the client side (A_c). Because the purpose of our proposed framework is to make a server unable to uniquely identify the *spot* in photos, lower A_s is desirable. At the same time, since clients should be able to identify the *spot* for getting correct *spot*-information, higher A_c is desirable. We employed nonlinear SVM with a RBF kernel as a recognition method on the server side since the server is expected to have rich computational resources for training complex recognizers, whereas we employed 1-nearest neighbor algorithm for a recognizer of the client side.

4.2   Results and Discussions

First we evaluated A_s without any feature transformation, which results in 99.5%. This result means that users' current location can be almost uniquely identified by the server and therefore supports the importance of privacy protection for this scenario. Next, we evaluated how A_s is degraded by the proposed method of feature transformation, where the set of *spots* S was divided into three subsets and each subset included three or four *spots*. Parameters α, β, γ, and ϵ was empirically set as $\alpha = 4, \beta = 7, \gamma = 5$, and $\epsilon = 0.01$. For comparison, we also used PCA and linear discriminant analysis (LDA) as a method of computing feature transformation matrix L and evaluated A_s in these two cases. Note that the dimensionality of the subspace, i.e., m in Sect. 3.3, is set as 8 in all cases. Figure 3 shows the result with a form of cumulative matching characteristic (CMC) curve; that is, the vertical axis means the probability that the correct *spot* ID is within the top λ candidates and the horizontal axis means the λ. A_s is identical to the score of $\lambda = 1$. In the cases of PCA and LDA, A_s is not degraded so much; it keeps more than 85% after the feature transformation. This indicates that PCA and LDA can hardly protect the users' current location. This is because PCA and LDA do not aim to decrease the class-separability of image recognizers or classifiers. In contrast, A_s falls to 45% with the proposed feature transformation method. This indicates that the proposed method can make the server unable to uniquely identify the *spot*.

Next, we evaluated the accuracy of *spot*-identification on the client side, i.e., A_c. Noting that A_c depends on the number of candidate *spots* sent from the server and the number of publicly available visual features per *spot*, we tried various settings of these parameters. Let q be the number of candidate *spots* and M be the number of publicly available features per *spot*. Figures 4, 5, and 6 show the result of the cases using PCA, LDA, and the proposed method, respectively. It is derived from these figures that the proposed method can be comparable to PCA and LDA with $q \geq 4$. This is because each subset includes at most four

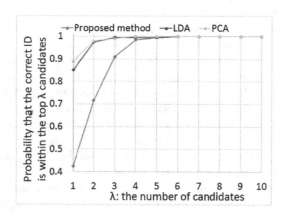

Fig. 3. Accuracy of spot-identification on the server side

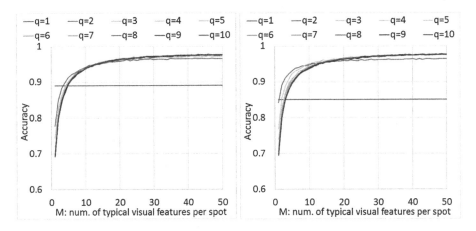

Fig. 4. Accuracy of spot-identification using PCA on the client side

Fig. 5. Accuracy of spot-identification using LDA on the client side

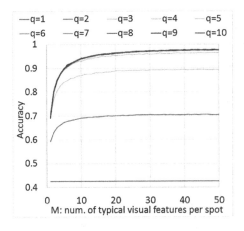

Fig. 6. Accuracy of spot-identification using the proposed method on the client side

' *spots* in this experiment. In the proposed method, a *spot* is hardly separable from the other *spots* in the same subset. Therefore q should be equal to or larger than the maximum number of *spots* in a single subset. (Conversely, the number of subsets in a partition for S, i.e., K, should be larger than N/q, where N is the number of *spots* in a *field*.) As far as this condition is satisfied, the proposed method can provide a good recognition power to the client side. As for the effect of parameter M, A_c becomes higher than 95% with $M > 20$ in the case of $q \geq 4$. Based on these results, we employed the settings of $q = 4$ and $M = 20$ in the evaluation described next.

Finally, we evaluated the effect of parameters α, β, γ, and ϵ on the performance. To this end, we first fixed three of these parameters and then evaluated A_s and A_c, varying the other parameter. Figure 7 shows the result. It is derived

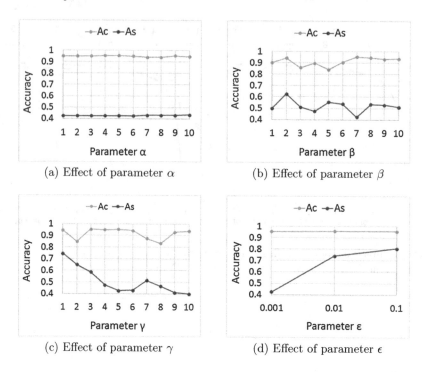

(a) Effect of parameter α (b) Effect of parameter β

(c) Effect of parameter γ (d) Effect of parameter ϵ

Fig. 7. effect of parameters α, β, γ, and ϵ on the accuracy of spot-identification

from Fig. 7 that change in parameters α and β do not give a clear effect to both A_s and A_c. On the other hand, parameters γ and ϵ give a significant effect to A_s and A_c; both A_s and A_c become lower with larger γ, and higher A_s and A_c are obtained with larger ϵ. Both of these two parameters are related to the last term of Eq. (1), i.e., $\gamma(\epsilon I_n + \Sigma_b(C_k))^{-1}$. If we employ larger γ, between-class variance for each subset C_k is more strongly minimized, so that the *spot*-separability becomes lower. This is also the case with smaller ϵ. These results indicate that we only have to take care of the ratio of γ to ϵ for parameter tuning in the proposed method.

5   Conclusion

In this paper, we assumed the following scenario of information services: users take a photo of a certain *spot* and send it to a server, who identifies the *spot* in the photo and returns the users some information about the identified *spot*. This kind of services can cause a privacy issue because image recognition results are sometimes privacy sensitive. To deal with the privacy issue, we proposed a novel framework for privacy-preserving image recognition in which the server cannot uniquely identify the recognition result but users can do so. We demonstrated the effectiveness of the proposed framework with several experimental results.

In fact, the current version of the proposed framework has a drawback; there is a possibility that the location history protected by the proposed framework is disclosed to the server using a spatial relationship between *spots* and a temporal relationship between queries sent from the same client. To deal with this drawback is an important future work.

This work was supported by JSPS KAKENHI Grant Numbers 16H06302 and 15H01686.

References

1. Deng, L., Yu, D.: Deep learning: methods and applications. Found. Trends Sig. Process. **7**(3–4), 197–387 (2014)
2. Zeng, Y., Chan, Y., Lin, T., Shih, M., Hsieh, P., Chao, G.: Scene feature recognition-enabled framework for mobile service information query system. In: Proceedings of the 17th International Conference on Human Interface and the Management of Information (2015)
3. Cavallaro, A., Steiger, O., Ebrahimi, T.: Semantic video analysis for adaptive content delivery and automatic description. IEEE Trans. Circ. Syst. Video Technol. **15**(10), 1200–1209 (2005)
4. Chinomi, K., Nitta, N., Ito, Y., Babaguchi, N.: PriSurv: privacy protected video surveillance system using adaptive visual abstraction. In: Proceedings of the 14th International Conference on MultiMedia Modeling (MMM2008), pp. 144–154 (2008)
5. Mitsugami, I., Mukunoki, M., Kawanishi, Y., Hattori, H., Minoh, M.: Privacy-protected camera for the sensing web. In: Proceedings of the International Conference on Information Processing and Management of Uncertainty in Knowledge-Based Systems (2010)
6. Frome, A., Cheung, G., Abdulkader, A., Zennaro, M., Wu, B., Bissacco, A., Adam, H., Neven, H., Vincent, L.: Large-scale privacy protection in Google Street View. In: Proceedings of the International Conference on Computer Vision, pp. 2373–2380 (2009)
7. Flores, A., Belongie, S.: Removing pedestrians from Google Street View images. In: Proceedings of the International Workshop on Mobile Vision, pp. 53–58 (2010)
8. Zhang, Y., Lu, Y., Nagahara, H., Taniguchi, R.: Anonymous camera for privacy protection. In: Proceedings of the 22nd International Conference on Pattern Recognition (2014)
9. Lu, W., Swaminathan, A., Varna, A.L., Wu, M.: Enabling search over encrypted multimedia databases. In: Proceedings of the SPIE Conference on Media Forensics and Security, vol. 7254, pp. 18–29 (2009)
10. Zhang, L., Jung, T., Feng, P., Liu, K., Li, X., Liu, Y.: PIC: enable large-scale privacy preserving content-based image search on cloud. In: Proceedings of the 44th International Conference on Parallel Processing (2015)
11. Chu, W., Chang, F.: A privacy-preserving bipartite graph matching framework for multimedia analysis and retrieval. In: Proceedings of the 5th ACM International Conference on Multimedia Retrieval, pp. 243–250 (2015)
12. Erkin, Z., Franz, M., Guajardo, J., Katzenbeisser, S., Lagendijk, I., Toft, T.: Privacy-preserving face recognition. In: Proceedings of the 9th International Symposium on Privacy Enhancing Technologies, pp. 235–253 (2009)

13. Sadeghi, A., Schneider, T., Wehrenberg, I.: Efficient privacy-preserving face recognition. In: Proceedings of the 12th International Conference on Information Security and Cryptology, pp. 229–244 (2009)
14. Bringer, J., Chabanne, H., Patey, A.: Privacy-preserving biometric identification using secure multiparty computation: an overview and recent trends. IEEE Sig. Process. Mag. **30**(2), 42–52 (2013)
15. Weng, L., Amsaleg, L., Morton, A., Marchand-Maillet, S.: A privacy-preserving framework for large-scale content-based information retrieval. IEEE Trans. Inf. Forensics Secur. **10**(1), 152–167 (2015)
16. Fanti, G., Finiasz, M., Friedland, G.: Toward efficient, privacy-aware media classification on public databases. In: Proceedings of the International Conference on Multimedia Retrieval (2014)
17. Liu, C., Shang, Z., Tangc, Y.Y.: An image classification method that considers privacy-preservation. Neurocomputing **208**(5), 80–98 (2016)
18. Jia, Y., Shelhamer, E., Donahue, J., Karayev, S., Long, J., Girshick, R., Guadarrama, S., Darrell, T.: Caffe: convolutional architecture for fast feature embedding. In: Proceedings of the 22nd ACM International Conference on Multimedia, pp. 675–678 (2014)

A Real-Time 3D Visual Singing Synthesis: From Appearance to Internal Articulators

Jun Yu[(⊠)]

Department of Automation, University of Science and Technology of China,
Hefei 230026, China
harryjun@ustc.edu.cn

Abstract. A facial animation system is proposed for visual singing synthesis. With a reconstructed 3D head mesh model, both finite element method and anatomical model are used to simulate articulatory deformation corresponding to each phoneme with musical note. Based on an articulatory song corpus, articulatory movements, phonemes and musical notes are trained simultaneously to obtain the visual co-articulation model by a context-dependent Hidden Markov Model. Articulatory animations corresponding to all phonemes are concatenated by visual co-articulation model to produce the song synchronized articulatory animation. Experimental results demonstrate the system can synthesize realistic song synchronized articulatory animation for increasing the human computer interaction capability objectively and subjectively.

Keywords: Articulatory animation · Song-to-articulator mapping

1 Introduction

Visual speech synthesis has numerous interesting applications [1, 2], and has been a hot topic recently. The existing efforts mostly focus on head mesh modeling [3], acquiring data to control the synchronized movements of articulators [4, 5], and using limited data to synthesize various articulatory motions during speech [6]. However, emerging applications require the visual speech synthesis system to have the singing ability for enhancing the diversity of human computer interaction (HCI).

For visual speech synthesis, Anderson et al. [7] used active appearance model (AAM) to obtain a high degree of realism. Wang et al. [8] and Liu and Ostermann [9] built the data-driven systems by unit selection. Wang et al. [10] proposed a high quality facial animation system that can be easily built based on affordable off-the-shelf components. Xu et al. [11] produced high quality lip synchronization for multiple languages. Hartholt et al. [12] offered a flexible framework for exploring a variety of different types of virtual human systems. Taylor et al. [13] redefined visemes as temporal units that describe distinctive speech movements of the visual speech articulators, capable of generating a dynamic, concatenative unit of visual speech. Wang et al. [2, 14, 15] built a system by applying 3D acoustical animation. Li et al. [16] synthesized 2D animations of lips and tongue for Mandarin Chinese.

There are two main differences between visual singing synthesis and visual speech synthesis [17]. Contextual factors which affect singing voice are different from those

L. Amsaleg et al. (Eds.): MMM 2017, Part I, LNCS 10132, pp. 53–64, 2017.
DOI: 10.1007/978-3-319-51811-4_5

used in speech synthesis. Another difference is there are time-lags between the start timings of musical notes and singing [18] for singing synthesis.

For introducing the singing ability to the visual speech synthesis community, we proposes a real-time musical score + lyric driven 3D facial animation system for the first time. A complete and accurate 3D head mesh model, including appearance and internal articulators, is reconstructed for fast modeling. It has hair and more articulators compared with the work in [2]. Combining with contextual factors which affect singing voice and the time-lags model, multi-stream Hidden Semi-Markov Model (HSMM) [19] is used to establish the mapping between song and articulatory trajectory by training them together. Whereas the visual speech synthesis method in [2] only trains the articulatory trajectory alone. A biomechanical model is built to deform the tongue model by given muscle activations directly by improving the method in Tang et al. [20] with the Moneey-Rivlin material.

2   Constructing 3D Head Mesh Model

The head mesh includes internal articulators and appearance. To achieve proper compromise between performance and realism, 3D volume mesh is adopted for the tongue mesh, while 3D surface meshes are adopted for others. Magnetic Resonance Imaging (MRI) data of a person (Fig. 1(a)) are captured for the internal articulators. The contour on each MRI slice is marked (Fig. 1(b)) to obtain the 3D surface meshes. The meshes obtained consist of teeth, tongue, palate, pharynx, mandible and oral cavity. Then the surface mesh (Fig. 1(c)) is discretized into a mesh of 20,181 tetrahedrons for creating the volume mesh. The appearance mesh (Fig. 1(d)) is obtained by matching the front and profile visible images of a person by the approach reported in [21].

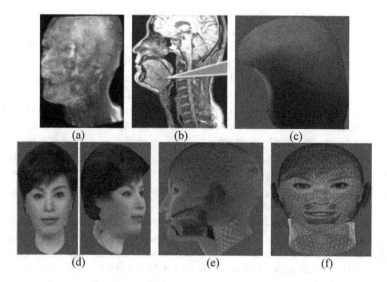

Fig. 1. (a) All MRI slices on the sagittal plane. (b) The MRI slice with marked tongue contour on the midsagittal plane. (c) Tongue mesh. (d) Skin mesh and hair mesh. (e)-(f) Head mesh.

The internal articulatory mesh and appearance mesh may be obtained from different persons; thus they need to be integrated together for forming a complete head mesh. Because the MRI data capturing process of internal articulators is time-consuming, it is only applied once, and the internal articulatory mesh is integrated with the appearance mesh. This is suitable for fast modeling because we only need to capture the limited visible images of a desired singer when changing the role for animation. The integration process is as follows. Several dominant vertices pairs of appearance mesh and internal articulatory mesh, such as lip top and tongue tip, are selected. The global motion parameters of internal articulatory mesh are obtained from the above dominant vertices pairs by the method reported in [22] as well as anatomical knowledge, which defines the relative position between different articulators [23]. Then $U_i = V_i - V_i^0$ (V_i^0, V_i are the coordinates of the ith dominant vertex of the internal articulatory mesh before and after displacing) are obtained by the global motion parameters, and used to construct an interpolation function by radial basis function.

Based on the solved radial basis function, the complete head mesh (Fig. 1(e)–(f)) can be obtained.

3   Song-To-Articulator Mapping

An articulatory song corpus (Table 1) is built by Electro-Magnetic Articulography (EMA). EMA records the 3D trajectories of articulatory movements (from sensors on Tongue Rear (TR), Tongue Blade (TB), Tongue Tip (TT), Lower Incisor (LI), Lower Lip (LL), Left Corner (LC), Right Corner (RC), Upper Lip (UL)) and audio data simultaneously. In the recorded corpus, 90% of data are used for training, and 10% of data are used for testing.

Table 1. Articulatory song corpus.

Singer	Songs	Sampling rate (kHz)		Length
3	468	Audio: 22	Articulators: 200	232 min

In the feature extraction stage, the extracted features include two parts. The first is the displacement vectors of EMA coils. The observation of the t th frame is $Y_t = \left[y_t^T, \Delta y_t^T, \Delta^2 y_t^T\right]^T$, y_t is the static feature, $\Delta y_t, \Delta^2 y_t$ are the dynamic features [24]. Then $y = [y_1^T, \cdots, y_t^T]^T$, $Y = [Y_1^T, \cdots, Y_t^T]^T$ are the static feature and observation sequences, and $Y = W_y \cdot y$. The second is the acoustic parameter x_t defined as the 24 orders of mel-cepstrum, and the definition of X is similar to above.

In the training stage, multi-stream HSMM is used to train y_t and x_t. It has the left-to-right topology and 5 emitting states, and the output distributions are represented by single Gaussian densities. Four types of contextual factors are extracted from the musical score and lyric. The first type is the phoneme: the preceding, current, and succeeding phonemes. The second type is the tone: the musical tones of the preceding, current, and succeeding musical notes. The third type is the duration: the durations of the preceding, current, and succeeding musical notes. The fourth type is the position: the positions of the preceding, current, and succeeding musical notes in the corresponding musical bar. The HSMM is trained by Baum Welch algorithm. Then context-dependent tri-phone models are obtained, and the decision tree-based clustering [25] is conducted to alleviate the sparse data problem. Finally, the model λ is obtained, and the decision-tree-clustered context-dependent time-lag models are obtained according to the method in [18].

In the synthesis stage, firstly, the l, namely phonemes, notes and durations, are extracted from input musical score and lyric. Secondly, the optimal state sequence of the HSMM and time-lags of musical notes are determined simultaneously as $\hat{q}, \hat{g} = \arg\max_{q,g} P(q, g|l, X, \lambda)$ by Viterbi alignment [26]. Finally, y is estimated:

$$\hat{y} = \arg\max_y P(Y|\hat{q}, \hat{g}, \lambda, l) \tag{1}$$

4   3D Facial Animation

4.1   Animation of Appearance

With respect to appearance, the anatomical model [27, 28] simulates facial motion by the contraction of muscles, the motion of skeleton, and the deformation of skin. It is suitable for describing facial motion intuitively [29]. Therefore, the proposed anatomical model for appearance includes three parts: skeleton, skin, muscle. The skeleton includes skull and jaw. The skull is generally passive, and the up-down rotation degree of the jaw is computed with the trajectory of LI. The skin is approximated by an elastic mesh [30], and connected with muscle by two classes of spring. The first class is used to simulate the elasticity of skin. The second class is used to ensure that the skin not to be split. The muscles are modeled by Waters model [28], in which linear and sheet muscles are used for tension and sphincter muscle is used for shrinkage. Therefore, the animation of appearance is synthesized as follows. Given the trajectories of LI, LL, LC, RC, UL, the force on the skin mesh is first computed, which includes the elasticity of springs between skin and muscle, the contraction of muscles, the drawing of jaw and the restriction of skull. The displacement of each vertex on the skin mesh is then obtained by the force.

4.2   Animation of Internal Articulators

With respect to internal articulators, the movements of the palate and upper teeth are generally passive, and the jaw and lower teeth only have the movements of up-down rotation. So they can be abbreviated, and we focus our attention on the modeling of the tongue whose motion is more complicated and non-rigid. The tongue is mostly composed of muscle and connective tissue [31], and most of them are nonhomogeneous. It is regarded as incompressible because it contains a high percentage of water.

For biomechanical modeling, soft tissues are usually assumed as a hyperelastic or viscoelastic material, and modeled by using a strain energy approach.

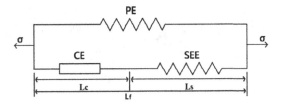

Fig. 2. Hill's three element muscle model.

The connective tissue is very thin and accounts for only a small part of tongue, hence can be assumed as isotropic. It is modeled as a Mooney-rivlin material [32], and the strain energy approach is given as:

$$U = U_I(I_1, I_2) + U_J(J) \tag{2}$$

where U_I reflects the nonlinear, isotropic, hyperelastic properties, U_J reflects the incompressibility. More details can be found in [32].

The muscle is endowed with a constitutive model which embodies the active and passive properties of muscle fibers [20]. The strain energy function for muscle is a sum of muscle fiber term and the term for connective tissues as:

$$U = U_I(I_1, I_2) + U_J(J) + U_f(\lambda_f, A) \tag{3}$$

where λ_f is the fiber stretch ratio in the along-fiber direction, A is the activation value vector of tongue muscles which can be computed from the trajectories of TR, TB and TT. $U_f(\lambda_f, A)$ represents the active and passive non-linear mechanical behaviors of the muscle fiber during contracting, and is modeled by Hill's three element model [33] (Fig. 2(a)), which is constituted of a contractile element (CE), a serial elastic element (SEE) and a parallel elastic element (PE).

According to above biomechanical model, FEM is applied to simulate the deformation of tongue mesh driven by the muscle activations.

5   Experiments

Experiments are conducted by a workstation with AMD Athlon™ II X4 640 3.01G, memory 2G, NVIDIA GT200.

Figure 3 shows the 3D facial animation results of seventh of music notes and Chinese phonemes. The system generates a virtual head which is able to produce realistic articulatory animation. The virtual head has a human-like appearance and motion, thus providing a solid foundation for visual singing synthesis.

The Mandarin Chinese lyric "Wo tao yan", which means "I hate" in English, consists of following Chinese phonemes: "w, o, t, ao, y, an". The song synchronized 3D articulatory animation driven by this lyric is produced as Fig. 4 shows. The first 3 visemes are displayed with appearance, while the others are displayed with transparent

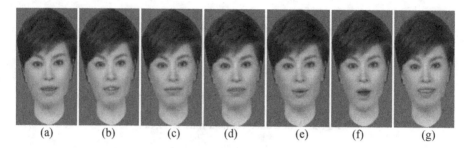

(a)   (b)   (c)   (d)   (e)   (f)   (g)

Fig. 3. (a) do. (b) re. (c) mi. (d) fa. (e) suo. (f) la. (g) xi.

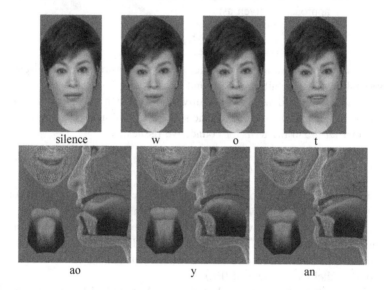

Fig. 4. Song synchronized 3D articulatory animation driven by a Mandarin Chinese lyric.

skin. We can see that the visual co-articulation model can produce song synchronized visemes corresponding to the visual effect of co-articulation in continuous song.

5.1   Objective and Perceptual Evaluations

Firstly, the accuracy of song-to-articulator mapping is evaluated objectively. In [2], a Hidden Markov Model (HMM)-based synthesis is performed with a Maximum Likelihood Parameter Generation algorithm for smoothing, and also used as the baseline here. The training data of the articulatory song corpus and HTK toolkit are used for training. There are 59 monophone models and 25,650 triphone models in our trained HSMM. It is difficult to obtain the absolute truth of real 3D articulatory movement. MRI has the drawbacks of poor real-time capability and high cost. X-ray technique has the unique advantage of real-time capability, but is harmful to the human health.

Consider the pros and cons, the Root Mean Square Error (*RMSE*) between articulatory trajectory synthesis results in EMA sensors positions and the EMA true displacement trajectories (ground truth) is used. Table 2 shows the RMSE results by the method in [2] and our method. As can be seen from them, our articulatory trajectory synthesis results outperform the results of the method in [2]. This is because that our method uses multi-stream HSMM to train musical sore, song and articulatory trajectory together, while the method in [2] only trains the articulatory trajectory alone.

Table 2. The accuracy comparison.

	Our method	The method in [2]
RMSE (milimeter)	2.077	2.802

Secondly, the animation of articulators is evaluated subjectively. Because the *Realism*, *Expressiveness* and *Coherence* of articulators are crucial to the human visual system, they are used as the questionnaire (Table 3) for evaluating human acceptance. The available image database [34] contains not only the videofluoroscopic images of X-ray, but also the front visible images of mouth. The front visible images of mouth are captured with the corresponding videofluoroscopic images simultaneously. Then the shapes and textures of articulators in the image database are chosen as the baseline for perceptual evaluation. The articulatory shape is detected by the AAM [7] plus manual intervention. Figure 5 illustrates the detection result.

Table 3. The questionnaire and mean scores of evaluation.

Construct	Question	Mean score
Realism	If the appearance of articulators is realistic	8.16
Expressiveness	If the articulators can display the visual signature of speech	8.31
Coherence	If the dynamics of articulatory animation are same as that in the images of X-ray	8.85

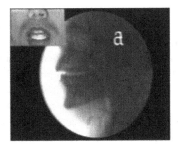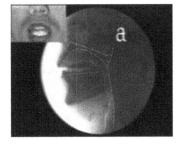

Fig. 5. The detection result (right image) of articulatory shape on one frame (left image) of the X-ray images when pronouncing vowel /a/.

The system performs 3D articulatory animation driven by text input, which is same as that accompanied with the image database, and shows the 29 Chinese students the corresponding articulatory movement (images + detected shape contours) in the image database simultaneously. Table 3 shows the mean scores after evaluation. The maximum is 10, while the minimum is 0. The participants can repeat the animation and films in the image database as many times as they want. After participants see the presentation, they are asked to fill in the questionnaire. The obtained scores are all greater than 8.0, indicating that the system is acceptable when audiences watching the animation results.

Table 4. The realism scores of several methods when using large training data.

	The proposed method	Anderson et al. [7]	Liu et al. [9]	Xu et al. [10]
Expression	Neutral	Neutral	Neutral	Neutral
Realism score	4.01	3.79	4.27	3.61

5.2   Comparison with Other Visual Text-to-Speech Systems

As indicated in [7], the method in [9] is the state-of-the-art visual text-to-speech system for neutral expression, and the method in [7] is the state-of-the-art visual text-to-speech system for large range expression. In addition, the method in [10] is a high quality facial animation system in which the co-articulation model is rule-based. We have implemented them. The methods in [7, 9] are trained on the training data of the video database in an articulatory speech corpus [34], in which the text are same as the lyrics in the articulatory song corpus used by the proposed method, while the method in [10] does not need training data. Our proposed method and the methods in [7, 9, 10] all synthesize the images of the same singer, and the synthesized images are scaled to a face region height of approximately 300 pixels. To make the comparison fair, the virtual head is kept static, and the same background is used for them. The definition of realism score in [7] is used: the maximum is 5, the minimum is 0.

The results in Table 4 show that the method in [9] is rated most realistic among the methods for neutral expression. The proposed method performs comparably to other methods in the neutral expression category.

However, when the data size for training is reduced largely, namely only one quarter training data are used, the performance in the Table 5 has a large difference from that in the Table 4. The performance of the proposed method does not deteriorate very much, and has higher realism scores. On the other hand, the performance of methods in [7, 9] degenerate to a large extent. This is because the proposed method constructs the head mesh model accurately, and uses the anatomical knowledge and mesh model to constrain the animation model. For example, the visemes are synthesized by the precise mesh model and anatomical knowledge, and thus not depend on the training data. Training data are only used to blend visemes by constructing visual co-articulation model. In other words, the synthesis results are less influenced by training data because only the sparse EMA data are used for training. Whereas, the methods in [7, 9] totally depend on training data, and need vast data to describe details, especially when the virtual head opens mouth. The performance of the method in [10] does not degenerate because of its independence on training data. However, the performance of the proposed method still outperforms that of the method in [10]. This is because the used co-articulation rules by the method in [16] are hardly to accurately model co-articulation in speech animation.

Table 5. The realism scores of several methods when using limited training data.

	The proposed method	Anderson et al. [7]	Liu et al. [9]	Xu et al. [10]
Expression	Neutral	Neutral	Neutral	Neutral
Realism score	3.83	3.27	3.58	3.61

5.3   Visual Singing Synthesis vs. Visual Speech Synthesis

To verify the effectiveness of singing ability in HCI, the subjective comparison between the visual singing synthesis system and a visual speech synthesis system is conducted. The object of the comparison is to decide if singing a lyric is more acceptable than speaking a lyric.

The technique for constructing the visual speech synthesis system for comparison is similar to that for the visual singing synthesis system, while there are three differences between them. The first one is the training data for the visual speech synthesis system are from an articulatory speech corpus [34], in which the text are same as the lyrics in the articulatory song corpus used for training the visual singing synthesis system. The second one is the contextual factors for training the visual speech synthesis system are only phoneme and duration, and the time-lags is not modeled. The third one is the input of the visual singing synthesis system is musical score + lyric, while the input of the visual speech synthesis system is only the lyric of song.

The animation results are evaluated by comparing with the articulatory movements of 34 singer perceptually. The virtual head is kept static, so the evaluation focus is the motion of lips. Several feature vertices are selected on the lip contour of 3D head mesh model, and the rendering results of feature vertices during animation form feature points trajectories, called trajectories 1. Similarly, the corresponding feature points in the detected lip contours of captured image sequence of singer form another feature points trajectories, called trajectories 2 (ground truth).

Because of the normalization problem and the problem of comparing males and females, it is hard to make the comparison objectively. To alleviate the first problem, firstly, trajectory 1 and trajectories 2 are aligned to have the same dimension. They are aligned in the time domain using the Dynamic Time Warping technique [35]. Secondly, the amplitudes of trajectories 1 are normalized according to those of trajectories 2. A_{max}, B_{max} are the max absolute amplitude values of trajectories 1 and trajectories 2. Then the amplitude value of ith frame, A_i, of trajectories 1 is normalized to $A_i = A_i \cdot (B_{max}/A_{max})$. To alleviate the second problem, the comparison between trajectories 1 and trajectories 2 is a perceptual evaluation by singers. Singers judge if the shapes of trajectories 1 are similar to those of ground truth. For example, if the numbers and location of local maxima and minima of above two curves are same.

Table 6 shows the mean opinion score (MOS) with a scale from 1 (poor) to 10 (good), and the standard deviation of opinion scores (SDOS). The obtained MOSs are all greater than 8.0, indicating that the system is acceptable when the virtual head is talking or singing.

Table 6. The mean evaluation scores.

Construct	Question	Visual singing synthesis		Visual speech synthesis	
		MOS	SDOS	MOS	SDOS
Realism	If the animation is realistic	8.46	0.36	8.46	0.36
Coherence	If the dynamics of animation are same as that of articulatory movement of singers	8.98	0.67	8.34	0.78
Naturalness of HCI	If the virtual head increases the naturalness of HCI	9.35	0.45	8.29	0.51

The visual singing synthesis system outperforms the visual speech synthesis system in the second construct *Coherence*. Because the contextual factors, i.e., tone and position, are not considered and the time-lags is not modeled in the visual speech synthesis system, participants think the visual synthesis results produced by the visual speech synthesis system are less synchronized to the musical characteristics than those produce by the visual singing synthesis system.

The visual singing synthesis system outperforms the visual speech synthesis system in the third construct *Naturalness of HCI*. This is because the input text of visual speech synthesis system is same as the input lyric of visual singing synthesis system, and participants think singing a lyric is more acceptable than speaking a lyric.

6 Conclusion

This paper proposes a real-time musical score + lyric driven facial animation system, which can produce not only the 3D facial animation for appearance, but also that for internal articulators. The data from visual images, MRI, anatomical characteristics and EMA are fused to obtain accurate song synchronized realistic 3D articulatory animation. The system can not only illustrate the visual differences among phonemes in a song, but also increase the human computer interaction capability of visual speech synthesis significantly, and quantitative improvements are demonstrated in objective evaluation and user studies, comparing with the output of different visual speech synthesis systems.

In future, diverse cues, such as video, will be added to drive the developed system for constructing a multi-modal HCI system. A practical hardware robot with the developed head and animation capacity will also be constructed.

Acknowledgement. This work is supported by the National Natural Science Foundation of China (Nos. 61572450, 61303150), the Open Project Program of the State KeyLab of CAD&CG, Zhejiang University (No. A1501), the Fundamental Research Funds for the Central Universities (WK2350000002), the Open Funding Project of State Key Laboratory of Virtual Reality Technology and Systems, Beihang University (No. BUAA-VR-16KF-12), the Open Funding Project of State Key Laboratory of Novel Software Technology, Nanjing University (No. KFKT2016B08).

References

1. Taylor, S.L., Mahler, M., Theobald, B.J., Matthews, I.: Dynamic units of visual speech. In: ACM/Eurographics Symposium on Computer Animation, pp. 245–250 (2012)
2. Wang, L., Chen, H., Li, S., Meng, H.M.: Phoneme-level articulatory animation in pronunciation training. Speech Commun. **54**(7), 845–856 (2012)
3. Badin, P., Elisei, F., Bailly, G., Tarabalka, Y.: An audiovisual talking head for augmented speech generation: models and animations based on a real speaker's articulatory data. In: Perales, F.J., Fisher, R.B. (eds.) AMDO 2008. LNCS, vol. 5098, pp. 132–143. Springer, Heidelberg (2008). doi:10.1007/978-3-540-70517-8_14
4. Fagel, S., Clemens, C.: An articulation model for audio-visual speech synthesis - determination, adjustment, evaluation. Speech Commun. **44**(1–4), 141–154 (2004)
5. Deng, A., Neumann, U.: Expressive speech animation synthesis with phoneme-level controls. Comput. Graph. **27**(8), 2096–2113 (2008)
6. Ma, J., Cole, R., Pellom, W., Ward, B.: Accurate automatic visible speech synthesis of arbitrary 3D models based on concatenation of diviseme motion capture data. Comput. Animat. Virtual Worlds **15**, 485–500 (2004)
7. Anderson, R., Stenger, B., Wan, V., et al.: Expressive visual text-to-speech using active appearance models. In: CVPR, pp. 146–152 (2013)
8. Wang, L., Han, W., Qian, X., Soong, F.: Photo-real lips synthesis with trajectory-guided sample selection. In: ISCASSW, pp. 1–6 (2010)
9. Liu K., Ostermann, J.: Realistic facial expression synthesis for an image based talking head. In: ICME, pp. 1–6 (2011)
10. Wang, A., Emmi, M., Faloutsos, P.: Assembling an expressive facial animation system. In: SIGGRAPH VGS, pp. 21–26 (2007)
11. Xu, Y., Feng, A.W., Marsella, S., Shapiro, A.: A practical and configurable lip sync method for games. In: SIGGRAPH CMG, pp. 84–89 (2013)
12. Hartholt, A., Traum, D., Marsella, Stacy, C., Shapiro, A., Stratou, G., Leuski, A., Morency, L.-P., Gratch, J.: All together now: introducing the virtual human toolkit. In: Aylett, R., Krenn, B., Pelachaud, C., Shimodaira, H. (eds.) IVA 2013. LNCS (LNAI), vol. 8108, pp. 368–381. Springer, Heidelberg (2013). doi:10.1007/978-3-642-40415-3_33
13. Taylor, S.L., Mahler, M., et al.: Dynamic units of visual speech. In: SCA, pp. 245–250 (2012)
14. Li, S., Wang, L., Qi, E.: The phoneme-level articulator dynamics for pronunciation animation. In: ICALP, pp. 283–286 (2011)
15. Wang, L., Chen, H., et al.: Evaluation of external and internal articulator dynamics for pronunciation learning. In: InterSpeech, pp. 2247–2250 (2009)
16. Li, H., Yang, M.H., Tao, J.H.: Speaker-independent lips and tongue visualization of vowels. In: ICASSP, pp. 8106–8110 (2013)

17. Oura, K., et al.: Pitch adaptive training for HMM-based singing voice synthesis. In: ICASSP, pp. 5377–5380 (2012)
18. Saino, K., et al.: An HMM-based singing voice synthesis system. In: Proceedings of InterSpeech, pp. 2274–2277 (2006)
19. Yu, S.Z.: Hidden semi-Markov model. Artif. Intell. **174**(2), 215–243 (2010)
20. Tang, C.Y., Zhang, G., Tsui, C.P.: A 3D skeletal muscle model coupled with active contraction of muscle fibres and hyperelastic behavior. J. Biomech. **42**(7), 865–872 (2009)
21. Wang, Z.F., Zheng, Z.G.: A region based stereo matching algorithm using cooperative optimization. In: Proceedings of CVPR, pp. 701–708 (2008)
22. Kampmann, M.: Automatic 3-D face mode adaption for model-based coding of videophone sequences. IEEE Trans. CSVT **12**(3), 172–182 (2002)
23. Ekman, P., Friesen, W.V.: Manual for the Facial Action Coding System. Psychologists Press, New York (1978)
24. Ling, Z.H., Richmond, K., Yamagishi, J., Wang, R.H.: Articulatory control of hmm-based parametric speech synthesis driven by phonetic knowledge. In: Proceedings of InterSpeech, pp. 573–576 (2008)
25. Youssef, B., Badin, P., Bailly, G., Heracleous, C.: Acoustic-to articulatory inversion using speech recognition and trajectory formation based on phoneme hidden markov models, In: Proceedings of Interspeech, pp. 2255–2258 (2009)
26. Tokuda, K., Kobayashi, T., Imai, S.: Speech parameter generation from hmm using dynamic features. In: Proceedings of ICASSP, pp. 660–663 (1995)
27. Waters, K.: A muscle model for animating three dimensional facial expression. Comput. Graph. **22**(4), 17–24 (1987)
28. Marcos, S., Gómez-García-Bermejob, J., Zalama, E,.: A realistic facial animation suitable for human-robot interfacing. In: ICIRS, , pp. 3810–3815 (2008)
29. Safakis, E., Neverov, I., Fedkiw, R.: Automatic determination of facial muscle activations from sparse motion. ACM Trans. Graph. **24**(3), 417–425 (2005)
30. Lee, Y.C., Terzopoulos, D., et al.: Realistic modeling for facial animation. In: SIGGRAPH, pp. 55–62 (1995)
31. Miyawaki, K.: A study on the musculature of the human tongue. Annu. Bull. Res. Inst. Logop. Phoniatr. **8**, 23–50 (1974)
32. Mooney, M.: A theory of large elastic deformation. J. Appl. Phys. **11**(9), 582–592 (1940)
33. Kojic, M., Mijailovic, S., Zdravkovic, N.: Modelling of muscle behaviour by the finite element method using Hill's three-element model. Int. J. Numer. Methods Eng. **43**(5), 941–953 (1998)
34. Yu, J., Li, A.J., Wang, Z.F.: Data-driven 3D visual pronunciation of Chinese IPA for language learning. In: Proceedings of OCOCOSDA, pp. 93–98 (2013)
35. Muller, M.: Information Retrieval for Music and Motion. Springer, Berlin (2007)

A Structural Coupled-Layer Tracking Method Based on Correlation Filters

Sheng Chen[1], Bin Liu[1(✉)], and Chang Wen Chen[2]

[1] CAS Key Laboratory of Electromagnetic Space Information, University of Science and Technology of China, Hefei, China
flowice@ustc.edu.cn
[2] Department of Computer Science and Engineering, University at Buffalo, State University of New York, Buffalo, USA

Abstract. A recent trend in visual tracking is to employ correlation filter based formulations for their high efficiency and superior performance. To deal with partial occlusion issue, part-based methods via correlation filters have been introduced to visual tracking and achieved promising results. However, these methods ignore the intrinsic relationships among local parts and do not consider the spatial structure inside the target. In this paper, we propose a coupled-layer tracking method based on correlation filters that resolves this problem by incorporating structural constraints between the global bounding box and local parts. In our method, the target state is optimized jointly in a unified objective function taking into account both appearance information of all parts and structural constraint between parts. In that way, our method can not only have the advantages of existing correlation filter trackers, such as high efficiency and robustness, and the ability to handle partial occlusion well due to part-based strategy, but also preserve object structure. Experimental results on a challenging benchmark dataset demonstrate that our proposed method outperforms state-of-art trackers.

Keywords: Visual tracking · Correlation filter · Structural constraint · Part-based

1 Introduction

Visual tracking is an important problem in computer vision and artificial intelligence. Though much progress has been made in recent years, there are still unsolved issues due to critical scenes such as illumination changes, background clutter, fast object motion changes, and occlusions.

Recently, tracking-by-detection methods have become popular in visual tracking for their superior performance [13,19,20,24]. In particular, correlation filter based tracking-by-detection methods have been proven to be able to achieve fairly high speed and robust tracking performance in relatively simple environments [1,4]. And a recent trend in tracking is to employ correlation filter based formulations. In [2], Henriques et al. introduced a kernelized correlation filter

© Springer International Publishing AG 2017
L. Amsaleg et al. (Eds.): MMM 2017, Part I, LNCS 10132, pp. 65–76, 2017.
DOI: 10.1007/978-3-319-51811-4_6

method which achieved promising performance with high speed. Scale adaptation of correlation filters [3] was presented by Danelljan et al. which made use of the scale-space pyramid representation. In [5,6], correlation filters were integrated with other tracking strategies and their tracking performances were appealing. However, what these trackers suffer from the general drawback of holistic models is that they cannot cope with partial occlusion issue.

To deal with partial occlusion, the part-based strategy were introduced to correlation filters [7,8,22,23] and made much progress. While in [7,8], the object parts were independently tracked by the KCF tracker [2] and a simple weighted voting strategy was used to denote the final target position without considering structural constraint among parts. Thus, the object parts can move freely and have different directions, which results in drifting of the tracker. In [23], both global and local appearance information were taken into consideration, but structural constraint among local parts was still ignored. Liu et al. [22] proposed a SCF appearance model which designed a complex objective function to learn optimal states of local parts and structural constraints among them jointly, but it might be time-consuming. Actually, for general objects, although the appearance may change drastically over time due to multiple challenges, the structure relationships inside the target remain stable. Besides, in [9–12], the experimental results have shown that spatial relationships among parts are helpful to improve the tracking performance. Therefore, incorporating geometric constraints into correlation filters helps enhancing the robustness of a tracker.

Motivated by the above observations, we propose a novel couple-layer tracker based on correlation filters, which represents the target with its global and local appearance and models the intrinsic relationships among them by hierarchical star structure. At the global layer, the target is represented by the target template to capture the holistic information of the target; At the local layer, the target is represented by patches containing partial informations of the target. Every local patch and the global bounding box are allocated with an associated correlation filter. And the global bounding box is connected with local parts by a hierarchical star model. We choose the KCF tracker [2] as our base tracker (or called base appearance model) due to its high efficiency and robustness. Firstly, the part detection is performed with corresponding base tracker for global bounding box and local patches individually. Then the optimal state of the target and parts is obtained by quantifying the detection score and the structural cost between global bounding box and local parts, and optimized jointly in a unified objective function. Thus, the global layer and local layer can benefit from each other, and the overall performance of the tracker is enhanced. According to the final tracking state, we update the appearance model and hierarchical star structure online. In summary, the proposed method not only handles partial occlusion well, but also preserves object structure which makes the tracker more robust.

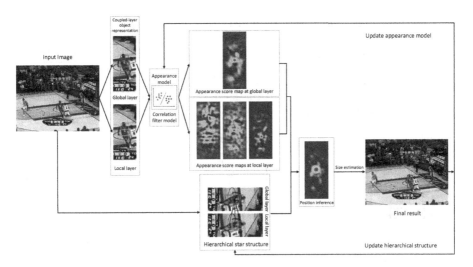

Fig. 1. Workflow of the proposed tracking algorithm

2 The Proposed Method

In this section, we first review the adopted base tracker (or called base appearance model), and then present the proposed coupled-layer object representation. Next, we incorporate the information from the global and local layers into a unified framework taking into account both the appearance score and structural cost to obtain the optimal positions of all parts. After that, we introduce a scale estimation method to obtain the final tracking state. Finally, the update strategy is performed to adapt to the change of the object. The workflow of the proposed algorithm is illustrated in Fig. 1.

2.1 The KCF Tracker

The KCF tracker [2] has drawn considerable attention for its impressive high efficiency and robustness. By considering the property of cyclic shifts, the correlation filter can learn from a relatively large number of training samples effectively and perform fast tracking in the Fourier domain under the tracking-by-detection framework.

The KCF tracker models the object appearance using a correlation filter w trained on an image patch x centred around the target with size $K \times L$. Taking advantages of the property of cyclic matrix and befitting padding, all the circular shifts of $x_{k,l}$, $(k,l) \in \{0, 1, ..., K-1\} \times \{0, 1, ..., L-1\}$, are considered as training samples. Each training sample $x_{k,l}$ is assigned with a soft label $y_{k,l}$, which is generated by a Gaussian function and takes a value of 1 for the centred target, and smoothly decays to 0 for any other shifts. The goal of training is to find a function $f(z) = w^T z$ that minimizes the following cost function

$$w = \min_{w} \sum_{k,l} |\langle \phi(x_{k,l}), w \rangle - y_{k,l}|^2 + \lambda_c ||w||^2 \tag{1}$$

where ϕ represents the mapping to the kernel space, and λ_c is a regularization parameter. The solution w can be expressed as $w = \sum_{k,l} \alpha(k,l) \phi(x_{k,l})$. Employing a kernel $\kappa(x, x') = \langle \phi(x), \phi(x') \rangle$, the coefficient α is calculated as

$$\alpha = \mathcal{F}^{-1} \left(\frac{\mathcal{F}(y)}{\mathcal{F}(\kappa(x,x)) + \lambda_c} \right) \tag{2}$$

where $y = \{y_{k,l} | (k,l) \in \{0,1,\ldots,K-1\} \times \{0,1,\ldots,L-1\}\}$, $\mathcal{F}$ and $\mathcal{F}^{-1}$ denote the Fourier transform and its inverse. During tracking, an image patch z with the search window size $K \times L$ is cropped out in the new frame. Then the score map $\hat{y}$ is calculated as

$$\hat{y} = \mathcal{F}^{-1}(\mathcal{F}(\alpha) \odot \mathcal{F}(\kappa(\hat{x}, z))) \tag{3}$$

here, $\odot$ is the element-wise product and $\hat{x}$ denotes the learned target appearance template. For the KCF tracker, the optimal position of the target is obtained by searching the maximal value of the score map $\hat{y}$.

2.2 Coupled-Layer Object Representation

Target Appearance Model. The target appearance is modeled at two different layers of granularity: the bounding box (global) layer and the partial (local) layer. The appearance model $M = (M^g, M^l)$ consists of two models: M^g at the global layer, M^l at the local layer. At the global layer, we use a bounding box to delimit the target. At the local layer, we divide the holistic bounding box into several patches to represent the target. For the bounding box at the global layer and each part at the local layer, an independent correlation filter is attached as its base tracker. Thus the appearance model M^g at the global layer and M^l at the local layer are denoted as

$$M^g = (s_0, \hat{x}_0, \alpha_0) \tag{4}$$

$$M^l = \{(s_i, \hat{x}_i, \alpha_i)\}_{i=1}^{N} \tag{5}$$

where $s_i = [c_i, w_i, h_i]$ is the tracking state, with c_i being the position, w_i and h_i being the width and height for ith part. $\hat{x}_i$ and α_i are the learned appearance template and the parameters of the correlation filter for ith part, respectively. N is the number of local parts.

Hierarchical Star Structure. To constrain the inference of the part states, we need a specific structural model to represent the relationships among parts. Here we choose a hierarchical star structural model due to its high-efficiency.

There are a number of parts at local layer and one bounding box at global layer. Thus we exploit a hierarchical star model to construct the geometric structure between the global layer and the local layer. In this framework, each part at the local layer is connected with the bounding box at the global layer via the edges in star model. We denote the edge between ith local part and the global bounding box as $e_{0,i} = c_0 - c_i$. The edges in the star model can be viewed as springs that represent spatial constraints between the local patches and the global bounding box.

Adaptive Weights for Parts. During tracking, different parts of the target may suffer from different problems such as occlusion, illumination change and background clutter, while base trackers for the bounding box and parts may not always cope with well. If we apply the same weight for each part to vote for final target state, the influence of falsely tracked parts may be unfairly emphasized. Given the detection result of each partial base tracker, the contribution made for final target state should be different from each part, i.e. higher weight signifies that this part is more reliable than others, and makes more contributions to the final state of target. To estimate the confidence level a patch can be tracked correctly, we adopt three different evaluation criteria as our confidence metric. Firstly, we introduce the maximum response value of score map (MR) to evaluate the fitness of that score maps. The higher MR value means the detected result fit in with the appearance model better. Secondly, we accept the peak-to-sidelobe ratio (PSR) [7] to measure the sharpness of the score map peak. The higher PSR value means more discriminative detection. Thirdly, we take the smooth constraint of score maps (SCCM) [8] as a metric to measure the temporal smoothness of score maps. For tracking problems, the temporal smoothness property is helpful for detecting whether the target is occluded, i.e. the lower SCCM value means that the detecting score is more stable temporally. The weight parameter is a combination of these criteria:

$$W_i^t = MR_i \left(PSR_i + \eta \frac{1}{SCCM_i} \right) \tag{6}$$

$$MR_i = \max \left(\hat{f}_{c_i}^t \right) \tag{7}$$

$$PSR_i = \frac{\max \left(\hat{f}_{c_i}^t \right) - \mu_i}{\sigma_i} \tag{8}$$

$$SCCM_i = \| \hat{f}_{c_i}^t - \hat{f}_{c_i}^{t-1} \oplus \Delta \|_2^2 \tag{9}$$

where $\hat{f}_{c_i}^{t-1}$ and $\hat{f}_{c_i}^t$ are the score maps of the ith part at frame $t-1$ and t, respectively. μ_i and σ_i are the mean and the standard deviation of the ith score map respectively. η is the balance parameter between sharpness and smoothness of score maps. $\oplus$ means a circular shift operation of the score map, and $\Delta = (\Delta x, \Delta y)$ denotes the relative shift vector of maximum value in score maps from frame $t-1$ and t. While Δx and Δy are relative shift values in the horizontal axis and vertical axis direction, respectively.

2.3   Target State Inference

Given the appearance score for coupled-layer of the target, the difficulty is how to combine the score maps to infer positions of different parts. Here we employ a unified scheme in which positions of all parts are optimized jointly in a score objective function taking into account both the appearance score and the hierarchical star structural constraint.

The configuration of all the parts in a new frame I is denoted as $C = \{c_0, c_1, \ldots, c_N\}$, where c_0 is the position of the bounding box in the appearance model M, $c_1, \ldots, c_N$ are corresponding positions of local patches in M. The score of a configuration C consists of two terms: (1) the appearance score which is the weighted sum of appearance scores of all parts, and (2) the deformation cost which measures how much a configuration compresses or stretches the springs between the local patches and the global bounding box (connected edges). The optimal configuration is calculated by maximizing its scores defined as

$$S\left(C; I\right) = S_{app} - \lambda_s S_{def} \tag{10}$$

where S_{app} is the appearance score, S_{def} is the deformation cost. And λ_s is the trade-off term between them.

The appearance score S_{app} is defined as

$$S_{app} = \bar{W}_0 S^g\left(I, c_0\right) + \sum_{i=1}^{N} \bar{W}_i S_i^l\left(I, c_i\right) \tag{11}$$

where $S^g\left(I, c_0\right)$ and $S_i^l\left(I, C_i\right)$ are appearance scores of the global bounding box at location c_0 and ith local patch at location c_i, respectively. While we obtain these appearance scores directly from score maps of detected results. And $\bar{W}_i$ is the normalized form of the weight parameter W_i for ith part.

The deformation cost S_{def} is defined as

$$S_{def} = \sum_{i=1}^{N} \bar{W}_i || \left(c_0 - c_i\right) - e_{0,i}||_2^2 \tag{12}$$

Finally, the unified objective function is presented as

$$S\left(C; I\right) = \bar{W}_0 S^g\left(I, c_0\right) + \sum_{i=1}^{N} \bar{W}_i (S_i^l\left(I, c_i\right) - \lambda || \left(c_0 - c_i\right) - e_{0,i}||_2^2) \tag{13}$$

The optimal configuration C^* is then given by

$$C^* = \arg\max_{\hat{C}} S\left(\hat{C}; I\right) \tag{14}$$

and can be solved by dynamic programming.

2.4 Scale Estimation

After obtaining optimal location of the target, the target size should be estimated in current frame. Similar to recent scale-adaptive state-of-the-art tracker (DSST) [3], we take a fast and effective way: learning a separate 1-dimensional correlation filter for the scale estimation.

Similar to 2-dimensional correlation filter model for the KCF tracker, the key idea of 1-dimensional correlation filter model for the scale estimation is to learn a discriminative classifier in the Fourier domain and estimate the scale of the target by searching the scale space for the scale which has the maximal response value. The difference is that the training example for updating the scale filter is calculated by extracting features using variable patch sizes centred around the target rather than using circular shifts. Let $w_0 \times h_0$ denote the target size in current frame and D be the size of the scale filter. For each $i \in \left\{ -\frac{D-1}{2}, \ldots, \frac{D-1}{2} \right\}$, we extract an image patch J_i of size $\varepsilon^i w_0 \times \varepsilon^i h_0$ centred around the target. Here, ε is the scale factor between feature layers. All D image patches are then resized to the template size for the feature extraction. The training and detection steps can be referred to [3] in detail. Finally, the optimal scale is obtained by searching the scale space for the scale with maximal detection score.

2.5 Update Strategy

After obtaining the optimal positions and scale for target and local parts, we need to update appearance model and structural constraint.

Appearance Model Update. During tracking, the object appearance will change because of a number of factors such as illumination change and occlusion. Typically, trackers using a holistic correlation filter model employ the forget-remember mechanism to update their filter coefficients: $\mathcal{F}(\alpha^t) = (1 - \beta_c)\mathcal{F}(\alpha^{t-1}) + \beta_c \mathcal{F}(\alpha)$, where β_c is the learning rate parameter. For part-based method, it is apparent that the model of an occluded or unreliable part should not be updated to avoid introducing errors. Thus we address these situations with an adaptive update strategy.

For the correlation filter of ith part in frame t, the update scheme is presented as

$$\mathcal{F}(\alpha_i^t) = \begin{cases} (1 - \beta_c)\mathcal{F}(\alpha_i^{t-1}) + \beta_c \mathcal{F}(\alpha_i) \,, PSR_i \geq ThreP, SCCM_i \leq ThreS \\ \mathcal{F}(\alpha_i^{t-1}) \,, otherwise \end{cases}$$

(15)

$$\hat{x}_i^t = \begin{cases} (1 - \beta_c)\hat{x}_i^{t-1} + \beta_c x_i \,, PSR_i \geq ThreP, SCCM_i \leq ThreS \\ \hat{x}_i^{t-1}, otherwise \end{cases}$$

(16)

where $ThreP$ and $ThreS$ are threshold parameters. As can be seen, these updates are performed when parts are reliable. And two predefined threshold parameters, $ThreP$ and $ThreS$ are used to judge whether a part is reliable as we found the two adopted confidence metrics, PSR and SCCM could reflect the reliability of a part to a large extent. While lower PSR means the detection result of corresponding part is less reliable, and higher SCCM indicates there is larger possibility that corresponding part suffer from heavy occlusion.

Structural Constraint Update. After obtaining optimal positions of all parts, the edge of star model is updated as $e_{0,i} = c_0^t - c_i^t$, where c_0^t and c_i^t are obtained optimal positions of the bounding box and ith part in current frame t, respectively.

What is more, we consider the hierarchical star model as spring mechanism which contracts and stretches within limited lengths. And it will not be able to recover when going beyond the limited lengths. Therefore, lengths of edges between the target center and local patches are relatively loose but have limited thresholds. When $\left| e_{0,i} - v \cdot e_{0,i}^{init} \right| \geq \rho \cdot [w_0, h_0]$ or c_i is out of the bounding box, we think ith part is broken, and then reallocate a new part in the bounding box according to $e_{0,i}^{init}$ to keep the number of local parts unchanged. $e_{0,i}$ and $e_{0,i}^{init}$ are the current edge and the initial edge between the target center and ith local patch, and v is the scale variation from the initial frame to current frame. ρ is a threshold parameter.

3   Experiments

3.1   Experimental Setting

We compare the proposed tracker on all 51 video sequences from OTB benchmark [18] with 8 state-of-the-art trackers: TLD [13], Struck [14], L1T [15], CSK [16], Frag [17], and three correlation-filter based trackers including the base tracker (KCF) [2], a part-based tracker (RPT) [7], and a scale estimation approach (DSST) [3]. Our approach is implemented in Matlab with HoG feature [21] extraction implemented in C++ and performed at about 16 fps on an Intel Core i7 machine.

In our experiment, we divide the object bounding box into three or four patches as local parts according to the initial aspect ratio of the object, as shown in Fig. 2. For parameter setting of the base tracker, the padding parameter is 3, the regularization parameter λ_c in Eq. (1) is 0.0001, and the learning rate β_c in Eq. (16) is set to 0.01. The balance parameter η in Eq. (6) is 3. The trade-off parameter λ_s in Eq. (10) is set to be 0.002. For the scale estimation step, the scale factor ε is set to 1.02 and $D = 33$. For update steps, we found that when PSR drops to around 8.0, it is an indication that the detection of corresponding part is not reliable. When the value of SCCM is larger than 0.6, it means that the parts may suffer heavy occlusion. So $ThreP$ and $ThreS$ are set to 8.0 and 0.6, respectively. And the threshold parameter ρ is set to 0.2.

Fig. 2. The initial local parts setting in first frame. (a) Left: when the aspect ratio is less than 0.5. (b) Center: when the aspect ratio is greater than 2. (c) Right: otherwise. Video sequences are from *singer*1, *suv* and *faceOcc*1 [17], respectively.

3.2 Quantitative Evaluation

Two common evaluation metrics are used here for quantitative comparison: central location error (CLE) and overlap Rate (OR). OR is defined as $OR = \frac{B_T \cap B_G}{B_T \cup B_G}$, where B_T is the bounding box generated by trackers and B_G is the ground-truth bounding box. CLE is defined as the Euclidean distance between the center locations of B_T and B_G. We adopt the precision plot and success plot suggested by [18]. The precision plot shows the percentage of successfully tracked frames vs. the CLE, which ranks the trackers as precision score at 20 pixels. The success plot depicts the percentage of successfully tracked frames vs. OR threshold, where Area Under The Curve (AUC) is used as metric for ranking.

As shown in Fig. 3, our proposed method ranks first and achieves the best performance in both the precision plot and success plot. Compared with the base tracker KCF with 0.742 precision ranking score and 0.516 success ranking score, our method has achieved over 13% and 20% improvements, respectively. And compared with the part-based formulation tracker based on correlation filters (RPT), our approach obtained better tracking performance by 3.3% at precision

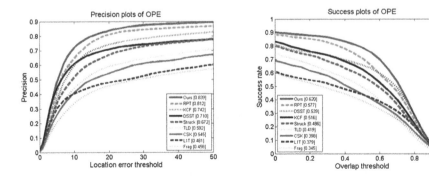

Fig. 3. The precision plot and success plot of the OTB benchmark

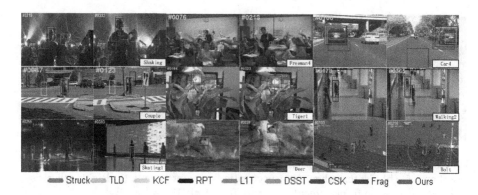

Fig. 4. Tracking results of 9 trackers on 9 challenging video sequences

ranking score and 7.6% at success ranking score. Overall, our proposed method outperforms those state-of-art trackers.

3.3 Qualitative Evaluation

In Fig. 4, we show some key frames of the comparison experiment on 9 challenging sequences between our proposed method and existing algorithms. The 9 video sequences contains various challenging factors including fast motion, scale and illumination variation, background clutter and occlusion. Overall, our method is robust against those challenging factors and performs best, and other methods have different problems. The tracking performances are discussed below according to different challenges.

Fast Motion. Due to fast motion, L1T, KCF, RPT methods drift in the *couple* sequence. In the *bolt* sequence, all existing trackers fail except the KCF tracker and our tracker. While in the *deer* sequence, TLD, KCF and Frag methods lose tracking. Comparatively, our tracker performs well throughout the situation of fast motion.

Scale and Illumination Variation. In the *car*4 and *walking*2 sequences, L1T, CSK and Frag methods fail due to scale variation. In the *skating*1 sequence, the illumination changes makes Struck, TLD, L1T and CSK methods fail. Comparatively speaking, our method handles scale and illumination variation problems well.

Background Clutter. In the *shaking* sequence, Struck, TLD, KCF, L1T, DSST and Frag methods suffer from the background clutter and drift far away from the target, while our method tracks the target successfully till the end.

Occlusion. In the *tiger* and *walking*2 sequences, Struck, TLD, L1T, DSST, CSK and Frag methods drift away when the target undergoes partial occlusion. In the *freeman*4 sequence, only our tracker can locate the target accurately, while all other methods fail. Tracking results of these sequences demonstrate our tracker is robust with the situation of partial occlusion.

4 Conclusion

In this paper, we have proposed a structural coupled-layer tracking method based on correlation filter. The proposed method takes part-based strategy into correlation filter tracker, and exploits the spatial structure inside the target. The optimal target state is obtained by incorporating both appearance scores and structural cost into a unified objective function. Thus, it not only has the advantages of existing correlation filter trackers, such as high efficiency and robustness, and the ability to handle partial occlusion well due to part-based strategy, but also can preserve object structure. Experimental results on challenging benchmark video sequences demonstrate the superiority of our proposed approach compared with 8 recent state-of-art trackers.

References

1. Bolme, D.S., Beveridge, J.R., Draper, B.A., Lui, Y.M.: Visual object tracking using adaptive correlation fillters. In: IEEE Conference on Computer Vision and Pattern Recognition, pp. 2544–2550 (2012)
2. Henriques, J.F., Caseiro, R., Martins, P., et al.: High-speed tracking with kernelized correlation filters. IEEE Trans. Pattern Anal. Mach. Intell. 37(3), 583–596 (2014)
3. Danelljan, M., Hger, G., Khan, F., et al.: Accurate scale estimation for robust visual tracking. In: British Machine Vision Conference (2014)
4. Rodriguez, A., Boddeti, V.N., Kumar, B.V.K.V., et al.: Maximum margin correlation filter: a new approach for localization and classification. IEEE Trans. Image Process. 22(2), 631–643 (2013)
5. Hong, Z., Chen, Z., Wang, C., et al.: MUlti-Store Tracker (MUSTer): a cognitive psychology inspired approach to object tracking. In: IEEE Conference on Computer Vision and Pattern Recognition, pp. 749–758 (2015)
6. Ma, C., Yang, X., Zhang, C., et al.: Long-term correlation tracking. In: IEEE Conference on Computer Vision and Pattern Recognition, pp. 5388–5396 (2015)
7. Li, Y., Zhu, J., Hoi, S.C.H.: Reliable patch trackers: robust visual tracking by exploiting reliable patches. In: IEEE Conference on Computer Vision and Pattern Recognition, pp. 353–361 (2015)
8. Liu, T., Wang, G., Yang, Q.: Real-time part-based visual tracking via adaptive correlation fillters. In: IEEE Conference on Computer Vision and Pattern Recognition, pp. 4902–4912 (2015)
9. Felzenszwalb, P.F., Girshick, R.B., McAllester, D., et al.: Object detection with discriminatively trained part-based models. IEEE Trans. Pattern Anal. Mach. Intell. 32(9), 1627–1645 (2010)
10. Zhang, L., Maaten, L.V.D.: Structure preserving object tracking. In: IEEE Conference on Computer Vision and Pattern Recognition, pp. 1838–1845 (2013)
11. Čehovin, L., Kristan, M., Leonardis, A.: An adaptive coupled-layer visual model for robust visual tracking. In: IEEE Conference on Computer Vision and Pattern Recognition, pp. 1363–1370 (2011)
12. Wang, J., Yu, N., Zhu, F., et al.: Multi-level visual tracking with hierarchical tree structural constraint. Neurocomputing 202, 1–11 (2016)
13. Kalal, Z., Mikolajczyk, K., Matas, J.: Tracking-learning-detection. IEEE Trans. Pattern Anal. Mach. Intell. 34(7), 1409–1422 (2012)

14. Hare, S., Saffari, A., Torr, P.H.S.: Struck: structured output tracking with kernels. In: IEEE Conference on Computer Vision and Pattern Recognition, pp. 263–270 (2011)
15. Bao, C., Wu, Y., Ling, H., et al.: Real time robust L1 tracker using accelerated proximal gradient approach. In: IEEE Conference on Computer Vision and Pattern Recognition, pp. 1830–1837 (2012)
16. Henriques, J.F., Caseiro, R., Martins, P., et al.: Exploiting the circulant structure of tracking-by-detection with kernels. In: European Conference on Computer Vision, pp. 702–715 (2012)
17. Adam, A., Rivlin, E., Shimshoni, I.: Robust fragments-based tracking using the integral histogram. In: IEEE Conference on Computer Vision and Pattern Recognition, pp. 798–805 (2006)
18. Wu, Y., Lim, J., Yang, M.H.: Online object tracking: a benchmark. In: IEEE Conference on Computer Vision and Pattern Recognition, pp. 2411–2418 (2013)
19. Grabner, H., Grabner, M., Bischof, H.: Real-time tracking via on-line boosting. In: The British Machine Vision Conference, vol. 1, no. 5, p. 6 (2006)
20. Babenko, B., Yang, M.H., Belongie, S.: Robust object tracking with online multiple instance learning. IEEE Trans. Pattern Anal. Mach. Intell. **33**, 1619–1632 (2011)
21. Dalal, N., Triggs, B.: Histograms of oriented gradients for human detection. In: IEEE Conference on Computer Vision and Pattern Recognition, pp. 886–893 (2005)
22. Liu, S., Zhang, T., Cao, X., et al.: Structural correlation filter for robust visual tracking. In: IEEE Conference on Computer Vision and Pattern Recognition, pp. 4312–4320 (2016)
23. Akin, O., Erdem, E., Erdem, A., et al.: Deformable part-based tracking by coupled global and local correlation filters. J. Vis. Commun. Image Represent. **38**, 763–774 (2016)
24. Wang, J., Fei, C., Zhuang, L., et al.: Part-based multi-graph ranking for visual tracking. In: IEEE International Conference on Image Processing, pp. 1714–1718 (2016)

Augmented Telemedicine Platform
for Real-Time Remote Medical Consultation

David Anton[(✉)], Gregorij Kurillo, Allen Y. Yang, and Ruzena Bajcsy

University of California at Berkeley, Berkeley, USA
{davidantonsaez,gregorij,yang,
bajcsy}@eecs.berkeley.edu

Abstract. Current telemedicine systems for remote medical consultation are based on decades old video-conferencing technology. Their primary role is to deliver video and voice communication between medical providers and to transmit vital signs of the patient. This technology, however, does not provide the expert physician with the same hands-on experience as when examining a patient in person. Virtual and Augmented Reality (VR and AR) on the other hand have the capacity to enhance the experience and communication between healthcare professionals in geographically distributed locations. By transmitting RGB+D video of the patient, the expert physician can interact with this real-time 3D representation in novel ways. Furthermore, the use of AR technology at the patient side has potential to improve communication by providing clear visual instructions to the caregiver. In this paper, we propose a framework for 3D real-time communication that combines interaction via VR and AR. We demonstrate the capabilities of our framework on a prototype system consisting of a depth camera, projector and 3D display. The system is used to analyze the network performance and data transmission quality of the multimodal streaming in a remote scenario.

Keywords: Telepresence · Virtual reality · Augmented reality · Real-time multimedia streaming · 3D interaction

1 Introduction

The use of Health Information Technology (HIT) has been promoted as having tremendous promise in improving the efficiency, cost-effectiveness, quality, and safety of medical care delivery [1]. Telemedicine and eHealth have been constantly evolving and are intended to serve as means to reduce costs and improve healthcare [2, 3]. Several technologies have been proposed to deliver care to rural and remote areas away from large medical centers, which are typically located in urban areas. Until recently, telemedicine has relied on video-conferencing technology to establish communication between health care professionals and patients [4]. Two-dimensional (2D) wireless video-conferencing has been proposed as a replacement for cell phone and radio consultation in emergency medicine. These implementations have relied on a single-viewpoint video acquisition and transmission [5, 6]. Single viewpoint however provides limited information on patient's condition (e.g. injury) as compared to in-person

© Springer International Publishing AG 2017
L. Amsaleg et al. (Eds.): MMM 2017, Part I, LNCS 10132, pp. 77–89, 2017.
DOI: 10.1007/978-3-319-51811-4_7

examination. The fast access to a medical expert is especially relevant in emergency situations, where intervention, in the so-called "golden hour" since occurrence of trauma, is critical for the efficiency of care [7]. Real-time telemedicine systems have only in recent years been adopted in emergency medicine for civilian [8, 9] and military use [10–12]. In 1995, Satava [13] proposed to use virtual reality in telemedicine based diagnostics and treatment to improve combat casualty care in battlefield scenarios. A decade later, Welch et al. [14] demonstrated a 3D telepresence system for real-time streaming of surgical procedures based on surface reconstruction from 8 cameras. The prototype system was intended to facilitate real-time or offline remote observation of medical procedures, with 3D visualization, audio, and vital signs. Their system however required complex hardware setup, dedicated network, and custom network protocols. In contrast to this work, we propose to use affordable, low-complexity, and off-the-shelf technology that can be easily deployed in a clinical setting. Our framework aims to use virtual reality (VR) and augmented reality (AR) to connect two remote locations for real-time remote medical consultation in emergency scenarios. Our approach makes use of standard and widely used peer-to-peer library WebRTC for real-time communication. We provide details of the implementation of the framework and demonstrate its capabilities on a prototype setup consisting of a depth camera, projector and 3D display. Finally, we present the network performance results from a set of data streaming experiments conducted between a medical institution and an external remote site.

2   Augmented Telemedicine Framework

The proposed augmented telemedicine framework consists of two geographically distributed modules that are connected over the network to facilitate interaction between a physician at the hospital and the first responder at the patient side. The module at the patient side (*AR Client*) is capable of capturing texture, 3D body surface, and vital signs of the patient (e.g., heart rate, blood pressure, etc.). Furthermore, the module provides augmented visual feedback by overlaying graphical annotations generated by the expert physician over the real environment. At the physician side (*VR Client*), VR technology is used to visualize streamed information in an interactive 3D environment which allows user to closely examine specific parts of the remote scene. This creates an informative and immersive experience while real-time nonverbal communication (pointing, annotations, vital signs monitoring, etc.) enhances the access and exchange of meaningful information between the first responder and the physician (see Fig. 1).

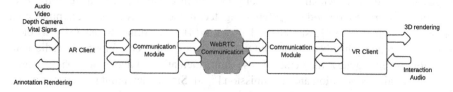

Fig. 1.   General overview of the framework modules and the data exchange pipeline

AR Client. This module is responsible for the data acquisition at the patient side and the rendering of augmented feedback sent from the remote location. A 3D camera is used to capture in real time 3D surface information of the area of interest (e.g., injury site). These data are transmitted to the *Communication Module* via *Named Pipes*, which facilitate inter-process communication in Microsoft Windows. In this way, each component of the framework is independent of the other in terms of the implementation. In the AR Client, three different pipes are used: RGB video, depth, and arbitrary (binary) data pipe (see Fig. 2). The client library is shared between the AR and VR clients and includes the following components. *Scene* and *Overlay Managers* are responsible for managing the list of objects and their location in the scene and the graphic user interface (GUI) components, respectively. The *Annotation Manager* provides methods to create different types of annotation and references their location in the scene. *Communication Manager* sends and receives scene updates between the two clients via the data pipe. Finally, the rendering component includes OpenGL-based implementation of the scene rendering using vertex buffer objects (VBOs).

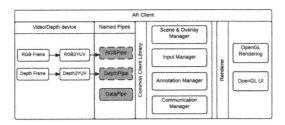

Fig. 2. Components of the AR Client

Communication Module. The Communication Module enables secure peer-to-peer connections between two geographically distributed clients and provides the infrastructure to stream audio, video, and binary data via standardized WebRTC protocol (Fig. 3). To transmit the data received from the client via Named Pipes, the following WebRTC channels are utilized: (1) two channels for RGB and depth combined in a single synchronized video stream, (2) data channel for binary data, and (3) audio channel for voice communication. As part of the standard implementation, the video frames are encoded using VP8 compression, the audio is encoded using Opus audio codec, and the binary data are encoded using *Protocol Buffers*.

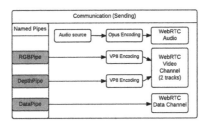

Fig. 3. Components of the communication module (for sending data)

VR Client. It is responsible for the rendering and data interaction at the physician side. The components of the VR Client are similar as those shown for the AR Client (see Fig. 2), with a different implementation of the rendering and the user interface modules. The VR Client receives remotely streamed data via the Communication Module, which decodes and redirects the data streams to the corresponding pipes. (Note that video, audio, depth and binary data encoding processes shown in Fig. 3 are reversed.) Afterwards the VR client renders the data from the remote site on a 3D display, allowing the physician to interact with the reconstruction and examine patient's data.

2.1   Real-Time Communication

One of the main goals and challenges in telepresence is achieving the real-time interaction between users. For this purpose, the Communication Module described in the previous section incorporates a real-time peer-to-peer communication library, based on WebRTC. WebRTC is a collection of standards, protocols, and APIs, which enables peer-to-peer audio, video, and data sharing in real time between peers in real-time. WebRTC manages different kinds of data in streams (video and audio) and data channels (binary or text data) while the streams are adapted in terms of resolution, bitrate or framerate to the available bandwidth and network state. Moreover, thanks to its implementation of secure communication protocols and platform independency, it is an ideal network framework for personal data and real-time interaction in remote locations as all the communications are encrypted by default [15].

2.2   Depth and Video Acquisition

Considering the variety of depth sensing devices available on the market (e.g., Microsoft Kinect, Intel RealSense, etc.), we have adopted a modular design that allows for transmission and visualization of RGB+D video independent of the input source. We have created an abstract frame grabber in WebRTC that retrieves the video frames from *Named Pipes* instead of a specific physical device. The user only needs to implement an appropriate device reader and encode the video and depth frames in 8-bit YUV format. While RGB video is easily converted to YUV color space, encoding the depth image directly to YUV would create undesirable artefacts in the mesh reconstruction due to video compression. To encode the depth image, we take advantage of the YUV frame structure, which consists of the 8-bit Y channel in the original resolution and two 8-bit U and V channels in one quarter of the original resolution. The depth map, reduced to 8-bits, is encoded into the Y channel while the UV channels are used to encode the numerical parameters needed to reconstruct the original depth map. To encode these parameters, we divide the UV channels into several horizontal strips and fill the pixels of each strip with the appropriate repeated values. Once the depth image is transmitted, it is possible to retrieve these parameters by averaging the pixel values of each horizontal segment and use them to restore the original depth map. The encoding solution is specific for each device depending on the range and quality of the depth maps. In Sect. 3.2.1, we present the details of our solution to encode Kinect depth frames in YUV 8-bit frames.

2.3 Custom Binary Data

Data channels are designed for the exchange of arbitrary data between peers. These channels can stream text (UTF8) or binary data. For the purposes of this project we stream binary messages encoded as ProtocolBuffers between two peers. Protocol buffers, developed by Google [16], provide fast and efficient binary serialization/ deserialization of data [17], while supporting human friendly XML-based data structure templates for several programming languages (e.g., Java, C++, and C#). For this framework, we created templates for four basic message categories: *Pointer messages*, *Annotation messages*, *Action messages*, and *Data messages*. *Pointer messages* define several modes to interact and highlight areas of the scene based on the stylus input. *Annotation messages* include attributes of the annotations that are overlaid over the patient's body. *Action messages* consist of control instructions to interact with the 3D scene (e.g., delete annotation, clear display, send file, etc.). Finally, *Data messages* incorporate data structures for streaming vital signals data obtained from an external sensor device (e.g., heartbeat, oxygen, etc.).

3 Architecture

Based on the AR/VR framework presented in the previous section, we implemented a prototype system shown in Fig. 4. The system consists of two distributed clients, a *patient site* where a caregiver assists a patient and the *expert site* where a doctor provides assistance. At the patient side, augmented feedback is achieved via the Microsoft Kinect Xbox One camera for real time 3D scanning and a projector to overlay graphical annotations over the patient's body. The expert site includes a virtual reality setup with a fully interactive 3D display zSpace (zSpace, Sunnyvale, CA) that depicts the remote scene in real-time and provides the tools for the provision of real-time feedback (see Fig. 4).

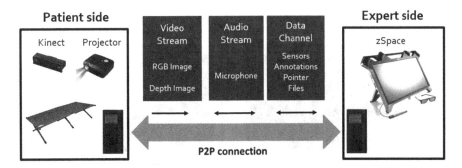

Fig. 4. Prototype architecture overview

3.1   Hardware

For the implementation of the prototype we used the following off-the-shelf devices that were integrated with our framework.

Kinect is a depth sensing camera developed by Microsoft that includes natural interaction capabilities. Kinect provides high resolution RGB (1920 × 1080 pixels) and depth video output (512 × 424 pixels) with 30 fps [18]. The depth map is obtained using time of flight (ToF) technology, allowing for the capture range between 0.7–6 m.

zSpace is 24″ high definition (1920 × 1080 pixels) stereoscopic display that allows for visualization of 3D data. A user has to wear a pair of lightweight circularly polarized glasses with small reflective markers, which are tracked to adjust the perspective to the user's viewpoint. The zSpace system includes a stylus for 3D interaction.

3.2   Patient Site Client

The Patient Site has been designed with portability and simplicity in mind. A desktop PC with Kinect camera and projector is used to acquire data and provide augmented feedback. The data acquisition and rendering is managed by the AR Client Module that connects to the Communication Module to establish the connection with the remote peer. To achieve alignment between the graphical annotations and the real-world geometry, the projector and Kinect are calibrated using a checkerboard and a software adapted for Kinect V2 based on the work presented in [19].

3.2.1   Kinect Depth Encoding

As mentioned earlier, WebRTC requires that the video streams are encoded as 8 bit YUV frames before applying the video compression. Since the depth map provided by the Kinect is 13-bit, it has to be first optimized for streaming. We developed a method to compress and encode depth maps as YUV frames compatible with WebRTC (see Fig. 5) as described in the following section. Using this approach, we reduce the depth information to a limited depth window in order to reduce the number of required bits.

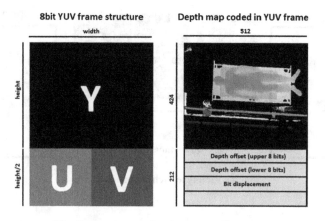

Fig. 5. Depth encoding in YUV 8 bit frames

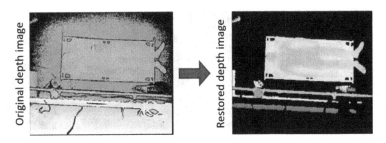

Fig. 6. Kinect depth encoding/decoding process

First, the depth range is constrained to a volume of interest (e.g. patient's body). 8-bits can provide a range of 256 unique depth values. In the presented setup, we limit the data to a depth window from 1300 mm to 2300 mm from the Kinect device. All the values out of range are set to 0. Next, we subtract 1296 from all non-zero values and we apply a 2-bit right shift, setting the range of values from 1 to 251. At this point the depth map can be encoded in the Y channel. To encode the numerical parameters, we divide the UV channels into 4 horizontal strips and fill the pixels of each strip with the appropriate repeated values. The depth offset (in this case 1296, in binary notation 00000101 00010000) is encoded into the first two strips, with the upper eight bits in the first strip (in this case 16) and the lower eight bits in the second line (in this case 5). In the third line, the 2-bit displacement value is encoded (in this case 2). All the parameters are multiplied by a factor of 10 to reduce the probability of retrieving incorrect values due to the noise in pixel intensity values introduced by the compression. Once encoded, this YUV frame is streamed through the network as a standard WebRTC video stream.

To restore the depth frame at the receiving side, the parameters for the reconstruction are first retrieved from the UV channels by averaging all the values of each horizontal strip. The depth image is then restored by applying the encoding steps in a reversed order. Furthermore, morphological operations are applied to remove some of the noise and the artifacts that may appear after the transmission (see Fig. 6). These are especially apparent on the edges where abrupt changes in depth occur. Note that this approach reduces the size of the depth map packet to about 5% of the original size; however, the depth range is limited to 1.0 m while the depth values lose 2 bits of precision when recovered.

3.3   Expert Site Client

Once the Kinect data stream is received and decoded, each frame is in real-time reconstructed into a textured 3D mesh for rendering on the zSpace display.

We have designed an intuitive and user friendly interface to interact with this 3D reconstruction and to provide access to the different interaction tools. The tools include: view manipulation, on-body annotations, "laser" pointer, region highlighting, and 3D linear and surface measurements (see Fig. 7). The annotations, highlights, and pointer

Fig. 7. 3D interface running in zSpace (left) and annotations projected on patient's body (right)

location are transmitted for augmented feedback to the patient side via the communication module as described previously.

4 Performance Evaluation

We have conducted a set of data streaming experiments to evaluate the performance of the communication framework. The network experiments were performed between a medical institution and an external remote site within an approximate distance of 110 km. The experimental sessions took place on the same day at different times (one-hour difference between each test) to include various network conditions. We compared the performance of data transmission with and without CPU overuse detection. The CPU overuse detection is an additional WebRTC option that regulates the video resolution taking into account also the local CPU usage, we analyze results of using this feature as it is mainly intended to preserve battery in mobile devices, and therefore it can be useful to activate it in a remote context. A second computer at the remote site hosted the WebRTC server application to initiate the communication between peers.

The tests consisted of streaming two 3 min videos that were recorded using Kinect Studio. Pre-recorded video was used in order to provide consistent data input. *Video1* was a recording of simulated interaction of a first responder with a patient lying on a stretcher. *Video2* was a recording of an upper extremity exercise for physical therapy. The videos were chosen based on envisioned scenarios where this technology could be used. Both videos included a continuous audio recording consisting of a person reading instructions. For all the tests the initial quality of multimedia data was as follows: video data (1920 × 1080 pixels), depth data (512 × 424 pixels), 8-bit audio and also binary messages were streamed periodically during each connection.

4.1 Metrics

The metrics used for this analysis were collected at both locations. Audio and video statistics were obtained from provided WebRTC stats report tool, while the data stream statistics were collected via implemented custom code inside the client application. The following metrics were recorded for each data stream type:

- *Audio and video*: Bytes sent/received, packets sent/received/lost, current delay, round-trip time, bitrate;
- *Video and depth*: Available send/receive bandwidth, target/actual encoding bit rate, frame height/width, frame height/width, frame rate received;
- *Data*: Packet timestamp, packet type, packet size, packets sent/received/lost.

4.2 Network Connectivity

Table 1 presents the average bandwidth used during the experiments. The results are divided depending on the CPU Overuse Detection (COD) feature. It can be seen that the average depth and video bandwidth consumed varies when COD is activated. Using COD reduces the bandwidth required to stream both video and depth frames. The results show that COD does not affect audio streaming. The tests streaming *Video1* without COD were on average the most bandwidth consuming with 3.6 Mbits/s.

Table 1. Average bandwidth used by the different streams (Mbit/s)

	Video	Depth	Audio	Data	Total
V1 NO COD	1.785	1.785	0.026	0.004	3.600
V1 COD	1.662	1.444	0.026	0.004	3.136
V2 NO COD	1.775	1.190	0.030	0.004	2.998
V2 COD	1.433	0.555	0.026	0.004	2.019

In both cases, *Video1* and *Video2*, the bandwidth consumed is reduced when applying COD. However, the resolution changes do not affect the bandwidth consumed as much as it would be expected by reducing resolution to one quarter of the original size. According to the data retrieved, this is due to the fact that WebRTC detects a similar bandwidth for multimedia streaming and thus, the target encoding bitrate remained similar.

Figures 8 and 9 show the average video and audio delays measured during the tests (labeled with V1 or V2 for *Video1* or *Video2*, followed by C or N for tests with or without COD). Two of the tests (V1N1 and V2C2) showed significantly higher video delays, probably due to higher network traffic during that time. The remaining tests had video delay between 50 and 150 ms, with similar delay values for both tracks, video

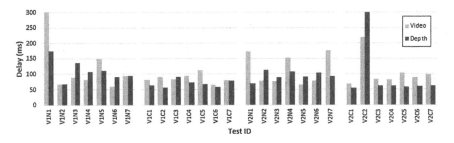

Fig. 8. Average video\depth delay (ms)

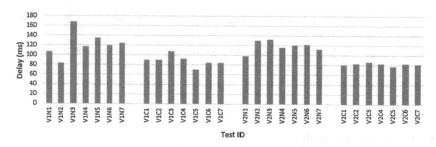

Fig. 9. Average audio delay (ms)

track and depth track. The average audio delay was similar during the tests, in a range of 70 and 150 ms which can be considered a range of values adequate for real-time multimedia communication. The minimum and maximum video delay registered during the tests were 10 ms and 1537 ms respectively, and the minimum and maximum audio delay were 66 ms and 639 ms respectively. Apparently, there is not a direct relation between video and audio delays, the results do not show delay spikes on both measures at the same time.

4.3   Video and Depth Image Quality

Video quality on the receiving end has been also measured on this tests, we recorded the video and depth frames received during each connection and these frames were compared with the original video and depth frames[1]. We have used two well-known standard measures to assess the quality of the video frames the PSNR and SSIM [20], this metrics intend to establish the quality of video images comparing their similarity. PSNR provides a quality measurement based on the squared error between original and processed images, on the other hand, SSIM measures structural similarity providing a metric from 0 to 1 that better reflects the human perception. Table 2 presents the results of the quality metrics on each test for video and depth frames.

The results show that video frames between the different tests and applying COD have very similar PSNR values, meaning that there is no significant reduction on the frame quality. SSIM, which measures the quality of an image from a perspective closer to the human perception gave values in the range 0.941 to 0.957 for *Video1* and from 0.9641 to 0.9731 for *Video2*, in both cases we obtained very high similarity with the original images. Both measures show that there is no a significant difference in video quality when using COD. The results for the depth frames are consistent with those obtained for video frames. Though the PSNR values of different images cannot compared, as they depend on the structure of the image, the higher PSNR values obtained for depth images it is due to the fact that after applying our compression algorithm the depth images are quite uniform in terms of pixel intensity as the depth range is limited. However, we can highlight that PSNR also shows no difference between the quality of the depth frames when using COD.

---

[1] RGB frames were reduced to 960 × 540 resolution.

Table 2. Streamed image quality measures (PSNR and SSIM)

	VIDEO1			VIDEO2		
	Video		Depth	Video		Depth
	PSNR (dB)	SSIM	PSNR (db)	PSNR (db)	SSIM	PSNR (db)
N1	31.052	0.948	44.509	32.760	0.973	44.107
C1	30.951	0.941	46.742	33.359	0.964	48.440
N2	31.140	0.948	45.754	33.244	0.973	47.685
C2	31.522	0.937	45.432	33.124	0.965	48.398
N3	30.690	0.957	44.639	33.258	0.970	47.603
C3	31.707	0.951	45.638	33.138	0.966	48.386
N4	31.023	0.948	45.420	33.051	0.972	47.555
C4	31.740	0.943	46.757	33.308	0.964	48.484
N5	31.195	0.947	44.932	33.116	0.970	47.599
C5	31.494	0.943	44.340	33.143	0.968	48.467
N6	31.193	0.945	45.429	32.929	0.969	47.621
C6	31.605	0.952	45.655	32.930	0.964	49.585
N7	31.262	0.947	44.816	33.206	0.973	47.602
C7	32.406	0.951	45.315	33.270	0.966	48.458

5 Conclusions

Providing experts with tools to remotely intervene and guide less experienced professionals has great potential in various applications, such as medicine, mechanical repair, aeronautics, military applications, archaeological excavation, education, and others. In this paper we focused on telemedicine by presenting the prototype that was designed based on the developed AR/VR real-time streaming framework. The developed framework was tested from network performance perspective by investigating the video quality, required bandwidth, and transmission delays. The results of the networking experiments show that our framework can provide the multimedia data at an interactive framerate with relatively small delays. Although the presented prototype was tested using Kinect camera and zSpace display the presented framework is designed to support different AR and VR technologies. On the patient side the system could use wearable as display devices such as Microsoft HoloLens to provide more immersive experience of the participants. In our future work we are planning to perform usability studies to investigate the user experience with this technology during a remote collaboration in a healthcare setting.

Acknowledgements. This work was partially supported by the following sources: National Science Foundation (NSF) grant #1427260, Office of Naval Research (ONR) grant #N00014-09-1-0230, and Siemens Fellowship (#20150859).

References

1. Goldzweig, C.L., Towfigh, A., Maglione, M., Shekelle, P.G.: Costs and benefits of health information technology: new trends from the literature. Health Aff. (Millwood) **28**, w282–w293 (2009)
2. Kvedar, J., Coye, M.J., Everett, W.: Connected health: a review of technologies and strategies to improve patient care with telemedicine and telehealth. Health Aff. (Millwood) **33**, 194–199 (2014)
3. Flodgren, G., Rachas, A., Farmer, A.J., Inzitari, M., Shepperd, S.: Interactive telemedicine: effects on professional practice and health care outcomes. The Cochrane Library (2015)
4. Augestad, K.M., Lindsetmo, R.O.: Overcoming distance: video-conferencing as a clinical and educational tool among surgeons. World J. Surg. **33**, 1356–1365 (2009)
5. Markarian, G., Mihaylova, L., Tsitserov, D.V., Zvikhachevskaya, A.: Video distribution techniques over WiMAX networks for m-health applications. IEEE Trans. Inf. Technol. Biomed. **16**, 24–30 (2012)
6. Pande, A., Mohapatra, P., Zambreno, J.: Securing multimedia content using joint compression and encryption. IEEE MultiMedia **20**, 50–61 (2013)
7. Latifi, R., Weinstein, R., Porter, J., Ziemba, M., Judkins, D., Ridings, D., Nassi, R., Valenzuela, T., Holcomb, M., Leyva, F.: Telemedicine and telepresence for trauma and emergency care management. Scand. J. Surg. **96**, 281–289 (2007)
8. Boniface, K.S., Shokoohi, H., Smith, E.R., Scantlebury, K.: Tele-ultrasound and paramedics: real-time remote physician guidance of the focused assessment with sonography for trauma examination. Am. J. Emerg. Med. **29**, 477–481 (2011)
9. Söderholm, H.M., Sonnenwald, D.H., Cairns, B., Manning, J.E., Welch, G.F., Fuchs, H.: The potential impact of 3D telepresence technology on task performance in emergency trauma care. In: Proceedings of the 2007 International ACM Conference on Supporting Group Work, pp. 79–88 (2007)
10. Garcia, P.: Telemedicine for the battlefield: present and future technologies. In: Rosen, J., Hannaford, B., Satava, R.M. (eds.) Surgical Robotics, pp. 33–68. Springer, Heidelberg (2011)
11. Irizarry, D., Wadman, M.C., Bernhagen, M.A., Miljkovic, N., Boedeker, B.H.: Using the battlefield telemedicine system (BTS) to train deployed medical personnel in complicated medical tasks-a proof of concept. In: MMVR, pp. 215–217 (2012)
12. Ling, G.S., Rhee, P., Ecklund, J.M.: Surgical innovations arising from the Iraq and Afghanistan wars. Annu. Rev. Med. **61**, 457–468 (2010)
13. Satava, R.M.: Virtual reality and telepresence for military medicine. Comput. Biol. Med. **25**, 229–236 (1995)
14. Welch, G., Fuchs, H., Cairns, B., Mayer-Patel, K., Sonnenwald, D.H., Ilie, A., Noland, M., Noel, V., Yang, H.: Improving, expanding and extending 3D telepresence. In: International Workshop on Advanced Information Processing for Ubiquitous Networks, 15th International Conference on Artifical Reality and Telexistence (ICAT 2005) (2005)
15. Jang-Jaccard, J., Nepal, S., Celler, B., Yan, B.: WebRTC-based video conferencing service for telehealth. Computing **98**, 169–193 (2016)
16. Protocol buffers (2016). https://developers.google.com/protocol-buffers
17. Maeda, K.: Performance evaluation of object serialization libraries in XML, JSON and binary formats. In: 2012 Second International Conference on Digital Information and Communication Technology and Its Applications (DICTAP), pp. 177–182. IEEE (2012)

18. Kinect for windows (2016). https://developer.microsoft.com/en-us/windows/kinect
19. Kinect projector toolkit (2016). https://github.com/genekogan/KinectProjectorToolkit
20. Hore, A., Ziou, D.: Image quality metrics: PSNR vs. SSIM. In: 20th International Conference on Pattern Recognition (ICPR), pp. 2366–2369. IEEE (2010)

Color Consistency for Photo Collections Without Gamut Problems

Qi-Chong Tian$^{(\boxtimes)}$ and Laurent D. Cohen

Université Paris-Dauphine, PSL Research University, CNRS, UMR 7534,
CEREMADE, 75016 Paris, France
{tian,cohen}@ceremade.dauphine.fr

Abstract. In this paper, we present a color consistency technique in order to make images in the same collection share the same color style and to avoid gamut problems. Some previous methods define simple global parameter-based models and use optimizing algorithms to obtain the unknown parameters, which usually cause gamut problems in bright and dark regions. Our method is based on the range-preserving histogram specification and can enforce images to share the same color style, without resulting in gamut problems. We divide the input images into two sets having respectively high visual quality and low visual quality. The high visual quality images are used to make color balance. And then the low visual quality images are color transferred using the previous corrected high quality images. Our experiments indicate that such histogram-based color correction method is better than the compared algorithm.

Keywords: Color consistency · Color correction · Histogram specification · Gamut preserving · Image quality assessment

1 Introduction

With the popularity of digital cameras, online photo albums and social networks, we can easily find thousands of photos of the same scenes, especially for the famous landmarks. And these kinds of photo collections have been applied in high level computer vision problems, including visual photo tourism [1], scene completion [2], landmark recognition [3], content-based image retrieval [4], photo stitching [5], and time-lapse [6]. However, these online images are usually taken with different capture devices [7], different illuminations, and different views, which are the reasons for the different color styles in the same photo collection.

In order to obtain better results of the above mentioned interesting computer vision applications, the color consistency or color correction is an essential pre-processing procedure before implementing the main algorithms. The most simple color correction method is the auto color adjustment technique, which is implemented in most commercial photo editing tools. However, this kind of method makes the color correction for each image independently, not considering the color distribution relations among all images in the same collection.

© Springer International Publishing AG 2017
L. Amsaleg et al. (Eds.): MMM 2017, Part I, LNCS 10132, pp. 90–101, 2017.
DOI: 10.1007/978-3-319-51811-4_8

Another kind of method is manually adjusting all images by professional photographers, which requires a significant amount of processing time for a large number of photos.

Recently, Park et al. [8] proposed a method to optimize color consistency in community photo collections depicting the same scenes. Their method is based on the assumption of a color correction model, which is a simple combination of white balance and gamma correction. Park et al.'s method is globally making color mapping in the three color channels respectively, and may cause gamut problems, since the dark pixel values are easily saturated to 0 and the bright pixel values are easily saturated to 255 in this simple global parameter-based model. Another disadvantage of their algorithm is that the final results are highly depending on the color distribution of all images. In other words, some low quality images with bad color distribution will affect the quality of other images in the same collection.

In order to avoid gamut problems [9,10] and reduce the effects from bad quality images, we present an automatic color consistency algorithm for photo collections based on image quality assessment and range-preserving histogram specification. In our method, input images are divided into two sets, namely high quality image set and low quality image set. And then the images having high quality will be used to produce corresponding color balanced images. At last, the images classified in the low quality image set will be color transferred with the corrected high quality images as the reference. These three stages can obtain color consistency among the whole set of images and not cause the disadvantages existing in Park et al.'s method.

The rest of this paper is organized as follows. The relevant previous work is summarized in Sect. 2. The overview of the proposed algorithm is described in Sect. 3 and the detailed description is given in Sect. 4. The experiments and comparisons are shown in Sect. 5. Finally, we give the conclusion and discuss the future work in Sect. 6.

2   Related Work

In this section, we summarize the previous related work and categorize these methods by the number of input images. Color image enhancement is an ancient image processing technique, which is normally applied to a single image based on the statistic information of the input image or some specialized filters. Color transfer is a technique of transferring the color style of a reference image to a test image. Color balance for image collections is a new and interesting problem, which can enforce thousands of images to share the same color style.

2.1   Color Enhancement for a Single Image

Color image enhancement is an ancient and challenging task in image processing. Histogram equalization [11] is a useful method for image enhancement. Sapiro and Caselles [12] proposed a PDE-based image histogram modification method, which was further extended to color image enhancement by many researchers [13–17]. The automatic color enhancement (ACE) algorithm [18] is a popular method, which has been used in many photo editing software. Recently, Yuan and Sun [19] proposed an automatic exposure correction method, which was based on region-level exposure evaluation and detail-preserving S-curve adjustment. This kind of algorithm operates on each image independently, so they can not obtain color consistency among all images in the same collection.

2.2 Color Transfer Between Two Images

Color transfer between two images has received significant attention in recent years. The goal of this kind of algorithm is transferring the color style of a reference image to a test image, which can enforce the corresponding two images to share the same color style. Reinhard et al. [20] proposed a simple statistics-based algorithm to transfer colors between two images, which was also extended by many researchers. Papadakis et al. [21] proposed a variational model for color image histogram transfer, which used the energy functional minimization to obtain the goal of transferring the image color style and maintaining the image geometry. Hristova et al. [22] presented a style-aware robust color transfer method, which was based on the style feature clustering and the local chromatic adaptation transform. Hwang et al. [23] proposed a geometry-based color transfer algorithm using probabilistic moving least squares. A more detailed review of this kind of algorithm can been seen in [24].

2.3 Color Balance for Image Collections

Automatic color consistency for photo collections firstly obtained attention in HaCohen et al. [25]. Their method used the non-rigid dense correspondence algorithm [26] to obtain matched features, and then constructed a graph with edges linking image pairs sharing the similar contents. Color consistency was achieved by globally optimizing three piece-wise quadratic curves over the whole graph, which can minimize the color difference between all correspondences. However, their method is highly depending on dense and accurate correspondences (high computational complexity), which makes this algorithm unsuitable for a large number of images in photo collections.

A more efficient and robust color consistency algorithm for photo collections was proposed by Park et al. [8]. They firstly recovered sparse point correspondences in the same image collection, and then constructed a simple color correction model which can be considered the combination of white balance and gamma correction. Then a robust low-rank matrix factorization method was used to estimate the unknown parameters of the defined model. However, this correction model does not consider gamut problems and negative effects resulting from low quality images in the same collection.

3   Overview

Given an input photo collection as a sequence of images $I = \{I_1, I_2, I_3, ..., I_n\}$ which describe the same scene with different views, different capture devices or different illustrations, our goal is to make these images share the same color style and to be color consistent in the shared contents.

Our method is inspired by [8], which has given a whole framework to minimizing color difference among all input images. However, we do not adopt the simple global parameter-based model to correct colors, which may result in gamut problems. The contributions of our work are classifying input images based on image quality metrics and finding suitable target histograms for the range-preserving histogram specification.

Figure 1 illustrates the overall framework of our color consistency algorithm, which consists of three main stages.

1. Image quality assessment: use non-reference measures to assess the image visual quality, and then construct two image sets having respectively high quality and low quality.

2. Color balance: use matched points between images having high quality to construct correspondences, and then obtain early corrected images with linear transforms. These early images can produce the target histograms. At last, the range-preserving histogram specification is applied to each image.

3. Color transfer: use matched points between low quality images and corrected high quality images to construct correspondences, and then obtain final corrected images with linear transforms (can produce target histograms) and the range-preserving histogram specification.

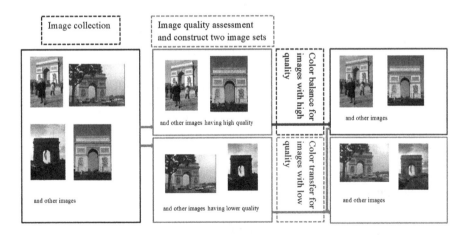

Fig. 1. The overall framework of the proposed method

4   Color Consistency without Gamut Problems

4.1   Image Quality Assessment by Non-reference Measures

As discussed in Sect. 1, the results of Park et al.'s method [8] depend on the color distribution of all images in the same collection. This kind of algorithm may obtain some bad results when there are many low quality images in the same image collection. So we firstly make quality assessment for input images, which can sort the images by their visual quality. Then we make color balance for the images ranking high quality. At last, this kind of color style obtained from high visual quality images will be propagated to the images which are ranked lower visual quality in the assessment procedure. This kind of processing can reduce bad effects from lower visual quality images.

Much effort [27, 28] has been made in recent years to design non-reference image quality assessment metrics. Panetta et al. [27] proposed a non-reference color image quality measurement combining the contrast, colorfulness and sharpness metrics. In other words, this measurement is a linear weighting of three terms assessing the contrast, color and edge information respectively. We simplify this assessment method and adopt the contrast and the sharpness metrics in our method. The measurement IQA (Image Quality Assessment) is written in Eq. (1).

$$IQA = w_1 * Contrast + w_2 * Sharpness \, , \tag{1}$$

where w_1 is the weighting of the contrast measure, w_2 is the weighting of the sharpness measure, $Contrast$ is defined in Eq. (2), $Sharpness$ is defined in Eq. (3).

$$Contrast = AME(I\_Y)$$

$$AME(I\_Y) = \frac{1}{k_1 * k_2} \sum_{l=1}^{k_1} \sum_{k=1}^{k_2} \left(log \left(\frac{I\_Y(max, k, l) + I\_Y(min, k, l)}{I\_Y(max, k, l) - I\_Y(min, k, l)} \right) \right)^{-0.5} \, , \tag{2}$$

where AME is the Michelson-Law measure of enhancement [29], I_Y is the values of Y channel of the input image in the YCbCr color space, the image is divided into $k_1 * k_2$ blocks, $I\_Y(min, k, l)$ is the minimum of the block (k, l) in I_Y, $I\_Y(max, k, l)$ is the maximum of the block (k, l) in I_Y, k and l are the row and column index of blocks in the image.

$$Sharpness = \sum_{c=1}^{3} \lambda_c EME(GrayEdge_c)$$

$$EME(GrayEdge) = \frac{2}{k_1 * k_2} \sum_{l=1}^{k1} \sum_{k=1}^{k2} log \left(\frac{GrayEdge(max, k, l)}{GrayEdge(min, k, l)} \right) \, , \tag{3}$$

where c is the index of the color channel, EME is the measure of enhancement [30], $GrayEdge_c$ is the edge map of the corresponding color channel, λ_c is the weighting of each color channel, the image is divided into $k_1 * k_2$ blocks.

$GrayEdge(min, k, l)$ and $GrayEdge(max, k, l)$ are respectively the minimum and maximum of the block (k, l) in $GrayEdge$, k and l are the row and column index of blocks in the image.

After this kind of quality assessment, we obtain the corresponding visual quality values of each image, which are written by $IQA = \{IQA_1, IQA_2, IQA_3, ..., IQA_n\}$. Then these images can be sorted by descending the values of IQA. Some best quality images $I_{high} = \{I_{high(1)}, I_{high(2)}, ..., I_{high(s)}\}$ will be chosen for make color balance in the next stage. s is the total number of images in the set I_{high}. The rest images are labeled by $I_{low} = \{I_{low(1)}, I_{low(2)}, ..., I_{low(t)}\}$, where $t = n - s$.

4.2   Color Balance Without Gamut Problems

In this stage, the goal is to make color balance for the image set $I_{high} = \{I_{high(1)}, I_{high(2)}, ..., I_{high(s)}\}$. We adopt the global color correction method, which is robust to feature matching errors and has low computational complexity. We do not define the simple correction models which were defined in [8, 25, 26]. For example, the correction model in [8] is the simulation of white balance and gamma correction, which is depending on only two parameters for each image and may result in gamut problems. Our color balance method consists of two parts, linear transforms to obtain target histograms and range-preserving histogram specifications to produce corrected images. The detailed algorithm is described in **Algorithm 1**.

Algorithm 1. Color Balance without Gamut Problems

1: **Input:** $I_{high} = \{I_{high(1)}, I_{high(2)}, ..., I_{high(s)}\}$.
2: **Extract SIFT features** in $I_{high(i)}$, where $i \in [1, 2, ..., s]$.
3: **Feature matching** between $I_{high(i)}$ and $I_{high(j)}$, where $i, j \in [1, 2, ..., s]$ and $i \neq j$, store matching results in a sparse Matrix M_1 with size $k \times s \times 3$, k is the number of matching points in the same scene.
4: **Correction for the sparse Matrix** M_1: find the non-zero values in each row of M_1, compute the mean of these non-zero values and replace the original non-zero values, which produces a new sparse Matrix M_2.
5: **Compute linear transforms** $T(i)$ for each image:$M_2(i) = M_1(i) \bullet T(i)$, the sizes of $M_1(i)$ and $M_2(i)$ both are $k \times 3$, the size of $T(i)$ is 3×3.
6: **Compute transformed images:** $I_{Trans(i)} = I_{high(i)} \bullet T(i)$.
7: **Histogram Specification** for $I_{high(i)}$ with the histogram of $I_{Trans(i)}$, the result is $I_{Correction(i)}$.
8: **Output:** $I_{Correction(i)}$.

For the step 7 in **Algorithm 1**, we combine the hue-preserving color image enhancement framework [9] and the fast exact histogram specification [15,16], which can avoid gamut problems. Given a color image IM, we firstly get the corresponding intensity image G using Eq. (4). Then the histogram specification will be applied to G, the result is $\hat{G}$. At last, we use the hue-preserving enhancement framework to obtain the final corrected color image $I\hat{M}$, which is computed using the Eq. (5) for the three color channels respectively.

$$G = 0.299 * I\_R + 0.587 * I\_G + 0.114 * I\_B , \qquad (4)$$

where I_R, I_G, I_B are respectively the red color channel, green color channel and blue color channel of the corresponding image IM.

$$
\begin{aligned}
\hat{im}(k) &= \frac{\hat{G}(k)}{G(k)}im(k), & if\ \frac{\hat{G}(k)}{G(k)} <= 1, \\
\hat{im}(k) &= \frac{255 - \hat{G}(k)}{255 - G(k)}\big(im(k) - G(k)\big) + \hat{G}(k), & if\ \frac{\hat{G}(k)}{G(k)} > 1,
\end{aligned} \qquad (5)
$$

where im is the corresponding color channel of the image IM, $\hat{im}$ is the corresponding color channel of the final corrected image $I\hat{M}$, k is the index of pixels in each gray image.

4.3   Color Transfer Without Gamut Problems

In the first stage, we divided the input image collection into two image sets, based on a non-reference image quality assessment method. The first image set has been color balanced in the previous stage. And the other image set having lower quality will be color corrected in this stage. We use the idea of color transfer or color propagation between two images to finish the task at this stage. The detailed algorithm is described in **Algorithm 2**.

Algorithm 2. Color Transfer without Gamut Problems

1: **Input:** $I_{low} = \{I_{low(1)}, I_{low(2)}, ..., I_{low(t)}\}$.
2: **Extract SIFT features** in $I_{low(j)}$, where $j \in [1, 2, ..., t]$.
3: **Feature matching** between $I_{low(j)}$ and $I_{high(i)}$, where $j \in [1, 2, ..., t]$, $i \in [1, 2, ..., s]$, $I_{high(h(j))}$ is the best corresponding image pair of $I_{low(j)}$, the corresponding values are stored in two Matrices M_j and $M_{h(j)}$.
4: **Compute linear transforms** $T(j)$ for each image:$M_{h(j)} = M_j \bullet T(j)$.
5: **Compute transformed images:**$I_{Trans(j)} = I_{low(j)} \bullet T(j)$.
6: **Histogram Specification** for $I_{low(j)}$ with the histogram of $I_{Trans(j)}$, the result is $I_{Correction(j)}$.
7: **Output:** $I_{Correction(j)}$.

For the step 6 in **Algorithm 2**, we adopt the same method described in the previous stage, which can avoid gamut problems.

5   Experiments

5.1   Image Datasets

We had run our algorithm and the comparison algorithm [8] on the landmark images download from FLICKR, which were also collected as a public test dataset on the project website (landmark3d.codeplex.com) [3]. This dataset was popularly used as a benchmark for evaluating and comparing different computer vision algorithms. In our experiments, four landmarks had been chosen to test color consistency algorithms. ARC TRIOMPHE, NOTRE DAME CATHEDRAL, STATUE OF LIBERTY, TREVI FOUNTAIN are the names of these landmarks. For each landmark, we chose 20 images having different color styles, which were captured with different cameras, different illuminations, different views by different photographers. Some examples will be shown in the following part.

5.2   Results

For the comparison algorithm, we use the original codes shared on the author's homepage and the parameters are the default values. Our method is implemented on the Matlab platform. Figures 2, 3, 4 and 5 show the results of Park et al.'s method [8] and our method. Since there is no efficient object evaluation metric to assess the performance of color consistency algorithms for photo collections, we compare the results with the subject method based on the color distribution and visual quality of corrected images.

Fig. 2. Results on the photo collection ARC TRIOMPHE, only four images are shown. The upper row shows the original images, the middle row shows the results by Park et al.'s method [8], the lower row shows our results. Red and yellow squares indicate gamut problems in Park et al.'s method. (Color figure online)

The results shown in Fig. 2 demonstrate that our approach can enforce the image collection to share a similar color style and not cause gamut problems.

Fig. 3. Results on the photo collection NOTRE DAME CATHEDRAL, only four images are shown. The upper row shows the original images, the middle row shows the results by Park et al.'s method [8], the lower row shows our results. Red squares indicate gamut problems in Park et al.'s method. (Color figure online)

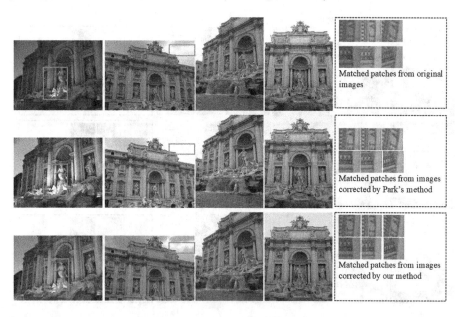

Fig. 4. Results on the photo collection TREVI FOUNTAIN, only four images are shown. The upper row shows the original images, the middle row shows the results by Park et al.'s method [8], the lower row shows our results. Red squares indicate gamut problems in Park et al.'s method. (Color figure online)

The matched patches from the original images and the corrected images can easily describe the color consistency after our method. The yellow squares indicate that the compared algorithm causes gamut problems in the dark regions. In the

Fig. 5. Results on the photo collection STATUE OF LIBERTY, only five images are shown. The upper row shows the original images, the middle row shows the results by Park et al.'s method [8], the lower row shows our results.

same way, the red squares indicate that it causes gamut problems in the bright regions. Our method do not result in this kind of disadvantage, since we adopt a range preserving exact histogram specification in the proposed algorithm.

In Fig. 3, the results of Park et al.'s method and our approach show these two methods both enforce the images to share a similar color style. However, our method outperforms the compared algorithm on the bright sky regions, which is easy to cause gamut problems in Park et al.'s method. The results shown in Fig. 4 are similar to the results of Fig. 3, which indicates that the two approaches both make color consistency well and Park et al.'s method can not correct pixels in bright and dark regions.

In Fig. 5, it is not easy to compare the results from the whole images on the left of this figure. However, we can know that our results are slightly better than the compared algorithm, from the re-sized matched patches on the right of the figure.

6   Conclusion

In this work, we have proposed a color consistency approach for image collections, based on range-preserving histogram specification. This method can enforce the same image collection to share the same color style and not to cause gamut problems. The main contributions of our work are reducing the effects of bad visual quality images to the whole image collection based on non-reference image quality assessment method, and finding suitable target histograms for range-preserving histogram specifications. The experiments demonstrate the proposed method outperforms the compared algorithm in terms of color consistency and avoiding gamut problems.

The proposed method is also based on a global model, since the global histogram specification is adopted. Generally speaking, global model may result in some color artifacts when the input images have some bright and dark regions. So we will focus on the local-based model in the future.

References

1. Snavely, N., Seitz, S.M., Szeliski, R.: Photo tourism: exploring photo collections in 3D. ACM Trans. Graph. (TOG) **25**, 835–846 (2006)
2. Hays, J., Efros, A.A.: Scene completion using millions of photographs. Commun. ACM **51**, 87–94 (2008)
3. Hao, Q., Cai, R., Li, Z., Zhang, L., Pang, Y., Wu, F.: 3D visual phrases for landmark recognition. In: The IEEE Conference on Computer Vision and Pattern Recognition (CVPR), pp. 3594–3601 (2012)
4. Lew, M.S., Sebe, N., Djeraba, C., Jain, R.: Content-based multimedia information retrieval: state of the art and challenges. ACM Trans. Multimedia Comput. Commun. Appl. (TOMM) **2**, 1–19 (2006)
5. Shan, Q., Curless, B., Furukawa, Y., Hernandez, C., Seitz, S.M.: Photo uncrop. In: Fleet, D., Pajdla, T., Schiele, B., Tuytelaars, T. (eds.) ECCV 2014. LNCS, vol. 8694, pp. 16–31. Springer, Heidelberg (2014). doi:10.1007/978-3-319-10599-4_2
6. Martin-Brualla, R., Gallup, D., Seitz, S.M.: Time-lapse mining from internet photos. ACM Trans. Graph. (TOG) **34**, Article No. 62 (2015)
7. Faridul, H.S., Stauder, J., Trmeau, A.: Illumination and device invariant image stitching. In: IEEE International Conference on Image Processing (ICIP), pp. 56–60 (2014)
8. Park, J., Tai, Y.W., Sinha, S.N., Kweon, I.S.: Efficient and robust color consistency for community photo collections. In: The IEEE Conference on Computer Vision and Pattern Recognition (CVPR), pp. 430–438 (2016)
9. Naik, S.K., Murthy, C.A.: Hue-preserving color image enhancement without gamut problem. IEEE Trans. Image Process. (TIP) **12**, 1591–1598 (2003)
10. Solomon, C., Breckon, T.: Fundamentals of Digital Image Processing: A Practical Approach with Examples in Matlab. Wiley, Hoboken (2011)
11. Russ, J.C.: The Image Processing Handbook, 6th edn. CRC Press, Boca Raton (2011)
12. Sapiro, G., Caselles, V.: Histogram modification via differential equations. J. Differ. Equ. **135**, 238–268 (1997)
13. Palma-Amestoy, R., Provenzi, E., Bertalmio, M., Caselles, V.: A perceptually inspired variational framework for color enhancement. IEEE Trans. Pattern Anal. Mach. Intell. (PAMI) **31**, 458–474 (2009)
14. Provenzi, E.: Perceptual color correction: a variational perspective. In: Trémeau, A., Schettini, R., Tominaga, S. (eds.) CCIW 2009. LNCS, vol. 5646, pp. 109–119. Springer, Heidelberg (2009). doi:10.1007/978-3-642-03265-3_12
15. Nikolova, M., Steidl, G.: Fast ordering algorithm for exact histogram specification. IEEE Trans. Image Process. (TIP) **23**, 5274–5283 (2014)

16. Nikolova, M., Steidl, G.: Fast hue and range preserving histogram specification: theory and new algorithms for color image enhancement. IEEE Trans. Image Process. (TIP) **23**, 4087–4100 (2014)
17. Pierre, F., Aujol, J.-F., Bugeau, A., Ta, V.-T.: Luminance-hue specification in the RGB space. In: Aujol, J.-F., Nikolova, M., Papadakis, N. (eds.) SSVM 2015. LNCS, vol. 9087, pp. 413–424. Springer, Heidelberg (2015). doi:10.1007/978-3-319-18461-6_33
18. Rizzi, A., Gatta, C., Marini, D.: A new algorithm for unsupervised global and local color correction. Pattern Recogn. Lett. **24**, 1663–1677 (2003)
19. Yuan, L., Sun, J.: Automatic exposure correction of consumer photographs. In: Fitzgibbon, A., Lazebnik, S., Perona, P., Sato, Y., Schmid, C. (eds.) ECCV 2012. LNCS, vol. 7575, pp. 771–785. Springer, Heidelberg (2012). doi:10.1007/978-3-642-33765-9_55
20. Reinhard, E., Ashikhmin, M., Gooch, B., Shirley, P.: Color transfer between images. IEEE Comput. Graph. Appl. **21**, 34–41 (2001)
21. Papadakis, N., Provenzi, E., Caselles, V.: A variational model for histogram transfer of color images. IEEE Trans. Image Process. (TIP) **20**, 1682–1695 (2011)
22. Hristova, H., Le Meur, O., Cozot, R., Bouatouch, K.: Style-aware robust color transfer. In: Proceedings of the workshop on Computational Aesthetics, pp. 67–77 (2015)
23. Hwang, Y., Lee, J.Y., Kweon, I.S., Kim, S.J.: Color transfer using probabilistic moving least squares. In: IEEE Conference on Computer Vision and Pattern Recognition (CVPR), pp. 3342–3349 (2014)
24. Faridul, H.S., Pouli, T., Chamaret, C., Stauder, J., Reinhard, E., Kuzovkin, D., Tremeau, A.: Colour mapping: a review of recent methods, extensions and applications. Comput. Graph. Forum **35**, 59–88 (2016)
25. HaCohen, Y., Shechtman, E., Goldman, D.B., Lischinski, D.: Optimizing color consistency in photo collections. ACM Trans. Graph. (TOG) **32**, Article No. 38 (2013)
26. HaCohen, Y., Shechtman, E., Goldman, D.B., Lischinski, D.: Non-rigid dense correspondence with applications for image enhancement. ACM Trans. Graph. (TOG) **30**, Article No. 70 (2011)
27. Panetta, K., Gao, C., Agaian, S.: No reference color image contrast and quality measures. IEEE Trans. Consum. Electron. **59**, 643–651 (2013)
28. Bringier, B., Richard, N., Larabi, M.C., Fernandez-Maloigne, C.: No-reference perceptual quality assessment of colour image. In: the 14th European Signal Processing Conference (EUSIPCO), pp. 1–5 (2006)
29. Agaian, S.S., Silver, B., Panetta, K.A.: Transform coefficient histogram-based image enhancement algorithms using contrast entropy. IEEE Trans. Image Process. (TIP) **16**, 741–758 (2007)
30. Agaian, S.S., Panetta, K., Grigoryan, A.M.: A new measure of image enhancement. In: IASTED International Conference on Signal Processing & Communication, pp. 19–22 (2000)

Comparison of Fine-Tuning and Extension Strategies for Deep Convolutional Neural Networks

Nikiforos Pittaras[1], Foteini Markatopoulou[1,2(✉)], Vasileios Mezaris[1], and Ioannis Patras[2]

[1] Information Technologies Institute (ITI), CERTH, 57001 Thermi, Greece
{npittaras,markatopoulou,bmezaris}@iti.gr
[2] Queen Mary University of London, Mile end Campus, London E14NS, UK
i.patras@qmul.ac.uk

Abstract. In this study we compare three different fine-tuning strategies in order to investigate the best way to transfer the parameters of popular deep convolutional neural networks that were trained for a visual annotation task on one dataset, to a new, considerably different dataset. We focus on the concept-based image/video annotation problem and use ImageNet as the source dataset, while the TRECVID SIN 2013 and PASCAL VOC-2012 classification datasets are used as the target datasets. A large set of experiments examines the effectiveness of three fine-tuning strategies on each of three different pre-trained DCNNs and each target dataset. The reported results give rise to guidelines for effectively fine-tuning a DCNN for concept-based visual annotation.

Keywords: Concept detection · Deep learning · Visual analysis

1 Introduction

Concept-based video annotation, also known as video concept detection, refers to the problem of annotating a video fragment (e.g., a keyframe) with one or more semantic concepts (e.g., "table", "dog") [14]. The state of the art approach to doing this is to train a deep convolutional neural network (DCNN) on a set of concepts [3,5,12]. Then, a test keyframe can be forward propagated by the DCNN, to be annotated with a set of concept labels. DCNN training requires the learning of millions of parameters, which means that a small-sized training set could easily over-fit the DCNN on the training data. It has been proven that the bottom layers of a DCNN learn rather generic features, useful for different domains, while the top layers are task-specific [18]. Transferring a pre-trained network in a new dataset by fine-tuning its parameters is a common strategy that can take advantage of the bottom generic layers and adjust the top layers to the target dataset and the new target concepts [2,5,18].

In this study we compare three fine-tuning methods in order to investigate the best way to transfer the parameters of a DCNN trained on a source dataset

© Springer International Publishing AG 2017
L. Amsaleg et al. (Eds.): MMM 2017, Part I, LNCS 10132, pp. 102–114, 2017.
DOI: 10.1007/978-3-319-51811-4_9

to a different target dataset that requires a different set of concepts. Although DCNN fine-tuning has been presented in previous studies [2,5,18], this is the first work, to our knowledge, that performs a large number of experimental comparisons considering three different pre-trained DCNNs, two different subsets of concepts for which the pre-trained networks will be fine-tuned, two different target datasets, and three fine-tuning strategies with many parameter evaluations for each of them, with the purpose of comparing these strategies. Experiments performed on the TRECVID 2013 SIN [10] and the PASCAL VOC-2012 classification [4] datasets show that increasing the depth of a pre-trained network with one more fully-connected layer and fine-tuning the rest of the layers on the target dataset can improve the network's concept detection accuracy compared to other fine-tuning approaches.

2 Related Work

Fine-tuning is a process where the weights of a pre-trained DCNN are used as the starting point for a new target training set and they are modified in order to adapt the pre-trained DCNN to the new target dataset. Then, the fine-tuned DCNN can be used either as feature generator, i.e., the output of one or more hidden layers is typically used as a global keyframe/image representation [13], or as standalone classifier that performs the final class label prediction directly.

Different DCNN-based transfer learning approaches have been successfully applied in many datasets. The most straight-forward approach replaces the classification layer of a pre-trained DCNN with a new output layer that corresponds to the categories that should be learned with respect to the target dataset, [2,5,18]. Generalizing this approach, the weights of the first K network layers can remain frozen, i.e., they are copied from the pre-trained DCNN and kept unchanged, and the rest of the layers (be it just the last one or more than one) are learned from scratch [1,9]. Alternatively, the copies of the first K layers could be allowed to adapt to the target dataset with a low learning rate. For example, [18] investigates which layers of Alexnet [7] are generic, i.e., can be directly transferred to a target domain, and which layers are dataset-specific. Furthermore, experiments in [18] show that fine-tuning the transferred layers of a network works better than freezing them. However, neither of these studies investigates how low the learning rate for the aforementioned layers should be, relative to the new layers, during fine-tuning. Other studies extend the pre-trained network by one or more fully connected layers, which seems to improve the above transfer learning strategies [1,8,9,15]. However, the optimal number of extension layers and the size of them has not been investigated before. Although fine-tuning has been applied in many studies, a complete understanding of what fine-tuning parameters (e.g., number/size of extension layers, learning rate) work better has not been extensively examined. Furthermore, a thorough comparison of all the available fine-tuning alternatives is yet to appear in the literature.

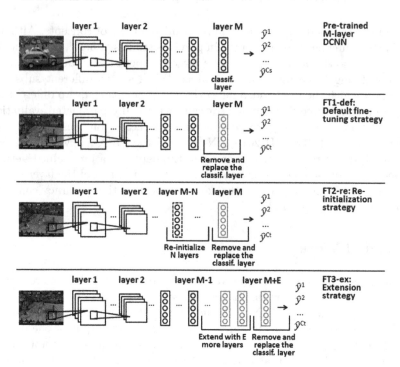

Fig. 1. Fine-tuning strategies outline.

3   Comparison of Fine-Tuning and Extension Strategies for DCNNs

Let D_s denote a pre-trained DCNN, trained on C_s categories using a source dataset, and D_t denote the target DCNN, fine-tuned on C_t categories of a different target dataset. In this section we present the three fine-tuning strategies (Fig. 1) that we compare for the problem of visual annotation, in order to effectively fine-tune DCNNs D_s that were trained on a large visual dataset for a new target video/image dataset. These three fine-tuning strategies are as follows:

- FT1-def: Default fine-tuning strategy: This is the typical strategy that modifies the last fully-connected layer of D_s to produce the desired number of outputs C_t, by replacing the last fully-connected layer with a new C_t-dimensional classification fully-connected layer.
- FT2-re: Re-initialization strategy: In this scenario, similar to FT1-def, the last fully-connected layer is replaced by a new C_t-dimensional classification layer. The weights of the last N layers, preceding the classification layer, are also re-initialized (i.e., reset and learned from scratch).
- FT3-ex: Extension strategy: Similar to the previous two strategies, the last fully-connected layer is replaced by a new C_t-dimensional classification fully-connected layer. Subsequently, the network is extended with E fully-connected

layers of size L that are placed on the bottom of the modified classification layer. These additional layers are initialized and trained from scratch during fine-tuning, at the same rate as the modified classification layer. One example of a modified network after the insertion of one extension layer for two popular DCNN architectures that were also used in our experimental study in Sect. 4, is presented in Fig. 2. Regarding the GoogLeNet architecture, which has two additional auxiliary classifiers, an extension layer was also inserted in each of them.

Each fine-tuned network D_t is used in two different ways to annotate new test keyframes/images with semantic concepts. (a) Direct classification: Each

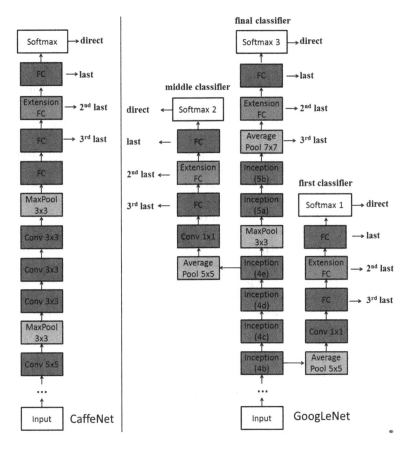

Fig. 2. A simplified illustration of the CaffeNet [7] (left) and GoogLeNet [16] (right) architectures used after insertion of one extension layer. Each of the inception layers of GoogLeNet consists of six convolution layers and one pooling layer. The figure also presents the direct output of each network and the output of the last three layers that were used as features w.r.t. FT3-ex strategy. Similarly, the corresponding layers were used for the FT1-def and FT2-re strategies.

test keyframe/image is forward propagated by D_t and the network's output is used as the final class distribution assigned to the keyframe/image. (b) D_t is used as feature generator: The training set is forward propagated by the network and the features extracted from one or more layers of D_t are used as feature vectors to subsequently train one supervised classifier (e.g., Logistic Regression) per concept. Then, each test keyframe/image is firstly described by the DCNN-based features and subsequently these features serve as input to the trained Logistic Regression classifiers.

4   Experimental Study

4.1   Datasets and Experimental Setup

The TRECVID SIN task 2013 [10] dataset and the PASCAL VOC-2012 [4] dataset were utilized to train and evaluate the compared fine-tuned networks. The TRECVID SIN dataset consists of low-resolution videos, segmented into video shots; each shot is represented by one keyframe. The dataset is divided into a training and a test set (approx. 600 and 200 h, respectively). The training set is partially annotated with 346 semantic concepts. The test set is evaluated on 38 concepts, i.e., a subset of the 346 concepts. The PASCAL VOC-2012 [4] dataset consists of images annotated with one object class label of the 20 available object classes. PASCAL VOC-2012 is divided into training, validation and test sets (consisting of 5717, 5823 and 10991 images, respectively). We used the training set to train the compared methods, and evaluated them on the validation set. We did not use the original test set because ground-truth annotations are not publicly available for it (the evaluation of a method on the test set is possible only through the evaluation server provided by the PASCAL VOC competition, submissions to which are restricted to two per week).

For each dataset we fine-tuned the following three pre-trained DCNNs: (i) CaffeNet-1k, the reference implementation of Alexnet [7] by Caffe [6], trained on 1000 ImageNet categories, (ii) GoogLeNet-1k [16], trained on the same 1000 ImageNet [11] categories and (iii) GoogLeNet-5k, trained using 5055 ImageNet [11] categories. Each of these networks was fine-tuned on the 345 TRECVID SIN concepts (i.e., all the available TRECVID SIN concepts, except for one which was discarded because only 5 positive samples are provided for it), which resulted to a training set of 244619 positive examples. CaffeNet-1k was also fine-tuned on a subset of 60 TRECVID SIN concepts. We refer to each of these fine-tuned networks as CaffeNet-1k-345-SIN, GoogLeNet-1k-345-SIN, GoogLeNet-5k-345-SIN and CaffeNet-1k-60-SIN, respectively. In addition, each of these original networks was fine-tuned on the positive examples of the PASCAL VOC-2012 training set. These networks are labeled as CaffeNet-1k-VOC, GoogLeNet-1k-VOC and GoogLeNet-5k-VOC, respectively.

In performing pre-trained DCNN fine-tuning, we compared the three fine-tuning strategies presented in Sect. 3. Specifically, in all cases we discarded and replaced the classification fully-connected (fc) layer of the utilized pre-trained network, with a 60-dimensional or a 345-dimensional fc classification layer for the 60 or 345 concepts of the TRECVID SIN dataset respectively, or with a 20-dimensional classification layer for the 20 object categories of the PASCAL VOC-2012 dataset. We examined two values for parameter N of the FT2-re strategy; we refer to each configuration as FT2-re1 (for $N = 1$) and FT2-re2 (for $N = 2$). The FT3-ex strategy was examined for two settings of network extensions $E \in \{1, 2\}$: i.e., extending the network by one or two fc layers, respectively, followed by ReLU (Rectified Linear Units) and Dropout layers. The size of each extension layer was examined for 7 different dimensions: $L \in \{64, 128, 256, 512, 1024, 2048, 4096\}$. We refer to these configurations as FT3-exE-L. The new layers' learning rate and momentum was set to 0.01 and 5e−4, whereas the mini-batch size was restricted by our hardware resources and set to 256 and 128 for the CaffeNet and GoogLeNet configurations, respectively.

For the purpose of evaluation, we then tested each fine-tuned network on the TRECVID SIN 2013 test set that consists of 112677 representative keyframes and 38 semantic concepts on the indexing problem; that is, given a concept, return the 2000 test keyframes that are more likely to represent it. In addition, we examined classification performance on the PASCAL VOC-2012 validation set, consisting of 5823 images and 20 object categories. We fine-tuned the total of 17 configurations times 7 networks on a Tesla K40 GPU, over a period of 2 months. All networks were trained and implemented in Caffe [6].

4.2 Preliminary Experiments for Parameter Selection

A set of preliminary experiments on the CaffeNet-1k-60-SIN and the FT1-def strategy was performed, in order to investigate how the learning rate of the pre-trained layers and the number of training epochs affect the performance of a fine-tuned network. Specifically, we partitioned the training set of the TRECVID SIN dataset into training and validation sets, which resulted to 71457 and 3007 keyframes, respectively. Momentum and weight decay were set to 0.9 and 5e−4, respectively. We examined learning rate values for the pre-trained layers equal to $LR_{pre} = k \times LR_{new}$, where $k \in \{0, 0.05, 0.075, 0.1, 0.25, 0.5, 0.75, 1\}$ and $LR_{new} = 0.01$ is the learning rate of the new classification layer that will be trained from scratch. A value of $k = 0$ keeps the pre-trained layers' weights "frozen", while the value of 1 makes them learn as fast as the new layers. To investigate the effect of the training epochs, each fine-tuning run was examined for a range of maximum epochs equal to: $\{0.25, 0.5, 1, 2, 4, 8, 16, 32\}$.

Table 1 presents the results w.r.t. the accuracy on the validation set, as this metric is implemented in the Caffe framework. We can observe that smaller learning rate values for the pre-trained layers and higher values for the training epochs improve accuracy. Consequently, we selected the best values of 0.1 and 32 for the learning rate multiplier and the maximum number of epochs, respectively, and kept them fixed for the rest of the experiments.

Table 1. Classification accuracy for the CaffeNet-1k-60-SIN and the FT1-def strategy for different values of the learning rate multiplier of the pre-trained layers (k), in the vertical axis, and different number of training epochs (e), in the horizontal axis. For each value of parameter e the best accuracy reached is underlined. The globally best accuracy is bold and underlined.

k/e	0.25	0.5	1	2	4	8	16	32
0.050	<u>0.348</u>	<u>0.362</u>	<u>0.402</u>	0.417	0.437	0.434	0.451	0.462
0.075	0.341	0.349	0.388	0.412	0.438	0.453	0.462	0.462
0.100	0.346	0.354	0.388	0.420	0.434	<u>0.455</u>	<u>0.463</u>	**<u>0.470</u>**
0.250	0.328	0.361	0.397	<u>0.421</u>	0.430	0.450	0.455	0.468
0.500	0.306	0.354	0.388	0.415	<u>0.439</u>	0.447	0.451	0.444
0.750	0.284	0.349	0.381	0.410	0.431	0.443	0.448	0.448
1.000	0.257	0.321	0.367	0.390	0.430	0.442	0.450	0.436

4.3 Main Findings of the Study

Table 2 presents the results, in terms of Mean Extended Inferred Average Precision (MXinfAP), of the CaffeNet-1k-60-SIN (left) and the CaffeNet-1k-345-SIN (right), for the three fine-tuning strategies of Sect. 3. In addition, Table 3 presents the MXinfAP of the GoogLeNet-1k-345-SIN (top) and GoogLeNet-5k-345-SIN (bottom). MXinfAP [17] is an approximation of the MAP, suitable for the TRECVID SIN partially annotated dataset. Similarly, Table 4 presents the results in terms of MAP of the CaffeNet-1k-VOC and Table 5 presents the MAP of the GoogLeNet-1k-VOC (top) and GoogLeNet-5k-VOC (bottom).

For each pair of utilized network and fine-tuning strategy we evaluate: (i) The direct output of the network (Tables 2, 3, 4 and 5: col. (a)). (ii) Logistic regression (LR) classifiers trained on DCNN-based features. Specifically, the output of each of the three last layers of each fine-tuned network was used as feature to train one LR model per concept (Tables 2, 3, 4 and 5: col. (b)–(d)). Furthermore, we present results for the late-fused output (arithmetic mean) of the LR classifiers built using the last three layers (Tables 2, 3, 4 and 5: col. (e)). For the GoogLeNet-based networks evaluations are also reported for the two auxiliary classifiers (Tables 3 and 5: col. (f)–(i)). The details of the two DCNN architectures mentioned above (CaffeNet, GoogLeNet) and the extracted features are also illustrated in Fig. 2. Based on the results reported in the aforementioned tables, we reach the following conclusions:

(a) According to Table 2, fine-tuning a pre-trained network on more concepts (going from 60 to 345) leads to better concept detection accuracy for all the fine-tuning strategies.

(b) Across all the networks and for both datasets, the FT3-ex strategy almost always outperforms the other two fine-tuning strategies (FT1-def, FT2-re) for specific (L, E) values.

Table 2. MXinfAP (%) for CaffeNet-1k-60-SIN (sub-table (A), left) and CaffeNet-1k-345-SIN (sub-table (B), right). For each sub-table, the best result per column is underlined. The globally best result per sub-table is bold and underlined.

Conf./layer	(A) CaffeNet-1k-60-SIN					(B) CaffeNet-1k-345-SIN				
	Direct	Last	2nd last	3rd last	Fused	Direct	Last	2nd last	3rd last	Fused
	(a)	(b)	(c)	(d)	(e)	(a)	(b)	(c)	(d)	(e)
FT1-def	16.86	23.60	22.93	20.86	24.74	18.26	24.10	23.51	20.84	25.21
FT2-re1	15.35	22.09	21.52	20.79	23.93	17.43	23.36	22.88	20.71	24.48
FT2-re2	13.97	19.76	17.49	18.46	20.16	15.74	22.52	19.50	20.81	23.12
FT3-ex1-64	<u>18.74</u>	23.22	22.80	23.46	24.69	<u>20.36</u>	22.89	21.44	22.92	24.79
FT3-ex1-128	18.49	24.45	<u>23.74</u>	23.09	**25.35**	20.30	24.21	22.96	22.92	25.26
FT3-ex1-256	17.25	24.02	23.23	23.29	25.04	19.79	24.72	<u>24.01</u>	23.00	25.63
FT3-ex1-512	17.01	<u>24.49</u>	23.10	23.13	25.27	19.64	25.23	23.99	23.46	**<u>25.94</u>**
FT3-ex1-1024	16.53	24.19	21.37	22.16	24.78	18.59	25.29	22.39	23.65	25.64
FT3-ex1-2048	15.81	24.40	20.97	22.75	24.77	18.22	25.25	22.03	22.54	25.45
FT3-ex1-4096	14.94	24.11	19.88	21.42	24.02	17.67	24.79	21.50	23.18	25.05
FT3-ex2-64	17.26	20.53	21.03	21.48	21.61	17.02	19.53	20.23	19.62	20.67
FT3-ex2-128	17.72	22.17	22.27	22.36	22.88	19.42	22.93	22.24	21.79	23.13
FT3-ex2-256	17.17	23.46	23.12	23.20	24.19	19.35	24.09	23.98	23.60	24.80
FT3-ex2-512	16.77	23.71	23.65	<u>23.72</u>	24.96	19.00	24.89	<u>24.01</u>	<u>23.90</u>	25.52
FT3-ex2-1024	16.18	24.14	22.88	22.78	24.82	18.48	25.08	23.66	23.07	25.37
FT3-ex2-2048	15.67	24.17	22.36	21.42	24.59	18.20	<u>25.55</u>	22.13	22.50	25.08
FT3-ex2-4096	16.34	24.12	21.66	20.33	24.59	17.64	25.23	22.08	21.94	25.07

(c) With respect to the direct output, FT3-ex1-64 and FT3-ex1-128 constitute the top-two methods for the TRECVID SIN dataset irrespective of the employed DCNN. On the other hand, FT3-ex1-2048 and FT3-ex1-4096 are the top-two methods for the PASCAL VOC-2012 dataset and the GoogLeNet-based networks, while FT3-ex1-512 and FT3-ex1-1024 are the best performing strategies for the CaffeNet network on the same dataset. That is, the FT3-ex strategy with one extension layer is always the best solution, but the optimal dimension of the extension layer varies, depending on the target domain dataset and the network architecture.

(d) The highest concept detection accuracy for each network is always reached when LR classifiers are trained on features extracted from the last and the second last fully connected layer for TRECVID SIN and PASCAL VOC-2012 dataset, respectively, using the FT3-ex strategy. That is, features extracted from the top layers are more accurate than layers positioned lower in the network, but the optimal layer varies, depending on the target domain dataset.

(e) DCNN-based features significantly outperform the direct output alternative in the vast majority of cases. However, in a few cases the direct network output works comparably well. The choice between the two approaches should

Table 3. MXinfAP (%) for GoogLeNet-1k-345-SIN (sub-table (A), top) and GoogLeNet-5k-345-SIN (sub-table (B), bottom). For each sub-table, the best result per column is underlined. The globally best result per sub-table is bold and underlined.

Conf./layer	Final classifier					Middle classifier		First classifier	
	Direct	Last	2nd last	3rd last	Fused	Direct	Fused	Direct	Fused
	(a)	(b)	(c)	(d)	(e)	(f)	(g)	(h)	(i)
(A) GoogleNet-1k-345-SIN									
FT1-def	21.67	28.15	28.29	-	29.05	<u>22.70</u>	28.62	<u>20.36</u>	25.95
FT2-re1	19.80	26.85	26.97	-	27.97	22.17	28.13	20.08	25.29
FT2-re2	19.32	26.90	26.71	-	27.65	21.61	27.60	19.40	25.01
FT3-ex1-64	24.18	27.46	25.77	28.59	28.60	22.08	26.27	18.07	22.49
FT3-ex1-128	<u>24.44</u>	28.90	28.10	28.82	30.09	22.65	27.38	20.03	24.53
FT3-ex1-256	23.93	29.48	28.61	28.64	30.23	22.04	27.54	19.87	25.00
FT3-ex1-512	23.63	29.82	28.34	28.11	**30.43**	22.15	28.19	19.94	25.33
FT3-ex1-1024	23.01	29.95	27.37	28.28	30.05	21.87	28.41	19.90	26.03
FT3-ex1-2048	23.34	30.21	27.05	28.80	30.42	21.41	<u>28.69</u>	19.93	26.18
FT3-ex1-4096	22.47	<u>30.43</u>	26.90	28.44	30.27	21.60	28.33	19.48	<u>26.59</u>
FT3-ex2-64	15.91	16.72	18.55	17.83	18.49	10.65	13.58	9.50	12.03
FT3-ex2-128	22.09	24.38	24.86	24.74	25.59	17.58	21.65	14.10	18.05
FT3-ex2-256	23.91	27.80	27.82	27.95	28.75	20.77	24.46	17.56	21.51
FT3-ex2-512	23.06	28.31	28.02	28.45	29.04	21.16	26.19	18.95	23.18
FT3-ex2-1024	22.77	29.01	<u>28.63</u>	<u>28.86</u>	29.92	20.58	26.64	18.87	24.15
FT3-ex2-2048	22.84	29.53	28.53	27.99	29.83	20.44	27.07	18.86	24.87
FT3-ex2-4096	22.36	29.90	27.85	27.49	29.93	20.72	27.32	19.01	25.44
(B) GoogleNet-5k-345-SIN									
FT1-def	22.45	29.60	<u>29.80</u>	-	30.58	23.08	29.41	<u>21.25</u>	26.00
FT2-re1	20.88	28.44	28.43	-	29.58	22.51	28.55	20.37	25.16
FT2-re2	19.08	27.21	27.17	-	28.02	21.73	28.44	20.07	25.74
FT3-ex1-64	25.48	28.86	26.86	29.22	29.62	23.30	28.37	20.20	24.47
FT3-ex1-128	<u>25.52</u>	29.75	28.66	29.57	30.60	<u>23.98</u>	28.82	20.87	25.38
FT3-ex1-256	24.79	30.16	28.99	<u>30.26</u>	31.11	23.62	29.56	21.06	26.32
FT3-ex1-512	24.28	30.86	29.26	29.68	31.47	23.54	29.86	20.71	26.32
FT3-ex1-1024	24.03	31.02	28.78	29.35	31.55	23.43	29.90	20.53	26.57
FT3-ex1-2048	23.37	31.02	27.24	29.37	31.02	23.29	<u>29.94</u>	20.56	<u>26.61</u>
FT3-ex1-4096	23.07	30.91	28.98	29.61	**31.57**	22.85	29.64	20.82	26.26
FT3-ex2-64	16.44	17.51	19.62	19.95	20.09	11.43	15.12	10.65	13.33
FT3-ex2-128	23.87	26.19	26.73	26.05	27.02	18.70	23.64	14.87	19.95
FT3-ex2-256	24.46	28.94	28.69	28.68	29.57	22.68	26.98	18.75	23.10
FT3-ex2-512	23.95	29.44	29.07	28.94	30.14	22.72	28.22	20.20	24.79
FT3-ex2-1024	23.41	30.03	28.80	29.54	30.63	22.79	29.10	19.74	25.68
FT3-ex2-2048	23.38	30.74	28.98	28.21	30.61	22.29	29.34	19.57	26.23
FT3-ex2-4096	23.07	<u>31.21</u>	28.94	27.98	30.93	22.11	29.40	19.64	26.11

Table 4. MAP % for CaffeNet-1k-VOC. For the FT-re strategy we trained the network with learning rate 10 times lower that of all the other cases. Otherwise, the network did not converge. The best result per column is underlined. The globally best result is bold and underlined.

Conf./layer	CaffeNet-1k-VOC				
	Direct	Last	2nd last	3rd last	Fused
	(a)	(b)	(c)	(d)	(e)
FT1-def	72.77	69.80	72.59	69.95	73.29
FT2-re1	8.22	8.22	8.22	27.87	27.57
FT2-re2	23.53	25.05	26.33	25.74	29.27
FT3-ex1-64	72.61	70.27	72.75	69.70	73.86
FT3-ex1-128	73.53	71.59	74.20	69.96	74.52
FT3-ex1-256	73.63	71.90	74.59	69.78	74.81
FT3-ex1-512	<u>73.84</u>	72.34	74.18	69.64	74.85
FT3-ex1-1024	73.76	72.42	72.49	69.91	74.48
FT3-ex1-2048	73.41	<u>72.59</u>	72.46	69.39	74.30
FT3-ex1-4096	73.04	71.14	74.17	69.44	74.70
FT3-ex2-64	51.42	43.82	59.64	64.63	61.55
FT3-ex2-128	62.33	58.36	67.95	71.16	69.24
FT3-ex2-256	67.97	64.64	72.64	73.52	73.04
FT3-ex2-512	70.89	68.38	74.94	<u>73.80</u>	75.04
FT3-ex2-1024	72.55	71.22	<u>75.28</u>	73.26	**<u>75.65</u>**
FT3-ex2-2048	73.02	72.37	73.80	72.31	75.29
FT3-ex2-4096	66.43	60.83	69.10	71.69	72.67

be based on the application that the DCNN will be used. E.g., real time applications' time and memory limitations would most probably render using DCNNs as feature extractors in conjunction with additional learning (LR or SVMs) prohibitive. Furthermore, we observe that the features extracted from the final classifier of GoogLeNet-based networks outperform the other two auxiliary classifiers, in most cases.

(f) Using DCNN layers' responses as feature vectors, on the one hand, FT3-ex1-512 is in the top-five methods irrespective of the employed DCNN, the extracted feature and the used dataset. Regarding the PASCAL VOC-2012 dataset this is always the case except for the features extracted from the third last layer of the CaffeNet network (Table 4: col. (d)). On the other hand, FT3-ex2-64 is always among the five worst fine-tuning methods. The rest of the FT3-ex configurations, present fluctuations of their performance across the different utilized DCNNs and DCNN-based features.

(g) Finally, it is better to combine features extracted from many layers; specifically, performing late fusion on the output of the LR classifiers trained with

Table 5. MAP % for GoogleNet-1k-VOC (sub-table (A), top) and GoogleNet-5k-VOC (sub-table (B), bottom). For each sub-table, the best result per column is underlined. The globally best result per sub-table is bold and underlined.

Conf./layer	Final classifier					Middle classifier		First classifier	
	Direct	Last	2nd last	3rd last	Fused	Direct	Fused	Direct	Fused
	(a)	(b)	(c)	(d)	(e)	(f)	(g)	(h)	(i)
(A) GoogleNet-1k-VOC									
FT1-def	82.00	85.83	86.31	-	87.69	80.71	84.01	78.93	79.01
FT2-re1	80.42	83.77	86.34	-	86.66	79.39	81.73	78.06	76.92
FT2-re2	77.34	77.43	82.51	-	82.31	75.94	77.46	74.28	73.01
FT3-ex1-64	78.98	80.29	85.19	85.76	86.11	76.07	81.93	71.46	76.58
FT3-ex1-128	80.14	83.86	87.74	86.22	87.64	77.89	83.52	73.64	78.00
FT3-ex1-256	81.11	84.97	88.04	86.46	87.95	78.95	83.96	75.00	79.01
FT3-ex1-512	81.42	85.41	87.70	86.57	**88.16**	79.54	84.35	76.03	79.39
FT3-ex1-1024	81.82	85.85	86.95	86.69	88.08	80.06	84.27	76.86	79.37
FT3-ex1-2048	82.18	86.36	86.31	86.81	87.99	80.70	84.38	77.33	79.08
FT3-ex1-4096	82.01	86.49	86.23	86.93	87.86	80.96	84.26	78.30	79.56
FT3-ex2-64	44.81	42.04	52.48	51.70	52.19	36.57	44.36	33.32	38.77
FT3-ex2-128	75.11	67.81	81.32	81.70	80.54	61.05	70.30	52.39	61.69
FT3-ex2-256	78.05	76.65	85.38	85.49	85.00	69.45	77.59	62.56	71.45
FT3-ex2-512	80.06	82.70	87.12	86.69	87.09	73.35	81.16	67.26	74.49
FT3-ex2-1024	81.26	84.32	86.75	86.39	87.39	76.46	82.53	70.19	76.38
FT3-ex2-2048	81.72	85.08	86.46	85.96	87.52	78.20	83.19	73.10	78.19
FT3-ex2-4096	81.71	84.82	86.14	85.05	87.00	78.93	83.14	74.08	77.80
(B) GoogleNet-5k-VOC									
FT1-def	82.39	86.75	86.74	-	88.01	81.10	84.25	78.96	79.06
FT2-re1	80.50	85.21	86.91	-	87.44	79.58	82.76	77.78	77.23
FT2-re2	77.73	78.81	83.13	-	83.11	75.28	77.34	71.99	69.65
FT3-ex1-64	79.74	82.86	86.41	86.26	86.92	76.36	82.72	72.32	77.51
FT3-ex1-128	80.47	85.50	88.26	86.56	88.12	78.57	84.12	74.01	78.76
FT3-ex1-256	81.43	85.81	88.33	86.73	**88.36**	79.31	84.48	75.29	79.12
FT3-ex1-512	81.65	85.91	87.84	86.90	88.33	79.99	84.76	76.25	79.69
FT3-ex1-1024	82.30	86.48	87.01	86.89	88.20	80.68	84.56	77.32	79.32
FT3-ex1-2048	82.51	86.93	86.80	86.96	88.23	81.15	84.51	77.97	79.62
FT3-ex1-4096	82.39	87.20	86.37	87.05	88.13	81.52	84.45	78.43	79.65
FT3-ex2-64	43.85	45.11	53.99	51.67	52.81	39.10	47.22	32.42	38.72
FT3-ex2-128	75.89	70.96	82.85	83.34	82.51	63.27	72.34	54.45	63.64
FT3-ex2-256	78.94	80.30	86.44	86.43	86.01	69.19	77.67	65.31	72.75
FT3-ex2-512	80.47	82.83	87.56	87.00	87.38	75.17	81.44	66.50	74.38
FT3-ex2-1024	81.47	84.54	86.81	86.53	87.58	76.99	82.85	71.09	76.74
FT3-ex2-2048	82.11	85.49	86.90	86.28	87.76	78.15	83.24	73.55	77.69
FT3-ex2-4096	80.50	83.83	85.82	84.71	86.64	77.49	81.79	74.66	78.21

each one of the last three fully connected layers almost always outperforms using a single such classifier irrespective of the employed network (Tables 2, 3, 4 and 5: col. (e)). The above conclusion was also reached for the auxiliary classifiers of GoogLeNet-based networks but for space-limitations we only present the fused output for each of these auxiliary classifiers (Tables 2, 3, 4 and 5: col. (g), (i)).

5   Conclusions

In this paper we presented a large comparative study of three fine-tuning strategies on three different pre-trained DCNNs and two different subsets of semantic concepts. Experiments performed on the TRECVID 2013 SIN dataset [10] and PASCAL VOC-2012 classification dataset [4] show that the method of increasing the depth of a pre-trained network with one fully-connected layer and fine-tuning the rest of the layers on the target dataset can improve the network's concept detection accuracy, compared to other fine-tuning approaches. Using layers' responses as feature vectors for a learning model such as logistic regression can lead to additional gains, compared to using the direct network's output, at an additional cost of computation time and memory.

Acknowledgements. This work was supported by the European Commission under contract H2020-687786 InVID.

References

1. Campos, V., Salvador, A., Giro-i Nieto, X., Jou, B.: Diving deep into sentiment: understanding fine-tuned CNNs for visual sentiment prediction. In: 1st International Workshop on Affect and Sentiment in Multimedia (ASM 2015), pp. 57–62. ACM, Brisbane (2015)
2. Chatfield, K., Simonyan, K., Vedaldi, A., Zisserman, A.: Return of the devil in the details: delving deep into convolutional nets. In: British Machine Vision Conference (2014)
3. Donahue, J., Jia, Y., Vinyals, O., Hoffman, J., Zhang, N., Tzeng, E., Darrell, T.: DeCAF: a deep convolutional activation feature for generic visual recognition (2013). CoRR abs/1310.1531
4. Everingham, M., Van Gool, L., Williams, C.K.I., Winn, J., Zisserman, A.: The PASCAL Visual Object Classes Challenge 2012 (VOC 2012) Results (2012)
5. Girshick, R., Donahue, J., Darrell, T., Malik, J.: Rich feature hierarchies for accurate object detection and semantic segmentation. In: Computer Vision and Pattern Recognition (CVPR 2014) (2014)
6. Jia, Y., Shelhamer, E., Donahue, J., Karayev, S., Long, J., Girshick, R., Guadarrama, S., Darrell, T.: Caffe: convolutional architecture for fast feature embedding (2014). arXiv preprint: arXiv:1408.5093
7. Krizhevsky, A., Ilya, S., Hinton, G.: ImageNet classification with deep convolutional neural networks. In: Advances in Neural Information Processing Systems (NIPS 2012), pp. 1097–1105. Curran Associates, Inc. (2012)

8. Markatopoulou, F., et al.: ITI-CERTH participation in TRECVID 2015. In: TRECVID 2015 Workshop. NIST, Gaithersburg (2015)
9. Oquab, M., Bottou, L., Laptev, I., Sivic, J.: Learning and transferring mid-level image representations using convolutional neural networks. In: Computer Vision and Pattern Recognition (CVPR 2014) (2014)
10. Over, P., et al.: TRECVID 2013 - an overview of the goals, tasks, data, evaluation mechanisms and metrics. In: TRECVID 2013. NIST, Gaithersburg (2013)
11. Russakovsky, O., Deng, J., Su, H., et al.: ImageNet large scale visual recognition challenge. Int. J. Comput. Vis. (IJCV) 115(3), 211–252 (2015)
12. Sermanet, P., Eigen, D., Zhang, X., Mathieu, M., Fergus, R., Lecun, Y.: Over-Feat: integrated recognition, localization and detection using convolutional networks (2014)
13. Simonyan, K., Zisserman, A.: Very deep convolutional networks for large-scale image recognition. arXiv technical report (2014)
14. Snoek, C.G.M., Worring, M.: Concept-based video retrieval. Found. Trends Inf. Retr. 2(4), 215–322 (2009)
15. Snoek, C., Fontijne, D., van de Sande, K.E., Stokman, H., et al.: Qualcomm Research and University of Amsterdam at TRECVID 2015: recognizing concepts, objects, and events in video. In: TRECVID 2015 Workshop. NIST, Gaithersburg (2015)
16. Szegedy, C., et al.: Going deeper with convolutions. In: Computer Vision and Pattern Recognition (CVPR 2015) (2015)
17. Yilmaz, E., Kanoulas, E., Aslam, J.A.: A simple and efficient sampling method for estimating AP and NDCG. In: 31st ACM SIGIR International Conference on Research and Development in Information Retrieval, pp. 603–610. ACM, USA (2008)
18. Yosinski, J., Clune, J., Bengio, Y., Lipson, H.: How transferable are features in deep neural networks? CoRR abs/1411.1792 (2014)

Describing Geographical Characteristics with Social Images

Huangjie Zheng[1,2]([envelope]), Jiangchao Yao[2], and Ya Zhang[2]

[1] SJTU-ParisTech Elite Institute of Technology, Shanghai Jiao Tong University,
Shanghai, China
[2] Cooperative Medianet Innovation Center,
Shanghai Jiao Tong University, Shanghai, China
{zhj865265,sunarker,ya_zhang}@sjtu.edu.cn

Abstract. Images play important roles in providing comprehensive understanding of our physical world. When thinking of a tourist city, one can immediately imagine pictures of its famous attractions. With the boom of social images, we attempt to explore the possibility of describing geographical characteristics of different regions. We here propose a Geographical Latent Attribute Model (GLAM) to mine regional characteristics from social images, which is expected to provide a comprehensive view of the regions. The model assumes that a geographical region consists of different "attributes" (e.g., infrastructures, attractions, events and activities) and "attributes" are interpreted by different image "clusters". Both "attributes" and image "clusters" are modeled as latent variables. The experimental analysis on a collection of 2.5M Flickr photos regarding Chinese provinces and cities has shown that the proposed model is promising in describing regional characteristics. Moreover, we demonstrate the usefulness of the proposed model for place recommendation.

Keywords: Geographic characteristics · Recommender systems · Latent variable models · Region description

1 Introduction

Geotagged images are pervasive, and they also provide an intuitive and objective view of our life. Thanks to these properties, images can easily reflect personal, regional, even social characteristics, and plenty of research works have been conducted with social images to facilitate people's life. Geographical analysis from social media has been widely investigated in the recent years. While most of existing studies focus their analysis on landmarks with the assumption that they are representative to regions [1–4], other perspectives such as local festivals

Y. Zhang—The work is partially supported by the High Technology Research and Development Program of China 2015AA015801, NSFC 61521062, STCSM 12DZ2272600.

L. Amsaleg et al. (Eds.): MMM 2017, Part I, LNCS 10132, pp. 115–126, 2017.
DOI: 10.1007/978-3-319-51811-4_10

and events could also be essential for profiling a region. We thus study the problem of forming comprehensive description of geographical characteristics from social media. With the description of geographical characteristics in one specific region, we could better recognize this region and boost a number of utilities such as tourist advertising, etc.

While some existing applications such as tourist recommendation and location retrieval could also extend to this problem [5–8], they mainly rely on the textual information, e.g., social tags. To our best knowledge, geotagged photos help understand intuitively a specific region and it can boost plenty of applications in several domains. For example, it is interesting that systems could generate a recommendation based on its understanding of images, which leads us free from taking effort to find a proper word for the description of the region. Therefore, since the goal is to understand a region from images, the challenge lies in how to map low level visual features to semantic characteristics.

Fig. 1. Motivation of the model. We assume that in every geographic area, people's life consists of several aspects, e.g. sports, music, etc. These aspects could be presented by several clusters, while clusters are formed by vast images.

In this paper, we propose a Geographical Latent Attribute Model (GLAM) to learn geographical characteristics from photo collections. We assume that each region consists of some latent "attributes" (considered as characteristics) and each "attribute" consists of image "clusters". The motivation of our model is illustrated in Fig. 1 using Beijing as an example. A city may be described by several aspects (e.g., historical buildings), and each aspect includes different image clusters (e.g., antiques, temples, sculptures). These clusters are summarized from images taken in Beijing. Following the idea of the generative model, we introduce corresponding latent variables to formalize this procedure. By learning the latent parameters, a comprehensive view about geographical regions is formed.

The major contributions of this paper could be summarized as follows:

- We propose a Geographical Latent Attribute Model (GLAM) to learn geographical characteristics from photo collections without utilizing any textual information.
- We validate the proposed model with 2.5M Flickr photos taken in China to demonstrate its effectiveness in both qualitative and quantitative ways.
- As one of the potential applications, a region recommendation strategy is proposed based on the similarity between region's characteristics and user's interest according to his/her photo album.

The rest of paper is organized as follows: In Sect. 2, we review the related work. Section 3 explains our model and its inference technique. The experiment results will be displayed in Sect. 4, and we conclude our paper in Sect. 5.

2   Related Work

Plenty of works have been conducted in geographical analysis. Ji, et al. [2] propose a hierarchical structure to mine city landmarks from view, scene and city layers. [9] analyzes the attribute at region level for region exploration and [10] handles the urban understanding with CNN. Livia Hollenstein and Ross S. Purves [11,12] focus on social media to find out how people generate their understanding for a city. Similarly, [1] extract the tags representing landmarks to better present and extract view of one region. In [3,4], the authors find the popular landmarks using mean shift.

This work is also related to several applications such as location retrieval, tourist recommendation, etc. [5] shows the same viewpoint that users are more interested in a geographic area than the precise GPS coordinate. Our work thus pay more effort into recommending users with a proper geographic area rather than location estimation with exact geographic coordinates. [6,7] give personalized tourist recommendation based on users' interest and their similarity, while our work focus more on the similarity between user's interest and geographic characteristics.

3   Model

3.1   Geographical Latent Attribute Model

The plate notation of GLAM is illustrated in Fig. 2. Assuming that we have M regions and each region has N_m images, we target to learn the regional attribute distributions $\{\theta_m\}_{m=1,...,M}$ from these images. We first use GoogLeNet to extract one D dimensional feature vector v_{mn} for each image. Then our problem could be formalized to learn $\{\theta_m\}_{m=1,...,M}$ from the feature collection $\{v_{11}, ..., v_{MN_M}\}$.

We transform this problem into a generative procedure and consider that each region has a distribution over characteristics and each characteristic has a distribution over clusters which are modeled by a series of Gaussian mixtures. Both "characteristic" and "cluster" are introduced as latent variables in this hierarchical structure and could be inferred by the observed variables $\{v_{11}, ..., v_{MN_M}\}$. The generative procedure is summarized as follows:

- Choose regional characteristic proportion $\theta_m \sim Dir(\alpha)$.
- Choose the characteristic of one image $i_{mn} \sim Multinomial(\theta_m)$.
- Choose the cluster $z_{mn} \sim Multinomial(\phi_{i_{mn}})$, where $i_{mn} \in \{1, 2, ..., K\}$.
- Choose each visual vector $v_{mn} \sim \mathcal{N}(\mu_{z_{mn}}, \sigma_{z_{mn}}\mathbf{I})$, where $z_{mn} \in \{1, 2, ..., K'\}$.

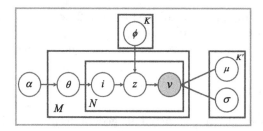

Fig. 2. The plate notation of GLAM

In our model, $\{(\mu_{k'}, \sigma_{k'})\}_{k'=1,...,K'}$ constitute the visual space and $\{\Phi_k\}_{k=1,...,K}$ are used to capture the characteristic-cluster distributions. Latent variables z_{mn} and i_{mn} are decided by v_{mn} and reversely affect the regional characteristic distribution θ_m. In short, we use a topic model structure to learn the high level concepts at the top layer and facilitate Gaussian mixture model to cluster low level visual features at the bottom layer.

3.2 Inference and Learning

In this part, we present our inference algorithm. The key inferential problem of our model is to compute the posterior distribution of latent variables given data as Eq. 1.

$$p(\theta, i, z | \alpha, \phi, \mu, \sigma, v) = \frac{p(\theta, i, z, v | \alpha, \phi, \mu, \sigma)}{p(v | \alpha, \phi, \mu, \sigma)} \qquad (1)$$

Above equation is intractable due to the non-integrable denominator and an alternative method, e.g., Gibbs sampling or variational approximation [13], could be employed. In this paper, we adopt a mean field variational bayes method [14] (variational EM) to deal with our model. Following its methodology, we assume that the variational distribution is defined as

$$q(\theta, i, z) = q(\theta|\gamma)q(i|\psi)q(z|\varPhi),$$

(2)

where γ is the Dirichlet parameter and ψ, $\varPhi$ are the multinomial parameters. With this specification, the latent variables could be approximated by minimizing the Kullback-Leibler (KL) divergence between Eqs. 1 and 2.

$$\arg\min\nolimits_{(\gamma,\psi,\varPhi)} D(q(\theta, \psi, \varPhi)|p(\theta, \psi, \varPhi))$$

(3)

By setting the derivative of free parameters γ, ψ, $\varPhi$ in Eq. 3 to zero, we obtain the following equations.

$$\varPhi_{mnk'} \propto \exp(\sum_k \psi_{ijk} \log \varPhi_{kk'}) \mathcal{N}(v_{ij}|\mu_{k'}, \sigma_{k'})$$

(4)

$$\psi_{ijk} \propto \exp(\Psi(\gamma_{ik})) \exp(\sum_{k'} \varPhi_{ijk'} \log \phi_{kk'})$$

(5)

$$\gamma_{ik} = \alpha_k + \sum_j \psi_{ijk}$$

(6)

The most frequent approach to estimate the model parameters is maximizing the likelihood of observed variables, i.e., $p(v|\alpha, \phi, \mu, \sigma)$. Although there is no analytical integral for this likelihood, Jensen's inequality could be used to get an adjustable lower bound.

$$
\begin{aligned}
&\ln p(v|\alpha, \phi, \mu, \sigma)) \\
&= \ln \int_\theta \sum_{i,z} p(v, \theta, i, z|\alpha, \phi, \mu, \sigma) d\theta \\
&= \ln \int_\theta \sum_{i,z} \frac{p(v, \theta, i, z|\alpha, \phi, \mu, \sigma)q(\theta, i, z)}{q(\theta, i, z)} d\theta \\
&\geqslant E_q(\ln p(v, \theta, i, z|\alpha, \phi, \mu, \sigma)) - E_q(\ln q(\theta, i, z)) \\
&\triangleq L(\alpha, \phi, \mu, \sigma)
\end{aligned}
$$

(7)

With previous optimal free parameters γ, ψ, $\varPhi$, we could maximize the lower bound L by setting the derivatives to zero with respect to the parameters ϕ, μ, σ respectively. Then, we have following solutions:

$$\phi_{kk'} \propto \sum_i \sum_j \psi_{ijk} \varPhi_{ijk'}$$

(8)

$$\mu_{k'} = \frac{\sum_i \sum_j \varPhi_{ijk'} v_{ij}}{\sum_i \sum_j \varPhi_{ijk'}}$$

(9)

$$\sigma_{k'} = \frac{\sum_i \sum_j \Phi_{ijk'} (\mu'_k - v_{ij})^{\mathrm{T}} (\mu'_k - v_{ij})}{D \sum_i \sum_j \Phi_{ijk'}} \tag{10}$$

And for Dirichlet prior α, we use Newton-Raphson method to update it like LDA [15]. Iterating the inference and parameter estimation procedure, we would gradually acquire the solution of our model.

4   Experimental Results

To validate GLAM for geographical analysis, we evaluate it on a Flickr dataset of 2.5M photos in both qualitative and quantitative ways. In addition, we show its potential to retrieve the regions of interest.

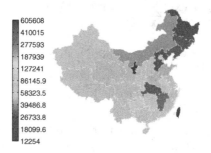

Fig. 3. The color map of data distribution in China. The warmer the color is, the more images are taken there. Taiwan possesses the most amount of data, while Ningxia possesses the least. The average amount in each province is about 85K.

4.1   Experimental Settings

We crawled 6.5M photos that had the GPS information in the YFCC100M dataset [16]. Then with the database of GADM[1], which is a database containing the boundary geo-coordinates of each administration region, we filter out the photos not taken in China and the 2.5M remaining photos are divided into 34 groups according to the administration regions as shown in Fig. 3. One feature vector is extracted for each image from the dropout layer (the second last layer) of GoogLeNet [17].

4.2   Quantitative Evaluation

In this section, we provide a quantitative evaluation for our GLAM model. The GLAM aims to find a better description for regions based on social images. As we know, textual content is good at delivering semantic information. Thus,

---

[1] https://www.gadm.org/.

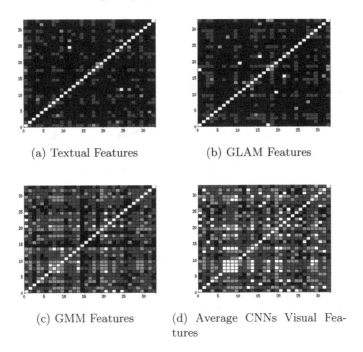

(a) Textual Features

(b) GLAM Features

(c) GMM Features

(d) Average CNNs Visual Features

Fig. 4. Region's similarity computed with different features. We can observe the results of text feature and our model are quite coherent, while the results of the others are difficult to determine the similar regions. Presented with $n = 20$, $K = 15$, $K' = 500$.

we employ the documents from the online tour guide "TravelChinaGuide"[2], the largest and most authoritative online tour operator in China, for comparison. Each document covers general introduction, facts, even life details for each region. We build topic models with LDA [15] from the textual document. The Euclidean distance between regions is computed based on the learned topic model. Similarly, we compute the distance between regions based on visual features learned by GLAM, Gaussian Mixture Model (GMM), and average visual features extracted directly from GoogLeNet. The corresponding distance matrix are shown in Fig. 4, where brighter colors mean higher similarity. It can be seen that our model presents more similar results as textual features, suggesting that our model generates a better semantic description for regions.

To test the effectiveness of our model, we employ the Kernel Canonical Correlation Analysis (KCCA) to compute the correlation between the distance matrix obtained from the textual feature and the other three types of visual features. As shown in Table 1, from textual feature we learn respectively 5, 10, 15, 20, 25 and 30 topics. Meanwhile, GLAM is severally trained with 200 and 500 clusters, and the number of characteristics K is set to 10, 15 and 20 respectively in the experiments. Distance matrix built from GMM and average visual features lead

---

[2] https://www.travelchinaguide.com/.

Table 1. Comparing the correlation between ground truth and the three types of features.

	$GLAM_{K'=200}$			GMM	$GLAM_{K'=500}$			GMM	θ_{avg}
	θ_{10}	θ_{15}	θ_{20}	$K'=200$	θ_{10}	θ_{15}	θ_{20}	$K'=500$	
$Text_{5topics}$	0.5548	0.5945	0.5835	0.3904	0.5910	0.6010	0.6192	0.3912	0.3484
$Text_{10topics}$	0.6191	0.6515	0.6568	0.3920	0.6310	0.6571	0.6780	0.4040	0.3726
$Text_{15topics}$	0.6764	0.7414	0.7251	0.4304	0.7021	0.7827	0.7574	0.4467	0.4038
$Text_{20topics}$	0.7550	0.8014	0.7842	0.5064	0.7704	0.8212	0.8195	0.5163	0.4595
$Text_{25topics}$	0.7253	0.7843	0.7725	0.4739	0.7502	0.8130	0.7982	0.4973	0.4510
$Text_{30topics}$	0.7181	0.7838	0.7670	0.4865	0.7446	0.8056	0.7941	0.4836	0.4477

to a weak correlation to that of textual feature, with the highest correlation at 0.52 and 0.46, respectively, while the highest correlation for GLAM is 0.82, confirming it has a higher similarity to textual features in terms of semantic region description. This superiority is due to that geographical characteristics is abstract and semantic, while GMM and CNN features lack the mechanism to model the semantic features, which makes them difficult to discover complex patterns.

4.3   Qualitative Evaluation

We illustrate here an example (Fig. 5). A region is described by its dominant characteristics and each characteristic is described by the corresponding top 5 clusters. Here we only present one set of experiment results for qualitative evaluation, where the number of characteristics and number of clusters are respectively set to 15 and 500 with the strongest correlation in Table 1. The rest of results can be accessed at: https://sites.google.com/site/geolatentim/.

Take Beijing and Shanghai, two famous cities in China as an example. As shown in Fig. 5, according to Beijing's characteristic distribution, the characteristic 11 dominates, which can be regarded as the main descriptor for Beijing. To interpret this characteristic, the top 5 representative clusters are picked out to describe it. We manually summarize these five clusters, which correspond to Chinese antique, Chinese tower, Chinese architecture, Chinese roof decoration and pedestrian street, indicating people in Beijing prefer a Chinese traditional atmosphere. This conclusion is well-aligned with Beijing because Beijing is the national center of Chinese history and culture and the historical sites are quite common. Similarly, we can see that Shanghai, the economic center of China, is a modern city with large population, as its characters are mainly described by skyscraper, city scene, urban night, modern traffic, and street scene with people crowd. Among all these regions[3], it is remarkable that some cities are

---

[3] To see other examples with different parameter sets, please go to our website: https://sites.google.com/site/geolatentim/.

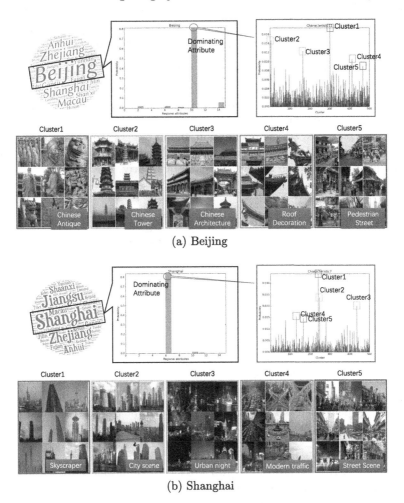

(a) Beijing

(b) Shanghai

Fig. 5. Analysis of the region "Beijing" and "Shanghai".

dominated by one single characteristic (e.g. Beijing, Shanghai) while others possess diverse characteristics (e.g. Sichuan, Shandong) because of geographical and cultural reasons.

4.4 City Recommendation

In this section, we introduce a strategy for region recommendation based on user's photo album. We evaluate the effectiveness of GLAM for recommendation with the Mean Reciprocal Rank metric (MRR).

A photo collection could reflect a user's interest since it contains snapshots of things that the user adores. Here we design a strategy based on the similarity between a user's interest and a region's geographical characteristics for recommendation. First, we compute an interest distribution θ_{new} for a photo collection by Eq. 6. Then, we measure the similarity between this distribution and a region's characteristics with the following distance metric:

$$d_i = ||\theta_i - \theta_{new}||_{i=1,...,M}^2$$

where θ_i is the characteristic distribution in the ith region. The smaller the distance is, the more similar the collection and the region are. The top 3 similar provinces are picked as a recommendation. In our experiments, we crawled additional photos with GPS information from Flickr community[4] (not included in our training data) for both quantitative and qualitative evaluation.

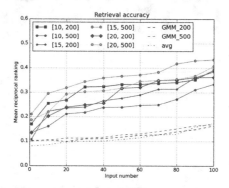

Fig. 6. The recommendation accuracy. In this figure, we can observe that the recommendation accuracy increases as the input number of images increases and GLAM features outperform than GMM and visual features.

For quantitative evaluation, according to the GPS information, we choose 100 images from a province to form a virtual album and the province is regarded as the label of this album. Then we input different amount, accumulating gradually until 100, of images for each album and compute the average MRR to show the recommendation accuracy. Figure 6 presents the average recommendation accuracy with different parameters. The best average MRR performance of GLAM region feature ($K = 15, K' = 500$) is over 40% when input number is more than 70, and according to the property of MRR, we can infer that the label region appears in the top 3 recommended regions, which provide us a reliable recommendation result. Compared with GMM features and visual features, they possess close performance when the input number is small. Nevertheless, it is clear that our model could better perform with more input images and outperform GMM feature and average CNNs visual features because more images could better cover the personal characteristics. For qualitative evaluation, we randomly pick several users, and in each user's photo collection, we randomly select 100 images to form test photo albums. Since the parameter set as 15 "attributes"

---

[4] https://www.flickr.com/.

and 500 "clusters" provide the best performance (Fig. 6), we here employ this parameter setting. Figure 7 present one example: the photo collection containing mostly nature scenes which present mountain and waterside. This indicates the owner of the photo collection may be a fan of traveling in nature. Our recommendation result shows Yunnan, Chongqing and Jiangxi, which are famous for their landscape. Browsing the photos in these regions, we observe the scenery is similar to the photo collection.

Fig. 7. The album recommendation. It is clear that the recommended regions possess the similar natural scene like the input ones.

5   Conclusion

In this paper, assuming "attributes" as the descriptors of regional characteristics, we have attempted to find the characteristic relevance of a region and use the high-relevant ones to describe this region. Meanwhile, representative clusters, formed by social images, are picked out to present the attributes of regions. The experiments on photos in China qualitatively and quantitatively demonstrate our model has the capacity to semantically describe a region with image content. Based on our model, the regional features could be extracted, from which the recommendation strategy profits to provide reliable results and outperform GMM features, as well as average CNNs features in the experiments. Therefore, our model is promising for plenty of applications and could be further developed in future work related to geographical characteristics.

Acknowledgments. The work is partially supported by the High Technology Research and Development Program of China 2015AA015801, NSFC 61521062, STCSM 12DZ2272600.

References

1. Kennedy, L.S., Naaman, M.: Generating diverse and representative image search results for landmarks. ACM, New York, April 2008
2. Ji, R., Xie, X., Yao, H., Ma, W.-Y.: Mining city landmarks from blogs by graph modeling. ACM, October 2009
3. Crandall, D.J., Backstrom, L., Huttenlocher, D., Kleinberg, J.: Mapping the world's photos. ACM, New York, April 2009
4. Crandall, D., Snavely, N.: Modeling people and places with internet photo collections. Commun. ACM **55**, 52–60 (2012)
5. Cao, L., Jie, Y., Luo, J., Huang, T.S.: Enhancing semantic and geographic annotation of web images via logistic canonical correlation regression. In: Proceedings of the 17th ACM International Conference on Multimedia, pp. 125–134. ACM (2009)
6. Clements, M., Serdyukov, P., de Vries, A.P., Marcel, J.T.: Reinders: using flickr geotags to predict user travel behaviour. ACM, New York (2010)
7. Popescu, A., Grefenstette, G.: Mining social media to create personalized recommendations for tourist visits. In: Proceedings of the 2nd International Conference on Computing for Geospatial Research & Applications, COM.Geo 2011, pp. 37:1–37:6. ACM, New York (2011)
8. Li, J., Qian, X., Lan, K., Qi, P., Sharma, A.: Improved image GPS location estimation by mining salient features. Image Commun. **38**(C), 141–150 (2015)
9. Fang, Q., Sang, J., Changsheng, X.: Giant: geo-informative attributes for location recognition and exploration. In: Proceedings of the 21st ACM International Conference on Multimedia, pp. 13–22. ACM (2013)
10. Porzi, L., Bulò, S.R., Lepri, B., Ricci, E.: Predicting and understanding urban perception with convolutional neural networks. In: Proceedings of the 23rd ACM International Conference on Multimedia, pp. 139–148. ACM (2015)
11. Hollenstein, L., Purves, R.: Exploring place through user-generated content: using Flickr tags to describe city cores. J. Spat. Inf. Sci. **2010**, 21–48 (2010)
12. Cranshaw, J., Schwartz, R., Hong, J.I., Sadeh, N.: The livehoods project: utilizing social media to understand the dynamics of a city. In: International AAAI Conference on Weblogs and Social Media, p. 58 (2012)
13. Blei, M.D.: Probabilistic topic models. Commun. ACM **55**(4), 77–84 (2012)
14. Xing, E.P., Jordan, M.I., Russell, S.: A generalized mean field algorithm for variational inference in exponential families. In: Proceedings of the Nineteenth Conference on Uncertainty in Artificial Intelligence, pp. 583–591. Morgan Kaufmann Publishers Inc. (2002)
15. Blei, M.D., Ng, A.Y., Jordan, M.I.: Latent dirichlet allocation. J. Mach. Learn. Res. **3**, 993–1022 (2003)
16. Thomee, B., Elizalde, B., Shamma, D.A., Ni, K., Friedland, G., Poland, D., Borth, D., Li, L.-J.: YFCC100M: the new data in multimedia research. Commun. ACM **59**(2), 64–73 (2016)
17. Szegedy, C., Liu, W., Jia, Y., Sermanet, P., Reed, S., Anguelov, D., Erhan, D., Vanhoucke, V., Rabinovich, A.: Going deeper with convolutions. In: 2015 IEEE Conference on Computer Vision and Pattern Recognition (CVPR), pp. 1–9. IEEE (2015)

Fine-Grained Image Recognition from Click-Through Logs Using Deep Siamese Network

Wu Feng and Dong Liu[⊠]

CAS Key Laboratory of Technology in Geo-Spatial Information
Processing and Application System, University of Science and Technology of China,
Hefei, China
fw092@mail.ustc.edu.cn, dongeliu@ustc.edu.cn

Abstract. Image recognition using deep network models has achieved remarkable progress in recent years. However, fine-grained recognition remains a big challenge due to the lack of large-scale well labeled dataset to train the network. In this paper, we study a deep network based method for fine-grained image recognition by utilizing the click-through logs from search engines. We use both click times and probability values to filter out the noise in click-through logs. Furthermore, we propose a deep siamese network model to fine-tune the classifier, emphasizing the subtle difference between different classes and tolerating the variation within the same class. Our method is evaluated by training with the Bing clickture-dog dataset and testing with the well labeled dog breed dataset. The results demonstrate great improvement achieved by our method compared with naive training.

Keywords: Click-through logs · Deep network · Dog breed recognition · Fine-grained recognition · Siamese network

1 Introduction

Deep network models have achieved remarkable progress in many aspects of computer vision, especially in image recognition since the pioneering work of Krizhevsky *et al.* [12]. The success of deep network-based image recognition is attributed to renewed network structure, efficient learning algorithm, as well as large-scale training data. Indeed, the amount and quality of available training data greatly restrict the capability of deep network models. Therefore, several large-scale well labeled datasets have been constructed to serve for image recognition, including the well-known ImageNet [4].

Inspired by the progress in *general* image recognition, *fine-grained* image recognition, which aims to classify images of the same category (e.g. dog) into subcategories (e.g. dog breeds), is receiving more and more attention of researches. Fine-grained classification is usually more difficult than coarse-grained classification, because it is challenging to identify the subtle difference

© Springer International Publishing AG 2017
L. Amsaleg et al. (Eds.): MMM 2017, Part I, LNCS 10132, pp. 127–138, 2017.
DOI: 10.1007/978-3-319-51811-4_11

between different subcategories and at the same time adapt to the variation within the same subcategory, the variation possibly due to change of viewpoint, illumination, occlusion, context, and so on. When using deep network models for fine-grained classification, another challenge is how to acquire large-scale well labeled datasets for training. Since fine-grained classification, such as dog breed recognition, may require domain specific knowledge, it is not a trivial task even for ordinary people, collecting manual labels is then difficult and expensive.

To address the problem of lacking well labeled data, researchers have proposed to utilize the *weakly* labeled data from the Internet media. There are already billions of images on the Internet, many of them have their related data, such as surrounding text (e.g. in webpages), user provided tags (e.g. in Flickr), and query words (e.g. in search engines). Collecting these data is much easier and cheaper than asking human to manually label each image. However, these are far from accurate labels of images, because some of them may not directly relate to the images, and some others may contain noise. Therefore, how to effectively utilize these weakly labeled data is an important problem, which has been studied for constructing high quality datasets [1], as well as for providing solutions to image classification [16,22] and image retrieval [5,9].

In this paper, we study a method for fine-grained image recognition by utilizing the click-through logs as weakly labeled data. Click-through logs recorded the query words input by users and the following clicked images. They represent the correlation between query words and images, which can serve for extracting the semantic information of images. And it is easy for search engine providers to record these logs. Since the queries raised by Internet users can be very diverse, click-through logs provide a good data source for fine-grained recognition of many different categories and subcategories. However, the click-through logs may contain severe noise, thus simply using the logs as training data turns out not effective.

We present a method to filter out the noise in the click-through logs, using both click times and probability values with pre-trained network. Moreover, we propose a deep siamese network model for fine-grained image recognition, which enhances the network capability in differentiating similar subcategories by using pairs of images to fine-tune. We evaluate our method by performing the task of dog breed recognition, with the Bing clickture-dog dataset [8] for training and the existing well labeled dog breed datasets for test. Experimental results demonstrate the superior performance achieved by our method compared with simply using the raw data for training.

The remainder of this paper is organized as follows. Section 2 presents related work. In Sect. 3, we give the details of our method, including the processing of noisy data and the proposed deep siamese network. Section 4 presents experimental results. Section 5 concludes our work.

2   Related Work

Using Noisy Data for Image Recognition. Deep learning has shown very impressive progress in image recognition, evolving from AlexNet [12], VGG [18],

GoogLeNet [19], to the very deep ResNet [7]. New modules are also proposed to be inserted into deep network models, such as bilinear model [15] and spatial transformer network [10]. However, these methods presume the usage of large-scale well labeled datasets, which are often expensive and time-consuming to obtain. Researchers then proposed to utilize the massive data from the Internet media, though such data are quite noisy. For example, Lu et al. [16] proposed to automatically construct the attribute vocabulary by mining latent topics from the click-through logs, where the mining problem was formulated as a matrix factorization, and further improved by weighted terms-based matrix factorization to address the extreme sparsity of the click-through matrix. Xiao et al. [22] introduced a general framework to train deep networks with only a limited number of clean labels and millions of noisy labels; they modeled the relationships between images, class labels and label noises with a probabilistic graphical model and further integrated it into an end-to-end deep learning system.

There was a grand challenge in the ICME 2016 conference about using click-through logs to perform dog breed recognition. Proposals for this challenge can be divided into two categories. The first category is to directly use the noisy data, such as [23], which proposed to use more dog related data from the full click-through logs and to reuse the pre-trained network model as much as possible. The second category is to utilize additional well labeled data, such as [14] that used high quality public dog datasets to enlarge the training data, and adopted edge enhancement and multi-scale training strategies; both [13,17] proposed to use well labeled datasets to train a dog detector and then use the detector to filter out noisy images in the click-through logs, and proposed to adopt multiple networks as an ensemble. The method in [13] was the winner of the grand challenge, but it was only slightly better than [23] which did not use external data. In this paper, we perform the same task without using external data and we achieve better results than [23].

Deep Siamese Networks. Most of the previous studies on image recognition adopt a singular network, yet siamese network, i.e. a network consisting of two copies of the same mapping, had been investigated in different contexts. Siamese network integrates feature learning and distance learning into an end-to-end network, which is shown to be useful in verification tasks. It was first proposed by Bromley et al. [2] for signature verification, where two identical subnets were used to extract features from two signatures and the cosine distance between two feature vectors was compared with label. Siamese network was also adopted for face verification [3,20] and dimensionality reduction of images [6]. Moreover, siamese network has been extended to include multiple copies of the same mapping, for example three subnets for processing triplets and learning to rank [21]. In this paper, we adopt siamese network to fine-tune the classifier and we propose a new loss function suitable for fine-grained image recognition.

3   Approach

Inspired by the abundance of click-through data and the great success of deep learning, we design an approach for building a deep network model for fine-grained image recognition from the click-through logs. Our approach has contributions in both data and model: first, we propose a method to filter out noise in click-through data; second, we propose a siamese network for better training for fine-grained recognition.

3.1   Data Denoising

In order to remove noise in click-through data, several works proposed to use a pre-trained classifier or detector to select the relevant images, where the pre-training adopted external clean data. But in this paper, we do not consider the adoption of external data, and thus may encounter the overfitting if we use solely the click-through data to pre-train then to select. To solve this problem, we propose to choose partial data for pre-training and then perform denoising with the pre-trained model.

In the pre-training step, we consider the click times of each image in the click-through logs to choose data. Intuitively, click times can reflect the relation between query and image, since most users tend to click the "correct" image after a specific query. We have verified this intuition with real data. As shown in Fig. 1, after sorting the images by their click times in descending order, it can be observed that the noisy images are mostly ranked behind. Thus in experiments, we choose the images with more click times for pre-training.

Next, we use the pre-trained network to filter out noise in click-through data. Given a query (class), its clicked images are input into the network and the output probability values corresponding to the class are compared with a predefined threshold. If the probability value is small, then the image is probably noise. Moreover, we empirically set the threshold according to click times, i.e. for images with more click times the threshold is lower, and vice versa.

3.2   Siamese Network for Fine-Grained Recognition

Most of works on image recognition adopt a singular network, where the predicted class is compared with the ground-truth label to determine loss. Essentially, singular network performs an "independent" prediction for each image without considering its relations to other images. For fine-grained image recognition, the difficulty is how to identify the subtle difference between different classes while also adapt to the variation within the same class. Therefore, comparing several images during learning may help. This idea has been verified in [21], where triplets of images were used for comparative learning for the purpose of ranking. In this paper, we follow the same idea but design a new network for image recognition.

(a) Before sorting by click times

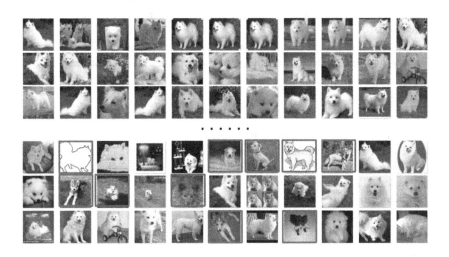

(b) After sorting by click times (and resizing)

Fig. 1. The clicked images corresponding to the dog breed *American Eskimo dog* before and after sorting by click times in descending order. Red blocks indicate noisy images that are not related to the dog breed as per human observation. (Color figure online)

Our proposed deep siamese network is depicted in Fig. 2. There are two identical subnets (a variant of VGG 19 [18]) that process two input images, respectively. After the softmax layer, one dimension of the predicted probability values is extracted, corresponding to the class of the first image. The extracted

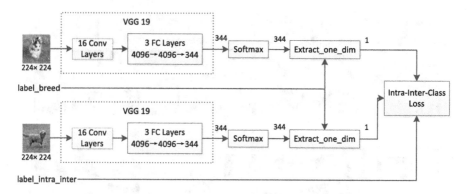

Fig. 2. Our proposed deep siamese network with intra-inter-class loss for fine-grained image recognition.

probability values of both images are then compared to determine loss. Specifically, if the two input images belong to the same class, we require their probability values to be similar, but we still tolerate some difference because the variation within the same class is possible. Similarly, if the two images belong to different classes, their probability values shall be dissimilar, but some similarity is tolerable because the two images indeed belong to the same superclass. Thus, we define a new loss function, namely intra-inter-class loss as follows,

$$\mathcal{L}(I_1, I_2) = \begin{cases} \max\left(0, t_1 - (p_{c_1}(I_1) - p_{c_1}(I_2))^2\right) & \text{if } c_1 \neq c_2 \\ \max\left(0, (p_{c_1}(I_1) - p_{c_1}(I_2))^2 - t_2\right) & \text{if } c_1 = c_2 \end{cases} \quad (1)$$

where I_1 and I_2 are two images in a pair, c_1 and c_2 are the corresponding classes of the two images, and $p_{c_1}()$ stands for extracting the probability value of the image corresponding to c_1, t_1 and t_2 are two thresholds representing the tolerance. This loss is minimized by the standard gradient descent using backpropagation. Note that the siamese network is only used for training. After training, either one of the two identical subnets is actually used in test.

4   Experimental Results

4.1   Settings

We use the Bing clickture-dog dataset provided by Microsoft [8] as the click-through data for training. This dataset consists of 344 queries (dog breeds) and about 95k images, each image has its content and the number of click times. It is a subset of the entire clickture dataset, which comes from the click-through logs of Bing search engine in one year. After inspection of the dataset, we observed much noise in this dataset, such as some images that are not related to dog at all, some images that are falsely labeled, and some images that include several different breeds of dogs. Besides, the amount of images under each dog breed varies significantly, from only one image per breed to several thousands of images per breed. Thus, training with this dataset is quite challenging.

Since the click-through data contain severe noise, we build another dataset from well labeled data to serve for test. Following the process in [23], the test set is union of the Stanford-Dog test set, the Columbia-Dog test set, and the ImageNet-Dog test set. Dog breeds are selected as those appear in the clickture-dog dataset, and images of the same breed are merged together. The test set include 129 dog breeds and 12358 images.

Our training process consists of three steps as shown in Fig. 3. First, we choose images based on click times to pre-train a network. Second, we use the pre-trained network to filter out noise and to select images for training. Third, we select images and construct image pairs to fine-tune the network using our proposed siamese network. All experiments were performed using the Caffe software [11]. More details are discussed as follows.

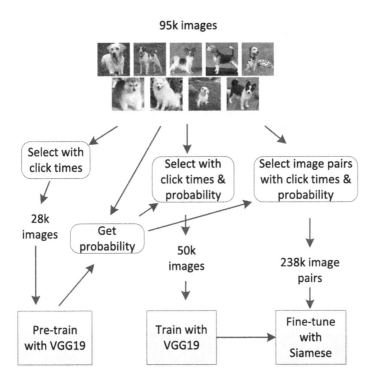

Fig. 3. Flowchart of the entire training process.

4.2   Baseline

To verify the effectiveness of our proposed method, we first set a baseline as a naive training without removing noise from the click-through data. Specifically, we use all the 95k images for training. The network is initialized with the VGG 19 model [18], which was pre-trained on ImageNet, but the final full-connection layer (fc8) is replaced with 344 neurons corresponding to the 344 dog breeds, and this layer is randomly initialized. The baseline achieves 53.68% top-1 accuracy and 79.92% top-5 accuracy, as shown in Fig. 5.

4.3   Pre-training

We use the criterion of click times to choose images for pre-training. The images of a same breed are sorted by click times and the first 80% images are chosen. However, if the amount of images in one breed is less than 100, then all images are chosen; at most 300 images are chosen for each breed. This is trying to balance the data of different classes. Finally, 28k images are chosen for pre-training.

The network is initialized in the same manner as that in baseline. After pre-training, we use the network for test, and the results are shown in Fig. 5, denoted by "Click." It achieves 66.41% top-1 accuracy, more than 12% higher than baseline, and 89.36% top-5 accuracy, more than 9% higher than baseline. Such results demonstrate the effectiveness of selecting images by click times.

4.4   Training

Then we use the pre-trained network to filter out noise in the click-through data. As shown in Fig. 4, the pre-trained network has a good discriminative ability, the images having very high probability values are probably correct samples and the images having very low probability values are mostly noise. We then utilize both click times and the probability values calculated by the pre-trained network to clean the data. Specifically, the images of a same breed are sorted by click times, for the first half, if an image has probability value higher than 0.3, it is selected; for the second half, if an image has probability value higher than 0.5, it is selected.

We use the selected data to train a VGG 19 network. The network is initialized in the same manner as that in baseline. And after training, the network achieves 67.51% top-1 accuracy and 89.77% top-5 accuracy, as shown in Fig. 5 denoted by "Prob." It outperforms the pre-trained network ("Click" in Fig. 5) and the baseline. Note that these three networks are initialized from the same status and trained with different data. Such results demonstrate the effectiveness of data denoising.

4.5   Fine-Tuning with Deep Siamese Network

Next, we utilize the proposed siamese network to further improve. For training siamese network, it is very important to carefully construct the pairwise training data. We use the images having more click times and higher probability values as candidates for pairs. Same-breed pairs and different-breed pairs are constructed separately, and the amounts of them are made comparable to make a balanced dataset.

| 0.962462 | 0.992127 | 0.689608 | 0.999716 | 0.99723 | 0.99999 | 0.999992 | 0.999838 | 0.613759 | 0.034775 | 0.95919 |
| 0.118604 | 0.999952 | 0.999253 | 0.832449 | 0.995671 | 0.804785 | 0.999755 | 0.999401 | 0.185536 | 0.772475 | 0.092379 |

(a) The dog breed *basenji*

| 0.903557 | 0.000146 | 0.008649 | 0.083307 | 0.986059 | 0.003993 | 0.000011 | 0.148574 | 0.017137 | 0.983142 | 0.202895 |
| 0.057141 | 0.002685 | 0.001138 | 0.005388 | 0.050676 | 0.995755 | 0.998531 | 0.842944 | 0.957932 | 0.81592 | 0.483269 |

(b) The dog breed *American Eskimo dog*

Fig. 4. The clicked images corresponding to the dog breeds *basenji* and *American Eskimo dog*, and their probability values calculated by the pre-trained model (refer to Fig. 3). Very high and very low probability values as well as corresponding images are indicated by blue and red blocks, respectively. (Color figure online)

The siamese network is initialized by setting its two subnets with the trained VGG 19 network achieved in the last step. After training, either subnet is used for test. The network finally achieves 68.77% top-1 accuracy and 89.77% top-5 accuracy, as shown in Fig. 5 denoted by "Siamese." It outperforms the other networks, especially the baseline by more than 15% in top-1 accuracy. Such results demonstrate the effectiveness of our proposed deep siamese network.

4.6   Comparisons and Analyses

As mentioned before, we perform the same task as that in [13,14,17,23], i.e. training with clickture-dog dataset and testing with well labeled data. However, external training data had been utilized in [13,14,17], but not used in our work and [23]. Therefore, we compare our results with those reported in [23], which are also shown in Fig. 5. It can be observed that our network "Siamese" outperforms the best two methods in [23] by a large margin, more than 5%. It further verifies the effectiveness of our method.

We also analyzed the top-1 accuracy of each dog breed to observe the per-class performance. Some accuracy values are shown in Fig. 6. It can be observed that our method outperforms the baseline for most of the dog breeds, and the

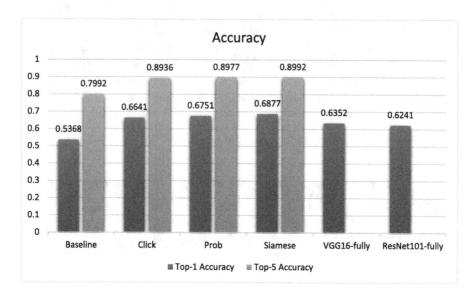

Fig. 5. Recognition (top-1 and top-5) accuracy values of different methods. The methods *VGG16-fully* and *ResNet101-fully* are the best two methods proposed in [23], which did not report top-5 accuracy.

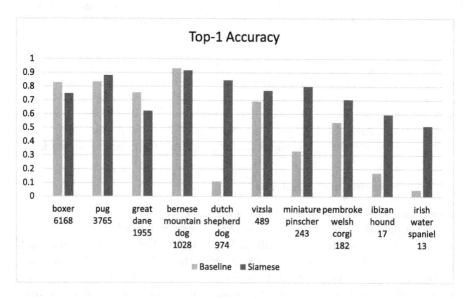

Fig. 6. Recognition (top-1) accuracy values of different dog breeds using baseline and our proposed method. The number beneath each dog breed is the amount of the clicked images of that breed.

improvement is often very significant if the amount of images in that dog breed is limited. However for several breeds, our method performs worse than baseline. We plan to investigate and try to improve the performance further in future work.

5 Conclusions

In this paper, we perform the task of dog breed recognition by learning from click-through logs. The training consists of three steps. We first pre-train a network with portion of images selected by click times. Second, we remove noise in click-through logs using the pre-trained network together with click times. Third, we fine-tune the network using our proposed deep siamese network. Experimental results demonstrate great improvement achieved by our method.

Acknowledgment. This work was supported by the National Program on Key Basic Research Projects (973 Program) under Grant 2015CB351803, by the Natural Science Foundation of China (NSFC) under Grants 61331017 and 61390512, and by the Fundamental Research Funds for the Central Universities under Grants WK2100060011 and WK3490000001.

References

1. Bai, Y., Yang, K., Yu, W., Xu, C., Ma, W.Y., Zhao, T.: Automatic image dataset construction from click-through logs using deep neural network. In: ACM Multimedia, pp. 441–450. ACM (2015)
2. Bromley, J., Bentz, J.W., Bottou, L., Guyon, I., LeCun, Y., Moore, C., Säckinger, E., Shah, R.: Signature verification using a "siamese" time delay neural network. Int. J. Pattern Recogn. Artif. Intell. **7**(04), 669–688 (1993)
3. Chopra, S., Hadsell, R., LeCun, Y.: Learning a similarity metric discriminatively, with application to face verification. In: CVPR, vol. 1, pp. 539–546. IEEE (2005)
4. Deng, J., Dong, W., Socher, R., Li, L.J., Li, K., Fei-Fei, L.: ImageNet: a large-scale hierarchical image database. In: CVPR, pp. 248–255. IEEE (2009)
5. Dong, J., Li, X., Liao, S., Xu, J., Xu, D., Du, X.: Image retrieval by cross-media relevance fusion. In: ACM Multimedia, pp. 173–176. ACM (2015)
6. Hadsell, R., Chopra, S., LeCun, Y.: Dimensionality reduction by learning an invariant mapping. In: CVPR, vol. 2, pp. 1735–1742. IEEE (2006)
7. He, K., Zhang, X., Ren, S., Sun, J.: Deep residual learning for image recognition. In: CVPR, pp. 770–778 (2016)
8. Hua, X.S., Yang, L., Wang, J., Wang, J., Ye, M., Wang, K., Rui, Y., Li, J.: Clickage: towards bridging semantic and intent gaps via mining click logs of search engines. In: ACM Multimedia, pp. 243–252. ACM (2013)
9. Hua, X.S., Ye, M., Li, J.: Mining knowledge from clicks: MSR-Bing image retrieval challenge. In: IEEE International Conference on Multimedia and Expo Workshops (ICMEW), pp. 1–4. IEEE (2014)
10. Jaderberg, M., Simonyan, K., Zisserman, A., et al.: Spatial transformer networks. In: NIPS, pp. 2017–2025 (2015)

11. Jia, Y., Shelhamer, E., Donahue, J., Karayev, S., Long, J., Girshick, R., Guadarrama, S., Darrell, T.: Caffe: convolutional architecture for fast feature embedding. In: ACM Multimedia, pp. 675–678. ACM (2014)
12. Krizhevsky, A., Sutskever, I., Hinton, G.E.: ImageNet classification with deep convolutional neural networks. In: NIPS, pp. 1097–1105 (2012)
13. Li, C., Song, Q., Wang, Y., Song, H., Kang, Q., Cheng, J., Lu, H.: Learning to recognition from bing clickture data. In: IEEE International Conference on Multimedia and Expo Workshops (ICMEW), pp. 1–4. IEEE (2016)
14. Li, W., Ke, C.: Ensemble deep neural networks for domain-specific image recognition. In: IEEE International Conference on Multimedia and Expo Workshops (ICMEW), pp. 1–4. IEEE (2016)
15. Lin, T.Y., RoyChowdhury, A., Maji, S.: Bilinear CNN models for fine-grained visual recognition. In: ICCV, pp. 1449–1457 (2015)
16. Lu, Y.J., Yang, L., Yang, K., Rui, Y.: Mining latent attributes from click-through logs for image recognition. IEEE Trans. Multimed. **17**(8), 1213–1224 (2015)
17. Ou, X., Wei, Z., Ling, H., Liu, S., Cao, X.: Deep multi-context network for fine-grained visual recognition. In: IEEE International Conference on Multimedia and Expo Workshops (ICMEW), pp. 1–4. IEEE (2016)
18. Simonyan, K., Zisserman, A.: Very deep convolutional networks for large-scale image recognition (2014). arXiv preprint: arXiv:1409.1556
19. Szegedy, C., Liu, W., Jia, Y., Sermanet, P., Reed, S., Anguelov, D., Erhan, D., Vanhoucke, V., Rabinovich, A.: Going deeper with convolutions. In: CVPR, pp. 1–9 (2015)
20. Taigman, Y., Yang, M., Ranzato, M., Wolf, L.: DeepFace: closing the gap to human-level performance in face verification. In: CVPR, pp. 1701–1708 (2014)
21. Wang, J., Song, Y., Leung, T., Rosenberg, C., Wang, J., Philbin, J., Chen, B., Wu, Y.: Learning fine-grained image similarity with deep ranking. In: CVPR, pp. 1386–1393 (2014)
22. Xiao, T., Xia, T., Yang, Y., Huang, C., Wang, X.: Learning from massive noisy labeled data for image classification. In: CVPR, pp. 2691–2699 (2015)
23. Xie, G., Yang, K., Bai, Y., Shang, M., Rui, Y., Lai, J.: Improve dog recognition by mining more information from both click-through logs and pre-trained models. In: IEEE International Conference on Multimedia and Expo Workshops (ICMEW), pp. 1–4. IEEE (2016)

Fully Convolutional Network with Superpixel Parsing for Fashion Web Image Segmentation

Lixuan Yang[1,2(✉)], Helena Rodriguez[2], Michel Cruanu[1], and Marin Ferecatu[1]

[1] Conservatoire National des Arts et Metiers,
292 Rue Saint-Martin, 75003 Paris, France
{lixuan.yang,michel.crucianu,marin.ferecatu}@cnam.fr
[2] Shopedia SAS, 16 Rue des Blancs Manteaux, 75004 Paris, France
helena.rodriguez@shopedia.fr

Abstract. In this paper we introduce a new method for extracting deformable clothing items from still images by extending the output of a Fully Convolutional Neural Network (FCN) to infer context from local units (superpixels). To achieve this we optimize an energy function, that combines the large scale structure of the image with the local low-level visual descriptions of superpixels, over the space of all possible pixel labellings. To assess our method we compare it to the unmodified FCN network used as a baseline, as well as to the well-known Paper Doll and Co-parsing methods for fashion images.

Keywords: Clothing extraction · Semantic segmentation · FCN · Superpixel parsing

1 Introduction and Related Work

Although the interest in developing dedicated search engines for fashion databases is several decades old [18], the field only started to develop with the recent massive proliferation of fashion web-stores and online retail shops. Indeed, more and more users expect online advertising to propose items that truly correspond to their expectations in terms of design, manufacturing and suitability. Localizing, extracting, describing and tracking fashion items during web browsing allows professionals to better understand the users' preferences and design web interfaces that make for them a better web shopping experience.

In this work we propose a method to extract, without user supervision, clothes and other fashion items from web images. This is an important building block in designing software systems that solve the problems inspired by the needs of professionals of online advertising and fashion media: present to the users relevant items from a database of clothes, based on the content of the web application they are consulting and its context of use. This goes far beyond the requirements of a search engine: the user is not asked to interact with any search interface or formulate a query, but instead she is accompanied by automatic suggestions presenting in a non-intrusive way a selection of products that are likely to interest her.

© Springer International Publishing AG 2017
L. Amsaleg et al. (Eds.): MMM 2017, Part I, LNCS 10132, pp. 139–151, 2017.
DOI: 10.1007/978-3-319-51811-4_12

1.1 Related Work

Many recent research efforts focusing on fashion images deal with a quite different use case, that of meta search engines federating and comparing several online shops. These efforts focus on improving existing search engines to help users find products that match their preferences while preserving the "browsing" aspect [5]. Online shops sometimes provide image tags for common visual attributes, such as color or pattern, but they usually form a proprietary, heterogeneous and non-standardized vocabulary, typically too small to characterize the visual diversity of desired clothing [7,30]. Moreover, in many cases users look for characteristics expressed through very subjective concepts to describe a style, a brand or a specific design. For this reason, recent research focused on the development of detection, recognition and search of fashion items based on visual characteristics [5,21,24].

One of the approaches models the target item based on attribute selection and high-level classification [6]. For example, in [7] the authors train attribute classifiers on fine-grained clothing styles, formulating image retrieval as a classification problem; they rank items that contain the same visual attributes as the input, which can be a list of words or an image. A similar idea is explored in [13], where a set of features such as color, texture, SIFT features and object outlines are used to determine similarity scores between pairs of images. In [3], the authors propose to extract low-level features in a pose-adaptive manner and combine complementary features for learning attribute classifiers by employing conditional random fields (CRF) to explore mutual dependencies between the attributes. To narrow the semantic gap between the low-level features of clothing items and the high-level categories, in [24] it is proposed to adopt mid-level clothing attributes (e.g., clothing category, color, pattern) as a bridge. More specifically, the clothing attributes are treated as latent variables in a latent Support Vector Machine (SVM) recommendation model. To address larger fine-grained clothing attributes, [4] introduced a novel double-path deep domain adaptation network for attribute prediction by modeling the data jointly from unconstrained photos and the images issued from large-scale online shopping stores. To be able to propose a set of items to specific users, [14] put forward a functional tensor factorization method to model the interactions between users and fashion items.

A second approach consists in using part-based models to compensate for the lack of pose estimation and to model complex interactions in deformable objects [11]. To predict human occupations, in [33] part-based models are employed to characterize complex details and variable appearances of human clothing on the automatically aligned patches of human body parts, using sparse coding and noise-tolerant capacities. A similar part-based model is proposed in [29], where image patches are described by a mixture of color and texture features. Parts are also employed in [26] to reduce the "feature gap" caused by human pose discrepancy, by employing a graph parsing technique to align human parts for cloth retrieval.

Another approach relies on segmentation and aggregation to select different cloth categories. In [16], articulated pose estimation is followed by an over-segmentation of the relevant parts. Then, clustering by appearance creates a reference frame that facilitates rapid classification without requiring an actual training step. Body joints are incorporated in [15] by estimating their prior distributions and then learning the cloth-joint co-occurrences of different cloth types as part of a conditional random field framework to segment the image into different clothes. A similar idea is proposed in [36], where a CRF is formulated by inter-object or inter-attribute compatibility. Simo-Serra et al. [31] have exploited another way to formulate a CRF by taking into account the garment's priors, 2D body pose condition and limb segments. The framework in [35] is based on bottom-up clothing parsing from semantic labels attached to each pixel. Local models of clothing items are learned on-the-fly from retrieved examples and parse mask predictions are transferred from these examples to the query image. Face detection is used in [37] to locate and track human faces in surveillance videos, then clothing is extracted by Voronoi partitioning to select seeds for region growing. For the video applications, [25] use SIFT Flow and superpixel matching to build correspondences across frames and exploit the cross-frame contexts to enhance human pose estimation. Also for human parsing and pose estimation, [8] proposed an unified framework to formulate the problem jointly via a tailored And-Or graph.

Recently, convolutional networks have started to achieve good performance in recognition and have also been deployed for semantic segmentation. For example, the fully convolutional network introduced in [27] improved the performance on the Pascal VOC database by using convolutional layers for pixel prediction and deconvolution layers for upsampling the result. DeepLab [22] improved on this by integrating a Markov Random Field into the upsampled network. In [39], conditional random fields were formulated as Recurrent Neural Networks and achieved good results on most classes of Pascal VOC [10]. Meanwhile, convolutional networks have also been used in fashion retrieval. Hadi et al. [17] employ three methods for street-to-shop retrieval, including two deep learning methods, and a deep similarity learning for the street and shop domains. In [34], the style retrieval uses Siamese CNN to transform features into a latent space. To achieve fast retrieval in a large scale dataset, [23] devised hashes-like representations learned by a latent layer added to the network during fine-tuning on the clothing dataset. Another emerging topic is *fashionability*: predict how fashionable a person looks. A conditional random field model that reasons about several fashionability factors by using deep network features was put forward in [32].

1.2   Outline of Our Approach

Our proposal aims to segment *precisely* the object of interest from the background (foreground separation, see Fig. 1(b)), a difficult problem without user interaction and without using an extensive training database. Indeed, to propose meaningful results in terms of high-level expectations (such as product style or design) we need to achieve a good description of the visual appearance. For this, it is clearly much better if the object is segmented to eliminate the effect of

(a) Original image   (b) Desired output (our result)

Fig. 1. Our goal is to produce a precise segmentation (extraction) of the fashion items as in (b).

mixing with the background and to let the description take into account the shape outline.

Most recent segmentation methods optimize a pixel-based objective function eventually followed by corrections relying on local edge behavior or segmentation templates. However, clothing items are deformable objects that can be encountered in a large variety of poses in real images. Extracting such objects requires an awareness of the object's global properties and context.

To take this into account, we extend the output of a Fully Convolutional Network (denoted by FCN in the following) by optimizing an objective function that iterates over all possible pixel-level labellings. The objective function considers the local adequacy with the class (label) under test, the global mid-scale structure of higher level units (superpixels), as well as the global smoothness of the labeling.

We test our method on the fashion image database CFPD [25] and we compare to the unmodified but fine-tuned FCN introduced in [27] that was shown to achieve state-of-the-art results on the Pascal VOC benchmark. We also compare to Co-parsing [25] and to the Paper-Doll framework [35] (see Sect. 1.1). We provide examples of successful segmentation, analyze difficult cases and evaluate each component of our framework. The rest of the paper is organized as follows. In Sect. 2 we give an outline of our proposal, followed by a detailed description of each component. An experimental validation including both quantitative and qualitative results is then presented in Sect. 3. We conclude the paper in Sect. 4 and provide perspectives for future work.

2   Our Proposal

Detecting clothes in images is a difficult problem because the objects are deformable, have large intraclass diversity and may appear against complex backgrounds. To extract objects under these difficult conditions and without user intervention, methods solely relying on optimizing a local criterion (or pixel classification based on local features) are unlikely to perform well. Some knowledge about the global shape of the class of objects to be extracted is necessary to help a local analysis converge to a correct object boundary. In this paper we use this intuition to develop a framework that takes into account the local/global duality to select the most likely object segmentation.

As baseline we employ the output of the fully convolutional network [27] that was shown to perform very well on the Pascal VOC benchmark. For every pixel in a test image, the FCN provides scores for all the labels (object classes) in the training set.

However, convolutional networks tends to produce a smooth output to preserve the network's generality. Thus, the resulting scores can predict the presence and rough position of objects but are less well suited to detect the exact outline of the objects. To address the localization challenge, some approaches use information from several convolution layers to better estimate the object boundaries [9,27]. Other approaches employ local representations, transforming the localization into a local optimization task [28].

In this paper we pursue an alternative path, by inserting before the softmax layer a fully connected conditional random field model. This idea has emerged recently [1,22,38]. The novelty of our proposal is that our model takes advantage of the middle-scale structure of the image by using superpixel co-localization. As we shall see, this helps the network recover object boundaries at a higher level of detail.

Let X be the test image, of size $W \times H$ pixels. For simplicity, we consider the pixels of the image are enumerated in linear fashion, from 1 to $N = W \times H$. The goal is to associate to each pixel one of the L classes (denoted by the labels $a_1, \ldots, a_L$) learnt by the FCN. Just before the output softmax layer, the FCN produces a vector of L scores for each pixel of the image. The FCN scores for the entire image are given by $F(X)$, the element of $F(X)$ corresponding to pixel i being the L-dimensional vector f_i. A labeling L of the test image consists in attaching a label l_i to each pixel $i \in 1 \ldots N$. The goal is then to maximize a likelihood function $P(L|X, F(X))$ over the space of all labellings L, or equivalently to minimize the corresponding energy:

$$\arg\min_{L} E(X) = \arg\min_{L}(-\log(P(L|X, F(X))))$$

The function $E(X)$ contains several terms aimed at preserving the predictive power of the FCN while improving the localization of the contours:

$$E(X) = \sum_{i=1}^{N} \lambda_{l_i}^{(f)} \theta_i^{(f)}(x_i) + \sum_{i=1}^{N} \lambda_{l_i}^{(r)} \theta_i^{(r)}(x_i) + \sum_{i,j=1, i>j}^{N} \theta_{ij}(x_i, x_j) \qquad (1)$$

Sums are over all the pixels of the image X. The first term, containing $\theta_i^{(f)}(x_i)$, encodes the degree of agreement between the produced output and the FCN scores, which is denoted by the superscript (f) in the equation. If only this

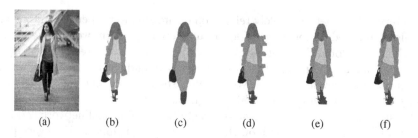

(a)   (b)   (c)   (d)   (e)   (f)

Fig. 2. Original image (a), ground truth (b) and the stages of our approach: (c) softmax FCN, (d) softmax with superpixels labeled by FCN, (e) superpixel parsing, (f) final result.

term were present, the output would be the same as the one provided by the unmodified FCN. To allow for more flexibility, the terms in the sum are weighted by the parameters $\lambda_{l_i}^{(f)}$, where l_i is the label assigned by L to the pixel i. The best values for these parameters, including $\lambda_{l_i}^{(r)}$ from the second term, are found by using a Nelder-Mead simplex method as described in [20]. The second term in Eq. 1 encodes the agreement between the proposed labeling and the visual descriptions of the middle-scale image content units (superpixels). Finally, the third term is a smoothness measure based on the fact that nearby pixels with similar low-level features are likely to belong to the same class. We now provide a detailed description of each of these terms. See also Fig. 2 for an illustration of the results of different stages of our work chain.

Convolutional Term. The FCN [27] takes as input an image of arbitrary size and produces an output of the same size. The classifier transforms fully connected layers into convolutional layers to output a spatial classification map. To make a dense prediction, a deconvolutional layer upsamples the coarse outputs to pixelwise outputs. It employs a skip architecture by combining the final prediction layer with lower layers with finer strides. The network is initialized by using a pre-trained model learnt on Pascal VOC [10] and then fine-tuned to the dataset employed here following a procedure that is similar to the one described in [2]. The output contains score predictions for each class and each pixel. The first term in Eq. 1 is given by the FCN pixel prediction: $\theta_i^{(f)}(x_i|l_i) = -\log f(i, l_i)$, where $f(i, l_i)$ is the FCN output for pixel i and label l_i.

Region/Superpixel Prediction. The second term in Eq. 1 encodes the level of agreement between the neighboring labels and the label of the current pixel. The image is first over-segmented in superpixels, following the idea that all pixels in a superpixel should be attributed the same label. This is reasonable because objects are in general delimited by physical contours, and are thus obtained as disjoint unions of superpixels. This procedure should thus improve the contour localization, which was one of the weaknesses of the original FCN. To extract the superpixels we use the well-known method from [12], also employed by other work on fashion retrieval [25]. For each superpixel we compute a single label, obtained by the Softmax procedure applied on the average of FCN class scores of the pixels in the region. Given a label l_i, let $N(l_i)$ be the set of all superpixels having this label. For a given superpixel S and for each possible label l_i we compute the agreement between the superpixels in $N(l_i)$ and S according to:

$$\phi(l_i|S) = \frac{1}{|N(l_i)|} \sum_{s \in N(l_i)} \frac{1}{1 + |h_S - h_s|^2} \cdot \frac{1}{1 + |p_S - p_s|^2} \tag{2}$$

where h_s denotes the low-level feature (see Sect. 3) of superpixel s, p_s is the barycenter of this superpixel and $\|\cdot\|$ the L_2 norm. The first component in Eq. 2 is larger for superpixels having similar low-level descriptions, implementing the idea that visually similar superpixels must share the same label. However, this behavior is weighted by the second term that is smaller for superpixels that are far away in the image. This filters the effect of similar superpixels that are far from S. The 1 is included in the denominator to protect against numerical instabilities.

If S_i is the superpixel to whom pixel i belongs and l_i is its candidate label, then the second term in Eq. 1 is computed using $\theta_i^{(r)}(x_i) = -\log \phi(l_i|S_i)$. To limit even further the influence of superpixels that are too far away from the candidate, we use the superpixels situated inside a circle of a given radius around the candidate. The best radius is likely to depend on the scale of the objects; in our case, by cross-validation we found a radius of 0.2 of the image size.

Smoothness Term. The third term in Eq. 1 implements a smoothing condition: two pixels are more likely to share the same label if they have similar low-level visual features and are not very distant in the image, corresponding to the idea that objects are localized units in an image. We found that the following formulation, proposed in [19], works well for our purpose:

$$\theta_{ij}(x_i, x_j) = -\log(g_{ij}(x_i, x_j))$$

where

$$g_{ij}(x_i, x_j) = (1 - \delta_{ij})(w_1 \exp(-\frac{\|p_i - p_j\|^2}{2\delta_\alpha^2} - \frac{\|h_i - h_j\|^2}{2\delta_\beta^2}) + w_2 \exp(-\frac{\|p_i - p_j\|}{2\delta_\gamma^2}))$$

where p_i is the position in the image of pixel i, h_i its visual description (see Sect. 3) and δ_{ij} the Kronecker delta. The first part is an appearance kernel inspired by the observation that nearby pixels that are visually similar are likely to be in the same class. The second part is a smoothness kernel that helps removing small isolated regions. The values of $w_{1,2}$ and $\delta_{\alpha,\beta,\gamma}$ are obtained by cross-validation.

Training. The FCN model is pre-trained on the Pascal VOC dataset [10] and then fine-tuned on the specific clothing dataset considered here. Fine-tuning is performed with a lower learning rate of 10^{-14} and a high momentum of 0.99.

For FCN-32s we employ 200K iterations, then 100K iterations for FCN-16s and FCN-8s.

The CRF parameters λ_f and λ_r are obtained by optimization of the F_1 score on the validation set. Given the size of the search space $(2 \times L)$, the Nelder-Mead simplex method [20] is employed.

3   Experimental Results

The proposed method is evaluated on the Colorful Fashion Parsing Data (CFPD), put forward in [25]. This clothing dataset consists of 2,682 images with 23 class labels. We employed the same training and test partitions suggested in [25]. We defined a validation set consisting of 100 randomly selected images from the initial training set. As visual features for describing the superpixels we use the concatenation of normalized RGB and HSV joint histograms, each having 10 bins.

To assess the performance of the proposed method, we perform two sets of experiments. First, a class-by-class comparison is performed on CFPD with the recent deep network FCN [27] method for semantic segmentation. Then, the global performance of our proposal is compared to one of the Co-parsing [25] fashion item annotation method. We eventually provide a qualitative evaluation of our proposal.

Class-by-Class Comparison. The FCN [27] is used as a baseline method in the recent work on object segmentation. The FCN has significantly improved state-of-the-art results in semantic segmentation and has an open implementation. This makes it a good candidate as a baseline and supports a class-by-class comparison. For the purpose of this evaluation, we also fine-tuned the FCN on our training images and employ the `argmax` output of the FCN-8s net.

In Table 1 we show a class-by-class comparison between the proposed method and the FCN (best results are in boldface). As performance measure we employ the average F_1 score over pixels, further averaged over all the testing images of each class. It can be seen that the proposed method performs significantly better on all the classes. While both segmentation methods are automatic (do not require any interaction on test images), these results speak in favor of better taking into account local image information into the algorithm. In our case this is achieved by parsing superpixels.

Global Comparison. We also have to compare our proposal to existing methods that were specifically developed for labelling or extracting fashion items from images. Two prominent frameworks are Paper Doll [35] and Co-parsing [25]. To validate our proposed new term, the algorithm without the superpixel term (second term in Eq. 1) was also tested (denoted "CRF w.o. superpixels" in Table 2). The authors of Paper Doll had introduced the Fashionista database of 685 images that was used to test annotation algorithms. However, this database only contains 456 training images, which is quite small for fine-tuning the FCN, and the

Table 1. Class-by-class comparison with the FCN [27] using the average F_1 score.

Class	Background	T-shirt	Bag	Belt	Blazer	Blouse	Coat	Dress
FCN	95.38	24.30	29.02	10.09	15.44	23.76	13.32	35.08
Our method	**96.67**	**28.73**	**34.83**	**12.87**	**18.80**	**26.94**	**16.48**	**39.44**
Class	Face	Hair	Hat	Jeans	Legging	Pants	Scarf	Shoe
FCN	46.03	45.11	17.02	26.99	20.77	24.94	11.34	35.40
Our method	**49.71**	**54.82**	**21.59**	**31.38**	**24.61**	**28.22**	**13.95**	**43.35**
Class	Shorts	Skin	Skirt	Socks	Stocking	Sunglass	Sweater	−
FCN	38.07	43.48	41.90	9.48	28.17	2.43	11.65	−
Our method	**46.35**	**51.35**	**48.23**	**10.91**	**34.79**	**2.75**	**13.39**	−

Table 2. Comparison of Paper Doll [35], Co-parsing [25], FCN [27] and our method.

Measure	Accuracy	FG. accuracy	Avg. precision	Avg. recall	Avg. F_1
Paper Doll [35]	82.79	44.08	49.20	32.00	32.66
Co-Parsing [25]	84.7	52.49	42.31	42.31	41.42
FCN [27]	86.09	50.62	47.06	51.13	40.29
CRF w.o. superpixels	86.77	49.87	50.21	49.64	40.45
Our method	**88.69**	**55.69**	**53.40**	**56.93**	**45.91**

classes are not exactly the same, so we did not evaluate on Fashionista but only on CFPD.

Table 2 presents a synthesis of the results obtained on CFPD by Paper Doll [35], Co-parsing [25], FCN [27], CRF w.o. superpixels and by the proposed method. Note that the results for Paper Doll come from [25]. Several performance measures are shown: the accuracy, the foreground accuracy, the average precision, the average recall and the average F_1 score, traditionally used for fashion segmentation evaluation. The measures are averaged over pixels, over all the testing images of each class and over classes. As seen from Table 1, the proposed method compares favorably to the other methods according to all the performance measures considered.

Qualitative Evaluation. In Fig. 3 we present some difficult but quite successful segmentations: (a, d) for clothes against confusing or cluttered background, (g, j) for deformed clothes (opened jacket) and (m, p) for small object extraction (shoes). Some parts of our results show that the ground truth is not perfect and an automatic segmentation method can do better. Figure 3 also shows examples where the proposed method is not perfect: when comparing to the ground truth, in (s) it failed to detect the sunglasses and in (v) it failed to detect the belt and the skin (neck). This reflects the lower F_1 score in Table 2 for sunglasses

T-shirt	Bag	Belt	Blazer	Blouse	Coat	Dress	Face	Hair	Hat	Jeans

| Legging | Pants | Scarf | Shoe | Shorts | Skin | Skirt | Socks | Stocking | SunGlass | Sweater |

Fig. 3. Qualitative evaluation: original images (a, d, g, j, m, p, s, v), ground truth (b,e, h, k, n, q, t, w) and associated segmentation results (c, f, i, l, o, r, u, x).

Fig. 4. Comparison with FCN: original image (first), ground truth (second), our method (third) and FCN (last).

and belt. Small objects are quite difficult to extract and may require a specific setting, including *e.g.* a higher penalty during training.

A visual comparison with the results of the FCN is also shown in Fig. 4. As hinted by the quantitative results, the FCN leads to an excessive smoothing and the segmented clothes include larger parts of external objects. This occurs on most of the images in the database, explaining the lower performance of FCN in Tables 1 and 2.

4 Conclusion

To extract clothing objects from web images, we propose to exploit both the output of a Fully Convolutional Neural Network (FCN) used for semantic segmentation and the superpixels obtained from local visual information. By bridging the high-level prediction provided by the deep network and the mid-level image description, the proposed method significantly improves contour localization. The proposed approach is validated by comparisons with the (fine-tuned) FCN alone and to the Co-parsing [25] method, arguably the current state-of-the-art in fashion item extraction.

The results can probably be further improved by the use of more training data and of refined visual features for the superpixels. To better extract small objects we intend to relate the penalty to the relative size of the objects. Also, we believe that some confusion is inevitable between specific classes, *e.g.* legging *vs.* pants or blouse *vs.* sweater, but this may not be a problem for subsequent similarity-based clothing retrieval if object segmentation is correct. The proposed method can easily be extended to other classes at relatively low cost, *i.e.* by manually annotating objects from these new classes to train the FCN and the CRF.

References

1. Bell, S., Upchurch, P., Snavely, N., Bala, K.: Material recognition in the wild with the materials in context database. In: IEEE Conference on Computer Vision and Pattern Recognition (CVPR), pp. 3479–3487 (2015)
2. Bengio, Y.: Deep learning of representations for unsupervised and transfer learning. In: JMLR W&CP: Proceedings of Unsupervised and Transfer Learning Challenge and Workshop, vol. 27, pp. 17–36 (2012)
3. Chen, H., Gallagher, A., Girod, B.: Describing clothing by semantic attributes. In: Fitzgibbon, A., Lazebnik, S., Perona, P., Sato, Y., Schmid, C. (eds.) ECCV 2012. LNCS, vol. 7574, pp. 609–623. Springer, Heidelberg (2012). doi:10.1007/978-3-642-33712-3_44
4. Chen, Q., Huang, J., Feris, R., Brown, L.M., Dong, J., Yan, S.: Deep domain adaptation for describing people based on fine-grained clothing attributes. In: CVPR, pp. 5315–5324. IEEE Computer Society, Boston (2015)
5. Datta, R., Joshi, D., Li, J., Wang, J.Z.: Image retrieval: ideas, influences, and trends of the new age. ACM Comput. Surv. **40**(2), 5:1–5:60 (2008)
6. Deselaers, T., Keysers, D., Ney, H.: Features for image retrieval: an experimental comparison. Inf. Retr. **11**(2), 77–107 (2008)

7. Di, W., Wah, C., Bhardwaj, A., Piramuthu, R., Sundaresan, N.: Style finder: fine-grained clothing style detection and retrieval. In: Proceedings of 2013 IEEE Conference on Computer Vision and Pattern Recognition Workshops, CVPRW 2013, pp. 8–13. IEEE Computer Society, Washington, DC (2013)
8. Dong, J., Chen, Q., Shen, X., Yang, J., Yan, S.: Towards unified human parsing and pose estimation. In: Proceedings of 2014 IEEE Conference on Computer Vision and Pattern Recognition, CVPR 2014, Washington, DC, USA, pp. 843–850 (2014)
9. Eigen, D., Fergus, R.: Predicting depth, surface normals and semantic labels with a common multi-scale convolutional architecture (2014). arXiv:abs/1411.4734
10. Everingham, M., Van Gool, L., Williams, C.K.I., Winn, J., Zisserman, A.: The Pascal visual object classes (VOC) challenge. Int. J. Comput. Vis. **88**(2), 303–338 (2010)
11. Felzenszwalb, P.F., Girshick, R.B., McAllester, D., Ramanan, D.: Object detection with discriminatively trained part-based models. IEEE Trans. Pattern Anal. Mach. Intell. **32**(9), 1627–1645 (2010)
12. Felzenszwalb, P.F., Huttenlocher, D.P.: Efficient graph-based image segmentation. Int. J. Comput. Vis. **59**(2), 167–181 (2004)
13. Hsu, E., Paz, C., Shen, S.: Clothing image retrieval for smarter shopping (Stanford project) (2011)
14. Hu, Y., Yi, X., Davis, L.S.: Collaborative fashion recommendation: a functional tensor factorization approach. In: Proceedings of 23rd ACM International Conference on Multimedia, MM 2015, pp. 129–138. ACM, New York (2015)
15. Jammalamadaka, N., Minocha, A., Singh, D., Jawahar, C.V.: Parsing clothes in unrestricted images. In: British Machine Vision Conference, BMVC 2013, Bristol, UK, 9–13 September 2013
16. Kalantidis, Y., Kennedy, L., Li, L.J.: Getting the look: clothing recognition and segmentation for automatic product suggestions in everyday photos. In: Proceedings of 3rd ACM Conference on International Conference on Multimedia Retrieval, ICMR 2013, pp. 105–112. ACM, New York (2013)
17. Kiapour, M.H., Han, X., Lazebnik, S., Berg, A.C., Berg, T.L.: Where to buy it: matching street clothing photos in online shops. In: Proceedings of 2015 IEEE International Conference on Computer Vision (ICCV), ICCV 2015, pp. 3343–3351. IEEE Computer Society, Washington, DC (2015)
18. King, I., Lau, T.K.: A feature-based image retrieval database for the fashion, textile, and clothing industry in Hong Kong. In: International Symposium on Multi-Technology Information Processing (ISMIP 1996), Hsin-Chu, Taiwan, pp. 233–240 (1996)
19. Krähenbühl, P., Koltun, V.: Efficient inference in fully connected CRFs with Gaussian edge potentials. In: Advances in Neural Information Processing Systems, vol. 24, pp. 109–117. Curran Associates Inc. (2011)
20. Lagarias, J.C., Reeds, J.A., Wright, M.H., Wright, P.E.: Convergence properties of the Nelder-Mead simplex method in low dimensions. SIAM J. Optim. **9**(1), 112–147 (1998)
21. Lew, M.S., Sebe, N., Djeraba, C., Jain, R.: Content-based multimedia information retrieval: state of the art and challenges. ACM Trans. Multimedia Comput. Commun. Appl. **2**(1), 1–19 (2006)
22. Chen, L.-C., George, P., Kokkinos, I., Murphy, K., Yuille, A.: Semantic image segmentation with deep convolutional nets and fully connected CRFs. In: International Conference on Learning Representations, San Diego, United States, May 2015

23. Lin, K., Yang, H.F., Liu, K.H., Hsiao, J.H., Chen, C.S.: Rapid clothing retrieval via deep learning of binary codes and hierarchical search. In: Proceedings of 5th ACM on International Conference on Multimedia Retrieval, New York, USA, pp. 499–502 (2015)

24. Liu, S., Feng, J., Song, Z., Zhang, T., Lu, H., Xu, C., Yan, S.: Hi, magic closet, tell me what to wear!. In: Proceedings of 20th ACM International Conference on Multimedia, MM 2012, pp. 619–628. ACM, New York (2012)

25. Liu, S., Liang, X., Liu, L., Lu, K., Lin, L., Yan, S.: Fashion parsing with video context. In: Proceedings of 22nd ACM International Conference on Multimedia, MM 2014, pp. 467–476. ACM, New York (2014)

26. Liu, S., Song, Z., Wang, M., Xu, C., Lu, H., Yan, S.: Street-to-shop: cross-scenario clothing retrieval via parts alignment and auxiliary set. In: Proceedings of 20th ACM International Conference on Multimedia, MM 2012, pp, 1335–1336. ACM, New York (2012)

27. Long, J., Shelhamer, E., Darrell, T.: Fully convolutional networks for semantic segmentation (2014). arXiv:abs/1411.4038

28. Mostajabi, M., Yadollahpour, P., Shakhnarovich, G.: Feedforward semantic segmentation with zoom-out features (2014). arXiv: abs/1412.0774

29. Nguyen, T.V., Liu, S., Ni, B., Tan, J., Rui, Y., Yan, S.: Sense beauty via face, dressing, and/or voice. In: Proceedings of 20th ACM International Conference on Multimedia, MM 2012, pp. 239–248. ACM, New York (2012)

30. Redi, M.: Novel methods for semantic and aesthetic multimedia retrieval. Ph.D. thesis, Université de Nice, Sophia Antipolis (2013)

31. Simo-Serra, E., Fidler, S., Moreno-Noguer, F., Urtasun, R.: A high performance CRF model for clothes parsing. In: Cremers, D., Reid, I., Saito, H., Yang, M.-H. (eds.) ACCV 2014. LNCS, vol. 9005, pp. 64–81. Springer, Heidelberg (2015). doi:10.1007/978-3-319-16811-1_5

32. Simo-Serra, E., Fidler, S., Moreno-Noguer, F., Urtasun, R.: Neuroaesthetics in fashion: modeling the perception of fashionability. In: CVPR (2015)

33. Song, Z., Wang, M., Hua, X.S., Yan, S.: Predicting occupation via human clothing and contexts. In: Proceedings of the 2011 International Conference on Computer Vision, ICCV 2011, pp. 1084–1091. IEEE Computer Society, Washington, DC (2011)

34. Veit, A., Kovacs, B., Bell, S., McAuley, J., Bala, K., Belongie, S.: Learning visual clothing style with heterogeneous dyadic co-occurrences. In: International Conference on Computer Vision (ICCV), Santiago, Chile (2015)

35. Yamaguchi, K., Hadi, K., Luis, E., Tamara, L.B.: Retrieving similar styles to parse clothing. IEEE TPAMI **37**, 1028–1040 (2015)

36. Yamaguchi, K., Okatani, T., Sudo, K., Murasaki, K., Taniguchi, Y.: Mix and match: joint model for clothing and attribute recognition. In: Proceedings of British Machine Vision Conference (BMVC), pp. 51.1–51.12. BMVA Press, September 2015

37. Yang, M., Yu, K.: Real-time clothing recognition in surveillance videos. In: ICIP, ICIP 2011, pp. 2937–2940. IEEE (2011)

38. Zhang, N., Donahue, J., Girshick, R.B., Darrell, T.: Part-based R-CNNs for fine-grained category detection (2014). arXiv: abs/1407.3867

39. Zheng, S., Jayasumana, S., Romera-Paredes, B., Vineet, V., Su, Z., Du, D., Huang, C., Torr, P.H.S.: Conditional random fields as recurrent neural networks (2015). arXiv: abs/1502.03240

Graph-Based Multimodal Music Mood Classification in Discriminative Latent Space

Feng Su[(⊠)] and Hao Xue

State Key Laboratory for Novel Software Technology,
Nanjing University, Nanjing 210023, China
suf@nju.edu.cn

Abstract. Automatic music mood classification is an important and challenging problem in the field of music information retrieval (MIR) and has attracted growing attention from variant research areas. In this paper, we proposed a novel multimodal method for music mood classification that exploits the complementarity of the lyrics and audio information of music to enhance the classification accuracy. We first extract descriptive sentence-level lyrics and audio features from the music. Then, we project the paired low-level features of two different modalities into a learned common discriminative latent space, which not only eliminates between modality heterogeneity, but also increases the discriminability of the resulting descriptions. On the basis of the latent representation of music, we employ a graph learning based multi-modal classification model for music mood, which takes the cross-modality similarity between local audio and lyrics descriptions of music into account for effective exploitation of correlations between different modalities. The acquired predictions of mood category for every sentence of music are then aggregated by a simple voting scheme. The effectiveness of the proposed method has been demonstrated in the experiments on a real dataset composed of more than 3,000 min of music and corresponding lyrics.

Keywords: Music mood classification · Multimodal · Graph learning · Locality Preserving Projection · Bag of sentences

1 Introduction

With the development of information and multimedia technologies and emerging on-line music services and applications nowadays, digital music has become widely available from various media. The vast amount of music calls for efficient methods for end users to index, retrieve and manage the music of interest. As an effective and commonly exploited property of one music piece for this objective, the mood category of music describes the inherent emotional meaning of the music, and has been recognized as an important criterion when people manage or seek for music. Accordingly, techniques for automated music mood classification have attracted increasing research interests in music information retrieval (MIR) field in recent years and yielded very promising results [6,18].

© Springer International Publishing AG 2017
L. Amsaleg et al. (Eds.): MMM 2017, Part I, LNCS 10132, pp. 152–163, 2017.
DOI: 10.1007/978-3-319-51811-4_13

For automatic categorizing the mood of one music piece, techniques exploiting the audio descriptions of music and effective machine learning schemes have demonstrated satisfying results in the previous research [2,8,11]. Due to the semantic gap between low-level audio features and high-level affective concepts of music, however, mood classification systems solely based on acoustic descriptions sometimes achieve suboptimal performances. On the other hand, as psychological studies [1] have shown, the lyrics of music convey exclusively part of semantic information (including emotional cues) of the music, which is independent to that embedded in the audio data. Accordingly, a few methods have been developed focusing on the exploitation of lyrics features of music and have also yielded encouraging results on music mood classification [4,5]. However, the sparsity of the lyrics data usually brings difficulties to lyrics-based descriptions in discriminating some complicated mood types such as relaxed. By combining both audio and lyrics information, furthermore, several methods [7,15,19] have shown the improved mood prediction performance by making use of inter-modality correlations of audio and lyrics data that could be distinctive for specific mood. Despite the progress that has been made, automatic music mood classification nevertheless remains a challenging task. The heterogeneity between audio and lyrics data spaces also increases the difficulty of effective exploitation of cross-modal correlations for mood classification.

In this paper, we discriminate four mood categories: *happy, angry, sad and relaxed* defined in the commonly employed Russell's circumplex model of affect [12]. We propose a novel multimodal music mood classification method based on cross-modal graph learning on the sentence-level latent representation of music, which effectively captures local correlations between audio and lyrics of the music. Figure 1 shows the main components of our method. The proposed method differs from most existing ones in two main aspects:

1. On basis of the sentence-level audio and lyrics features extracted from the music, we further project these paired low-level features of two different modalities into a learned common discriminative latent space, which not only eliminates between modality heterogeneity, but also increases the discriminability of the resulting descriptions and in certain degree narrows down the semantic gap between the low-level audio/textual features and the high-level concept of music mood.
2. We propose a graph learning based mood classification scheme, which allows measuring the cross-modality similarity of local audio and lyrics descriptions of music in the homogeneous common latent space, so as to more effectively exploit the correlations between two modalities.

The rest of the paper is organized as follows. Section 2 introduces the audio and lyrics feature representation of the music and the further learning of the common latent description space. Section 3 describes the graph learning based mood classification algorithm. Section 4 presents the experimental results, and final remarks are given in Sect. 5.

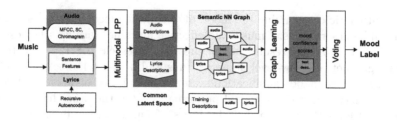

Fig. 1. Block diagram of the proposed method.

2   Music Representation

In this work, we assume both the audio track and the textual lyrics of one music clip are available. For the lyrics, we further assume that the time stamps of each lyrics sentence are available, as provided in the common lyrics format 'LRC'. Then, instead of the commonly used *bag-of-words* holistic model of music, we adopt the *bag of sentence-level descriptions* representation scheme for music proposed in our previous work [17], which captures the local correlations between audio and lyrics descriptions on the sentence scale.

2.1   Lyrics Description

Lyrics are semantically rich and expressive for music mood and naturally have a sentence-oriented structure. Given the time stamps of lyrics sentences, we divide the lyrics into a series of textual segments (a.k.a. sentences), which, as in [17], are depicted by fixed-dimensional continuous descriptions via following steps:

1. We first represent each lyrics word as a 50-dimensional continuous vector of parameters, called *distributed representation of word*, which is learned by a semi-supervised neural language model [16].
2. We then convert one lyrics sentence composed of varying number of words to a single feature vector of fixed dimensionality using the *recursive Autoencoder (RAE)* model adopted in [14], in which adjacent words/phrases in the sentence are iteratively combined and represented as the vectors of activation values at the hidden layer of the Autoencoders in the RAE tree in a bottom-up manner. Finally, the values of the hidden units of the topmost Autoencoder at the root of the RAE tree constitute the description vector for the input lyrics sentence.

Similar to [17], before extraction of lyrics descriptions, we filter the lyrics to retain words with higher discriminability on mood categories and replace those rarely occurring but semantically discriminative words with their synonymous peers for better descriptiveness of the lyrics description.

2.2   Audio Description

We exploit three popular audio features in Music Information Retrieval (MIR) research to capture variant aspects of acoustic characteristics of music for mood analysis: (1) the *Spectral Contrast* depicting the low-level relative spectral distribution of music, (2) the *Mel-frequency Cepstral Coefficients (MFCCs)* that describe the perceptual timbre characteristic of music (and especially the vocal property of singer) by short-time spectrum envelope, and (3) the *Chromagram* feature (a.k.a. Pitch Class Profile), which is independent of timbre, loudness and dynamics of music and robust to ambient noises or percussive sounds.

Specifically, we extract the first 20 MFCCs, the 7-dim spectral contrast and the 12-dim chromagram features from one 23 ms analysis window with 50% overlapping between two successive hops, which are then concatenated to form a 39-dim composite feature. We then compute the mean and variance of the composite feature within the temporal span of a sentence, and use the resulting 78-dim description as the audio feature of the lyrics sentence.

2.3   Common Discriminative Latent Space for Audio and Lyrics

Due to the heterogeneity between the audio and the lyrics feature spaces, previous multi-modal fusion methods usually adopt either the feature concatenation or the decision level fusion scheme to combine multiple modalities, which cannot make full use of the correlation between different modalities. In this work, given the pair of low-level audio and lyrics features that are temporally aligned at the sentence level, we learn a common discriminative latent space with a multimodal extension of the Locality Preserving Projection (LPP) algorithm [3] and then project the original features of two different modalities into this common description space, which not only eliminates between modality heterogeneity, but also increases the discriminability of resulting descriptions.

LPP is a linear dimensionality reduction algorithm particularly designed for preserving the local structure of the original data in projection, which allows acquiring similar search result for nearest neighbours in the original and the new feature spaces. The original objective function of LPP is:

$$\arg\min tr(L_t^T X L_p X^T L_t) \qquad s.t. \quad L_t^T X D X^T L_t = 1 \tag{1}$$

where, L_t is the linear transformation matrix. X is the feature matrix of samples, whose row represents one sample. D is a diagonal matrix whose entries are column sums of the distance weight matrix W of training samples. $L_p = D - W$ is the Laplacian matrix.

To extend LPP to couple the projection directions of two different modalities, i.e., projecting the paired features of two modalities of the same object into a discriminative common latent space in which samples of the same class are united and those from different classes are separated, the objective function of LPP is modified to combine two modalities and incorporate an extra term for the covariance between the exemplars of different modalities, as argued in [13]:

$$\arg\min tr(L_{t_1}^T X_1 L_{p_1} X_1^T L_{t_1} + L_{t_2}^T X_2 L_{p_2} X_2^T L_{t_2} - L_{t_1}^T Z_1 Z_2^T L_{t_2})$$
$$s.t. \quad L_{t_1}^T X_1 D_1 X_1^T L_{t_1} + L_{t_2}^T X_2 D_2 X_2^T L_{t_2} = 1 \tag{2}$$

where, in this work, the i-th row of matrix $Z_{k=1,2}$ is the mean of the samples of the modality k (either the audio or the lyrics) that belong to the mood category i. The resulting linear transformation matrices L_{t_1} and L_{t_2} that minimize Eq. (2) are then used to project the audio and lyrics features from their original description space to the common latent space, respectively.

As LPP enforces the closeness of the projected feature samples of the same class in the learned latent space, consequently, the latent features of both audio and lyrics modalities of the same music sample will be located closely to each other in the common latent space, which makes it possible to exploit the data of one modality to compensate for the weakness (such as unstructuredness) of data of the other modality, which helps to improve the classification accuracy.

3   Graph-Based Mood Classification

On basis of the sentence-level latent descriptions of the music in both audio and lyrics modalities, we exploit a graph learning based music classification model that propagates the mood category labels of the nearest neighbouring training descriptions to the description of one test music sample, which are then aggregated to classify the music to one of the predefined music mood categories.

3.1   Graph of Multimodal Sentence Descriptions

We employ a graph structure to capture the intra-modality and cross-modality similarity between sentence-level descriptions of music in the latent space. Each latent sentence description is represented as a node in the graph, so that the audio and lyrics descriptions of one music sentence correspond to two separate graph nodes. Owing to the homogeneity of the common latent description space, the similarity between two nodes (i.e., sentence descriptions) of *different modalities* in the graph can be measured and is used for computing the weight of the edge connecting them, which captures their cross-modality similarity in the latent description space. We show the advantage of this scheme by comparing it experimentally with the scheme that only the similarity between nodes of same modality is considered.

Specifically, we first divide the set of training sentence descriptions T of both modalities into several semantic groups $\{T_k \subset T\}_{k \in [1..c]}$ (c is the number of mood categories), each consisting of descriptions of the same mood category k. Then, for each sentence description S of a test music, we localize the neighbourhood $N(S)$ of S as a refined subset of training descriptions and correspondingly construct a *semantic nearest neighbour graph* χ_S specific to S on basis of T_k by:

1. For each mood category k, we find the K nearest neighbours of S, which may belong to different modalities, in the corresponding semantic group T_k, resulting in c training sample subsets $\{T_{S,k}\}_{k \in [1..c]}$, each of which consists of the semantic neighbours of S for one mood category.

2. We then construct the S's neighbourhood $N(S) = \bigcup_{k \in [1..c]} T_{S,k}$ and the graph χ_S consisting of nodes $\{x_1, \cdots, x_m\}$, in which the first $m - 1$ nodes are S's neighbours in $N(S)$ and the m-th node is S itself.

We further define a *weight matrix* $W \in R^{m \times m}$ for the semantic nearest neighbour graph χ_S, whose element W_{ij} indicates the pairwise weight of the edge connecting two nodes i and j in χ_S, which is computed as follows:

$$W_{ij} = \begin{cases} L(x_i, x_j), & i \neq j \\ 0, & i = j \end{cases} \tag{3}$$

where, $L(x_i, x_j) = \exp\left(-\frac{1}{2\sigma_L^2} D(x_i, x_j)^2\right)$ is a Gaussian kernel function measuring the similarity of two descriptions x_i and x_j, $D(x_i, x_j)$ is the Euclidean distance between two descriptions, and σ_L is the scale hyperparameter.

3.2 Graph Learning on K Nearest Semantic Neighbours

Based on the semantic nearest neighbour graph χ_S of a test sentence description S, we compute the confidence score for S being of a specific mood category by label propagation on χ_S. We define a $m \times c$ objective matrix $Y = [y_1, y_2, \cdots, y_m]^T = [y_{(1)}, y_{(2)}, \cdots, y_{(c)}]$ containing known category labels for the training sentence descriptions $x_{i \in [1..m-1]}$, in which $Y_{i,k} = 1$ if x_i is of class k, otherwise $Y_{i,k} = 0$. We then exploit the graph learning algorithm to estimate a $m \times c$ scoring matrix $P = [p_1, p_2, \cdots, p_m]^T = [p_{(1)}, p_{(2)}, \cdots, p_{(c)}]$, which approaches Y and in the meanwhile has node labels varying smoothly on the graph. The matrix element $P_{i,k}$ is the confidence score that the sentence description x_i is classified to mood category k, and once learned, p_m gives the predicted mood category labels for the test sentence description S.

For label propagation, we seek an optimal labelling of mood category $p^*_{(l)}$ for $l = 1, \cdots, c$ on both training and test sentence descriptions by minimizing the following objective function as in [20]:

$$E(P) = E_z(P) + \mu E_f(P) \tag{4}$$

$$E_z(P) = \frac{1}{2} \sum_{i=1}^{m} \sum_{j=1}^{m} W_{ij} \left\| p_i / \sqrt{D_{ii}} - p_j / \sqrt{D_{jj}} \right\|^2 \tag{5}$$

$$E_f(P) = \sum_{i=1}^{m} \left\| p_i - y_i \right\|^2 \tag{6}$$

where, the term $E_z(P)$ enforces the smoothness of mood category labels on the graph so that nearby nodes (i.e., similar descriptions) should be assigned similar category labels, while $E_f(P)$ is a loss function to penalize the deviation of the predicted mood categories from the given categories of the training descriptions, D is a diagonal matrix with $D_{ii} = \sum_{j=1}^{m} W_{ij}$, and $\mu > 0$ is the regularization parameter.

We employ an iterative way to obtain the optimal P minimizing $E(P)$, as described by Eq. (7):

$$P(t+1) = \alpha D^{-1/2} W D^{-1/2} P(t) + (1-\alpha)Y \tag{7}$$

where, t is the number of iteration, $\alpha = \frac{1}{1+\mu}$. We use the nearest neighbour method to initialize the mood category label of the test sentence description in $P(0)$ for its effectiveness and simplicity.

Upon convergence, we compute the final confidence score $F(S,k)$ of the test sentence S being classified to mood category k as a weighed combination of $P_{m,k}$ and the average similarity $Q(S,k)$ between S and its semantic neighbourhood $T_{S,k}$ of mood category k as defined in Sect. 3.1:

$$F(S,k) = \lambda P_{m,k} + (1-\lambda)Q(S,k) \tag{8}$$
$$Q(S,k) = \log \frac{1}{|T_{S,k}|} \sum_{S' \in T_{S,k}} L(S,S') \tag{9}$$

where $L(S,S')$ is the Gaussian based similarity of descriptions described before. We found $\lambda = 0.4$ yields the best result in the experiment.

3.3 Music Mood Prediction

Given the confidence scores of every sentence description of the test music on each potential mood category, we aggregate them to obtain the holistic mood category of the music. Let $\mathbf{q}_i^a(k)$ and $\mathbf{q}_i^t(k)$ be the mood category confidence vector computed on the semantic nearest neighbour graph of the audio and lyrics descriptions of the test sentence i respectively, and N_s be the total number of sentences in the test music. We compute two class histograms $H^a(k)$ and $H^t(k)$ ($k = 1..c$, c is the number of mood categories), one for each modality, by accumulating votes cast by every sentences:

$$H^a(k) = \sum_{i=1}^{N_s} \mathbf{q}_i^a(k) \qquad H^t(k) = \sum_{i=1}^{N_s} \mathbf{q}_i^t(k) \tag{10}$$

Next, we combine the two modality-specific histograms to the holistic mood class confidence distribution:

$$H(k) = \beta H^a(k) + (1-\beta)H^t(k) \tag{11}$$

where, β controls the relative importance of each modality in mood prediction. Finally, the maxima over all $H(k)$ ($k = 1..c$) gives the estimate for the potential mood category of the test music.

4 Results

We evaluate the proposed music mood classification method using the dataset [9] adopted in [17], which is composed of music samples whose mood category labels are derived from social tags on the music recommendation website `last.fm`. The total length of audio data of the dataset is about $3,300\,min$ and the lyrics consist of over $34,000$ sentences and $215,000$ words. For each of four mood categories, 100 music samples are used for training with the rest being used for testing. In the experiment, we set the $K = 20$ (the number of nearest semantic neighbours of one specific modality in the graph) in Sect. 3.2, the dimensionality of the latent description space to 25, and $\beta = 0.6$ in Eq. (11), unless otherwise explicitly stated.

4.1   Evaluation of Latent Discriminative Description of Music

We first evaluate the proposed latent representation of music by comparing it with the original audio/lyrics features. Note that, for the latter, due to the heterogeneity between audio and lyrics modalities, we have to look for the K nearest neighbours of a test description in the training descriptions of the same modality, rather than cross-modal searching in the latent space. Table 1 shows the comparison result, which reveals the superiority of the latent representation for more thoroughly exploitation of inter-modal correlation of data.

In Table 2, we further compare the accuracy by exploiting each modality (either audio or lyrics) solely in the proposed model, on basis of either the latent representation or the original representation (marked by $*$). Note 'A+L*' in the table denotes the concatenation of original audio and lyrics features. The result clearly shows, fusion of the two modalities for mood classification *averagely* outperforms any uni-modal configurations. Moreover, compared to the original space, the discriminability of the lyrics modality is averagely significantly

Table 1. Classification accuracy (%) by original and latent representation of music

Model	Mood category				Avg.
	Happy	Angry	Sad	Relaxed	
Original	55.1	82.4	60.6	**37.6**	58.9
Latent	**57.0**	**90.5**	**61.6**	34.7	**61.0**

Table 2. Classification accuracy (%) by audio, lyrics and fusion of both modalities

Model	Mood category				Avg.
	Happy	Angry	Sad	Relaxed	
Audio	42.1	79.7	45.5	**39.6**	51.7
Audio*	49.5	73.0	49.5	38.6	52.7
Lyrics	49.5	83.8	43.4	25.7	50.6
Lyrics*	43.9	73.0	46.5	18.8	45.5
A+L*	48.6	73.0	50.5	39.6	52.9
Proposed	**57.0**	**90.5**	**61.6**	34.7	**61.0**

enhanced in the latent space at the cost of small drop of discriminability of the audio modality, which demonstrates the effect of the coupled learning of multimodal latent representation.

On the other hand, it can also be seen in Table 2 that not all mood categories can be improved by modality fusion. Specifically, the relaxed mood is better classified by the audio descriptions than exploiting descriptions of both modalities. The potential reason for this result is that, compared to the audio descriptions, the lyrics of the relaxed mood are much less discriminative, as revealed by the significant gap between the classification accuracies by each modality solely. As the result, the low discriminability of the lyrics decreases the accuracy of localizing cross-modal nearest neighbours of an audio/lyrics graph node of the relaxed mood as well as the label propagation in the graph-based classification model, e.g. more nodes of other mood categories could be incorporated into the graph.

Table 3 shows the change of accuracy when different dimensionality of the discriminative latent space is used for the description of the music. The proposed method achieves the peak performance around the dimensionality of 25.

4.2   Evaluation of Graph-Based Classification Model

To inspect the impact of the number (K) of nearest neighbours considered in constructing the graph on the final classification accuracy, in Table 4 we compare the proposed model's performances with different values of K. It can be seen that, as K is increased, the accuracy of the proposed method is also constantly improved and reaches the peak around $K = 20$. When further increasing K such as beyond 25, however, the accuracy begins to decline, which indicates more training samples of music that have less correlation with the input sample have been incorporated in the semantic nearest neighbour graph, leading to the drop of classification performance based on nearest neighbours.

Figure 2 further shows the change of classification accuracy by assigning different relative importance ($\beta \in [0, 1]$ in Eq. (11)) to the two modalities in fusion of categorical votes cast by the test audio and lyrics descriptions via graph learning. For comparison, we also show the accuracy by exploiting the original

Table 3. Classification accuracy (%) by exploiting different dimensionality of the latent description space

Dimensionality	10	15	20	25	30	35	40
Accuracy	57.0	58.1	60.4	**61.0**	60.4	58.3	57.9

Table 4. Classification accuracy (%) using different number (K) of nearest neighbours in graph

K	5	10	15	20	25	30	40	100
Accuracy	56.6	59.8	60.5	**61.0**	60.5	60.3	60.2	59.8

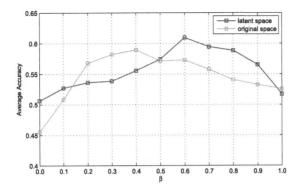

Fig. 2. Classification accuracy by assigning different relative importance β in Eq. (11) to the audio and lyrics modality in fusion of their votes for mood categories.

feature space within the classification framework. It can be seen that, the proposed graph-based classification model on basis of latent descriptions achieved the peak accuracy at $\beta = 0.6$, which indicates that the audio modality plays a relatively more important role in classification than the lyrics modality. Comparatively, the peak performance was achieved by the model at $\beta = 0.4$ in the original feature space. The result shows that, in the latent description space, the discriminability of the audio modality becomes more prominent in contrast to the lyrics modality owing to the higher structuredness of the audio data than the lyrics, which effectively enhances the classification accuracy.

4.3 Comparison with Other Methods

We compared the classification accuracy by our method on the four music mood categories with some other methods in Table 5, which were evaluated on the same experiment dataset aforementioned. The BSTI method [4] uses the bag-of-words (BOW) model with stemming and *tf-idf* weighting for generating lyrics features, which are then concatenated with 63 Marsyas spectral features for mood classification with SVM. The EFFC (Early Fusion by Feature Concatenation) method [19] concatenates textual features (BOW and PLSA) and audio features (MFCC and spectral centroid/moment/roughness), which are then classified by a SVM predictor. The LFSM (Late Fusion by Subtask Merging) method [19] uses one SVM trained with the audio features to classify the arousal aspect of emotion and another SVM trained with the textual features to classify the valence respectively, and then merges the two classification output for predicting the final emotion category. The MMP-SVM method [10], which achieves the top accuracy in MIREX 2012 music mood classification test, uses a total of 312-dim audio features extracted from three frameworks - Marsyas, MIR Toolbox and PsySound3 and employs a SVM classifier. The BOS-HF method [17] adopts a sentence-level representation of music and uses a Hough forest to aggregate the votes on mood category cast by temporally coupled sentence features of both audio and lyrics

Table 5. Classification accuracy (%) by different methods in comparison

Method	Mood category				Avg.
	Happy	Angry	Sad	Relaxed	
BSTI [4]	50.4	64.9	39.4	35.6	47.6
EFFC [19]	49.5	67.6	45.5	41.6	51.1
LFSM [19]	45.8	68.9	53.5	44.6	53.2
MMP-SVM [10]	47.7	81.1	38.4	**51.5**	54.7
BOS-HF [17]	53.3	89.2	51.5	44.6	59.7
Random forest	49.1	82.4	44.4	48.0	56.0
Proposed	**57.0**	**90.5**	**61.6**	34.7	**61.0**

modalities for mood classification. Furthermore, to evaluate the effectiveness of the proposed graph learning classification framework, we implemented a random forest based method that adopts the same latent sentence-level representation of music and exploits a random forest classifier to predict the mood category for an input sentence description.

The comparison results show the effectiveness of the proposed model, which achieves the averagely highest classification accuracy among all unimodal and multimodal (via feature concatenation or decision fusion) representative methods being compared. On the other hand, as observed earlier in Table 2, the high bias of discriminability to the audio modality for the relaxed mood causes significant drop of performance of the proposed graph-based method on that mood.

5   Conclusions

We propose a novel multimodal approach for music mood classification based on cross-modal graph learning on the sentence-level latent representation of music, which effectively exploited the local correlations between audio and lyrics descriptions of the music. The main contribution of the paper is in two aspects: (1) We project the original low-level audio and lyrics features of music into a latent discriminatively learned common description space for enhanced discriminability and homogeneity. (2) We employ a graph learning based classification model, which takes the cross-modality similarity between local descriptions of music into account for effective exploitation of correlations between different modalities. The experiment results demonstrate the effectiveness of the proposed method.

Acknowledgments. Research supported by the National Science Foundation of China under Grant Nos. 61003113, 61672273 and 61321491.

References

1. Besson, M., Faita, F., Peretz, I., Bonnel, A.M., Requin, J.: Singing in the brain: independence of lyrics and tunes. Psychol. Sci. **9**(6), 494–498 (1998)
2. Fu, Z., Lu, G., Ting, K.M., Zhang, D.: A survey of audio-based music classification and annotation. IEEE Trans. Multimedia **13**(2), 303–319 (2011)
3. He, X., Yan, S., Hu, Y., Niyogi, P., Zhang, H.J.: Face recognition using Laplacian faces. IEEE TPAMI **27**(3), 328–340 (2005)
4. Hu, X., Downie, J.S., Ehmann, A.F.: Lyric text mining in music mood classification. In: ISMIR 2009, pp. 411–416 (2009)
5. Hu, Y., Ogihara, M.: Identifying accuracy of social tags by using clustering representations of song lyrics. In: ICMLA 2012, vol. 1, pp. 582–585, December 2012
6. Kim, S., E.M., Migneco, R., Youngmoo, E.: Music emotion recognition: a state of the art review. In: ISMIR 2010, pp. 255–266 (2010)
7. Laurier, C., Grivolla, J., Herrera, P.: Multimodal music mood classification using audio and lyrics. In: ICMLA 2008, pp. 688–693 (2008)
8. Lu, L., Liu, D., Zhang, H.J.: Automatic mood detection and tracking of music audio signals. IEEE TASLP **14**(1), 5–18 (2006)
9. Music mood classification dataset. http://cs.nju.edu.cn/sufeng/data/musicmood. htm
10. Panda, R., Paiva, R.P.: Mirex 2012: mood classification tasks submission (2012)
11. Ren, J.M., Wu, M.J., Jang, J.S.R.: Automatic music mood classification based on timbre and modulation features. IEEE Trans. Affect. Comput. **6**(3), 236–246 (2015)
12. Russell, J.A.: A circumplex model of affect. J. Pers. Soc. Psychol. **39**(6), 1161–1178 (1980)
13. Sharma, A., Kumar, A., III, H.D., Jacobs, D.W.: Generalized multiview analysis: a discriminative latent space. In: CVPR 2012, pp. 2160–2167, June 2012
14. Socher, R., Pennington, J., Huang, E.H., Ng, A.Y., Manning, C.D.: Semi-supervised recursive autoencoders for predicting sentiment distributions. In: EMNLP 2011, pp. 151–161 (2011)
15. Su, D., Fung, P., Auguin, N.: Multimodal music emotion classification using AdaBoost with decision stumps. In: ICASSP 2013, pp. 3447–3451 (2013)
16. Turian, J., Ratinov, L., Bengio, Y.: Word representations: a simple and general method for semi-supervised learning. In: 48th Annual Meeting of the Association for Computational Linguistics, pp. 384–394 (2010)
17. Xue, H., Xue, L., Su, F.: Multimodal music mood classification by fusion of audio and lyrics. In: He, X., Luo, S., Tao, D., Xu, C., Yang, J., Hasan, M.A. (eds.) MMM 2015. LNCS, vol. 8936, pp. 26–37. Springer, Heidelberg (2015). doi:10.1007/978-3-319-14442-9_3
18. Yang, Y.H., Chen, H.H.: Machine recognition of music emotion: a review. ACM Trans. Intell. Syst. Technol. **3**(3), 40:1–40:30 (2012)
19. Yang, Y.-H., Lin, Y.-C., Cheng, H.-T., Liao, I.-B., Ho, Y.-C., Chen, H.H.: Toward multi-modal music emotion classification. In: Huang, Y.-M.R., Xu, C., Cheng, K.-S., Yang, J.-F.K., Swamy, M.N.S., Li, S., Ding, J.-W. (eds.) PCM 2008. LNCS, vol. 5353, pp. 70–79. Springer, Heidelberg (2008). doi:10.1007/978-3-540-89796-5_8
20. Zhou, D., Bousquet, O., Lal, T.N., Weston, J., Scholkopf, B.: Learning with local and global consistency. Adv. Neural Inf. Process. Syst. **16**(16), 321–328 (2004)

Joint Face Detection and Initialization for Face Alignment

Zhiwei Wang$^{(\boxtimes)}$ and Xin Yang

School of Electronic Information and Communications,
Huazhong University of Science and Technology, Wuhan, China
{zhiweiwang,xinyang2014}@hust.edu.cn

Abstract. This paper presents a joint face detection and initialization method for cascaded face alignment. Unlike existing methods which consider face detection and initialization as separate steps, we concurrently obtain a bounding box and initial facial landmarks (i.e. shape) in one step, yielding better accuracy and efficiency. Specifically, each image region is represented using shape-indexed features [6] derived from different head poses. A multipose face detector is trained: regions whose shapes are roughly aligned with faces can have a good feature representation and are utilized as positive samples, otherwise are considered as negative samples. During the face detection phase, initial landmarks can be explicitly placed on the detected faces according to the corresponding shape-indexed features. To accelerate our method, an ultrafast face proposal method based on face probability map (FPM) and boosted classifiers. Experimental results on public datasets demonstrate superior efficiency and robustness to existing initialization schemes and great accuracy improvement for the cascaded face alignment.

Keywords: Face detection · Face alignment · Initialization

1 Introduction

Face alignment, *a.k.a.* facial landmark localization, plays a very important role in many computer vision applications, such as face recognition [7,29], facial animation [5] and face synthesis/morphing [14,17]. Among several face alignment methods [4,6,9,16,18,19,24] proposed during the past decades, cascaded face alignment [4,6,19,24] has been demonstrated to be one of the best approaches due to its high efficiency and accuracy. Cascaded face alignment typically starts from an initial shape, *e.g.*, mean shape of training samples, and progressively updates the shape through sequentially trained regressors until the shape is aligned with the actual face.

However, cascaded face alignment approaches are usually initialization dependent, i.e. the final output of a cascaded face alignment system might change if a different starting shape to the same input image is provided. In particular, if the initial shape is beyond the convergence of the shape model, the discrepancy

© Springer International Publishing AG 2017
L. Amsaleg et al. (Eds.): MMM 2017, Part I, LNCS 10132, pp. 164–175, 2017.
DOI: 10.1007/978-3-319-51811-4_14

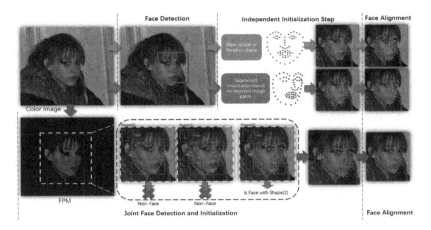

Fig. 1. Illustration of our joint face detection and initialization method (path in red color) vs. the supervised initialization method [26] (path in green color) and conventional unsupervised initialization schemes (path in orange color). (Color figure online)

between the initial shape and actual shape could not be completely rectified by subsequent iterations in the cascade, yielding a wrong alignment result, as shown by the alignment result indicated in orange in Fig. 1. To improve the robustness, some methods [6,19] proposed to initialize facial landmarks using a mean shape within the face bounding box or multiple randomly selected shapes from the training set. However, for faces with large pose variations, these initialization schemes can hardly guarantee a good performance for all cases. Recently, several supervised solutions have been employed in [25–27] to predict a reasonable initial shape. However, these supervised initializations are usually integrated into the framework as an independent module between face detection and alignment, which brings extra computation cost.

In this paper, we propose a new supervised initialization scheme called joint face detection and initialization, which is capable to explicitly place an initial facial shape during the face detection process. Our work is motivated by the fact that head poses, which can be roughly estimated by performing multipose face detection on 2D images, can guide facial landmarks initialization [26]. However, most existing multipose face detection methods [11,13,15,23,31] are either formulated as a classification problem without providing an explicit face shape or are too computational expensive so that are unable to run real-time on mobile devices. In this work, we introduce a multipose face detector which can explicitly output facial landmarks efficiently in addition to face bounding boxes. Specifically, our face detection is based on shape-indexed features extracted from patches around facial landmarks: several face shapes are overlaid on a face region, and only the shape which is aligned with the face (or closer to the actual shape) could provide an accurate face representation (as shown in Fig. 1). To make our joint method more efficient, an ultrafast face proposal method based on a two-level classifier is proposed. At the first level, we convert a color image to a FPM via a Naive Bayesian Model. Each pixel of a FPM represents its likelihood to being belong to a face. At the second level, a boosted classifier based on FPM is employed to eliminate most non-face regions. Comparing to previous initialization schemes which use mean shape or random shapes, our method could guarantee a more reliable and accurate initialization. In addition, our method combines face detection and initialization in one step and share the same face feature for detection, initialization and even for the subsequent alignment tasks, eliminating the redundant computation cost for extracting different features in different stages, which are usually required for some supervised initialization schemes [26] that consider face detection, initialization and alignment as independent modules.

In summary, we make the following contributions:

- We propose a new joint face detection and initialization method, which outperforms other initialization schemes in terms of robustness and efficiency.
- We propose an ultrafast and robust face proposal method to further speed up the joint detection and initialization process. The proposed face proposal method significantly excludes non-face regions so that the following joint face detection and initialization can focus on a small set of candidate regions, facilitating the real-time performance of the overall system (i.e. 32 fps on a single thread CPU of a x86 computer).

1.1   Related Work

Cascaded face alignment has been demonstrated to be one of the best methods for face alignment among emourous amount of approaches [6,9,10,16,18,19,24]. However, most if not all cascaded face alignment methods [6,19] are initialization dependent. In general, existing initialization schemes employed in the cascaded framework can be divided into three groups: *random, mean pose and supervised*.

Random initialization scheme such as [4,6] uses a face shape randomly selected from a training set as an initial shape for a candidate facial region. Each initialization is used to run the face algorithm independently and all alignment results based on different initialization are averaged as the final prediction. Random way may make poor estimations a little bit better, but could also make the good become worse in all possibility and increase the runtime linearly.

Mean pose initialization scheme like [2,19,24] uses the average shape of the training samples as an initial shape with the assumption that the shape variations are relatively small. However, this assumption does not always hold especially for faces with large pose variations [30].

Supervised initialization scheme proposes to predict a roughly aligned initial shape from a given face image. Yang *et al.* [25] performed sparse facial landmarks estimation based on an additional regression forest model and use the estimation result for the initialization. However, this method is too costly for real-time performance. Similarly the authors in [26] proposed a ConvNet framework to estimate head pose, which is then used to generate a reasonable initial shape according to detected face bounding box's size and location. However, a small variation of the location and size of the detected bounding boxes greatly affects the performance of the supervised initialization, degrading the robustness of the initialization scheme.

Some methods proposed to jointly address the tasks of face detection and face alignment. Zhu and Ramanan [31] trained a mixture of deformable part models to capture large face variations under different viewpoints and expressions. The model is comprehensive and could detect faces, estimate head poses and facial points at the same time. However, the model is too slow to meet the real-time requirement. Recently, Chen *et al.* [8] used a boosted cascade structure and simple features to jointly detect and align faces. Its motivation is the observation that accurate face alignment is helpful to distinguish faces/non-faces. However, this method still utilizes mean pose for initialization, which is sensitive to large pose variations as well.

2   Method

The framework of our joint method consists of two major steps as shown in Fig. 1: (1) ultrafast face proposal based on a two-level classifier to significantly eliminate non-face regions (Sect. 2.1); (2) joint face detection and initialization based on a multipose classifier to further detect face in the remaining face-like regions and explicitly place a roughly aligned shape on the detected face (Sect. 2.2). In the following, we provide details of each step.

2.1   Two-Level Classifier Based Face Proposal

Face probability map is a mixture of skin probability map (SPM), eyes probability map (EPM) and mouth probability map (MPM). All three probability maps are obtained by applying the Bayes rule to each pixel for a given image. Specifically, SPM is constructed by a learned Naive Bayesian Model (the first-level classifier) in the $YCbCr$ color space, in which pixels belong to human skin forms a relatively tight cluster even for different races. We collect dozens of images containing human skin and label all skin pixels to build a two dimensional histogram of Cb-Cr chromaticity of skin colors, $h_{skin}(Cb, Cr)$. Similarly, we build another histogram of the entire pixels, $h_{total}(Cb, Cr)$. By using Bayes rule, the probability of skin for a given (Cb, Cr) color vector is described as:

$$p_{skin}(Cb, Cr) = \frac{p(Cb, Cr|skin) \times p(skin)}{p(Cb, Cr)} \qquad (1)$$

$p(Cb, Cr|skin)$ and $p(Cb, Cr)$ can be easily gotten from $h_{skin}(Cb, Cr)$ and $h_{total}(Cb, Cr)$ respectively. And $p(skin)$ is approximated by the fraction of observed skin-like pixels as $p(skin) \cong N_{skin}/N_{total}$, where N_{skin} and N_{total} are sum over Cb and Cr of $h_{skin}(Cb, Cr)$ and $h_{total}(Cb, Cr)$ respectively. Similarly, we can construct an EPM and an MPM.

We observe that three probability maps provide complementary information of faces and combining three maps can better suppress background clutters and enhance face regions comparing to using only one map, as shown in Fig. 2.

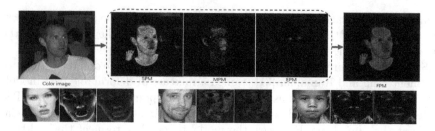

Fig. 2. The first row shows that FPM is a mixture of three probability maps. The second row gives some examples. (Color figure online)

Given a color image, we define the face likelihood for each pixel at position (i, j) a combination of three values as Eq. 2:

$$\mathbf{P}_{face}(i,j) = (p_{skin}(i,j), p_{eyes}(i,j), p_{mouth}(i,j))^T \quad (2)$$

which can be calculated by Eq. 3:

$$\mathbf{P}_{face}(i,j) = ((p_{skin}, p_{eyes}, p_{mouth})^T | Cb(i,j), Cr(i,j)) \quad (3)$$

At the second level, we implement a boosted classifier based on Local Binary Pattern (LBP) features, which are extracted from the converted FPM. When a FPM is provided, we extract three sets of LBP features on its three channels respectively and cascade them as the final LBP features. Such features based on FPM are more reliable than those based on only SPM or intensity values, for that inaccuracy occasionally caused by errors in one probability map could be compensated by the other two, still providing a face-like pattern as shown in Fig. 2.

To demonstrate the efficiency and effectiveness of our face proposal method, we collect 446 single face contained color images and compare the detection rate, false positive rate and runtime of our method with conventional Viola-Jones face detection methods relying on LBP or Haar features extracted from gray scale images, denoted as GRAY+LBP and GRAY + HAAR, respectively. Our face proposal method utilizing LBP and Haar features are denoted as FPM+LBP and FPM+HAAR respectively. Table 1 shows that by utilizing the FPM the

Table 1. Comparison between our FPM based face proposal and gray image based face proposal.

Method	True positives	False positives	Time (ms/face)
GRAY+LBP	381	7968	2.65
FPM+LBP	395	2524	4.66
GRAY+HAAR	415	8294	14.29
FPM+HAAR	423	7147	15.20

second-level classifier can emphasize face information while deemphasizing non-face information, which enables our face proposal method to remove most non-face regions at a ultrahigh speed. Instead of having our joint method search the whole image, we just input the small set of remaining face-like regions, which ensures a real-time performance of our joint method.

2.2   Joint Face Detection and Initialization

Given face-like regions generated by our face proposal method, we propose a multipose face detector to further detect and initialize faces. Our face detector is based on a cascaded face detection framework and shape-indexed features. Moreover, we take a multiple poses strategy for extracting shape-indexed features, making our detector a multipose face detector and explicitly produce initial shapes.

Shape-Indexed Feature Based Cascaded Face Detection. Without loss of generality, the classification score in the general cascade face detection can be written as

$$f^N = \sum_{i=1}^{N} C^i(\phi(x)). \tag{4}$$

Each C^i is a weak classifier and $\phi(x)$ is a feature vector extracted from an image window x. To test x, the weak classifiers are sequentially evaluated and the window is rejected immediately whenever $f^i < \theta^i$ for any $i = 1, 2, ..., N$, where θ^i is the bias threshold. Therefore, the cascade detection is very fast as most negative image windows are rejected after evaluating only a few weak classifiers.

We utilize the shape-indexed features [6] instead of the Haar-like features, based on the reason that the shape-indexed features can explicitly provide shapes. For efficient detection, we use simple shape-indexed pixel-difference features, i.e., the intensity difference of two pixels in the image. During learning, a face shape is overlaid on the image and pixels are indexed by it instead of indexed in global coordinate system (the upper left corner of the picture is the coordinate origin). Specifically, we index a pixel by the local coordinate with respect to its closest facial landmark in the normalized face. Such features present more invariance to the geometric variations in the face shapes and are extremely cheap to compute.

When shape-indexed features are involved, the operation of extracting features in Eq. 4, $\phi(x)$, can be rewritten as $\phi(x, S)$, where S indicates to a given face shape, often represented as a vector of landmark locations, i.e., $S = (l_1, ..., l_k, ..., l_K) \in R^{2K}$, where K is the number of landmarks. $l_k \in R^2$ is the 2D coordinates of the k-th landmark. Therefore, Eq. 4 should be rewritten as

$$f^N = \sum_{i=1}^{N} C^i(\phi(x, S)). \tag{5}$$

In practice, however, the variable S remains unknown or even non-existent for non-face images. The easiest solution is to take the mean shape as S and integrate an operation of face alignment to estimate the actual shape just like Chen *et al.* did in [8]. However, it is essentially an unsupervised initialization method as we discussed in Sect. 1.1 and brings no help for cascaded face alignment.

Joint Detection and Initialization Method. In stead of extracting features by single shape (i.e. mean shape), we introduce a multiple poses features extracting strategy. More specifically, we first cluster face shapes in trainset based on the Hausdorff distance [12] and obtain multiple mean shapes as templates by averaging shapes in every classes. Given an image, we extract shape-indexed features repeatedly relative to different templates. Only the template and the image in the same class share a similar pose, which makes the extracted shape-indexed features a good face representation. Therefore, if template and image are from the same class, features will be labeled positive, otherwise they are labeled negative, which is visualized in Fig. 3(b).

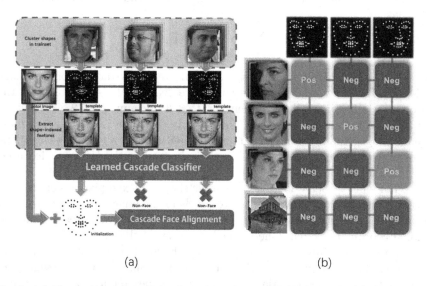

(a)   (b)

Fig. 3. (a) Illustration of our joint face detection and initialization; (b) Definition of positive and negative samples.

Given the two sets of positives and negatives, we train a random forest as a cascade classifier using the RealBoost algorithm, where every decision trees are cascaded and each one acts as a weak classifier corresponding to the variable C in Eq. 5. To learn each decision tree, we use a similar strategy as in [8]: in the split test for an internal node, we minimize the weighted binary entropy for classification. The weight w_i for each sample i is calculated before learning a decision tree as:

$$w_i = e^{-y_i f_i}. \tag{6}$$

where $y_i \in \{-1, 1\}$ is the face/non-face label and f_i is the current classification score. In each tree leaf node, the classification score is computed as

$$\frac{1}{2} \ln \left(\frac{\sum_{\{i \in leaf \cap y_i = 1\}} w_i}{\sum_{\{i \in leaf \cap y_i = -1\}} w_i} \right). \tag{7}$$

where the enumerator and denominator are the sum of positive and negative samples' weight in the leaf node, respectively.

To generate a feature during split test for an internal node, we randomly choose up two pixels in the global coordinate system firstly, then index the two pixels with their corresponding closest landmarks and take the difference of the two shape-indexed pixels as the feature.

During testing, we extract shape-indexed features by each template on a testing image. If the extracted feature vector get the highest score from the learned cascade classifier, it means that the corresponding template is closer to the actual shape than other templates, and can be considered as an initialization for the following face alignment. The complete pipeline of our joint face detection and initialization model is illustrated in Fig. 3(a).

3   Evaluation

3.1   Face Detection

In this section, we compare our joint method with the Viola-Jones detector [22] which is widely used as the first step in many face alignment systems. Although the HELEN dataset provided by [20] is usually used for evaluation of face alignment, we first test on it for two reasons: (1) we want to certify that our joint method is a better alternative for face alignment than Viola-Jones; (2) in the HELEN, the proportion of face in the picture is relatively large, which conforms to many face alignment required scenarios like facial AR or Pule Measurement from facial videos, where images or videos are captured by a frontal camera. We generate face bounding box from annotated facial points as groundtruth and use the precision-recall curves and the 50% overlapping criteria for evaluation. The results in Fig. 4(a) show our method has a better performance than Viola-Jones, indicating that our method is a better alternative in face alignment required scenarios.

Without loss of fairness, we also evaluate on the FDDB dataset, which is a popular and famous face detection dataset and benchmark. We use the same evaluation criteria as on HELEN. The results in Fig. 4(b) show our detector outperforms Viola Jones detector by a large margin. It is worth noting that only 2811 images and their mirror version from the training set of HELEN and LFPW provided by [20] are used for training in our joint method.

Moreover, our joint method achieves 32.48 fps detecting faces larger than 30×30 in a 640×480 resolution video, using single thread on a 3.6 GHz CPU, compared to 33.09 fps for OpenCV Viola-Jones detector, which is known to be optimized for speed.

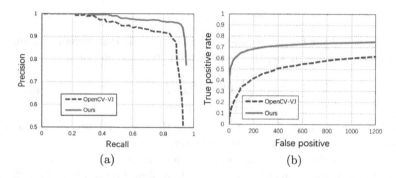

Fig. 4. (a) Comparison with Viola-Jones detector on HELEN dataset; (b) Comparison with Viola-Jones detector on FDDB dataset.

3.2 Face Alignment

We implement local binary features based face alignment method (LBF) [19] with the same settings as LBF-fast and use initial shapes generated by our joint method. Evaluations are performed on three widely used benchmark datasets: (1) *LFPW* originally contains 1100 training and 300 test images. However due to some invalid URLs, we only employ the 811 training and 224 test images provided by [20]. (2) *HELEN* contains 2000 training and 330 test images (provided by [20]). (3) *300 W* [20] is short for 300 Faces in-the-Wild. For the 300 W, we follow the partition of recent methods [19,26,30] to set up the experiments. More specifically, we use 3148 training images from AFW (337 images), HELEN training set (2000 images) and LFPW training set (811 images) and their mirror version for training, and use 689 test images from HELEN test set (330 images), LFPW test set (224 images) and iBug (135 images) for evaluation. All faces we used are annotated with 68 facial landmarks.

We first clarify how many poses is reasonable for our joint method. We test our joint method based LBF-fast on 300 W using one pose (mean shape), three poses and five poses respectively, and the result in the form of Cumulative Error Distribution (CED) is shown in Fig. 5(a). It can be observed that three poses can significantly improve the performance in comparison with only using mean shape (reduce the normalised error by 10.6%), and using five poses has a very close performance to three poses (reduce only 0.2 error). This indicates three poses are enough for our joint method considering both efficiency and accuracy.

We further compare our joint method based LBF-fast with recent state of the art methods (CCNF [3], CFAN [27], DRMF [1], GNDPM [21], IFA [2], LBF [19], SDM [24], TCDCN [28], PCPR [4], CFSS [30], Zhu and Ramanan [31]). For missed faces during detection in our method, we use prescribed face bounding boxes provided by [20] and mean shape for initialization for a fair comparison. The averaged error comparative results are shown in Table 2. It can be observed that our joint method based LBF-fast outperforms all previous methods. Especially, results on 300 W show that our joint method brings a large improvement on the original LBF-fast. To compare the results with literatures reporting CED

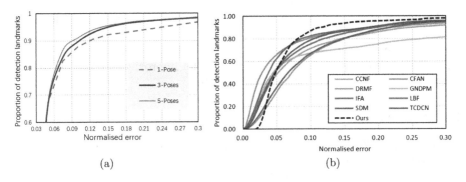

(a)   (b)

Fig. 5. (a) Comparison using different number of poses; (b) Comparison with recent methods on 300 W.

Table 2. Results on three public datasets: LFPW, HELEN and 300 W with 68-point annotation. Ours outperforms all previous methods and especially brings a large improvement on the original LBF-fast on 300 W dataset (best performance in bold).

Method	LFPW (68-pts)	HELEN (68-pts)	300 W (68-pts)
Zhu and Ramanan [31]	8.29	8.16	10.2
DRMF [1]	6.57	6.70	9.22
RCPR [4]	6.56	5.93	8.35
SDM [24]	5.67	5.50	7.5
GNDPM [21]	5.92	5.69	-
CFAN [27]	5.44	5.33	-
LBF [19]	-	-	6.32
LBF-fast [19]	-	-	7.37
CFSS [30]	4.87	4.63	5.76
Ours	**4.05**	**4.13**	**5.55**

performance, we plot CED curve for various methods in Fig. 5(b). Again, our joint method based LBF-fast achieves a better performance benefiting from the reliable initializations.

4   Conclusion

In this work, we present a joint face detection and initialization method for cascade face alignment. Our method first enhances facial regions by computing the likelihood of locating on a face (i.e. skin, eyes or mouth) for each pixel. Such process can be implemented ultrafast via lookup tables and meanwhile can significantly remove distractions from background and in turn improve the robustness and distinctiveness of face representation. A cascaded classifier is then built based on the face probability map to exclude most non-face regions.

In the second step, a multipose face detector based on shape-indexed features relative to face shapes from different poses is applied to the remaining candidate regions. Experimental results demonstrate that the initial shape generated by our method are more reliable than mean shape for improving the face alignment performance with around 25% error decreasing.

References

1. Asthana, A., Zafeiriou, S., Cheng, S., Pantic, M.: Robust discriminative response map fitting with constrained local models. In: Proceedings of the IEEE Conference on Computer Vision and Pattern Recognition, pp. 3444–3451 (2013)
2. Asthana, A., Zafeiriou, S., Cheng, S., Pantic, M.: Incremental face alignment in the wild. In: Proceedings of the IEEE Conference on Computer Vision and Pattern Recognition, pp. 1859–1866 (2014)
3. Baltrušaitis, T., Robinson, P., Morency, L.-P.: Continuous conditional neural fields for structured regression. In: Fleet, D., Pajdla, T., Schiele, B., Tuytelaars, T. (eds.) ECCV 2014. LNCS, vol. 8692, pp. 593–608. Springer, Heidelberg (2014). doi:10.1007/978-3-319-10593-2_39
4. Burgos-Artizzu, X., Perona, P., Dollár, P.: Robust face landmark estimation under occlusion. In: Proceedings of the IEEE International Conference on Computer Vision, pp. 1513–1520 (2013)
5. Cao, C., Weng, Y., Lin, S., Zhou, K.: 3D shape regression for real-time facial animation. ACM Trans. Graph. (TOG) **32**(4), 41 (2013)
6. Cao, X., Wei, Y., Wen, F., Sun, J.: Face alignment by explicit shape regression. Int. J. Comput. Vis. **107**(2), 177–190 (2014)
7. Chen, C., Dantcheva, A., Ross, A.: Automatic facial makeup detection with application in face recognition. In: 2013 International Conference on Biometrics (ICB), pp. 1–8. IEEE (2013)
8. Chen, D., Ren, S., Wei, Y., Cao, X., Sun, J.: Joint cascade face detection and alignment. In: Fleet, D., Pajdla, T., Schiele, B., Tuytelaars, T. (eds.) ECCV 2014. LNCS, vol. 8694, pp. 109–122. Springer, Heidelberg (2014). doi:10.1007/978-3-319-10599-4_8
9. Cootes, T.F., Edwards, G.J., Taylor, C.J.: Active appearance models. IEEE Trans. Pattern Anal. Mach. Intell. **6**, 681–685 (2001)
10. Cootes, T.F., Taylor, C.J., Cooper, D.H., Graham, J.: Active shape models-their training and application. Comput. Vis. Image Underst. **61**(1), 38–59 (1995)
11. Huang, C., Ai, H., Li, Y., Lao, S.: High-performance rotation invariant multiview face detection. IEEE Trans. Pattern Anal. Mach. Intell. **29**(4), 671–686 (2007)
12. Huttenlocher, D.P., Klanderman, G.A., Rucklidge, W.J.: Comparing images using the hausdorff distance. IEEE Trans. Pattern Anal. Mach. Intell. **15**(9), 850–863 (1993)
13. Jones, M., Viola, P.: Fast multi-view face detection. Mitsubishi Electr. Res. Lab TR-20003-96 **3**, 14 (2003)
14. Kemelmacher-Shlizerman, I., Suwajanakorn, S., Seitz, S.M.: Illumination-aware age progression. In: 2014 IEEE Conference on Computer Vision and Pattern Recognition (CVPR), pp. 3334–3341. IEEE (2014)
15. Li, S.Z., Zhu, L., Zhang, Z.Q., Blake, A., Zhang, H.J., Shum, H.: Statistical learning of multi-view face detection. In: Heyden, A., Sparr, G., Nielsen, M., Johansen, P. (eds.) ECCV 2002. LNCS, vol. 2353, pp. 67–81. Springer, Heidelberg (2002). doi:10.1007/3-540-47979-1_5

16. Liang, L., Xiao, R., Wen, F., Sun, J.: Face alignment via component-based discriminative search. In: Forsyth, D., Torr, P., Zisserman, A. (eds.) ECCV 2008. LNCS, vol. 5303, pp. 72–85. Springer, Heidelberg (2008). doi:10.1007/978-3-540-88688-4_6

17. Liu, L., Xing, J., Liu, S., Xu, H., Zhou, X., Yan, S.: Wow! you are so beautiful today!. ACM Trans. Multimedia Comput. Commun. Appl. (TOMM) 11(1s), 20 (2014)

18. Matthews, I., Baker, S.: Active appearance models revisited. Int. J. Comput. Vis. 60(2), 135–164 (2004)

19. Ren, S., Cao, X., Wei, Y., Sun, J.: Face alignment at 3000 FPS via regressing local binary features. In: Proceedings of the IEEE Conference on Computer Vision and Pattern Recognition, pp. 1685–1692 (2014)

20. Sagonas, C., Tzimiropoulos, G., Zafeiriou, S., Pantic, M.: 300 faces in-the-wild challenge: the first facial landmark localization challenge. In: Proceedings of the IEEE International Conference on Computer Vision Workshops, pp. 397–403 (2013)

21. Tzimiropoulos, G., Pantic, M.: Gauss-Newton deformable part models for face alignment in-the-wild. In: Proceedings of the IEEE Conference on Computer Vision and Pattern Recognition, pp. 1851–1858 (2014)

22. Viola, P., Jones, M.J.: Robust real-time face detection. Int. J. Comput. Vis. 57(2), 137–154 (2004)

23. Wu, B., Ai, H., Huang, C., Lao, S.: Fast rotation invariant multi-view face detection based on real Adaboost. In: Proceedings of 2004 Sixth IEEE International Conference on Automatic Face and Gesture Recognition, pp. 79–84. IEEE (2004)

24. Xiong, X., Torre, F.: Supervised descent method and its applications to face alignment. In: Proceedings of the IEEE conference on computer vision and pattern recognition, pp. 532–539 (2013)

25. Yang, H., He, X., Jia, X., Patras, I.: Robust face alignment under occlusion via regional predictive power estimation. IEEE Trans. Image Process. 24(8), 2393–2403 (2015)

26. Yang, H., Mou, W., Zhang, Y., Patras, I., Gunes, H., Robinson, P.: Face alignment assisted by head pose estimation. In: Xie, X., Jones, M.W., Tam, G.K.L. (eds.) Proceedings of the British Machine Vision Conference (BMVC), pp. 130.1–130.13. BMVA Press. https://dx.doi.org/10.5244/C.29.130

27. Zhang, J., Shan, S., Kan, M., Chen, X.: Coarse-to-fine auto-encoder networks (CFAN) for real-time face alignment. In: Fleet, D., Pajdla, T., Schiele, B., Tuytelaars, T. (eds.) ECCV 2014. LNCS, vol. 8690, pp. 1–16. Springer, Heidelberg (2014). doi:10.1007/978-3-319-10605-2_1

28. Zhang, Z., Luo, P., Loy, C.C., Tang, X.: Facial landmark detection by deep multi-task learning. In: Fleet, D., Pajdla, T., Schiele, B., Tuytelaars, T. (eds.) ECCV 2014. LNCS, vol. 8694, pp. 94–108. Springer, Heidelberg (2014). doi:10.1007/978-3-319-10599-4_7

29. Zhao, W., Chellappa, R., Phillips, P.J., Rosenfeld, A.: Face recognition: a literature survey. ACM Comput. Surv. (CSUR) 35(4), 399–458 (2003)

30. Zhu, S., Li, C., Change Loy, C., Tang, X.: Face alignment by coarse-to-fine shape searching. In: Proceedings of the IEEE Conference on Computer Vision and Pattern Recognition, pp. 4998–5006 (2015)

31. Zhu, X., Ramanan, D.: Face detection, pose estimation, and landmark localization in the wild. In: 2012 IEEE Conference on Computer Vision and Pattern Recognition (CVPR), pp. 2879–2886. IEEE (2012)

Large-Scale Product Classification via Spatial Attention Based CNN Learning and Multi-class Regression

Shanshan Ai[1,2], Caiyan Jia[1(✉)], and Zhineng Chen[2]

[1] School of Computer and Information Technology and Beijing Key Lab
of Traffic Data Analysis and Mining, Beijing Jiaotong University, Beijing, China
{15120384, cyjia}@bjtu.edu.cn
[2] Institute of Automation, Chinese Academy of Sciences, Beijing, China
zhineng.chen@ia.ac.cn

Abstract. Large-scale product classification is an essential technique for better product understanding. It can provide support to online retailers from a number of aspects. This paper discusses the CNN based product classification with the existence of class hierarchy. A SaCNN-MCR method is developed to settle this task. It decomposes the classification into two stages. Firstly, a spatial attention based CNN model that directly classifies a product to leaf classes is proposed. Compared with traditional CNNs, the proposed model focuses more on product region rather than the whole image. Secondly, the outputted CNN score together with class hierarchy clues are jointly optimized by employing a multi-class regression (MCR) based refinement, which provides another kind of data fitting that further benefits the classification. Experiments on nearly one million real-world product images show that, based on the two innovations, SaCNN-MCR steadily improves the classification performance over CNN models without these modules. Moreover, it is demonstrated that CNN features characterize product images much better than traditional features, whose classification performance outperforms those of the traditional features by a large margin.

Keywords: Product classification · Convolutional neural network · Spatial attention learning · Multi-class regression

1 Introduction

With the prevalence of eCommerce and advancement of multimedia techniques, visual-based protocols are playing an increasingly important role in online business activities such as online shopping, advertising, etc. A typical example is *Taobao*, one of the most popular online retailers in China. It has launched a functionality named *Pailitao*[1] recently, which enables users to take a photo of an interested product by using camera on mobile devices. As a response, a number of highly similar products are

---

[1] http://pailitao.com/.

© Springer International Publishing AG 2017
L. Amsaleg et al. (Eds.): MMM 2017, Part I, LNCS 10132, pp. 176–188, 2017.
DOI: 10.1007/978-3-319-51811-4_15

instantly returned. Similar efforts are also observed in *CamFind*[2], *Google Goggles*[3], etc. Nowadays, it is a clear trend that multimedia content analysis is becoming a new driving force for better online shopping experience.

This work focuses on automatic product classification, one of the essential means for better understanding and analyzing of product images. It tries to classify a product image to a certain class, which is selected from a number of predefined classes (e.g., long sleeved sweater, running shoes) given by online retailers. It provides support to several interesting applications in these stores including: first, classification serves as a fundamental step for a number of online retailer services such as product annotation, recommendation and search [1–8, 13, 14]. Better classification performance always corresponds to higher quality of the services and thus better user experience. Second, classification is beneficial to the management of online retailers. For instance, cases that people upload a product but assigned a wrong class label, either unintentionally or deliberately, is not rare in online retailers. With the predicted labels, mistakenly labeled products can be found out much easier and rectified by manual intervention. Besides, statistics about the products could also be more accurate if rectified.

There have been several studies undergone for the task of automatic product classification [1–4], due to the great economic benefits. Despite with these promising efforts, we argue that studies on this topic are still confronted with two fundamental problems when adopted to real-world online retailers. The first problem is the class coverage, famous online retailers like *Taobao* and *eBay* sell products covering a wide range of classes. However, existing studies are mostly focused on a subset of them. For example, shoes [1], clothes [4], grocery [2], or evaluated on tens of classes only [3]. Directly extending these methods to real-world retailers, which have hundreds of categories with various products, is not straightforward. The second problem is about methods employed. Existing studies mostly relied on handcrafted features and traditional models. The performance of large-scale visual classification has been significantly boosted by recent advancements on deep convolution neural networks (CNN) [5–7, 12]. However, efforts of applying CNN models to generic product classification are still limited. Moreover, product categories are always organized in a hierarchical way in existing stores. The mechanism remains unclear that how to combine CNN with the class hierarchy to further enhance the classification performance. The task of automatic product classification is not well investigated particularly under large-scale realistic scenarios.

Motivated by this, in this work we study the problem of large-scale product classification in real-world corpuses that depict product images and categories as realistic as possible. Specifically, we propose a SaCNN-MCR method that implements the CNN based product classification with the existence of the class hierarchy. The proposed method is featured by two major stages, spatial attention based CNN learning and multi-class regression (MCR) based refinement. The first stage learns a spatial attention based CNN model that directly classifies a product image to leaf classes. To this end, we first implement the task using AlexNet, a well-recognized CNN structure [5].

---

[2] http://camfindapp.com/.

[3] https://play.google.com/store/apps/details?id=com.google.android.apps.unveil&hl=en.

The CNN model is pre-trained with ILSVRC12 training set and fine-tuning using hundreds of thousands of real-world product images. Since it is common that only a part of image is related to the product, a spatial attention layer is stacked behind the convolutional layers, instructing the network to pay more attention to the product region. The structure is termed as SaCNN in this work. In the second stage, we further propose a MCR based refinement. It takes the outputted CNN score as input, and outputs a series of new scores jointly refined based on the CNN score and class hierarchy clue, aiming at further improving the classification performance by performing data fitting from another angle. We have conducted extensive experiments on a public dataset containing nearly one million real-world product images[4]. Evaluations show that both the spatial attention learning and MCR based regression leads to steady performance improvement over CNN models without the two modules. Moreover, it is also shown that CNN features characterize product images much better than traditional handcrafted features, which produce significantly better classification performance.

The work has three main contributions. (1) We give a comprehensive study to the task of product image classification by using CNN, where CNN with different configurations are evaluated. Moreover, comparisons in both feature and model aspects are also carried out. To our knowledge, it is the first attempt of applying the CNN model to large-scale generic product classification. (2) We propose a meaningful modification to AlexNet, where a spatial attention layer is inserted between the convolutional layers and fully connected layers, such that the network could focus more on the product region rather than the whole image and benefit the classification. (3) We propose a MCR based refinement to jointly utilize the CNN prediction scores and the class hierarchy. As a result, data are further fitted from a different aspect and performance gains are observed.

2 Related Work

Image classification is a longstanding research task in the multimedia community. Product image classification is a specific scenario in this branch. It is closely related to large-scale image classification, which mainly focuses on classify a wide range of natural images, and fine-grained product classification, which further classifies products belonging to a certain category like clothes to more fine-level sub-categories. Both of them have received extensive research efforts in the past years [1–7].

Recently, the task of large-scale image classification has been significantly boosted by the advancement in CNN. For example, the top-5 test error rate on the classification task of the ILSVRC competition has dropped to 15.3% in 2012 by AlexNet [5], to 6.67% in 2014 by GoogLeNet [6], and to 3.57% in 2015 by ResNet [7]. As a comparison, the lowest error rate of the non-CNN methods is 26.2% in ILSVRC12, which has been surpassed by a large margin. On the other hand, existing methods on fine-grained product classification largely depends on the data to be classified. Different

---

[4] https://tianchi.shuju.aliyun.com/competition/introduction.htm?spm=5176.100066.333.8.
0JtHPC&raceId=231510.

methods are developed for different products. For example in [1], the authors proposed the Circle & Search system. It associates several attributes with each kind of shoes and implement the shoe classification and retrieval in an attribute-aware manner. In [2], George *et al.* presented a pre-exemplar multi-label product classification method, targeting simultaneously product recognition and localization. The method is validated on a grocery dataset. Deep models are also considered in [4], where a FashionNet is developed to learn clothing features by jointly predicting clothing attributes and landmarks, and then to generate the learned features. In [3], Chai *et al.* use the Campana-Keogh descriptor to predict the product labels of top ranked product images and re-rank them. However, the method only has been validated on 100 k images of 25 categories without hierarchy.

Compared with the above two kinds of classification tasks, generic product classification is less studied to date. However, it is shown that classification has served as a building block for many online retailer services. For example, knowing the class label is helpful for product retrieval and recommendation [1–4]. Despite with significant value, how to deploy a product classification service for typical online retailers is still unclear, which motivates us to investigate this work.

3   The Proposed Method

3.1   Overview

Given a product image I with groundtruth label c_i, the objective of product classification aims at finding a class label c_t, which satisfies

$$c_t = \arg\max_{c \in R} f_c(I) \tag{1}$$

where c is a certain category and R is the union of product categories predefined by online retailers. In our scenario, R is structured as a two-level class hierarchy (e.g., coat – long sleeved sweater). $f_c(I)$ denotes the prediction score that product I classifies as class c. If $c_i = c_t$, it is said a correct classification, otherwise it is an incorrect classification.

We propose a SaCNN-MCR method to implement the task. As depicted in Fig. 1, the method consists of two major stages. In the first stage, a SaCNN structure is developed. It inserts a spatial attention layer between the convolutional layers and fully

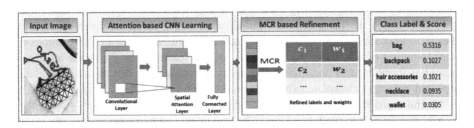

Fig. 1. An illustrative flowchart of the proposed SaCNN-MCR method

connected layers, instructing the network to learn from the product region rather than the whole image, or saying, performing the spatial attention based CNN learning. In the second stage, since typical online retailers use a two-level class hierarchy to organize products, we take the prediction score of SaCNN, termed as $\hat{f}_c(I)$, on both root and leaf levels as the inputs and design a MCR based refinement to adjust the prediction by utilizing the class hierarchy. The refined score, i.e., $f_c(I)$, is used to decide the final leaf class label. We will elaborate the two stages in detail as follows.

3.2   CNN with Spatial Attention Layer

CNN is recognized as the most powerful model for its superior performance in large-scale image classification. Therefore, it is a straightforward idea to extend it to large-scale product classification. To this end, we use AlexNet, an 8-layer deep convolutional neural network, as the classification model. The model structure is inspired by the organization of the animal visual cortex. It is suitable to deal with visual signals.

AlexNet is pre-trained by ILSVRC12 training set. Since the training images mainly composed of natural images rather than product images, adaption is essential for good performance. In the implementation, to classify products to leaf classes, the number of neurons in output layer is set to the number of leaf classes. We use hundreds of thousands of product images to fine-tuning the parameters. To ensure a good convergence, learning rate of the output layer and the rest layers are set to 0.01 and 0.001, respectively. The obtained model is termed as the fine-tuning AlexNet.

As a comparison, we also train the network from scratch and initialize the parameters following methods described in [5], using the same training dataset. Top-1 classification accuracy of the obtained model is 0.6782, while the fine-tuning AlexNet achieves 0.7210. It clearly verifies that fine-tuning is a better choice. Therefore, we take the fine-tuning AlexNet as the basic model and abbreviate it as AlexNet in the rest of the paper.

The fine-tuning adaption already could result a decent performance. Nevertheless, it is observed that for many product images, the product region only occupies a part of the image. Besides, the product region is somehow prominent in the image in order to quickly attract the user's interest. Therefore, similar to [8, 15], we propose to integrate spatial attention learning with CNN to enhance the classification performance.

In the implementation, we add a spatial attention layer behind conv5, i.e., the fifth convolutional layer, which outputs 256 feature maps of size 13 * 13. To ensure adding the attention layer without affecting structure of the rest network, the attention layer consists of the following building blocks. First, a convolutional layer with kernel size of 1 * 1 is added. Its parameters are initialized with the Gaussian distribution. The layer is followed by a sigmoid activation to rectify the values to (0,1). A common weight template of size 13 * 13 is learned for all the feature maps. Then, the attention learning is formulated as performing element wise product between the conv5 feature maps and the weight template, which also output 256 feature maps of size 13 * 13.

Several exemplars of the spatial attention learning are depicted in Fig. 2, where the top row corresponds to product images. The middle and bottom rows are feature maps before and after the attention learning, respectively. Brightness of the feature maps

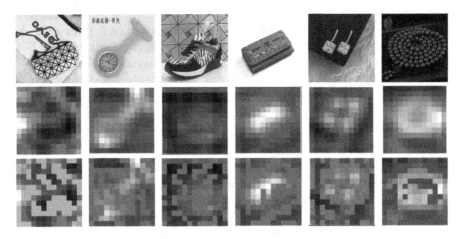

Fig. 2. Typical feature maps before (middle) and after (bottom) the spatial attention learning.

indicates the importance of the corresponding position in the image. It is observed that more product regions are likely to be highlighted after spatial attention learning, showing that the spatial attention learning indeed instructs the network to focus more on the product itself. The modified network and the outputted prediction score are termed as SaCNN and $\hat{f}_c(I)$, respectively.

3.3   Multi-class Regression Based Refinement

In SaCNN the classes are regarded as independent to each other. However, online retailers typically use a class hierarchy to organize products. There are intrinsic relationships between the classes. For example, if a product is classified as a *coat* (a root class), it is less likely to be classified as a *running shoe* (a leaf class). The hierarchy presents a lot of priori that can be utilized to further model the classification process, which have not been exploited in the SaCNN learning.

Motivated by this, we propose a multi-class regression (MCR) based refinement to further adjust the classification score on top of $\hat{f}_c(I)$, aiming at leveraging the hierarchical relationships to rectify the previous wrongly classified samples. Our intuition is that the class hierarchy carries valuable clues for better classification. However, they have not been fully explored in the SaCNN learning stage. To utilize these clues, let $S(c) \subseteq R$ be a set of classes in the hierarchy (including both root and leaf classes) that semantically related to leaf class c, e.g., classes with visually similar products, $\hat{f}^c(I)$ be the score vector corresponding to $S(c)$, we can formulate the refinement by MCR as

$$f_c(I) = P\left(c \mid \hat{f}^c(I), W\right) = \frac{exp\left(w_c^T \hat{f}^c(I)\right)}{\Sigma_{t \in S(c)} \exp\left(w_t^T \hat{f}^c(I)\right)} \tag{2}$$

where w_c is the weight vector for leaf class c. $W = [w_{c_1}, w_{c_2}, \ldots, w_{c_n}]$ be the complete weight matrix, where n is the number of leaf classes. Equation 2 gives $f_c(I)$, the prediction score that image I classifies as class c, which is calculated based on both SaCNN scores on semantically related classes and weight parameters.

Given a set of samples $X = \{x_1, x_2, \cdots\}$ with labels $C = \{c_1, c_2, \cdots\}$, the optimal weight matrix of W, denoted as W^*, can be obtained by

$$W^* = \arg\max_W - \sum_{x_i \in X} \log P\left(c_i \mid W, \hat{f}(x_i),\right) + \lambda \|W\|^2 \tag{3}$$

where $\hat{f}(x_i)$ the score vector corresponding to R. Equation 3 is referred to as L2-regularized MCR, which can be efficiently solved by Quasi-Newton method [9]. With the weights, $f_c(I)$ can be obtained accordingly by Eq. 2.

An interesting issue in Eq. 3 is how to determine $S(c)$. Intuitively, setting $S(c) = R$ is not a wise scheme, as many classes are not semantically related (e.g., shoe and book), which is likely to introduce noises. On the contrary, setting $S(c) = \{c\}$ is also meaningless as it totally ignores the class hierarchy. Therefore, we propose a heuristic to determine $S(c)$. Specifically, a SaCNN that classifies products to root categories is implemented. Given an image I, we pick out all root classes and leaf classes of the top-K scored root categories as the $S(c)$. The reason of incorporating all root classes is twofold. First, classification scores on root classes are obtained from an independent SaCNN. They are complementary to the leaf scores. Second, by considering semantically related classes only, the irrelevant noises are suppressed to a large extent. We empirically set K to 3 in this paper. Preliminary results show that the proposed heuristic leads to 0.1–0.2 absolute performance gains in top-1 accuracy compared with the scheme of setting $S(c) = R$.

Selecting a subset of classes as $S(c)$ also triggers an engineering problem when solving Eq. 3, the length of $\hat{f}^c(I)$ is varying across classes. To address this problem, we fix the length of $\hat{f}^c(I)$ to be $|R|$, i.e., with the same size as $\hat{f}(x_i)$. Alternatively, we set the dimensions corresponding to $c \notin S(c)$ to value zero when optimizing Eq. 3.

It also should be noted that similar to [10], we use the unnormalized prediction score rather than the class posteriors returned by the softmax layer as $\hat{f}_c(I)$, which approximate the differences between classes more realistic.

4 Experiments and Results

4.1 Dataset and Settings

We use a public dataset containing 971,467 real-world product images for evaluation. The dataset is first released by the large-scale image retrieval competition from the *Tianchi big data contest*. The images, their classes and attribute labels (a few of them might be inaccurate) are all given. All images are picked out from taobao.com. It thus exhibits typical challenges in large-scale visual classification such as variants in product location, capturing view, illumination, etc. The class hierarchy is also the same as taobao.com. It is a two-level hierarchy consists of 10 root classes and 547 leaf classes.

Although released for image retrieval competition originally, the dataset is also suitable for evaluating large-scale product classification.

We perform a series of pre-processing on the dataset. Since images are not uniformly distributed across leaf classes, as many as 99,351 and as few as 1. To ensure the evaluation reliability, we filter out classes with less than 1000 samples. As a result, we obtain 10 root classes and 129 leaf classes with 894,571 images in total, which compose the dataset evaluated in this work. We randomly split the images to training, testing and validation sets class-by-class following the fixed partition ratios of 70%, 20% and 10%. To increase stability of the experiment, the testing and validation sets are mixed and then randomly split for 5 times, such that we get 5 sets of results for each trained model. As the evaluation metric, similar to the ILSVRC competition, the top-1 and top-5 accuracies are adopted.

4.2 Feature and Model Comparison

Since little attention has been paid to the task of generic product image classification previously, we carry out two experiments to compare different methods regarding feature and model aspects, aiming at giving an overview of the advantages of CNN over the traditional solutions in the task.

Feature: GIST V.S. CNN-FC7. GIST is a global scene feature well-recognized for years [11]. By using GIST feature, each image can be represented by a 512 dimensional feature vector. As a comparison, we extract the fc7 feature from the SaCNN, which is a 4096 dimensional feature vector. We use a weighted KNN as the classification model. The model is featured by taking two practical factors into account, i.e., imbalanced sample distribution across classes, and the varying distance between neighbor images. Concretely, given an image I, let image I_i be one of its k nearest neighbors, we define the weight w_i as

$$w_i = w_{i_1} \cdot w_{i_2} = \frac{n_i}{\log_2 N_i} \cdot \sum_{j=1}^{n_i} \frac{2}{1 + e^{d_j}} \qquad (4)$$

where w_{i_1} and w_{i_2} are the weights contributed from the category and distance, respectively. n_i is the number of k nearest neighbors with class label i, d_j is the distance between image I and the jth sample in class i among the k nearest neighbors, N_i is the number of images in the whole training set with class label i. By using Eq. 4, a neighbor image with small distance and coming from a category with a limited number of training samples will be assigned with a large weight.

Since the experiment targets to make a comparison between traditional feature and CNN feature, with the KNN model above, the two methods are denoted as GIST-KNN and SaCNN-KNN, respectively. The performance is shown in Table 1. As anticipated, SaCNN-KNN performs much better than GIST-KNN. When k equals to 19 and 25, GIST-KNN and SaCNN-KNN achieve their best top-1 accuracy of 0.3311 and 0.7261. SaCNN-KNN has surpassed GIST-KNN over 119.3% in this case, which no doubly shows the advantage of the feature extracted from CNN models.

Table 1. Top-1 accuracy of different KNN based methods with varying k

	$k = 5$	$k = 9$	$k = 13$	$k = 17$	$k = 19$	$k = 21$	$k = 25$
GIST-KNN	0.3063	0.3251	0.3301	0.3311	0.3311	0.3303	0.3302
SaCNN-KNN	0.7071	0.7183	0.7228	0.7241	0.7250	0.7254	0.7261

Classification Model: KNN V.S. CNN. We then carry out experiments to compare models, namely KNN and CNN. The first one is represented by the SaCNN-KNN above, while the second one is the SaCNN, which is a purely deep based end-to-end method that classifies a testing image to a leaf class with the largest posteriors returned by the last soft-max layer. The top-1 and top-5 accuracies of SaCNN and SaCNN-KNN are 0.7264 and 0.9420, and 0.7269 and 0.9166, respectively. When observing the top-1 accuracy, the two methods perform quite similar. However, their performances on top-5 accuracy are quite different. It can be interpreted as: although the simple model (i.e., KNN) could achieve decent performance by learning from big data, the complex model (i.e., CNN) is more powerful in big data learning. It could grasp information more precisely from the sophisticated and unstructured visual signals, leading to general better performance.

4.3   SaCNN-MCR Results

In this section, we gives the performance of CNN models with different configurations, and stepwisely tests the performance gains obtained from the proposed spatial attention learning and MCR based refinement, aiming at presenting a comprehensive evaluation of the CNN based solutions in the task.

We first list the top-1 and top-5 accuracies of the four CNN related models mentioned in this work, as shown in Table 2. They are AlexNet, SaCNN that applies spatial attention learning on top of AlexNet, SaCNN-KNN that uses the 4096-d fc7 feature vector as the image descriptor, and employ the weighted KNN as the classification model, as described in Sect. 4.2, and SaCNN-MCR that further employ a MCR based refinement on top of SaCNN. It is seen that both spatial attention learning and MCR based refinement steadily improves both the top-1 and top-5 accuracies, resulting in performance gains of 1.24 and 0.55 in the two terms on average, respectively. In Fig. 3, we have presented three rows of classification examples that wrongly classified by a

Table 2. Performance of different methods on the five test rounds. The numbers on the left and right of the separator are top-1 and top-5 accuracies, respectively.

	AlexNet	SaCNN	SaCNN-KNN	SaCNN-MCR
1	0.7211/0.9408	0.7260/0.9414	0.7261/0.9162	0.7325/0.9462
2	0.7208/0.9412	0.7265/0.9422	0.7276/0.9168	0.7334/**0.9468**
3	0.7219/0.9407	0.7272/0.9420	0.7274/0.9162	**0.7347**/0.9462
4	0.7206/0.9409	0.7260/0.9419	0.7261/0.9167	0.7331/0.9467
5	0.7207/0.9413	0.7265/0.9423	0.7273/0.9169	0.7331/**0.9468**
Average	0.7210/0.9410	0.7264/0.9420	0.7269/0.9166	**0.7334/0.9465**

Fig. 3. Classification examples of different methods. Row(a)/(b)/(c) gives examples that wrongly classified by AlexNet/SaCNN/SaCNN-KNN but correctly classified by SaCNN/SaCNN-KNN/SaCNN-MCR. Conversely classified examples in each row are marked in red. (Color figure online)

method but correctly classified by another method, or conversely (such ones are marked in red). By incorporating the spatial attention learning, several clothes and shoes with complex background (in row (a)) are successfully rectified. As a result, it leads to an improvement gain of 0.54 in top-1 accuracy on average. Moreover, by further integrating the MCR based refinement, the class hierarchy takes effects such that noisy predictions are also suppressed to some extent (in row (c)), which results in another improvement gain of 0.7 in top-1 accuracy on average.

In Fig. 4, we further give a few classification examples of the SaCNN-MCR method. As can be seen, SaCNN-MCR could make reliable classifications with the existence of complex background, as well as on artificial images such as picture in picture and splice picture, all of which are rather challenging samples. The performance indicates SaCNN-MCR is indeed appealing in classifying product images. It is also observed that the method works less stable with the existence of large words and multiple objects. On one hand, words are often less relevant to the product class. It is likely to be interpreted as noises eventually for many classes after the CNN learning, resulting in low classification performance. On the other hand, simultaneously

Fig. 4. Classification examples of SaCNN-MCR. Error classified ones are marked in green. (Color figure online)

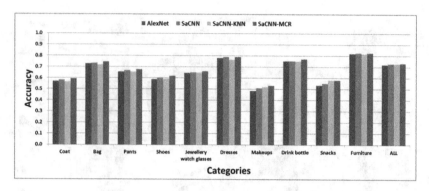

Fig. 5. Performance of different methods on the ten root categories.

presenting several similar products on an image is also likely to confuse the classification system. Perhaps more samples are essential for addressing these problems. We will also try to relieve these issues by investigating more advanced feature pooling and/or gating techniques.

To inspecting the results from a macro point of view, we group the leaf-level classification results to the 10 root categories. Figure 5 shows the results, from which the main observations are:

- SaCNN-MCR has the best performance on the 10 root categories among the four compared methods, indicating that the spatial attention learning and MCR based refinement lead to more accurate feature representation compared to other methods. What's more, the advantage of SaCNN-MCR is more prominent when viewed from the root level.
- When compared among the methods, SaCNN-KNN performs well on makeup and snacks but worse on other categories. This could be explained as products in the two categories are generally rigid objects with small deformation. It would be easier to find quite similar or exactly the same product in a large dataset. In these cases, SaCNN-KNN is likely to perform well and obtain relatively good results.
- When looking into via the root classes, furniture, dresses and bag generally get better classification performance than other classes, due to their relative small semantic gap. On the contrary, the performance of makeup and snacks are somehow low. It's mainly due to products in those categories are often small in size, and with larger variants in quantity, illumination and capturing view, all of which are open issues for a visual-based classification protocols.

5   Conclusion and Future Work

We have proposed the SaCNN-MCR method for large-scale product image classification. Compared with AlexNet, it introduced a spatial attention layer to the network for attention learning, and proposed to utilize the class hierarchy by a MCR based refinement. The experiments conducted on a real-world product dataset with nearly one million images basically validate our proposal, from which significant performance improvements are observed when compared with traditional GIST feature. Moreover, steady improvement gains are also obtained with the incorporation of both spatial attention learning and MCR based refinement. The proposed method, though achieve the top performance among the compared methods, is still limited by the fact that the introduced learning and refinement are conducted separately. The two stages could not benefit each other in a more compact way. Thus, future work includes the exploitation of more advanced CNN models that could encode both the visual and class hierarchy information effectively. We are also interested in applying CNN models to efficiently picking out the wrongly labeled products in online retailers.

Acknowledgements. This research is supported by National Nature Science Foundation of China (Grant No. 61303175, 61473030).

References

1. Lu, S., Mei, T., Wang, J., Zhang, J., Wang, Z., Li, S.: Exploratory product image search with circle-to-search interaction. IEEE Trans. Circuits Syst. Video Technol. **25**(7), 1190–1202 (2015)
2. George, M., Floerkemeier, C.: Recognizing products: a per-exemplar multi-label image classification approach. In: Fleet, D., Pajdla, T., Schiele, B., Tuytelaars, T. (eds.) ECCV 2014. LNCS, vol. 8690, pp. 440–455. Springer, Heidelberg (2014). doi:10.1007/978-3-319-10605-2_29
3. Chai, L., Qin, Z., Zhang, H., Guo, J., Shelton, C.R.: Re-ranking using compression-based distance measure for content-based commercial product image retrieval. In: International Conference on Image Processing (2012)
4. Liu, Z., Luo, P., Qiu, S., Wang, X., Tang, X.: DeepFashion: powering robust clothes recognition and retrieval with rich annotations. In: Computer Vision and Pattern Recognition (2016)
5. Krizhevsky, A., Sutskever, I., Hinton, G.E.: ImageNet classification with deep convolutional neural networks. In: Neural Information Processing Systems (2012)
6. Szegedy, C., Liu, W., Jia, Y., Sermanet, P., Reed, S., Anguelov, D., Rabinovich, A.: Going deeper with convolutions. In: Computer Vision and Pattern Recognition (2015)
7. He, K., Zhang, X., Ren, S., Sun, J.: Deep residual learning for image recognition. In: Computer Vision and Pattern Recognition (2015)
8. Xu, K., Ba, J., Kiros, R., Cho, K., Courville, A., Salakhutdinov, R., Bengio, Y.: Show, attend and tell: neural image caption generation with visual attention. In: International Conference on Machine Learning (2015)
9. Zhu, S.A., Wei, X.Y., Ngo, C.W.: Collaborative error reduction for hierarchical classification. Comput. Vis. Image Underst. **124**, 79–90 (2014)
10. Simonyan, K., Vedaldi, A., Zisserman, A.: Deep inside convolutional networks: visualising image classification models and saliency maps (2014). arXiv:1312.6034v2 [cs.CV]
11. Oliva, A., Torralba, A.: Modeling the shape of the scene: a holistic representation of the spatial envelope. Int. J. Comput. Vis. **42**(3), 145–175 (2001)

12. Bell, S., Bala, K.: Learning visual similarity for product design with convolutional neural networks. ACM Trans. Graph. **34**(4), 98 (2015)
13. Murthy, V., Maji, S., Manmatha, R.: Automatic image annotation using deep learning representations. In: International Conference on Multimedia Retrieval (2015)
14. Li, G., Wang, M., Lu, Z., Hong, R., Chua, T.: In-video product annotation with web information mining. ACM Trans. Multimed. Comput. Commun. Appl. (TOMCCAP) **8**(4), 1–19 (2012)
15. Itti, L., Koch, C., Niebur, E.: A model of saliency-based visual attention for rapid scene analysis. IEEE Trans. Pattern Anal. Mach. Intell. **20**(11), 1254–1259 (1998)

Learning Features Robust to Image Variations with Siamese Networks for Facial Expression Recognition

Wissam J. Baddar$^{(\boxtimes)}$, Dae Hoe Kim, and Yong Man Ro

Image and Video Systems Lab, KAIST, Daejeon, Korea
{wisam.baddar,dhkim10,ymro}@kaist.ac.kr

Abstract. This paper proposes a computationally efficient method for learning features robust to image variations for facial expression recognition (FER). The proposed method minimizes the feature difference between an image under a variable image variation and a corresponding target image with the best image conditions for FER (i.e. frontal face image with uniform illumination). This is achieved by regulating the objective function during the learning process where a Siamese network is employed. At the test stage, the learned network parameters are transferred to a convolutional neural network (CNN) with which the features robust to image variations can be obtained. Experiments have been conducted on the Multi-PIE dataset to evaluate the proposed method under a large number of variations including pose and illumination. The results show that the proposed method improves the FER performance under different variations without requiring extra computational complexity.

Keywords: Deep learning · Convolutional neural networks · Siamese-network · Facial expression recognition

1 Introduction

Recently, researchers in the computer vision and human-computer interaction have been increasingly investigating automatic face expression recognition (FER), due to its vast potential applications [1]. Nonetheless, achieving accurate FER from expressive face images is still a challenging problem. This is mainly attributed to the multitude of inter-personal variations, as well as variations in the imaging conditions, such as pose or illumination variations [2, 3].

Previous FER methods try to automatically identify the facial expressions from a set of basic expressions (i.e., anger, disgust, fear, happiness, sadness and surprise) that are found globally among humans [2, 4]. According to the feature representation utilized, FER methods can be divided into geometric-based methods, appearance-based methods or even a fusion of both methods [2, 3]. Geometric-based methods utilize features such as, shape and locations of facial components, or relationships between pairs of facial landmarks. Although geometric methods have shown sufficient FER performance [5], they suffer from limitations caused by facial landmark misalignments due to detection or tracking errors under various imaging conditions [6, 7]. On the other

© Springer International Publishing AG 2017
L. Amsaleg et al. (Eds.): MMM 2017, Part I, LNCS 10132, pp. 189–200, 2017.
DOI: 10.1007/978-3-319-51811-4_16

hand, appearance-based methods focus on capturing detailed textural changes that occur in the facial expressive image such as wrinkles and bulges [3, 8]. Most of appearance based methods exploit hand-crafted features (such as the local binary patterns [9] or Gabor filters [10]) extracted from the whole face, or from local patches around facial components [8]. However, hand-crafted features have been found to lack the generalization flexibility to spontaneous facial expressive images in the wild (i.e., expressive face images captured under uncontrolled environments) [11].

The recent success in deep learning in various computer vision tasks has influenced researchers to investigate the utilization of deep learning techniques in FER [11–18]. A FER method influenced by the action units concept has been introduced in [12, 13], where textural features (namely, micro-action-patterns) were learned from local patches of expressive face images. Subsequently, the learned features were clustered and passed to a restricted Boltzmann machine to learn a spatial feature representation of the expression. In [15], appearance and geometric feature representations were learned from expressive face sequences through a fusion of a convolutional neural network (CNN) and a deep neural network. In [14], the authors proposed adopting deeper CNN models by utilizing the concept of inception layers, to improve the FER performance in expressive face images. Although these methods achieved impressive improvement in FER performance in terms of FER accuracy compared to previous methods, these method did not address the image variation problem directly. The authors of [18] proposed utilizing hierarchical committee of deep CNNs to improve the FER accuracy. However, the improvement in the FER performance was obtained at a very computationally burdensome cost, as it requires the evaluation of a number of deep CNN models and aggregates the decision on a hierarchy of networks.

This paper proposes a computationally efficient feature learning method to reduce the effect of image variations on FER performance. To that end, we proposes an objective function that reduces the feature distance between an image with a variable image variation and a target image with the best image conditions of the same expression. To implement this objective term, a Siamese network structure is utilized at the training (feature learning) stage. As a result, the expression class separability of the learned feature is improved at different image variations. The main contributions of this paper can be summarized as follow:

1. The proposed method introduces a new objective function that minimizes the classification error while simultaneously minimizing the features difference between expressive face images with variable image variations and corresponding target image with best image conditions (i.e., frontal face pose with uniform illumination). To calculate the variation regularization term of the objective function, a Siamese structure is employed during the training stage. After the learning is complete, the model parameters are transferred to a conventional CNN which can extract features robust to image variations from the probe images.

2. FER for interactive applications requires robust and computationally efficient classification models. While Deep models have shown robustness towards various imaging conditions [19–21], they can be computational expensive [19] and difficult to train with limited training data [21]. On the other hand, the proposed method achieves FER robust to image variations, while it maintains low computational

complexity. This can be achieved by transferring the learned network parameters to a conventional CNN structure at the test stage.

Extensive and comprehensive experiments have been conducted on the Multi-PIE [22] dataset, which contains a large number of illumination and pose variations. The experimental results demonstrate that the proposed method outperforms the conventional CNN learning method, in terms of the FER accuracy. Moreover, the proposed method learns features that are more robust to different image variations including illumination and pose.

The reminder of this paper is organized as follows. In Sect. 2, the proposed method for learning features robust to image variations with Siamese Networks for FER is described. Section 3, discusses the experimental design and the results. Finally, conclusions are drawn in Sect. 4.

2 Methodology

Figure 1 shows an overview of the proposed method for learning features robust to image variations with Siamese networks for FER. As shown in the figure, the learning stage of the proposed method is composed of a Siamese CNN network with two objective function terms. The first term of the objective function is responsible for minimizing the expression class classification error. The second term of the objective function is responsible for minimizing the feature difference between two CNN networks which, respectively, have variably image variations and target best image conditions. As a result, the learned feature is more robust to image variations. After the learning process is complete, the learned network parameters are transferred to a CNN with the same structure such that, the learned features robust to image variations are obtained from the probe images. Note that, the features obtained from the probe image

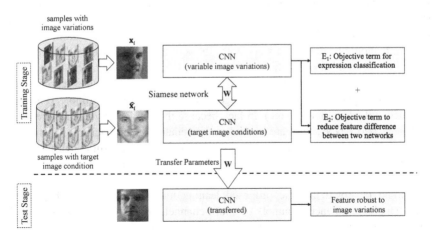

Fig. 1. Overview of the proposed method for learning features robust to image variations with Siamese networks for FER

are robust to image variations without requiring additional computational complexity compared to a conventional CNN.

2.1 The Proposed Method for Learning Features Robust to Image Variations with Siamese Networks

The recognition performance of learned features could vary depending on the objective function [23]. Thus, designing objective functions that can cater for the given problem is important to determine the performance of the network. In this paper, features robust to image variations are learned with an objective function consisting of two objective terms. The first term (E_1) of the objective function is responsible for recognizing the expression by minimizing the expression classification error. To that end, a softmax loss function is readily utilized as:

$$E_1 = -\sum_i \sum_c y_i^c \log \hat{y}_i^c, \qquad (1)$$

where c is the expression class index, y_i^c is the expression ground truth of the i-th sample (1 if c is the correct class and 0 otherwise), and $\hat{y}_i^c$ is the predicted probability that the sample belongs to the class c.

To mitigate the effect of image variations on the FER performance, the term E_2 is devised as:

$$E_2 = \frac{1}{2N} \sum_i \left\| f(\mathbf{x}_i^c) - f(\hat{\mathbf{x}}_i^c) \right\|_2^2, \qquad (2)$$

where $\mathbf{x}_i^c$ is the i-th training sample of the c-th expression class with any image variation, and $\hat{\mathbf{x}}_i^c$ is the target sample with best imaging condition, that corresponds to the i-th training sample $\mathbf{x}_i$, with the same expression class c. $f(\mathbf{x}_i^c)$ is the feature at the last fully connected layer of the variable image variations CNN (fc1 layer in Fig. 3) obtained from sample $\mathbf{x}_i$, and $f(\hat{\mathbf{x}}_i^c)$ is the feature at the last fully connected layer of the best image conditions CNN of the Siamese network that is obtained from the target sample $\hat{\mathbf{x}}_i$. It should be noted that since both CNNs of the Siamese network have shared network parameters. Thus, by minimizing E_2, the feature representation learned from any image variation ($f(\mathbf{x}_i^c)$) is enforced to be similar to the feature representation ($f(\hat{\mathbf{x}}_i^c)$) of the target facial expressive image with the best image conditions (uniformly illuminated and frontal face image). In fact, because the network parameters are shared between the two CNNs, they are continuously updated in the back propagation and converge to the parameters where features $f(\mathbf{x}_i)$ and $f(\hat{\mathbf{x}}_i)$ most similar in the feature space. As a result the learned feature is robust to image variations, which in turn improves the expression class classification.

It should be noted that the proposed leaning method can work with grayscale images (single luminance channel) or multi-channel images (e.g. colored images), simply by changing the input channels of the network. This is applicable because the proposed method updates the feature representation rather than the image characteristics.

2.2 Facial Expression Recognition

At the test stage, the input to the network should only be the probe image taken from any image variation. Therefore, after the learning processes is complete, the network parameters are transferred from the variable image variations CNN (refer to Fig. 1) to a test CNN with an identical architecture. A softmax classifier is attached to the last fully connected layer of the test CNN (fc1 layer in Fig. 3) in order to perform the FER task. The predicted class is obtained by feed-forwarding the probe image to the test CNN. Note that, because the network parameters of the test CNN were transferred from the variable image variations CNN in the training, the test CNN obtains features robust to image variations without the need for extra computational cost to reduce the effect of image variations such as pose or illumination.

3 Experiments

3.1 Experimental Setup

To evaluate the proposed method for learning features robust to image variations, experiments were conducted on the Multi-PIE [22] dataset. The dataset consists of 6 expressions (i.e., neutral, smile, surprise, squint, disgust and scream) recorded from 337 subjects over four sessions. Each expression was recorded from 15 viewpoints and 20 illuminations variations (labeled with an illumination index from 0 to 19). In the experiments, expressive images from the 20 illumination and 5 viewpoints ($\pm 30°$, $\pm 15°$ and $0°$) were collected from 318 subjects. As a result, 126,500 expressive face images such that 100 variations of each facial expression image were included (20 illuminations $\times$ and 5 head poses). Note that, the face regions were automatically aligned according to the eye centers, cropped and resized to 64×64 pixels. To that end, the landmark detection method described in [24] was utilized. Figure 2(a) and (b) show examples of the different illumination variations at two different pose variations ($-30°$ and $0°$ viewpoints) of the same expression. Moreover, an example of the image variations (100 variations) and the target image with best image conditions (i.e., frontal pose with uniform illumination) that has been used during the training of the proposed method are shown in Fig. 2(c).

All the evaluations in the experiments were conducted by a five-fold subject-independent cross-validation, such that subjects in the test set were excluded from the training set. To implement the experimental network structures, Caffe library [25] was utilized. As shown in Fig. 3 the Siamese network structure proposed for learning the features robust to image variations is composed of two identical networks with shared parameters. Each network is comprised of three convolutional layers, each of which is followed by a max-pooling layer. The convolution filter sizes were defined as $(5 \times 5 \times 64)$, $(3 \times 3 \times 128)$ and $(3 \times 3 \times 256)$ for layers conv1, conv2 and conv3, respectively. For the activation function at each layer, the rectified linear unit (ReLU) [26] was utilized. The last layer of each of the CNNs is a fully connected layer (fc1) that consisted of 300 nodes. It should be noted that each channel of the input images was normalized to have a zero mean and a unit variance. For the test network and the baseline comparative network, a CNN structure was constructed to have a CNN

(a)   Example of image illumination variations at the 0° viewpoint (frontal pose).

(b)   Example of image illumination variations at the -30° viewpoint (pose).

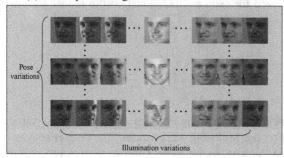

(c)   Example of the variations used at the training stage, with the corresponding target
best image condition (i.e., frontal pose with uniform illumination)

Fig. 2. Example images of the Multi-PIE dataset used in this paper. Note that in (a) and (b) the images are arranged according to the illumination index value provided by the Multi-PIE dataset; i.e., the first row consists of 10 images with illumination index 0 to 9 and the second row consists of 10 images with illumination index 10 to 19. (c) An example of the variations and the corresponding target best conditions used at the training stage of the experiments.

structure identical to the variable imaging variations CNN shown in Fig. 3. At the test stage, a softmax classifier was attached to fc1 to perform the FER classification task.

3.2   Performance Comparison with Baseline CNN

In this section, the effectiveness of the proposed method for learning features robust to image variations is measured in terms of the FER accuracy. As a reference performance, a baseline CNN structure (a CNN structure identical to the variable image variation CNN shown in Fig. 3) was trained and tested, via 5-fold cross validation, with the best image conditions (i.e., frontal face images with uniform illumination). Because the best image condition images only constitute 1% of the utilized dataset, augmentation was performed during the training to avoid overfitting. As a result, the FER accuracy at the best image conditions was 85.07%, which was used as the reference performance for our experiments.

To measure the effect of the proposed learning method on the FER accuracy, a direct comparison was performed between two models learned with the full dataset, including all illumination and pose variations. The first model was a baseline CNN

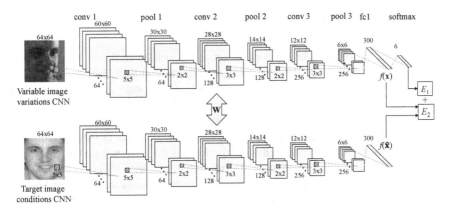

Fig. 3. Structure of the Siamese network implementation used in the experiments

structure, which represents a conventional CNN structure, which was learned using only the expression classification objective term (E_1). The second model was obtained with the proposed feature learning method, by transferring the learned CNN parameters from a Siamese network trained with the full dataset, as described in Sect. 2.

Table 1 shows a comparison between the proposed methods, the baseline CNN and the reference performance (obtained at the best image conditions). The results indicate that even when a conventional CNN (represented by the baseline CNN) was trained with variable image variations, the performance of the CNN is sub-optimal compared to the reference performance. On the other hand, the proposed method can efficiently learn facial expression features, regardless of the image variations. As a result, the proposed method learns features that are robust to those image variations. In fact, the propose method was able to outperform the reference performance (obtained at the best image conditions). This can be attributed to the fact that, when learning the reference performance CNN, the network can learn features related to the subject appearance which can hamper the FER performance. On contrary, the second objective function term (E_2) in the proposed method, reduces the intra-class variations between the best image conditions and the variable imaging condition. For example, the subject appearance is not preserved across poses, so the effect of subject appearance is also reduced, when the learning utilizes data with pose variations. This results in the improvement of the overall performance even at the best image conditions (refer to result in Sect. 3.3).

Table 1. Comparison with baseline CNN FER accuracy

Method	ACC (%)
Reference performance[a]	85.07
Baseline CNN	83.83
Proposed method	**87.34**

[a]The reference performance was obtained by training and testing the baseline CNN with the best image conditions

3.3   Measuring the Effect of the Proposed Method on Feature Robustness Towards Pose Variations

In this section, we measure the effect of the proposed method on the feature robustness towards pose variations. To that end, the models learned in Sect. 3.2 were utilized to obtain the FER accuracy at each pose separately. Figure 4(a) shows a comparison between the obtained accuracies at each pose for both models (the proposed method and the baseline CNN). Note that at all the pose variations, the proposed method shows an improved performance compared to the baseline performance. In fact, regardless of the pose variation, the proposed method outperformed the reference performance FER accuracy (85.07%), which was trained and tested on best image conditions (as described in Sect. 3.2). Moreover, the proposed method results in a more stable performance where, the standard deviation of the FER accuracy across all poses was is reduced to 0.8149, compared to 1.5336 in the baseline CNN learned with the same training dataset.

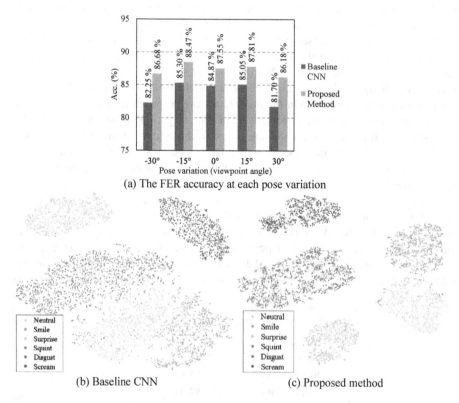

(a) The FER accuracy at each pose variation

(b) Baseline CNN   (c) Proposed method

Fig. 4. Effect of proposed method on pose variations robustness. (a) FER accuracy at every pose variation. (b) and (c) are visualizations of the features space using t-SNE for features learned by the baseline CNN and the proposed method, respectively. Best viewed in color. (Color figure online)

To qualitatively evaluate the effect of proposed method on the learned feature space, t-distributed stochastic neighbor embedding (t-SNE) [27] was utilized. For visualizing the t-SNE, features obtained from uniformly illuminated images (i.e. Illumination index 7 in the dataset) at the 5 different poses were plotted as shown Fig. 4(b) and (c). As seen in the figure, the features at different pose variations of the same expression are less scattered and more compactly presented compared to the baseline CNN. This reduces the classification confusion resulting in improved FER performance.

3.4 Measuring the Effect of the Proposed Method on Feature Robustness Towards Illumination Variations

To measure the effect of the proposed method on the feature robustness towards illumination variations, the FER accuracy of the learned model was obtained at each illumination variation individually. Figure 5(a) presents a radar chart that compares the

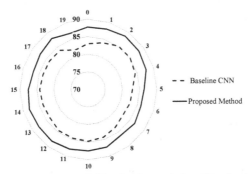

(a)! The FER accuracy at each illumination variation (illumination index 0 to 19)

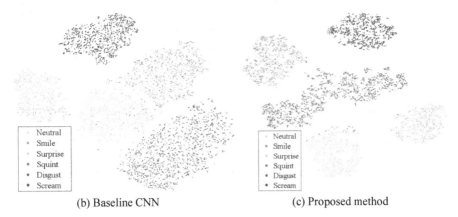

(b) Baseline CNN   (c) Proposed method

Fig. 5. Effect of proposed method on illumination variations robustness. (a) A radar chart of the FER accuracy at each illumination variation. (b) and (c) are visualization of the features space using t-SNE for features learned by the baseline CNN and the proposed method, respectively. Best viewed in color. (Color figure online)

FER accuracy of the proposed method and the baseline CNN at each illumination variation. The center of the chart represents the lowest FER accuracy and the outmost contour represents the highest FER accuracy. Each vertex in the chart represents a separate illumination variation, at which the FER accuracy was obtained. The indexes of the illumination variations correspond to the Multi-Pie dataset illumination index (refer to Fig. 2(a)). As seen in the figure, the proposed method outperforms baseline CNN at all illumination variations. Moreover, the FER accuracy at each illumination variation was improved compared to the reference performance FER accuracy (85.07%), which was trained and tested on best image conditions (as described in Sect. 3.2).

We further visualize the effect of the proposed method on the learned feature space of the proposed method compared to the baseline CNN. This time, the t-SNE was obtained by fixing the pose to the frontal pose (0 ° view point) and using the 20 illumination variations. Due to the large number of samples, 25% of the samples were randomly selected and plotted in Fig. 5(a) and (b). As the figure shows, the intra-class variation was reduced while the interclass distance was increased. As a results, the proposed method improved the robustness of the learned features against different illumination variations.

4 Conclusions

This paper proposed a new method for learning features robust to image variations for FER. The proposed method is computationally efficient, as it does not require any added computational complexity to the baseline conventional CNN. At the training stage, the proposed method minimizes the feature difference between an expressive face image with variable image variation and a corresponding target expressive face image with best image conditions (i.e. frontal face with uniform illumination). This was achieved by a supervised learning strategy, which involves the feature of the best image condition images in the objective function, to regulate the learning process with Siamese networks. At the test stage, the learned network parameters were transferred to a conventional CNN structure, such that the learned robust feature was obtained from probe images without requiring additional computational complexity. Extensive and comprehensive experiments were conducted on the Multi-PIE dataset. The experimental results showed that the proposed method result in improved FER performance under a large number of image variations. For future work, more evaluations of the proposed method will be conducted on additional datasets, containing acted and spontaneous facial expression in the wild.

Acknowledgement. This work was supported by the National Research Foundation of Korea (NRF) grant funded by the Korea government (MSIP) (No. 2015R1A2A2A01005724).

References

1. Cohn, J.F., Ekman, P.: Measuring facial action. In: The New Handbook of Methods in Nonverbal Behavior Research, pp. 9–64 (2005)
2. Tian, Y.-L., Kanade, T., Cohn, J.F.: Facial expression analysis. In: Tian, Y.-L., Kanade, T., Cohn, J.F. (eds.) Handbook of Face Recognition, pp. 247–275. Springer, New York (2005)
3. Zeng, Z., Pantic, M., Roisman, G.I., Huang, T.S.: A survey of affect recognition methods: Audio, visual, and spontaneous expressions. Pattern Anal. Mach. Intell. **31**, 39–58 (2009)
4. Ekman, P., Friesen, W.V.: Constants across cultures in the face and emotion. J. Pers. Soc. Psychol. **17**, 124 (1971)
5. Valstar, M.F., Patras, I., Pantic, M.: Facial action unit detection using probabilistic actively learned support vector machines on tracked facial point data. In: 2005 IEEE Computer Society Conference on Computer Vision and Pattern Recognition (CVPR 2005)-Workshops, pp. 76–76. IEEE (2005)
6. Ramirez Rivera, A., Rojas Castillo, J., Chae, O.: Local directional number pattern for face analysis: Face and expression recognition. IEEE Trans. Image Process. **22**, 1740–1752 (2013)
7. Lee, S.H., Baddar, W.J., Ro, Y.M.: Collaborative expression representation using peak expression and intra class variation face images for practical subject-independent emotion recognition in videos. Pattern Recogn. **54**, 52–67 (2016)
8. Jiang, B., Valstar, M., Martinez, B., Pantic, M.: A dynamic appearance descriptor approach to facial actions temporal modeling. IEEE Trans. Cybern. **44**, 161–174 (2014)
9. Mäenpää, T.: The local binary pattern approach to texture analysis: extensions and applications. Oulun yliopisto (2003)
10. Zhang, L., Tjondronegoro, D., Chandran, V.: Random Gabor based templates for facial expression recognition in images with facial occlusion. Neurocomputing **145**, 451–464 (2014)
11. Liu, M., Li, S., Shan, S., Wang, R., Chen, X.: Deeply learning deformable facial action parts model for dynamic expression analysis. In: Cremers, D., Reid, I., Saito, H., Yang, M.-H. (eds.) ACCV 2014. LNCS, vol. 9006, pp. 143–157. Springer, Heidelberg (2015). doi:10.1007/978-3-319-16817-3_10
12. Liu, M., Li, S., Shan, S., Chen, X.: AU-aware deep networks for facial expression recognition. In: 2013 10th IEEE International Conference and Workshops on Automatic Face and Gesture Recognition (FG), pp. 1–6. IEEE (2013)
13. Liu, M., Li, S., Shan, S., Chen, X.: AU-inspired deep networks for facial expression feature learning. Neurocomputing **159**, 126–136 (2015)
14. Mollahosseini, A., Chan, D., Mahoor, M.H.: going deeper in facial expression recognition using deep neural networks (2015). arXiv preprint arXiv:1511.04110
15. Jung, H., Lee, S., Yim, J., Park, S., Kim, J.: Joint fine-tuning in deep neural networks for facial expression recognition. In: Proceedings of IEEE International Conference on Computer Vision, pp. 2983–2991 (2015)
16. Khorrami, P., Paine, T., Huang, T.: Do deep neural networks learn facial action units when doing expression recognition? In: Proceedings of IEEE International Conference on Computer Vision Workshops, pp. 19–27 (2015)
17. Gupta, O., Raviv, D., Raskar, R.: Deep video gesture recognition using illumination invariants (2016). arXiv preprint arXiv:1603.06531
18. Kim, B.-K., Roh, J., Dong, S.-Y., Lee, S.-Y.: Hierarchical committee of deep convolutional neural networks for robust facial expression recognition. J. Multimodal User Interfaces **10**, 1–17 (2016)

19. Ba, J., Caruana, R.: Do deep nets really need to be deep? In: Advances in Neural Information Processing Systems, pp. 2654–2662 (2014)
20. Szegedy, C., Liu, W., Jia, Y., Sermanet, P., Reed, S., Anguelov, D., Erhan, D., Vanhoucke, V., Rabinovich, A.: Going deeper with convolutions (2014). arXiv preprint arXiv:1409.4842
21. He, K., Zhang, X., Ren, S., Sun, J.: Deep residual learning for image recognition (2015). arXiv preprint arXiv:1512.03385
22. Gross, R., Matthews, I., Cohn, J., Kanade, T., Baker, S.: Multi-pie. Image Vis. Comput. **28**, 807–813 (2010)
23. Wan, L., Zeiler, M., Zhang, S., Cun, Y.L., Fergus, R.: Regularization of neural networks using dropconnect. In: Proceedings of 30th International Conference on Machine Learning (ICML-13), pp. 1058–1066. (2013)
24. Asthana, A., Zafeiriou, S., Cheng, S., Pantic, M.: Incremental face alignment in the wild. In: 2014 IEEE Conference on Computer Vision and Pattern Recognition (CVPR), pp. 1859–1866. IEEE (2014)
25. Jia, Y., Shelhamer, E., Donahue, J., Karayev, S., Long, J., Girshick, R., Guadarrama, S., Darrell, T.: Caffe: convolutional architecture for fast feature embedding. In: Proceedings of ACM International Conference on Multimedia, pp. 675–678. ACM (2014)
26. Nair, V., Hinton, G.E.: Rectified linear units improve restricted boltzmann machines. In: Proceedings of 27th International Conference on Machine Learning (ICML-10), pp. 807–814. (2010)
27. Van der Maaten, L., Hinton, G.: Visualizing data using t-SNE. J. Mach. Learn. Res. **9**, 85 (2008)

M3LH: Multi-modal Multi-label Hashing for Large Scale Data Search

Guan-Qun Yang, Xin-Shun Xu$^{(\boxtimes)}$, Shanqing Guo, and Xiao-Lin Wang

School of Computer Science and Technology, Shandong University, Jinan, China
yguanqun@163.com, {xuxinshun,guoshanqing,xlwang}@sdu.edu.cn

Abstract. Recently, hashing based technique has attracted much attention in media search community. In many applications, data have multiple modalities and multiple labels. Many hashing methods have been proposed for multi-modal data; however, they seldom consider the scenario of multiple labels or only use such information to build a simple similarity matrix, e.g., the corresponding value is 1 when two samples share at least one same label. Apparently, such methods cannot make full use of the information contained in multiple labels. Thus, a model is expected to have good performance if it can make full use of information in multi-modal and multi-label data. Motivated by this, in this paper, we propose a new method, multi-modal multi-label hashing-M3LH, which can not only work on multi-modal data, but also make full use of information contained in multiple labels. Specifically, in M3LH, we assume every label is associated with a binary code in Hamming space, and the binary code of a sample can be generated by combining the binary codes of its labels. While minimizing the Hamming distance between similar pairs and maximizing the Hamming distance between dissimilar pairs, we also learn a project matrix which can be used to generate binary codes for out-of-samples. Experimental results on three widely used data sets show that M3LH outperforms or is comparable to several state-of-the-art hashing methods.

Keywords: Hashing · Multi-modal · Multi-label · Cross-modal retrieval

1 Introduction

With an explosive growth of data, Approximate Nearest Neighbor (ANN) search plays a fundamental role in data mining, computer vision, database and information retrieval. Many techniques have been developed for ANN search. For example, tree based approaches can make the search very efficiently by storing data with special data structure. However, for high-dimensional data, their performance may degrade; in addition, they have to use much memory to store data structure.

Recently, hashing-based approximate nearest neighbor search has garnered considerable interest, which can embed high-dimensional data into compact binary codes where the similarity in the original space is preserved in the new

© Springer International Publishing AG 2017
L. Amsaleg et al. (Eds.): MMM 2017, Part I, LNCS 10132, pp. 201–213, 2017.
DOI: 10.1007/978-3-319-51811-4_17

space (Hamming space) [13,15,16,22]. Such methods have a constant or sub-linear time complexity for search due to the efficiency of pairwise comparison with Hamming distance. Moreover, the storage cost can also be reduced greatly [17,19–21,23]. Correspondingly, many hashing methods have been proposed and applied to many applications, e.g., Locality-Sensitive Hashing (LSH) [3], Spectral Hashing (SH) [18], Semi-Supervised Sequential Projection Learning for Hashing (S3PLH) [14], Iterative Quantization (ITQ) [6] and Least Square Regularized Spectral Hashing (LSSPH) [26], K-means Hashing (KMH) [7], etc.

Most early hashing methods focus on unimodal data, which means that all data samples only contain one type of modality. However, in many scenarios, data could have multiple modalities. For example, an image is usually associated with tags or texts. Thus, how to search multimodal data has become a hot topic in these years. Many multimodal hashing methods have been proposed. Generally, existing multi-modal hashing methods can be divided into two categories: supervised and unsupervised.

Unsupervised hashing methods do not use supervised information, e.g., tags, labels and semantic similarity between samples, etc. For example, Inter-Media Hashing (IMH) [12] is one of representative methods, which introduces inter-view and intra-view consistency to learn linear hash functions for mapping features in different views into a common Hamming space. Collective Matrix Factorization Hashing for Multi-modal Data (CMFH) [4] is another unsupervised hashing method that uses collective matrix factorization to learn latent factor from different modalities. Latent Semantic Sparse Hashing (LSSH) [25] learns latent semantic features for multi-modal with sparse coding and matrix factorization, then maps them to a joint space and generates unified hash codes.

Supervised hashing method usually have better performance than unsupervised ones because of using supervised information, e.g., labels or semantic similarity, etc. Many supervised hashing methods have been proposed for multi-modal data. For example, Cross-View Hashing (CVH) [9] extends the single-view Spectral Hashing [18]; it learns hash functions via minimizing the similarity-weighted Hamming distances between hash codes of training data. Semantic Correlation Maximization (SCM) [24] integrates semantic labels into the hashing learning procedure via maximizing semantic correlations. Semantics-Preserving Hashing (SePH) [10] standardizes all Hamming distances by transforming each into a probability that depends on all others.

As mentioned above, many hashing methods have been proposed for multi-modal data. In some scenarios, one sample contains multiple labels, i.e., multi-label data. Most existing hashing methods do not make full use of multi-label information. For example, most of them only set two samples as similar when they share at least one label. However, the degree of similarity between two samples should be different if they share different number of labels. Apparently, if a model could consider this and make full use of the information contained in multiple labels, it is expected to obtain better performance.

Motivated by this, in this paper, we propose a new supervised hashing method for multi-modal multi-label data, i.e., M3LH, which can not only deal with multi-modal data, but also make full use of multi-label information. Specifically, we assign each label a binary code in the Hamming space as an anchor [5,8]. Then the hashing code of each instance should be close to its corresponding anchor. At the same time, we assume that the hashing code of a sample is a linear combination of its nearest neighbors. Thus, the hashing code of a sample is a combination of the binary codes of its labels. In addition, while maximizing the intra-class similarity and minimizing inter-class similarity, we also learn a project matrix which can generate binary codes for out-of-samples. Extensive experiments are conducted on three data sets to test the performance of the proposed method. The results demonstrate that M3LH outperforms or is comparable to state-of-the-art methods.

2   Proposed Method

2.1   Notation and Problem Definition

Without loss of generality, in this paper, we suppose that each data sample has two modalities, e.g., image and text. Then, we suppose that $O = \{o_i\}_{i=1}^n$ is a training set containing n instances, where $o_i = (x_i^{(1)}, x_i^{(2)})$ is the i-th instance, $x_i^{(1)} \in \mathbb{R}^{d_1}$ is a d_1-dimensional image feature, and $x_i^{(2)} \in \mathbb{R}^{d_2}$ is a d_2-dimensional text feature, (usually $d_1 \neq d_2$). Correspondingly, $X^{(1)} \in \mathbb{R}^{d_1 \times n}$ and $X^{(2)} \in \mathbb{R}^{d_2 \times n}$ are data matrices of two modalities. In addition, $L = [l_1, ..., l_n] \in \{-1, 1\}^{c \times n}$ denotes the label matrix of all training samples, where $l_{ij} = 1$ if o_j contains the i-th label and -1 otherwise. We further assume all data are zero-centered which is usually adopted in existing hashing algorithms. Given the code length k, the problem is to learn the unified hash codes $b_i \in \{-1, 1\}^k$ for o_i, $i = 1, 2, ..., n$.

2.2   Objective Function

As mentioned previously, to generate unified hash codes, we first define an anchor matrix $A = [a_1, ..., a_c] \in \{-1, 1\}^{k \times c}$, where each anchor represents a prototype of one label. Then, we further assume that the code b_i of o_i is a linear combination of a_j, $j = 1, 2, ..., c$, and the coefficient is its label l_i, i.e., $b_i = sign(Al_i)$. In addition, the binary codes should preserve the similarity of samples in original space.

From the other point of view, we further define the view-specific hash function as follows.

$$f_t(x^{(t)}) = W^{(t)} x^{(t)} \tag{1}$$

where $W^{(t)} \in \mathbb{R}^{k \times d_t}$ is projection matrix, $t = 1, 2$. Then, the inner-product of binary codes can be used to measure the similarity of two data points in Hamming space, i.e.,

$$<f_t(x_i^{(t)}), f_t(x_j^{(t)})> = (W^{(t)} x_i^{(t)})^T W^{(t)} x_j^{(t)} \tag{2}$$

Once we have the above assumption and definition, the problem becomes how to obtain the anchor matrix A and the projection matrices $W^{(t)}$.

For simplicity of representation, we can further combine the representation of different modalities into a unified form. For example, we can define the training data matrix as follows:

$$X = \begin{bmatrix} X^{(1)} & 0 \\ 0 & X^{(2)} \end{bmatrix} \tag{3}$$

Correspondingly, the unified projection matrix and label matrix can be defined as $W = [W^{(1)}, W^{(2)}]$ and $L' = [L, L]$, respectively. Then, the projection function becomes:

$$f(X) = WX \tag{4}$$

where $W \in \mathbb{R}^{k \times (d_1 + d2)}$, $X \in \mathbb{R}^{(d_1 + d2) \times 2n}$, $L' \in \{-1, 1\}^{c \times 2n}$.

In addition, the projection should preserve the similarity of data points, which means that data pairs in the new subspace should have large inner-product if they are similar in original space, and small inner-product if they are dissimilar. Then, we have the following sub-optimization problem.

$$\max \sum_{i=1}^{2n} \sum_{j=1}^{2n} S_{ij} x_i^T W^T W x_j \tag{5}$$

where S is an affinity matrix which can be constructed from labels L', e.g., $S_{ij} = 1$ if the corresponding data pairs share at least one same label, and $S_{ij} = -1$ otherwise. In addition, the above problem can be rewritten in a matrix form, i.e.,

$$\max tr(WXSX^T W^T) \tag{6}$$

In addition, according to the definition of A, a_i must be similar to the binary codes of all data points that contain the i-th label. This means the following equation should also be maximized:

$$\max \sum_{i=1}^{c} \sum_{j=1}^{2n} l'_{ij} a_i^T W x_j \tag{7}$$

which can also be rewritten in a matrix form, i.e.,

$$\max tr(AL'X^T W^T) \tag{8}$$

By maximizing the above equation, the similarity between Wx_j and its corresponding anchor is maximized.

Note that the norm of projected vector will highly influence the results. To reduce such influence, we should minimize the following term.

$$\min_W \|WX\|_F^2 \tag{9}$$

By combining Eqs. (6), (8) and (9) together, the final objective function is given as:

$$G(W, A) = \|WX\|_F^2 - \alpha tr(WXSX^T W^T) - \beta tr(AL'X^T W^T) + \lambda \|W\|_F^2 \tag{10}$$

where α, β and λ are tradeoff parameters, $\|W\|_F^2$ is a regularization term to avoid overfitting.

3   Optimization

In order to tackle the optimization problem in Eq. (10), we adopt an alternating procedure to find W and A. In addition, we also propose a method to initialize W and A. These steps are described in the following paragraphs.

Initializing W and A

Usually, the initialized values of parameters play an important role on the quality of solutions or convergence speed. In addition, the anchors should be different so that similar data pairs should be close and dissimilar pairs should be far. Here, we propose a method to initialize W and A. Specifically, we first initialize W by minimizing the first two terms of Eq. (10).

$$\begin{aligned} G(W) &= \|WX\|_F^2 - \alpha tr(WXSX^TW^T) \\ &= tr(WXX^TW^T) - \alpha tr(WXSX^TW^T) \end{aligned} \tag{11}$$

$$s.t.\ WW^T = I$$

where the constraint $WW^T = I$ is added to make the problem easy for optimization, and avoid the situation that W is zero. The above optimization problem can be rewritten as:

$$\min tr(W^TJW)$$
$$s.t.\ W^TW = I \tag{12}$$

where $J = XX^T - \alpha XSX^T$. Apparently, we can obtain optimal W by computing the eigenvalue decomposition of J.

After getting W, we can initialize anchor A by solving the following optimization problem:

$$\max tr(AL'X^TW^T) \tag{13}$$

We can find that the above equation is maximized when A_{ij} has the same sign as $(WXL'^T)_{ij}$. Thus, we have

$$A = sign(WXL'^T) \tag{14}$$

Once we have initialized values of W and A, we then adopt an alternating optimization procedure to iteratively update them.

Fixing A and Updating W

When A is fixed, we have the following optimization problem.

$$G(W) = \|WX\|_F^2 - \alpha tr(WXSX^TW^T) - \beta tr(AL'X^TW^T) + \lambda\|W\|_F^2 \tag{15}$$

Algorithm 1. M3LH: Multi-modal Multi-label Hashing

Input:

Training set $X^{(1)}$ and $X^{(2)}$, label matrix L; code length k; iteration number N; parameters, α, β and λ.

Output:

Binary codes B, projection matrix $W^{(1)}$ and $W^{(2)}$.

1: Solve Eq. (12) to initialize W using eigenvalue decomposition.
2: Using Eq. (13) to initialize A.
3: **for** i=1 to N **do**
4:    Update W using Eq. (17)
5:    Update A using Eq. (19)
6: **end for**
7: $B = sign(AL)$

Apparently, we can get the optimal solution by setting the gradient of Eq. (15) with respect to W to zero, i.e.,

$$\frac{\partial G}{\partial W} = 2(WXX^T - \alpha WXSX^T - \beta AL'X^T + \lambda W) = 0 \qquad (16)$$

Then we have

$$W = \beta(XX^T - \alpha XSX^T + \lambda I)^{-1}AL'X^T \qquad (17)$$

Fixing W and Updating A

When W is fixed, the optimization problem becomes

$$G(A) = -\beta tr(AL'X^TW^T) \qquad (18)$$

As mentioned previously, this equation is maximized when A_{ij} has the same sign as $(WXL'^T)_{ij}$. Thus, we have

$$A = sign(WXL'^T) \qquad (19)$$

To get the final solution, we just need to repeat the above two steps and iteratively update W, and A until it converges or reaches the max number of iterations. Finally, to have an overall look of the proposed algorithm, we summarized it in Algorithm 1.

4   Out-of-Sample Extension

M3LH can easily generate binary codes for objects in training data set. For example, it can use $b_i = sign(Al_i)$ to generate a unified binary code for object o_i because different modalities of one object share the same labels. A good hashing method should be able to generate hash codes for any data points besides training data, e.g., those samples in test set. Apparently, M3LH can also generate hash code for any out-of-sample points with the following equation.

$$b_i^{(t)} = sign(W^{(t)}x_i^{(t)}) \tag{20}$$

where $x_i^{(t)}$ is the t-th modality of a query sample, $b_i^{(t)}$ is its corresponding hash code, $W^{(t)}$ is the project matrix learned from training set.

5 Experiments

We conduct extensive experiments on three data sets to test the performance of the proposed method-M3LH. We also compare it with seven state-of-the-art hashing methods for multi-modal data.

5.1 Datasets

NUS-WIDE: It is a web image database containing 269,648 images with tags [2]. Each instance is annotated with at least one of 81 labels. We only use ten largest concepts and the corresponding 186,577 images. For each instance, image is represented by 500-dimension SIFT feature vector and text is represented by index vectors of the most frequent 1000 tags. Each pair is annotated by at least one of 10 labels. Pairs are supposed to be similar if they share at least one same label. Here we take 2,000 instances as training set and 1,000 instances as query set.

MIRFlickr 25000: It contains 25,000 instances collected from Flickr, each being an image with its associated textual tags [1]. For each instance, image is represented by 150-dimension edge histogram vector and text is represented by 500-dimension feature vector derived from PCA on its binary tagging vector. There are 24 labels and each instance has at least one label. Here we take 5,000 instances as training set, and 500 instances as retrieval set.

Wiki: It consists 2866 instances collected from wikipedia, each instance is an image-text pair [11]. For one instance, image is represented by 128-dimension SIFT feature vector and text is represented by 10-dimension topic vector. It has 10 classes and each instance is annotated with one label. We use 75% of pairs as training set, and the remaining 25% as the query set.

5.2 Baseline Methods

Our method is compared with seven state-of-the-art hashing methods, i.e., CVH [9], IMH [12], SCM-orth [24], SCM-seq [24], LSSH [25], CMFH [4] and SePH [10], which can be classified into unsupervised hashing methods and supervised hashing methods. Their parameters are carefully tuned according to the schemes in those references.

5.3   Evaluation Metric

We use the widely used mean Average Precision (mAP) to measure the performance of all multi-modal hashing methods. It is the mean of average precision (AP), and defined as:

$$mAP = \frac{1}{Q} \sum_{i=1}^{Q} AP(q_i) \tag{21}$$

where Q is the number of queries, and AP is defined as:

$$AP(q) = \frac{1}{L} \sum_{r=1}^{R} P_q(r)\delta(r) \tag{22}$$

where L is the number of true retrieved instances of q, and $P_q(R)$ denotes the precision of the top r retrieved points of query q, $\delta(r)$ is an indicator function which equals 1 if the rth result is a true neighbour of the query and 0 otherwise. In our experiment, ground-truth is defined as those pairs who share at least one label.

We also use precision-recall curves to measure the performance. It can be obtained by varying the Hamming radius of the retrieved points and evaluating the precision, recall and the number of retrieved points.

5.4   Other Implementation Details

To further improve the performance M3LH, we adopt a nonlinear embedding scheme to map original training samples into m-dimension subspace, i.e.,

$$\phi(x) = [e^{-\frac{\|x - x_1\|}{\sigma}}, ..., e^{-\frac{\|x - x_m\|}{\sigma}}] \tag{23}$$

where $\{x_j\}_{j=1}^{m}$ are the randomly selected m anchor points from the training samples and σ is the kernel width.

In addition, the parameters in M3LH are selected based on a validation procedure. Specifically, in all experiments, they take the following values: $\alpha = 1$, $\beta = 100$, and $\lambda = 0.00005$.

5.5   Results and Discussions

We report the cross-modal retrieval performance of M3LH and all compared methods, i.e., the performance of retrieving text database with image query and that of retrieving image database with text query.

Results on NUS-WIDE

The mAP values of M3LH and all baselines on NUS-WIDE are shown in Table 1. We can observe that M3LH obtains the best results on all cases. In addition, we can also find that the results of "Text Query v.s Image Database" are much better than those of "Image Query v.s Text Database". The reasonable explanation

Table 1. mAP comparison on NUS-WIDE.

Task	Method	8 bits	16 bits	32 bits	64 bits	128 bits
Image Query v.s. Text Database	CVH	0.3373	0.3368	0.3365	0.3378	0.3384
	IMH	0.3437	0.3419	0.3405	0.3392	0.3387
	LSSH	0.3670	0.3648	0.3597	0.3671	0.3658
	SCM-orth	0.3992	0.3789	0.3669	0.3594	0.3560
	SCM-seq	0.4998	0.5252	0.5261	0.5328	0.5307
	CMFH	0.3596	0.3632	0.3652	0.3630	0.3603
	SePH	0.5528	0.5697	0.5705	0.5597	0.5841
	M3LH	**0.5754**	**0.5885**	**0.5946**	**0.5985**	**0.6014**
Text Query v.s. Image Database	CVH	0.3378	0.3370	0.3368	0.3394	0.3405
	IMH	0.3458	0.3429	0.3440	0.3414	0.3406
	LSSH	0.3950	0.4016	0.4022	0.4066	0.4018
	SCM-orth	0.4040	0.3780	0.3626	0.3510	0.3463
	SCM-seq	0.4722	0.4993	0.5071	0.5134	0.5174
	CMFH	0.3544	0.3541	0.3570	0.3513	0.3522
	SePH	0.6538	0.6520	0.6771	0.6836	0.6885
	M3LH	**0.6675**	**0.6729**	**0.6844**	**0.6898**	**0.6932**

is that text can better describe the content of the image-text pair than image. We can further observe that the performance of some methods degrade with increasing the length of binary code, e.g., CVH and SCM-orth. The reason is that such methods have orthogonality constraints on the projection to make each bits uncorrelated. However, these constraints may lead to additional problems. For instance, their first few projects may have high variance and the corresponding bits are discriminative. However, with the code length increasing, the binary codes may be dominated by low variance bits. Thus, their performance may degrade.

The precision of top-N retrieved data points and precision-recall curves are shown in Fig. 1. To save space, we only plot the curves of the case with 32 bits. The first two subfigures are the top-N retrieval precision curves. The precision-recall curves are plotted in the left two subfigures. From this figure, we can

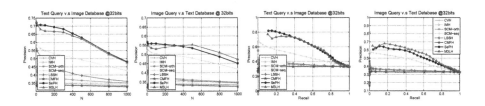

Fig. 1. Top-N retrieval precision and precision-recall curves on NUS-WIDE

observe that M3LH generally outperforms or is comparable to other compared methods.

Results on MIRFlickr

The mAP results of all methods on MIRFlickr are shown in Table 2. The precision of top-N retrieved data points and precision-recall curves are plotted in Fig. 2. From Table 2 and Fig. 2, we can observe the similar results as those on NUS-WIDE. Generally, M3LH outperforms or is comparable to compared state-of-the-art methods.

Results on Wiki

Note that our proposed M3LH can also work on single-label data set. To test its performance on single-label data set, we conduct experiments on Wiki, in which one instance is only associated with one label. The mAP values of M3LH and

Table 2. mAP comparison on MIRFlickr.

Task	Method	8 bits	16 bits	32 bits	64 bits	128 bits
Image Query v.s. Text Database	CVH	0.5785	0.5739	0.5699	0.5673	0.5648
	IMH	0.5656	0.5680	0.5686	0.5677	0.5659
	LSSH	0.5709	0.5736	0.5709	0.5747	0.5778
	SCM-orth	0.6021	0.5876	0.5750	0.5690	0.5662
	SCM-seq	0.6121	0.6280	0.6357	0.6473	0.6526
	CMFH	0.5831	0.5791	0.5795	0.5801	0.5796
	SePH	0.6765	0.6825	0.6896	0.6905	0.6938
	M3LH	**0.6918**	**0.7003**	**0.7077**	**0.7077**	**0.7110**
Text Query v.s. Image Database	CVH	0.5786	0.5736	0.5699	0.5674	0.5658
	IMH	0.5647	0.5669	0.5687	0.5672	0.5665
	LSSH	0.5877	0.5836	0.5892	0.5917	0.5929
	SCM-orth	0.5964	0.5862	0.5742	0.5677	0.5651
	SCM-seq	0.6034	0.6149	0.6214	0.6285	0.6358
	CMFH	0.5899	0.5902	0.5883	0.5869	0.5890
	SePH	0.7216	0.7342	0.7424	0.7514	0.7557
	M3LH	**0.7385**	**0.7599**	**0.7675**	**0.7721**	**0.7741**

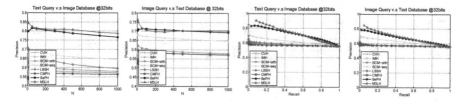

Fig. 2. Top-N retrieval precision and precision-recall curves on MIRFlickr

Table 3. mAP comparison on Wiki.

Task	Method	8 bits	16 bits	32 bits	64 bits	128 bits
Image Query v.s. Text Database	CVH	0.1893	0.1692	0.1611	0.1550	0.1552
	IMH	0.1620	0.1655	0.1637	0.1655	0.1688
	LSSH	0.1900	0.2168	0.2126	0.2041	0.2036
	SCM-orth	0.1763	0.1596	0.1440	0.1361	0.1353
	SCM-seq	0.2125	0.2341	0.2410	0.2427	0.2546
	CMFH	0.2249	0.2118	0.2314	0.2319	0.2391
	SePH	0.2539	0.2915	0.2968	0.3047	0.3135
	M3LH	**0.3033**	**0.3385**	**0.3567**	**0.3647**	**0.3688**
Text Query v.s. Image Database	CVH	0.1863	0.1547	0.1382	0.1249	0.1213
	IMH	0.1320	0.1377	0.1401	0.1324	0.1320
	LSSH	0.4279	0.4980	0.5218	0.5278	0.5318
	SCM-orth	0.1795	0.1509	0.1362	0.1278	0.1238
	SCM-seq	0.2013	0.2257	0.2459	0.2460	0.2544
	CMFH	0.4848	0.4977	0.5252	0.5210	0.5409
	SePH	0.5842	0.6496	0.6555	0.6607	0.6737
	M3LH	**0.7082**	**0.7317**	**0.7490**	**0.7636**	**0.7709**

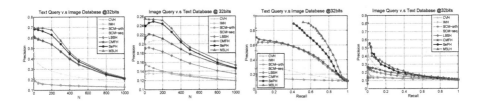

Fig. 3. Top-N retrieval precision and precision-recall curves on Wiki

all baseline methods on Wiki are reported in Table 3. From this table, we can find that M3LH outperforms all other methods on all cases. The precision-recall curves and top-N retrieval precision on Wiki are plotted in Fig. 3. From Fig. 3, we can find that we can observe that M3LH generally outperforms other compared methods. These results on Wiki confirm that M3LH can also perform well on single-label data.

6 Conclusions

In this paper, we propose a supervised hashing method for multi-modal and multi-label data, i.e., M3LH. It can not only work on multi-modal data, but also make full use of multi-label information. Specifically, it assumes every label is associated with a binary code in Hamming space, and the binary code of a sample can be generated by combining the binary codes of its labels. By minimizing the Hamming distances between similar pairs and maximizing the Hamming distances between dissimilar pairs, it also learns a project matrix which can be used to generate binary codes for out-of-samples. Experimental results on three data sets show that M3LH outperforms or is comparable to state-of-the-art multi-modal hashing methods.

Acknowledgements. This work was partially supported by National Natural Science Foundation of China (61173068, 61573212, 91546203), Program for New Century Excellent Talents in University of the Ministry of Education, Independent Innovation Foundation of Shandong Province (2014CGZH1106), Key Research and Development Program of Shandong Province (2016GGX101044, 2015GGE27033).

References

1. http://press.liacs.nl/mirflickr/
2. Chua, T.S., Tang, J., Hong, R., Li, H., Luo, Z., Zheng, Y.: NUS-WIDE: a real-world web image database from National University of Singapore. In: Proceedings of CIVR, article no. 48 (2009)
3. Datar, M., Immorlica, N., Indyk, P., Mirrokni, V.S.: Locality-sensitive hashing scheme based on p-stable distributions. In: Proceedings of SCG, pp. 253–262 (2004)
4. Ding, G., Guo, Y., Zhou, J.: Collective matrix factorization hashing for multimodal data. In: Proceedings of CVPR, pp. 2075–2082 (2014)
5. Ding, K., Huo, C., Fan, B., Pan, C.: kNN hashing with factorized neighborhood representation. In: Proceedings of ICCV, pp. 1098–1106 (2015)
6. Gong, Y., Lazebnik, S.: Iterative quantization: a procrustean approach to learning binary codes. In: Proceedings of CVPR, pp. 817–824 (2011)
7. He, K., Wen, F., Sun, J.: K-means hashing: an affinity-preserving quantization method for learning binary compact codes. In: Proceedings of CVPR, pp. 2938–2945 (2013)
8. Kim, S., Choi, S.: Multi-view anchor graph hashing. In: Proceedings of ICASSP, pp. 3123–3127 (2013)
9. Kumar, S., Udupa, R.: Learning hash functions for cross-view similarity search. In: IJCAI Proceedings-International Joint Conference on Artificial Intelligence, vol. 22, p. 1360 (2011)
10. Lin, Z., Ding, G., Hu, M., Wang, J.: Semantics-preserving hashing for cross-view retrieval. In: Proceedings of CVPR, pp. 3864–3872 (2015)
11. Rasiwasia, N., Costa Pereira, J., Coviello, E., Doyle, G., Lanckriet, G.R., Levy, R., Vasconcelos, N.: A new approach to cross-modal multimedia retrieval. In: Proceedings of MM, pp. 251–260 (2010)
12. Song, J., Yang, Y., Yang, Y., Huang, Z., Shen, H.T.: Inter-media hashing for large-scale retrieval from heterogeneous data sources. In: Proceedings of SIGMOD, pp. 785–796 (2013)
13. Tang, J., Li, Z., Wang, M., Zhao, R.: Neighborhood discriminant hashing for large-scale image retrieval. IEEE Trans. Image Process. **24**(9), 2827–2840 (2015)
14. Wang, J., Kumar, S., Chang, S.F.: Sequential projection learning for hashing with compact codes. In: Proceedings of ICML, pp. 1127–1134 (2010)
15. Wang, J., Xu, X.S., Guo, S., Cui, L., Wang, X.: Linear unsupervised hashing for ANN search in Euclidean space. Neurocomputing **171**(c), 283–292 (2016)

16. Wang, S.S., Huang, Z., Xu, X.S.: A multi-label least-squares hashing for scalable image search. In: Proceedings of SDM, pp. 954–962 (2015)
17. Weinberger, K.Q., Saul, L.K.: Distance metric learning for large margin nearest neighbor classification. J. Mach. Learn. Res. **10**, 207–244 (2009)
18. Weiss, Y., Torralba, A., Fergus, R.: Spectral hashing. In: NIPS, pp. 1753–1760 (2009)
19. Wu, B., Yang, Q., Zheng, W.-S., Wang, Y., Wang, J.: Quantized correlation hashing for fast cross-modal search. In: Proceedings of IJCAI, pp. 3946–3952 (2015)
20. Xu, X.-S.: Dictionary learning based hashing for cross-modal retrieval. In: Proceedings of MM, pp. 177–181 (2016)
21. Yan, T.-K., Xu, X.-S., Guo, S., Huang, Z., Wang, X.-L.: Supervised robust discrete multimodal hashing for cross-media retrieval. In: Proceedings of CIKM, pp. 1271–1280 (2016)
22. Yang, Y., Shen, F., Shen, H.T., Li, H., Li, X.: Robust discrete spectral hashing for large-scale image semantic indexing. IEEE Trans. Big Data **1**(4), 162–171 (2015)
23. Yang, Y., Zha, Z.-J., Gao, Y., Zhu, X., Chua, T.-S.: Exploiting web images for robust semantic video indexing via sample-specific loss. IEEE Trans. Multimed. **16**(6), 1677–1689 (2014)
24. Zhang, D., Li, W.J.: Large-scale supervised multimodal hashing with semantic correlation maximization. In: Proceedings of AAAI, pp. 2177–2183 (2014)
25. Zhou, J., Ding, G., Guo, Y.: Latent semantic sparse hashing for cross-modal similarity search. In: Proceedings of SIGIR, pp. 415–424 (2014)
26. Zou, F., Liu, C., Ling, H., Feng, H., Yan, L., Li, D.: Least square regularized spectral hashing for similarity search. Signal Process. **93**(8), 2265–2273 (2013)

Model-Based 3D Scene Reconstruction Using a Moving RGB-D Camera

Shyi-Chyi Cheng[1(✉)], Jui-Yuan Su[1,2], Jing-Min Chen[1],
and Jun-Wei Hsieh[1]

[1] Department of Computer Science and Engineering,
National Taiwan Ocean University, 1, Peining Rd., Keelung, Taiwan
{csc,2035700,105570031,shieh}@mail.ntou.edu.tw
[2] Department of New Media and Communications Administration,
Ming Chuan University, 250 Sec. 5, Zhong Shan North Road, Taipei, Taiwan
rysu@mail.mcu.edu.tw

Abstract. This paper presents a scalable model-based approach for 3D scene reconstruction using a moving RGB-D camera. The proposed approach enhances the accuracy of pose estimation due to exploiting the rich information in the multi-channel RGB-D image data. Our approach has lots of advantages on the reconstruction quality of the 3D scene as compared with the conventional approaches using sparse features for pose estimation. The pre-learned image-based 3D model provides multiple templates for sampled views of the model, which are used to estimate the poses of the frames in the input RGB-D video without the need of a priori internal and external camera parameters. Through template-to-frame registration, the reconstructed 3D scene can be loaded in an augmented reality (AR) environment to facilitate displaying, interaction, and rendering of an image-based AR application. Finally, we verify the ability of the established reconstruction system on publicly available benchmark datasets, and compare it with the sate-of-the-art pose estimation algorithms. The results indicate that our approach outperforms the compared methods on the accuracy of pose estimation.

Keywords: Image-based 3D model · Multiple view templates · Iterative closed point · Template-to-frame registration · Augmented reality

1 Introduction

The ability to quickly reconstruct three-dimensional (3D) models from an image sequence is a primary step in many research areas including robotics, augmented reality (AR), computer vision, geodesy, remote sensing, and 3D printing. Researchers in remote sensing provide two traditional 3D reconstruction techniques including airborne image photogrammetry [1] and light detection and ranging (LiDAR) [2]. Both

This work was supported in part by Minister of Science and Technology, Taiwan under Grant Numbers MOST 105-2221-E-019-034-MY2 and 105-2218-E-019-001.

L. Amsaleg et al. (Eds.): MMM 2017, Part I, LNCS 10132, pp. 214–225, 2017.
DOI: 10.1007/978-3-319-51811-4_18

approaches reconstruct high quality 3D models [1], nevertheless, their acquiring cost is essentially high. To tackle the difficulty, researchers in computer vision have introduced many image-based 3D modeling techniques such as Simultaneous Location and Mapping (SLAM) [3], Multi-view Stereo (MVS) [4–6], Photo Tourism [7], and ARC 3D Webservices [8]. These techniques reconstruct 3D models without requiring photogrammetry information.

Image-based 3D reconstruction is predicated on the assumption that the observed imagery is generated jointly by the pose of a 3D scene. Recently, with the depth or combined color and depth imagery from a low-cost Kinect-like sensor as input data, the-state-of-the-art approaches focus on the development of novel methods that can infer scale and pose of the scene more accurately even in the environment of cluttered background. Template-based approach is an efficient method that exploits both depth and color images to capture the appearance and 3D shape of the object in a set of templates covering different views of an object [9]. Because the pose parameters of each template is given, it provides a coarse estimation of the pose of the test frame. To take full advantage of substantial parts of the RGB-D data, the dense SLAM can be followed to accurately estimate the poses in the remainder frames [10, 11].

The performance of template-based 3D model reconstruction is extremely good, however still has some disadvantages [9]: (1) the online-learned templates are often spotty in the coverage of viewpoints; (2) the pose output by template matching is only approximately correct since a template covers a range of views around its viewpoint; (3) the presence of false positives. An image-based 3D model can be used to deal with these difficulties. Note that many methods have been developed to quickly create high quality 3D models [12–14]. In industrial applications, a detailed 3D model often exist before the real object is even not existed. This implies that to require a pre-learned 3D model is not a disadvantage in 3D reconstruction.

In this work, the image-based 3D model represents a 3D scene as a set of RGB-D templates which correspond to different viewpoints of the scene by projecting the 3D points into 2D image planes using different pose parameters. We come up with the frame-to-model scene reconstruction by accurately matching every input frame against these templates in order to find the most similar template which provides the coarse pose estimate for the input frame. Next, taking the different level of 3D noise into account, we propose an Iterative Closet Point (ICP) framework which embeds the generalized Procrustes analysis [15] to decrease the interference of high noise on the accurate calculation of a projection matrix describing how a 3D model projects onto a set of input frames. At last, we evaluate our system on an open RGB-D benchmark provided by the Technical University of Munich [16]. All of our work is based on the open source project Point Cloud Library (PCL) [17]. It provides some visualization functions of 3D point cloud process and realizes the basic function of ICP algorithm.

2 Notation and Preliminaries

In this section, for the sake of illustration, we briefly summarize some relevant mathematical concepts and notation including the camera model, the representation of 3D rigid body transformation and the volumetric representation of 3D object model

[11, 12]. Let $p_c = (x_c, y_c, z_c, 1)^t$ be the homogeneous coordinates of a 3D point in the camera coordinate system. The pinhole camera model projects the 3D point into the pixel $x = (x, y, 1)^t$ in an image plane using the following equation:

$$x = \pi(p_c) = \left(\frac{f_x x_c + c_x}{z(x)}, \frac{f_y y_c + c_x}{z(x)}, 1 \right)^t. \tag{1}$$

where f_x and f_y are the x and y direction focal length, respectively; c_x and c_y are the principal point offsets; $z(x)$ is the depth value of the 2D point x. Obviously, given the depth value $z(x)$, the inverse projection function $\pi^{-1}(\cdot)$ projects a 2D point x back to the 3D point p_c:

$$p_c = \pi^{-1}(x, z(x)) = z(x) \left(\frac{x - c_x}{f_x}, \frac{y - c_y}{f_y}, 1 \right)^t. \tag{2}$$

Based on the theory of three-dimensional special orthogonal (SO(3)) Lie group, the pose of the classical rigid body can be described by the six-degree-of-freedom (6DOF) camera motion model which constitutes an orthogonal rotation matrix $R_{3\times3} \in SO(3)$ and a translation vector $t_{3\times1} \in \mathbb{R}^3$. The 6DOF camera motion model is also used to define the rigid body transformation in $\mathbb{R}^3$ as a 4×4 matrix $M_{4\times4} = \begin{bmatrix} R_{3\times3} & t_{3\times3} \\ 0 & 1 \end{bmatrix}$. The rigid body transformations in $\mathbb{R}^3$ form a smooth manifold and therefore are also a Lie group. The group operator is the matrix left-multiplication. Thus, a 3D point p_c in the camera coordinate system can be transformed into a 3D point $p_g = (x_g, y_g, z_g, 1)^t$ in the global coordinate system by matrix left-multiplication: $p_g = \psi(M_{4\times4}, p_c) = M_{4\times4}p_c$. Inversely, the inverse function $\psi^{-1}(\cdot)$ transformation p_g back to p_c:

$$p_c = \psi^{-1}(M_{4\times4}^{-1}, p_g) = M_{4\times4}^{-1} \begin{bmatrix} x_g \\ y_g \\ z_g \\ 1 \end{bmatrix} = \begin{bmatrix} R_{3\times3}^{-1} & -R_{3\times3}^{-1}t_{3\times1} \\ 0 & 1 \end{bmatrix} \begin{bmatrix} x_g \\ y_g \\ z_g \\ 1 \end{bmatrix}. \tag{3}$$

Estimating the pose of a rigid object means to determine the rigid object motion in the 3D space from 2D images. Researchers in 3D reconstruction have shown that the signed distance function (SDF) which contains the distances to the surface for each voxel can represent the geometry of a 3D scene as a volume. Accordingly, in this paper, we integrate dense template depth images into a SDF volume (SDFV) to represent the geometry and texture of the reconstructed 3D object model.

The fixed scale 3D volume is sub-divided uniformly into 3D grid of voxels in which the values are integrated from multiple 2D RGB-D images. At each time step, an RGB-D camera outputs two registered images, i.e. a color image $I_{RGB} : \mathbb{R}^2 \rightarrow \mathbb{R}^3$ and a depth image $I_D : \mathbb{R}^2 \rightarrow \mathbb{R}$. Given any point $p_g \in \mathbb{R}^3$, the SDF function $v : \mathbb{R}^3 \rightarrow \mathbb{R}$ returns the signed distance from p_g to the surface of an object. For each pixel $x = (x,y)$, we have its depth $z(x) = I_D(x,y)$. Using (2), we can reconstruct the corresponding 3D point $p_c = \pi^{-1}(x, z(x))$ in the camera coordinate system. Again, we can transform this point to the 3D point $p_g' = \psi(M_{4\times4}, p_c')$ in the global coordinate system. To query the SDF, we can read out the distance of p_g from the surface. Given that the SDF and the camera pose is correct, the distance of p_g is zero when p_g is on the boundary of the reconstructed object.

To follow the SDF volume integration process proposed in [11], we can reconstruct a 3D scene represented with a SDFV using a set of template depth images. Let $M_i = (R_i, t_i)$, $i = 1,\ldots, k$ be the global camera pose of the i-th depth image used to reconstruct the SDFV V. A single voxel center v can be projected into the depth map at $x = \pi(\psi^{-1}(M_i^{-1}, v))$. For each voxel v in the SDFV, we maintain a voxel value $f(v)$ and a weighted value $w(v)$. We re-project the pixel x with its depth value $z(x)$ in the i-th depth image back to measurement voxel $p_g = \psi(M_i, \pi^{-1}(x, z(x)))$. The signed distance between voxels v and p_g is determined by $d = ||v - t_i|| - ||p_g - t_i||$. Then this distance is truncated as the increment SDF value

$$\Delta f = \begin{cases} \dfrac{\min_{(|d|,\gamma)}}{\gamma} & d \geq 0 \\ \dfrac{-\min_{(|d|,\gamma)}}{\gamma} & d < 0 \end{cases} \tag{4}$$

where γ is the truncated distance threshold. Using the i-th depth image, the values of $w_i(v)$ and $f_i(v)$ of each v are updated using linear weighted summation as follow

$$w_i(v) = \min(w_{i-1}(v) + \Delta w, w_{\max}) \tag{5}$$

$$f_i(v) = \frac{w_i(v)f_{i-1}(v) + \Delta w \Delta f}{w_{i-1}(v) + \Delta w} \tag{6}$$

where $w_{\max} = 128$ is the maximal weight and $\Delta w = 1$ is the weight increment. Note that the voxel value can be extended to be a multi-channel RGB-D vector in order to synthesize the registered template depth and RGB images that sample the possible appearances of the scene covering the whole pose range.

3 Approach

We represent the geometry of the 3D model using a SDFV with each voxel storing a voxel value and a weight value. Given a RGB-D video with the target scene stored in each frame, we follow an iterative approach consists of three major steps: (1) to search a suitable template that samples the possible appearance of the model in the current frame using the proposed pose shape Hough model; (2) to estimate the camera pose of the input frame using the template; (3) to fine tune the coarse pose parameters by an Interactive Closest Point (ICP) algorithm [15]. Then the camera pose of the input frames is used to reconstruct the SDFV of the target model. Figure 1 shows the flow chart to reconstruct the target 3D model in a RGB-D video. The approach is divided into two phases: the pose shape modeling and the 3D model reconstruction.

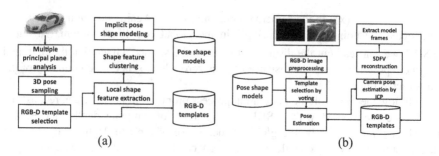

(a)   (b)

Fig. 1. The workflow of the proposed 3D model reconstruction on RGB-D video: (a) the pose shape modeling; (b) the reconstruction approach.

3.1   The Pose Sampling for Template Selection

With a pose sampling process, the first step of the implicit pose shape modeling is to automatically build a set of RGB-D templates for a CAD 3D model which is manually prepared in advance. To construct a complete image-based 3D model, the number of templates should be large enough to cover all pose ranges though this results in high computational complexity of model reconstruction. Thus, the pose sampling process aims at balancing the tradeoff between the coverage of the scene for reliability and the number of template for efficiency. To achieve the goal, using the multiple principal plane (MPP) analysis proposed in our previous work [18], the initial 3D model is first recursively approximated by a set of surfaces which segment the input model into multiple surfaces. With the centers of the surfaces as the vertices of the simplified model, Fig. 2 shows that the vertices of the resulting simplified model give us then the two out-of-plane rotations for the sampled pose with respect to the model center. For each surface S, we define a vector $\vec{n}_S = p_S - p_c$ which is originated from model center p_c to p_S, i.e., the center of S. As shown in Fig. 2(e), given a scaling factor s, the point defined by $v_S = p_c + s\vec{n}_S$ is the virtual camera center used to select the template that covers the pose defined by S. Then a set of viewpoints $\Phi = \{v_i\}_{i=1}^m$ can be computed if the MPP analysis returns m surfaces for the 3D model.

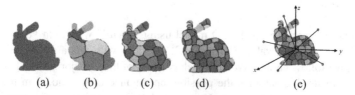

(a)   (b)   (c)   (d)   (e)

Fig. 2. Model simplification results for Stanford bunny using the proposed multiple principal plane analysis with different numbers of surfaces: (a) the original model; (b)–(d) are the segmentation results with 10, 50, and 100 surfaces, respectively; (e) outer red vertices are the viewpoints which represent the virtual camera centers used to select templates. (Color figure online)

Let $\Gamma = \{T_i\}_{i=1}^n$ be the set of templates to construct the image-based 3D model. By fixing the value of the scaling factor s mentioned above, we can select a set of templates from Γ, which are used to estimate the frames' camera pose in the input RGB-D video. For each training template T_i, we use the inverse projection $\pi^{-1}(\cdot)$ to project the 2D center point x_i of T_i back to the 3D point $p_i^{(c)} = \pi^{-1}(x_i, z(x_i))$ in the camera coordinate system. This computes a set of template camera centers $\Omega = \{p_i^{(c)}\}_{i=1}^n$ using the camera parameters and depth information associated with the templates in Γ. The template selection problem can then be formulated as the problem of establishing the one-to-one point correspondence between point sets Φ and $\tilde{\Omega} \subset \Omega$, where $\tilde{\Omega}$ is the set of templates selected to reconstruct the image-based 3D model.

For each viewpoint v_i, the final step of the 3D point correspondence algorithm is to compute the line L that passes both v_i and p_c. The point $p_i^{(c)}$ in $(\Omega - \tilde{\Omega})$ of the smallest distance to L is then selected to be the correspondence point of v_i. To collect all correspondence points of viewpoints, the point set $\tilde{\Omega}$ that determines the templates of the 3D object model is finally constructed. The 3D points in $\tilde{\Omega}$ are then defined as the new viewpoints with the corresponding templates to reconstruct the 3D object using the SDFV representation.

3.2 The Pose Shape Modeling

The model-based template search relies two different features: *SIFT* descriptors, computed from the key-points of the template color image T_{RGB}; and surface normal vectors computed from the template depth map by considering the 3D surrounding neighboring points of the key-points. Let $x_i = (x_i, y_i, 1)^t$ be the homogeneous coordinate of a pixel in the template depth map T_D. Each pixel in T_D is computed by the inverse projection function: $p_i^{(c)} = \pi^{-1}(x_i, z(x_i))$, where $z(x_i)$ is the depth value of x_i. The surface normal (*SN*) vector is simply computed using back-projected points:

$$\vec{n}_i^{(c)} = \left(p_i^{(c)}(x+1, y) - p_i^{(c)}(x, y)\right) \times \left(p_i^{(c)}(x, y+1) - p_i^{(c)}(x, y)\right). \tag{7}$$

Since the camera pose $M_{4x4} = [R_{3x3}|t_{3x3}]$ of template T_D is given, a vertex and a surface normal vector can be converted into global coordinates $(p_i^{(g)}, \vec{n}_i^{(g)}) = (M_{4\times4}p_i^{(c)}, R_{3\times3}\vec{n}_i^{(c)})$.

In this work, we only compute the surface normal vectors of edges in the template depth image. Although the surface normal vectors of the key-points can be used to estimate the poses of frames in an input RGB-D video, we want to keep among the remaindering ones. This is because the number of key-points is often small when the target objects to be detected are texture less.

Let the appearance of the training templates be represented by a set of descriptors of key-points $\left\{K_i = (F_i^{(SIFT)}, F_i^{(SN)}, v_i)\right\}$ where $F_i^{(SIFT)}$ and $F_i^{(SN)}$ are the i-th key-point features, computed from the template RGB and depth images, respectively and v_i is the pose label, i.e. the identification of template T_i. Based on the key-point features of the training templates, one step towards learning the appearance variability of a 3D model is to build a visual codebook to characterize the pose shapes. As the basic representation for our approach, for the collection of key-points from all training template objects, we introduce the pose shape Hough model (*PSHM*), which is a two-directional codebook with two types of code words, shown in Fig. 3. The system uses an efficient k-means clustering [19] to cluster features of training features regardless of their pose labels. Notice that we perform the k-means clustering twice, each for individual feature type, to generate the two-directional codebook. More concisely, each cluster of the *PSHM* is a key-point list which is indexed by two types of features, the *SIFT* and *SN* codewords. For each key-point cluster KC_{ij}, we also maintain an inverting list to retrieve relevant training template objects. Without loss of generality, we assume each training key-point only belongs to one key-point cluster in the *PSHM*.

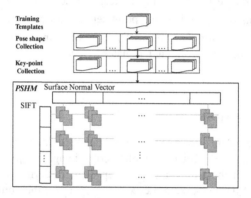

Fig. 3. The pose shape Hough model (*PSHM*) for a key-point collection: a two-directional codebook.

3.3   The Template Selection with Hough Voting

In the 3D reconstruction phase, the input RGB-D video is first pre-processed to extract key-frames from the video. This reduces the complexity of 3D reconstruction. Next, the *SIFT* features and *SN* vectors of key-points in each key-frame are extracted for the further template selection.

For every key-point K at (K_x, K_y) in the current key-frame I, the features of K are matched against the codewords of the *PSHM* to search for the nearest cluster in which the training key-points K's are considered as the similar key-points. The distance of a key-point pair (K, K') is obtained from the following equation:

$$\varepsilon(K, K') = w_1||F_K^{(SIFT)} - F_{K'}^{(SIFT)}|| + w_2||F_K^{(SN)} - F_{K'}^{(SN)}||, w_1 + w_2 = 1 \quad (8)$$

where w_1 (w_2) is the weighting of *SIFT* (*SN*) feature and $||.||$ is the Euclidean distance. The similarity measure for key-point pair (K, K') is then obtained by $S(K, K') = \frac{2}{1 + e^{\xi \varepsilon(K, K')}}$, where ξ is a positive constant, *e.g.*, $\xi = 2$ to regulate the key-point distance. Let $v^{(K')}$ denote the pose type of K', which is stored in advance in the *PSHM*. To determine the pose type of I, the local similarity measurement for (K, K') casts a vote on the one-dimensional array $A(v)^{(new)} = A(v)^{(old)} + S(K, K')$, where $v = v^{(K')}$ is the estimated pose type. Note that a key-point pair of its similarity value less than a pre-defined threshold, *i.e.*, 0.8, is excluded from casting a vote on the Hough-voting matrices to avoid generating spurious peaks. Obviously, the peak in A group key-points of the same pose type which aligns the corresponding template with current key-frame I.

3.4 3D Reconstruction with Pose Estimation

The pose parameters associated with a template T of a 3D model provides a coarse pose estimate of the detected object in a key-frame of the input RGB-D video. In this work, the Interactive Closest Point (ICP) algorithm [15] is used to refine the pose estimation of the detected object by aligning the 3D model surface with the depth map. In the current depth image, we consider the pixels that lie on the object projection according to the pose estimate.

Let $x_i = (x_i, y_i, 1)^t$ be the homogeneous coordinate of a pixel of the object projection in the current depth image I_D. Using the camera parameters of the initial pose estimate, each pixel in I_D is computed by the inverse projection function: $p_i^{(c)} = \pi^{-1}(x_i, z(x_i))$, where $z(x_i)$ is the depth value of x_i. Similar to template depth image, we can also compute the normal vector $\vec{n}_i^{(c)}$ of x_i. Since the camera pose $M_{4x4} = [R_{3x3} | t_{3x3}]$ of the template image T provides the initial pose estimate for I_D, the vertex $p_i^{(c)}$ and the surface normal vector $\vec{n}_i^{(c)}$ of x_i can be converted into global coordinates: $(p_i^{(g)}, \vec{n}_i^{(g)}) = (M_{4\times4} p_i^{(c)}, R_{3\times3} \vec{n}_i^{(c)})$. This process computes the vertex map V_I and surface normal map N_I for the current key-frame I. Similarly, in the global coordinates, we can compute the point $p_i^{'(g)}$ and the surface normal vector $\vec{n}_i^{'(g)}$ of the pixel x_i' in the depth image of T. This process also constructs the template reference model $T_r = (V_T, N_T)$.

Second we propose an ICP algorithm to estimate the transformation matrix by aligning the current frame map with the template reference model $T_r = (V_T, N_T)$. The goal of each iteration of the ICP is to find an optimized transformation matrix ϕ^* such that the summation of point-to-plane error is minimum:

$$\phi^* = \arg \min_{\phi} \sum_{x_i \in I} [\vec{n}_i^{'(g)} \cdot (\phi \cdot p_i^{(g)} - p_i^{'(g)})]^2. \tag{9}$$

Because the initial pose is closed to the final solution, the nonlinear matrix ϕ can be approximated to a linear one by limiting the rotation angle. So this nonlinear optimization problem is easily transferred to linear problem by linear iteration. Then the accurate current camera pose is the coarse predicted pose M multiplied by the transformation matrix ϕ^*: M = ϕ^*M. As mentioned above, the final 3D object can be reconstructed using the pose parameters in every key-frame of the input RGB-D video.

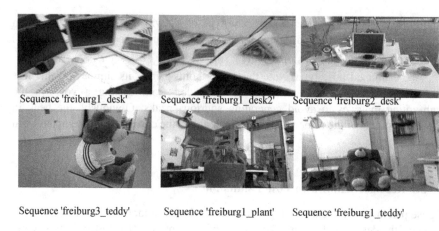

Sequence 'freiburg1_desk'   Sequence 'freiburg1_desk2'   Sequence 'freiburg2_desk'

Sequence 'freiburg3_teddy'   Sequence 'freiburg1_plant'   Sequence 'freiburg1_teddy'

Fig. 4. Example frames of test datasets.

4   Experimental Results

The system is implemented on a PC with Intel Core i5-4670 3.4 GHz CPU. The RGB-D benchmark provided by the Technical University of Munich [16] is used verify the effectiveness of our approach in terms of the pose estimation accuracy. In this benchmark, several datasets including color and depth sequences captured with a single RGB-D camera are provided. The accurate ground-truth of each frame's camera pose is also provided, which is measured by an external motion capture system. The frames of each dataset are separated into two disjoint sets: a template dataset and a test dataset.

To compare the performance of our approach with other state-of-the-arts methods in pose estimation [20–23], six typical datasets are selected as the test samples. In [20, 21], the compared approaches are model-based methods which is similar to our approach. In [22] and [23], the authors proposed an RGB-D SLAM system. Some post-processing methods like global optimization are used to modify the global camera trajectory in these methods. Figure 4 shows an example of RGB frame in each dataset.

Table 1. Camera pose error for datasets with groundtruth from [16] in terms of RMSE of the relative error. The value of k is the stride to select the key-frames from the test datasets.

Sequence	Our approach				Multi-resolution map	RGB-D SLAM
	$k = 1$	$k = 5$	$k = 10$	$k = 20$	[20]	[23]
freiburg1 desk2	0.049	0.071	0.052	**0.033**	0.060	0.102
freiburg1 desk	0.041	0.045	0.034	**0.033**	0.044	0.049
freiburg1 plant	0.041	0.050	0.068	**0.033**	0.036	0.142
freiburg1 teddy	0.057	0.090	0.076	**0.052**	0.061	0.138
freiburg2 desk	0.031	0.030	0.030	**0.027**	0.091	0.143

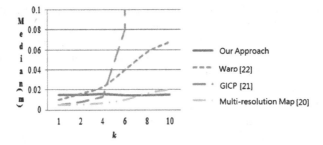

Fig. 5. Performance comparison in terms of median of relative pose error using different numbers of key-frames to construct a 3D scene.

Table 1 summarizes the experiment results. The first column is the dataset name. The rest columns are the results for the root mean square error (RMSE) of the relative errors. The relative error is the translational drift in m/s between estimated pose and ground-truth. The parameter k is the stride to select the key-frames from the template

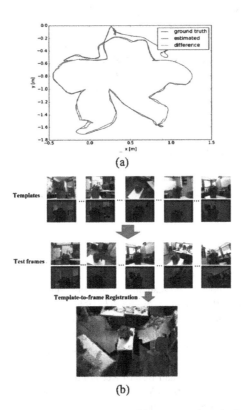

(a)

(b)

Fig. 6. (a) An example of the 2D trajectory error of the proposed method using the test sequence 'freiburg1_plant'. (b) The reconstructed 3D scene using the proposed template-to-frame registration technique.

datasets. The larger value of k implies the less templates are selected in the training sequences to build up the 3D model for reconstructing the target scene or object in the testing phase. Since a large amount of redundancy could exist in two consecutive frames of each sequence, we can actually use very few frames to reconstruct a 3D model. With the usage of less templates also highly reduces the time complexity to estimate the pose of each frame in order to achieve the goal of 3D object reconstruction in real-time. As shown in Fig. 5, compared with the state-of-the-arts, the proposed method is not sensitive to the value of k. Accordingly, the proposed method can use very few key-frames to reconstruct a 3D scene without scarifying the accuracy of pose estimation. Figure 6(a) shows an example of the 2D trajectory error of the proposed method using the test sequence 'freiburg1_plant'; the reconstructed 3D scene is also shown in Fig. 6(b).

5   Conclusion

In this paper, we have presented a model-based scalable 3D scene reconstruction system. The contributions of the proposed method include (1) to start from a 3D CAD model, the multiple principal plane analysis facilitates the searches of the template images to construct a high quality image-based 3D model; (2) the usage of a pose shape Hough model speeds up the searches of model templates to estimate the pose of a test frame; (3) the initial pose estimate of a test frame provided by the corresponding template is optimized by an ICP algorithm. In our experiments on publicly available datasets, we show that our approach gives the lowest trajectory error on the whole and outperforms the state-of-the-art methods.

In the next step we plan to explore a large scale AR interaction system based on this RGB-D based pose estimation. Furthermore, based on the reconstruction models, we want to extend our approach to 3D object detection, segmentation, reconstruction and printing.

References

1. Wolf, P.R., Dewitt, B.A.: Elements of Photogrammetry: with Applications in GIS. McGraw-Hill, New York (2000)
2. Ackermann, F.: Airborne laser scanning – present status and further expectations. ISPRS J. Photogramm. Remote Sens. **54**, 64–67 (1999)
3. Davison, A., Reid, I., Molton, N., Stasse, O.: MonoSLAM: real-time single camera SLAM. IEEE Trans. Pattern Anal. Mach. Intell. **29**(6), 1052–1067 (2007)
4. Seitz, S.M., Curless, B., Diebel, J., Scharstein, D., Szeliski, R.: A Comparison and evaluation of multi-view stereo reconstruction algorithms. In: IEEE Computer Society Conference on Computer Vision and Pattern Recognition (2006)
5. Furukawa, Y., Curless, B., Seitz, S.M., Szeliski, R.: Towards internet-scale multi-view stereo. In: IEEE Computer Society Conference on Computer Vision and Pattern Recognition (2010)

6. Furukawa, Y., Ponce, J.: Accurate, dense, and robust multi-view stereopsis. IEEE Trans. Pattern Anal. Mach. Intell. **32**(8), 1362–1376 (2010)
7. Snavely, N., Seitz, S.M., Szeliski, R.: Modeling the world from internet photo collections. Int. J. Comput. Vis. **80**, 189–210 (2008)
8. Vergauwen, M., Van Gool, L.: Web-based 3D reconstruction service. Mach. Vis. Appl. **17** (6), 411–426 (2006)
9. Hinterstoisser, S., Lepetit, V., Ilic, S., Holzer, S., Bradski, G., Konolige, K., Navab, N.: Model based training, detection and pose estimation of texture-less 3D objects in heavily cluttered scenes. In: Lee, K.M., Matsushita, Y., Rehg, J.M., Hu, Z. (eds.) ACCV 2012. LNCS, vol. 7724, pp. 548–562. Springer, Heidelberg (2013). doi:10.1007/978-3-642-37331-2_42
10. Kerl, C., Sturm, J., Cremers, D.: Robust odometry estimation for RGB-D cameras. In: International Conference on Robotics and Automation (ICRA), pp. 3748–3754 (2013)
11. Li, J.N., Wang, L.H., Li, Y., Zhang, J.F., Li, D.X., Zhang, M.: Local optimized and scalable frame-to-model SLAM. Multimed. Tools Appl. **75**, 8675–8694 (2016)
12. Newcombe, R.A., Izadi, S., Hilliges, O., Molyneaux, D., Kim, D., Davison, A.J., Kohli, P., Shotton, J., Hodges, S., Fitzgibbon, A.: KinectFusion: real-time dense surface mapping and tracking. In: ISMAR (2011)
13. Tong, J., Zhou, J., Liu, L., Pan, Z., Yan, H.: Scanning 3D full human bodies using kinects. IEEE Trans. Vis. Comput. Graph. **18**(4), 643–650 (2012)
14. Alexiadis, D.S., Zarpalas, D., Daras, P.: Real-time, full 3-D reconstruction of moving foreground objects from multiple consumer depth cameras. IEEE Trans. Multimed. **15**(2), 339–358 (2013)
15. Toldo, R., Beinat, A., Crosilla, F.: Global registration of multiple point clouds embedding the generalized procrustes analysis into an ICP framework. In: Proceedings of 3DPVT Conference, pp. 17–20 (2010)
16. Sturm, J., Engelhard, N., Endres, F., Burgard, W., Cremers D.: A benchmark for the evaluation of RGB-D SLAM systems. In: 2012 IEEE/RSJ International Conference on Intelligent Robots and Systems (IROS), pp. 573–580 (2012)
17. Rusu, R.B., Cousins, S.: 3D is here: point cloud library (PCL). In: IEEE International Conference on Robotics and Automation (ICRA), pp. 1–4 (2011)
18. Cheng, S.C., Kuo, C.T., Wu, D.C.: A novel 3D mesh compression using mesh segmentation with multiple principal plane analysis. Pattern Recogn. **43**(1), 267–279 (2010)
19. Kanungo, T., Mount, D.M., Netanyahu, N.S., Piatko, C.D., Silverman, R., Wu, A.Y.: An efficient k-means clustering algorithm: analysis and implementation. IEEE Trans. Pattern Anal. Mach. Intell. **24**, 881–892 (2002)
20. Jörg Stückler, J., Behnke, S.: Multi-resolution surfel maps for efficient dense 3D modeling and tracking. J. Vis. Commun. Image Represent. **25**(1), 137–147 (2014)
21. Segal, A., Haehnel, D., Thrun, S.: Generalized-ICP. In: Proceedings of Robotics: Science and Systems (RSS) Conference (2009)
22. Steinbruecker, F., Sturm, J., Cremers, D.: Real-time visual odometry from dense RGB-D images. In: Proceedings of Workshop on Live Dense Reconstruction with Moving Cameras at ICCV, pp. 719–722 (2011)
23. Endres, F., Hess, J., Engelhard, N., Sturm, J., Cremers, D., Burgard, W.: An evaluation of the RGB-D SLAM system. In: Proceedings of the IEEE International Conference on Robotics and Automation (ICRA) (2012)

Modeling User Performance for Moving Target Selection with a Delayed Mouse

Mark Claypool[1]([✉]), Ragnhild Eg[2], and Kjetil Raaen[2]

[1] Computer Science, Worcester Polytechnic Institute, Worcester, MA 01720, USA
claypool@cs.wpi.edu
[2] School of Art, Communication and Technology,
Westerdals – Oslo, Christian Krohgs gate 32, 0186 Oslo, Norway
{egrag,raakje}@westerdals.no

Abstract. The growth in networking and cloud services provides opportunities to host multimedia on remote servers, but also brings challenges to developers who must deal with added delays that degrade interactivity. A fundamental action for many computer-based multimedia applications is selecting a moving target with the mouse. While previous research has modeled both moving target selection and target selection with delay, there have not been models of moving target selection with delay. Our work presents a user study that measures the effects of delay and target speed on the time to select a moving target with a mouse, with analysis of trends and derivation of a model. The analysis shows delay and speed impact target selection time exponentially and that selection time is well-represented by a model with three terms - two with exponential relationships for delay and speed and one an important interaction term.

1 Introduction

Interactive multimedia applications vary in purpose and content, but some user actions are nearly ubiquitous. One such action is using a mouse or similar device to select a target in an interface, a task common from Web browsers and spreadsheets to video editors and computer games.

Although target selection is commonly in use already, target selection with delay has not been thoroughly explored or modeled. When performing target selection with a mouse, there is always delay between moving and clicking the physical mouse and the corresponding rendered results on the screen. While the delay due to the local device is often negligible, when used in ever more popular cloud computing, all actions are rendered on a cloud server, accumulating delays from the local client, network and remote server [1]. These delays may affect the interaction in at least two ways: (1) reducing user performance and, thereby, productivity, and (2) degrading the quality of experience and causing frustration.

Fitts' early seminal work in target selection [2] provided a robust model (Fitts' Law) for the time it takes a user to select a target of a specified size at a specified distance. Extensions to this work expanded the model to two dimensions (such as for a computer mouse) [7] and moving targets [6]. While effective

© Springer International Publishing AG 2017
L. Amsaleg et al. (Eds.): MMM 2017, Part I, LNCS 10132, pp. 226–237, 2017.
DOI: 10.1007/978-3-319-51811-4_19

for modeling target selection, the models fail to incorporate delays present in Internet multimedia applications. A model that does incorporate delay [5] concentrates on much higher delays than are typically encountered on the Internet and does not combine delay with moving target selection. Ideally, real-time, multimedia developers would have a model as far reaching and robust as Fitts' Law, but incorporating moving targets, common to many multimedia applications (e.g., computer games) and delay, present in all networked multimedia applications. Our work takes a first step towards providing such a model.

We design and implement a game that isolates the fundamental action of selecting a moving target with a mouse and allows for control of the delay between the user input and the rendered action. With our game, task performance is the time it takes the user to hit the target and quality of experience is the user's subjective opinion of the game's responsiveness. The game is central to a user study of over 80 total participants, studying added delays ranging from 0 to 400 ms and target speeds ranging from 150 to 1550 px/s.

Analysis of the results shows the time to select a moving target with a mouse increases exponentially with both delay and target speed – this is in contrast to earlier works that showed a linear relationship with delay [5] and a linear relationship with speed [6]. The combination of delay and speed results in higher times than either alone would suggest. As such, we propose a model for the time to select a moving target with delay that uses exponential terms for both delay and target speed and has a combined interaction between them. Our model fits our experimental data well and provides a basis for future models and multimedia development. Brief analysis of the quality of experience shows a marked decay in perceived responsiveness with delay.

The rest of this paper is organized as follows: Sect. 2 describes our methodology, including game development and user study; Sect. 3 analyzes the user study results for trends and presents our derived model; and Sect. 4 summarizes our conclusions and presents possible future work.

2 Methodology

To investigate how delay affects users when selecting a moving target with a mouse, we deployed the following methodology: (1) Design and develop a game that isolates the target selection action with controlled delay (Sect. 2.1); (2) Conduct user studies with the game to evaluate the impact of delay on moving target selection, measuring game performance and perceived game responsiveness (Sect. 2.2); and (3) Analyze the user study results and develop an analytic model for moving target selection with delay (Sect. 3).

2.1 Application for Experiments

In order to avoid the monotony that often occurs during behavioral experiments, we designed the experimental task as a game that requires both temporal and

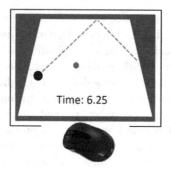

Fig. 1. Puck Hunt. Players select moving target (black puck) with mouse cursor (red ball), with controlled delay and varied target speeds. (Color figure online)

spatial precision isolated in mouse interactions. In the game *Puck Hunt*[1], the objective is simple – move the cursor and click on the target as quickly as possible. To minimize the inherent local delay, Puck Hunt is written in C++ using OpenGL with the Angel 2D[2] game engine (Fig. 1).

Puck Hunt consists of a series of short rounds; each round starts with a randomly placed large, black puck (the target) and a smaller red ball (the mouse cursor). The puck bounces around in a space confined by the screen according to simple 2d physics. The user moves the mouse to control the ball, and attempts to "hit" the puck by placing the ball over the puck and clicking the mouse button. Once the user has successfully hit the puck, the puck disappears and a notification pops up telling the user to prepare for the next round. Upon pressing space, a new round starts, with the puck at a new starting position and velocity. User performance is defined by the time it takes to complete a round, shown by a timer that counts up from zero. Puck Hunt adds a configurable amount of delay to the mouse input, both moving and clicking, and uses a configurable set of puck speeds to control game difficulty. Since rounds with fast pucks and high delays are quite difficult, Puck Hunt has an upper time limit of 30 s to avoid situations where a user may just give up. So a recorded score of 30 or higher indicates the game task was not completed the task for that round.

Our experiments had two different versions of the game at separate university locations – Worcester Polytechnic Institute (WPI) and Westerdals-Oslo (WO) – the sole difference being the speeds of the puck. While the differences in the puck speeds at each university was an inadvertent side effect of the game engine on a Mac (at WO) versus Windows (at WPI), the results do provide for analysis of a wider speed spread than would the same speeds at each university. Both game versions use 3 possible speeds varied randomly between rounds, for a total of 6 speed levels total, shown in Table 1(a). Effectively, these speeds create different levels of difficulty.

---

[1] A pun on the classic game *Duck Hunt* (Nintendo, 1984).
[2] http://angel2d.com/.

Table 1. Independent variables for user study.

Speed (pixels/second)	
WPI: Slow	150
WPI: Medium	300
WPI: Fast	450
WO: Slow	550
WO: Medium	1100
WO: Fast	1550

(a) Target speeds.

Delay (milliseconds)
0,   25,   50,   75
100, 125, 150, 175
200
300
400

(b) Added mouse input delays.

Both game versions add a controlled amount of delay selected from a set of 11 possible values, shown in Table 1(b). The delay is added to all mouse movements and button clicks for the duration of the round. Each delay and speed combination is presented to each user 5 times to control for variation, and all combinations and repetitions are shuffled so as to appear in a random order and prevent learning effects.

Every 30 rounds, the game stops for a minimum of 20 s to allow the user to rest his/her visual focus and regain concentration, displayed as a countdown timer via a popup window.

Puck Hunt runs in fullscreen mode at 1080 p resolution (1920×1080 pixels). The puck is 100 pixels in diameter and the red ball (the mouse cursor) is 25 pixels in diameter.

2.2   User Studies

To test a wide range of delay and speed combinations, experiments were run at our two locations in parallel, Worcester Polytechnic Institute (WPI) and Westerdals-Oslo (WO). Conditions and procedures for the participants were matched as closely as possible, but practicalities necessitate some differences.

At WPI, 32 participants were solicited through advertising via university email lists. Incentives included a \$25 gift card raffle for participating and a \$25 gift card for the user with the highest score. Game development students that participated also received 1 extra point on their final exams. At WO, 52 participants were recruited by approaching students on campus and asking them if they wanted to take part in a research experiment. A bag of candy was offered as incentive.

The experiments were conducted in dedicated computer labs, running up to four participants at a time at WPI, and up to five participants at a time at WO. Once participants were seated at their respective computers, they listened to a scripted brief outlining the study and signed a consent form. Next, participants were asked to make themselves comfortable by adjusting chair height and monitor angle/tilt so as to be looking at the center of the screen. At WO, participants could also adjust desk height. Importantly, participants were encouraged to shift the mouse to whichever hand they preferred. Participants then completed an online questionnaire that assessed potential visual impairments, handedness, computer competence and gaming experience; Table 2 summarizes the most relevant information.

Table 2. Summary of key demographics. Standard deviations in parentheses for mean values. Mean ability is from 1 (low) to 5 (high).

Uni.	Mean age	Gender	Mouse hand	Self-reported mean ability	
				PC	PC game
WPI	20.9 (1.9)	23 ♂, 8 ♀, 1 ?	32 right	3.5 (1.3)	3.5 (1.3)
WO	23.9 (3.5)	45 ♂, 8 ♀	52 right, 1 left	4.1 (1.9)	3.8 (1.0)

Upon completion of the questionnaire, play commenced immediately. The first two rounds were practice trials to familiarize participants with the game and controls – these results were not recorded. Play then proceeded through 5 iterations of shuffled combinations of all the delays Table 1(b) and puck (target) speeds Table 1(a), where WPI used only the slower 3 speeds, and WO only the faster 3 speeds. The forced break was introduced every 30 rounds.

To assess user perception of the experience, participants were asked to "rate the quality of responsiveness of the last round" (1 to 5) once for every combination of delay and puck speed (in total, 33 times).

In total, each participant played 165 recorded rounds, taking a total of 15–30 min.

Note, the delays in Table 1(b) added by Puck Hunt are in addition to any base delays inherent in the computer system. Since such base delays have been shown to be significant [8], we measured the base delay for mouse actions in each lab. At WPI, we used a Blur-busters type technique[3] whereby an instrumented mouse and a high-speed camera were used to precisely measure the delay between a mouse click and visual output. The measurement method was repeated 5 times, resulting in a mean base delay of 100 ms, which is added to all subsequent delay values in the WPI data. At WO, we used an alternate technique whereby a mouse button and a photosensor connected to an oscilloscope provided precise readings of mouse input delay. Again, the measurement method was repeated 5 times, resulting in a mean base delay of 20 ms which is added to all subsequent delay values in the WO data. While the techniques differed at the two universities because of available hardware (e.g., an oscilloscope, a high-speed camera), given that both techniques have milliseconds of precision, the difference in the reported base delays is likely not due to technique.

---

[3] http://www.blurbusters.com/gsync/preview2/.

3   Analysis

This section first analyzes user performance in selecting a moving target using a mouse with delay (Sect. 3.1), derives our model (Sect. 3.2) and provides a brief look at quality of experience (Sect. 3.3).

3.1   User Performance

Participant performance is defined by the time taken to hit the puck (i.e., select the moving target with a mouse). Figure 2 depicts two graphs of selection time versus delay. The x-axes correspond the input delay and the y-axes the median time to select the puck, recorded for all participants. The two graphs use the same data, where the right-hand graph applies logscale to both axes. There are six trend-lines in each graph, one for the three puck speeds tested in the two game versions. From the graphs, there are two clear trends: (1) the time to hit the puck increases with delay, a trend that appears exponential; and (2) the time to hit the puck increases with puck speed, but only for the three fastest pucks – speed has little effect for the three slowest pucks.

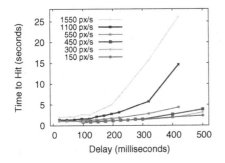 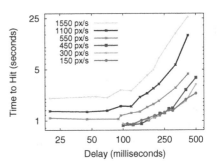

Fig. 2. Median hit time across input delays. Trend-lines are in order of decreasing puck speed, top to bottom. Same data in both graphs, right is logscale.

While the median times illustrate how moving target selection time is affected by delay, modeling from median values does not provide expected (mean) values and is not comparable to other research that models expected values.

Unfortunately, the mean values recorded for fast pucks at high delays are skewed by the 30 s time limit per round. This skew is illustrated by the cumulative distribution functions (CDFs) for hit times across delay in Fig. 3 (left). The x-axis represents the time to hit the puck in seconds and the y-axis the cumulative distribution. Again, the six trend-lines correspond to the puck speeds tested, with the slowest speeds on top and to the left. In this graph, the plotted distributions are only for total delays of about 400 ms.[4] The distributions show

---

[4] Because of differences in local latencies, total delays are 410 ms for the slowest puck speeds and 425 ms for the fastest puck speeds.

clear separation for all trend-lines, with the slower pucks closer together than the faster pucks. The three fastest puck speeds all have maximum scores at 30 s and the two fastest puck speeds have a significant portion of the distribution at 30 s.

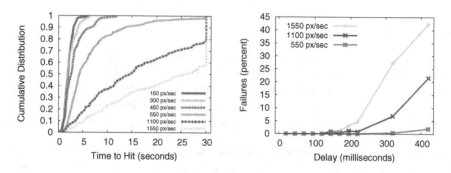

Fig. 3. Distribution of hit time (left). Percentage of failure to hit (right).

While all participants hit the puck within the 30 second time limit for the three slowest puck speeds, this was not the case for the three fastest speeds. Figure 3 (right) illustrates the rate of failure for the three fastest puck speeds. The x-axis corresponds to input delay and the y-axis the percentage of failures (based on the fraction of rounds that participants failed to hit the puck within 30 s). The three trend-lines represent the three fastest speeds, 550 pixels/second (px/s), 1100 px/s and 1550 px/s. For delays under 100 ms, users always hit the puck within 30 s. However, for delays of 125 ms and above, increasingly more users failed to hit the pucks moving at 1100 and 1550 px/s. At the highest added delays, 300 and 400 ms, high puck speeds led to fairly high rates of failure, with a maximum failure percent of more than 40% for the 1550 px/s puck at 400 ms delay. This illustrates that the combination of speed and delay can severely impact selecting a moving target with delay for a range of difficulty (speed) levels.

In order to obtain representative means for puck hit times, specifically for fast speeds and high delays, we model the CDFs of the three fastest puck speeds using only scores below 30 s. This way, the model can be extended past the 30 second limit to provide for a full distribution range and allowing for derivation of a mean. Because CDFs naturally trend to 1, we fit[5] a logarithmic regression model to data points below 30 s for the rounds that had failures and then integrated to yield the means. Table 3 summarizes the results, showing good fits for the analytic models (R^2 0.93 or above).

For illustration, Fig. 4 depicts the CDFs of the 1550 px/s puck speed with the models. The x-axis corresponds to the hit time and the y-axis the cumulative distribution. The six trend-lines correspond to the input delay, beginning with

---

[5] Using R, https://www.r-project.org/.

Table 3. Analytic models for CDFs for speed and delay combinations with failures.

Speed (px/s)	Delay (ms)	Equation	R^2	Mean (ms)
550	420	$0.39 \cdot ln(x) - 0.11$	0.96	6.2
1100	320	$0.34 \cdot ln(x) - 0.09$	0.98	8.2
1100	420	$0.27 \cdot ln(x) - 0.20$	0.94	23.1
1550	195	$0.29 \cdot ln(x) + 0.05$	1.00	7.4
1550	220	$0.30 \cdot ln(x) - 0.07$	0.97	10.3
1550	320	$0.25 \cdot ln(x) - 0.17$	0.94	30.0
1550	420	$0.21 \cdot ln(x) - 0.22$	0.93	66.8

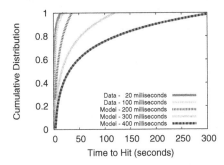

Fig. 4. Distribution of hit time. Trend-lines are in order of increasing delay, left to right. Delays of 220 ms and above are derived from models in Table 3.

the lowest delays on top and to the left. The distribution builds on users' mean data for delays of 20 and 120 ms, while data for 220, 320 and 420 ms delays are derived from the models in Table 3.

Using the mean values both computed and modeled, Fig. 5 presents graphs for mean hit time versus delay. In both graphs, the x-axes correspond to input

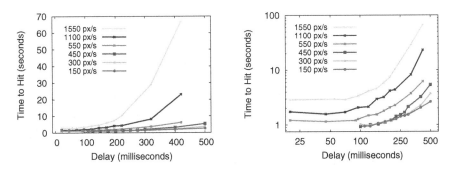

Fig. 5. Mean time to hit puck versus delay. Trend-lines are in order of decreasing puck speed, top to bottom. Same data in both graphs, right is logscale.

delay and the y-axes to the time to hit the puck. The two graphs are based on the same data, with logscale applied to the right-hand graph. The six trend-lines belong to each of the six puck speeds. From the graphs, similar trends hold as for the median graphs (Fig. 2) – an exponential growth in the mean time to hit a puck with delay. Also similar, the mean times show no difference for the three slowest pucks and for delays less than 175 ms. An additional observation relates to the flatness of the mean distributions for short delays – specifically, there is a near-absence of an increase in targeting time for the three fastest pucks with delays below 100 ms. This observation suggests that moving target selection is only impacted by delays once delays exceed 100 ms.

In Fig. 6, the same data are clustered by speed. Here, the x-axis defines the speed, in pixels per second, and the y-axis the time to hit the puck. Both axes are logscale. The six trend-lines in each graph represent six different delay values, ranging from 100 to 420 ms.[6] Delays of 100 and 120 ms correspond to approximately the same total delays, same as 200 and 220 ms and 400 and 420 ms (the discrepancies are attributed to the base delays); the pairs are plotted with the same colors. The graph shows a visual trend with gradually increasing target selection time as puck speeds increase (the left side of the graph), followed by sharper increase for the longer delays (on the right). The overall trend appears exponential with speed.

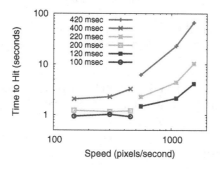

Fig. 6. Mean time to hit puck versus speed. Trend-lines are in order of decreasing delay, top to bottom.

3.2   Model

While trends in user performance with delay provide valuable insights, an analytic model illustrating the relationships can be used by developers and researchers for designing and implementing more effective interactive systems and applications in the presence of delay. Hence, we model the mean time to select a moving target with a delayed mouse.

---

[6] Not all delay values studied are shown in the graph to keep it readable.

Based on the previous analysis (e.g., Fig. 5), there is a clear upward trend in mean time to select a moving target with increased delay, a trend that appears exponential. The trends for mean time to select a moving target with increased speed are less clear, being flat for low delays and speeds (e.g., Fig. 6), but overall also appears exponential. Moreover, previous work has suggested an exponential relationship with speed in the time to select a moving target without delay [3,4]. The greater increase in time to select a moving target at fast speeds and high delays suggests an interaction between speed and delay.

Thus, we propose modeling the time to select a moving target with a mouse (T) with exponential functions for delay (D) and speed (S), as well as an interaction term:

$$T = k_1 + k_2 \cdot e^D + k_3 \cdot e^S + k_4 \cdot e^D \cdot e^S \tag{1}$$

where k_1, k_2, k_3 and k_4 are constants determined empirically through user study. We standardized our user study data[7] using the values in Table 4.

Table 4. User study data values used for standardization.

Factor	Mean	StDev
Delay	206 ms	122
Speed	683 px/s	488

Fitting a regression model to the standardized data (See footnote 5) yields an adjusted R^2 0.99, F-stat 3151 and $p < 2.511 \times 10^{-16}$, and the simplified final model for the time to select a moving target with a delayed mouse (T):

$$T = 1 - 0.7e^d - 0.5e^s + 2e^d e^s \tag{2}$$

where d and s are the standardized delay (D) and speed (S), respectively – $d = \frac{D-206}{122}$ and $s = \frac{S-683}{488}$.

3.3 Quality of Experience (QoE)

Our presented model for moving target selection with a delayed mouse provides insights into how user performance deteriorates with delayed visual feedback, but does not capture the user's quality of experience (QoE) with a delayed system. The user study experiments gathered QoE in the form of users' opinions on the responsiveness from 1 (worst) to 5 (best), which we analyzed and then graphed in Fig. 7. The x-axis corresponds to input delay and the y-axis the QoE. As before, the six trend-lines correspond to the puck speeds. The QoE decreases with increasing delay values without a clear trend with puck speed. These data could indicate a user's QoE for responsiveness is predominantly affected by the mouse delay and not the moving target's speed. However, given the variance in the trend-lines, further study is warranted.

---

[7] Subtracting the mean and dividing by the standard deviation.

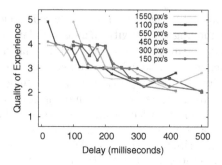

Fig. 7. QoE versus delay.

4 Conclusion

Many multimedia applications involve target selection with a mouse, and in many cases (e.g., computer games) the targets are moving. With the growth in networking and cloud services, applications are increasingly hosted on servers, adding significant delay between user input and rendered results. In order to build systems that accommodate delay, multimedia designers benefit from improved understanding of human performance with delay. This work presents a user study and analysis that provides the foundation for a model of target selection in the presence of delay.

We design and develop a custom game that focuses on moving target selection while allowing explicit control of input delay. The game is the basis for experiments with over 80 participants, providing over 32k game sessions with target speeds from 150–1550 px/s and total delays from 20–500 ms.

Our findings reveal a rise in target selection time that is exponential with both delay and target speed. Slow speeds (below 550 px/s) and low delays (below 100 ms total) have little impact on target selection time. However, delay and speed added to the difficulty of the selection task. Fast speeds (550+ px/s) yielded longer selection times at all delays, while high delays (100+ ms total) extended selection times at all speeds. Moreover, an interaction between speed and delay demonstrates the difficulty in selecting a target that is both severely delayed and moving fast and takes longer than the combination of speed and delay would suggest. Preliminary analysis of QoE indicates that ratings for responsiveness may only be affected by delay, independent of target speed, with further confirmation required.

Based on the trends, we model the time to select a moving target with a delayed mouse with additive exponential terms for delay and speed and a multiplicative interaction term. This last term is critical for accuracy as models tried without delay and speed interaction fit poorly. As proposed, the model has an excellent fit (adjusted R^2 of 0.99) and should be relatively simple to use in multimedia development and assessment.

Applying this work to practical scenarios can validate our model, while delving deeper into user backgrounds, such as age or gaming experience, may provide potential predictors to target selection performance. Future investigations could examine other forms of interactions, input devices, or measures of performance. In some cases, QoE is as important as performance, possibly warranting a model that includes both performance and QoE, as dependent on delay and task difficulty.

References

1. Chen, K.T., Chang, Y.C., Hsu, H.J., Chen, D.Y., Huang, C.Y., Hsu, C.H.: On the quality of service of cloud gaming systems. IEEE Trans. Multimed. **26**(2), 480–495 (2014)
2. Fitts, P.M.: The information capacity of the human motor system in controlling the amplitude of movement. J. Exp. Psychol. **47**(6), 381 (1954)
3. Hajri, A.A., Fels, S., Miller, G., Ilich, M.: Moving target selection in 2D graphical user interfaces. In: Campos, P., Graham, N., Jorge, J., Nunes, N., Palanque, P., Winckler, M. (eds.) INTERACT 2011. LNCS, vol. 6947, pp. 141–161. Springer, Heidelberg (2011). doi:10.1007/978-3-642-23771-3_12
4. Hoffmann, E.: Capture of moving targets: a modification of Fitts' law. Taylor Francis Ergon. **34**(2), 211–220 (1991)
5. Hoffmann, E.: Fitts' law with transmission delay. Taylor Francis Ergon. **35**(1), 37–48 (1992)
6. Jagacinski, R., Repperger, D., Ward, S., Moran, M.: A test of Fitts' law with moving targets. J. Hum. Factors Ergon. Soc. **22**(2), 225–233 (1980)
7. MacKenzie, I., Buxton, W.: Extending Fitts' law to two-dimensional tasks. In: Proceedings of ACM CHI Human Factors in Computing Systems (1992)
8. Raaen, K., Petlund, A.: How much delay is there really in current games? In: Proceedings of the 6th ACM Multimedia Systems Conference (2015)

Multi-attribute Based Fire Detection in Diverse Surveillance Videos

Shuangqun Li, Wu Liu, Huadong Ma$^{(\boxtimes)}$, and Huiyuan Fu

Beijing Key Laboratory of Intelligent Telecommunications Software and Multimedia,
Beijing University of Posts and Telecommunications, Beijing 100876, China
shuangqunli@hotmail.com, {liuwu,mhd,fhy}@bupt.edu.cn

Abstract. Fire detection, as an immediate response of fire accident to avoid great disaster, has attracted many researchers' focuses. However, the existing methods cannot effectively exploit the comprehensive attribute of fire to give satisfying accuracy. In this paper, we design a multi-attribute based fire detection system which combines the fire's color, geometric, and motion attributes to accurately detect the fire in complicated surveillance videos. For geometric attribute, we propose a descriptor of shape variation by combining contour moment and line detection. Furthermore, to utilize fire's instantaneous motion character, we design a dense optical flow based descriptor as fire's motion attribute. Finally, we build a fire detection video dataset as the benchmark, which contains 305 fire and non-fire videos, with 135 very challenging negative samples for fire detection. Experimental results on this benchmark demonstrate that the proposed approach greatly outperforms the state-of-the-art method with 92.30% accuracy and only 8.33% false positives.

Keywords: Fire detection · Geometric attribute · Motion estimation · Contour moment

1 Introduction

Fire is one of the most harmful disasters, which seriously threatens human life and property security in the world. Therefore, the technique of automatic fire detection has attracted massive interests from academic to industrial communities [1]. Some examples of the fire detection applications are listed in Fig. 1. Undoubtedly, these challenging applications depend greatly on the accuracy of the fire detection algorithm. However, the fire detection in real-world scenario still faces the following unique challenges: (1) many natural objects have similar appearances with fire, e.g. the sun, artificial lights, light reflections, moving red objects, which can be often mistakenly detected as fire; (2) modeling of the chaotic and complex nature of the fire phenomenon in video surveillance is difficult [2], but very important for accurate fire detection.

As the above challenges, to the best of our knowledge, the existing fire detection techniques [1,3–5] can seldom achieve satisfactory detection results.

© Springer International Publishing AG 2017
L. Amsaleg et al. (Eds.): MMM 2017, Part I, LNCS 10132, pp. 238–250, 2017.
DOI: 10.1007/978-3-319-51811-4_20

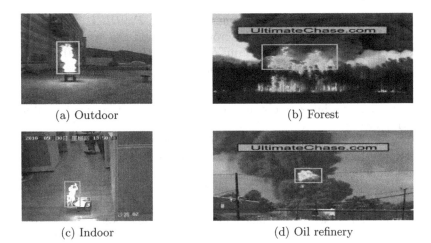

(a) Outdoor   (b) Forest

(c) Indoor   (d) Oil refinery

Fig. 1. Examples of our proposed fire detection methods. Given an input video, our algorithm can detect whether fire is present. Typical scenes: (a) Outdoor; (b) Forest; (c) Indoor; (d) Oil refinery. (Color figure online)

These methods mainly rely on color and shape variation of fire, which are particularly sensitive to the changes in brightness, burning material, and weather conditions. Furthermore, it is hard to distinguish between fire and fire-color objects. Although the motion attribute has been used to detect fire [1,6], the formalization of motion attribute is also very difficult. For example, Foggia et al. used Scale-invariant Feature Transform (SIFT) matching [7] to calculate the motion information. However, it is difficult to find the suitable SIFT points on the fire. Differently, except color and shape attributes, we mainly focus on modeling the contour variation and motion attribute of fire, which can exploit the complementary nature of fire's static and dynamic characters to meet the needs of fire detection applications.

In this paper, we design a Multi-Attribute based Fire Detection (MAFD) system, which combines the fire's color, geometry, and motion attributes to give accurate fire detection in forest, indoor, outdoor, and vehicular environments. Based on YUV color space, we firstly extract color feature. Furthermore, we extract the geometric attribute of fire, which can describe the significant variation of fire appearance using contour moment. In addition, as it is observed that there is no long line segment in the contour of fire, so we detect the line in the contour as a part of geometric descriptor. The proposed geometric descriptor is robust to the changes in brightness, burning material, and weather conditions, which can also help to discriminate fire and nonfire objects. More importantly, to utilize the disordered motion character of fire, we propose a novel descriptor adopting a bag-of-words (BoW) approach to represent the valid directions of optical flow extracted from the moving object. The proposed motion descriptor can explicitly capture instantaneous motions of fire with a much lower feature dimension. Meanwhile, it is also able to well represent the inside and boundary motions of fire. Finally, experiments on the dataset *FireDB* (153 fire and 152 nonfire videos) demonstrate that the proposed method can greatly outperform the state-of-the-art method [1] in terms of accuracy. In summary, the contributions of this paper can be concluded as follows:

(1) We design a fire detection system, which combines fire's color, geometric, and motion attributes to obtain accurate fire detection results in real-world environments;

(2) We propose two novel descriptors for characterizing the geometric and motion attributes of fire, in which one represents the outline of fire using contour moment and line detection, and the other captures the instantaneous motion of fire with dense optical flow to well represent the inside and boundary motions of fire;

(3) We build a more wide and challenging dataset *FireDB* for fire detection. The comprehensive evaluation on this dataset demonstrates that the proposed method can significantly outperform the state-of-the-art method.

2   Related Works

According to the used features, the existing fire detection methods can be divided into three types: color-based [3–5,15,16], shape-based [6,8,9,18–20], and motion-based [1,10,21,22].

The color-based methods can reliably detect fire on condition that fires generated by common combustibles (e.g. wood, fabric) have similar colors. For example, Qi *et al.* combined the RGB color and HSV saturation to distinguish non-fire moving objects from uncontrolled fires [3]. However, these approaches are particularly sensitive to the changes in brightness, thus causing a low accuracy due to the presence of shadows or different tonalities of the red. This problem can be mitigated by switching to a YUV color space. For example, Çelik *et al.* experimentally defined six rules in the YUV space to effectively separate the luminance from the chrominance [4]. Ko *et al.* applied a probabilistic approach based on YUV color space for the thresholding of potential fire pixels [5]. In particular, this paper exploits the combination of six different rules to model the color of fire. However, except the disturbance of light and background, fires generated by different materials also have different colors. Therefore, the color based methods cannot well detect fire in the complicated environments.

The shape-based methods distinguish fire from other fire-colored objects based on the assumption that fire randomly changes its shape in surveillance video. For example, Borges and Izquierdo used area size and boundary roughness for analyzing the shape variation of fire regions [6]. Foggia *et al.* applied the ratio between the perimeter and the area of fire for evaluating shape variation [1]. Demi proposed a spatio-temporal intensity moment which integrated the two basic functions of edge detection and contour tracking [8]. Sun *et al.* presented a novel approach to non-rigid objects contour tracking based on a supervised level set model (SLSM) [9]. Inspired by the above methods, we exploit geometric attribute descriptor for characterizing the large variation of fire appearance.

As for the motion-based methods, many motion descriptors have been proposed in [1,10,21,22]. More specifically, Foggia *et al.* proposed a descriptor based on a BoW model to represent fire motion [1]. In addition, Mueller *et al.* put forward a set of motion features based on optical flow to distinguish the disordered motion of fire and the structured motion of other objects [10]. Töreyin *et al.* estimated flames flicker by analyzing the motion of candidate regions in wavelet domain [11]. In contrast, we propose a novel motion attribute descriptor using the weighted normalized entropy of the optical-flow orientation to characterize both the inside and boundary motions of fire, which provides the significant advantage of lower feature dimension and accurate description.

In summary, compared with the above methods, we aim to provide fire detection method which combines fire's color, geometric, and motion attributes. Furthermore, we propose two novel descriptors to characterize the geometric and motion attributes to explicitly capture fire's instantaneous motion and achieve robust detection results. As a result, the proposed MAFD outperforms the state-of-the-art method in terms of accuracy and false positives.

3 Our Method

Figure 2 shows the architecture of the proposed method MAFD. Initially, we apply the background subtraction method based on the Gaussian Mixture Model (GMM) [12] to detect moving objects in surveillance video. Then a morphological operator is applied to eliminate small noise. For the foreground areas, we label them as independent objects (blobs). After that, three different classifiers respectively based on color, geometric, and motion attributes have been introduced to classify these objects. Finally, the final decisions are taken by the multi-classifier system. In order to avoid depending on sampling rates between two consecutive frames, we collect video frames with 100 ms intervals feeding into the three classifiers. Next, we will describe each part in detail.

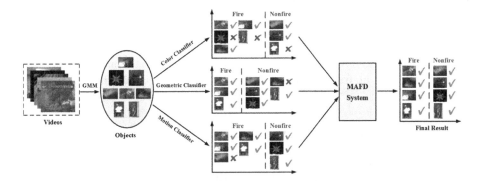

Fig. 2. The overview of our proposed multi-attribute based fire detection system. (Color figure online)

3.1   Color Attribute Based Classifier

This classifier firstly extracts the color attribute from the YUV color space. Here we use YUV because it can separate the luminance from the chrominance, which is less sensitive to the changes in brightness. In particular, to model the color attribute of fire, we combine six different rules, denoted as $r_1^c, \cdots, r_6^c$, in this classifier [4].

Specifically, according to the observation, for fire object, its value in red channel is bigger than green (i.e., r_1^c), as well as the green is greater than the blue (i.e., r_2^c). On the other hand, the red channel value of fire pixel is higher than the mean red channel value. In the YUV space, it implies that for a fire pixel, the value in the Y and V channel is bigger than the mean value in Y and V channel respectively (i.e., r_3^c and r_5^c), while the value in U channel is lower than the mean value in U channel in the frame (i.e., r_4^c). Finally, fire pixels are characterized by a considerable difference between U and V channel value (i.e., r_6^c).

As a result, the color attribute based classifier C_{CA} is taken by combining $r_1^c, \cdots, r_6^c$ in Eq. (1),

$$C_{CA} = \begin{cases} F & if \ r_1^c \wedge r_2^c \wedge r_3^c \wedge r_4^c \wedge r_5^c \wedge r_6^c \\ \overline{F} & otherwise \end{cases}, \tag{1}$$

if all rules are satisfied, then the object is assigned to fire class.

3.2   Geometric Attribute Based Classifier

As mentioned before, only depending on color attribute is not enough to discriminate fire from other fire-color objects. Through our careful observation, we have discovered that the shape of fire always changes very quickly in the video. Therefore, a new contour variation descriptor is introduced to characterize the large variation of fire shapes between two frames. We firstly extract the exterior contour of object i at frame t, then use Hu moment invariants (i.e., the invariance of translation, scale change, mirroring, and rotation) to represent the exterior contours, which are denoted as $Hu_t^i = \{h_t^1, \cdots, h_t^7\}$.

Next, the contour variation is estimated by using the contour difference D_t^i between object i at frame t and the corresponding object at frame $(t-1)$ in Eq. (2),

$$D_t^i = \sum_{k=1}^{7} \left| sign(h_t^k) \cdot \log \left| h_t^k \right| - sign(h_{t-1}^k) \cdot \log \left| h_{t-1}^k \right| \right|, \tag{2}$$

where $sign(\cdot)$ is the sign function, $log(\cdot)$ is the common logarithm.

Furthermore, we have discovered that in most cases: fire has an irregular shape — long line segments are not always appeared in the contours of fire. In particular, the shape of fire in burst also changes quickly in irregular way and is very hard to generate long line segments. Hence, it is feasible to filter out the contours of nonfire video with the rationality of long line segments. The rationality R_t^i of long line segments in the contours of object i is defined as a logical value. If line segments are all shorter than the maximal edge of object i at frame t, R_t^i is set to true, otherwise R_t^i is false.

According to the above observations, the geometric attribute based classifier C_{GA} is finally taken by combining D_t^i with R_t^i. If D_t^i is higher than a given threshold τ_g, R_t^i at frame t and R_{t-1}^i at frame $(t-1)$ are all true, then the object i is assigned to fire class with Eq. (3).

$$C_{GA} = \begin{cases} F & if \ D_t^i > \tau_g \wedge R_t^i \wedge R_{t-1}^i \\ \overline{F} & otherwise \end{cases}. \tag{3}$$

In the implementation, τ_g is set to 3 according to the experiments.

3.3   Motion Attribute Based Classifier

Except color and geometric attributes, motion information of fire can also help us to discriminate fire and nonfire objects. Due to the wind or burning material, the motion of fire is chaotic and unpredictable. By contrast, the rigid or articulated object always moves in limited directions in a while. To effectively discriminate these objects, we extract optical flow and represent the motion attribute as bag of motion words model. The detailed steps are described as follows:

(1) Optical Flow Feature Extraction: Different from [1], we use an efficient dense optical flow model [13] to compute the motion amplitude and orientation of each pixel in object i. For each pixel (x, y) in object i at frame t, the optical flow field is described by $(u_t^i(x, y), v_t^i(x, y))$, where the two composing channel $u_t^i(x, y)$ and $v_t^i(x, y)$ denotes the horizontal and vertical velocity respectively. The size of the optical flow field is the same as object i. For each pixel (x, y) in object i at frame t, its magnitude $L_t^i(x, y)$ and orientation $O_t^i(x, y)$ can be computed via the following equations,

$$L_t^i(x,y) = \sqrt{u_t^i(x,y)^2 + v_t^i(x,y)^2}, \qquad O_t^i(x,y) = \tan^{-1}\left(\frac{v_t^i(x,y)}{u_t^i(x,y)}\right). \tag{4}$$

(2) Motion Descriptor Extraction: To efficiently distinguish the disordered and regular motion, we evenly divide the angle space $O_t^i(x, y)$ into $|S|$ sectors. $|S|$ has been experimentally set to 18 in this paper. Thus each sector covers $20°$, which represents a good tradeoff between a suitable representation of the motion and the immunity to the noise.

First of all, in order to obtain the rapidly changing pixels (valid pixels) between two frames, we filter the pixels of object i using an adaptive mask $M_t^i(x, y)$, which is composed of 0 and 1 (the magnitude is larger than its mean), and its size is the same as object i. The $O_t^i(x, y)$ is an important clue to discriminate fire and nonfire objects. For each object, the orientation histogram with 18 bins is obtained by counting the total number of valid pixels located in each bin, which are denoted as $H_t^i(k), k = 1, \cdots, |S|$.

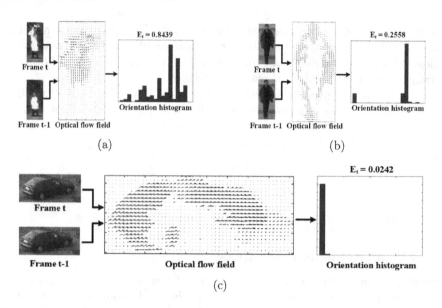

Fig. 3. Optical flow field and histograms of orientation for computing motion attribute. (a) Fire. (b) Pedestrian. (c) Vehicle.

(3) Classification: As the examples shown in Fig. 3, we can build the motion attribute based classification according to the follow assumptions: For fire object shown in Fig. 3(a), its motion is disordered and spreads around the fire region, hence its orientation histogram is dispersed. Differently, for non-fire objects like human and car shown in Fig. 3(b) and (c), their motion is regular and distributes on the boundary, hence its orientation histogram is concentrated in a few bins. Based on the above assumptions, we propose the weighted normalized entropy E_t^i of the optical-flow orientation to represent the homogeneity of the histogram. For object i, the E_t^i is computed by using the following equation,

$$E_t^i = -\frac{N_t^i}{A_t^i} \frac{1}{\log_2 |S|} \sum_{k=1}^{|S|} \left(\frac{H_t^i(k)}{N_t^i} \cdot \log_2 \left(\frac{H_t^i(k)}{N_t^i} \right) \right), \tag{5}$$

where $N_t^i = \sum_{k=1}^{|S|} H_t^i(k)$ is the total number of valid pixels in object i, A_t^i denotes the area of the object i. The weight $\frac{N_t^i}{A_t^i}$ enhances the entropy of fire object, and reduces the entropy of nonfire object. Hence, the E_t^i has more distinguishing ability for fire detection.

As the orientation histogram of fire is dispersed, its entropy is large, as shown in Fig. 3(a). Conversely, the entropy of rigid or articulated object with concentrated histogram is small, as shown in Fig. 3(b) and (c). The property provides a clue to classify: if E_t^i is greater than a given threshold τ_{min} and less than another given threshold τ_{max}, the object is classified as fire.

$$C_{MA} = \begin{cases} F & if \ \tau_{min} < E_t^i < \tau_{max} \\ \overline{F} & otherwise \end{cases}. \tag{6}$$

In our experiments, τ_{min} and τ_{max} is set to 0.6 and 0.95 respectively. In particular, τ_{max} is used for the detection of fire-colored moving objects with flicker effect, such as car lights, reflections, neon lamps and so on.

3.4 Multi-classifier System

The main factor determining the performance of the multi-classifier system is the combination rule. It has been proved that one of the most robust combination rules is the weighted voting rule [14,17].

In a formal way, the generic k-th classifier, $k \in \{CA, GA, MA\}$, assigns the class $c_k(i)$ to object i. The vote of the generic class m can be defined as

$$\delta_{mk}(i) = \begin{cases} 1 & c_k(i) = m \\ 0 & otherwise \end{cases}. \tag{7}$$

Furthermore, the weight w_k is dynamically evaluated by a Bayesian formulation to obtain the highest recognition rate. Given the classification matrix $C^{(k)}$ computed by the k-th classifier on the training step, w_k can be determined by Eq. (8),

$$w_k(m) = P(i \in m | c_k(i) = m) = \frac{C_{mm}^{(k)}}{\sum\limits_{r=1}^{M} C_{rm}^{(k)}}, \tag{8}$$

where M is the number of classes.

The classification results are acquired by maximizing the reliability of the multi-classifier system in recognizing a particular class. In detail, the reliability $\psi(m)$ that the object i belongs to the class m is computed by a weighted sum of the votes,

$$\psi(m) = \frac{\sum_{k \in \{CA,GA,MA\}} \delta_{mk}(i) \cdot w_k(m)}{\sum_{k \in \{CA,GA,MA\}} w_k(m)}. \tag{9}$$

4 Experimental Results

4.1 The Proposed FireDB Dataset

Although fire detection is a very important research topic in computer vision, until now there is no benchmark dataset in this field. To enhance the community in the fire detection research and well evaluate the proposed method, we build a fire detection video dataset named *FireDB* in this paper. The dataset contains 153 fire videos and 152 non-fire videos which are collected from the public video

(a) Outdoor Fire   (b) Outdoor Fire   (c) Indoor Fire

(d) Forest Fire   (e) Outdoor Nonfire   (f) Indoor Nonfire

Fig. 4. Examples of different fire and nonfire scenarios in the *FireDB* dataset. (Color figure online)

sharing dataset. In these videos, there are 33 forest scenes, 103 indoor scenes and 169 outdoor scenes. Besides, it also contains 135 very challenging negative samples such as car lights, human with red clothes, red vehicles, the flame of special materials, flashing red lights, moving smoke, illumination changing, and reflections. Some examples of different scenarios in *FireDB* dataset are shown in Fig. 4. Each video is automatically segmented into many clips in our methods. A video is set to "fire video" if it contains at least one fire clip.

In the experiments, we compare nine different methods on the collected *FireDB* dataset, including Foggia's method implemented in rigidly accordance with the literature [1]. The dataset has been partitioned into two parts: 80% as training and 20% as testing.

4.2   Performance Evaluation

We use the Accuracy, False Positive Rates (FPR), and False Negative Rates (FNR) to evaluate the performance of the proposed MAFD method in this paper. An overview of the performance achieved on *FireDB* are summarized in Table 1.

4.3   Comparison with Other Methods

First of all, we find that the proposed geometric attribute performs obviously better than traditional shape variation based features in accuracy, which demonstrates that it can more effectively capture the discriminative power of shape variation. From the results, we also find that traditional shape variation has high false alarm rate in some complex scenes, such as occlusion of multiple objects, settled fire. Differently, the contour moment invariants and long line detection used in geometric attribute can distinguish these scenes. Compared to the motion estimation in [1], the proposed motion attribute can also obviously improve the

Table 1. Comparison of the fire detection accuracies on the FireDB dataset.

Typology	Method	Accuracy	FPR	FNR
Single classifier	(CA) [4]	84.85%	17.42%	15.15%
	(SV) [1]	61.01%	51.67%	15.15%
	(GA)	72.61%	40.56%	5.03%
	(ME) [1]	66.67%	41.67%	24.24%
	(MA)	76.52%	46.33%	0%
Combined classifier	CA + GA	88.86%	14.64%	12.73%
	CA + MA	90.24%	13.34%	12.85%
	CA + SV + ME [1]	89.96%	11.67%	9.09%
	CA + GA + MA (MAFD)	**92.30%**	**8.33%**	9.09%

detection accuracy by using the global motion information. The reason is that motion estimation uses the SIFT matching to compute the motion information but the suitable SIFT points are very difficult to be detected in the fire. Conversely, the dense optical flow in motion attribute is easier and more accurate to be obtained. Finally, we find that the proposed method MAFD achieves the highest accuracy (92.30%) by combining the color attribute, geometric attribute and motion attribute with the weighted voting rule, which is nearly 2.4% higher than the state-of-the-art method [1].

4.4 Comparison of Different Attribute

Furthermore, we provide a more detailed analysis about the contribution of color attribute, geometric attribute and motion attribute based classifiers on the entire dataset. As shown in Table 1, among the three classifiers employed in this paper, the most important one is still color attribute, which achieves a promising performance on *FireDB*. However, if we only use color attribute for fire detection, we can not correctly discriminate fire from nonfire objects in some special cases. Therefore, we apply the combination of geometric attribute and motion attribute for detecting fire object in these special cases. Our method achieves correct detection results as well, as shown in the first two rows of Fig. 5. Foggia's method [1] achieves incorrect detection results, because motion estimation and shape variation are difficult to effectively characterize the large variation of fire appearance and motion attribute. Compared with geometric attribute and motion attribute, we can find that geometric attribute is not good at detecting the fire with large occlusion, like forest and smoke. Differently, it is feasible to detect fire by using motion attribute in these cases, which are shown in the third row of Fig. 5. Therefore, geometric attribute and motion attribute are able to reduce false positives introduced by color attribute. The consideration is confirmed by filtering false alarms, such as human with red clothes, red vehicles, car lights, reflections. Meanwhile, geometric attribute and motion attribute can also reduce false negatives introduced by color attribute, which are confirmed

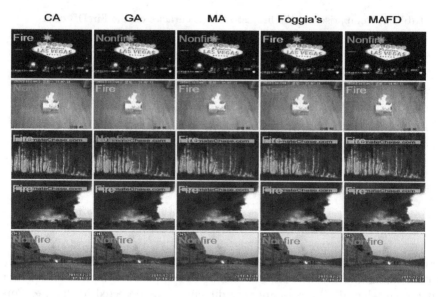

Fig. 5. Typical examples of fire detection results from different methods. (Color figure online)

by correctly identifying the fires caused by special materials, such as gasoline, heptane, and so on. For the common fire, our method and Foggia's method [1] can work well, which are shown in the fourth row of Fig. 5.

However, our approach may yield a few failure cases based on *FireDB*, as shown in the final row of Fig. 5. For some videos that record the burning of special materials (such as gasoline, heptane) from a long distance (i.e. exceeding 30 m), our approach cannot correctly detect the fire objects. Note that in these cases, characterizing the geometric attribute and motion attribute of fire are essential for classifying because the color attribute of flame violates the six rules in Sect. 3. Nevertheless, as the fire areas are too small, the geometric attribute and motion attribute cannot work well. In view of these special cases, our approach still has room for future improvements.

5   Conclusions

In this paper, we design a multi-attribute based fire detection system which combines the color, geometric, and motion attributes with the weighted voting rule to improve the overall fire detection accuracy. In the system, we propose two novel descriptors to characterize the geometric and motion attributes of fire. In addition, a more wide and challenging fire detection dataset is built to evaluate the proposed system. Experimental results on the dataset demonstrate that the proposed approach greatly outperforms the state-of-the-art method in terms of accuracy and false positives. In the future works, we will attempt to apply more effective learning method, such as deep learning, to further improve the fire detection results.

Acknowledgment. This work is supported by the Funds for Creative Research Groups of China (No. 61421061), the Beijing Training Project for the Leading Talents in S&T (ljrc 201502), the National Natural Science Foundation of China (No. 61602049, 61402048), the CCF-Tencent Open Research Fund (No. AGR20160113).

References

1. Foggia, P., Saggese, A., Vento, M.: Real-time fire detection for video-surveillance applications using a combination of experts based on color, shape, and motion. IEEE Trans. Circuits Syst. Video Technol. **25**(9), 1545–1556 (2015)
2. Dimitropoulos, K., Barmpoutis, P., Grammalidis, N.: Spatio-temporal flame modeling and dynamic texture analysis for automatic video-based fire detection. IEEE Trans. Circuits Syst. Video Technol. **25**(2), 339–351 (2015)
3. Qi, X., Ebert, J.: A computer vision-based method for fire detection in color videos. Int. J. Imaging Robot. **2**(S09), 22–34 (2009)
4. Çelik, T., Demirel, H.: Fire detection in video sequences using a generic color model. Fire Saf. J. **44**(2), 147–158 (2009)
5. Ko, B.C., Cheong, K.H., Nam, J.Y.: Fire detection based on vision sensor and support vector machines. Fire Saf. J. **44**(3), 322–329 (2009)
6. Borges, P.V.K., Izquierdo, E.: A probabilistic approach for vision-based fire detection in videos. IEEE Trans. Circuits Syst. Video Technol. **20**(5), 721–731 (2010)
7. Lowe, D.G.: Distinctive image features from scale-invariant keypoints. Int. J. Comput. Vis. **60**(2), 91–110 (2004)
8. Demi, M.: Contour tracking with a spatio-temporal intensity moment. IEEE Trans. Pattern Anal. Mach. Intell. **38**(6), 1141–1154 (2016)
9. Sun, X., Yao, H., Zhang, S., Li, D.: Non-rigid object contour tracking via a novel supervised level set model. IEEE Trans. Image Process. **24**(11), 3386–3399 (2015)
10. Mueller, M., Karasev, P., Kolesov, I., Tannenbaum, A.: Optical flow estimation for flame detection in videos. IEEE Trans. Image Process. **22**(7), 2786–2797 (2013)
11. Töreyin, B.U., Dedeoglu, Y., Güdükbay, U., Çetin, A.E.: Computer vision based method for real-time fire and flame detection. Pattern Recogn. Lett. **27**(1), 49–58 (2006)
12. Liang, C., Juang, C.: Moving object classification using a combination of static appearance features and spatial and temporal entropy values of optical flows. IEEE Trans. Intell. Transp. Syst. **16**(6), 3453–3464 (2015)
13. Sun, D., Roth, S., Black, M.J.: A quantitative analysis of current practices in optical flow estimation and the principles behind them. Int. J. Comput. Vis. **106**(2), 115–137 (2014)
14. Kittler, J.: Combining classifiers: a theoretical framework. Pattern Anal. Appl. **1**(1), 18–27 (1998)
15. Liu, W., Mei, T., Zhang, Y., Che, C., Luo, J.: Multi-task deep visual-semantic embedding for video thumbnail selection. In: CVPR, pp. 3707–3715 (2015)
16. Liu, W., Mei, T., Zhang, Y.: Instant mobile video search with layered audio-video indexing and progressive transmission. IEEE Trans. Multimed. **16**(8), 2242–2255 (2014)

17. Chu, L., Zhang, Y., Li, G., Wang, S., Zhang, W., Huang, Q.: Effective multimodality fusion framework for cross-media topic detection. IEEE Trans. Circuits Syst. Video Technol. **26**(3), 556–569 (2016)
18. Zeng, C., Ma, H.: Robust head-shoulder detection by PCA-based multilevel HOG-LBP detector for people counting. In: ICPR, pp. 2069–2072 (2010)
19. Chu, L., Jiang, S., Huang, Q.: Fast common visual pattern detection via radiate geometric model. In: ICIP, pp. 2465–2468 (2011)
20. Nie, W., Liu, A., Gao, Z., Su, Y.: Clique-graph matching by preserving global & local structure. In: CVPR, pp. 4503–4510 (2015)
21. Liu, A., Su, Y., Nie, W., Kankanhalli, M.: Hierarchical clustering multi-task learning for joint human action grouping and recognition. IEEE Trans. Pattern Anal. Mach. Intell. **38**(6), 1 (2016)
22. Gan, C., Wang, N., Yang, Y., Yeung, D.Y., Hauptmann, A.G.: DevNet: a deep event network for multimedia event detection and evidence recounting. In: CVPR, pp. 2568–2577 (2015)

Near-Duplicate Video Retrieval by Aggregating Intermediate CNN Layers

Giorgos Kordopatis-Zilos[1,2]($\boxtimes$), Symeon Papadopoulos[1], Ioannis Patras[2], and Yiannis Kompatsiaris[1]

[1] Information Technologies Institute, CERTH, Thessaloniki, Greece
{georgekordopatis,papadop,ikom}@iti.gr
[2] Queen Mary University of London, Mile end Campus, London E14NS, UK
i.patras@qmul.ac.uk

Abstract. The problem of Near-Duplicate Video Retrieval (NDVR) has attracted increasing interest due to the huge growth of video content on the Web, which is characterized by high degree of near duplicity. This calls for efficient NDVR approaches. Motivated by the outstanding performance of *Convolutional Neural Networks* (CNNs) over a wide variety of computer vision problems, we leverage intermediate CNN features in a novel global video representation by means of a layer-based feature aggregation scheme. We perform extensive experiments on the widely used CC_WEB_VIDEO dataset, evaluating three popular deep architectures (AlexNet, VGGNet, GoogLeNet) and demonstrating that the proposed approach exhibits superior performance over the state-of-the-art, achieving a mean Average Precision (mAP) score of 0.976.

Keywords: Near-duplicate · Video retrieval · CNNs · Bag of keyframes

1 Introduction

Near-duplicate video retrieval (NDVR) is a research topic of increasing interest in recent years. It is considered essential in a variety of applications that involve video retrieval, indexing and management, video recommendation and search, copy detection and copyright protection. The exponential growth of the Web is accompanied by a proportional increase of video content, typically posted and shared through social media platforms. At the moment, YouTube reports more than one billion users and approximately 500 h of video content is uploaded every minute[1]. This fact renders the NDVR problem extremely important.

NDVR is defined in various ways among the multimedia research community as pointed in [12]. Here, we adopt the definition of Wu et al. [21]: **near-duplicate videos are considered to be identical or close to exact duplicate of each other**, but different in terms of file format, encoding parameters, photometric variations (color, lighting changes), editing operations (caption, logo and border insertion), different lengths, and other modifications.

---

[1] https://www.youtube.com/yt/press/statistics.html (accessed on August 2016).

© Springer International Publishing AG 2017
L. Amsaleg et al. (Eds.): MMM 2017, Part I, LNCS 10132, pp. 251–263, 2017.
DOI: 10.1007/978-3-319-51811-4_21

Motivated by the excellent performance of Convolutional Neural Networks (CNNs) on many computer vision problems, such as image classification, retrieval and object detection, in this paper we propose using intermediate convolutional layers to construct features for NDVR. Although CNN features have been recently used for video retrieval [14,22], it is the first time that **intermediate CNN layers are exploited for NDVR**. A first use of these layers was recently presented on the problem of image retrieval [13,23].

To make use of intermediate convolutional layers for NDVR, we extract layer-level feature descriptors by applying max pooling to the activations of each convolutional layer. In addition, **we propose two layer aggregation techniques,** a first by concatenating the layer vectors in a single vector, and a second by computing layer-specific codebooks and aggregating the resulting bag-of-words representations. Furthermore, **we evaluate three popular deep architectures** [10,16,19] in combination with both layer aggregation schemes by means of a thorough experimental study on an established NDVR dataset (CC_WEB_VIDEO [21]), and **we demonstrate the superior performance of the proposed approach over five state-of-the-art methods**. In particular, the best configuration of the proposed approach achieves a mean Average Precision (mAP) score of 0.976, i.e. a clear improvement over the already high mAP of 0.958 achieved by the Multiple Feature Hashing and Pattern-based approaches of Song et al. [18] and Chou et al. [3], respectively.

2   Related Work

NDVR is a very challenging task, which has attracted increasing research interest in recent years. Liu et al. [12] provide a survey with detailed overviews of the NDVR research problem and a number of recent approaches. These are typically classified based on the level of matching performed to determine the near-duplicate videos: video-level, frame-level and hybrid-level matching.

Video-Level Matching: Here, videos are represented with a global signature such as an aggregate feature vector, a fingerprint or a hash code. Huang et al. [6] proposed a video representation model called Bounded Coordinate System (BCS) which extends Principal Component Analysis (PCA). In [18], Song et al. present an approach for Multiple Feature Hashing (MFH) based on a supervised method that uses multiple image features and learns a group of hash functions that map the video keyframes into the Hamming space. The video signatures are generated by the combination of the keyframe hash codes and they constitute the video representation in the dataset.

Frame-Level Matching: Near-duplicate videos are determined by the comparison between individual frames or sequences of the candidate videos. Douze et al. [4] detect local points of interest, extract the SIFT [11] and CS-LBP [5] descriptors, and create a visual codebook for hamming embedding. Using post-filtering, they verify retrieved matches with spatiotemporal constrains. In [15], Shang et al. introduce compact spatio-temporal features to represent videos and construct a modified inverted file index. The spatio-temporal features are extracted using a feature selection and w-shingling scheme. Cai et al. [2] presented a large-scale approach by applying a scalable K-means clustering technique to learn a visual vocabulary on the color correlograms of a training set of images and using inverted file indexing for fast retrieval of candidate videos.

Hybrid-Level Matching: A typical such approach is [21], where Wu et al. apply a hierarchical filter-and-refine scheme to cluster and filter out near-duplicate videos. When a video cannot be clearly classified as novel or near-duplicate, they apply an expensive local feature-based NDVR scheme. In a more recent approach [3], Chou et al. filter the non near-duplicate videos with a pattern-based indexing tree and rank candidate videos with m-pattern-based dynamic programming and time-shift m-pattern similarity.

The well-known TRECVID copy detection task [9] is also a specific case of NDVR. However, in the TRECVID copy detection task, the duplicates are artificially generated by applying standard transformations, whereas in case of NDVR duplicates correspond to real content.

3   Approach Overview

The proposed NDVR approach leverages features produced by the intermediate CNN layers of deep architectures (Subsect. 3.1) and introduces a layer-based aggregation scheme for deriving a bag-of-word representation for each video (Subsect. 3.2). The bag-of-words representations of videos are stored in an efficient inverted file index, while video retrieval is carried out based on cosine similarity between *tf-idf* weighted versions of the extracted vectors (Subsect. 3.3).

3.1   CNN Based Feature Extraction

In some recent research works [13,23], pre-trained CNN models are adopted to extract visual features from intermediate convolutional layers. These features are computed through the forward propagation of an image over the CNN network and the use of an aggregation function (e.g., VLAD encoding [7], max/average pooling) on every convolutional layer.

We experiment with three deep network architectures: AlexNet [10], VGGNet [16] and GoogLeNet [19]. All three architectures receive images of size 224×224 as input. For all experiments, input images are resized to fit these dimensions.

To extract frame descriptors, we are following the process of [23]. A pre-trained CNN network $\mathcal{C}$ is employed, with a total number of L convolutional layers, denoted as $\mathcal{L}^1, \mathcal{L}^2, ..., \mathcal{L}^L$. Forward propagating an image I through $\mathcal{C}$ generates a total of L feature maps, denoted as $\mathcal{M}^l \in \mathbb{R}^{n_d^l \times n_d^l \times c^l}$ ($l = 1, ..., L$), where $n_d^l \times n_d^l$ is the dimension of every channel for convolutional layer $\mathcal{L}^l$ (which depends on the size of the input image) and c^l is the total number of channels.

To extract a single descriptor vector from every layer, an aggregation function is applied on the above feature maps. In particular, we apply max pooling on every channel of feature map $\mathcal{M}^l$ to extract a single value. The extraction process is formulated in Eq. 1.

$$v^l(i) = \max \mathcal{M}^l(\cdot, \cdot, i), \quad i = \{1, 2, ..., c^l\} \tag{1}$$

where layer vector v^l is a c^l-dimensional vector that is derived from max pooling on every channel of feature map $\mathcal{M}^l$. The layer vectors are L2-normalized to unit length after their extraction.

Table 1. Total number of CNN channels per layer used by the proposed approach for the three selected deep architectures.

(a) AlexNet		(b) VGGNet		(c) GoogLeNet	
Layer $\mathcal{L}^l$	c^l-dim	Layer $\mathcal{L}^l$	c^l-dim	Layer $\mathcal{L}^l$	c^l-dim
conv1	96	conv2_1	128	Inception 3a	256
conv2	256	conv2_2	128	Inception 3b	480
conv3	384	conv3_1	256	Inception 4a	512
conv4	384	conv3_2	256	Inception 4b	512
conv5	256	conv3_3	256	Inception 4c	512
total	1376	conv4_1	512	Inception 4d	528
		conv4_2	512	Inception 4e	832
		conv4_3	512	Inception 5a	832
		conv5_1	512	Inception 5b	1024
		conv5_2	512	total	5488
		conv5_3	512		
		total	4096		

Table 1 depicts the employed CNN architectures and the number of channels in the respective convolutional layers. We extract image descriptors only from the activations in intermediate layers, since we aim to construct a visual representation that preserves local structure in different scales. The fully-connected layer activations are not used. A positive side-effect of this decision is that the resulting descriptor is compact, reducing the total processing time and storage requirements. For the VGGNet and GoogLeNet architectures, we do not use the initial layer activations as features, since those layers are expected to capture very primitive image features (e.g. edges, corners, etc.) that could lead to false matches. For the extraction of the above descriptors we use the Caffe framework [8], which provides pre-trained models on ImageNet for all three CNN networks[2].

3.2   Feature Aggregation

We then follow two alternative feature aggregation schemes (i.e. ways of aggregating features from layers into a single descriptor for the whole frame): (a)

---

[2] https://github.com/BVLC/caffe/wiki/Model-Zoo.

vector aggregation and (b) **layer** aggregation. The outcome of both schemes is a frame-level histogram H_f that is considered as the representation of a frame. Finally, a video-level histogram H_v is derived from the respective keyframe representations by plain summing. Figure 1 gives an overview of the two schemes.

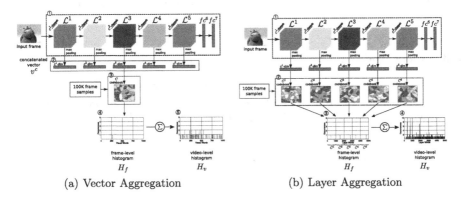

(a) Vector Aggregation   (b) Layer Aggregation

Fig. 1. The two aggregation schemes and the final video representation.

Vector Aggregation. A bag-of-words scheme is applied on the vector v^c resulting from the concatenation of individual layer features to generate a codebook of K visual words, denoted as $C_K = \{t_1, t_2, ..., t_K\}$. The selection of K has critical impact on the performance of the approach and it is considered a system parameter, which is further explored in Sect. 5. Having generated the visual codebook, every video keyframe is assigned to the nearest visual word. Accordingly, every frame f with feature descriptor v_f^c is aggregated to the nearest visual word $t_f = NN(v_f^c)$, hence its H_f contains only a single visual word.

Layer Aggregation. To preserve the structural information of intermediate CNN layers $\mathcal{L}$, we generate L layer-specific codebooks of K words (denoted as $C_K^l = \{t_1^l, t_2^l, ..., t_K^l\}, l = 1, ..., L$), which we then use to extract separate bag-of-words representations (one per layer). The layer vectors v_f^l of frame f are mapped to the nearest layer words $t_f^l = NN(v_f^l)$, $(l = 1, 2, ..., L)$. In contrast to the previous scheme, every frame f is represented by a frame-level histogram H_f that results from the concatenation of the individual layer-specific histograms, therefore comprising L words instead of a single one.

In both schemes, the visual codebooks are generated using scalable K-Means++ [1] on a sample of 100 K randomly selected video frames. The Apache Spark[3] implementation of the algorithm is used for efficiency and scalability.

---

[3] http://spark.apache.org (accessed on August 2016).

Keyframe to Video Aggregation. The final video representation is generated using the bag-of-words histograms of its keyframes. Given a video d with $|F|$ keyframes, $F = \{f_1, f_2, ..., f_F\}$, its video-level histogram H_v is derived by summing the histogram vectors corresponding to its keyframes, i.e. $H_v = \sum_{f_i \in F} H_{f_i}$. Note that for the two aggregation schemes, histograms of different sizes are generated. In the first case, the total number of visual words is K, whereas in the second case it is $K \cdot L$.

3.3 Video Indexing and Querying

We use *tf-idf* weighting to calculate the similarity between two video histograms. The *tf-idf* weights are computed for every visual word in every video in collection D_b based on $w_{td} = n_{td} \cdot \log |D_b| / n_t$, where w_{td} is the weight of word t in video d, n_{td} and n_t are the number of occurrences of word t in video d and the entire collection respectively, while $|D_b|$ is the number of videos in the collection. The former factor of the equation is called *term frequency* (tf) and the latter *inverted document frequency* (idf). The calculation of weights take place in the offline part of the method, i.e. they are not recalculated for every new query.

The feature extraction and aggregation steps for a query video q are the same as the ones described above. Once the final histogram H_v^q is extracted from q, an inverted file indexing structure [17] is used for fast retrieval of videos that have at least a common visual word with the query video. Then, all these videos are ranked in descending order based on their cosine similarity with the query video, computed using the corresponding *tf-idf* representations.

4 Evaluation

4.1 Dataset

Experiments were performed on the CC_WEB_VIDEO dataset [21], which is available by the research groups of City University of Hong Kong and Carnegie Mellon University. The collection consists of a sample of videos retrieved by submitting 24 popular text queries to popular video sharing websites (i.e. YouTube, Google Video, and Yahoo! Video). For every query, a set of video clips was collected and the most popular video was considered to be the query video. Subsequently, all videos in the video set retrieved by the query were manually annotated based on their near-duplicate relation to the query video. Table 2 depicts the types of near-duplicate types and their annotation. In the present work, all videos annotated with any symbol but **X** are considered near-duplicates. The dataset contains a total of 13,129 videos consisting of 397,965 keyframes.

All experiments were carried out on a system with Intel(R) Core(TM) i7-4770K CPU at 3.50 GHz, 16 GB RAM, NVIDIA GTX 980 GPU and 64-bit Ubuntu 14.04 operating system.

Table 2. Type of transformation.

Annotation	Transformation
E	Exactly duplicate
S	Similar video
V	Different version
M	Major change
L	Long version
X	Dissimilar video

4.2 Evaluation Metrics

To measure detection accuracy, we employ the interpolated precision-recall (PR) curve. Precision is determined as the fraction of retrieved videos that are relevant to the query, while recall is the fraction of the total relevant videos that are retrieved. We further use mean average precision (mAP) as defined in [21] and in Eq. 2, where n is the number of relevant videos to the query video, and r_i is the rank of the i-th retrieved relevant video.

$$AP = \frac{1}{n} \sum_{i=0}^{n} \frac{i}{r_i} \qquad (2)$$

4.3 Competing Approaches

In Sect. 5.4, we compare the proposed approach with five widely used content-based NDVR approaches.

Color Histograms (CH) - Wu et al. [21] generated a global video representation based on the color histograms of keyframes. The color histogram is a concatenation of 18 bins for Hue, 3 bins for Saturation, and 3 bins for Value, resulting in a 24-dimensional vector representation for every keyframe. The global video signature is the normalized color histogram over all keyframes in the video.

Auto Color Correlograms (ACC) - Cai et al. [2] used uniform sampling to extract one frame per second for the input video. The auto-color correlograms of each frame are computed and aggregated based on a visual codebook generated from a training set of video frames. The retrieval of near-duplicate videos is performed using tf-idf weighted cosine similarity over the visual word histograms of a query and a dataset video.

Local Structure (LS) - Wu et al. [21] combined global signatures and local features in a hierarchical method. Color signatures are employed to detect near-duplicate videos with high confidence and to filter out very dissimilar videos. For the rest of videos, a local feature based method was developed, which compares the keyframes in a sliding window using their local features (PCA-SIFT).

Multiple Feature Hashing (MFH) - Song et al. [18] exploited multiple image features to learn a group of hash functions that project the video keyframes into the Hamming space. The combination of the keyframe hash codes generates a video signature which constitutes the video representation in the dataset. Hamming distance is employed to determine similarity between candidate videos.

Pattern-based approach (PPT) - Chou et al. [3] built a pattern-based indexing tree (PI-tree) based on a sequence of symbols encoded from keyframes, which facilitates the efficient retrieval of candidate videos. They used m-pattern-based dynamic programming (mPDP) and time-shift m-pattern similarity (TPS) to determine video similarity.

5 Experiments

5.1 Impact of CNN Architecture and Vocabulary Size

In this section, we study the performance of the proposed approach in the CC_WEB_VIDEO dataset in relation to the underlying CNN architecture and the size of the visual vocabulary.

Regarding the first aspect, three CNN architectures are tested: AlexNet, VGGNet and GoogLeNet, with both aggregation schemes implemented using $K = 1000$ words. Figure 2 illustrates the PR curves of the different CNN architectures with the two aggregation schemes. Layer-based aggregation runs outperform vector-based ones for every architecture. GoogLeNet achieves the best results for the vector-based aggregation experiments with a precision close to 100% up to a 70% recall. For recall values in the range 80%–100%, all three architectures have similar results. For the layer-based aggregation scheme, all three architectures exhibit near perfect performance up to 75% recall.

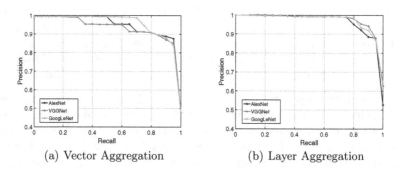

(a) Vector Aggregation   (b) Layer Aggregation

Fig. 2. Precision-recall curve of the proposed approach based on three CNN architectures and for the two aggregation schemes.

Similar conclusions are obtained from the analysis of mAP achieved using different CNN architectures, as depicted in Table 3. For the vector-based aggregation experiments, GoogLeNet achieved the best performance with a mAP of

0.958, and VGGNet the worst (mAP = 0.937). On the other hand, when using the layer-based aggregation scheme, the best mAP score (0.976) was based on VGG-Net. The lowest, yet competitive results in the case of layer-based aggregation, are obtained for AlexNet (mAP = 0.969).

Table 3. mAP per CNN architecture and aggregation scheme.

Method	K	AlexNet	VGGNet	GoogLeNet
Vector aggregation	1000	0.951	0.937	**0.958**
	10,000	0.879	0.886	0.857
Layer aggregation	1000	0.969	**0.976**	0.974
	10,000	0.948	0.959	0.958

To study the impact of vocabulary size, we compare the two schemes when used with $K = 1000$ and $K = 10,000$ (Table 3). Results reveal that the performance of vector-based aggregation for $K = 10,000$ is significantly lower compared to the case when $K = 1000$ words are used. It appears that the vector-based aggregation suffers considerably more from the increase of K compared to the layer-based aggregation, which appears to be less sensitive to this parameter. Due to this fact, we did not consider to use the same amount of visual words for the vector-based and the layer-based aggregation, since the performance gap between the two types of aggregation with the same number of visual words would be much more pronounced.

5.2 Performance Using Individual Layers

We also assessed the retrieval capability of every layer for the three tested CNN architectures. Figure 3 depicts the mAP of the approach using only a selected layer vector. In the AlexNet and VGGNet architectures, the mAP of the first layers are quite low and as we are moving to deeper layers, the retrieval performance improves. In both cases, there are several layers that exceed the performance of the vector-based aggregation scheme. This indicates that it is better to extract

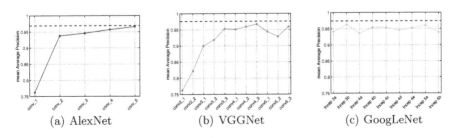

(a) AlexNet   (b) VGGNet   (c) GoogLeNet

Fig. 3. mAP of every layer for the three architectures.

the feature descriptors only from one layer than concatenating all layers in a single vector. However, no single layer overpasses the performance of the layer-based aggregation scheme, displayed with a dashed line. In GoogLeNet, the first layer (Inception 3a) is already deep enough to achieve competitive performance. In this case, the performance for all layers fluctuates between 0.935 and 0.960.

5.3   Performance per Query

Here, we analyze the performance of the best vector-based aggregation instance (GoogLeNet) with the best layer-based aggregation instance (VGGNet) on different queries. Figure 4 displays their Average Precision per query. Layer aggregation outperforms vector aggregation for every single query. However, both approaches fail in the difficult queries of the dataset, namely query 18 (Bus uncle) and query 22 (Numa Gary). The major factor leading to errors is that both videos have relatively low resolution/quality and the candidate videos are heavily edited, which leads to a significant number of relevant videos not to be retrieved at all (i.e. many false negatives). Nevertheless, CNN-L leads to considerably better results in both queries in comparison to CNN-V (Fig. 4).

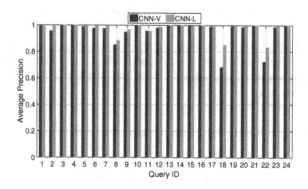

Fig. 4. Average precision/query for GoogLeNet (CNN-V) and VGGNet (CNN-L).

Table 4. Comparison between our approach and existing approaches.

Method	CH	ACC	LS	MFH	PPT	CNN-V	CNN-L
mAP	0.892	0.944	0.952	0.958	0.958	0.958	**0.976**

5.4   Comparison Against Existing NDVR Approaches

For comparing the performance of our approach with the five NDVR approaches from the literature, we select the same runs as in the previous section. The numeric data for the interpolated PR curves of the CH, LS and PPT methods

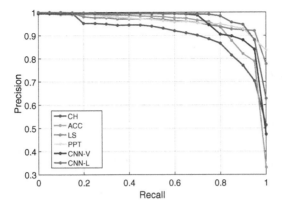

Fig. 5. Precision-recall curve comparison of two versions of the proposed approach with four state-of-the-art methods.

on the CC_WEB_VIDEO dataset were provided by the authors of [3,21], respectively. For the ACC method, we developed our own implementation, which was fine-tuned on the dataset.

Figure 5 illustrates the PR curves of the compared approaches. CNN-L outperforms all other methods up to 90% of recall, at which point the LS and PPT methods start outperforming it. Additionally, CNN-V is at the same level with CNN-L up to 70%, after which it starts performing worse. It is noteworthy that the approaches based on the bag-of-word scheme have low precision at high values of recall (>90%). In terms of mAP, both versions of the proposed approach are competitive in comparison to the state-of-the-art, as attested by Table 4. CNN-L achieves the best score (mAP = 0.976), followed by CNN-V, MFH and PPT (mAP = 0.958).

6 Conclusions and Future Work

We presented a new video-level representation for Near-Duplicate Video Retrieval, which leverages the effectiveness of CNN features and a newly introduced layer-based aggregation scheme that exhibited considerably improved performance over five popular approaches on the CC_WEB_VIDEO dataset in terms of Precision-Recall and mAP.

In the future, we plan to apply the necessary modifications to our method to exploit the use of generic C3D features [20]. Furthermore, we are going to conduct more comprehensive evaluations of the method using more challenging datasets, and we will also assess the applicability of the approach on the problem of Partial Duplicate Video Retrieval (PDVR).

Acknowledgement. This work is supported by the InVID project, partially funded by the European Commission under contract numbers 687786.

References

1. Bahmani, B., Moseley, B., Vattani, A., Kumar, R., Vassilvitskii, S.: Scalable k-means++. Proc. VLDB Endow. **5**(7), 622–633 (2012)
2. Cai, Y., Yang, L., Ping, W., Wang, F., Mei, T., Hua, X.S., Li, S.: Million-scale near-duplicate video retrieval system. In: Proceedings of the 19th ACM International Conference on Multimedia, pp. 837–838 (2011)
3. Chou, C.L., Chen, H.T., Lee, S.Y.: Pattern-based near-duplicate video retrieval and localization on web-scale videos. IEEE Trans. Multimed. **17**(3), 382–395 (2015)
4. Douze, M., Jegou, H., Schmid, C.: An image-based approach to video copy detection with spatio-temporal post-filtering. IEEE Trans. Multimed. **12**(4), 257–266 (2010)
5. Heikkila, M., Pietikainen, M., Schmid, C.: Description of interest regions with local binary patterns. Pattern Recogn. **42**(3), 425–436 (2009)
6. Huang, Z., Shen, H.T., Shao, J., Zhou, X., Cui, B.: Bounded coordinate system indexing for real-time video clip search. ACM Trans. Inf. Syst. **27**(3), 17 (2009)
7. Jègou, H., Douze, M., Schmid, C., Prez, P.: Aggregating local descriptors into a compact image representation. In: Proceedings of the IEEE Conference on Computer Vision and Pattern Recognition, pp. 3304–3311 (2010)
8. Jia, Y., Shelhamer, E., Donahue, J., Karayev, S., Long, J., Girshick, R., Guadarrama, S., Darrell, T.: Caffe: convolutional architecture for fast feature embedding. In: Proceedings of the 22nd ACM International Conference on Multimedia, pp. 675–678 (2014)
9. Kraaij, W., Awad, G.: TRECVID 2011 content-based copy detection: task overview. In: Online Proceedings of TRECVid 2010 (2011)
10. Krizhevsky, A., Sutskever, I., Hinton, G.E.: ImageNet classification with deep convolutional neural networks. In: Advances in Neural Information Processing Systems, pp. 1097–1105 (2012)
11. Lowe, D.: Distinctive image features from scale-invariant keypoints. Int. J. Comput. Vis. **60**(2), 91–110 (2004)
12. Liu, J., Huang, Z., Cai, H., Shen, H.T., Ngo, C.W., Wang, W.: Near-duplicate video retrieval: current research and future trends. ACM Comput. Surv. **45**(4), 44 (2013)
13. Ng, J.Y.H., Yang, F., Davis, L.S.: Exploiting local features from deep networks for image retrieval. In: Proceedings of the IEEE CVPR Workshops, pp. 53–61 (2015)
14. Razavian, A.S., Azizpour, H., Sullivan, J., Carlsson, S.: CNN features off-the-shelf: an astounding baseline for recognition. In: Proceedings of the IEEE CVPR Workshops, pp. 806–813 (2014)
15. Shang, L., Yang, L., Wang, F., Chan, K.P., Hua, X.S.: Real-time large scale near-duplicate web video retrieval. In: Proceedings of the 18th ACM International Conference on Multimedia, pp. 531–540 (2010)
16. Simonyan, K., Zisserman, A.: Very deep convolutional networks for large-scale image recognition (2014). arXiv preprint: arXiv:1409.1556
17. Sivic, J., Zisserman, A.: Video Google: a text retrieval approach to object matching in videos. In: Proceedings of Ninth IEEE International Conference on Computer Vision, pp. 1470–1477 (2003)
18. Song, J., Yang, Y., Huang, Z., Shen, H.T., Luo, J.: Effective multiple feature hashing for large-scale near-duplicate video retrieval. IEEE Trans. Multimed. **15**(8), 1997–2008 (2013)

19. Szegedy, C., Liu, W., Jia, Y., Sermanet, P., Reed, S., Anguelov, D., Erhan, D., Vanhoucke, V., Rabinovich, A.: Going deeper with convolutions. In: Proceedings of the IEEE Conference on Computer Vision and Pattern Recognition, pp. 1–9 (2015)
20. Tran, D., Bourdev, L., Fergus, R., Torresani, L., Paluri, M.: Learning spatiotemporal features with 3D convolutional networks. In: Proceedings of the IEEE International Conference on Computer Vision, pp. 4489–4497 (2015)
21. Wu, X., Hauptmann, A.G., Ngo, C.W.: Practical elimination of near-duplicates from web video search. In: Proceedings of the 15th ACM International Conference on Multimedia, pp. 218–227 (2007)
22. Xu, Z., Yang, Y., Hauptmann, A.G.: A discriminative CNN video representation for event detection. In: Proceedings of the IEEE Conference on Computer Vision and Pattern Recognition, pp. 1798–1807 (2014)
23. Zheng, L., Zhao, Y., Wang, S., Wang, J., Tian, Q.: Good practice in CNN feature transfer (2016). arXiv preprint: arXiv:1604.00133

No-Reference Image Quality Assessment Based on Internal Generative Mechanism

Xinchun Qian, Wengang Zhou[⊠], and Houqiang Li

CAS Key Laboratory of Technology in Geo-Spatial Information Processing and
Application System, University of Science and Technology of China, Hefei, China
qxc@mail.ustc.edu.cn, {zhwg,lihq}@ustc.edu.cn

Abstract. No-reference (NR) image quality assessment (IQA) aims to
measure the visual quality of a distorted image without access to its non-
distorted reference image. Recent neuroscience research indicates that
human visual system (HVS) perceives and understands perceptual signals with an internal generative mechanism (IGM). Based on the IGM,
we propose a novel and effective no-reference IQA framework in this
paper. First, we decompose an image into an orderly part and a disorderly one using a computational prediction model. Then we extract the
joint statistics of two local contrast features from the orderly part and
local binary pattern (LBP) based structural distributions from the other
part, respectively. And finally, two groups of features extracted from the
complementary parts are combined to train a regression model for image
quality estimation. Extensive experiments on some standard databases
validate that the proposed IQA method shows highly competitive performance to state-of-the-art NR-IQA ones. Moreover, the proposed metric
also demonstrates its effectiveness on the multiply-distorted images.

Keywords: No-reference image quality assessment · Internal generative
mechanism · Joint statistics · Local binary pattern

1 Introduction

With the exponential growth of image and video data that is being captured,
stored, transmitted and received, image quality assessment (IQA) has gained
tremendous significance as a research field in the past several years [14].

Perceptual image quality assessment can be classified into two fundamental
groups: subjective assessment by humans, and objective assessment by designed
algorithms [31]. Subjective assessment is time-consuming and very expensive.
Thus, it is not realistic in many applications. The purpose of objective IQA
is to automatically predict the image quality that is consistent with human
perception [1]. Hence, only objective IQA will be discussed in the rest of the
paper.

According to the existence of reference images, IQA algorithms can be categorized into three classes: full-reference IQA (FR-IQA), reduced-reference IQA
(RR-IQA), and no-reference IQA (NR-IQA). FR-IQA approaches [19,23,32]

© Springer International Publishing AG 2017
L. Amsaleg et al. (Eds.): MMM 2017, Part I, LNCS 10132, pp. 264–276, 2017.
DOI: 10.1007/978-3-319-51811-4_22

require the original reference images to predict perceptual quality of a distorted image. And for RR-IQA, partial information of the reference image is necessary for prediction. In contrast, there is no need of any reference information for NR-IQA.

NR-IQA methods [11,18,22,27,34] have been extensively studied, since the original images are unavailable in most cases. Many existing NR-IQA metrics are based on Natural Scene Statistics (NSS). NSS-based approaches depend on an assumption that natural image scenes contain certain statistical properties that will be altered by the existence of distortion. In [15,16], Moorthy and Bovik extracted NSS features in the wavelet domain, and support vector machine (SVM) was used for model training and final quality estimation. In [18], Saad *et al.* trained a probabilistic model based on contrast and statistical features in the DCT domain. And an NSS-based NR-IQA algorithm that works in the spatial domain was proposed in [13]. In addition to NSS-based methods, many training-based NR-IQA approaches were also proposed. Ye and Doermann constructed a visual codebook consisting of Gabor-filter-based local features in [29]. This codebook was further utilized to yield an estimation of quality. And in [30], Ye *et al.* utilized raw-image-patches extracted from a set of unlabelled images to learn a dictionary in an unsupervised manner.

Recent proposed Bayesian brain theory [3,8] in neuroscience shows that the brain works with an internal generative mechanism (IGM) [3,4,24] for visual information perception and understanding [24]. The brain works as an inference system that actively predicts the visual sensation and avoids the residual disorder [8,24]. Thus in this work, we propose a novel and effective NR-IQA model based on IGM. We utilize a Bayesian prediction model [8,24] to decompose an input image into orderly and disorderly parts. For the orderly part, we extract the joint statistics of two local contrast features. And for the disorderly part, we extract LBP distributions as features. At last, we combine all features extracted from the two parts for further model learning.

The rest of this paper is organized as follows. In Sect. 2, we introduce our framework in details. Section 3 conducts extensive experiments on benchmark databases to validate the proposed NR-IQA model. Finally, the paper is concluded in Sect. 4.

2 The Proposed No-Reference IQA Model

In this section, we will introduce the proposed NR-IQA model. Our image quality estimation framework is shown in Fig. 1. An input image is firstly decomposed into orderly and disorderly parts with an autoregressive (AR) model [4,35] based on IGM [3,8]. Then we extract the joint statistics of two local contrast features, i.e., the gradient magnitude (GM) map and the Laplacian of Gaussian (LOG) response [27], from the orderly content. And we utilize LBP distributions [17] to encode structural information from the disorderly one. At last, two groups of features mentioned above are combined for the final image quality prediction.

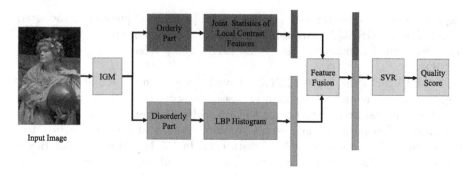

Fig. 1. The framework of the proposed no-reference image quality assessment model based on IGM.

2.1   IGM Based Image Decomposition

We follow the Bayesian prediction based autoregressive (AR) model used in [24, 26] for image content inference and further image decomposition. According to recent human brain research [3,8], human visual system (HVS) works as an active inference system with an internal generative mechanism (IGM) [3,4,24]. And a Bayesian brain theory is introduced to imitate the inference procedure of the IGM [3,8]. The essential of the Bayesian brain theory is a Bayesian probabilistic model that optimizes the input scene by minimizing the prediction error [24,25].

For instance, given an input image, the value of a pixel x is predicted by maximizing the conditional probability $p(x|\mathcal{X})$ for error minimization according to the correlation between the central pixel x and its neighbors $\mathcal{X}$ [24,25]. By further analyzing the relationship between x and its neighbors x_i in $\mathcal{X}$, we take the mutual information $I(x; x_i)$ as the AR coefficient. Then we build an AR model to predict the value of the pixel x [24,35],

$$f'(x) = \sum_{x_i \in \mathcal{X}} \mathcal{C}_i f(x_i) + \epsilon,\tag{1}$$

where $f'(x)$ is the predicted value of pixel x, $f(x_i)$ is the value of its neighbor x_i, $\mathcal{C}_i = I(x; x_i) / \sum_k I(x; x_k)$ is the normalized coefficient, and ϵ is the white noise.

With Eq. 1, the visual contents of an input scene are actively predicted. As shown in Fig. 2, an input image $\mathcal{I}$ is decomposed into an orderly part $\mathcal{I}_o$, which is the predicted portion, and a disorderly one $\mathcal{I}_d$, which is the residual information (i.e., prediction error $\mathcal{I} - \mathcal{I}_o$). Since the two parts contain different visual information and have different impacts towards the perceptual quality, we extract features from them, respectively.

2.2   The Joint Statistics Extraction

We adopt the joint statistics of the gradient magnitude (GM) map and the Laplacian of Gaussian (LOG) response [27] for the orderly part $\mathcal{I}_o$, considering

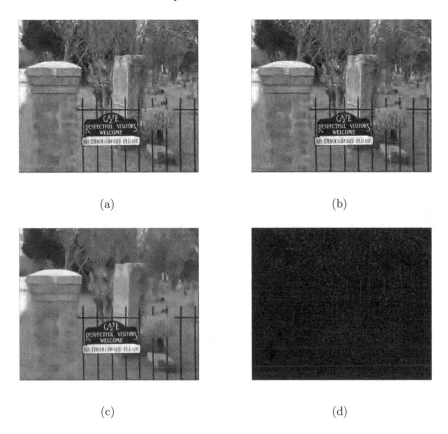

(a)   (b)

(c)   (d)

Fig. 2. Image decomposition with the AR model. (a) Is an image contaminated by JPEG2k distortion (from the commonly used LIVE database [20]), (b) Is the input grayscale image $\mathcal{I}$ simply converted from (a), (c) Is the orderly part $\mathcal{I}_o$ of $\mathcal{I}$, and (d) Is the disorderly one $\mathcal{I}_d$ of $\mathcal{I}$.

that image local contrast features convey significant structural information. In this work, we just follow the implementation steps proposed in [27] to extract four joint statistical features. However, these features are extracted not from the original image but from its orderly part.

For $\mathcal{I}_o$, its GM map $\mathcal{G}_{I_o}$ and LOG map $\mathcal{L}_{I_o}$ can be easily computed, respectively. Both GM and LOG operations could reduce a certain amount of of image spatial redundancies, but some correlations between neighboring pixels will still remain. Devisive normalization models have been developed to further reduce local correlations [12,27]. We normalize the GM and LOG coefficients jointly to obtain more robust image representations. As in [27], let

$$\mathcal{F}_{I_o}(i,j) = \sqrt{\mathcal{G}_{I_o}^2(i,j) + \mathcal{L}_{I_o}^2(i,j)}. \tag{2}$$

Then we compute a locally adaptive normalization factor at each location (i, j):

$$\mathcal{N}_{I_o}(i, j) = \sqrt{\sum\sum_{(l,k)\in\Omega_{i,j}} w(l, k)\mathcal{F}_{I_o}^2(l, k)}, \tag{3}$$

where $\Omega_{i,j}$ is a local window centered at (i, j), $w(l, k)$ are positive weights satisfying $\Sigma_{l,k}w(l, k) = 1$ [27]. The GM and LOG feature maps are further normalized as:

$$\bar{\mathcal{G}}_{I_o} = \mathcal{G}_{I_o}/(\mathcal{N}_{I_o} + C), \tag{4}$$

$$\bar{\mathcal{L}}_{I_o} = \mathcal{L}_{I_o}/(\mathcal{N}_{I_o} + C), \tag{5}$$

where C is a small positive constant to avoid numerical instability. As in [27], we call the above normalization step joint adaptive normalization (JAN). After employing JAN, $\bar{\mathcal{G}}_{I_o}$ is quantized into M levels $\{g_1, g_2, ..., g_M\}$ and $\bar{\mathcal{L}}_{I_o}$ into N levels $\{l_1, l_2, ..., l_N\}$. Now, we can get the joint empirical probability function of $\bar{\mathcal{G}}_{I_o}$ and $\bar{\mathcal{L}}_{I_o}$

$$\mathcal{K}_{m,n} = P(\bar{\mathcal{G}}_{I_o} = g_m, \bar{\mathcal{L}}_{I_o} = l_n), m = 1, ..., M; n = 1, ..., N. \tag{6}$$

According to [27], the marginal probability functions of $\bar{\mathcal{G}}_{I_o}$ and $\bar{\mathcal{L}}_{I_o}$, denoted by P_G and P_L, are more straightforward choices:

$$\begin{cases} P_G(\bar{\mathcal{G}}_{I_o} = g_m) = \sum_{n=1}^{N}\mathcal{K}_{m,n} \\ P_L(\bar{\mathcal{L}}_{I_o} = l_n) = \sum_{m=1}^{M}\mathcal{K}_{m,n}. \end{cases} \tag{7}$$

Besides the marginal probability functions P_G and P_L, the dependencies between GM and LOG are also used [27]. First, we define the following index to measure the dependency:

$$D_{m,n} = \frac{\mathcal{K}_{m,n}}{P_G(\bar{\mathcal{G}}_{I_o} = g_m) \times P_L(\bar{\mathcal{L}}_{I_o} = l_n)}. \tag{8}$$

Then define the following measure of the overall dependency of each specific value $\bar{\mathcal{G}}_{I_o} = g_m$ against all possible values of $\bar{\mathcal{L}}_{I_o}$:

$$Q_G(\bar{\mathcal{G}}_{I_o} = g_m) = P_G(\bar{\mathcal{G}}_{I_o} = g_m) \times \frac{1}{N}\sum_{n=1}^{N} D_{m,n}. \tag{9}$$

And similarly:

$$Q_L(\bar{\mathcal{L}}_{I_o} = l_n) = P_L(\bar{\mathcal{L}}_{I_o} = l_n) \times \frac{1}{M}\sum_{m=1}^{M} D_{m,n}. \tag{10}$$

In summary, we extract the joint statistical features P_G, P_L, Q_G and Q_L from the orderly part $\mathcal{I}_o$ to build the NR-IQA model.

2.3 LBP Based Structure Extraction

Since the local binary pattern (LBP) [17] can be considered as the binary approximation of the statistics of local structure primitives in the early stage of HVS [33], we choose to extract LBP histograms from the disorderly part $\mathcal{I}_d$ to represent its structural information. The LBP value of a pixel x_c is computed as the difference with its circularly symmetric neighborhood x_i [17]:

$$LBP_{P,R} = \sum_{i=0}^{P-1} s(f(x_i) - f(x_c))2^i, \tag{11}$$

$$s(f(x_i) - f(x_c)) = \begin{cases} 1, & f(x_i) - f(x_c) > 0 \\ 0, & f(x_i) - f(x_c) < 0, \end{cases} \tag{12}$$

where P is the number of neighbors, R is the radius of the neighborhood, $f(x_c)$ is the value of the central pixel x_c and $f(x_i)$ is the value of its neighbor x_i.

To reduce the effect of rotation, rotation invariant local binary pattern $LBP_{P,R}^{ri}$ can be defined as [17]:

$$LBP_{P,R}^{ri} = min\{ROR(LBP_{P,R}, i) \mid i = 0, 1, ..., P - 1\}, \tag{13}$$

where ri is short for rotation invariant, and $ROR(x, i)$ performs a circular bit-wise right shift on the P-bit number x i times. $LBP_{P,R}^{ri}$ quantifies the occurrence statistics of individual rotation invariant patterns corresponding to certain structural information in the image. Rotation invariant LBP histograms are extracted from the disorderly part $\mathcal{I}_d$ for further model learning.

2.4 Image Quality Estimation

After all features extracted, we concatenate them for final image quality prediction. In this work, support vector regression (SVR) [21] is employed to learn the regression model. Given training data $\{(\boldsymbol{x_1}, y_1), (\boldsymbol{x_2}, y_2), ..., (\boldsymbol{x_n}, y_n)\}$, where $\boldsymbol{x_i}$ is the feature vector and y_i is the target output, we aim to find a mapping from feature space to perceptual quality measure. SVR model can be represented as [21]:

$$\min_{\boldsymbol{\omega}, b, \boldsymbol{\xi}, \boldsymbol{\xi}^*} \quad \frac{1}{2}\boldsymbol{\omega}^T\boldsymbol{\omega} + C\{\sum_{i=1}^{n}\xi_i + \sum_{i=1}^{n}\xi_i^*\}$$

$$s.t. \quad \begin{cases} \boldsymbol{\omega}^T\phi(\boldsymbol{x_i}) + b - y_i \leq \epsilon + \xi_i, \\ y_i - \boldsymbol{\omega}^T\phi(\boldsymbol{x_i}) - b \leq \epsilon + \xi_i^*, \\ \xi_i, \xi_i^* \geq 0, \ i = 1, 2, ..., n, \end{cases} \tag{14}$$

where $\boldsymbol{\omega}$ is the weight vector, b is the bias parameter, $\boldsymbol{\xi}$ and $\boldsymbol{\xi}^*$ are the slack variables, C is a hyper parameter and $K(\boldsymbol{x_i}, \boldsymbol{x_j}) = \phi(\boldsymbol{x_i})^T\phi(\boldsymbol{x_j})$ is the kernel function. The radial base function (RBF) kernel $K(\boldsymbol{x_i}, \boldsymbol{x_j}) = exp(-\gamma||\boldsymbol{x_i} - \boldsymbol{x_j}||^2)$ is adopted as the kernel function in our work, where γ is the parameter to control precision. After training the SVR model, the output of regression is the predicted image quality score.

3　Experimental Results and Discussions

3.1　Experimental Protocol

The performance of a NR-IQA metric is evaluated by measuring the correlation between the predicted image quality scores and the subjective scores provided by the databases. Three widely used image quality assessment databases, including LIVE [20], CSIQ [9] and LIVE multiple distortion (LIVEMD) [7], are employed in our experiments.

The **LIVE** database [20] is the most commonly used database for the evalution of IQA algorithms. It consists of 779 distorted images generated from 29 reference images by processing them with 5 types of distortions - JPEG2k, JPEG, white Gaussian noise (WN), Gaussian blurring (BLUR) and fast fading channel distortion (FF), at 5 to 6 levels. A different mean opinion score (DMOS) associated with each distorted image is provided and is generally in the range [0,100], where higher DMOS indicates lower quality. The consistency experiments are conducted on the LIVE database.

The **CSIQ** database [9] consists of 30 reference images and their distorted versions with 6 different types of distortions at 4 to 5 different levels. DMOS is provided for each distorted image. And the DMOS is in the range [0,1]. We employed the CISQ database to demonstrate the database independence test, where we train on the LIVE database and test on the CSIQ database.

The **LIVEMD** database [7] consists of two subsets. The first subset includes 15 reference images and 215 degraded versions distorted by BLUR&JPEG (Gaussian blurring followed by JPEG compression). The second subset includes the same 15 reference images and 215 degraded versions distorted by BLUR&WN (Gaussian blurring followed by white Gaussian noise). DMOS is also provided in LIVEMD database. We evaluated the performance of the proposed model on multiply-distorted images in the LIVEMD database.

Two commonly used quantitative criterions [2] are used to evaluate the performance of the objective quality assessment models. The first criterion is the Spearman rank ordered correlation coefficient (SROCC) between the objective quality predictions and subjective DMOS values, and it is related to prediction monotonicity of a model. The second one is the Person's linear correlation coefficient (LCC) which is considered as a measure of prediction accuracy of a model. Before computing these criterions, a nonlinear regression is conducted between the subjective and objective scores [20]:

$$f(x) = \beta_1 \left(\frac{1}{2} - \frac{1}{1 + exp(\beta_2(x - \beta_3))} \right) + \beta_4 x + \beta_5, \tag{15}$$

where $\beta_i, i = 1, 2, ..., 5$ are the parameters to be fitted. Both higher SROCC and LCC values indicate better IQA performance.

3.2　Implementation Details

In image decomposition procedure, $\mathcal{X}$ is set as a 21×21 surrounding region as what did in [25]. As in [27], we set $M = 10$ and $N = 10$, and thus all the joint statistical features P_G, P_L, Q_G and Q_L are of dimension 10. And as discussed in [17], LBP can obtain a good texture classification performance with 16 neighbors in the radius of 2, and multiresolution information can boost the performance. Thus, we employ two operators of varying (P, R), i.e. (16,2) and (8,1), to extract rotation invariant LBP features from $\mathcal{I}_d$. The extracted features for the SVR model learning are 4192 ($10 \times 4 + 4116 + 36$) dimensions after concatenation.

3.3   Consistency Experiment on the LIVE Database

In this part, thorough experiments on the LIVE database are conducted to validate how the objective assessment corresponds to human evaluation. By default, all results reported in this section are obtained by 1000 train-test iterations with randomly selected 80% of the reference images and their associated distorted versions as training set and the remaining 20% of the reference images and their distorted versions as testing set [30]. We use the realigned DMOS scores [20] and report results only on the distorted images.

Tables 1 and 2 show our experimental results. The first four columns show results from distortion-specific (DS) experiments. The last columns in Tables 1 and 2 are obtained by performing train-test procedures on the images with all five types of distortions, i.e., non-distortion-specific (NDS) experiments. For comparison, the results of several previous state-of-the-art FR-IQA and NR-IQA models are also listed, including PSNR, SSIM [23], FSIM [32], BIQI [15], DIIVINE [16], BLIINDS-II [18], BRISQUE [13], M3 [27] and CORNIA [30]. The results of those models are just taken from the original or related papers. The LCC results of M3 on five individual distortions are not reported in its original paper [27], so we omit them in Table 2. The best two NR-IQA models are shown in boldface.

Table 1. SROCC comparison on LIVE database. Italicized are FR-IQA models. The best two NR-IQA models are shown in boldface.

SROCC	JP2K	JPEG	WN	BLUR	FF	ALL
PSNR	0.870	0.885	0.942	0.763	0.874	0.866
SSIM	0.939	0.946	0.964	0.907	0.941	0.913
PSIM	0.970	0.981	0.967	0.972	0.949	0.964
BIQI	0.800	0.891	0.951	0.846	0.707	0.820
DIIVINE	0.913	0.910	**0.984**	0.921	0.863	0.916
BLIIDNS-II	0.929	0.942	0.969	0.923	0.889	0.931
BRISQUE	0.914	**0.965**	0.979	0.951	0.877	0.940
M3	0.928	**0.966**	**0.985**	0.940	0.901	**0.951**
CORNIA	**0.943**	0.955	0.976	**0.969**	**0.906**	0.942
Proposed	**0.934**	0.956	0.959	**0.961**	**0.962**	**0.958**

Table 2. LCC comparison on LIVE database. Italicized are FR-IQA models. The best two NR-IQA models are shown in boldface.

LCC	JP2K	JPEG	WN	BLUR	FF	ALL
PSNR	0.873	0.876	0.926	0.779	0.870	0.856
SSIM	0.921	0.955	0.982	0.893	0.939	0.906
PSIM	0.910	0.985	0.976	0.978	0.912	0.960
BIQI	0.809	0.901	0.954	0.830	0.733	0.821
DIIVINE	0.922	0.921	**0.988**	0.923	0.888	0.917
BLIINDS-II	0.935	**0.968**	0.980	0.938	0.896	0.930
BRISQUE	0.923	**0.973**	0.985	0.951	0.903	0.942
M3	-	-	-	-	-	**0.955**
CORNIA	**0.951**	0.965	**0.987**	0.968	0.917	0.935
Proposed	**0.940**	0.966	0.965	**0.969**	0.967	**0.960**

As shown in Tables 1 and 2, the proposed NR IQA method achieves promising results in both NDS experiments and DS experiments. In NDS experiments, our approach outperforms those state-of-the-art BIQA methods for the overall evaluation. In DS experiments, the proposed method obtains promising results compared to those NR-IQA and FR-IQA methods on all the five distortions, especially on JPEG2k, Gaussian blurring (BLUR) and fast fading channel distortion (FF).

3.4 Database Independence Experiment

To verify the generalization ability of our approach, we train our model on the LIVE database and test them on the CSIQ database. We use the CSIQ database for the database independence test because there is no overlap between images in the CSIQ database and images in the LIVE database. We only report results on the four distortions - JPEG2k, JPEG, WN and BLUR - that are present in both LIVE database and the CSIQ database.

Table 3 shows our results. For comparison, we also report results of some state-of-the-art NR-IQA models, including BIQI, DIIVINE, QAC [28], CORNIA, BRISQUE, and ILNIQE [31]. We can observe that our proposed approach outperforms the compared NR-IQA methods except BRISQUE. Actually, the performance of the proposed metric and BRISQUE are very close. The database independence test demonstrates the strong generalization ability of the proposed method.

3.5 Performance on Multiply-Distorted Images

Compared with IQA models dedicated to single distortion types, IQA models for multiply-distorted images are of more significance. Thus, in addition to the above

Table 3. SROCC and LCC comparisons on CSIQ database (only images of four distortions shared by CSIQ and LIVE database are tested) with models trained on LIVE. The best two models are shown in boldface.

Models	SROCC	LCC
BIQI	0.8115	0.8476
DIIVINE	0.8760	0.8983
QAC	0.8416	0.8736
BRISQUE	**0.9099**	**0.9278**
CORNIA	0.8930	0.9175
ILNIQE	0.8885	0.9173
Proposed	**0.9005**	**0.9262**

mentioned experiments conducted on singly-distorted images, we also evaluate our model on multiply-distorted images in LIVEMD database. Results of the proposed model and some related models - PSNR, SSIM, BIQI, BRISQUE, SHANIA [10], SISBLIM [6], GAO [5] - are shown in Table 4. We can easily conclude from Table 4 that our proposed approach beat almost all state-of-the-art methods in all items except SROCC of GAO on BLUR&WN subset.

Table 4. SROCC and LCC comparisons on LIVEMD database. Italicized are FR-IQA models. The best two NR-IQA models are shown in boldface.

Models	BLUR&JPEG		BLUR&WN	
	SROCC	LCC	SROCC	LCC
PSNR	0.6514	0.6890	0.6378	0.7218
SSIM	0.7443	0.7615	0.7022	0.7473
BIQI	0.6542	0.6398	0.4902	0.6634
BRISQUE	0.7902	0.8454	0.2992	0.2868
SHANIA	0.8014	0.7629	0.7528	0.7073
SISBLIM	0.8749	0.8651	0.8802	0.8795
GAO	**0.8939**	**0.9265**	**0.8983**	**0.9015**
Proposed	**0.9021**	**0.9353**	**0.8949**	**0.9221**

4 Conclusion

In this paper, we propose a no-reference image quality assessment model based on the internal generative mechanism. We decompose an input image into orderly portion and disorderly portion with the Bayesian prediction model. And we extract four joint statistical features of GM map and LOG response from the orderly portion and LBP distributions as features the disorderly. The proposed model achieves state-of-the-art performance on three benchmark IQA databases. Consistency experiment on LIVE database validates the effectiveness of the proposed model. Database independence test demonstrates that the novel method has robust generalization ability. And experiment on LIVEMD shows that the proposed method can tackle multiply-distorted images.

Acknowledgment. This work was supported in part to Prof. Houqiang Li by 973 Program under Contract 2015CB351803, Natural Science Foundation of China (NSFC) under Contract 61390514 and Contract 61325009, and in part to Dr. Wengang Zhou by the Fundamental Research Funds for the Central Universities.

References

1. Chen, Z., Jiang, T., Tian, Y.: Quality assessment for comparing image enhancement algorithms. In: 2014 IEEE Conference on Computer Vision and Pattern Recognition, pp. 3003–3010. IEEE (2014)
2. Corriveau, P., Webster, A.: The video quality experts group: evaluates objective methods of video image quality assessment. In: 140th SMPTE Technical Conference and Exhibit, pp. 1–8, October 1998
3. Friston, K.: The free-energy principle: a unified brain theory? Nat. Rev. Neurosci. **11**(2), 127–138 (2010)
4. Gao, D., Han, S., Vasconcelos, N.: Discriminant saliency, the detection of suspicious coincidences, and applications to visual recognition. IEEE Trans. Pattern Anal. Mach. Intell. **31**(6), 989–1005 (2009)
5. Gao, F., Tao, D., Gao, X., Li, X.: Learning to rank for blind image quality assessment. IEEE Trans. Neural Netw. Learn. Syst. **26**(10), 2275–2290 (2015)
6. Gu, K., Zhai, G., Yang, X., Zhang, W.: Hybrid no-reference quality metric for singly and multiply distorted images. IEEE Trans. Broadcast. **60**(3), 555–567 (2014)
7. Jayaraman, D., Mittal, A., Moorthy, A.K., Bovik, A.C.: Objective quality assessment of multiply distorted images. In: 2012 Conference Record of 46th Asilomar Conference on Signals, Systems and Computers (ASILOMAR), pp. 1693–1697, November 2012
8. Knill, D.C., Pouget, A.: The Bayesian brain: the role of uncertainty in neural coding and computation. Trends Neurosci. **27**(12), 712–719 (2004)
9. Larson, E.C., Chandler, D.M.: Most apparent distortion: full-reference image quality assessment and the role of strategy. J. Electron. Imaging **19**(1), 011006 (2010)
10. Li, Y., Po, L.M., Xu, X., Feng, L.: No-reference image quality assessment using statistical characterization in the shearlet domain. Sig. Process.: Image Commun. **29**(7), 748–759 (2014)
11. Lu, Q., Zhou, W., Li, H.: A no-reference image sharpness metric based on structural information using sparse representation. Inf. Sci. **369**, 334–346 (2016)
12. Lyu, S., Simoncelli, E.P.: Nonlinear image representation using divisive normalization. In: IEEE Conference on Computer Vision and Pattern Recognition, CVPR 2008, pp. 1–8, June 2008
13. Mittal, A., Moorthy, A.K., Bovik, A.C.: No-reference image quality assessment in the spatial domain. IEEE Trans. Image Process. **21**(12), 4695–4708 (2012)

14. Mittal, A., Moorthy, A.K., Bovik, A.C.: No-reference approaches to image and video quality assessment. Multimedia Qual. Exp. (QoE): Curr. Status Future Requir., Ch. 5, 99–121 (2015)
15. Moorthy, A.K., Bovik, A.C.: A two-step framework for constructing blind image quality indices. IEEE Sig. Process. Lett. **17**(5), 513–516 (2010)
16. Moorthy, A.K., Bovik, A.C.: Blind image quality assessment: from natural scene statistics to perceptual quality. IEEE Trans. Image Process. **20**(12), 3350–3364 (2011)
17. Ojala, T., Pietikainen, M., Maenpaa, T.: Multiresolution gray-scale and rotation invariant texture classification with local binary patterns. IEEE Trans. Pattern Anal. Mach. Intell. **24**(7), 971–987 (2002)
18. Saad, M.A., Bovik, A.C., Charrier, C.: Blind image quality assessment: a natural scene statistics approach in the DCT domain. IEEE Trans. Image Process. **21**(8), 3339–3352 (2012)
19. Sheikh, H.R., Bovik, A.C., de Veciana, G.: An information fidelity criterion for image quality assessment using natural scene statistics. IEEE Trans. Image Process. **14**(12), 2117–2128 (2005)
20. Sheikh, H.R., Sabir, M.F., Bovik, A.C.: A statistical evaluation of recent full reference image quality assessment algorithms. IEEE Trans. Image Process. **15**(11), 3440–3451 (2006)
21. Smola, A.J., Schölkopf, B.: A tutorial on support vector regression. Stat. Comput. **14**(3), 199–222 (2004)
22. Sun, C., Li, H., Li, W.: No-reference image quality assessment based on global and local content perception. In: Visual Communications and Image Processing (VCIP), November 2016
23. Wang, Z., Bovik, A.C., Sheikh, H.R., Simoncelli, E.P.: Image quality assessment: from error visibility to structural similarity. IEEE Trans. Image Process. **13**(4), 600–612 (2004)
24. Wu, J., Lin, W., Shi, G., Liu, A.: Perceptual quality metric with internal generative mechanism. IEEE Trans. Image Process. **22**(1), 43–54 (2013)
25. Wu, J., Lin, W., Shi, G., Wang, X., Li, F.: Pattern masking estimation in image with structural uncertainty. IEEE Trans. Image Process. **22**(12), 4892–4904 (2013)
26. Wu, J., Lin, W., Shi, G., Xu, L.: Reduced-reference image quality assessment with local binary structural pattern. In: 2014 IEEE International Symposium on Circuits and Systems (ISCAS), pp. 898–901, June 2014
27. Xue, W., Mou, X., Zhang, L., Bovik, A.C., Feng, X.: Blind image quality assessment using joint statistics of gradient magnitude and Laplacian features. IEEE Trans. Image Process. **23**(11), 4850–4862 (2014)
28. Xue, W., Zhang, L., Mou, X.: Learning without human scores for blind image quality assessment. In: 2013 IEEE Conference on Computer Vision and Pattern Recognition (CVPR), pp. 995–1002, June 2013
29. Ye, P., Doermann, D.: No-reference image quality assessment using visual codebooks. IEEE Trans. Image Process. **21**(7), 3129–3138 (2012)
30. Ye, P., Kumar, J., Kang, L., Doermann, D.: Unsupervised feature learning framework for no-reference image quality assessment. In: 2012 IEEE Conference on Computer Vision and Pattern Recognition (CVPR), pp. 1098–1105, June 2012
31. Zhang, L., Zhang, L., Bovik, A.C.: A feature-enriched completely blind image quality evaluator. IEEE Trans. Image Process. **24**(8), 2579–2591 (2015)
32. Zhang, L., Zhang, L., Mou, X., Zhang, D.: FSIM: a feature similarity index for image quality assessment. IEEE Trans. Image Process. **20**(8), 2378–2386 (2011)

33. Zhang, M., Muramatsu, C., Zhou, X., Hara, T., Fujita, H.: Blind image quality assessment using the joint statistics of generalized local binary pattern. IEEE Sig. Process. Lett. **22**(2), 207–210 (2015)
34. Zhang, P., Zhou, W., Wu, L., Li, H.: SOM: semantic obviousness metric for image quality assessment. In: 2015 IEEE Conference on Computer Vision and Pattern Recognition (CVPR), pp. 2394–2402, June 2015
35. Zhang, X., Wu, X.: Image interpolation by adaptive 2-D autoregressive modeling and soft-decision estimation. IEEE Trans. Image Process. **17**(6), 887–896 (2008)

On the Exploration of Convolutional Fusion Networks for Visual Recognition

Yu Liu$^{(\boxtimes)}$, Yanming Guo, and Michael S. Lew

LIACS Media Lab, Leiden University, Leiden, The Netherlands
{y.liu,y.guo,m.s.lew}@liacs.leidenuniv.nl

Abstract. Despite recent advances in multi-scale deep representations, their limitations are attributed to expensive parameters and weak fusion modules. Hence, we propose an efficient approach to fuse multi-scale deep representations, called convolutional fusion networks (CFN). Owing to using 1×1 convolution and global average pooling, CFN can efficiently generate the side branches while adding few parameters. In addition, we present a locally-connected fusion module, which can learn adaptive weights for the side branches and form a discriminatively fused feature. CFN models trained on the CIFAR and ImageNet datasets demonstrate remarkable improvements over the plain CNNs. Furthermore, we generalize CFN to three new tasks, including scene recognition, fine-grained recognition and image retrieval. Our experiments show that it can obtain consistent improvements towards the transferring tasks.

Keywords: Multi-scale deep representations · Locally-connected fusion module · Transferring deep features · Visual recognition

1 Introduction

Since their repeated success in ImageNet classification [10,16,27,29], deep convolutional neural networks (CNNs) have contributed much to computer vision and the wider research community around it. CNN features can be used for many visual recognition tasks, and obtain top-tier performance [2]. Moreover, some works [1,21,34] begin capturing complementary features from intermediate layers. However, their methods mainly make use of one off-the-shelf model trained on the ImageNet dataset [26], but not to train a new network that can integrate intermediate layers. Instead, recent work [33] trains a multi-scale architecture for scene recognition at the expense of increasing algorithm complexity.

Hence, we propose to train an efficient fusion architecture to integrate intermediate layers for visual recognition. Our architecture is called convolutional fusion networks (CFN), which mainly consists of three characteristics: (1) *Efficient side outputs*: we add few parameters to generate new side branches due to using efficient 1×1 convolution and global average pooling [20]. (2) *Early fusion and late prediction*: in contrast to [33], we present an "early fusion and late prediction" strategy. It can not only reduce the number of parameters, but also

ⓒ Springer International Publishing AG 2017
L. Amsaleg et al. (Eds.): MMM 2017, Part I, LNCS 10132, pp. 277–289, 2017.
DOI: 10.1007/978-3-319-51811-4_23

produce a richer image representation. (3) *Locally-connected fusion*: in the fusion module, we propose making use of a locally-connected layer to learn adaptive weights (importance) for the side outputs. To the best of our knowledge, this is the first attempt to apply a locally-connected layer to a fusion module.

In a nutshell, our contributions can be summarized as follows. First, an efficient fusion architecture is presented to provide promising insights into efficiently exploiting multi-scale deep features. Second, we train CFN on the CIFAR and ImageNet 2012 datasets, and evaluate its efficiency and effectiveness. Experimental results demonstrate the superiority of CFN over the plain CNN. Third, we generalize the CFN model to other new tasks, including scene recognition, fine-grained recognition and image retrieval. Results show that CFN can consistently achieve significant improvements on these transferring tasks.

2   Related Work

In this section, we summarize existing approaches that focus on intermediate layers in the following three aspects.

Employment of Intermediate Layers. In CNNs, intermediate layers can capture complimentary information to the top-most layers. For example, Ng et al. [35] employed features from different intermediate layers and encoded them with VLAD scheme. Similarly, Cimpoi et al. [5] and Wei et al. [31] made use of Fisher Vectors to encode intermediate activations. Moreover, Liu et al. [21] and Babenko and Lempitsky [3] aggregated several intermediate activations and generated a more discriminative and expensive image descriptor. Based on intermediate layers, these methods are able to achieve promising performance on their tasks, as compared to using the fully-connected layers.

Intermediate Supervision. Considering the importance of intermediate layers, Lee et al. [18] proposed the deeply supervised nets, which imposed additional supervision to guide the intermediate layers earlier, rather than the standard approach of only supervising the final prediction. Similarly, GoogLeNet [29] created two extra branches from the intermediate layers and supervised them jointly. However, these approaches do not explicitly fuse the outputs of intermediate layers.

Multi-scale Fusion (or Skip Connections). To incorporate intermediate outputs explicitly during training, multi-scale fusion is presented to train multi-scale deep neural networks [22,32,33]. A similar work in [33] built a DAG-CNNs model that summed up the multi-scale predictions from intermediate layers. However, DAG-CNNs required processing a large number of additional parameters. In addition, its fusion module (i.e. sum-pooling) failed to consider the importance of side branches. However, our CFN can learn adaptive weights for fusing side branches, while adding few parameters.

3   Proposed Approach

In this section, we introduce the architecture of CFN and its training procedure.

3.1 Architecture

Similar to [10,20,29] we use 1×1 convolutional layer and global average pooling at the top layers, To reduce the number of parameters in a plain CNN model. Based on a plain CNN, we develop our convolutional fusion networks, as illustrated in Fig. 1. Overall, our CFN manly consists of three characteristics that will be described in the following.

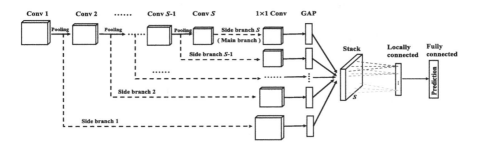

Fig. 1. The general pipeline of convolutional fusion networks (best viewed in zoom in). The side branches start from the pooling layers and consist of a 1×1 convolution and global average pooling. All side outputs are then stacked together. A locally-connected layer is used to adaptively learn adaptive weights for the side outputs. Finally, the fusion feature is fed to the following fully-connected layer that is used to make a final prediction.

Efficient Side Outputs. Instead of using the fully-connected layers, CFN efficiently generates the side branches from the intermediate layers while adding few parameters. First, the side branches are grown from the pooling layers by inserting 1×1 convolution layers like the main branch. All 1×1 convolutional layers must have the same number of channels so that they can be integrated together. Then, global average pooling is performed over the 1×1 convolutional maps to obtain one-dimensional feature vector, called GAP feature here. Notably, we can also consider the full depth main branch as a side branch.

Assume that there are S of side branches in total and the last side branch (i.e. S-th) indicates the main branch. We notate $h_{i,j}^{(s)}$ as the input of 1×1 convolution in the s-th side branch, where $s = 1, 2, \ldots, S$ and (i, j) is the spatial location across feature maps. As 1×1 convolution has K of channels, its output associated with the k-th kernel, denoted as $f_{i,j,k}^{(s)}$, where $k = 1, \ldots, K$. Next, let $H^{(s)}$ and $W^{(s)}$ be the height and width of features maps derived from the s-th 1×1 convolution. Thereby, global average pooling performed over the feature map $f_k^{(s)}$ is calculated by

$$g_k^{(s)} = \frac{1}{H^{(s)} W^{(s)}} \sum_{i=1}^{H^{(s)}} \sum_{j=1}^{W^{(s)}} f_{i,j,k}^{(s)}, \tag{1}$$

where $g_k^{(s)}$ is the k-th element in the s-th GAP feature vector. Thus, we can notate $g^{(s)} = [g_1^{(s)}, \ldots, g_K^{(s)}]$, a $1 \times K$ dimensional vector, as the whole GAP feature from the s-th side branch. Recall that $g^{(S)}$ represents the GAP feature from the full depth main branch.

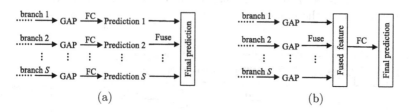

Fig. 2. Comparison between EPLF and EFLP. (a) The pipeline of EPLF strategy; (b) The pipeline of EFLP strategy.

Early Fusion and Late Prediction. Considering when to fuse the side branches, related work [32,33] used an "early prediction and late fusion" (EPLF) strategy. In contrast to EPLF [33], in which a couple of FC layers are added, we present an opposite strategy called "early fusion and late prediction" (EFLP). EFLP can fuse the GAP features from the side outputs and obtain a fused feature. Then, one fully-connected layer following the fused feature is used to estimate the final prediction. Figure 2 shows the comparison between EPLF and EFLP. As compared to EPLF, EFLP consumes less parameters due to using only one fully-connected layer. We assume that each fully-connected layer has C units that correspond to the number of object categories in the dataset. The fusion module has W_{fuse} of parameters. Quantitatively, we can compare the parameters (i.e. weights and bias) between EFLP and EPLF by

$$W_{EFLP} = S(C+1) + W_{fuse} < W_{EPLF} = SK(C+1) + W_{fuse}. \tag{2}$$

More importantly, the fused feature in EFLP can be extracted as a richer image representation, compared with the widely-used fc6 and fc7 [16,27]. The fused feature could be transferred from generic to specific vision recognition tasks. However, EPLF fails to specify which feature can serve as a good representation. Additionally, EFLP can achieve the same accuracy as EPLF, though EPLF consumes more parameters.

Locally-Connected Fusion. Another significant component in CFN is its fusing the branches based on a locally-connected (LC) layer. Owing to its no-sharing filters over spatial dimensions, LC layer can learn different weights in each local field [9]. We aim to make use of a LC layer to learn adaptive weights (or importance) for the side branches, and to generate the fused feature. As we know, this is the first attempt to apply a locally-connected layer to a fusion module. The detail computation are introduced as follows.

At first, we stack GAP features together (from $g^{(1)}$ to $g^{(S)}$), and form a stack layer G with size of $1 \times K \times S$, see Fig. 1. The s-th feature map of G is $g^{(s)}$. Then, one LC layer which has K of no-sharing filters is convolved over G. Each filter has $1 \times 1 \times S$ kernel size. As a result, LC can learn adaptive weights for different elements in the GAP features. Here, the fused feature convolved by LC also has $1 \times K$ shape, denoted as $g^{(f)}$. Each element in $g^{(f)}$ can be computed via

$$g_i^{(f)} = \sigma \left(\sum_{j=1}^{S} W_{i,j}^{(f)} \cdot g_i^{(j)} + b_i^{(f)} \right), \tag{3}$$

where $i = 1, 2, \ldots, K$; σ indicates the activation function (i.e. ReLU). $W_{i,j}^{(f)}$ and $b_i^{(f)}$ represent the weights and bias. The number of parameters in the LC fusion is $K \times (S+1)$. These additional parameters benefit adaptive fusion while do not need any manual tuning.

To be clear, Fig. 3 compares LC fusion with other simple fusion methods. In Fig. 3(a), the sum-pooling fusion simply sums up the side outputs without learning any weights. In Fig. 3(b), the convolution fusion can learn only one sharing filter over the whole spatial dimensions (as drawn in the same blue color). On the contrary, LC can learn independent weights over each local field (i.e. $1 \times 1 \times S$ size), as drawn in different colors in Fig. 3(c). Although LC fusion has a little more parameters than the sum-pooling and convolution fusion, these parameters are nearly negligible as compared to the whole network parameters.

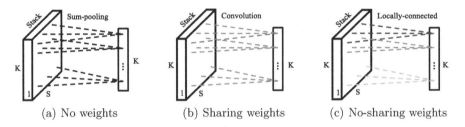

(a) No weights   (b) Sharing weights   (c) No-sharing weights

Fig. 3. Comparison of three fusion modules (best viewed in color). Left: Sum-pooling fusion has no weights; Middle: Convolution fusion learns sharing weights over spatial positions, as drawn in the same color; Right: Locally-connected fusion learns no-sharing weights over spatial positions, as drawn in different colors. To learn element-wise weights, we use 1×1 local field.

3.2   Training

Since CFN has efficient forward propagation and backward propagation, it can maintain the ease of training as similar to CNN. Assume that W indicates the set of all parameters learned in the CFN (including the LC fusion weights), and $\mathcal{L}$ is the total loss cost during training. To minimize the total loss, the partial derivative of the loss with respect to any weight will be recursively computed by the chain rule during the backward propagation [6]. Since the main components in our CFN model are the side branches, we will induce the detail computations of their partial derivatives. For notational simplicity, we consider each image independently in the following.

First, we compute the gradient of the loss cost with respect to the outputs of the side branches. As an example of the s-th side branch, the gradient of $\mathcal{L}$ with respect to the side output $g^{(s)}$ can be formulated as below

$$\frac{\partial \mathcal{L}}{\partial g^{(s)}} = \frac{\partial \mathcal{L}}{\partial g^{(f)}} \cdot \frac{\partial g^{(f)}}{\partial g^{(s)}}, s = 1, 2, \ldots, S. \tag{4}$$

Second, we formulate the gradient of $\mathcal{L}$ with respect to the inputs of the side branches. We notate $a^{(s)}$ as the input of the s-th side branch. As depicted in Fig. 1, $a^{(s)}$ represents the pooling layer. Note that the input of the main branch, denoted as $a^{(S)}$, refers to the last convolutional layer (i.e. conv S). We can observe that the gradient of $a^{(s)}$ depends on several related branches. For example in Fig. 1, the gradient of $a^{(1)}$ is influenced by S of branches; the gradient of $a^{(2)}$ need to consider the results from the 2-th to S-th branch; but the gradient of $a^{(S)}$ is updated by only the main branch. Mathematically, the gradient of $\mathcal{L}$ with respect to the side input $a^{(s)}$ can be computed as follows:

$$\frac{\partial \mathcal{L}}{\partial a^{(s)}} = \sum_{i=s}^{S} \frac{\partial \mathcal{L}}{\partial g^{(i)}} \cdot \frac{\partial g^{(i)}}{\partial a^{(i)}}, \tag{5}$$

where i indexes the related branch that contributes to the gradient of $a^{(s)}$. We then need to sum up the gradients from these related branches. Like [16], we employ standard stochastic gradient descent (SGD) algorithm with mini-batch to train the whole network.

3.3   Discussion

To present more insights into CFN, we compare CFN with other related models.

Relationship with CNN. Normally, a plain CNN only estimates a final prediction based on the topmost layer, as a result, the effects of intermediate layers on the prediction are implicit and indirect. In contrast, CFN can connect the intermediate layers using side branches, and deliver their effects on the final prediction jointly. Hence, CFN can take advantage of intermediate layers explicitly and directly.

Relationship with DSN. DSN [18] adds extra supervision to intermediate layers for earlier guidance. However, CFN that still uses one supervision towards the final prediction aims to generate a fused and richer feature. In a nutshell, DSN focuses on "loss fusion", but CFN instead focuses on "feature fusion".

Relationship with ResNet. ResNet [10] addresses the vanishing gradient problem by adding "linear" shortcut connections. CFN has three main differences as compared to ResNet: (1) The side branches in CFN are not shortcut connections. They start from a pooling layer and end in a fusion module together. (2) In contrast to adding a "linear" branch, we still use ReLU in each side branch. (3) The output of a fusion module is fed to the final prediction. As mentioned in the ResNet work, when the network is not overly deep (e.g. 11 or 18 layers), ResNet may obtain little improvement over a plain CNN. However, CFN can obtain some considerable improvements as compared to CNN. In contrast to increasing the depth, CFN can serve as an alternative to improving the discriminative capacity of not-very-deep models. This explains the usefulness and effectiveness of CFN.

4   Experiments

First, we trained CFN models on the CIFAR-10/100 [15] and ImageNet 2012 [26]. Then, we transferred the trained ImageNet model to three new tasks, including scene recognition, fine-grained recognition and image retrieval. We conducted all experiments using the Caffe framework [13] with a NVIDIA TITAN X card.

4.1   CIFAR Dataset

Both CIFAR-10 [15] and CIFAR-100 [15] consist of 50,000 training images and 10,000 testing images. But they define 10 and 100 object categories, respectively. We preprocessed their RGB images by global contrast normalization [7]. We built a plain CNN that consists of seven convolutional layers and one fully-connected layer. The first six convolutional layers have 3×3 kernel size, but the seventh one is 1×1 convolution. Global average pooling locates between the last convolutional layer and the fully-connected layer. Based on the plain CNN, we developed the CFN counterpart as illustrated in Fig. 4.

Overall, we use the same hyper-parameters to train CNN and CFN. We use a weight decay of 0.0001, a momentum of 0.9, and a mini-batch size of 100. The learning rate is initialized with 0.1 and is divided by 10 after 10×10^4 iterations. The whole training will be terminated after 12×10^4 iterations. As for CFN, the initialized weights in LC fusion is set with $1/S$ (S is the number of branches).

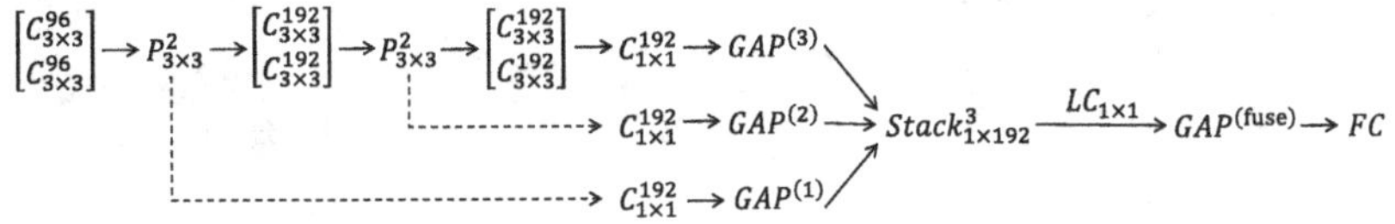

$$\begin{bmatrix} C_{3\times3}^{96} \\ C_{3\times3}^{96} \end{bmatrix} \rightarrow P_{3\times3}^{2} \rightarrow \begin{bmatrix} C_{3\times3}^{192} \\ C_{3\times3}^{192} \end{bmatrix} \rightarrow P_{3\times3}^{2} \rightarrow \begin{bmatrix} C_{3\times3}^{192} \\ C_{3\times3}^{192} \end{bmatrix} \rightarrow C_{1\times1}^{192} \rightarrow GAP^{(3)}$$
$$\dashrightarrow C_{1\times1}^{192} \rightarrow GAP^{(2)} \rightarrow Stack_{1\times192}^{3} \xrightarrow{LC_{1\times1}} GAP^{(fuse)} \rightarrow FC$$
$$\dashrightarrow C_{1\times1}^{192} \rightarrow GAP^{(1)}$$

Fig. 4. Illustration of CFN built for CIFAR dataset. For the convolutional layers, the right lower numbers indicate the kernel size; the right upper number indicates the number of channels.

Table 1. Test error (%) on CIFAR-10/100 dataset (without data augmentation).

Model	#parameters	CIFAR-10	CIFAR-100
CNN	1.287M (basic)	9.28	31.89
CNN-Sum	1.287M + 0.074M (extra branches) + 0 (fusion)	8.84	31.42
CNN-Conv	1.287M + 0.074M (extra branches) + 4 (fusion)	8.68	31.16
CFN	1.287M + 0.074M (extra branches) + 768 (fusion)	**8.27**	**30.68**

Results. Table 1 shows the results on CIFAR-10 test set. We can analyze the results from the following aspects: (1) Compared with the plain CNN, CFN achieves about 1.01% and 1.21% improvement on CIFAR-10 and CIFAR-100, respectively. (2) In order to demonstrate the advantage of LC fusion, we also implement the sum-pooling fusion and convolution fusion, denoted as CNN-Sum and CNN-Conv. We can see that LC fusion used in CFN outperforms both CNN-Sum and CNN-Conv. (3) We compute the number of parameters in each model. Importantly, the additional parameters for extra side branches and LC fusion are significantly fewer than the number of basic parameters. Although LC fusion uses a little more parameters for fusing branches, these parameters are nearly negligible for a deep network. To reflect the efficiency, we also compare the training time between CNN and CFN. For example on CIFAR-10, CNN and CFN consumes 1.67 and 2.08 hours, respectively.

In Fig. 5, we visualize and compare the learned feature maps in CNN and CFN. We select ten images from CIFAR-10 dataset. We extract the feature maps in the 1×1 convolutional layer and visualize the top-4 maps (we rank the feature maps by averaging their own activations). One can observe that CFN can learn

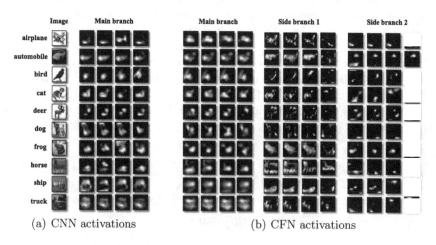

(a) CNN activations   (b) CFN activations

Fig. 5. Illustration of features activations of CIFAR-10 images. (a) For CNN, we visualize its 1×1 convolutional layer in the main branch. (b) For CFN, the 1×1 convolutional maps from the main branch and two side branches are shown.

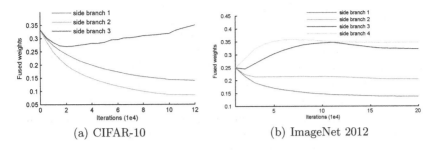

(a) CIFAR-10   (b) ImageNet 2012

Fig. 6. Illustration of adaptive weights of the side branches learned in the LC fusion. The top branches have larger weights than the bottom branches.

complementary clues in the side branches to the full depth main branch. For example, the side-output 1 mainly learns the boundaries or shapes around the objects. The side-output 2 focuses on some semantic "parts" that fire strong near the objects. Furthermore, Fig. 6(a) shows the adaptive weights learned in the LC fusion. The side branch 3 (main branch) plays a core role, but other side branches are also complementary to the full depth main branch.

Comparison with the State-of-the-Art. Table 2 compares the results on CIFAR datasets. Overall, CFN can obtain comparative results and outperform recent not-very-deep state-of-the-art models. It is worth mentioning that some work intends to push the results using much deeper networks [10] and large data augmentation [8]. In contrast to purely pushing the results, our aim is to demonstrate the advantage of fusing intermediate layers. Thus we only use a not-overly deep model and standard data augmentation [18]. We believe that Adapting CFN to a very deep model will be an interesting future work.

Table 2. Test error on CIFAR-10/100 to compare CFN with recent state-of-the-art. A superscripted * indicates to use the standard data augmentation [18].

Method	CIFAR-10	CIFAR-10*	CIFAR-100
Maxout networks [7]	11.68%	9.38%	38.57%
NIN [20]	10.41%	8.81%	35.68%
DSN [18]	9.69%	7.97%	34.54%
ALL-CNN [28]	9.08%	7.25%	33.71%
R-CNN [19]	8.69%	7.09%	31.75%
NIN + SReLU [14]	8.41%	6.98%	31.10%
CNN (baseline)	9.28%	7.34%	31.89%
CFN (ours)	**8.27%**	**6.77%**	**30.68%**

4.2   ImageNet 2012

We developed a basic 11-layer plain CNN (i.e. CNN-11) whose channels of feature maps range from 64 to 1024. Based on this CNN, we built its CFN counterpart (i.e. CFN-11) as illustrated in Fig. 7. We create three extra side branches from the pooling layers (excluding the first pooling layer). Following existing literature [10, 16,27,29], we use a weight decay of 0.0001, and a momentum of 0.9 and a mini-batch size of 64. Batch normalization (BN) [11] is used after each convolution. The learning rate starts from 0.01 and decreases to 0.001 at 10×10^4 iterations, and to 0.0001 at 15×10^4 iterations. The whole training will be terminated after 20×10^4 iterations. LC weights are initialized with 0.25 due to four side branches in total.

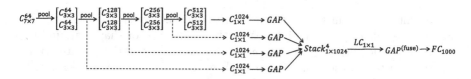

Fig. 7. The architecture of CFN built for ImageNet classification.

Results. Table 3 compares the results on the validation set. First, CNN-11 can achieve competitive results as compared to AlexNet [16], however, it consumes much fewer parameters (∼6.3 millions) than Alexnet (∼60 millions). Second, CFN-11 obtains about 1% improvement over CNN-11, while adding few parameters (∼0.5 millions). It verifies the efficiency of fusing multi-scale deep representations. Furthermore, we reproduce the DSN [18] and ResNet [10] models based on the plain CNN-11. As a result, CFN-11 can achieve better accuracy than DSN-11 and ResNet-11. For such a not-overly deep network, CFN can serve as an alternative to improving the discriminative capacity of CNNs, instead of increasing the depth like ResNet. Moreover, to test the generalization of CFN to deeper networks, we build a 19-layer model following the principle of 11-layer model. Likewise, CFN-19 outperforms CNN-19 by a consistent improvement as seen in Table 3.

Similar to CIFAR-10, Fig. 6(b) shows the adaptive weights learned in the LC fusion. We can see that the top branches (i.e. 3 and 4) have larger weights than the bottom branches (i.e. 1 and 2). In Fig. 8, we illustrate and compare the feature maps in the side branches.

Table 3. Error rates (%) on the ImageNet 2012 validation set.

Method	AlexNet	CNN-11	DSN-11	ResNet-11	CFN-11	CNN-19	CFN-19
Top-1	42.90	43.11	42.24	43.02	**41.96**	36.99	**35.47**
Top-5	19.80	19.91	19.24	19.85	**19.09**	14.74	**13.93**

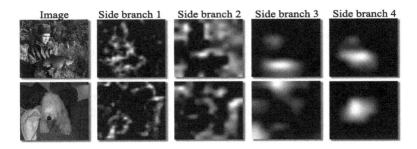

Fig. 8. Illustration of the feature activations of images of the ImageNet dataset.

4.3 Transferring Fused Feature to New Tasks

To evaluate the generalization of CFN, we transferred the trained ImageNet model to three new tasks: scene recognition, fine-grained recognition and image retrieval. Each task is evaluated on two datasets: Scene-15 [17] and Indoor-67 [25], Flower [23] and Bird [30], and Holidays [12] and UKB [24]. For AlexNet, the fc7 layer is used as a baseline; For CNN-11, we extract the result of global average pooling as another baseline; For CFN-11, the fused feature is extracted to represent images. For scene and fine-grained recognition, we use linear SVM [4] to compute the classification accuracy. For image retrieval, we use KNN to compute the mAP on Holidays and N-S score on UKB. Table 4 reports the evaluation results on six datasets. We can see that CFN-11 obtains consistent improvement performance on all datasets. Interestingly, their gains are more remarkable than those in ImageNet. It reveals that learning multi-scale deep representations are beneficial for diverse vision recognition problems. In addition, fine-tuning the model on the target datasets will further improve the results.

Table 4. Results on transferring the ImageNet model to three target tasks.

| Method | Dim | Scene recognition | | Fine-grained recognition | | Image retrieval | |
		Scene 15	Indoor 67	Flower	Bird	Holidays	UKB
AlexNet [16]	4096	83.99	58.28	78.68	45.79	76.77	3.45
CNN-11	1024	84.32	60.45	76.79	45.98	78.33	3.47
CFN-11	1024	**86.83**	**62.24**	**82.57**	**48.12**	**80.32**	**3.54**

5 Conclusions

We proposed efficient convolutional fusion networks (CFN) by adding few parameters. It can serve as an alternative to improving recognition accuracy instead of increasing the depth. Experiments on the CIFAR and ImageNet datasets demonstrate the superiority of CFN over the plain CNN. Additionally, CFN outperforms not-very-deep state-of-the-art models by considerable gains. Moreover, we verified its significant generalization while transferring CFN to three new tasks. In future work, we will evaluate CFN with much deeper neural networks.

Acknowledgments. This work was supported mainly by the LIACS Media Lab at Leiden University and in part by the China Scholarship Council. We would like to thank NVIDIA for the donation of GPU cards.

References

1. Agrawal, P., Girshick, R., Malik, J.: Analyzing the performance of multilayer neural networks for object recognition. In: Fleet, D., Pajdla, T., Schiele, B., Tuytelaars, T. (eds.) ECCV 2014. LNCS, vol. 8695, pp. 329–344. Springer, Heidelberg (2014). doi:10.1007/978-3-319-10584-0_22
2. Azizpour, H., Razavian, A.S., Sullivan, J., Maki, A., Carlsson, S.: From generic to specific deep representation for visual recognition. In: CVPR, DeepVision Workshop (2015)
3. Babenko, A., Lempitsky, V.S.: Aggregating deep convolutional features for image retrieval. In: ICCV (2015)
4. Chang, C.C., Lin, C.J.: LIBSVM: A library for support vector machines. ACM Trans. Intell. Syst. Technol. **2**, 27:1–27:27 (2011)
5. Cimpoi, M., Maji, S., Vedaldi, A.: Deep filter banks for texture recognition and segmentation. In: CVPR (2015)
6. Cun, L., Boser, B., Denker, J.S., Henderson, D., Howard, R.E., Hubbard, W., Jackel, L.D.: Handwritten digit recognition with a back-propagation network. In: NIPS (1990)
7. Goodfellow, I.J., Warde-Farley, D., Mirza, M., Courville, A.C., Bengio, Y.: Maxout networks. In: ICML (2013)
8. Graham, B.: Fractional max-pooling (2014). CoRR arXiv:abs/1412.6071
9. Gregor, K., LeCun, Y.: Emergence of complex-like cells in a temporal product network with local receptive fields (2010). CoRR arXiv:abs/1006.0448
10. He, K., Zhang, X., Ren, S., Sun, J.: Deep residual learning for image recognition. In: CVPR (2016)
11. Ioffe, S., Szegedy, C.: Batch normalization: accelerating deep network training by reducing internal covariate shift. In: ICML (2015)
12. Jegou, H., Douze, M., Schmid, C.: Hamming embedding and weak geometric consistency for large scale image search. In: Forsyth, D., Torr, P., Zisserman, A. (eds.) ECCV 2008. LNCS, vol. 5302, pp. 304–317. Springer, Heidelberg (2008). doi:10.1007/978-3-540-88682-2_24
13. Jia, Y., Shelhamer, E., Donahue, J., Karayev, S., Long, J., Girshick, R., Guadarrama, S., Darrell, T.: Caffe: convolutional architecture for fast feature embedding. In: ACM Multimedia (2014)
14. Jin, X., Xu, C., Feng, J., Wei, Y., Xiong, J., Yan, S.: Deep learning with S-shaped rectified linear activation units. In: AAAI (2016)
15. Krizhevsky, A.: Learning multiple layers of features from tiny images. Master's thesis, Department of Computer Science, University of Toronto (2009)
16. Krizhevsky, A., Sutskever, I., Hinton, G.E.: ImageNet classification with deep convolutional neural networks. In: NIPS (2012)

17. Lazebnik, S., Schmid, C., Ponce, J.: Beyond bags of features: spatial pyramid matching for recognizing natural scene categories. In: CVPR (2006)
18. Lee, C., Xie, S., Gallagher, P., Zhang, Z., Tu, Z.: Deeply-supervised nets. In: AISTATS (2015)
19. Liang, M., Hu, X.: Recurrent convolutional neural network for object recognition. In: CVPR (2015)
20. Lin, M., Chen, Q., Yan, S.: Network in network. In: ICLR (2014)
21. Liu, L., Shen, C., van den Hengel, A.: The treasure beneath convolutional layers: cross convolutional layer pooling for image classification. In: CVPR (2015)
22. Long, J., Shelhamer, E., Darrell, T.: Fully convolutional networks for semantic segmentation. In: CVPR (2015)
23. Nilsback, M.E., Zisserman, A.: Automated flower classification over a large number of classes. In: Indian Conference on Computer Vision, Graphics and Image Processing (2008)
24. Nister, D., Stewenius, H.: Scalable recognition with a vocabulary tree. In: CVPR (2006)
25. Quattoni, A., Torralba, A.: Recognizing indoor scenes. In: CVPR (2009)
26. Russakovsky, O., Deng, J., Su, H., Krause, J., Satheesh, S., Ma, S., Huang, Z., Karpathy, A., Khosla, A., Bernstein, M., Berg, A.C., Li, F.-F.: ImageNet large scale visual recognition challenge. IJCV **115**, 1–42 (2015)
27. Simonyan, K., Zisserman, A.: Very deep convolutional networks for large-scale image recognition. In: ICLR (2015)
28. Springenberg, J.T., Dosovitskiy, A., Brox, T., Riedmiller, M.: Striving for simplicity: the all convolutional net. In: ICLR (2015)
29. Szegedy, C., Liu, W., Jia, Y., Sermanet, P., Reed, S., Anguelov, D., Erhan, D., Vanhoucke, V., Rabinovich, A.: Going deeper with convolutions. In: CVPR (2015)
30. Wah, C., Branson, S., Welinder, P., Perona, P., Belongie, S.: The Caltech-UCSD Birds-200-2011 dataset. Technical report CNS-TR-2011-001 (2011)
31. Wei, X.S., Gao, B.B., Wu, J.: Deep spatial pyramid ensemble for cultural event recognition. In: ICCV Workshops (2015)
32. Xie, S., Tu, Z.: Holistically-nested edge detection. In: ICCV (2015)
33. Yang, S., Ramanan, D.: Multi-scale recognition with DAG-CNNs. In: ICCV (2015)
34. Yoo, D., Park, S., Lee, J.Y., Kweon, I.S.: Multi-scale pyramid pooling for deep convolutional representation. In: CVPR, Deep Vision Workshop (2015)
35. Yue-Hei Ng, J., Yang, F., Davis, L.S.: Exploiting local features from deep networks for image retrieval. In: CVPR, Deep Vision Workshops, June 2015

Phase Fourier Reconstruction for Anomaly Detection on Metal Surface Using Salient Irregularity

Tzu-Yi Hung[1][✉], Sriram Vaikundam[1],
Vidhya Natarajan[1], and Liang-Tien Chia[2]

[1] Rolls-Royce@NTU Corporate Lab, Singapore, Singapore
e090043@e.ntu.edu.sg
[2] Nanyang Technological University, Singapore, Singapore

Abstract. In this paper, we propose a Phase Fourier Reconstruction (PFR) approach for anomaly detection on metal surfaces using salient irregularities. To get salient irregularity with images captured from an automatic visual inspection (AVI) system using different lighting settings, we first trained a classifier for image selection as only dark images are utilized for anomaly detection. By doing so, surface details, part design, and boundaries between foreground/background become indistinct, but anomaly regions are highlighted because of diffuse reflection caused by rough surfaces. Then PFR is applied so that regular patterns and homogeneous regions are further de-emphasized, and simultaneously, anomaly areas are distinct and located. Different from existing phase-based methods which require substantial texture information, our PFR works on both textual and non-textual images. Unlike existing template matching methods which require prior knowledge of defect-free patterns, our PFR is an unsupervised approach which detects anomalies using a single image. Experimental results on anomaly detection clearly demonstrate the effectiveness of the proposed method which outperforms several well-designed methods [8, 12, 15, 16, 18, 19] with a running time of less than 0.01 seconds per image.

Keywords: Anomaly detection · Defect detection · Saliency · Unsupervised · Automatic visual inspection · Metal surface

1 Introduction

Automatic Visual Inspection (AVI) is one of the critical industrial applications which uses computer vision technology to assist human inspection in an effective and efficient manner. In such a system, anomaly detection is one of the challenging tasks with high demands [1,2,10]. For example, many issues have been discussed on electronic chips and integrated circuits for defect detection in images [2,4,7,14].

In the literature, anomaly detection can be categorized into three types. The first type is to detect local textual irregularities on textured surfaces [1,11], as

© Springer International Publishing AG 2017
L. Amsaleg et al. (Eds.): MMM 2017, Part I, LNCS 10132, pp. 290–302, 2017.
DOI: 10.1007/978-3-319-51811-4_24

an example shown in Fig. 1(a). Many works on texture features and textural analysis have been discussed and investigated [17].

The second type is to detect abnormalities among defect-free examples, which is a typical application applied for die and chip inspections, as shown in Fig. 1(b). Template matching is one of the traditional solutions discussed in this category [2,9,15], where anomalies are found by computing similarities between a test image and images in an anomaly-free set. In other words, prior knowledge such as a pre-collected template dataset is required.

(a)   (b)   (c)

Fig. 1. Examples of different image sources. (a) a texture-based image [1], (b) an image with single lighting condition [2], and (c) an image set with multiple lighting conditions. Note that dark images at the left two columns are adjusted with 40% increase of brightness and contrast to be visible in the paper.

The third type is to detect abnormalities using images of metallic components captured at pre-defined viewpoints and lighting conditions. Figure 1(c) shows two examples of how the changes in lighting directions influence the reflectivity of regions with and without anomalies on a metal surface. Note that dark images at the left two columns are adjusted to be visible in the paper. Between these two examples, the anomaly at the top row can be easily found in the bright images, but the one at the bottom is visually distinct in the dark images since it shines. Detection of such anomalies is a challenging task in a real-world application of AVI, but to our best knowledge, this area of research is yet to be explored.

In this paper, we are working on images from the third category for anomaly detection where not only different characteristics of anomalies are considered, but shapes of metallic objects are also taken into account which makes this task more challenging. Figure 2 lists five examples of anomalies, which from left

Fig. 2. Example of anomalies. From left to right are melt, plus metal, scratch, scuff and shadowing.

to right are melt, plus metal, scratch, scuff, and shadowing. Some of them are visually noticeable in bright images such as melt, plus metal and shadowing, but some of them shine in dark images such as scratch and scuff. Although certain anomalies may only be visually perceptible under certain lightings, we observed that because of diffuse reflection, those anomalies which are visually noticeable in bright images also appear irregularly salient in dark images when compared to its surroundings such as metal surface and shape boundary. Therefore, we propose to use PFR on dark images where details of smooth regions are reduced, but rough defect areas are highlighted. Our method outperforms several well-designed methods including saliency-based approaches, template matching, and keypoint detection schemes. Moreover, the processing time is less than 0.01 seconds per image, and no prior knowledge is required.

The rest of the paper is organized as follows. Section 2 introduces related work. Section 3 describes the proposed anomaly detection method. Section 4 shows the experimental results, and Sect. 5 concludes the paper.

2   Related Work

Most methods detect anomalies using images whose anomaly regions are clearly visible. Vaikundam et al. [15] proposed to detect anomalies on a residue image which is obtained by a template matching approach. More specifically, they firstly performed a SIFT keypoint matching to find the best-match anomaly-free image compared to the given input and obtained a residue image between an anomaly-free image and an anomaly one. Then the SIFT keypoint detection method is applied again to detect regions of interest in the residue image. This method works well on those anomalies that are visually noticeable in bright images. However, it is hard to perform template matching using dark images since component shape and design, and foreground/background boundary are obscure. Moreover, template matching may be time-consuming which highly depends on the size of the template set.

On the other hand, Bai et al. [2] proposed to use phase spectrum followed by a template matching method. Firstly, phase-only Fourier transform (POFT) is used to highlight potential anomalies on a collection of similar test images which is formed as an input matrix image for anomaly detection. Then a template matching step is adopted to finely compare local discrepancies between test images and anomaly-free templates, especially for those potential locations. This method needs several images to form a regular and repeated pattern when using POFT; otherwise, the method would fail for anomaly detection. Further, they only considered anomalies shown in the bright images. Instead, in our work, we apply PFR on a single dark image for anomaly detection without template matching.

Phase-based Fourier methods have been used for saliency detection and defect detection [1,3,5,6,13] such as automated surface inspection for directional textures [13], fabric defect detection [3], saliency detection on natural images [6] and videos [5], and texture-based surface defect detection [1]. The main difference is that they applied phase-based Fourier methods on bright images and tried to make input images into patterns.

3   Anomaly Detection on Metal Surface

In this paper, we propose an anomaly detection method for AVI using PFR. Figure 3 shows the framework. We firstly apply image selection to automatically select dark images as input to PFR for anomaly detection. The following subsections describe details of the proposed method.

Fig. 3. Framework of the proposed anomaly detection method.

3.1   Salient Irregularities in Dark Images

Most existing methods for anomaly detection aim to detect their targets using clearly visible images as mentioned earlier. Instead, we focus on salient irregularities in dark images. The key idea is the majority of image pixels on the metal surface are dark in such images, in particular for those smooth and homogenous areas; therefore the details of the surface become uniformly indistinct caused by specular reflection. On the other hand, irregular or abnormal regions become salient because of diffuse reflection on uneven surfaces. Figure 4 shows an illustration of specular and diffuse reflections, respectively. In dark images, the diffuse reflection makes anomaly salient comparing to its smooth surrounding metal surface. For example, in Fig. 5(a), a melt anomaly can be seen clearly in addition to details of the surface, shape of the component, and boundaries of the foreground and background, etc. Although the melt anomaly is visually salient, it is challenging to be detected as other details are also evident, in particular for regions of focus. On the contrary, in Fig. 5(b), it is too dark to see every detail except for salient irregularities from the rough surface caused by the diffuse

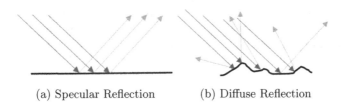

(a) Specular Reflection   (b) Diffuse Reflection

Fig. 4. Example of reflections. (a) Specular reflection on smooth surface, and (b) diffuse reflection on rough surface.

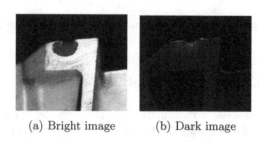

(a) Bright image   (b) Dark image

Fig. 5. Example of an anomaly under different lighting conditions.

reflection. This diffusion effect emphasizes anomaly regions while eliminating object details. Therefore, in this paper, we aim at detecting different anomalies such as melt, plus metal, scratch and scuff using dark images based on their salient irregularities.

3.2   Automatic Image Selection

To get such dark images, we manually labeled a small collection of images according to their lighting conditions and trained a classifier using support vector machine (SVM). There are two advantages to using this method for automatic image selection. First, this is a simple and intelligent method for selecting images. Second, it is easy to be applied to all test samples without considering their shape and the ratio of background and foreground. More specifically, a small subset of dark, gray and bright images are chosen for training. In our experiments, we use 220, 70 and 74 images, accordingly. Then a multi-class SVM model is used to learn a dark/gray/light classifier, and only dark image outputs are adopted for anomaly detection.

3.3   Phase Fourier Reconstruction (PFR)

Although shape and design of the component, and boundaries of foreground and background are still visible in the selected dark images, they are de-emphasized due to lower pixel intensities on these smooth surfaces. Instead, anomaly region is emphasized because of the diffuse reflection on rough surfaces, as shown in Fig. 5(b). In this section, we proposed to use PFR for anomaly detection in such dark images. The proposed PFR removes regular patterns and homogeneous regions, but highlights salient regions which may likely be an anomaly. After PFR, a 2D Gaussian filter and a thresholding method are applied to generate the corresponding saliency map. The key idea of the PFR is as follows. First, we transform an image from its spatial domain to the frequency domain represented by magnitude and phase using Fourier transform. Then, the image is reconstructed back to its spatial domain using phase with unity magnitude. Corresponding color information is removed from the image by setting the magnitude to unity, and distinct irregularities such as edge boundaries are highlighted due to the phase effect. More specifically, given an image f with size of $N \times M$, the two-dimensional discrete Fourier transform can be defined as follows:

$$F(u,v) = \sum_{x=0}^{N-1} \sum_{y=0}^{M-1} f(x,y) e^{-j2\pi(\frac{ux}{N} + \frac{vy}{M})} \tag{1}$$

where $f(x,y)$ is the pixel in the spatial domain, $F(u,v)$ is the point in the Fourier domain, and u and v are spatial frequencies in x and y directions, respectively. We can rewrite the equation to separate the real and imaginary parts:

$$F(u,v) = F_R(u,v) + jF_I(u,v) \tag{2}$$

where $F_R(u,v)$ is the real part of $F(u,v)$, and $F_I(u,v)$ is the imaginary part of F(u,v) in the complex domain. It can also be expressed in polar form in terms of magnitude and phase:

$$F(u,v) = |F(u,v)|e^{j\angle F(u,v)} \tag{3}$$

where $|F(u,v)|$ and $\angle F(u,v)$ are magnitude and phase, respectively, which can be calculated using $F_R(u,v)$ and $F_I(u,v)$ as below:

$$|F(u,v)| = \sqrt{F_R(u,v)^2 + F_I(u,v)^2} \tag{4}$$

$$\angle F(u,v) = \arctan \frac{F_I(u,v)}{F_R(u,v)} \tag{5}$$

The PFR can be performed by providing the inverse Fourier transform of $F(u,v)'$ using the following equation for image reconstruction:

$$F(u,v)' = e^{j\angle F(u,v)} \tag{6}$$

where the magnitude is set to one, $|F(u,v)| = 1$, and the original phase is retained. The reconstructed image is denoted by $f(x,y)'$. Then a 2D Gaussian filter is applied on the reconstructed image $f(x,y)'$ to get a smoother image $g(x,y)$ with reduced noise. The standard deviation of the Gaussian distribution is equal to 3 in the experiment. Finally, a thresholding method is used to get the saliency map s(x,y):

$$s(x,y) - \begin{cases} 255 & \text{if } g(x,y) \geq \mu + C\sigma \\ 0 & \text{otherwise} \end{cases} \tag{7}$$

where μ is the mean value of g(x,y), σ is the standard deviation of g(x,y) and C is a constant which is set to 6 in our experiment. This means that only the pixel whose intensity is much larger than its mean is picked as anomaly. We show the PFR method in Algorithm 1.

Algorithm 1. Phase Fourier Reconstruction for Anomaly Detection

---

Require: Given a dark image f
Ensure: Saliency map s for anomaly detection
 1: Fourier Transform. Get F(u,v) from f(x,y) using Eqs. (1) and (3).
 2: Remove Magnitude. Get F(u,v)' using Eq. (6).
 3: Phase Reconstruction. Get f'(x,y) using inverse Fourier transform on F(u,v)'.
 4: Smoothing. Get g(x,y) using 2D Gaussian filter on f(x,y)'.
 5: Thresholding. Get s(x,y) from g(x,y) using Eq. (7).

---

4   Experimental Results

In this section, we analyze and evaluate the proposed PFR method for anomaly detection. We compare our method with four well-designed saliency detection methods [12,16,18,19], a keypoint detection approach [8], and a template matching method [15]. The data used for this experiment is a subset of the actual data set which was acquired as a part of the Automated Visual Inspection process with about 18,600 images of metallic components captured at pre-defined viewpoints and lighting conditions.[1] Each image has a resolution of 2448 × 2050 pixels. All methods are applied to 13 pairs of bright and dark images obtained using a trained SVM classifier for image selection with different anomalies on metal surfaces. The visualization of the resulting saliency maps for these bright and dark images are shown in Figs. 6 and 7, respectively.

4.1   Comparison Between Dark and Bright Images Using PFR

The detection results of the proposed PFR method applied to the bright images are shown in Fig. 6(b). When applying PFR on the bright images, PFR fails in several cases as the details of the components are also salient along with the anomalies. Hence, no prominent peaks can be found after statistical thresholding. On the contrary, PFR works well on dark images because we utilize salient irregularities caused by diffuse reflection so that the details on smooth metal surfaces become visually obscure. This pattern-like effect makes anomalies salient among those regularities, as shown in Fig. 7(b). Comparing these results to the ground truth in Fig. 7(a), it can be observed that the proposed PFR method detects all the anomalies successfully. It is worth mentioning that while all dark images are adjusted for visualization, some of the anomalies still cannot be seen clearly. In such cases, please refer to the corresponding images in Fig. 6(a).

4.2   Comparison with Saliency Detection Methods

In this section, we compare our results with four well-established methods for saliency detection. They are geodesic saliency (GS) [16], manifold ranking (MR) [18], saliency filter (SF) [12] and saliency optimization (SO) [19]. Results are

---

[1] Limited disclosure due to confidentiality reasons.

Average F1 score	38.1	13.4	14.5	10.1	9.3	5.7	40.3
% of images with F1 scores	53.8	46.2	69.2	46.2	84.6	100	53.8

(a) Input   (b) PFR   (c) GS [16]   (d) MR [18]   (e) SF [12]   (f) SO [19]   (g) KD [8]   (h) TM [15]
            (our method)

Fig. 6. Saliency maps from different anomaly detection methods applying on bright images. (a) Bright images with bounding boxes indicating locations of anomalies, (b) results from the proposed PFR, (c) Geodesic Saliency (GS) [16], (d) Manifold Ranking (MR) [18], (e) Saliency Filter (SF) [12], (f) Saliency Optimization (SO) [19], (g) Keypoint Detection (KD) [8], and (h) Template Matching (TM) [15].

Average F1 score	68.9	13.5	14.0	46.1	38.4	61.6	22.5
% of images with F1 scores	100.0	53.8	61.5	38.5	38.5	38.5	7.7

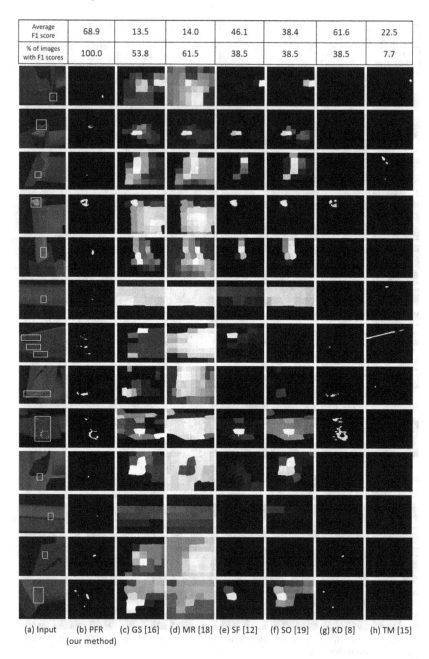

(a) Input   (b) PFR   (c) GS [16]   (d) MR [18]   (e) SF [12]   (f) SO [19]   (g) KD [8]   (h) TM [15]
             (our method)

Fig. 7. Saliency maps from different anomaly detection methods applying on dark images. (a) Dark images with bounding boxes indicating locations of anomalies (images are adjusted for visualization), (b) results from the proposed PFR, (c) Geodesic Saliency (GS) [16], (d) Manifold Ranking (MR) [18], (e) Saliency Filter (SF) [12], (f) Saliency Optimization (SO) [19], (g) Keypoint Detection (KD) [8], and (h) Template Matching (TM) [15].

shown in columns (c), (d), (e) and (f) of Figs. 6 and 7. Different from our application, these methods are designed for detecting saliency maps using visually distinct images. From the visualization, we can see that these saliency methods fail to work well in both bright and dark cases. Most of them separate foreground and background or detect regions in focus. This is because they use background/foreground information such as background connectivity [16,19], background and foreground cues [18] and contrast based filtering [12] to extract the saliency map of an image. On the other hand, in our images, we have a large portion of the component with silver metal surface and black background. Only a small region contains anomalies, so it is not surprising that these algorithms miss the anomaly regions although some of them are visually salient.

4.3   Comparison with a Keypoint Detection Method

In this section we compare our method with a popular keypoint detection (KD) method, SIFT [8], to evaluate the performance. Figures 6(g) and 7(g) show the results using SIFT on the bright and dark images, respectively. SIFT keypoints perform well on those anomalies that shine in the dark images due to reflection; However, it cannot detect those irregularities which do not shine. On the other hand, when applying SIFT on the bright images, it detects those focus regions or component shapes instead of anomalies. Instead, the proposed PFR can detect most of the anomalies in the dark images well.

4.4   Comparison with a Template Matching Method

We compare our method with a template matching (TM) method [15] which is also designed for anomaly detection using images captured from AVI. Results are shown in both Figs. 6(h) and 7(h), respectively. We can see that the template matching method performs well on some of the bright images where those anomalies are visually salient. However, it fails to detect anomalies on dark images. The main reason is that finding the best-match template requires viewpoint matching which could be difficult to perform in the dark.

4.5   Evaluation

In this section, we calculate the F1 score to evaluate the performance of different methods. The equation of the F1 score is defined as follows:

$$F1 = 2 \times \frac{Precision \times Recall}{Precision + Recall} \tag{8}$$

where precision is the percentage of detected results that are correct and recall is the percentage of ground truth that is detected. The F1 score has a range from a worst score of 0 to a best score of 100. It should be noted that the F1 score is undefined when the Precision + Recall = 0 which occurs in images where only false alarms are detected, or when the Precision is undefined which occurs in images whose saliency maps are NULL without any detection. Hence, the average F1 score is computed using the sum of valid F1 scores dividing by the number of test images considered. Individual results are listed at the top of each corresponding column in Figs. 6 and 7, respectively. Figure 8 shows four quadrants separating the average F1 score and the percentage of images considered into high and low regions for comparison. Only PFR applied on dark images can achieve a high F1 score with 100% of test images considered as shown in Quadrant I. Besides, the processing time per image is less than 0.01 seconds. While 40% of anomalies in the dark images are detected using Keypoint Detection (KD), as shown in Quadrant II, the remaining images are too dark to be detected. Although Salient Filter (SF) and Salient Optimization (SO) detect some anomalies on dark images similar to KD, their average F1 scores are lower and below 50% as shown in Quadrant III as they have more false alarms or missed detections. It is also worth mentioning that, Template Matching (TM) and PFR applied on bright images can detect some anomalies as well; however, similar to SF and SO, they also detect more false alarms or missed detections as shown in Quadrant IV.

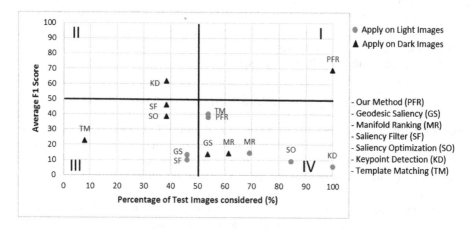

Fig. 8. Results between average F1 score and percentage of images with F1 scores from different approaches applying on bright and dark images

5   Conclusion

In this paper, we have proposed a phase Fourier reconstruction method (PFR) for anomaly detection on metal surface which utilizes the characteristics of salient irregularities. Experiments on anomalies clearly demonstrate that our method outperforms the well-designed saliency detection methods, the template matching approach, and the keypoint detection method. We have shown the effectiveness and efficiency of the proposed method which is applicable for real industrial applications.

Acknowledgement. This work was conducted within Rolls-Royce@NTU Corporate Lab with the support of National Research Foundation under the CorpLab@University scheme.

References

1. Aiger, D., Talbot, H.: The phase only transform for unsupervised surface defect detection. In: IEEE Conference on Computer Vision and Pattern Recognition (CVPR), pp. 295–302, June 2010
2. Bai, X., Fang, Y., Lin, W., Wang, L., Ju, B.F.: Saliency-based defect detection in industrial images by using phase spectrum. IEEE Trans. Ind. Inform. **10**(4), 2135–2145 (2014)
3. Chan, C.H., Pang, G.K.H.: Fabric defect detection by fourier analysis. IEEE Trans. Ind. Appl. **36**(5), 1267–1276 (2000)
4. Gao, H., Ding, C., Song, C., Mei, J.: Automated inspection of e-shaped magnetic core elements using K-tSL-center clustering and active shape models. IEEE Trans. Ind. Inform. **9**(3), 1782–1789 (2013)
5. Guo, C., Ma, Q., Zhang, L.: Spatio-temporal saliency detection using phase spectrum of quaternion fourier transform. In: IEEE Conference on Computer Vision and Pattern Recognition (CVPR), June 2008
6. Hou, X., Zhang, L.: Saliency detection: a spectral residual approach. In: IEEE Conference on Computer Vision and Pattern Recognition (CVPR), June 2007
7. Li, W.C., Tsai, D.M.: Defect inspection in low-contrast LCD images using hough transform-based nonstationary line detection. IEEE Trans. Ind. Inform. **7**(1), 136–147 (2011)
8. Lowe, D.G.: Distinctive image features from scale-invariant keypoints. Int. J. Comput. Vis. **60**(2), 91–110 (2004)
9. Mattoccia, S., Tombari, F., Stefano, L.D.: Efficient template matching for multichannel images. Pattern Recogn. Lett. **32**(5), 694–700 (2011)
10. Neogi, N., Mohanta, D.K., Dutta, P.K.: Review of vision-based steel surface inspection systems. J. Image Video Process. **2014**(1), 1–19 (2014)
11. Ngan, H.Y., Pang, G.K., Yung, N.H.: Automated fabric defect detection review. Image Vis. Comput. **29**(7), 442–458 (2011)
12. Perazzi, F., Krhenbhl, P., Pritch, Y., Hornung, A.: Saliency filters: contrast based filtering for salient region detection. In: IEEE Conference on Computer Vision and Pattern Recognition (CVPR), pp. 733–740, June 2012
13. Tsai, D.M., Hsieh, C.Y.: Automated surface inspection for directional textures. Image Video Comput. **18**(1), 49–62 (1999)
14. Tsai, D.M., Wu, S.C., Chiu, W.Y.: Defect detection in solar modules using ICA basis images. IEEE Trans. Ind. Inform. **9**(1), 122–131 (2013)
15. Vaikundam, S., Hung, T.Y., Chiat, L.T.: Anomaly region detection and localization in metal surface inspection. In: IEEE Conference on International Conference of Image Processing (ICIP), September 2016
16. Wei, Y., Wen, F., Zhu, W., Sun, J.: Geodesic saliency using background priors. In: Fitzgibbon, A., Lazebnik, S., Perona, P., Sato, Y., Schmid, C. (eds.) ECCV 2012. LNCS, vol. 7574, pp. 29–42. Springer, Heidelberg (2012). doi:10.1007/978-3-642-33712-3_3
17. Xie, X.: A review of recent advances in surface defect detection using texture analysis techniques. Electron. Lett. Comput. Vis. Image Anal. **7**(3), 1–22 (2008)

18. Yang, C., Zhang, L., Lu, H., Ruan, X., Yang, M.H.: Saliency detection via graph-based manifold ranking. In: IEEE Conference on Computer Vision and Pattern Recognition (CVPR), pp. 3166–3173, June 2013
19. Zhu, W., Liang, S., Wei, Y., Sun, J.: Saliency optimization from robust background detection. In: IEEE Conference on Computer Vision and Pattern Recognition (CVPR), pp. 2814–2821, June 2014

ReMagicMirror: Action Learning Using Human Reenactment with the Mirror Metaphor

Fabian Lorenzo Dayrit[1]([✉]), Ryosuke Kimura[3], Yuta Nakashima[1],
Ambrosio Blanco[2], Hiroshi Kawasaki[3], Katsushi Ikeuchi[2], Tomokazu Sato[1],
and Naokazu Yokoya[1]

[1] Nara Institute of Science and Technology,
Takayamacho 8916-5, Ikoma, Nara 630-0101, Japan
fabian-d@is.naist.jp
[2] Microsoft Research Asia, Building 2, No. 5 Dan Ling Street,
Haidian District, Beijing 100080, China
[3] Kagoshima University, Korimoto 1-21-24, Kagoshima 890-8580, Japan

Abstract. We propose ReMagicMirror, a system to help people learn actions (e.g., martial arts, dances). We first capture the motions of a teacher performing the action to learn, using two RGB-D cameras. Next, we fit a parametric human body model to the depth data and texture it using the color data, reconstructing the teacher's motion and appearance. The learner is then shown the ReMagicMirror system, which acts as a mirror. We overlay the teacher's reconstructed body on top of this mirror in an augmented reality fashion. The learner is able to intuitively manipulate the reconstruction's viewpoint by simply rotating her body, allowing for easy comparisons between the learner and the teacher. We perform a user study to evaluate our system's ease of use, effectiveness, quality, and appeal.

Keywords: 3D human reconstruction · Human reenactment · RGB-D sensors

1 Introduction

When people want to learn an action, e.g., a martial arts performance or a dance, one of the most intuitive ways is to watch a teacher performing the action and imitate it. This can be done in person, e.g., how most martial arts are traditionally learned, or from a recording, such as a video of teacher performing the action. Both ways have advantages and disadvantages: imitating a teacher in-person is limited by the availability of the teacher regarding time and place, but the learner is free to observe the action from any point of view. In contrast, the video may be watched anytime and anywhere, but is limited to the original capturing point of view, which may pose problems for difficult or hard-to-understand actions.

The original version of this chapter was revised: An acknowledgement has been added. The erratum to the chapter is available at DOI: 10.1007/978-3-319-51811-4_60

© Springer International Publishing AG 2017
L. Amsaleg et al. (Eds.): MMM 2017, Part I, LNCS 10132, pp. 303–315, 2017.
DOI: 10.1007/978-3-319-51811-4_25

There are several technical remedies that try to provide convenience for learners without spoiling the capability to view the action from an arbitrary viewpoint as in-person learning does. Most of them make use of augmented reality (AR), which is the technique of visualizing virtual objects in the real world. Several systems [11,12] provide visual guidance for tasks by overlaying, e.g., arrows and labels on key objects. We especially draw inspiration, however, from systems that allow easy comparison with the learner's own body by using a mirror [1,4].

Fig. 1. Our ReMagicMirror system. The learner is mirrored on the left in the screen, and the reenactment of the teacher is shown on the right.

This paper proposes a system that is able to be replayed like a video and allows viewing from an arbitrary viewpoint, combining these with the mirror metaphor to help learners comprehend difficult actions. Our system is centered around *reenactments* of motion, i.e. a sequence of novel views of a person in motion, and the system is called ReMagicMirror (Reenactment-based Magic Mirror). This system captures and displays reenactments of the teachers' motions in order to aid learner comprehension. It consists of a large screen that mirrors the learner, upon which the system then overlays the teacher's reenactment as shown in Fig. 1. The reenactment should be rendered from an intuitive viewpoint for the learner, i.e., matching the learner's own body orientation. To view the action from the side, for example, the learner has only to turn his/her own body to the side, and by doing so can easily see the differences between his/her side view and the teacher's. This makes it easy to compare two actions.

Our system acquires the reenactments by fitting a parametric model to two RGB-D sequences: First, we capture the teacher's motion sequence using two RGB-D cameras, acquiring the entire shape of the teacher. Next, the system generates a 3D mesh sequence by fitting a parametric model, such as [3,6], to the scans. Finally, the motion sequence is overlaid on top of a screen mirroring our learner (Fig. 1).

The main contributions of this paper are summarized as follows:

1. A novel method to synthesize views of humans in motion, called reenactments, using two RGB-D streams.
2. ReMagicMirror, an end-to-end reenactment display system, that helps learners by overlaying a mirror of them with teachers' reenactments with easily-controlled viewpoints.
3. User study results to show how well the proposed reenactment display system helps learners.

2 Related Works

Our system has its roots in two main fields of research: human shape reconstruction and augmented reality for learning.

2.1 Human Shape Reconstruction

For our system, we render novel views of the teacher in motion by first reconstructing the teacher's shape and motion from two depth cameras, rendering these from the desired viewpoint, and finally displaying it to the learner. Shape reconstruction is an active research field and techniques here are mainly divided into *model-free* and *model-based* approaches, referring to the usage or non-usage of a human shape model.

The model-free approach requires no prior data on human shape and makes no assumptions on the person captured. Most recent methods of this type employ variants of the signed distance field technique, which is basically a registration problem of multiple depth maps and represents 3D shape using the zero-level iso-surface of the signed distance field. This approach was originally designed for rigid scenes, and one of the more well-known examples is KinectFusion [15]. This approach was later extended to handle non-rigid objects by describing the deformation of objects with transformations of signed distance field [9,13,14]. These methods can generate surprisingly high quality 3D shapes, but may lack tracking stability with regards to, e.g., occlusions.

In contrast to the model-free approach the *model-based* approach uses prior knowledge on the object to be captured to facilitate entire object reconstruction, and most existing methods that take this approach are designed for human body reconstruction. Most methods of this type use a parametric model of human body shape, such as SCAPE [3] and TenBo [6]. These methods in particular describe plausible body shapes using pose and shape parameters that control the human body's attributes like weight, height, etc.

Existing methods that adopt the model-based approach basically fit the parametric model to a point cloud of depth observations. For example, Weiss et al. [17] proposed a system using SCAPE, where the fitting process is initialized with skeletal tracking results. Bogo et al. [5] use SCAPE with several modifications including multi-resolution mesh fitting and using displacement maps for finer details. Due to the modification of multiple resolution meshes, their system no longer relies on skeletal tracking, which is error-prone.

Since we know we are targeting humans, our system uses a model-based approach for stability. We fit a TenBo model [6] to our input depth sequences, applying temporal constraints for smoothness. This has the following advantages.

1. Our input depth maps give no guarantee that they cover the entire body of the teacher due to, e.g., self-occlusion. Fitting a body model to the visible regions gives us a plausible estimate of the unobserved ones.
2. Our fitting process is constrained in two ways to increase stability. Firstly, since our system is designed for capturing a single person, i.e., a teacher, we keep the same shape parameters of the TenBo model for the entire sequence. Also, we can assume that each frame in a sequence is captured a few milliseconds (at most 33 ms for a Kinect) after the previous one, which let us add temporal smoothness constraints to the fitting process.
3. The results of fitting include pose parameters, from which we can derive the relative camera view. We use this during the viewing stage in order to display the most appropriate viewpoint to the learner.

2.2 Augmented Reality for Learning

Using AR technology for learning has become a popular topic in recent years. AR visualizes virtual objects in the real world, which can range from placing labels on top of real objects to rendering entire virtual objects in real environments. Several systems implement AR techniques for the purpose of learning. For example, Henderson and Feiner [11] proposed a system to help users learn a procedure of actions by attaching a sequence of 3D arrows to key objects. Another system [12] generates virtual targets for rehabilitating users who have had a stroke. Dayrit et al. [8] proposed a similar system that renders a teacher through a handheld device, such as a tablet, using AR technology. The main technical difference of our work is that we adopt a parametric 3D human shape model to improve visual quality.

The mirror metaphor in particular is well-suited to AR, as it is instantly understandable. Mirracle [4] is a system for anatomy education that simulates a mirror and projects virtual organs in the appropriate place on top of the learner's body in the mirror. Here, the mirror metaphor allows users to easily learn about their own bodies. Another system employing the mirror metaphor, YouMove [1], projects a teacher's motions as stick figures on a half-silvered mirror, allowing learners to align their body with the stick figures in the mirror. In their user study, participants were asked to imitate motions either from videos or using the system, and those using the system were noticeably closer to the teacher's motion. We thus believe that using the mirror metaphor to let learners compare their own motion to the virtual teacher's motion will be helpful for action learning.

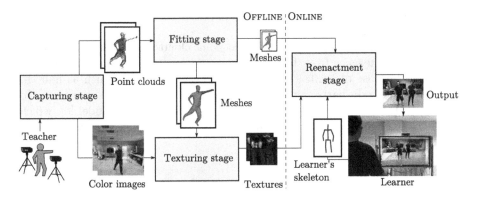

Fig. 2. System overview.

3   ReMagicMirror System

Figure 2 shows an overview of our ReMagicMirror system. The system has three offline stages and one online one. The offline portion captures and builds the reenactment, and is composed of the capturing, fitting, and texturing stage. In the capturing stage, a teacher captures his/her action with two RGB-D sensors with known relative poses. Our system merges the point clouds in the two RGB-D streams from the two sensors, feeding the merged point clouds into the fitting stage. Since the point clouds usually have unobserved regions due to, e.g., occlusion, we fit the parametric 3D human model [6] to each merged point cloud to obtain a mesh sequence. The texturing stage extracts the texture applied to each triangle of the every frame from RGB images. In the online reenactment stage, our system presents, to the learner, his/her flipped image for the mirror metaphor. The system also provides the teacher's reenactment as shown in Fig. 1 so that the learner can easily imitate the teacher's action. For each playback, the learner can intuitively determine the orientation of the reenactment, observing it from a desired direction. The following sections detail each stage.

3.1   Capturing Stage

In the capturing stage, our system records an action of the teacher using a pair of RGB-D sensors facing each other. The relative pose between these two sensors is calculated, and they are manually synchronized. Since we require the depth and color pixels belonging to the teacher, separate from the background, we extract the teacher's region using such a method as [16]. After extracting the teacher's region, we regain the 3D position of each depth pixel to form a point cloud. We merge the two point clouds from the pair of sensors using the relative pose calculated above.

We denote the f-th frame point cloud with N_f points, by

$$Z_f = \{\mathbf{z}_{fn} | n = 1, \ldots, N_f\}, \tag{1}$$

and the RGB images from first and second sensors as I_f^1 and I_f^2, respectively.

3.2 Fitting Stage

Figure 3(a, top) shows examples of merged point clouds. Generally, even though we capture the teacher from both his front and back, the point cloud can be incomplete because of occlusion or difficult-to-capture regions such as hair. In addition, some body parts can partially be out of the sensors' field of view. To reconstruct the complete shape of his body, we fit the TenBo model [6], which is a state-of-the-art statistical human shape model, to each point cloud.

(a)   (b)

Fig. 3. (a) Top: Example input point clouds. Middle: Examples of fit meshes. Bottom: Textured meshes. (b) Segmented reference mesh, front and back. Each color in (b) represents one of the 13 body parts: head, shoulders, upper arms, lower arms, torso, abdomen, upper legs, and lower legs. (Color figure online)

Training a statistical human shape model, usually requires a large amount of registered meshes of multiple subjects in various poses. We used the MPII dataset [10], which contains over 500 registered meshes. For stable fitting, we selected a mesh and reduced the number of vertices in it from 6,449 to 502 using the quadric edge collapse decimation algorithm [7]. From here we treat the decimated mesh as the reference.

This decimation is transferred to all other meshes in the dataset as they are registered, i.e., we keep the same vertices in a mesh as the reference and use the edges in the reference instead of the original ones. We refer to the reference as

$$M_X = \{X, E\}, \tag{2}$$

where $X = \{\mathbf{x}_j | j = 1, \ldots, J\}$, $\mathbf{x}_j$ being the j-th vertex, and E contains the pairs of vertex indices that form the edges of the reference. The TenBo model also requires segmenting the mesh into body parts so that each body part is not subjected to excessive deformation. Instead of using an automatic approach, such as [2], we manually segmented the mesh as in Fig. 3(b).

The TenBo model, like other parametric shape models such as [3], regresses a deformation matrix of each triangle in the reference given the body part poses Θ and shape parameter $\mathbf{v}$, where $\Theta = \{\boldsymbol{\theta}_l | l = 1, \ldots, L\}$ is a set of rotation representations for all body parts. Letting $\mathbf{D}_k(\Theta, \mathbf{v})$ be the deformation matrix and $\mathbf{R}(\boldsymbol{\theta}_l)$ be the rotation matrix obtained from $\boldsymbol{\theta}_l \in \Theta$ for body part l, the deformed triangle k's edges, which are called triangle vectors, $\triangle\mathbf{y}_{k1}$ and $\triangle\mathbf{y}_{k2}$ can be given by

$$\triangle\mathbf{y}_{k1} = \mathbf{R}(\boldsymbol{\theta}_l)\mathbf{D}_k(\Theta, \mathbf{v})\triangle\mathbf{x}_{k1}$$
$$\triangle\mathbf{y}_{k2} = \mathbf{R}(\boldsymbol{\theta}_l)\mathbf{D}_k(\Theta, \mathbf{v})\triangle\mathbf{x}_{k2},$$

where $\triangle\mathbf{x}_{km} = \mathbf{x}_{km} - \mathbf{x}_{k0}$ and $\mathbf{x}_{km}$ $(m = 0, 1, 2)$ is in X and forms a triangle of the mesh. In the above equation, l is the body part that triangle k belongs to.

The fitting algorithm tries to find the body part poses Θ and the shape parameter $\mathbf{v}$. We modify the fitting algorithm in [6] to take advantage of the temporal continuity of meshes in successive frames. More specifically, we apply an additional smoothness term for the pose parameters that penalizes pose differences between adjacent frames, as well as modifying the shape parameter fitting to simultaneously take multiple frames into account. The optimization involves three terms: the model error term $\mathcal{M}$, the point cloud error term $\mathcal{P}$, and the temporal pose smoothness term $\mathcal{R}$.

The model error term penalizes the difference between the TenBo model-based body shape prediction and the deformed mesh Y_f in frame f. $\triangle\mathbf{y}_{fkt}$ is triangle vector $t \in \{1, 2\}$ of triangle k in frame f, the term is given by

$$\mathcal{M}(Y_f, \Theta_f, \mathbf{v}) = \sum_{k=1}^{K} \sum_{t} \|\mathbf{R}(\boldsymbol{\theta}_{fl})\mathbf{D}_k(\Theta_f, \mathbf{v})\triangle\mathbf{x}_{kt} - \triangle\mathbf{y}_{fkt}\|^2. \quad (3)$$

The point cloud error term $\mathcal{P}$ for frame f is the difference between the deformed mesh Y_f and the point cloud Z_f. As there are no explicit correspondences between the deformed mesh and the point cloud, we first use the rigid iterative closest point (ICP) algorithm to bring the mesh into rough alignment, then assign correspondences by nearest neighbor. Using $\tilde{\mathbf{y}}_f(\mathbf{z}_{fn})$ as the nearest vertex in Y_f to point cloud point $\mathbf{z}_{fn}$, the point cloud error term is

$$\mathcal{P}(Y_f) = \sum_{n} \|\tilde{\mathbf{y}}_f(\mathbf{z}_{fn}) - \mathbf{z}_{fn}\|^2. \quad (4)$$

The pose smoothness term $\mathcal{R}$ for frame f penalizes large differences in pose between frames. Due to our assumption of fitting depth image sequences, we do not want subsequent frames to vary wildly. This term increases fitting robustness. The term is defined as the sum of squared Frobenius norms:

$$\mathcal{R}(\Theta_f, \Theta_{f+1}) = \sum_{l} \|\mathbf{R}(\boldsymbol{\theta}_{fl}) - \mathbf{R}(\boldsymbol{\theta}_{(f+1)l})\|_{\text{fro}}^2. \quad (5)$$

The final meshes $M_{Y,f} = \{Y_f, E\}$ can be found by minimizing the following objective with respect to Y_f and Θ_f for $f = 1, \ldots, F$ as well as $\mathbf{v}$:

$$\sum_{f=1}^{F} [\mathcal{M}(Y_f, \Theta_f, \mathbf{v}) + w_z \mathcal{P}(Y_f)] + w_r \sum_{f=1}^{F-1} \mathcal{R}(\Theta_f, \Theta_{(f+1)}). \tag{6}$$

We cannot handle all frames at once because of memory requirements. We instead use a sliding window of three frames at a time with the second and third frames' parameters being updated (frames 1 and 2 are independently minimized). The minimization is done using coordinate descent. In each iteration, we first minimize with respect to Y_f, and then Θ_f. Assuming that shape parameter $\mathbf{v}$ does not change along with the sequence, we deal with $\mathbf{v}$ only in the minimization for frame 1, and use the value for the rest of minimization processes. Figure 3(a, middle) shows examples of fit meshes.

3.3 Texturing Stage

Our system extracts textures from RGB images I_f^1 and I_f^2 from the first and second sensors using $M_{Y,f}$ $(f = 1, \ldots, F)$. For each triangle in frame f, we project its vertices $\mathbf{y}_{km}$ to I_f^1 and I_f^2. Since the image region corresponding to a triangle may not necessarily be visible (e.g., an arm may be occluding the body), we must detect and handle such regions.

To do this, we generate a depth map of $M_{Y,f}$ for each sensor that captures I_f^1 and I_f^2, and project a vertex to them. If the depth component of one of the vertices in a triangle is inconsistent with the corresponding depth value by a threshold T, we deem the triangle not visible. If the triangle is not visible from both sensors, we use the averaged texture calculated over corresponding visible triangles in the entire sequence. Figure 3(a, bottom) shows some examples of textured meshes. We create a 1024×1024 texture per frame.

3.4 Reenactment Stage

In the reenactment stage, the system reenacts the captured action and presents it to the learner through our interface with the mirror metaphor. This section describes reenactment generation and the interface in detail.

Figure 1 shows the configuration of our system's learning interface. The interface has one RGB-D sensor to capture the learner and the environment as well as a screen to present the captured live video stream from the sensor and the reenactment of the teacher. The RGB image in the live video stream is flipped before it is presented to the learner so that it appears like a mirror. Note that the image is not a true mirror image as the RGB-D sensor is on top of the screen. We however consider it similar enough to the learner's mental model of a mirror.

One key aspect of our system is that it can present the teacher's action from any direction that the learner wants. For this, we use a skeleton tracker (e.g., [16]) to obtain the learner's shoulders' position and compute the learner's direction. After a fixed amount of time, the system fixes the rotation of the teacher's reenactment and starts playing the action.

4   Evaluation

To implement our system, we used two Microsoft Kinect v2s as our RGB-D sensors. We used Kinect v2 SDK for extracting the teacher's region in depth maps and for skeleton tracking. The fitting stage is implemented on a Windows PC with 3.20 GHz CPU and 32 GB memory. Optimization process (Eq. (6)) takes around 5 min per frame. We use $w_z = 1$, $w_r = 0.05$, and $T = 10$ cm. For the reenactment stage, the screen is 165×97 cm. The system was implemented on a Windows PC with 3.40 GHz CPU and 8 GB memory. It runs at 20 FPS.

We conducted an objective evaluation to demonstrate how well our system helps users learn actions and a survey to subjectively evaluate our system in terms of ease of use, effectiveness, graphics quality, and appeal.

4.1   Objective Evaluation of Effectiveness

We compared the system against the process of learning by imitating a video. We recorded four Taekwondo actions (A, B, C, and D) for this purpose, ranging from 4–12 s long[1]. We divided the actions into two groups: Group 1, consisting of actions A and B, where the teacher mainly faced forward, and group 2, consisting of actions C and D, with no restriction. Users learned one action from each group using the system, and the other with the video.

For this evaluation, we recruited 14 users with ages ranging from 20–30, with 3 female and 11 male users. The process of learning an action is as follows: First, we show a video of the action to the user. Next, we establish a baseline by having the user perform the action and recording it, while the video plays again. After that, the user learns the action by practicing it over and over. The practice is accompanied either with a video of the action looping repeatedly, or with our system looping the reenactment repeatedly. For our system, the user can freely change the viewing direction before every repetition. Finally, we test the user's learning by playing the video or the reenactment one last time and recording, comparing it to the baseline.

We measured the error by recording the users' motion using a Kinect v2. Since we play the video or the reenactment at the same time that the users perform the action, we are able to match body pose frames up one to one and compare each frame directly. We compare body part orientations, normalizing all orientations relative to the spine.

Figure 4 summarizes the results of our experiment. For all sequences, those using our system were able to follow our teacher's motions more closely compared to the pre-test and those learning from a video. In fact, those learning from the

---

[1] The videos of the actions may be viewed at https://db.tt/qupIZ91a.

video barely changed from the pre-test. We consider that this is due to the fact that the user is not able to see their mistakes, while our system makes it easy to do so, allowing users to adjust their motions to better copy the teacher's by observing the teacher from desired directions.

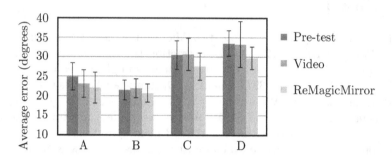

Fig. 4. Average error in degrees per joint, per frame, between the user and the teacher, for action sequences A, B, C, and D.

4.2 Survey

We asked the same users to try out 2 other reenactment methods: the untextured full mesh, and the skeleton of the teacher (Fig. 5). Finally, our users answered a survey consisting of 8 questions with the goal of evaluating the system's perceived ease of use, effectiveness, quality, and appeal (Fig. 6).

Fig. 5. (a) Textured full mesh reenactment. (b) Untextured full mesh reenactment. (c) Teacher skeleton reenactment.

Table 1 summarizes our users' responses. Most users preferred the reenactment with a fully textured mesh for all questions, even for the equivalent video questions. This means that users found our system easy to use, effective at helping them learn actions, having high output quality, and most would use a similar system given the chance. Many users also appreciated the mirroring as it was more difficult to tell left from right by watching the video.

Part 1. For each reenactment (Full mesh with full textures, untextured full mesh, skeletons only) and the video:

Q1 Was the reenactment/video comprehensible?
Q2 Was it easy to learn the motions using the system/video?

Part 2. For each reenactment (Full mesh with full textures, untextured full mesh, skeletons only):

Q3 Did the reenactment have good quality?
Q4 Did the reenactment resemble the original video?
Q5 Was it easy to manipulate the viewpoint to your desired one?
Q6 Were the differences between yourself and the reenactment clear?
Q7 Did changing the viewpoint help you learn the action?
Q8 Would you use this system in the future?

Fig. 6. Questions asked in our user study. Users answered from 1 (strongly disagree) to 5 (strongly agree).

Table 1. Users' averaged answers for the survey in Fig. 6, for full mesh with full textures (R1), untextured full mesh (R2), skeletons only (R3), and video (V). Users answered from 1 (strongly disagree) to 5 (strongly agree).

	R1	R2	R3	V
Q1	**4.11 ± 0.66**	3.86 ± 0.77	2.29 ± 1.07	3.86 ± 0.77
Q2	**4.29 ± 0.73**	3.71 ± 0.91	2.29 ± 0.83	2.93 ± 0.83
Q3	**4.00 ± 0.68**	3.71 ± 0.99	2.64 ± 1.22	—
Q4	**4.50 ± 0.65**	3.64 ± 1.08	2.50 ± 1.16	—
Q5	**3.93 ± 1.00**	**3.93 ± 1.00**	3.14 ± 1.29	—
Q6	**4.07 ± 1.21**	3.50 ± 1.22	2.29 ± 1.33	—
Q7	**4.00 ± 0.96**	3.79 ± 0.89	2.93 ± 1.14	—
Q8	**4.43 ± 0.85**	3.57 ± 1.02	2.00 ± 1.11	—

5 Conclusion

We have proposed and implemented an augmented reality system for helping users learn actions. The actions are performed by a teacher, and the system reconstructs the body and motion of the teacher using two RGB-D sensors. Using the reconstruction, the system overlays *reenactments*, which are novel views of the actions, onto a screen which also mirrors the learner. Learners are then are able to control the viewpoint intuitively by moving their own body. We conducted a user study, and found that this system allows for easy comparisons between learner and teacher, and users were able to perform more accurate motions using the system than with video. They appreciated the ability to intuitively control the point of view while comparing motions, which to our knowledge is unique to our system at the time of writing. In general, our users preferred learning using the system over watching a video.

From here, we have several possible avenues of improvement. One way is to further develop the application, for example by developing an automatic feedback system. Another way is to make capturing easier, for example by using only a single RGB-D sensor. Finally, the texture quality can also be improved by using a higher-resolution mesh.

Acknowledgement. This work is supported by MSR CORE 11/12 Project.

References

1. Anderson, F., Grossman, T., Matejka, J., Fitzmaurice, G.: YouMove: enhancing movement training with an augmented reality mirror. In: Proceedings of ACM Symposium on User Interface Software and Technology, pp. 311–320 (2013)
2. Anguelov, D., Koller, D., Pang, H.C., Srinivasan, P., Thrun, S.: Recovering articulated object models from 3D range data. In: Proceedings of Conference on Uncertainty in Artificial Intelligence, pp. 18–26 (2004)
3. Anguelov, D., Srinivasan, P., Koller, D., Thrun, S., Rodgers, J., Davis, J.: SCAPE: shape completion and animation of people. ACM Trans. Graph. **24**, 408–416 (2005)
4. Blum, T., Kleeberger, V., Bichlmeier, C., Navab, N.: mirracle: An augmented reality magic mirror system for anatomy education. In: Proceedings of IEEE Virtual Reality Workshops, pp. 115–116 (2012)
5. Bogo, F., Black, M.J., Loper, M., Romero, J.: Detailed full-body reconstructions of moving people from monocular RGB-D sequences. In: Proceedings of IEEE International Conference on Computer Vision, pp. 2300–2308 (2015)
6. Chen, Y., Liu, Z., Zhang, Z.: Tensor-based human body modeling. In: Proceedings of IEEE Computer Society Conference on Computer Vision and Pattern Recognition, pp. 105–112 (2013)
7. Cignoni, P., Callieri, M., Corsini, M., Dellepiane, M., Ganovelli, F., Ranzuglia, G.: Meshlab: an open-source mesh processing tool. In: Eurographics Italian Chapter Conference, pp. 129–136 (2008)
8. Dayrit, F.L., Nakashima, Y., Sato, T., Yokoya, N.: Increasing pose comprehension through augmented reality reenactment. Multimedia Tools Appl. 1–22 (2015). doi:10.1007/s11042-015-3116-1
9. Dou, M., Taylor, J., Fuchs, H., Fitzgibbon, A., Izadi, S.: 3D scanning deformable objects with a single RGBD sensor. In: Proceedings of IEEE Computer Society Conference on Computer Vision and Pattern Recognition, pp. 493–501 (2015)
10. Hasler, N., Stoll, C., Sunkel, M., Rosenhahn, B., Seidel, H.P.: A statistical model of human pose and body shape. Comput. Graph. Forum **28**, 337–346 (2009)
11. Henderson, S.J., Feiner, S.K.: Augmented reality in the psychomotor phase of a procedural task. In: Proceedings of IEEE International Symposium on Mixed and Augmented Reality, pp. 191–200 (2011)
12. Hondori, H.M., Khademi, M., Dodakian, L., Cramer, S.C., Lopes, C.V.: A spatial augmented reality rehab system for post-stroke hand rehabilitation. In: Proceedings of Medicine Meets Virtual Reality Conference, pp. 279–285 (2013)
13. Innmann, M., Zollhöfer, M., Nießner, M., Theobalt, C., Stamminger, M.: VolumeDeform: real-time volumetric non-rigid reconstruction. In: Leibe, B., Matas, J., Sebe, N., Welling, M. (eds.) ECCV 2016. LNCS, vol. 9912, pp. 362–379. Springer, Cham (2016). doi:10.1007/978-3-319-46484-8_22

14. Newcombe, R., Fox, D., Seitz, S.: DynamicFusion: reconstruction and tracking of non-rigid scenes in real-time. In: Proceedings of IEEE Computer Society Conference on Computer Vision and Pattern Recognition, pp. 343–352 (2015)
15. Newcombe, R.A., Izadi, S., Hilliges, O., Molyneaux, D., David Kim, A.J.D., Kohli, P., Shotton, J., Hodges, S., Fitzgibbon, A.: KinectFusion: real-time dense surface mapping and tracking. In: Proceedings of IEEE International Symposium on Mixed and Augmented Reality, pp. 127–136 (2011)
16. Shotton, J., Sharp, T., Kipman, A., Fitzgibbon, A., Finocchio, M., Blake, A., Cook, M., Moore, R.: Real-time human pose recognition in parts from single depth images. Commun. ACM **56**, 116–124 (2013)
17. Weiss, A., Hirshberg, D., Black, M.J.: Home 3D body scans from noisy image and range data. In: Proceedings of IEEE International Conference on Computer Vision, pp. 1951–1958 (2011)

Robust Image Classification via Low-Rank Double Dictionary Learning

Yi Rong[1,2], Shengwu Xiong[1(✉)], and Yongsheng Gao[2]

[1] School of Computer Science and Technology,
Wuhan University of Technology, Wuhan, China
r.yi@griffith.edu.au, xiongsw@whut.edu.cn
[2] School of Engineering, Griffith University, Brisbane, Australia
yongsheng.gao@griffith.edu.au

Abstract. In recent years, dictionary learning has been widely used in various image classification applications. However, how to construct an effective dictionary for robust image classification task, in which both the training and the testing image samples are corrupted, is still an open problem. To address this, we propose a novel low-rank double dictionary learning (LRD^{2}L) method. Unlike traditional dictionary learning methods, LRD^{2}L simultaneously learns three components from training data: (1) a low-rank class-specific sub-dictionary for each class to capture the most discriminative features owned by each class, (2) a low-rank class-shared dictionary which models the common patterns shared by different classes and (3) a sparse error container to fit the noises in data. As a result, the class-specific information, the class-shared information and the noises contained in data are separated from each other. Therefore, the dictionaries learned by LRD^{2}L are noiseless, and the class-specific sub-dictionary of each class can be more discriminative. Also since the common features across different classes, which are essential to the reconstruction of image samples, are preserved in class-shared dictionary, LRD^{2}L has a powerful reconstructive capability for newly coming testing samples. Experimental results on three public available datasets reveal the effectiveness and the superiority of our approach compared to the state-of-the-art dictionary learning methods.

Keywords: Low-rank dictionary learning · Class-specific dictionary · Class-shared dictionary · Robust image classification

1 Introduction

Image classification has been an active topic in the areas of machine learning and pattern recognition, due to its widely use in real-world applications, such as computational forensics, face recognition and medical diagnosis [1–3]. However, the images in the real-world applications are usually corrupted. The image corruptions, such as illuminations and disguises in face images, pose variants and sparse pixel noises in object images, will make the task of image classification more challenging.

In recent years, sparse representation based methods have led to the state-of-the-art results in image analysis. The core idea of sparse representation is to encode the test

© Springer International Publishing AG 2017
L. Amsaleg et al. (Eds.): MMM 2017, Part I, LNCS 10132, pp. 316–328, 2017.
DOI: 10.1007/978-3-319-51811-4_26

image as a linear combination of a few atoms chosen from a given dictionary [4, 5]. The dictionary plays an important role in the sparse coding process, and how to construct an effective dictionary is a key issue to the sparse representation based methods. Wright et al. [6] proposed a sparse representation based classifier (SRC) method, which is based on the data self-expressive property and seeks the sparse representation of the test image in terms of all the training samples. SRC has achieved promising results in many applications. However, directly using original training samples as dictionary may not fully exploit the discriminative information hidden in the training samples. The classification performance of SRC will significantly degrade, when there are not sufficient training data available.

Different from taking the training set as the dictionary, many currently proposed approaches choose to learn dictionaries from the training samples. K-SVD method [7] was proposed to learn an over-complete dictionary from training samples by solving the sparse coding problem and updating dictionary atoms iteratively. Based on the K-SVD method, Zhang and Li [8] proposed a discriminative K-SVD (DK-SVD) method for face recognition by integrating a classification error term into objective function. Jiang et al. [9] proposed a label consistent K-SVD (LC-KSVD) algorithm by introducing a binary code matrix to force the samples from the same classes to have similar representations, which makes the dictionary more discriminative. Ramirez et al. [10] proposed to learn a class-specified sub-dictionary for each class and introduce a structured incoherence regularization term to make sub-dictionaries independent. By imposing the Fisher criterion on the sparse coding coefficients, Yang et al. [11] proposed a Fisher discrimination dictionary learning (FDDL) method to make the coding coefficients have small within-class scatter and big between-class scatter. Kong and Wang [12] proposed to learn a particular dictionary (called particularity) for each class, and a common pattern pool (called commonality) shared by all the classes. In this way, the shared information of different sub-dictionaries are separated, and thus the particularity can be more compact and discriminative. Recently, Gu et al. [13] proposed a projective dictionary pair learning (DPL) method which simultaneously learns an analysis dictionary and a synthesis dictionary for pattern classification. Although the methods above can work well under the situation that the image data are noiseless, and even can handle the corruptions existing in the testing samples. Unfortunately, they cannot generalize well when training and testing samples are both badly corrupted. If training data are corrupted grossly, the corruptions will be introduced into the learned dictionary, resulting in a degraded classification performance.

Recently, much attention has been drawn to the low-rank matrix recovery theory, which has exhibited excellent performance for handling large image corruptions. By integrating low-rank regularization on each sub-dictionary, Ma et al. [14] proposed to learn discriminative low-rank dictionary for sparse representation (DLRD_SR) for robust image classification. Li et al. [15] proposed a discriminative dictionary learning with low-rank regularization ($D^2L^2R^2$) method by applying the Fisher discriminant function on the sparse coding coefficients, to make the representation coefficients more discriminative. Through rank minimization, the sparse noises are separated from the training samples, and dictionary atoms are updated to reconstruct the recovered samples, such that the dictionary atoms can be more noiseless. However, due to the low-rank regularization on each sub-dictionary, certain common information across different classes are removed. Thus, the dictionaries learned by these methods will have a weak representative power for the testing samples.

In this paper, we particularly focus on the robust image classification problem, in which both training and testing samples are corrupted. Such problem is challenging, because that the image samples often have large intra-class variants and small inter-class differences, due to the image corruptions. Inspired by [12], we find that an image in robust image classification task often contains three types of information (see examples of face recognition task in Fig. 1): (1) *Class-specific information*, which are the most discriminative features owned by one class. (2) *Class-shared information*, which are the common patterns shared by different classes. These information are essential to the representation of images but do not contribute to the discriminability of different classes. For example, in face recognition task, face images from different classes often share same illumination, expression and disguise variants. Such shared information are not helpful to distinguish different classes, but without them, images with similar variants cannot be well represented. (3) *Sparse noises*, such as pixel corruptions, which are useless for image classification and often enlarge the intra-class differences. If we can separate the class-specific information from the others and construct a dictionary to capture such information, this dictionary will be more noiseless, and the sparse representation on such dictionary will be more discriminative.

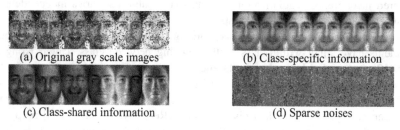

| (a) Original gray scale images | (b) Class-specific information |
| (c) Class-shared information | (d) Sparse noises |

Fig. 1. Take face recognition task as example. (b), (c) and (d) are the class-specific information, the class-shared information and the sparse noise in the (a) original images, respectively.

Based on above observation, we propose a novel dictionary learning method, called low-rank double dictionary learning (LRD^{2}L), to separate these three types of information from each other. More specifically, given a training samples set with labels, we propose to learn a low-rank *class-specific sub-dictionary* for each class, which captures the most discriminative features of the corresponding class. Simultaneously, a low-rank *class-shared dictionary* is constructed across all classes to represent the common patterns shared by different classes. With the help of the class-shared dictionary, the proposed method will have a powerful reconstructive capability for new testing samples. A sparse error term is also introduced to approximate the sparse noises contained in the samples. Since the noises are separated from the samples through the error term, the learned dictionaries can be more noiseless. The main contributions of this paper are summarized as follows:

1. Based on the low-rank matrix recovery theory, we propose a novel low-rank double dictionary learning (LRD^{2}L) method, to address the problem of robust image classification, with both the training and the testing data contain corruptions. By separating the sparse noises from the training samples through the error term, the learned class-specific sub-dictionaries and class-shared dictionary can be noiseless. Therefore, the proposed method is robust to extreme noises and gross corruptions.
2. Different from the existing low-rank dictionary learning methods, LRD^{2}L has a strong representative capability for newly coming testing samples. That is because the common patterns in the training data are preserved in the class-shared dictionary. The testing samples with the similar common patterns will be reasonably represented by the combination of the class-specific and class-shared dictionaries.
3. An alternative optimization algorithm is designed to effectively solve the optimization problem of the proposed method. Experimental evaluations on three public available datasets demonstrate the effectiveness and robustness of LRD^{2}L.

The remainder of this paper is organized as follows: In Sect. 2, we introduce the novel low-rank double dictionary learning (LRD^{2}L) model in detail. Section 3 presents the alternative algorithm to solve the corresponding optimization problem of the proposed method. Experimental results on three public databases under different experimental settings are reported in Sect. 4 to demonstrate the effectiveness of our method. Finally, the paper concludes in Sect. 5.

2   The Proposed LRD^{2}L Method

Given a training sample set $Y = [Y_1, Y_2, \ldots, Y_C] \in \Re^{d \times N}$, which consists of N training samples from C different classes. $Y_i \in \Re^{d \times N_i}$ denotes the sub-matrix which consists of all the training samples from the i-th class. d denotes the dimension of samples and N_i is the number of the training samples from the i-th class, which satisfies $N = \sum_{i=1}^{C} N_i$. The goal of LRD^{2}L is to simultaneously construct a class-specific dictionary $A = [A_1, A_2, \ldots, A_C] \in \Re^{d \times m_A}$ and a class-shared dictionary $B \in \Re^{d \times m_B}$, to represent the class-specific and the class-shared information of the data, respectively. m_A and m_B denote the dictionary sizes of A and B. A_i is a sub-dictionary associated with the i-th class. Therefore, we can represent the training samples from each class as follow:

$$Y_i = AX_i + BZ_i + E_i, \text{ for } i = 1, 2, \ldots, C, \tag{1}$$

where $X_i \in \Re^{m_A \times N_i}$ and $Z_i \in \Re^{m_B \times N_i}$ are the representation coefficient matrix of Y_i over dictionary A and B, respectively. $E_i \in \Re^{d \times N_i}$ is an error term to approximate the sparse noises in Y_i, and the sparse error term of the whole training samples can be denoted by $E = [E_1, E_2, \ldots, E_C] \in \Re^{d \times N}$. In order to design each term in (1) appropriately, and separate three types of information in training samples correctly, we analyze the properties of A, B and E as follows.

- If the image data are clean (i.e., only containing the class-specific information of its class), the images belonging to the same class tend to be drawn from the same

subspace, while the images from different classes are drawn from different subspaces. Thus, the image samples of the same class are usually highly correlated, whereas the image samples from different classes are independent. Therefore, the class-specific sub-dictionary A_i, which captures the class-specific information in training samples from the i-th class ($i = 1, 2, \ldots, C$), is expected to be low rank.
- The class-shared information represent the common patterns shared by all classes. Since the common features of different classes often have coherences or even share same atoms, the class-shared components of different classes are highly linearly correlated. Therefore, the class-shared dictionary B, which captures the class-shared information of all training samples, should reasonably be a low rank matrix.
- Because the sparse noise often contaminate a relatively small portion of the whole image and statistically uncorrelated between different images [6], the error term of each class E_i should be a sparse matrix.

Based on above analyses and sparse representation theory, we propose the following model for our low-rank double dictionary learning:

$$\min_{A_i, X_i, B, Z_i, E_i} \|A_i\|_* + \|B\|_* + \alpha(\|X_i\|_1 + \|Z_i\|_1) + \beta\|E_i\|_1 \\ s.t.\ Y_i = AX_i + BZ_i + E_i \text{ for } i = 1, 2, \ldots, C, \tag{2}$$

where α and β are positive-valued parameters that balance the sparsity of the coding coefficient matrices (i.e., X_i and Z_i) and the weight of the error term, respectively. $\|\cdot\|_*$ is the nuclear norm of a matrix, i.e., the sum of the singular values of this matrix. $\|A_i\|_*$ and $\|B\|_*$ enforce each class-specific sub-dictionary A_i and the class-shared dictionary B to be low-rank. $\|\cdot\|_1$ is the l_1-norm of a matrix, and the l_1-norm regularization $\|E_i\|_1$ is used to promote the sparseness of the error term E_i.

To further enhance the discrimination of the learned dictionaries, firstly, since A_i is the sub-dictionary associated with the i-th class, it is supposed to well represent the class-specific component of the i-th class. Therefore, by rewriting the coefficient matrix X_i as $X_i = [X_{i1}; X_{i2}; \ldots; X_{iC}]$, where X_{ij} denotes the coefficients of Y_i corresponding to A_j, we can have the constraint $Y_i = A_i X_{ii} + BZ_i + E_i$. Secondly, the class-specific components of different classes should be incoherent, therefore for Y_j, the coefficients X_{ji} ($i \neq j$) are expected to be zero matrices. To this end, an incoherence term $R(A_i) = \sum_{j=1, j \neq i}^{C} \|A_i X_{ji}\|_F^2$ ($i = 1, 2, \ldots, C$) is introduced into the objective function of the proposed model. The minimization of such term makes the correlation between $\bar{Y}_j$ and $A_i(i \neq j)$ as small as possible. Considering the both two factors, the objective function of Eq. (2) is further improved for LRD^{2}L as follow

$$\min_{A_i, X_i, B, Z_i, E_i} \|A_i\|_* + \|B\|_* + \alpha(\|X_i\|_1 + \|Z_i\|_1) + \beta\|E_i\|_1 + \lambda R(A_i) \\ s.t.\ Y_i = AX_i + BZ_i + E_i, \quad Y_i = A_i X_{ii} + BZ_i + E_i, \tag{3}$$

where $\lambda > 0$ is a parameter that controls the contribution of the incoherence term. Equation (3) is the overall objective function of our LRD^{2}L model, and we will present the optimization algorithm of solving this problem in Sect. 3.

3 Optimization Procedure

In this section, to solve the optimization problem in (3), we propose an effective optimization algorithm, by dividing the problem (3) into three sub-problems and solve them iteratively.

1. Updating the coding coefficients X_i and Z_i class-by-class by fixing the dictionaries A, B and other coefficients X_j and Z_j ($i \neq j$).
2. Updating the class-specific sub-dictionary A_i class-by-class by fixing other variables. The corresponding coefficients X_{ii} should also be updated to meet the constraint $Y_i = A_i X_{ii} + B Z_i + E_i$.
3. Updating the class-shared dictionary B and the corresponding coefficients Z iteratively to satisfy the constraint $Y_i = A X_i + B Z_i + E_i$.

Similar to [14], the error term E_i is updated in each sub-problem, and because the weights of the error E_i should be adjusted differently for the constraints $Y_i = A_i X_{ii} + B Z_i + E_i$ and $Y_i = A X_i + B Z_i + E_i$, the parameter β in the second sub-problem is set differently with the other two sub-problems.

3.1 Updating Coding Coefficients X_i and Z_i

Suppose that the dictionaries A and B are given, the coding coefficients X_i and Z_i are updated class by class. When calculating X_i and Z_i, all other coefficients X_j and Z_j ($i \neq j$) are fixed. Thus, ignore the terms that unrelated with X_i and Z_i, the problem (3) is reduced to a sparse coding problem, which can be formulated as follow:

$$\min_{X_i, Z_i, E_i} \alpha \big(\|X_i\|_1 + \|Z_i\|_1 \big) + \beta_1 \|E_i\|_1 \ s.t. \quad Y_i = A X_i + B Z_i + E_i. \tag{4}$$

Note that $\|[X_i; Z_i]\|_1 = \|X_i\|_1 + \|Z_i\|_1$ and the constraint can be rewritten as $Y_i = [A, B][X_i; Z_i] + E_i$. Therefore, by defining $P_i = [X_i; Z_i]$, $D = [A, B]$, and introducing an auxiliary variable H, the problem (4) is converted to the equivalent problem:

$$\min_{P_i, E_i, H} \alpha \|H\|_1 + \beta_1 \|E_i\|_1 \quad s.t. \quad Y_i = D P_i + E_i, \ P_i = H. \tag{5}$$

The above problem can be solved efficiently by the Augmented Lagrange Multipliers (ALM) [16] method, which minimizes the corresponding augmented Lagrange function of problem (5) as

$$\min_{P_i, E_i, H} \alpha \|H\|_1 + \beta_1 \|E_i\|_1 + \langle T_1, Y_i - D P_i - E_i \rangle + \langle T_2, P_i - H \rangle$$
$$+ \frac{\mu}{2} \Big(\|Y_i - D P_i - E_i\|_F^2 + \|P_i - H\|_F^2 \Big) \tag{6}$$

where T_1 and T_2 are Lagrange multipliers and $\mu > 0$ is a positive penalty parameter. $\langle A, B \rangle = Tr(A^T B)$ is the sum of the diagonal elements of the matrix $A^T B$, and A^T is the transpose of the matrix A.

3.2   Updating Class-Specific Sub-dictionary A_i

With the learned coefficients X_i, the sub-dictionary A_i is updated class by class, and the corresponding coefficients X_{ii} is also updated to meet the constraint $Y_i = A_iX_{ii} + BZ_i + E_i$. Then, the second sub-problem is converted to the following problem:

$$\min_{A_i,X_{ii},E_i} \|A_i\|_* + \alpha\|X_{ii}\|_1 + \beta_2\|E_i\|_1 + \lambda R(A_i) \qquad (7)$$
$$s.t.\ Y_i = A_iX_{ii} + BZ_i + E_i.$$

For mathematical brevity, we define $Y_A = Y_i - BZ_i$. Two auxiliary variables J and S are introduced and the problem (7) becomes to the equivalent optimization problem:

$$\min_{A_i,J,X_{ii},S,E_i} \|J\|_* + \alpha\|S\|_1 + \beta_2\|E_i\|_1 + \lambda R(A_i) \qquad (8)$$
$$s.t.\ Y_A = A_iX_{ii} + E_i,\quad A_i = J,\quad X_{ii} = S.$$

Problem (8) can be solved by solving the following Augmented Lagrange Multiplier problem through the ALM method:

$$\min_{A_i,J,X_{ii},S,E_i} \|J\|_* + \alpha\|S\|_1 + \beta_2\|E_i\|_1 + \lambda R(A_i)$$
$$+ \langle T_1, Y_A - A_iX_{ii} - E_i\rangle + \langle T_2, A_i - J\rangle + \langle T_3, X_{ii} - S\rangle \qquad (9)$$
$$+ \tfrac{\mu}{2}\left(\|Y_A - A_iX_{ii} - E_i\|_F^2 + \|A_i - J\|_F^2 + \|X_{ii} - S\|_F^2\right),$$

where T_1, T_2 and T_3 are Lagrange multipliers and $\mu > 0$ is a positive penalty parameter. Same as the existing dictionary learning methods [11, 12, 14], we also require that the column of the learned dictionary has a unit length, thus we add the normalization operation after updating J and A_i.

3.3   Updating Class-Shared Dictionary B

When all the class-specific sub-dictionaries are updated, we attempt to learn the class-shared dictionary B with all the other variables fixed. Hence, the objective function in (3) is reduced to

$$\min_{B,Z_i,E_i} \|B\|_* + \alpha\|Z_i\|_1 + \beta_1\|E_i\|_1\ s.t.\ Y_i = AX_i + BZ_i + E_i. \qquad (10)$$

Different from the class-specific sub-dictionary A_i that associated with only one class, the class-shared dictionary B is related to all the classes. Therefore, when updating B, we need to consider all the relationships corresponding to different classes together. Therefore, by summing up the objective function in Eq. (10) of all the classes, the optimization problem (10) can be converted to the following optimization problem:

$$\min_{B,[Z_1,Z_2,\ldots,Z_C],[E_1,E_2,\ldots,E_C]} \|B\|_* + \alpha \|[Z_1,Z_2,\ldots,Z_C]\|_1 + \beta_1 \|[E_1,E_2,\ldots,E_C]\|_1$$
$$s.t. [Y_1,Y_2,\ldots,Y_C] = A[X_1,X_2,\ldots,X_C] + B[Z_1,Z_2,\ldots,Z_C] + [E_1,E_2,\ldots,E_C].$$

$$(11)$$

Since $X = [X_1,X_2,\ldots,X_C]$, $Z = [Z_1,Z_2,\ldots,Z_C]$ and $E = [E_1,E_2,\ldots,E_C]$, the Eq. (11) can be reformulated as

$$\min_{B,Z,E} \|B\|_* + \alpha \|Z\|_1 + \beta_1 \|E\|_1 \ s.t. \ Y = AX + BZ + E. \tag{12}$$

We can observe that the objective function of problem (12) has the same form with the problem (7), except the incoherence term $R(A_i)$ in Eq. (7). Therefore, by setting the parameter λ to zero, the problem (12) can also be solved by using the same optimization strategy as problem (7), through the ALM method.

So far, the algorithms for the three sub-problems have been presented. The optimization procedures of (4), (7) and (12) need to iterate a few times to get the solution of LRD²L. The complete algorithm of LRD²L is summarized in Algorithm 1.

Algorithm 1. The Complete Algorithm of LRD²L

Input: Data matrix Y, initial dictionary matrix $D = [A, B]$, parameters $\alpha, \beta_1, \beta_2, \lambda$.
while *the maximal iteration number is not reached* **do**
    1. Fix the others and update X_i and Z_i class by class, by solving the problem (4);
    2. Fix the others and update A_i and X_{ii} class by class, by solving the problem (7);
    3. Fix the others and update B and Z by solving the problem (12);
end
Output: Class-specific dictionary A, class-shared dictionary B, coefficient matrix X, coefficient matrix Z and error matrix E.

---

4 Experimental Results

In this section, we evaluate the effectiveness and the robustness of the proposed LRD²L method on three public available datasets, under several experimental settings. The performance of our approach is compared with six state-of-the-art works, including sparse representation classifier (SRC) [6], fisher discrimination dictionary learning (FDDL) [11], label consistent KSVD [9] version 1 (LC-KSVD1) and version 2 (LC-KSVD2), DL-COPAR [12] and discriminative low-rank dictionary for sparse representation (DLRD_SR) [14]. Same as [17, 18], the image samples used in this paper are normalized to a unit length. (i.e., the ℓ_2-norm of each image vector equals to one). There are four parameters α, λ, β_1 and β_2 needed to be tuned in LRD²L, however, we find that the changes on α and λ would not affect the classification results very much. Therefore, we set $\alpha = 0.1$, $\lambda = 1$ for all the experiments in this paper. β_1, β_2

and the parameters in the competing algorithms are tuned manually for each experiment setting to get the best classification performance.

4.1   Face Recognition with Pixel Corruptions

The Extended YaleB [19] dataset consists of 2414 near frontal face images from 38 individuals, with each individual having around 59–64 images. The images are captured under various laboratory-controlled illumination conditions. For each subject, we randomly select 30 face images to compose the training samples set, and the remaining images are used as testing samples. All the images are manually cropped and normalized to the size of 32 × 32 pixels.

To evaluate the robustness to image corruption of different methods, a certain percentage of pixels (from 0% to 40%) randomly selected from each training and testing image are replaced with noise uniformly distributed over $[0, V_{max}]$, where V_{max} is the maximal pixel value in the image. For our approach, the parameters are set as $\beta_1 = 0.08$, $\beta_2 = 0.055$. The number of atoms of each class-specific sub-dictionary and the class-shared dictionary are set to 15 and 200, respectively. For fair comparison, the dictionary sizes of the competing methods are set to 760 (20 atoms for each class), we also set the size of the common feature dictionary in DL-COPAR to 20. SRC uses all the training samples as dictionary. All experiments are repeated 5 times and the average classification accuracies are reported in Table 1.

Table 1. Classification results of different methods on Extended YaleB dataset.

Methods	Classification accuracy (%) with different percentage of noises				
	0%	10%	20%	30%	40%
SRC	97.80%	80.46%	67.19%	50.94%	36.19%
FDDL	97.93%	76.22%	61.93%	44.51%	27.94%
LC-KSVD1	96.22%	83.64%	70.86%	54.87%	36.34%
LC-KSVD2	96.83%	83.67%	70.75%	55.65%	36.86%
DL-COPAR	97.25%	85.71%	73.63%	57.61%	40.74%
DLRD_SR	98.12%	92.07%	85.71%	74.41%	54.16%
Proposed LRD^{2}L	**99.43%**	**98.50%**	**96.88%**	**86.68%**	**71.65%**

From the Table 1, it can be seen that LRD^{2}L consistently obtains the highest classification accuracies under different pixel corruption conditions. With the percentage of pixel corruption increases, the performances of other competing methods are remarkably degraded, whereas the classification accuracies of LRD^{2}L decrease much more slowly. For example, when the percentage of pixel corruption increases from 0% to 20%, the noises have little influence on the performance of LRD^{2}L (decreasing from 99.43% to 96.88%). However, the classification accuracies of the other methods suffers a dramatically decrease, in the worst case, the performance of FDDL drops from 97.93% to 61.93%. This demonstrates that capturing the low-rankness property of dictionaries and introducing an error term, to model the noises in the images, make our approach robust to the pixel corruptions existing in the images.

4.2 Face Recognition with Occlusion

The AR dataset [20] contains more than 4000 frontal view face images of 126 subjects. For each subject, there are 26 images captured in two different sessions with a two-week interval, under different illumination conditions, expression changes and particular disguises. To evaluate the effectiveness of different methods in dealing with the occlusions both in training and testing samples, following the experimental settings in [21, 22], we conduct the experiments under three different scenarios:

- *Sunglasses:* For each subject, all seven undisguised images and one randomly selected image with sunglasses from session 1 are used to construct the training set. The testing set consists of seven undisguised images from session 2 and all remaining images with sunglasses. Thus, there are 8 training samples and 12 testing samples for each individual. The sunglasses cover about 20% of the face image.
- *Scarf:* Replace images with sunglasses in the above scenario by images with scarf. The scarf cover about 40% of the face image.
- *Mixed:* For each subject, all seven undisguised images, one randomly selected image with sunglasses and one randomly selected image with scarf from session 1 are chosen as training samples. Remaining images are used to compose the testing set. Thus, there are also 9 training samples and 17 testing samples for each subject.

In all above experiments, the parameters of our approach is fixed as $\beta_1 = 0.2$, $\beta_2 = 0.01$. The dictionary sizes of the class-specific sub-dictionary and the class-shared dictionary are set to 5 and 200, respectively. All the experiments repeat 5 times. Table 2 reports the mean classification results of different approaches under three scenarios.

Table 2. Classification results of different methods on AR dataset.

Methods	Classification accuracy (%) under different scenarios		
	Sunglass	Scarf	Mixed
SRC	89.83%	89.17%	88.59%
FDDL	90.08%	89.58%	89.47%
LC-KSVD1	87.75%	86.53%	86.24%
LC-KSVD2	89.36%	88.81%	87.76%
DL-COPAR	90.08%	89.17%	88.24%
DLRD_SR	94.08%	92.58%	91.53%
Proposed LRD^{2}L	**95.28%**	**94.23%**	**93.59%**

Again, our approach obtains the best classification results. More specifically, LRD^{2}L outperforms DLRD_SR by 1.20% for sunglasses scenario, 1.65% for scarf scenario and 2.06% for mixed scenario. Compared with other competing methods, LRD^{2}L achieves at least 5.20%, 4.65% and 4.12% improvements for three scenarios respectively. This suggests that the low-rank regularizations on the dictionaries enhance the robustness of LRD^{2}L to disguise occlusions, and compared to DLRD_SR, learning the class-shared dictionary makes our approach more discriminative.

4.3   Object Recognition with Pose Variants

The COIL-20 object dataset [23] consists of 1440 grayscale images from 20 objects with various poses. Each object contains 72 images, which are captured in equally spaced views, i.e., for every 5 degree, one image are taken. In this experiment, same as [24, 25], 10 images per class are randomly selected as training samples, while the remaining images compose the testing sample set. All the images are manually cropped and resized to 32×32 pixels. For our approach, the parameters are set as $\beta_1 = 0.07$, $\beta_2 = 0.04$. The size of each class-specific sub-dictionary and the class-shared dictionary are set to 6 and 80, respectively. For the competing methods, the size of each sub-dictionary is set to 10. For DL-COPAR, the common pattern pool size is set to 10. The experiment is repeated 5 times for each method, and the average classification accuracies of different approaches are reported in Table 3.

Table 3. Classification results of different methods on COIL-20 dataset.

Methods	Classification accuracy	Methods	Classification accuracy
SRC	87.50%	DL-COPAR	90.42%
FDDL	87.98%	DLRD_SR	89.63%
LC-KSVD1	90.68%	Proposed LRD^{2}L	**92.05%**
LC-KSVD2	90.87%		

As shown in Table 3, on COIL-20 object dataset, DLRD_SR does not work well and only achieves 89.63% classification rate, which is lower than DL-COPAR and LC-KSVD. This maybe because that, without sufficient training samples, the pose variants in data makes the low-rankness property of each class unobvious. And due to the low-rank regularization on each sub-dictionary, a portion of class-shared information are removed with sparse noises. By learning the class-shared dictionary to capture the pose variants, LRD^{2}L achieves the highest classification accuracy of 92.05%, which outperforms competing methods with the gains over 1.18%.

5 Conclusion

In this paper, we propose a novel dictionary learning model, namely low-rank double dictionary learning (LRD^{2}L), to learn robust and discriminative dictionary from corrupted data, for robust image classification task. Unlike traditional dictionary learning methods, besides learning a low-rank class-specific sub-dictionary for each class, our method also learns a low-rank class-shared dictionary, to represent the common features shared by all classes, and a sparse error term to approximate the sparse noises in data. By separating the class-specific information from the other information, the class-specific sub-dictionary only captures the most discriminative features of each class, which makes the proposed LRD^{2}L model more robust and discriminative. By capturing the common features, which are essential to the representation of image samples, through the class-shared dictionary, the reconstructive capability of LRD^{2}L is enhanced. Experimental results on three public datasets reveal the effectiveness and the superiority of LRD^{2}L compared to the state-of-the-art methods.

References

1. Yang, M., Van Gool, L., Kong, H.: Sparse variation dictionary learning for face recognition with a single training sample per person. In: ICCV (2013)
2. Yang, M., Zhang, L., Yang, J., Zhang, D.: Metaface learning for sparse representation based face recognition. In: ICIP (2010)
3. Li, S., Yin, H., Fang, L., Member, S.: Group-sparse representation with dictionary learning for medical image denoising and fusion. IEEE Trans. Biomed. Eng. **59**(12), 3450–3459 (2012)
4. Rubinstein, R., Bruckstein, A.M., Elad, M.: Dictionaries for sparse representation modeling. Proc. IEEE **98**(6), 1045–1057 (2010)
5. Wright, B.J., Mairal, J., Sapiro, G., Huang, T.S., Yan, S.: Sparse representation for computer vision and pattern recognition. Proc. IEEE **98**(6), 1031–1044 (2010)
6. Wright, J., Yang, A.Y., Ganesh, A., Sastry, S.S.: Robust face recognition via sparse representation. IEEE Trans. Pattern Anal. Mach. Intell. **31**(2), 210–227 (2009)
7. Aharon, M., Elad, M., Bruckstein, A.: K-SVD: an algorithm for designing overcomplete dictionaries for sparse representation. IEEE Trans. Signal Process. **54**(11), 4311–4322 (2006)
8. Zhang, Q., Li, B.: Discriminative K-SVD for dictionary learning in face recognition. In: CVPR (2010)
9. Jiang, Z., Lin, Z., Davis, L.S.: Label consistent K-SVD: learning a discriminative dictionary for recognition. IEEE Trans. Pattern Anal. Mach. Intell. **35**(11), 2651–2664 (2013)
10. Ramirez, I., Sprechmann, P., Sapiro, G.: Classification and clustering via dictionary learning with structured incoherence and shared features. In: CVPR (2010)
11. Yang, M., Zhang, D., Feng, X.: Fisher discrimination dictionary learning for sparse representation. In: ICCV (2011)
12. Kong, S., Wang, D.: A dictionary learning approach for classification: separating the particularity and the commonality. In: Fitzgibbon, A., Lazebnik, S., Perona, P., Sato, Y., Schmid, C. (eds.) ECCV 2012. LNCS, vol. 7572, pp. 186–199. Springer, Heidelberg (2012). doi:10.1007/978-3-642-33718-5_14

13. Gu, S., Zhang, L., Zuo, W., Feng, X.: Projective dictionary pair learning for pattern classification. In: NIPS (2014)
14. Ma, L., Wang, C., Xiao, B., Zhou, W.: Sparse representation for face recognition based on discriminative low-rank dictionary learning. In: CVPR (2012)
15. Li, L., Li, S., Fu, Y.: Learning low-rank and discriminative dictionary for image classification. Image Vis. Comput. **32**(10), 814–823 (2014)
16. Lin, Z., Liu, R., Su, Z.: Linearized alternating direction method with adaptive penalty for low-rank representation. In: NIPS (2011)
17. Zhuang, L., Gao, S., Tang, J., Wang, J.: Constructing a nonnegative low-rank and sparse graph with data-adaptive features. IEEE Trans. Image Process. **24**(11), 3717–3728 (2015)
18. Li, S., Fu, Y.: Low-rank coding with b-matching constraint for semi-supervised classification. In: IJCAI (2013)
19. Georghiades, A.S., Belhumeur, P.N., Kriegman, D.J.: From few to many: illumination cone models for face recognition under variable lighting and pose. IEEE Trans. Pattern Anal. Mach. Intell. **23**(6), 643–660 (2001)
20. Martinez, A.M.: The AR face database. CVC Technical report (1998)
21. Chen, C.F., Wei, C.P., Wang, Y.C.F.: Low-rank matrix recovery with structural incoherence for robust face recognition. In: CVPR (2012)
22. Zhang, Y., Jiang, Z., Davis, L.S., Park, C.: Learning structured low-rank representations for image classification. In: CVPR (2013)
23. Nene, S.A., Nayar, S.K., Murase, H.: Columbia Object Image Library (COIL-20). Technical report No. CUCS-006-96 (1996)
24. Wang, S., Fu, Y.: Locality-constrained discriminative learning and coding. In: CVPR Workshops (2015)
25. Li, S., Fu, Y.: Learning robust and discriminative subspace with low-rank constraints. IEEE Trans. Neural Netw. Learn. Syst. **27**, 2160–2173 (2015)

Robust Scene Text Detection for Multi-script Languages Using Deep Learning

Ruo-Ze Liu[1], Xin Sun[1], Hailiang Xu[1], Palaiahnakote Shivakumara[2],
Feng Su[1], Tong Lu[1(✉)], and Ruoyu Yang[1]

[1] National Key Lab for Novel Software Technology,
Nanjing University, Nanjing, China
liuruoze@163.com, mfl533042@smail.nju.edu.cn,
xhl_student@163.com, {suf,lutong,yangry}@nju.edu.cn
[2] Faculty of Computer Science and Information Technology,
University of Malaya, Kuala Lumpur, Malaysia
hudempsk@yahoo.com

Abstract. Text detection in natural images has been a high demand for a lot real-life applications such as image retrieval and self-navigation. This work deals with the problem of robust text detection especially for multi-script in natural scene images. Unlike the existing works that consider multi-script characters as groups of text fragments, we consider them as non-connected components. Specifically, we firstly propose a novel representation named Linked Extremal Regions (LER) to extract full characters instead of fragments of scene characters. Secondly, we propose a two-stage convolution neural networks for discriminating multi-script texts in clutter background images for more robust text detection. Experimental results on three well-known datasets, namely, ICDAR 2011, 2013 and MSRA-TD500, demonstrate that the proposed method outperforms the state-of-the-art methods, and is also language independent.

Keywords: Linked extremal regions · Scene text detection · Multi-script

1 Introduction

As a high-level kind of visual content of images, scene text now plays an important role in content-oriented applications such as image/video retrieval, automatic driving or navigation, and visual object recognition. In the past years, a number of scene detection methods have thus been explored, which can be categorized into two main classes consisting of *connected component* based approaches and *sliding window* based approaches. The connected components methods contain two main types, namely, Maximally Stable Extremal Regions (MSER) [1] and Stroke Width Transform (SWT) [2]. SWT based methods [2, 4] used stroke width information to detect texts efficiently. However, they are sensitive to noises and low contrast. MSER based methods [3, 5, 9] are robust to blur, noise, low contrast and illumination variations; however, they depend on the assumption that characters are composed of connected

© Springer International Publishing AG 2017
L. Amsaleg et al. (Eds.): MMM 2017, Part I, LNCS 10132, pp. 329–340, 2017.
DOI: 10.1007/978-3-319-51811-4_27

components, which results in poor detection quality on the cases of multi-script texts or texts composed of broken stokes. Sliding window methods [8] do not depend on the assumption that characters are connected components; however, they are often slow and not suitable for images having multi-oriented texts.

Nowadays, MSER based methods have become much popular in scene text detection because they are robust to scale size, low resolution or illumination variations and complex background. Despite this, MSER based methods are capable of handling multi- orientation texts. However, they report poor results for multi-script characters or characters with non-connected components as shown in Fig. 1(a)–(d). For example, as shown in Fig. 1(d), the digit character "2" is composed of several non-connected components due to poor image quality, which brings a lot difficulties for MSER based methods to detect. Here we refer "non-connected components" as the unconnected components that belong to the same character.

Fig. 1. Samples of non-connected components and their binary images.

Another problem that MSER based methods face is that they treat multi-script characters as groups of fragments. A text fragment here is a connected component extracted by the MSER algorithm. It has two disadvantages. Firstly, in the procedure of classification, it is hard to discriminate text fragments from clutters in background. As shown in Fig. 2(a), the fragments that constitute characters are hardly to be discriminated from noises in background even by humans. Secondly, text fragments are often very different in their geometry sizes, therefore it is hard to use this information for text grouping, which results in the rise of difficulties of text grouping.

text fragments   clutters   non-text   text

(a)   (b)

Fig. 2. Text fragments with clutters in background (a) and non-texts/texts (b).

To overcome the limitations of MSER based methods, in this paper, we propose a novel concept called Linked Extremal Regions (LER) to represent multi-script characters as non-connected components. Motivated by the fact that fragments of character components have uniform color, we exploit the link information of ERs (Extremal Regions) in the form of a tree structure to group them. This results in full characters rather than text fragments given by MSER. Therefore, the contributions of the proposed method are in three-folds: (1) Proposing a new LER representation for detecting character components from complex background, which helps a classifier to predict correct characters. (2) Since the above step outputs full characters, synthetic characters can be used to enlarge training datasets. (3) The difficulty of text grouping is also lightened by using geometry information. The difference of the outputs of the proposed LER and the state-of-the-art MSER can be seen in Fig. 3.

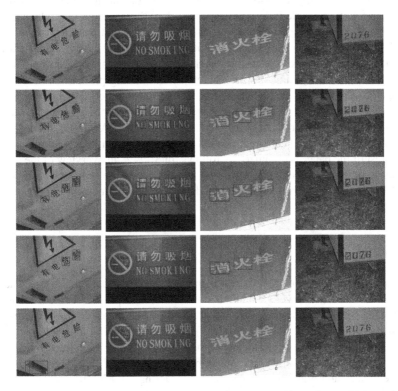

Fig. 3. The first row shows four source images, while the second row gives MSER results. For comparison, the third row illustrates the results of the proposed LER, where red rectangles are different from MSER results. The fourth row shows only the different boxes that LER outputs. Note that false positives are removed for better viewing in the above four rows. For completeness, the fifth row shows some of these false detections introduced by LER. (Color figure online)

Additionally, we propose a two-stage convolution neural network framework for handling the difficulties in discriminating multi-script texts from clutters in complex backgrounds. We find that the difficulties of discriminating texts from background are brought by the ambiguity in deciding either text or non-text. As seen in Fig. 2(b), symbols like road signs are considered as non-texts in most cases. Thus the decision boundary of text and non-text is hard to learn. Previous works use hand-crafted features [3] or deep learning models [5] to solve this problem. However, these approaches have two disadvantages. Firstly, the approaches using shallow models and hand-crafted features may not represent decision boundaries well, which results in a poor quality or accuracy. Secondly, the approaches using deep learning models such as convolution neural networks (CNN) have the power of representation, but due to the ambiguity of the task discussed in this paper, a CNN model may fall into over-fitting and thus reduce the performance for detection.

A two-stage CNN framework is proposed to solve this problem. Inspired by the cascade learning in face detection [13], the key idea of our approach is to divide a complex task to two simple tasks and conquer each of them respectively. One task is to distinguish clutters with symbols such as characters, signs and patterns, which can be seen as the searching of clear edges. The other one is to distinguish characters from other shapes, which can be seen as the searching of regular shapes. Decision boundary of each task is thus clear and simple. Therefore, the training process of these models converge quickly and the generation ability will be much better. By this way, we can build a robust and effective multi-script text detection system. Experiments on three well-known datasets such as ICDAR 2011, 2013 and MSRA-TD500 show that our system is robust in some challenging cases and capable of handling multiple scripts.

The rest of the paper is organized as follows. In Sect. 2, an overview of the existing methods is presented. Section 3 introduces the details of the proposed method, and Sect. 4 shows our experiment results to validate the proposed method. Conclusions are drawn in Sect. 5.

2   Related Work

It is noted from literature on text detection in natural scene images that a lot scene text detection methods are developed based on MSER concepts for text detection. For example, Yao et al. [11] proposed a method based on MSER for detecting multi-script texts in English and Chinese. Yin et al. [9] designed a pruning algorithm to extract MSER using the strategy of minimizing regularized variations. Then the candidates are grouped into texts using single-link clustering. Kang et al. [12] proposed a method to combine MSER and higher order correlation clustering (HOCC) to generate text line candidates. Based on the previous work, Yin et al. [10] proposed a modified system using MSER and adaptive clustering. From the above review of the existing methods, it is found that the methods consider multi-script as a group of text fragments and use MSER to extract those fragments. As discussed in the previous section, the methods which are developed based on MSER may not perform well for multi-script images especially for character classification and text grouping.

There are other methods proposed in literature which do not use MSER concepts for text detection in natural scene images. For example, Epshtein et al. [2] proposed stroke width transform (SWT) for detecting texts in natural scenes. The method also has the ability to handle multi-script images. However, the performance of the method depends on edge detection given by Canny of the input image. Similarly, sliding window based methods are also proposed for text detection in multi-script images. However, the methods are said to be inefficient and computationally expensive. Zhang et al. [8] proposed a symmetry based method to detect texts in natural scenes. However, it is not clear whether this method works well for multi-script images.

3 Proposed Method

Figure 4 shows the flow of the proposed method, where one can see LER is proposed to extract character candidates, then false character candidates are filtered by exploring a two-stage CNN classifier, and finally character candidates are grouped as a text line using a seed growing method.

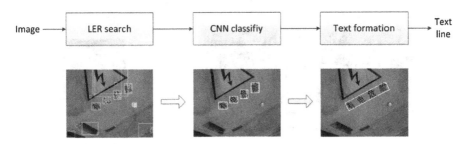

Fig. 4. The main process of the proposed method.

3.1 Linked Extremal Region

For an input image, I, a binary image $B_t(p)$ can be obtained by threshold t as

$$B_t(p) = \begin{cases} 1 & \text{if } I(p) \geq t, \\ 0 & \text{if } I(p) < t, \end{cases} \tag{1}$$

where $t \in [0, 255]$, and p denotes the position of a point in the image. An extremal region R at threshold t is defined by a connected component in the binary image $B_t(p)$ as follows

$$\forall q \in R, \forall p \in boundary(R) \rightarrow B_t(q) \geq B_t(p), \tag{2}$$

where q is another point in the image.

Extremal Regions (ERs) are obtained by constructing an ER tree, where the levels are determined with a certain threshold, and nodes at different levels are considered as ERs in the binary image. The relationship between each ER is represented by edges between nodes in the ER tree. For the region R_i which is the parent of another region R_j in the tree, it can be denoted as

$$\forall q \in R_j, \rightarrow q \in R_i \tag{3}$$

The proposed method considers the relationship between ERs represented by nodes as the "link" on different levels in the ER tree, and hence the method is named as LER as follows:

Linked Extremal Regions (LERs) are a group of regions, which are defined as

$$Q = \{\forall R_a, R_b \in Q \mid lv(R_a) = lv(R_b),\ \underset{\Delta}{argmin}(p_\Delta(R_a) = p_\Delta(R_b)) < t\}, \qquad (4)$$

where $lv(R)$ denotes the threshold level of R and $p_\Delta(R)$ denotes the parent of R in the ER tree at threshold level $(lv(R) + \Delta)$. And t is a parameter, which decides the depth of search paths. An example of LER can be seen in Fig. 5.

Fig. 5. Regions in the ER tree (left) and LER sample (right).

Distance Feature. LER exploits the link feature of ERs in the ER tree, and it captures the structure of split characters effectively. To filter some non-character structures and better extract split characters, distance feature is employed. The distance to all the ERs in LER is defined as

$$\|\text{centroid}(R_a) - \text{centroid}(R_b)\|_2 \le r(R_a) + r(R_b), \qquad (5)$$

where $r(R)$ denotes the radius of the minimum enclosing circle of region R.

LER search is an algorithm to extract LERs. Let $C_1, C_2 \ldots C_N$ represent the set of ERs on each level of the ER tree, C_1 be the root of the tree, $L_1, L_2 \ldots L_N$ represent the set of LERs on each level of the ER tree. LER search process can thus be described as follows:

1. Set $t = 0$, and search for the ERs set Ct on each level of the tree from top to bottom.
2. Lt = $\emptyset$ if the search set Ct is not the root of the ER tree. For each LER q in Lt−1, let qc = children (q).
3. If *stable* (q, qc) < m, then Ct = Ct \ *er* (qc) and Lt = Lt $\cup$ {qc}.
4. Use link and distance features to find LER set Ls in Ct. Then Lt = Lt $\cup$ Ls.
5. Set L t−1 = Lt and $t = t + 1$. If $t = N$, exit the search, else return back to step **2**.

Here *er* (q) denotes the group of regions that q contains. Children (q) is defined as the group of regions that are the children of *er* (q). And *stable* (q, q_c) is defined as

$$stable(q, q_c) = \frac{|q| - |q_c|}{|q|}, \qquad (6)$$

where $|q|$ denotes the number of points in q. And m is a parameter, which decides the threshold of stability.

It is noted that when the size of regions of LER is 1, the LER is regarded as an extremal region. As a result, the outputs of LER search contains both connected components and non-connected components. After LER search, we remove duplicates to filter those duplicate components as follows:

Remove Duplicates: Let C_{er} represent the set of remained components in the tree. For all the elements p in C_{er}, set *enable* (p) = false.

1. If $C_{er} = \varnothing$, exit. Else get a new element q from C_{er}, and create a new set $A = \varnothing$.
2. Set $q_p = $ parent (q). If *stable* $(q, q_p) < m$, go to stage 3. Else go to stage 4.
3. $A = A \cup \{q\}$ and set $q = q_p$. Then go to step 2.
4. Let r be the element which is in the middle of the positions of A. Set *enable* (r) = true. Then $C_{er} = C_{er} \setminus A$. Go to stage 1.

The above process outputs valid LER regions which are considered as the input for the proposed two stage CNN network.

3.2   Two-Stage CNN Framework

Since the considered problem is relatively complex, the LER method alone still may not be sufficient to achieve better results. As a result, one can expect false positives from LER. To alleviate this problem, we exploit a two-stage convolution neural network (Two-Stage CNN) to build a new classifier to filter out such false positives.

Figure 6 shows that the proposed framework consists of two stages. The first state of CNN is proposed to classify symbols such as characters and signs from the non-text created by complex background of images based on the fact that symbols exhibit regular shapes and fine edges compared to non-text components. The output of the first

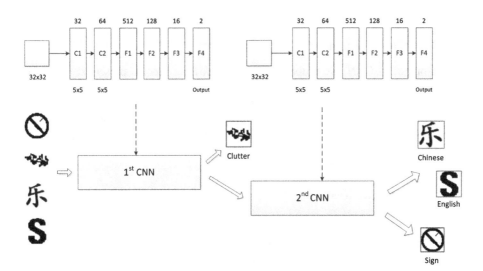

Fig. 6. The two-stage CNN framework.

stage is passed to the second stage to classify further symbols into characters and signs based on repeated patterns as shown in Fig. 6. Therefore, we can conclude that the proposed method combines two stage in a new fashion to achieve better results for complex issues.

In this work, we use LeNet-5 for constructing CNN with modifications. The proposed CNN considers ER regions of size 32×32 for learning. To ease the implementation difficulties, in this work, the size of regions given by the previous step is converted to standard size of 32×32. In addition, the two convolution layers with kernels of size 5×5 and mean pooling layers are proposed, which have the feature map size of 32 and 64, respectively. In this way, we propose three fully connected layers and one output layer of sizes 512, 128, and 64, respectively. Note that the activation functions of these layers are all rectifier linear units. The size of the first stage CNN output layers is 2, which divide the candidates into symbols and noise components. The size of the second stage CNN output layers is also 2, which labels symbol candidates as text and non-text. In this work, we consider characters for grouping into text lines and ignore signs.

The details for training is as follows. For the first stage, we choose 35000 clutter images (noise components) in background, which are considered as false samples. 5000 signs that have repeated patterns are collected as true samples. 10000 character images are collected as the ground-truth. The proposed method generates synthetic images about 5000 samples, which involve random rotations. In addition, for training our multi-script classifier, we use synthetic fonts to generate 15000 Chinese characters. Further, altogether 70000 samples are trained by deep learning tools to generate the first stage CNN. The same process is adopted for the second stage CNN. Here, the stochastic gradient descent algorithm is used as the training algorithm. The base learning rate is set to 0.005. We use a momentum of 0.9 and weight decay is set to 0.0005. The batch size of one iteration is set to 64, while the max number of iterations is set to 30000.

3.3   Text Line Detection

For the output of CNN, we propose to use the growing process presented in [6] for detecting seed candidates. The seed growing process classifies the candidates which get high scores by the character classifier as the seeds, which are then used to find other non-seeds (scored low). Furthermore, the proposed method uses clustering to group character candidates to form text lines. Finally, to improve text detection results, we use a random forest text classifier to remove false positives.

4   Experimental Results

To evaluate the proposed method, we use standard dataset, namely, ICDAR 2013, ICDAR 2011 and MSRA-TD500, and respective evaluation scheme in this work. In order to show the effectiveness of the proposed LER comparing to MSER, we consider

376 non-connected components from ICDAR and MSRA-TD500 datasets for experimentation. The following similarity is estimated to calculate recall measure.

$$Similarty(R_g, R_d) = \frac{Area(R_g) \cap Area(R_d)}{Area(R_g) \cup Area(R_d)}, \tag{7}$$

where R_g denotes the bounding box of ground truth of each character, and R_d denotes the bounding box of a detected character. If the similarity score is greater than 0.7, it is considered as the correct count.

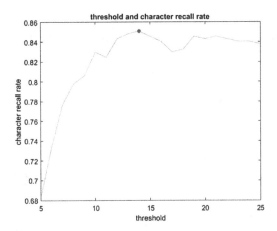

Fig. 7. Threshold and character recall rate

To determine suitable values for threshold t in the proposed LER algorithm, we conduct experiments on data chosen randomly from the datasets to calculate recall measure as shown in Fig. 7. Figure 7 shows that red color marked at 14 is considered as the value of the threshold t because after 14 the recall is decreasing as threshold increases. In the same way, we determined the threshold of stability m as 0.25.

It is observed from Table 1 that recall of the proposed LER is better than that of MSER. Therefore, one can conclude that the proposed LER is effective in detecting texts in scene images.

For calculating measures, we use the same the number of training and testing data in the ICDAR robust competition and MSRA-TD500 dataset. Sample qualitative results of the proposed method are shown in Figs. 8 and 9 for ICDAR and MSRA-TD500 dataset, respectively, where we can see the proposed method detects

Table 1. Character-level recall rate for non-connected components.

Algorithm	Character-level recall rate (%)
MSER [1]	57.6%
Proposed LER	89.3%

(a) art body style   (b) reflex light   (c) complex background

(d) low contrast   (e) broken stroke   (f) low resolution

Fig. 8. Detection results on ICDAR dataset.

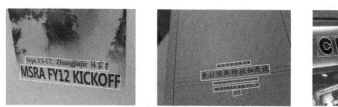

Fig. 9. Detection results on MSRA-TD500 dataset.

texts well for different challenges, such as low contrast, complex background, art body style, illumination effect and different orientation posed by datasets.

Quantitative results of the proposed and the existing methods are reported in Tables 2 and 3 for ICDAR 2011 and ICDAR 2013, respectively. It can be seen that results of the proposed method on ICDAR 2011 achieves the best at recall (78%) among all the methods. And the proposed method is also the best at F-measure (80%). The proposed method for ICDAR 2013 is the best at recall (79%) among all the methods. Though the proposed method does not score the best at precision (85%) compared to other methods, it is the best at F-measure (82%).

Table 2. Comparison on ICDAR 2011 dataset.

Methods	Precision	Recall	F-measure
Our method	0.81	0.78	0.80
Zhang et al. [8]	0.84	0.76	0.80
Huang et al. [5]	0.88	0.71	0.78
Yin et al. [9]	0.86	0.68	0.76
Neumann and Matas [4]	0.79	0.66	0.72

Table 3. Comparison on ICDAR 2013 dataset.

Methods	Precision	Recall	F-measure
Our method	0.85	0.79	0.82
Sung et al. [7]	0.88	0.74	0.80
Zhang et al. [8]	0.88	0.74	0.80
Yin et al. [9]	0.88	0.66	0.76

Similarly, the quantitative results of the proposed and the existing methods on MSRA-TD500 dataset are reported in Table 4, which shows that the proposed method is the best at recall (65%) compared to the other methods. Besides, the proposed method is the second best at F-measure (69%) compared to the other methods. This shows that the proposed method is language and orientation independent as shown in Fig. 9. The main reason to poor results of the proposed method is that the proposed method is sensitive to too low contrast and isolated characters as shown in Fig. 10.

Table 4. Comparison on MSRA-TD500 dataset.

Methods	Precision	Recall	F-measure
Yin et al. [10]	**0.81**	0.63	0.71
Our method	0.74	**0.65**	0.69
Kang et al. [12]	0.71	0.62	0.66
Yin et al. [9]	0.71	0.61	0.66
Yao et al. [11]	0.63	0.63	0.60

Fig. 10. Detection failures.

5 Conclusion

In this paper, we have proposed a novel representation called Linked Extremal Regions (LER) to extract full characters irrespective of orientation and script in natural scene images. To eliminate false text candidates, we have explored a two-stage convolution neural networks to tackle of challenges posed by multi-script texts, complex background and noisy components. Experimental results on three standard datasets demonstrate the effectiveness of this method. In our future work, we will focus on improve the accuracy of low contrast and isolated characters.

Acknowledgments. The work described in this paper was supported by the Natural Science Foundation of China under Grant Nos. 61672273, 61272218 and 61321491, the Science Foundation for Distinguished Young Scholars of Jiangsu under Grant No. BK20160021.

References

1. Matas, J., Chum, O., Urban, M., Pajdla, T.: Robust wide baseline stereo from maximally stable extremal regions. In: British Machine Vision Conference, pp. 384–393 (2002)
2. Epshtein, B., Ofek, E., Wexler, Y.: Detecting text in natural scenes with stroke width transform. In: IEEE Conference on Computer Vision and Pattern Recognition, pp. 2963–2970 (2010)
3. Neumann, L., Matas, J.: Real-time scene text localization and recognition. In: IEEE Conference on Computer Vision and Pattern Recognition, pp. 3538–3545 (2012)
4. Neumann, L., Matas, J.: Scene text localization and recognition with oriented stroke detection. In: IEEE International Conference on Computer Vision, pp. 97–104 (2013)
5. Huang, W., Qiao, Yu., Tang, X.: Robust scene text detection with convolution neural network induced MSER trees. In: Fleet, D., Pajdla, T., Schiele, B., Tuytelaars, T. (eds.) ECCV 2014. LNCS, vol. 8692, pp. 497–511. Springer, Heidelberg (2014). doi:10.1007/978-3-319-10593-2_33
6. Xu, H., Su, F.: A robust hierarchical detection method for scene text based on convolutional neural networks. In: IEEE International Conference on Multimedia and Expo, pp. 1–6 (2015)
7. Sung, M.C., Jun, B., Cho, H., Kim, D.: Scene text detection with robust character candidate extraction method. In: International Conference on Document Analysis and Recognition, pp. 426–430 (2015)
8. Zhang, Z., Shen, W., Yao, C., Bai, X.: Symmetry-based text line detection in natural scenes. In: IEEE Conference on Computer Vision and Pattern Recognition, pp. 4321–4329 (2015)
9. Yin, X.C., Yin, X., Huang, K., Hao, H.: Robust text detection in natural scene images. IEEE Trans. Pattern Anal. Mach. Intell. (TPAMI) **36**(5), 970–983 (2014)
10. Yin, X.C., Pei, W.Y., Zhang, J., Hao, H.: Multi-orientation scene text detection with adaptive clustering. IEEE Trans. Pattern Anal. Mach. Intell. (TPAMI) **37**(9), 1930–1937 (2015)
11. Yao, C., Bai, X., Liu, W., Tu, Z.: Detecting texts of arbitrary orientations in natural images. In: IEEE Conference on Computer Vision and Pattern Recognition, pp. 1083–1090 (2012)
12. Kang, L., Li, Y., Doermann, D.: Orientation robust text line detection in natural images. In: IEEE Conference on Computer Vision and Pattern Recognition, pp. 4034–4041 (2014)
13. Ren, S., Cao, X., Wei, Y., Sun, J.: Face alignment at 3000 FPS via regressing local binary features. In: IEEE Conference on Computer Vision and Pattern Recognition, pp. 1685–1692 (2014)

Robust Visual Tracking Based on Multi-channel Compressive Features

Jianqiang Xu and Yao Lu[(⊠)]

Beijing Key Laboratory of Intelligent Information Technology,
School of Computer Science and Technology,
Beijing Institute of Technology, Beijing 100081, China
{xujq,vis_yl}@bit.edu.cn

Abstract. Tracking-by-detection approaches show good performance in visual tracking, which often train discriminative classifiers to separate tracking target from their surrounding background. As we know, an effective and efficient image feature plays an important role for realizing an outstanding tracker. The excellent image feature can separate the tracking object and the background more easily. Besides, the feature should effectively adapt to many boring factors such as illumination changes, appearance changes, shape variations, and partial or full occlusions, etc. To this end, in this paper, we present a novel multi-channel compressive feature, which combine rich information from multiple channels, and then project it into a low-dimension compressive feature space. After that, we designed a new visual tracker based on the multi-channel compressive features. At last, extensive comparative experiments conducted on a series of challenging sequences demonstrate that our tracker outperforms most of state-of-the-art tracking approaches, which also proves that our multi-channel compressive feature is effective and efficient.

Keywords: Compressive features · Multi-channel compressive features · Multi-channel compressive tracking · Bayesian classifier

1 Introduction

Object tracking has been studied for several decades, and much progress has been made in recent years, see [1] for a review of tracking algorithms. Usually, the existed algorithms can be divided into two categories: generative algorithms [2–5] and discriminative algorithms [6, 9–11]. Generative tracking algorithms typically learn a model to represent the tracking object and use the learned model to match the image region with minimal reconstruction error. The IVT algorithm [2] proposed to learn an appearance model online to adapt the appearance change of the target, which laid a good foundation for many popular trackers. The Frag method [3] generated the appearance model based on integral histograms of object fragments, so it can resist minor deformation. To deal with large variation of appearance and motion, Kwon and Lee [4] combined multiple motion models and observation into a modified particle filtering framework. Oron et al.

© Springer International Publishing AG 2017
L. Amsaleg et al. (Eds.): MMM 2017, Part I, LNCS 10132, pp. 341–352, 2017.
DOI: 10.1007/978-3-319-51811-4_28

[5] added a spatial configuration of pixels into their appearance model, and the resulting joint model is more robust to rigid and nonrigid deformations.

While the discriminative algorithms often convert the tracking problem into a binary classification task, its aim is determining the decision boundary for separating the target object from the background. Zhang et al. [6] proposed an effective and efficient tracking algorithm based on compressive features extracted in the compressed domain. To deal with drifting problem, Stalder et al. [10] proposed an online semi-supervised approach which trains the classifier by only labeling the samples at the first frame while leaving the samples at the other frames unlabeled. MIL [9] drew multiple samples and put them into positive bag and negative bag and then trained classifier based on these bags. BSBT [11] present a multiple classifier system for model-free tracking. The tasks of detection, recognition, and tracking are split into separate classifiers to simplify each classification task.

The rest of this paper is organized as follows. We first introduce the compressive feature proposed by Zhang et al. in Sect. 2. Then the proposed novel multi-channel compressive features are detailed in Sect. 3. In Sect. 4, we designed a new MCT tracker (multi-channel compressive tracking) based on our feature. The experimental results are presented in Sect. 5 with comparisons to state-of-the-art methods on challenging sequences. At last, Sect. 6 concludes the paper.

2   Review of Compressive Features

In computer vision field, we often face the problem of "Curse of Dimensionality" due to that the original dimension of image feature is very high. To deal this problem, Zhang et al. [6] proposed compressive features according to compressive sensing theory that shows us when the dimension of the feature space is sufficiently high; these features can be projected into a randomly chosen low-dimensional space, which contains enough information to reconstruct the original high-dimensional features.

For a single-channel image sample of size $M \times N$, we can represent its original features with $x \subset \mathbb{R}^m (m = M \times N)$, the compressed feature is $y \subset \mathbb{R}^n (n \ll m)$, then we can get

$$y = Px \tag{1}$$

where $P \in \mathbb{R}^{n \times m}$ is a measurement matrix. When the measurement matrix satisfies the Johnson-Lindenstrauss lemma and the RIP (restricted isometry property) condition, we can reconstruct x with minimum error from y with high probability. Ideally, we hope the projection from $x \subset \mathbb{R}^m$ to $y \subset \mathbb{R}^n$ is a stable embedding process and those salient information of the signal can be preserved.

A typical measurement matrix satisfying the RIP condition is the random Gaussian matrix, i.e., zero mean and unit variance matrix. However, the common measurement matrix is dense, when m is large, the cost of memory and computation is also very high. So Zhang et al. adopted a very sparse random measurement matrix instead of the common measurement matrix. The very sparse random measurement is defined as

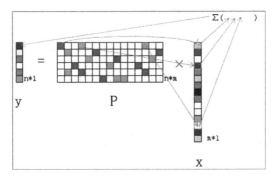

Fig. 1. Compress a high-dimensional vector x to a low-dimensional vector y. (Color figure online)

$$p_{rj} = \sqrt{s} \times \begin{cases} 1 & \textit{with probablity } \frac{1}{2s} \\ 0 & \textit{with probablity } 1 - \frac{1}{s} \\ -1 & \textit{with probablity } \frac{1}{2s} \end{cases} \tag{2}$$

The process of projecting a feature from high-dimension space into low-dimension space is shown in Fig. 1.

In the Fig. 1, the red boxes represent +1, the green boxes represent −1, and the other white boxes represent 0. According to the definition, there are at most m/s elements are nonzero in each row of the matrix. In this paper, we set $s = m/16$, and $m = M \times N$. So there are at most 16 elements with nonzero values in each row of the matrix. The merit of matrix P is that it is very sparse and only needs store those nonzero values with very light cost.

The top right corner of Fig. 1 shows how the compressed feature is computed. The first element of y is the cumulative sum of the nonzero value (1 or −1) of the first row of matrix P multiply the corresponding elements of x. The rest elements of y can be obtained in the same manner. So, we can get the compressed feature by this equation

$$y_i = \sum_{j=1}^{d} x_{rand(m)} (-1)^{rand(2)} \tag{3}$$

where $d \le m/s$, $i = 1, 2, \cdots, n$. Function $rand(m)$ is random function, it returns an integer value between 1 and m. The x is the original high-dimension feature, which is generated by convolving rectangle filters with the image. Those rectangle filters are defined by

$$F_{w,h}(x, y) = \begin{cases} 1, & 1 \le x \le w, 1 \le y \le h \\ 0, & \textit{otherwise} \end{cases} \tag{4}$$

where w and h are the width and height of the rectangle filter, respectively. For space limitation, please see [6] for more detail.

3 Multi-channel Compressive Feature

The compressive features described above obtain information from one-channel of the image, so it only preserves limited information of the images. Dollar et al. [7] showed that the features coming from multi-channel are more effective than those coming from one-channel. So if we combine the features coming from multi-channel into together, the combined features will be more effective. Inspired by this idea, we expanded the compressive feature into multi-channel image in this section. The procedure of the multi-channel compressive feature is shown in Fig. 2.

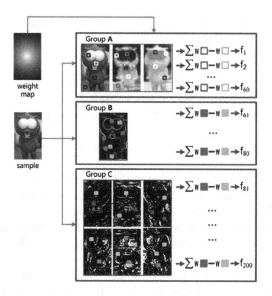

Fig. 2. Multi-channel compressive features. (Color figure online)

We can see from the Fig. 2 that there are total 10 image channels which are divided into three groups. In the first group, there are three image channels coming from LUV color space. The second group contains the gradient image channel only. In the third group, there are six image channels; each image channel is a gradient direction image. Here we divide $360°$ into six directions, so we can get six gradient direction image channels.

Suppose the length of our multi-channel compressive feature is 200, i.e. the final feature has 200 elements. Then we need get the first $3 * 20 = 60$ elements in the first group and $1 * 20 = 20$ elements in the second group and $6 * 20 = 120$ elements in the third group. At last, the features obtained in each group are concatenated together to form the final multi-channel compressive feature. For an element in a group, we need draw randomly at most 16 (there are at most 16 nonzero elements in each row of the matrix P) the basic units (rectangle with size $5 * 5$ in this paper), then the values of those basic units combine together to form the element. All elements should be calculated according to this method in each group. Figure 2 shows how the features are computed.

In our algorithm, we use rectangle features with size 5 * 5 as the basic unit. It is common that the value of the rectangle features is the sum of all pixels which fall within the rectangle area. A disadvantage of this calculation method is that the neighboring features are highly relevant, and this correlation could weaken the discrimination ability of the classifier. So we proposed a new method to generate the rectangle feature. To be specific, we accumulate the sum of those pixels that fall on the border of the rectangle box only. We name the new feature as hollow rectangle features. Figure 3 shows the merits of hollow rectangle features.

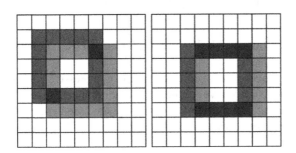

Fig. 3. The hollow rectangle features (Color figure online)

In Fig. 3, there are two neighboring hollow rectangle features which are showed in red color and green color respectively. In the left picture, a sample of hollow rectangle feature (red color) moves down to the right one pixel position and form the second hollow rectangle feature (green color). In this condition, there are only two pixels that are overlapped (showed in blue color). The overlap ratio is one eighth. But when it comes to the common rectangle features, the overlap ratio is 16/25. In the right picture, the hollow rectangle feature moves right one pixel, now, the overlap ratio is 8/16, while the overlap ratio of common rectangle feature is 20/25. The smaller overlap ratio means that the correlation of the features is fewer, so that the hollow rectangle features own more powerful discrimination ability of separating the tracking object from the surrounding background. But we adopted the hollow rectangle features only in the LUV color channels. In the other channels related to the gradient image, we still adopted the common rectangle features. The reason is that the gradient image and gradient direction image contain little information, and there are nonzero values near the boundary contour, and the values in most areas of the image are zero. In this case, the common rectangle features are more effective.

In most visual tracking algorithms, rectangle box is used to locate the tracking object. Nevertheless, due to the shape of the target is varied; the rectangle box often contains some background of the image. For example, in Fig. 2, there are parts of yellow table and blue book in the sample image. So we designed a weighted mechanism for these basic units. The weight of basic unit is related to the distance between the object center and the basic unit center. Suppose the center of the target is o and the center of the basic unit is m, $\vec{l}_o$ and $\vec{l}_m$ are location vectors of coordinate o and m respectively, then the weight of the basic unit m is defined as

$$w_m = e^{-c\left\|\vec{l}_o - \vec{l}_m\right\|_2^2} \tag{5}$$

where $\|\cdot\|_2$ is the norm of vector, and c is normalization constant. The weight w_m is in line with the two-dimensional Gaussian distribution. We can see an example of weight map at the upper-left corner of Fig. 2. The smaller the distance between the basic unit and the target center, the greater weight the basic unit gets.

4   Tracking Based on Multi-channel Compressive Features

In this section, we proposed a novel multi-channel compressive tracking (MCT) based on the Multi-channel compressive features. Our algorithm belongs to the discriminative algorithms. The multi-channel compressive features are used to separate the target from the surrounding background via a naive Bayesian classifier. The structure of our tracker is shown in Fig. 4.

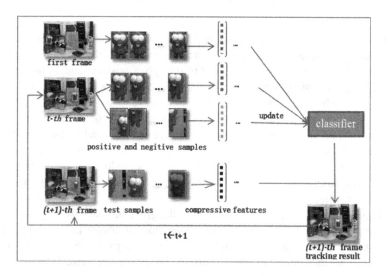

Fig. 4. The structure of our tracker

In our tracker, the initial position of the tracking target is given by human or other object detection algorithm. We use $[x, y, w, h]$ to represent the location of the object, where (x, y) is the location of the upper left corner of the object box and (w, h) are the width and the height of the object box respectively. At the very begin of the tracking, we generate the positive and the negative samples in the first frame, then these samples are used to training the initial classifier. From the second frame, the negative samples are obtained in the *t-th* frame (the current frame), but the positive samples used to train the classifier are those positive samples in the first frame plus the positive samples in the *t-th* frame. When the *(t + 1)-th* is coming, the algorithm first select the test samples in the neighboring position of the target position in the *t-th* frame, then these test samples are send into the classifier, at last, the most likely tracking result is obtained. The procedure of tracking is repeated until to the end.

The classifier we used in our tracker is based on the native Bayesian model which is essentially a conditional probability classification algorithm based on the independence assumption. Given a class label is $y \in \{0, 1\}$, and the multi-channel compressive feature of image sample is $v = (x_1, x_2, \cdots, x_n)$, the classifier computes the probability of the sample v belonging to each class, i.e., $p(y = 0|v)$ and $p(y = 1|v)$. Then the class label corresponding max probability value is regard as the final classification result. The formula of the classifier is

$$H(v) = \log \frac{p(y = 1|v)}{p(y = 0|v)} = \log \frac{p(v|y = 1) p(y = 1)}{p(v|y = 0) p(y = 0)} \tag{6}$$

when $H > 0$, the sample v belongs to calss **1**, and when $H < 0$, the sample v belongs to calss **0**. We suppose uniform prior, $p(y = 1) = p(y = 0)$ and the elements in v are independent of each other, we can get

$$H(v) = \log \frac{p(v|y = 1)}{p(v|y = 0)} = \sum_{i=1}^{n} \log \frac{p(x_i|y = 1)}{p(x_i|y = 0)} = \sum_{i=1}^{n} h_i \tag{7}$$

Here we can see that the classifier $H(v)$ is accumulated sum of a series of weaken classifier h_i which is constructed on each single feature. The conditional distributions $p(x_i|y = 1)$ and $p(x_i|y = 0)$ in the classifier $H(v)$ are assumed to be Gaussian distribution with four parameters $(\mu_i^1, \sigma_i^1, \mu_i^0, \sigma_i^0)$,

$$p(x_i|y = 1) \sim N(\mu_i^1, \sigma_i^1) \tag{8}$$

$$p(x_i|y = 0) \sim N(\mu_i^0, \sigma_i^0) \tag{9}$$

where $\mu_i^1 (\mu_i^0)$ and $\sigma_i^1 (\sigma_i^0)$ are the mean and the standard deviation of the positive (negative) class. In addition, the state and background of the target often change gradually in tracking, so the parameters of the classifier need be updated constantly to adapt to the changes. the updating procedure can be described as

$$\mu^1 \leftarrow r\mu^1 + (1 - r)\mu_i^1$$
$$\sigma^1 \leftarrow \sqrt{r(\sigma^1)^2 + (1 - r)(\sigma_i^1)^2 + r(1 - r)(\mu^1 - \mu_i^1)^2} \tag{10}$$

where $0 \leq r \leq 1$ is a learning parameter. And the definition of $(\mu_i^1, \sigma_i^1, \mu_i^0, \sigma_i^0)$ please refer to [6].

5 Experiments

In this section, we compare our MCT tracker with the other six state-of-the-art algorithms on 30 challenging videos. These sequences contain a variety of challenging factors, such as illumination changes, background cluster, heavy occlusion, non-rigid deformation, motion blur, etc. The algorithms we compared with are some representative discriminative algorithms which are realized in the similar manner to our tracker, including MIL [9], SemiT [10], CT [6], BSBT [11] and OAB [13]. In addition, we also compare our MCT tracker with Frag [3] algorithm which is a typical generative algorithm. For fair comparison, we use the source or binary codes provided by the authors with tuned parameters for best performance.

5.1 Experimental Setup

The parameters used in our MCT algorithm are fixed in all the experiments. The basic unit is a rectangle box with size 5 * 5 pixels. The dimensionality of the multi-channel compressive feature is set to $n = 200$, and each element of the compressive features is composed of 4–16 basic units. Given a target location at the current frame, the search radius for drawing positive samples is set to $\alpha = 4$ which generates 45 positive samples. When we generate negative samples, the inner and outer radiuses are 8 and 35 pixels respectively. When a new frame is coming, we draw test samples within the circle whose radius is 20 pixels. At last, the learning parameter r is set to 0.85 and all algorithm parameters are fixed for all the experiments.

5.2 Experimental Results

We use the popular Pascal VOC overlap ratio [12] as the evaluation criteria to evaluate the performance of our tracker and the other trackers. The VOC overlap ratio is defined as $R = area(B_t \cap B_g)/area(B_t \cup B_g)$, where B_t is the tracking bounding box and B_g is the ground truth bounding box. The $area(x)$ is a function that computes the area of the parameter x. The successful frame is indicated when the intersection of the ground truth bounding box and the tracking bounding box is not less than half of the union of the ground truth bounding box and the tracking bounding box. It should be emphasized that BSBT and SemiT algorithms have no tracking result on some frames, so we set the VOC overlap ratio is zero for these frames. Table 1 show the tracking results in terms of success rate, where the best result is shown in red bold fonts and blue fonts indicate the second best ones. From Table 1 we can see that our MCT tracker achieves the best or second best performance compared with the other algorithms for most of sequences.

Performance in Illumination Change. The tracking results for the trellis and fish sequences are shown in Fig. 5. These two videos including severe illumination change. Most of the features used in our MCT tracker is relate to the gradient and gradient direction, so our tracker is more robust than the other trackers. In the fish video, the global illumination change has little impact on the tracking results, so we get very good

Table 1. The Pascal VOC overlap ratio. The best result is shown in red bold fonts.

Sequence	CT	SemiT	OAB	MIL	BSBT	Frag	MCT
carDark	0.00	0.93	**0.95**	0.18	0.49	0.25	**1.00**
david	**0.43**	0.20	0.15	0.23	0.27	0.12	**0.62**
david2	0.00	0.56	0.25	0.32	**0.66**	0.30	**1.00**
sylvester	**0.83**	0.59	0.68	0.55	0.63	0.68	**0.89**
trellis	0.35	0.20	0.18	0.24	0.23	**0.36**	**0.83**
fish	**0.89**	0.12	0.04	0.39	0.70	0.55	**1.00**
mhyang	0.73	0.77	**0.96**	0.39	0.73	0.72	**1.00**
shaking	0.04	0.08	0.01	**0.23**	0.03	0.07	**0.98**
singer2	0.01	0.08	0.03	**0.48**	0.09	0.20	**0.99**
boy	0.69	0.46	**0.99**	0.39	0.65	0.46	**0.97**
dudek	0.85	0.60	0.80	0.86	**0.94**	0.59	**0.90**
crossing	**0.98**	0.88	0.83	**0.98**	0.11	0.38	0.96
football1	0.08	0.15	0.24	**0.78**	0.34	0.34	**0.62**
doll	0.53	0.15	**0.66**	0.43	0.21	0.66	**0.73**
girl	0.18	0.55	**0.94**	0.29	0.46	0.54	**0.58**
walking	0.50	0.28	0.48	**0.54**	0.17	0.51	**0.55**
fleetface	**0.64**	0.38	**0.68**	0.54	0.59	0.45	0.59
david3	0.35	0.18	0.34	0.68	0.37	**0.81**	**0.75**
jumping	0.01	0.06	0.05	0.48	0.13	**0.85**	**0.95**
carScale	**0.45**	0.44	0.44	0.45	0.43	0.44	**0.45**
skiing	0.07	0.05	**0.10**	0.07	0.04	0.04	**0.43**
dog1	0.65	0.59	**0.65**	**0.65**	0.60	0.62	0.64
suv	0.23	0.48	**0.76**	0.13	0.68	0.71	**0.72**
mountainBike	0.17	0.59	**0.91**	0.57	0.32	0.14	**1.00**
woman	0.16	0.36	**0.61**	0.19	0.16	0.18	**0.91**
faceocc1	0.85	0.91	0.91	0.76	**1.00**	1.00	**1.00**
faceocc2	0.74	0.57	0.75	**0.94**	0.76	0.75	**0.99**
basketball	0.26	0.03	0.02	0.28	0.11	**0.70**	**0.93**
football	**0.78**	0.23	0.36	0.74	0.29	**0.92**	0.64
subway	0.78	0.38	0.22	**0.80**	0.21	0.58	**0.93**

performance on this video. For the trellis video, though much local light change exists, most of the tracking results are satisfactory.

In the comparison algorithm, Frag algorithm uses histogram feature which cannot deal with light change well, so it resulting in serious drift. CT and MIL algorithms adopt features generated from gray image, which limits the represented ability of the

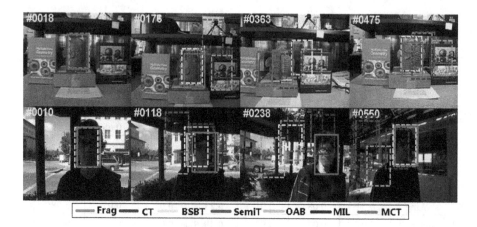

Fig. 5. The tracking results on trellis and fish sequences.

trackers, so CT and MIL algorithms perform not very well. OAB algorithm uses the positive samples coming from the last frame to train the classifier. When the illumination changes, the tracking result is not accurate, then drifting is coming soon. BSBT has no online learning procedure; it is only store the appearance of the first frame. When the light has great change, BSBT cannot find a similar object in many frames.

Performance in Deformation. The tracking results for the basketball and singer2 sequences are shown in Fig. 6. Lots of tracking targets are human person, which often include much of pose change. The basketball and singer2 sequences are good samples to show that. In our algorithm, we adopt weight mechanism to highlight the importance of central region, so the movements of hands and feet have little impact on our tracker. In addition, the multi-channels feature comprises much color and border information, that make our tracking results more stable. Just as the tracking result in Fig. 6 show that out tracker has found the right target in most time.

Fig. 6. The tracking results on basketball and singer2 sequences.

In the comparison algorithm, Frag algorithm vote by many small patches. In these two video, most patches on the head and body which are stable than hands and feet vote correctly, so Frag algorithm has good performance in these videos. The feature used by CT is not robust to deformation, and it lost the target eventually. MIL algorithm also performs well in these two videos; the reason is that feature used in MIL adopts weight mechanism similar to the MCT algorithm. All the other algorithms have poor performance in these videos.

Performance in Rotation. The tracking results for the Girl and mountain-bike sequences are shown in Fig. 7. There are a lot of in-plane and out-plane rotation. In these two videos, our MCT and OAB algorithms are superior to the other methods. OAB adopts Adaboost algorithm that can find the most discriminative features. Though MIL has a similar boosting mechanism, but it is based bag of positive and negative feature, so it is not very precise, when the target rotates, MIL cannot find suitable feature to discriminate the target. In the Girl video, when the girl turn her head to the back, the appearance change greatly, so most trackers cannot track the target correctly, besides MCT and OAB.

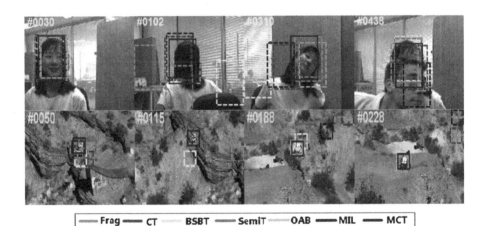

Fig. 7. The tracking results on girl and mountainBike sequences.

6 Conclusion

In this paper, we propose a novel multi-channel compressive feature, which make full use of the information coming from multiple image channels. There are total 10 image channel we used in this paper, including three color channels, one gradient channel and six gradient direction channels. These channels are divided into three groups, i.e. color channel group comprising L, U and V color channel, gradient group comprising the gradient channel, and the gradient direction group comprising six gradient direction channels. Within one group, the feature is across the channels, i.e., the basic units may be generated randomly at every channel in the group. Then, the features of three groups are concatenated together to form the original high-dimension vectors. At last, the high-dimension features are projected into low-dimension space by the measurement matrix. After that, we designed a new tracker based on the multi-channel compressive features. At the last of this paper, we conducted extensive experiments on some various challenging sequences, and experimental results show that our algorithm outperforms the other state-of-the-art algorithms.

References

1. Wu, Y., Lim, J., Yang, M.H.: Object tracking benchmark. IEEE Trans. Pattern Anal. Mach. Intell. **37**(9), 1834–1848 (2015)
2. Ross, D.A., Lim, J., Lin, R.S., et al.: Incremental learning for robust visual tracking. Int. J. Comput. Vis. **77**(1–3), 125–141 (2008)
3. Adam, A., Rivlin, E., Shimshoni, I.: Robust fragments-based tracking using the integral histogram. In: 2006 IEEE Computer Society Conference on Computer Vision and Pattern Recognition, pp. 798–805 (2006)
4. Kwon, J., Lee, K.M.: Visual tracking decomposition. In: IEEE Conference on Computer Vision and Pattern Recognition, pp. 1269–1276 (2010)
5. Oron, S., Bar-Hillel, A., Dan, L., et al.: Locally orderless tracking. Int. J. Comput. Vis. **111** (2), 1940–1947 (2012)
6. Zhang, K., Zhang, L., Yang, M.H.: Fast compressive tracking. IEEE Trans. Pattern Anal. Mach. Intell. **36**(10), 2002–2015 (2014)
7. Dollar, P., Tu, Z., Perona, P., Belongie, S.: Integral channel features. In: British Machine Vision Conference (2009)
8. Ng, A., Jordan, M.: On discriminative vs. generative classifiers: a comparison of logistic regression and naive Bayes. In: Advances in Neural Information Processing Systems, pp. 841–848 (2002)
9. Babenko, B., Yang, M.H., Belongie, S.: Robust object tracking with online multiple instance learning. IEEE Trans. Pattern Anal. Mach. Intell. **33**(8), 1619–1632 (2011)
10. Grabner, H., Leistner, C., Bischof, H.: Semi-supervised on-line boosting for robust tracking. In: Forsyth, D., Torr, P., Zisserman, A. (eds.) ECCV 2008. LNCS, vol. 5302, pp. 234–247. Springer, Heidelberg (2008). doi:10.1007/978-3-540-88682-2_19
11. Sato, K., Sekiguchi, S., Fukumori, T., et al.: Beyond semi-supervised tracking: tracking should be as simple as detection, but not simpler than recognition. In: IEEE International Conference on Computer Vision Workshops, pp. 1409–1416 (2009)
12. Everingham, M., Van Gool, L., Williams, C., Winn, J., Zisserman, A.: The pascal visual object classes (VOC) challenge. Int. J. Comput. Vis. **88**(2), 303–338 (2010)
13. Grabner, H., Grabner, M., Bischof, H.: Real-time tracking via on-line boosting. In: British Machine Vision Conference 2006, Edinburgh, UK, pp. 47–56, September 2006

Single Image Super-Resolution with a Parameter Economic Residual-Like Convolutional Neural Network

Ze Yang[1], Kai Zhang[2], Yudong Liang[1], and Jinjun Wang[1(✉)]

[1] Institute of Artificial Intelligence and Robotics, Xi'an Jiaotong University,
Xi'an 710049, Shaanxi, China
jinjun@mail.xjtu.edu.cn
[2] Harbin Institute of Technology, Harbin, China

Abstract. Recent years have witnessed great success of convolutional neural network (CNN) for various problems both in low and high level visions. Especially noteworthy is the residual network which was originally proposed to handle high-level vision problems and enjoys several merits. This paper aims to extend the merits of residual network, such as skip connection induced fast training, for a typical low-level vision problem, i.e., single image super-resolution. In general, the two main challenges of existing deep CNN for supper-resolution lie in the gradient exploding/vanishing problem and large amount of parameters or computational cost as CNN goes deeper. Correspondingly, the skip connections or identity mapping shortcuts are utilized to avoid gradient exploding/vanishing problem. To tackle with the second problem, a parameter economic CNN architecture which has carefully designed width, depth and skip connections was proposed. Experimental results have demonstrated that the proposed CNN model can not only achieve state-of-the-art PSNR and SSIM results for single image super-resolution but also produce visually pleasant results.

Keywords: Super-resolution · Deep residual-like convolutional neural network · Skip connections · The mount of parameters

1 Introduction

Single image super-resolution (SISR) which aims to recover a high-resolution (HR) image from the corresponding low-resolution (LR) image is a practical technique [11,21,22] due to its high value in various fields. Typically, it is very challenging to restore the missing pixels from an LR observation since the number of pixels to be estimated in the HR image is usually much larger than that in the given LR input. Generally, SISR techniques can be roughly divided into three categories: the interpolation methods, the reconstruction methods [8] and the example based methods [4,20].

© Springer International Publishing AG 2017
L. Amsaleg et al. (Eds.): MMM 2017, Part I, LNCS 10132, pp. 353–364, 2017.
DOI: 10.1007/978-3-319-51811-4_29

Most of the recent SISR methods fall into the example based methods which try to learn prior knowledge from LR and HR pairs, thus alleviating the ill-posedness of SISR. Representative methods include neighbor embedding regression [1,16,17], random forest [13,14] and deep convolutional neural network (CNN) [2,3,9,12].

Among the above techniques, deep learning techniques especially deep CNN have largely promoted the state-of-the-art performances in SISR area. Dong *et al.* [2] proposed a deep convolutional neural network termed SRCNN with three convolutional layers for image super-resolution. By learning an end-to-end mappings from LR to HR images, SRCNN extracts more discriminative features than handcrafted ones. Later, Dong *et al.* [3] extended SRCNN with larger filter size and filter numbers while keeping the depth of CNN fixed to further improve the performance. They have found that deeper model was hard to train and they failed to boost the performance by increasing the depth. Such findings indicate that deeper model is not suitable for image super-resolution, which is counter-intuitive as deeper model have been proved more effective in many tasks [5,6,15]. Instead of directly predicting the HR output, Kim *et al.* [9] proposed a very deep CNN (VDSR) with depth up to 20 to predict the residual image. VDSR surpasses SRCNN with a large margin which mainly benefits from two aspects: deeper architecture and predicting high frequency of images only which is called residual learning by [9].

As demonstrated in [9], the SR results have been improved as VDSR goes deeper. Although VDSR has achieved impressive results, the plain structure of VDSR which simply stacks layers hampers the convergence of deeper architectures due to the gradient exploding/vanishing problem. It would not bring any improvement as the network goes deeper. Fortunately, the residual network [5,6] has successfully addressed this issue. As a result, different from VDSR, this paper has designed a novel very deep residual-like convolutional neural network whose architecture is shown in Fig. 1. As LR image and target HR image is highly correlated, predicting high frequency of the image only is a kind of residual learning which largely lower the price for training. Thus, a totally residual-like deep CNN will fully take advantage of the correlations between LR and HR images.

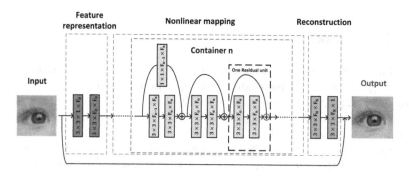

Fig. 1. The architecture of our residual-like model.

Moreover, skip connections or identity mapping shortcuts in residual-like deep CNN would alleviate gradient vanishing/exploding problem when the network becomes increasingly deeper.

While very deep CNN model would increase the model capacity, on the other hand, it would introduce a huge amount of parameters which is sometimes unacceptable for limited hardware. Thus, a computational economic architecture is essential for real word applications. In this paper, the 'shape' of deep CNN has been explored to largely reduce the amount of parameters. The 'shape' of deep CNN refers to all the filter size and numbers of each layer which decides featuremap size and numbers of each layer to form a global shapes. With a residual-like architecture and economic shape design, the proposed model can not only achieve state-of-the-art PSNR and SSIM results for single image super-resolution but also produce visually pleasant results.

2   Related Works

In the pioneer work by Freeman *et al.* [4], the co-occurrence priors were proposed that similar LR local structures often relate to similar HR local information. From LR and corresponding HR images, LR and HR examples (patches or sub images) could be extracted to form training database. The mappings from LR to HR examples call for accurate regression methods to be applied.

Since the work of SRCNN [2], deep CNNs have refreshed the state-of-the-art performances in super-resolution area. Dong *et al.* [3] elaborated the filter size and filter numbers for their three layers of SRCNN which further improved the performance. Wang *et al.* [18] incorporated the sparse coding prior into CNN architecture design based on the learned iterative shrinkage and thresholding algorithm (LISTA). With sparsity prior modeling, the performance boosted even with a model of smaller size compared with SRCNN.

Currently, the best performance and deepest CNN architecture for image super-resolution was VDSR with 20 convolutional layers proposed by Kim et al. [9], which largely accelerated the speed of training and outperformed SRCNN presented by Dong et al. [3]. To ensure the fast convergence of deep CNN and avoid gradient vanishing or exploding, a much larger learning rate for training was cooperated with adjustable gradient clipping in VDSR training. VDSR is inspired by the merits of VGG net which attempts to train a thin deep network. However, this kind of plain networks are not easy to be optimized when it goes even deeper as demonstrated by He *et al.* [5,6].

The difficulties of training deeper plain networks were carefully analyzed by He *et al.* [5,6]. It has been observed that the testing accuracy even the training accuracy becomes saturated then degrades rapidly as plain networks goes deeper, which is called as the degradation problem [5]. This degradation is caused by the difficulties of training other than overfitting. It has been demonstrated that learning a residual function is much more easier than learning the original prediction function with very deep CNN. Residual networks with a surprising depth were designed for image classification problems with skip connections or identity mapping shortcuts. Later, a detailed analysis on the mechanisms of identity mapping in deep residual networks and a new residual unit design has been represented in [6].

After largely easing the difficulties of training much deeper CNN with residual functions by shortcuts or skip connections, the huge amount of parameters is still a big problem for computational resources and storage. He *et al.* [5,6] attempts to alleviate the problem by bottleneck architectures. The bottleneck architectures first utilized 1×1 convolutions to reduce the dimensions, then after some operations, 1×1 convolutions are applied again to increase the dimensions. With such a residual unit design, the amount of parameters was largely reduced. Thus, the shape of CNN could be potentially explored to reduce the parameters while maintain the performances. In the meanwhile, contextual information is very important for image super-resolution, such residual unit design may give a negative effort to the SR results.

With a carefully design and exploration of the shape of the network, a novel residual-like deep model is proposed for image super-resolution task.

3   A Parameter Economic Residual-like Deep Model for Image Super-Resolution

Following the example based methods, HR examples I^h and LR examples I^l are extracted from HR images I^H and LR images I^L respectively. The degeneration process of LR images I^L from the corresponding HR images I^H could be considered as the following blurring process related with blur kernel G and downsampling process $\downarrow_s$ with a a scale factor s as in Eq. (1).

$$I^L = (I^H \otimes G) \downarrow_s, \tag{1}$$

3.1   Residual-like Deep Model

Our residual-like deep CNN for image super-resolution is an end-to-end mapping model which tries to predict HR versions from LR input ones. There are three sub-networks in our deep CNN to perform three steps: feature representation, nonlinear mapping, reconstruction.

The feature representation sub-network extracts discriminative features from the LR input images, while nonlinear mapping part maps the LR feature representations into HR feature representations. Reconstruction part restores the HR images from HR feature representation. Feature representation apply plain network stacking convolutional and ReLU layers as shown in Fig. 1 and reconstruction use convolutional layers. The main body of our model, nonlinear mapping part consists of residual-like units which eases the difficulties of training.

Typical units of our residual-like deep CNN are shown in Fig. 2. As residual unit with 2 layers and 3 layers worked well for image super-resolution problem, those two kinds of units are applied in the experiments. When featuremap dimensions change, the identity shortcut becomes a projection to change feature dimensions. The second right and rightmost are one unit of residual net for image

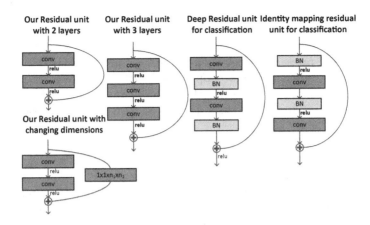

Fig. 2. The architectures of different residual units.

classification problems proposed by He *et al.* in [5,6] respectively. The architectures of our residual functions are composed of convolutional, ReLU layers and shortcuts, which is very different. Batch normalization units are discarded and deployments are different. Similar with VDSR, small convolutional filter size as 3×3 has been applied. Shortcuts or skip connections which are identity mappings are realized by element-wise additions. As this element-wise addition increases very little computations, our feed-forward deep CNN has a similar computational complexity with VDSR. Assuming the input as x^k of kth residual unit, the residual functions have the following form:

$$x^{k+1} = x^k + f(\theta^k, x^k), \tag{2}$$

where θ^k are the parameters of kth residual unit. As predicting high frequency can boost the performances and convergence speeds of deep CNN [9], a simple Euclidean loss function is adopted to approximate the high frequencies of examples

$$loss = \frac{1}{2n} \sum_{i=1}^{n} \|F(\Theta, I_i^l) - (I_i^h - I_i^l)\|^2 \tag{3}$$

where n is the number of patch pairs (I^l, I^h), $F(\Theta, I^l)$ denotes the predictions of our residual-like deep CNN with parameter Θ. Our residual-like deep CNN is composed of several **Containers** which have certain number of residual units. For succinctness, the filter numbers keep the same in each single container. The architectures of our residual-like deep CNN will be described as a sequence of the filter numbers $(N1_{k1}, N2_k, \cdots)$ in containers. If subscript k exists for N_k, it means there are k residual units with each having a filter number of N.

Stochastic gradient descent (SGD) with the standard back-propagation [10] is applied to train our residual-like deep CNN. In particular, the parameter is updated as Eq. (4), where m denotes the momentum parameter with a value of 0.9 and η is the learning rate.

$$\triangle_{i+1} = m \cdot \triangle_i + \eta \cdot \frac{\partial loss}{\partial \theta_i}, \quad \theta_{i+1} = \theta_i + \triangle_{i+1} \tag{4}$$

High learning rates are expected to boost training with faster and better convergency. Adjustable gradient clipping [9] is utilized to keep learning rates high while at the same time to prevent the net from gradient exploding problems. Gradients $\frac{\partial Loss}{\partial \theta_i}$ are clipped into the range of $[-\frac{\tau}{\eta}, \frac{\tau}{\eta}]$, where τ is a constant value.

3.2   Economic Design for the Proposed Model

In this paper, the 'shape' of deep CNN has been explored to reduce the amount of parameters. The 'shape' of deep CNN is determined by all the filter size and numbers of each layer. Thin but small filter size works well with padding which leads to larger receptive field as network goes deep, in specific, 3×3 filter size has been applied. Next, filter numbers and the combinations of filter numbers will be discussed.

Exploring the Shape of the Architecture. Inspired by the bottle-neck architecture [6], this paper supposes that changing the shape of the architecture may maintain the performance while largely reduces the computational parameters. Instead of applying 1×1 convolutions as bottle-neck architecture, the 3×3 convolutions are applied as image SR process largely depends on the contextual information in local neighbor areas. The impacts of the featuremaps number in each layers on performance are carefully explored in the following fashions: increase monotonically, decrease monotonically, increase monotonically then decrease monotonically, decrease monotonically then increase monotonically. The experiments demonstrate that some of different economic designs have achieved comparable performance which largely reduce the parameters. This will be further discussed in the experiments part. In comparison with our residual-like CNN, the performances of VDSR with different shapes have more variations. This proves our residual-like architecture are more robust to the shape of CNN.

Training with Multiple Upscaling Factors. It has been pointed out that it is feasible to train a deep CNN for different upscaling factors [9]. Training datasets for different specified upscaling factors are combined together to enable our residual-like deep CNN to handle multiple upscaling factors, as images across different scales share some common structures and textures. Parameters are shared across different predefined upscaling factors which further dispense with the trouble of retaining different models for different upscaling factors.

4   Experiments

In this section, we conducted a series of experiments to demonstrate the performance of the proposed method against the state-of-the-art SISR methods.

The same 291 training images applied by VDSR were utilized for training, including 91 images proposed in Yang *et al.* [20] and 200 natural images from Berkeley Segmentation Dataset (BSD). For testing, four datasets were investigated: 'Set5' and 'Set14' [2,16], 'Urban100' [7] and 'BSD100' [16,19].

The size of example is set as 41×41 and the batch size is set as 64. Momentum and weight decay parameters are fixed as 0.9 and 0.0001 respectively. Multi-scale training is applied in all of the following experiments. Weight initialization methods [5,6] were applied. Learning rate was initially set to 0.1 and then decreased by a factor of 10 every 30 epochs. All these setting ensures us to make a fair comparison with the competing approaches including VDSR method.

4.1 Comparisons with the State-of-the-art Methods

Table 1 shows the quantitative comparisons with A+ [17], RFL [14], SelfEx [7], SRCNN [2] and VDSR [9]. Visual results were also demonstrated to give intuitive assessment. In Table 1, two architectures of our residual-like deep CNN with different depth have been investigated, denoted as **R-basic** and **R-deep** respectively. R-basic $(16_3, 32_3, 64_3)$ has 22 layers, while R-deep $(16_3, 32_3, 64_3, 128_3, 256_3)$ has 34 layers. Deeper and larger model R-deep has achieved the best performance compared with other methods in most cases and comparable results in other situation.

R-basic outperform the other methods except VDSR. However, the performances of VDSR (20 layers) have not been obtained by us. For example, the average PSNR of VDSR for Set5 and Set14 are 37.32 dB and 32.89 dB respectively. Assisted with the missing tricks, the performance of our model is expected to be further boosted. In Fig. 3, the PSNR against training epochs has been compared among R-basic, R-deep, and VDSR trained by us. Deeper and larger model R-deep outperformed VDSR at very beginning with a large margin. Although R-basic contained much less parameters, R-basic model has obtained comparable performances with VDSR.

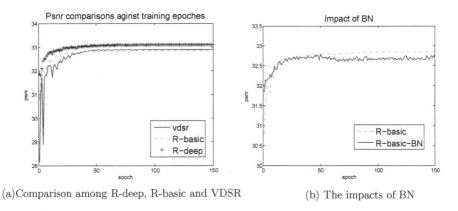

(a)Comparison among R-deep, R-basic and VDSR   (b) The impacts of BN

Fig. 3. Comparison of test psnr of Set 14 against training epoch among (a) R-deep, R-basic and VDSR, (b) our R-basic with and without Batch Normalization (BN).

In Fig. 4, all the compared results are obtained by the released code of the authors. Visual pleasing restorations have been achieved by our model. Results of our method contain more authentic texture and more clear details compared with other methods such as the texture of the zebra head. Our method have provided less artifacts, *e.g.* all the other methods except ours have restored obvious artifacts at the location of book. Shaper edges have appeared in our restorations which have represented visually more pleasing results.

Impacts of Batch Normalization on SISR. It seems adding batch normalization (BN) operations has hampered further improvement when more epochs have been performed. Normalizing input distribution of mini-batch to suppress data shifting has been proved powerful and largely accelerated the training convergency speed. It also enable deeper architecture and larger learning rates to be utilized in other tasks. However, whiten input and output of the intermediate layer may not be suitable for image super-resolution task which need precise output. Another suspects may be regularization effects of BN has not been fully exploited as the training set of Fig. 3(b) is still limited in contrast with ImageNet. As larger learning rates were enabled by gradient clipping methods, the benefits of BN for leaning rates are alleviated. The experiments with impacts of BN on SR would be elaborated in the future.

4.2   Amount of Parameters

For R-basic model, there are 22 convolutional layers and 0.3M(322721) parameters accumulated by the amounts of corresponding weights and bias. For R-deep model, 34 convolutional layers and 5M(4975905) parameters are applied. The compared VDSR in Table 1 is 20 layers and has 0.7M(664704) parameters. Although R-deep has more parameters, our R-deep model is still acceptable which can be efficiently trained with single GPU.

Table 1. Comparison in different datasets and with different scales.

Dataset	Scale	Bicubic PSNR/SSIM	A+[17] PSNR/SSIM	RFL[14] PSNR/SSIM	SelfEx[7] PSNR/SSIM	SRCNN[2] PSNR/SSIM	VDSR[9] PSNR/SSIM	R-deep PSNR/SSIM	R-basic PSNR/SSIM
Set5	x2	33.66/0.9299	36.54/0.9544	36.54/0.9537	36.49/0.9537	36.66/0.9542	**37.53/0.9587**	37.51/0.9587	37.27/0.9577
	x3	30.39/0.8682	32.58/0.9088	32.43/0.9057	32.58/0.9093	32.75/0.9090	33.66/0.9213	**33.72/0.9215**	33.43/0.9190
	x4	28.42/0.8104	30.28/0.8603	30.14/0.8548	30.31/0.8619	30.48/0.8628	31.35/**0.8838**	**31.37/0.8838**	31.15/0.8796
Set14	x2	30.24/0.8688	32.28/0.9056	32.26/0.9040	32.22/0.9034	32.42/0.9063	33.03/0.9124	**33.10/0.9131**	32.86/0.9113
	x3	27.55/0.7742	29.13/0.8188	29.05/0.8164	29.16/0.8196	29.28/0.8209	29.77/0.8314	**29.80/0.8317**	29.67/0.8297
	x4	26.00/0.7027	27.32/0.7491	27.24/0.7451	27.40/0.7518	27.49/0.7503	28.01/0.7674	**28.06/0.7681**	27.90/0.7648
BSD100	x2	29.56/0.8431	31.21/0.8863	31.16/0.8840	31.18/0.8855	31.36/0.8879	31.90/0.8960	**31.91/0.8961**	31.76/0.8940
	x3	27.21/0.7385	28.29/0.7835	28.22/0.7806	28.29/0.7840	28.41/0.7863	28.82/0.7976	**28.83/0.7980**	28.73/0.7954
	x4	25.96/0.6675	26.82/0.7087	26.75/0.7054	26.84/0.7106	26.90/0.7101	**27.29/0.7251**	27.27/0.7248	27.19/0.7221
Urban100	x2	26.88/0.8403	29.20/0.8938	29.11/0.8904	29.54/0.8967	29.50/0.8946	30.76/0.9140	**30.88/0.9150**	30.47/0.9100
	x3	24.46/0.7349	26.03/0.7973	25.86/0.7900	26.44/0.8088	26.24/0.7989	27.14/0.8279	**27.17/0.8283**	26.92/0.8208
	x4	23.14/0.6577	24.32/0.7183	24.19/0.7096	24.79/0.7374	24.52/0.7221	25.18/0.7524	**25.22/0.7537**	25.02/0.7452

4.3   The Deeper the Better, the Wider the Better

R-deep model perform much better than R-basic. Next, ablations of our system would be evaluated to unpack this performance gain. Two factors, width which is relate to filter numbers and depth of our model would be analyzed.

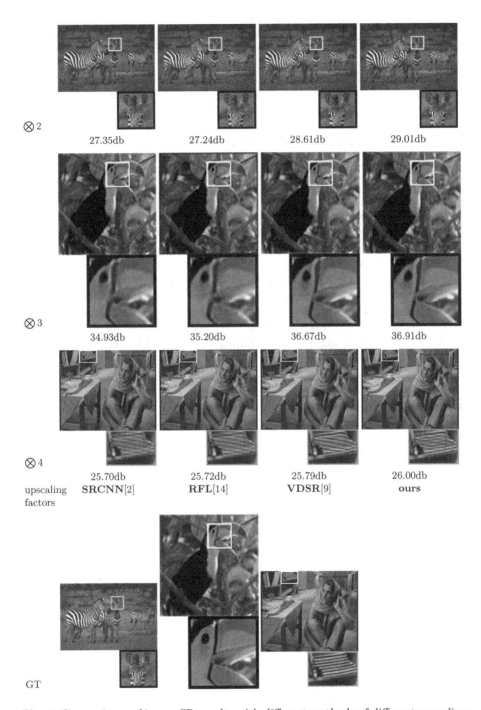

Fig. 4. Comparisons of image SR results with different methods of different upscaling factors

First, 20 layer VDSR has been added with identity shortcuts to form a residual-like net, $R(64_8)$. The performance of $R(64_8)$ is roughly the same as VDSR in Table 2. The shortcuts have very little impacts on the descriptive power.

Table 2. PSNR comparison between our residual-like CNN and VDSR trained by us

	Set5	Set14	BSD100	Urban100
$R(64_8)$	37.28	32.91	31.72	30.45
VDSR	37.32	32.89	31.77	30.51

Second, fixed the depth of model, simply broadened the width will improve the performance, *e.g.*, $R(16_3, 32_3, 64_3)$ vs $R(32_3, 64_3, 128_3)$, $R(4_3, 8_3, 16_3, 32_3, 64_3)$ vs $R(16_3, 32_3, 64_3, 128_3, 256_3)$ (Table 3).

Third, the deeper the architecture, the better the performance. Adding one more residual unit, *e.g.*, $R(16, 32, 64, 128)$ vs $R(32, 64, 128)$ will improve the performance. Certainly, the depth should be no more than certain number to avoid the overfitting problem as the training data is limited. Within this limit, the deeper the better. Our residual-like unit eases the training difficulties which enables a deeper CNN architecture to improve the situation. On the other side, plain deep CNN VDSR with a same depth as our residual-like R-deep model can not converge well and the restorations deteriorate. Another attempt to facilitate deeper net which is the economic design to solve the problem of parameter amounts will be discussed next.

Table 3. PSNR by the residual-like model of different depths and width with a magnification factors 2 in Set14.

	$R(16_3, 32_3, 64_3)$	$R(32_3, 64_3, 128_3)$	$R(16_3, 32_3, 64_3, 128_3)$	$R(16_3, 32_3, 64_3, 128_3, 256_3)$	$R(4_3, 8_3, 16_3, 32_3, 64_3)$
PSNR(dB)	32.85	32.96	33.00	33.10	32.91

Table 4. Performance by different residual-like models which have different shapes with a magnification factor 2 in Set14.

Residual	$R(16_4, 32_4, 64_4)$	$R(64_4, 32_4, 16_4)$	$R(16_2, 32_2, 64_2, 64_2, 32_2, 16_2)$	$R(64_2, 32_2, 16_2, 16_2, 32_2, 64_2)$
PSNR(dB)	32.91	32.85	32.94	32.89

4.4   Economic Design

The performances of different architectures which have different shapes have been investigated for our residual-like net and VDSR counterpart. To be specific, there are 28 layers as 6 residual **containers** stack and each **container** contains 2 residual units (2 layers). It can be calculated as $28 = 2 + 6 \times 2 \times 2 + 2$, where feature

First, 20 layer VDSR has been added with identity shortcuts to form a residual-like net, $R(64_8)$. The performance of $R(64_8)$ is roughly the same as VDSR in Table 2. The shortcuts have very little impacts on the descriptive power.

Table 2. PSNR comparison between our residual-like CNN and VDSR trained by us

	Set5	Set14	BSD100	Urban100
$R(64_8)$	37.28	32.91	31.72	30.45
VDSR	37.32	32.89	31.77	30.51

Second, fixed the depth of model, simply broadened the width will improve the performance, $e.g.$, $R(16_3, 32_3, 64_3)$ vs $R(32_3, 64_3, 128_3)$, $R(4_3, 8_3, 16_3, 32_3, 64_3)$ vs $R(16_3, 32_3, 64_3, 128_3, 256_3)$ (Table 3).

Third, the deeper the architecture, the better the performance. Adding one more residual unit, $e.g.$, $R(16, 32, 64, 128)$ vs $R(32, 64, 128)$ will improve the performance. Certainly, the depth should be no more than certain number to avoid the overfitting problem as the training data is limited. Within this limit, the deeper the better. Our residual-like unit eases the training difficulties which enables a deeper CNN architecture to improve the situation. On the other side, plain deep CNN VDSR with a same depth as our residual-like R-deep model can not converge well and the restorations deteriorate. Another attempt to facilitate deeper net which is the economic design to solve the problem of parameter amounts will be discussed next.

Table 3. PSNR by the residual-like model of different depths and width with a magnification factors 2 in Set14.

	$R(16_3, 32_3, 64_3)$	$R(32_3, 64_3, 128_3)$	$R(16_3, 32_3, 64_3, 128_3)$	$R(16_3, 32_3, 64_3, 128_3, 256_3)$	$R(4_3, 8_3, 16_3, 32_3, 64_3)$
PSNR(dB)	32.85	32.96	33.00	33.10	32.91

Table 4. Performance by different residual-like models which have different shapes with a magnification factor 2 in Set14.

Residual	$R(16_4, 32_4, 64_4)$	$R(64_4, 32_4, 16_4)$	$R(16_2, 32_2, 64_2, 64_2, 32_2, 16_2)$	$R(64_2, 32_2, 16_2, 16_2, 32_2, 64_2)$
PSNR(dB)	32.91	32.85	32.94	32.89

4.4 Economic Design

The performances of different architectures which have different shapes have been investigated for our residual-like net and VDSR counterpart. To be specific, there are 28 layers as 6 residual **containers** stack and each **container** contains 2 residual units (2 layers). It can be calculated as $28 = 2 + 6 \times 2 \times 2 + 2$, where feature

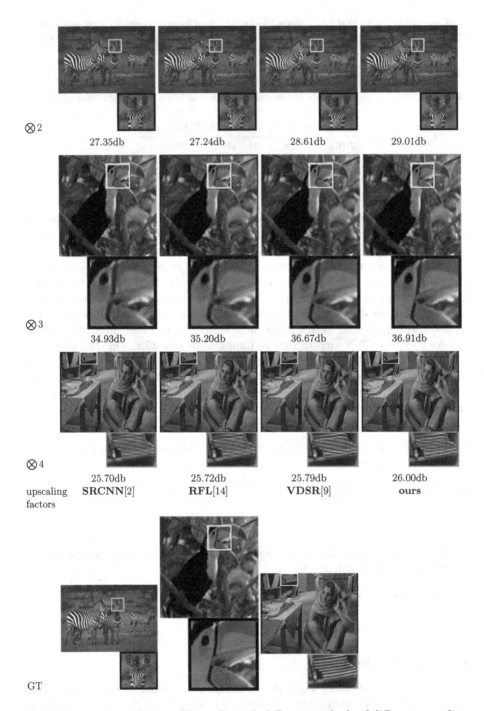

Fig. 4. Comparisons of image SR results with different methods of different upscaling factors

representation sub-network and reconstruction sub-network each have 2 layers. For VDSR, 12 layers VDSR have been explored. For residual-like architecture, different architectures have achieved comparable results (Tables 4 and 5). Analysis about the reasons and more detailed experiments will be investigated in the future.

Table 5. Performance by VDSR which have different shapes with a magnification factor 2 in Set14.

VDSR	$(8_2, 16_2, 64_2)$	$(64_2, 16_2, 8_2)$	$(8, 16, 64, 64, 16, 8)$	$(64, 16, 8, 8, 16, 64)$	(64_{16})
PSNR(dB)	32.68	32.59	32.66	32.50	32.85

5   Conclusion

In this paper, a novel residual-like deep CNN which takes advantage of skip connections or identity mapping shortcuts in avoiding gradient exploding/vanishing problem was proposed for single image super-resolution. In particular, the shape of CNN has been carefully designed such that a very deep convolutional neural network with much fewer parameters can produce even better performance. Experimental results have demonstrated that the proposed method can not only achieve state-of-the-art PSNR and SSIM results for single image super-resolution but also produce visually pleasant results.

Acknowledgments. This work is partially supported by National Science Foundation of China under Grant No. 61473219.

References

1. Chang, H., Yeung, D.Y., Xiong, Y.: Super-resolution through neighbor embedding. In: Proceedings of the 2004 IEEE Computer Society Conference on Computer Vision and Pattern Recognition, CVPR 2004, vol. 1, pp. 1–275. IEEE (2004)
2. Dong, C., Loy, C.C., He, K., Tang, X.: Learning a deep convolutional network for image super-resolution. In: Fleet, D., Pajdla, T., Schiele, B., Tuytelaars, T. (eds.) ECCV 2014. LNCS, vol. 8692, pp. 184–199. Springer, Heidelberg (2014). doi:10.1007/978-3-319-10593-2_13
3. Dong, C., Loy, C.C., He, K., Tang, X.: Image super-resolution using deep convolutional networks. IEEE Trans. Pattern Anal. Mach. Intell. **38**(2), 295–307 (2016)
4. Freeman, W.T., Pasztor, E.C., Carmichael, O.T.: Learning low-level vision. Int. J. Comput. Vis. **40**(1), 25–47 (2000)
5. He, K., Zhang, X., Ren, S., Sun, J.: Deep residual learning for image recognition. arXiv preprint arXiv:1512.03385 (2015)
6. He, K., Zhang, X., Ren, S., Sun, J.: Identity mappings in deep residual networks. arXiv preprint arXiv:1603.05027 (2016)

7. Huang, J.B., Singh, A., Ahuja, N.: Single image super-resolution from transformed self-exemplars. In: 2015 IEEE Conference on Computer Vision and Pattern Recognition (CVPR), pp. 5197–5206. IEEE (2015)
8. Irani, M., Peleg, S.: Motion analysis for image enhancement: resolution, occlusion, and transparency. J. Vis. Commun. Image Represent. 4(4), 324–335 (1993)
9. Kim, J., Lee, J.K., Lee, K.M.: Accurate image super-resolution using very deep convolutional networks. In: Proceedings of the IEEE Conference on Computer Vision and Pattern Recognition (2016)
10. Krizhevsky, A., Sutskever, I., Hinton, G.E.: Imagenet classification with deep convolutional neural networks. In: Advances in neural information processing systems, pp. 1097–1105 (2012)
11. Liang, Y., Wang, J., Zhang, S., Gong, Y.: Incorporating image degeneration modeling with multitask learning for image super-resolution. In: 2015 IEEE International Conference on Image Processing (ICIP), pp. 2110–2114. IEEE (2015)
12. Liang, Y., Wang, J., Zhou, S., Gong, Y., Zheng, N.: Incorporating image priors with deep convolutional neural networks for image super-resolution. Neurocomputing 194, 340–347 (2016)
13. Salvador, J., Pérez-Pellitero, E.: Naive bayes super-resolution forest. In: Proceedings of the IEEE International Conference on Computer Vision, pp. 325–333 (2015)
14. Schulter, S., Leistner, C., Bischof, H.: Fast and accurate image upscaling with super-resolution forests. In: Proceedings of the IEEE Conference on Computer Vision and Pattern Recognition, pp. 3791–3799 (2015)
15. Simonyan, K., Zisserman, A.: Very deep convolutional networks for large-scale image recognition. arXiv preprint arXiv:1409.1556 (2014)
16. Timofte, R., De, V., Gool, L.V.: Anchored neighborhood regression for fast example-based super-resolution. In: 2013 IEEE International Conference on Computer Vision (ICCV), pp. 1920–1927. IEEE (2013)
17. Timofte, R., De Smet, V., Van Gool, L.: A+: adjusted anchored neighborhood regression for fast super-resolution. In: Cremers, D., Reid, I., Saito, H., Yang, M.-H. (eds.) ACCV 2014. LNCS, vol. 9006, pp. 111–126. Springer, Heidelberg (2015). doi:10.1007/978-3-319-16817-3_8
18. Wang, Z., Liu, D., Yang, J., Han, W., Huang, T.: Deep networks for image super-resolution with sparse prior. In: Proceedings of the IEEE International Conference on Computer Vision, pp. 370–378 (2015)
19. Yang, C.-Y., Ma, C., Yang, M.-H.: Single-image super-resolution: a benchmark. In: Fleet, D., Pajdla, T., Schiele, B., Tuytelaars, T. (eds.) ECCV 2014. LNCS, vol. 8692, pp. 372–386. Springer, Heidelberg (2014). doi:10.1007/978-3-319-10593-2_25
20. Yang, J., Wright, J., Huang, T., Ma, Y.: Image super-resolution as sparse representation of raw image patches. In: 2008 IEEE Conference on Computer Vision and Pattern Recognition, CVPR 2008, pp. 1–8. IEEE (2008)
21. Zhang, K., Wang, B., Zuo, W., Zhang, H., Zhang, L.: Joint learning of multiple regressors for single image super-resolution. IEEE Signal Process. Lett. 23(1), 102–106 (2016)
22. Zhang, K., Zhou, X., Zhang, H., Zuo, W.: Revisiting single image super-resolution under internet environment: blur kernels and reconstruction algorithms. In: Ho, Y.-S., Sang, J., Ro, Y.M., Kim, J., Wu, F. (eds.) PCM 2015. LNCS, vol. 9314, pp. 677–687. Springer, Heidelberg (2015). doi:10.1007/978-3-319-24075-6_65

Spatio-Temporal VLAD Encoding for Human Action Recognition in Videos

Ionut C. Duta[1($\boxtimes$)], Bogdan Ionescu[2], Kiyoharu Aizawa[3], and Nicu Sebe[1]

[1] University of Trento, Trento, Italy
{ionutcosmin.duta,niculae.sebe}@unitn.it
[2] University Politehnica of Bucharest, Bucharest, Romania
bionescu@imag.pub.ro
[3] University of Tokyo, Tokyo, Japan
aizawa@hal.t.u-tokyo.ac.jp

Abstract. Encoding is one of the key factors for building an effective video representation. In the recent works, super vector-based encoding approaches are highlighted as one of the most powerful representation generators. Vector of Locally Aggregated Descriptors (VLAD) is one of the most widely used super vector methods. However, one of the limitations of VLAD encoding is the lack of spatial information captured from the data. This is critical, especially when dealing with video information. In this work, we propose Spatio-temporal VLAD (ST-VLAD), an extended encoding method which incorporates spatio-temporal information within the encoding process. This is carried out by proposing a video division and extracting specific information over the feature group of each video split. Experimental validation is performed using both hand-crafted and deep features. Our pipeline for action recognition with the proposed encoding method obtains state-of-the-art performance over three challenging datasets: HMDB51 (67.6%), UCF50 (97.8%) and UCF101 (91.5%).

Keywords: Action recognition · Video classification · Feature encoding · Spatio-temporal VLAD (ST-VLAD)

1 Introduction

Action recognition has become an important research area in computer vision and multimedia due to its huge pool of potential applications, such as automatic video analysis, video indexing and retrieval, video surveillance, and virtual reality. The most popular framework for action recognition is the Bag of Visual Words (BoVW) with its variations [16,32,34]. The BoVW pipeline contains three main steps: feature extraction and feature encoding followed by classification. In addition to these, there are several pre- and post-processing steps, such as Principal Component Analysis (PCA) with feature decorrelation or different normalizations, which can improve the performance of the system.

© Springer International Publishing AG 2017
L. Amsaleg et al. (Eds.): MMM 2017, Part I, LNCS 10132, pp. 365–378, 2017.
DOI: 10.1007/978-3-319-51811-4_30

In what concerns the feature extraction techniques, the recent approaches based on convolutional neural networks (ConvNets) [4,12,20,25,26,29,37,38] have proven to obtain very competitive results compared to traditional hand-crafted features such as Histogram of Oriented Gradients (HOG) [5,16], Histogram of Optical Flow (HOF) [16] and Motion Boundary Histograms (MBH) [6]. Regarding the encoding step, we can notice that in the recent approaches, super-vector based encoding methods, such as Fisher Vector [23] and VLAD [11] are presented as state-of-the-art approaches for the encoding step of action recognition tasks [18,22,30,31,34].

In this work we extend VLAD encoding method by considering feature localization within the video. There are many precursors who focus on improving VLAD representation, as this is an efficient super vector-based encoding method with very competitive results in many tasks. The work in [18] proposes to use Random Forest in a pruned version for the trees to build the vocabulary and then additionally concatenate second-order information, similar as in Fisher Vectors (FV) [23]. The work [2] uses intra-normalization to improve VLAD performance. The authors propose to L2 normalize the aggregated residuals within each VLAD block to suppress the negative effect of too large values within the vector. In our approach we use average pooling which deals with this issue. The work [14] proposes an extension to Spatial Fisher Vector (SFV) which computes per visual word the mean and variance of the 3D spatio-temporal location of the assigned features. VLAD encoding method can be viewed as a simplification of FV, which keeps only first-order statistics and performs hard assignment.

In [1], the RootSIFT normalization is proposed to improve the performance of the framework for object retrieval, by computing square root of the values to reduce the influence of large bin values. As recommended by [34] we also use this normalization for hand-crafted features. The works [16,17] consider Spatial Pyramid approach to capture the information about features location, however, the scalability is an issue for this method, as it increases considerably the size of the final representation and it is not feasible for more than four video divisions. In [8] the performance of VLAD is boosted by using a double assignment approach. The authors of [21] suggest improving VLAD by concatenating the second- and third-order statistics, and using supervised dictionary learning. Different from the above works, our method focuses on efficient capture the spatio-temporal information directly within the encoding process.

The encoding method which provides the final video representation is crucially important for the performance of an action recognition system as influences directly the classifier predicted class. The information regarding the location within the video and the feature grouping based on this can provide an additional useful source of information for the performance of the system. Besides the effectiveness of the super vector-based encoding method VLAD, the information regarding the spatio-temporal locations of the features is not considered. In this paper we tackle this limitation by proposing an encoding approach which incorporates the spatio-temporal information into the encoding step. We present Spatio-temporal VLAD (ST-VLAD), an encoding approach which aims to divide the video in different parts and performs one pooling based on the specific group of features regarding their location within the video, the other pooling is executed based on the similarity of the features. This provides an important additional information which contributes on improving the overall system performance.

The reminder of the paper is as following. Section 2 introduces the preposed encoding method. In Sect. 3 is presented local deep feature extraction pipeline. The experimental evaluation and the final results are outlined in Sect. 4. Finally, Sect. 5 concludes this work.

2   Proposed ST-VLAD Encoding Method

In this section we present our approach for video features encoding, Spatio-temporal VLAD (ST-VLAD). The goal of our method is to provide a better video representation by taking into consideration the spatio-temporal information of the features. VLAD is a super vector-based encoding method proposed initially in [11]. We embed in VLAD encoding method the information regarding the position of the features within the video. To include spatio-temporal information into the encoding step, apart from the features of a videos, we also retain their positioning within the video. To each feature we associate a position p:

$$p = (\bar{x}, \bar{y}, \bar{t}); \ \bar{x} = \frac{x}{h}, \ \bar{y} = \frac{y}{w}, \ \bar{t} = \frac{t}{\#fr} \tag{1}$$

where h, w and $\#fr$ represent the height, width and the number of frames of the video respectively. Therefore, $\bar{x}$, $\bar{y}$, $\bar{t}$ correspond to the normalized x, y, t position with respect to the video. This normalization guarantees that the position values range between the same interval [0 1] for any video input.

We initially learn a codebook with the size of $k1$ visual words with k-means $C = \{c_1, c_2, ..., c_{k1}\}$, which are the means for each cluster. The codebook C is learnt from a subset of features extracted from a subset of videos. In parallel we learn another codebook of $k2$ visual words with k-means $PC = \{pc_1, pc_2, ..., pc_{k2}\}$, which represents the points codebook. The codebook PC is computed from the location information of the features used for the first codebook C. This is an automatic way to propose a spatio-temporal video $k2$ divisions, which is independent from the feature extraction algorithm. Therefore, can be applied without any modification to any type of method for establishing the region of the feature extraction, such as dense [31, 36] or at specific locations [34].

Figure 1 presents an illustration of our pipeline for feature encoding with the path that a given video traverses to obtain its final representation, which serves as the input for a classifier. First, the video information is represented by computing certain content features. Then we aim at performing two hard assignments using the codebooks obtained before. The first assignment is performed as in standard VLAD, where each local video feature x_j from the set $X = \{x_1, x_2, ..., x_n\} \in R^{n \times d}$ is assigned to its nearest visual word (d is the feature dimensionality). After computing the residuals (the difference between the

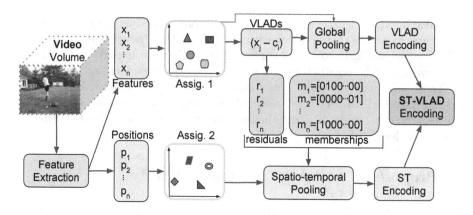

Fig. 1. The ST-VLAD framework for features encoding.

local feature and its assigned centroid), which we can call also the result VLADs vectors, we perform two actions. The first action is to perform a global pooling over the residuals for each centroid. Instead of doing sum pooling as in the standard VLAD, we perform average pooling over the residuals of each visual word:

$$v_i = \frac{1}{N_i} \sum_{j=1}^{N_i} (x_j - c_i) \tag{2}$$

where N_i is the number of features assigned to the cluster c_i. This division by the number of descriptors, that switches sum pooling to average pooling, is a very simple technique to deal with the problem of burstiness when some parts of VLAD vector can dominate the entire representation and may influence negatively the performance of the classifier. In the rest of the paper we refer to VLAD under this version. The concatenation of all v_i gives the video representation of VLAD encoding, which is a vector with size $k1 \times d$.

The second action is to save the residuals, therefore, to the local set of features $X \in R^{n \times d}$ is associated with the residuals $R = \{r_1, r_2, ..., r_n\} \in R^{n \times d}$. With the objective of preserving the associated similarity-based cluster information, we retain, together with residuals, the information regarding their centroid membership. We represent the membership information by a vector m with the size equal to the number of visual words $k1$, where all the elements are zero, except one value (which is equal to 1) that is located on the position corresponding to the associated centroid. For instance $m = [0100...00]$ maps the membership feature information to the second centroid of the codebook C.

At the bottom part of Fig. 1 is represented the encoding path where we consider the position of the extracted features $P = \{p_1, p_2, ..., p_n\} \in R^{n \times 3}$. We perform a second assignment based on the feature location within the video. This step proposes $k2$ splits of the obtained residuals. We perform the spatio-temporal pooling over each proposed division of the video by doing average pooling of the residuals part and sum pooling over the memberships. We obtain a new video representation, spatio-temporal encoding (ST encoding) by concatenating all the pooling results over the $k2$ divisions. The size of the resulted vector of ST encoding is $k2 \times (d + k1)$.

After we obtain the representations of VLAD and ST encoding, we concatenate them in a vector which represents the final representation for a video. The size of the resulted final representation is $k1 \times d + k2 \times (d + k1)$. Besides the additional information provided, this proposed representation has the advantage of the scalability over the number of video divisions. For instance, if we consider other methods such as Spatial Pyramid (SP) [16,17], for a division into 3 parts, the final video description is the concatenation of the resulted encoding for $(3 + 1)$ times ($+1$ is for the whole video). If we apply VLAD with the classic 256 clusters and the feature dimensionality is 54 then the size of final representation resulted with SP is 55,296 ($256 \times 54 \times 4$), while our ST-VLAD generates a final representation with the size 14,754 $((256 \times 54) + (3 \times (256 + 54)))$. If we increase the number of video divisions, for instance to 32, then the final vector size generated by SP explodes to more than 456K, which makes it very demanding for the computational resources, while the size of ST-VLAD is still reasonable (23,744). We make the code for ST-VLAD publicly available[1].

3   Local Deep Feature Extraction

This section presents the framework to extract deep local features for a given video. The approaches based on convolutional networks (ConvNets) [4,12,20, 25,26,29,37,38] obtained very competitive results over traditional hand-crafted features. In this work we also consider deep features resulted from ConvNets. Based on the fact that videos contain two main source of information, appearance and motion information, we consider also two ConvNets for capturing separately each information type. The pipeline for deep feature extraction is similar to [8].

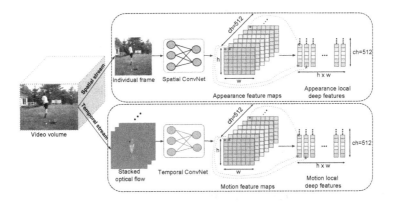

Fig. 2. Two-stream deep feature extraction pipeline for action recognition.

---

[1] http://disi.unitn.it/~duta/software.html.

For capturing the appearance information in our spatial stream we use the VGG ConvNet in [26]. This is a very deep network with 19 layers, including 16 convolutional layers and 3 fully connected layers. Together with a very high depth, VGG19 is characterized by a smaller size of the convolutional filters i.e., 3×3 and the stride is only 1 pixel. These enable the network to explore finergrained details from the feature maps. This network is trained on the ImageNet dataset [7], with state-of-the-art results for image classification. The used pipeline for the deep feature extraction is depicted in Fig. 2.

The input of VGG19 ConvNet is an image of spatial size of 224×224 with three channels for the color information. After we extract the individual frames from a video, we accordingly resize them for the request input of the network. For each individual frame we take the output of the last convolutional layer with spatial information, pool5, of the spatial ConvNet. Our choice for the convolutional layer is motivated by the fact that the deeper layers provide high discriminative information and by taking a layer with spatial information we can extract local deep features for each frame of the video. The output of pool5 is the feature maps with a spatial size of 7×7 and 512 channels. For extracting local deep feature from feature maps we take each spatial location from all channels and concatenate them to create the local deep features with 512 dimensions. Hence, from each frame we obtain $7 \times 7 = 49$ local deep features and each feature is a 512 dimensional vector. Therefore, for each video we obtain in total $\#frames \times (7 \times 7)$ local deep features.

To capture the motion information we use the re-trained network in [37]. This deep network, also VGG, is initially proposed in [26] and contains 16 layers. The authors of [37] re-trained the VGG ConvNet for a new task with new input data using several good practices for the network re-training, such as pre-training to initialize the network, smaller learning rate, more data augmentation techniques and high dropout ratio. The VGG ConvNet is re-trained for action recognition task using the UCF101 dataset [28]. The input for the temporal ConvNet is 10-stacked optical flow fields, therefore, in total there are 20-staked optical flow images as one input for the network. To extract optical flow fields we use the OpenCV implementation of TVL1 algorithm [39]. For the temporal ConvNet we also take the output of the last convolutional layer with structure information: pool5. The pool5 layer has the spatial size of feature maps of 7×7 and 512 channels. The final local deep features for an input are obtained by concatenating the values from each spatial location along all the channels, resulting in 49 local features for an input. The total number of local deep features for a video for the temporal ConvNet are $(\#frames - 9) \times 7 \times 7$. In the case of deep local features, the normalized positions, needed for ST-VLAD, are extracted based on the localization on the feature maps.

4   Experimental Evaluation

This part shows the experimental evaluation of our method, ST-VLAD, to create a final representation for a video. We validate our approach using both hand-crafted and deep features. The pipeline for extraction of deep local features is presented in the previous section, which provides us two features for the spatial and temporal stream: Spatial ConvNet (SCN) and Temporal ConvNet (TCN). Authors of [37] provide three networks re-trained for each split of UCF101 dataset. For the local deep feature extraction with temporal network another detail is that for the datasets UCF50 and HMDB51 we use the re-trained network from [37] for split1 and for UCF101 we accordingly use the re-trained networks for each split.

As hand-crafted features, we use Improved Dense Trajectories (iDT) approach with the code provided by the authors [34], keeping the default parameter settings recommended to extract four different descriptors: HOG, HOF, MBHx and MBHy. This is one of the state-of-the-art hand-crafted approaches for feature extraction. iDT is an improved version of [32] which removes the trajectories that are generated by camera motion. Each of these four descriptors is extracted along all valid trajectories and the resulted dimensionality is 96 for HOG, MBHx and MBHy, and 108 for HOF.

As default setting we use k-means for generating the codebook from 500 K random selected features. Before encoding step, we apply only for hand-crafted features the RootSIFT normalization [1] similar to [34], then for both hand-drafted and deep features we perform PCA to reduce the dimensionality by a factor of two and decorrelate the features. Therefore, the feature size of HOG, MBHx and MBHy is 48, for HOF is 54, and for the resulted deep feature size for SCN and TCN is 256.

After feature encoding we apply Power Normalization (PN) followed by L2-normalization ($||sign(\mathrm{x})|\mathrm{x}|^{\alpha}||$, where $0 \leq \alpha \leq 1$ is the normalization parameter). For all experiments, we fix $\alpha = 0.1$ for hand-crafted features and $\alpha = 0.5$ for deep features. For the classification step we use a linear one-vs-all SVM with $C = 100$. It is very important when different sources of information are combined to have values between the same range. Otherwise, the feature components with a bigger absolute value will dominate the vector representation and will weight more for the classifier, thus, may influence negatively the performance. The PN has a positive effect in reducing the picks within the vector, and α controls the level of penalization, by giving a smaller α, the large values are shrinked more. We combine different source of information, for instance the sum pooling over the membership information may give big values, therefore it is necessarily to balance the values within the final video representation vector. The reason of applying a different normalization level α depending on the feature type is related with the initial VLAD values before the concatenation with ST representation. For instance, the resulted values for VLAD representation of any iDT feature can range between $[-0.5\ 0.5]$ but for the deep features can range between $[-200\ 200]$, and ST representation can come with values until around 1500 (from the sum pooling of membership information). Therefore the absolute values resulted for deep features are more balanced after concatenation and we apply only PN with $\alpha = 0.5$, but for iDT as the values are less balanced, it is necessarily a bigger penalization of the vector picks, thus, we perform PN with smaller α of 0.1.

4.1   Datasets

We evaluate our framework on three of the most popular and challenging datasets for action recognition: HMDB51 [15], UCF50 [24] and UCF101 [28].

The HMDB51 dataset [15] contains 51 action categories, with a total of 6,766 video clips. It is one of the most challenging dataset with realistic settings. We use the original non-stabilized videos, and we follow the original protocol using three train-test splits [15]. We report average accuracy over the three splits as performance measure.

In total the UCF50 dataset [24] contains 6,618 realistic videos taken from YouTube. There are 50 human action categories mutually exclusive. The videos are split into 25 predefined groups. We follow the recommended standard procedure and perform leave-one-group-out cross validation and report average classification accuracy over all 25 folds.

The UCF101 dataset [28] is a widely adopted benchmark for action recognition, consisting in 13,320 realistic videos, which are divided in 25 groups for each action category. This dataset contains 101 action classes and there are at least 100 video clips for each class. We follow for evaluation the recommended default three training/testing splits. We report the average recognition accuracy over these three splits.

4.2   Parameter Tuning

In this set of experiments we evaluate different sizes of the vocabularies (C and PC) on the HMBD51 dataset. First, we tested different numbers of visual words ($k1$) for VLAD codebook. Figure 3a represents the graph evolution for VLAD performance when the codebook size varies. By increasing the codebook size the performance of VLAD is improved, especially for iDT features, where we can notice a consistent improvement. The impact of increasing the codebook size for the deep features is less steep, this is probably due to the fact that in general for a video, the number of generated iDT features is higher than in the case of deep features. While the codebook size of 512 gives the best results, we choose VLAD with the codebook size of 256 as a standard size of the vocabulary, since this is the best trade-off between the performance and the size of the resulted final representation.

Figure 3b illustrates the graph with the results for our proposed encoding method, ST-VLAD (Spatio-temporal VLAD). For this set of experiments we fix the size of $k1$ for VLAD to standard value of 256 visual words and we perform the tuning of $k2$, for the spatio-temporal divisions. The "0" value on the graph for the codebook size is for showing the results only for VLAD (with 256 visual words) without the contribution of the spatio-temporal encoding. By increasing the codebook size, the final results for all considered features continue to increase the performance up to the value of 32 video divisions, except for HOG and MBHx for which the accuracy continues to grow. Considering this and also taking into account the trade-off between accuracy and the final video representation size, we

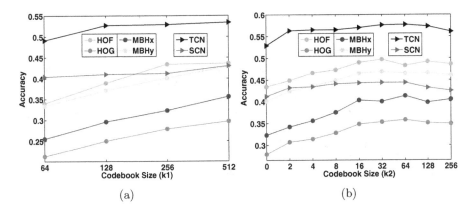

Fig. 3. Evaluation of the vocabulary size: (a) for VLAD, (b) for ST-VLAD.

fix the codebook size PC to 32 for all features in the next experiments. Therefore, we have the default settings for ST-VLAD of 256 for $k1$ and 32 for $k2$.

The performance of all iDT features has a continuous significant gain until $k2 = 32$, the algorithm behind iDT approach removes a consistent number of trajectories. Therefore, the features are not extracted anymore densely, instead they are extracted only at some positions, and by using an encoding method that provides information about the grouping of trajectories, brings an important source of information for the final representation.

4.3 Comparison to Baseline

The exact numbers of the comparison of our approach, ST-VLAD, with the baseline over all three considered datasets, can be visualized in Table 1. The parameters of ST-VLAD are fixed in the subsection above, and for the direct comparison with the baseline we take VLAD with the standard codebook sizes of 256 and 512. The ST-VLAD encoding approach outperforms VLAD by a large margin over all 6 features for each dataset.

The performance growth is prominent, especially on the HMDB51 dataset, which is one of the most challenging video collections for action recognition, where there is a considerable room for improvement. For instance, on HOF and HOG features, by adding the spatio-temporal information to VLAD256 boosts the performance with 6.3 and 7.4 percentage points respectively. Increasing the vocabulary size for VLAD to 512 may improve the performance, however, our approach still outperforms VLAD512 by a large margin, besides the fact that VLAD512 generates a higher size for the final video representation than our proposed approach. For instance, the size of the video representation using VLAD512 in the case of HOF descriptor is 27,648, though, our representation is 23,744; and in the case of deep local features (SCN or TCN) the video representation size for VLAD512 is 131,072, nevertheless, ST-VLAD representation size is 81,920.

Table 1. The performance (mean accuracy %) comparison of our encoding approach, ST-VLAD, to the baseline.

| | HMDB51 (%) | | | | | | UCF50 (%) | | | | | |
	HOF	HOG	MBHx	MBHy	SCN	TCN	HOF	HOG	MBHx	MBHy	SCN	TCN
VLAD256	43.4	27.9	32.3	40.1	41.2	52.9	82.9	70.3	78.7	82.7	84.5	96.0
VLAD512	43.6	29.7	35.6	42.9	43.0	53.4	84.0	72.3	80.4	84.3	84.2	95.8
ST-VLAD	**49.7**	**35.3**	**40.0**	**46.9**	**44.4**	**57.6**	**85.6**	**77.5**	**82.3**	**85.1**	**85.5**	**96.8**

| | UCF101 (%) | | | | | |
	HOF	HOG	MBHx	MBHy	SCN	TCN
VLAD256	70.8	58.2	66.3	69.5	75.9	85.1
VLAD512	72.9	59.6	69.2	72.0	75.4	85.5
ST-VLAD	**75.0**	**66.0**	**71.6**	**73.7**	**78.0**	**86.3**

Regarding the computational cost, we randomly sampled 500 videos form HMDB51 dataset, and report the number of frames per second for each encoding method, choosing as feature type TCN for the measurements which obtains the best performance over all the other features. Despite that our method outperforms by a large margin VLAD512 in terms of performance, ST-VALD is not more demanding for the computational cost. In fact, ST-VLAD is slightly faster, being able to process 1,233 frames/sec, while VLAD512 runs at a frame rate of 1,059. Obviously, the fastest method is VLAD256, with 1,582 frames/sec.

4.4 Comparison to State-of-the-art

For the comparison with state-of-the-art we evaluate the performance of our approach, ST-VLAD, by using three feature combinations. The first one is the combination of HOF, HOG, MBHx and MBHy to obtain the improved Dense Trajectories representation (iDT), the second one is to combine both deep features (SCN and TCN) to get the two-stream representation (2St). The last one is to combine all the 6 features for obtaining the final representation (iDT+2St).

Table 2. The final feature fusion results for ST-VLAD.

| | HMDB51 (%) | | | UCF50 (%) | | | UCF101 (%) | | |
	iDT	2St	iDT+2St	iDT	2St	iDT+2St	iDT	2St	iDT+2St
Early	58.4	61.2	66.5	90.6	97.6	97.1	82.9	90.2	91.3
sLate	56.3	59.2	64.6	89.8	97.1	95.6	81.0	88.7	88.0
wLate	56.7	60.1	65.5	90.0	97.6	97.6	81.2	88.9	89.9
sDouble	57.4	60.8	66.0	90.1	97.4	96.1	81.9	89.2	88.7
wDouble	**59.0**	**62.1**	**67.6**	**90.7**	**97.8**	**97.9**	**83.0**	**90.2**	**91.5**

For those three feature combination we tested both early and late fusion approaches. Table 2 presents the feature fusion results for all three datasets. For early fusion we concatenate for each specific combination the resulted individual video representation for each feature. Then we apply again the normalization, but only L2 norm, and this time we apply it over the resulted vector of the concatenation of the features. In the end we give the final vector as input to the linear classifier to get the predicted classes.

Late fusion is performed in two manners. For the first one we just make the sum (sLate) of the classifier output over each feature. For the second one we perform a wighted sum (wLate), where we give different weights for each feature representation classifier output, and then we perform the sum. The weights combination are tuned by taking values between 0 and 1 with the step 0.1. In addition to early and late fusion, we perform a double fusion where to the sum of the classifier output of the individual feature representation we add also the classifier output of their early fusion. Similar as for late fusion, we have sum double fusion (sDouble) and weighted sum double fusion (wDouble), the difference from late fusion is that in this case there is one additional classifier output (from the early fusion) which contributes to the sum or the wighted sum tuning.

From the Table 2 we can see that in general early fusion performs better than late fusion. Double fusion combines the benefits of early and late fusion, boosting the performance and obtains the best results for all three datasets. In general the combination of hand-crafted features (iDT) with deep features (2St) raises the performance, in special on more challenging datasets such as HMDB51, while on the less challenging ones such as UCF50, the hand-crafted features does not bring significant contribution.

Table 3. Comparison to the state-of-the-art.

HMDB51 (%)		UCF50 (%)		UCF101 (%)	
Jain et al. [10]	52.1	Kliper-Gross et al. [13]	72.7	Karpathy et al. [12]	65.4
Zhu et al. [40]	54.0	Solmaz et al. [27]	73.7	Wang et al. [35]	85.9
Oneata et al. [19]	54.8	Reddy et al. [24]	76.9	Wang et al. [33]	86.0
Park et al. [20]	56.2	Uijlings et al. [31]	81.8	Peng et al. [22]	87.9
Wang et al. [34]	57.2	Wang et al. [32]	85.6	Simonyan et al. [25]	88.0
Sun et al. [29]	59.1	Wang et al. [34]	91.2	Sun et al. [29]	88.1
Simonyan et al. [25]	59.4	Wang et al. [33]	91.7	Ng et al. [38]	88.6
Wang et al. [33]	60.1	Peng et al. [22]	92.3	Park et al [20]	89.1
Peng et al. [22]	61.1	Ballas et al. [3]	92.8	Bilen et al. [4]	89.1
Bilen et al. [4]	65.2	Duta et al. [9]	93.0	Wang et al. [37]	**91.4**
ST-VLAD	**67.6**	ST-VLAD	**97.8**	ST-VLAD	**91.5**

The comparison of our method, ST-VLAD, with the state-of-the-art is presented in Table 3. We obtain state-of-the-art results on UCF50 with a performance of 97.8% and with 91.5% for UCF101. The outstanding performance of our method on the challenging dataset HMDB51 of 67.6% outperforms the state-of-the-art, including on the very recent works based on deep learning, such as [4,20,29], showing the effectiveness of our representation.

5 Conclusion

This paper introduces Spatio-temporal VLAD (ST-VLAD), an encoding method which captures the spatio-temporal information within the encoding process. In the experimental part we show that our approach of grouping the features and the specific information extraction brings significant additional information to the standard representation. Our pipeline obtains state-of-the-art results on three of the most challenging datasets for action recognition: HMDB51, UCF50 and UCF101. For the future work we will focus more also on capturing the relationships between the features from the same group.

Acknowledgement. This work has been supported by the EC FP7 project xLiMe.

References

1. Arandjelović, R., Zisserman, A.: Three things everyone should know to improve object retrieval. In: CVPR (2012)
2. Arandjelovic, R., Zisserman, A.: All about VLAD. In: CVPR (2013)
3. Ballas, N., Yang, Y., Lan, Z.Z., Delezoide, B., Prêteux, F., Hauptmann, A.: Space-time robust representation for action recognition. In: ICCV (2013)
4. Bilen, H., Fernando, B., Gavves, E., Vedaldi, A., Gould, S.: Dynamic image networks for action recognition. In: CVPR (2016)
5. Dalal, N., Triggs, B.: Histograms of oriented gradients for human detection. In: CVPR (2005)
6. Dalal, N., Triggs, B., Schmid, C.: Human detection using oriented histograms of flow and appearance. In: Leonardis, A., Bischof, H., Pinz, A. (eds.) ECCV 2006. LNCS, vol. 3952, pp. 428–441. Springer, Heidelberg (2006). doi:10.1007/11744047_33
7. Deng, J., Dong, W., Socher, R., Li, L.J., Li, K., Fei-Fei, L.: Imagenet: a large-scale hierarchical image database. In: CVPR (2009)
8. Duta, I.C., Nguyen, T.A., Aizawa, K., Ionescu, B., Sebe, N.: Boosting VLAD with double assignment using deep features for action recognition in videos. In: ICPR (2016)
9. Duta, I.C., Uijlings, J.R.R., Nguyen, T.A., Aizawa, K., Hauptmann, A.G., Ionescu, B., Sebe, N.: Histograms of motion gradients for real-time video classification. In: CBMI (2016)
10. Jain, M., Jégou, H., Bouthemy, P.: Better exploiting motion for better action recognition. In: CVPR (2013)
11. Jégou, H., Perronnin, F., Douze, M., Sanchez, J., Perez, P., Schmid, C.: Aggregating local image descriptors into compact codes. TPAMI **34**(9), 1704–1716 (2012)
12. Karpathy, A., Toderici, G., Shetty, S., Leung, T., Sukthankar, R., Fei-Fei, L.: Large-scale video classification with convolutional neural networks. In: CVPR (2014)

13. Kliper-Gross, O., Gurovich, Y., Hassner, T., Wolf, L.: Motion interchange patterns for action recognition in unconstrained videos. In: Fitzgibbon, A., Lazebnik, S., Perona, P., Sato, Y., Schmid, C. (eds.) ECCV 2012. LNCS, vol. 7577, pp. 256–269. Springer, Heidelberg (2012). doi:10.1007/978-3-642-33783-3_19

14. Krapac, J., Verbeek, J., Jurie, F.: Modeling spatial layout with fisher vectors for image categorization. In: ICCV (2011)

15. Kuehne, H., Jhuang, H., Garrote, E., Poggio, T., Serre, T.: HMDB: a large video database for human motion recognition. In: ICCV (2011)

16. Laptev, I., Marszałek, M., Schmid, C., Rozenfeld, B.: Learning realistic human actions from movies. In: CVPR (2008)

17. Lazebnik, S., Schmid, C., Ponce, J.: Beyond bags of features: spatial pyramid matching for recognizing natural scene categories. In: CVPR (2006)

18. Mironică, I., Duţă, I.C., Ionescu, B., Sebe, N.: A modified vector of locally aggregated descriptors approach for fast video classification. Multimedia Tools and Applications (2016, in press)

19. Oneata, D., Verbeek, J., Schmid, C.: Action and event recognition with fisher vectors on a compact feature set. In: ICCV (2013)

20. Park, E., Han, X., Berg, T.L., Berg, A.C.: Combining multiple sources of knowledge in deep CNNs for action recognition. In: WACV (2016)

21. Peng, X., Wang, L., Qiao, Y., Peng, Q.: Boosting VLAD with supervised dictionary learning and high-order statistics. In: Fleet, D., Pajdla, T., Schiele, B., Tuytelaars, T. (eds.) ECCV 2014. LNCS, vol. 8691, pp. 660–674. Springer, Heidelberg (2014). doi:10.1007/978-3-319-10578-9_43

22. Peng, X., Wang, L., Wang, X., Qiao, Y.: Bag of visual words and fusion methods for action recognition: comprehensive study and good practice. arXiv:1405.4506 (2014)

23. Perronnin, F., Sánchez, J., Mensink, T.: Improving the fisher kernel for large-scale image classification. In: Daniilidis, K., Maragos, P., Paragios, N. (eds.) ECCV 2010. LNCS, vol. 6314, pp. 143–156. Springer, Heidelberg (2010). doi:10.1007/978-3-642-15561-1_11

24. Reddy, K.K., Shah, M.: Recognizing 50 human action categories of web videos. Mach. Vis. Appl. **24**(5), 971–981 (2013)

25. Simonyan, K., Zisserman, A.: Two-stream convolutional networks for action recognition in videos. In: NIPS (2014)

26. Simonyan, K., Zisserman, A.: Very deep convolutional networks for large-scale image recognition. arXiv preprint arXiv:1409.1556 (2014)

27. Solmaz, B., Assari, S.M., Shah, M.: Classifying web videos using a global video descriptor. Mach. Vis. Appl. **24**, 1473–1485 (2013)

28. Soomro, K., Zamir, A.R., Shah, M.: UCF101: a dataset of 101 human actions classes from videos in the wild. arXiv preprint arXiv:1212.0402 (2012)

29. Sun, L., Jia, K., Yeung, D.Y., Shi, B.E.: Human action recognition using factorized spatio-temporal convolutional networks. In: ICCV (2015)

30. Uijlings, J.R.R., Duta, I.C., Rostamzadeh, N., Sebe, N.: Realtime video classification using dense HOF/HOG. In: ICMR (2014)

31. Uijlings, J.R.R., Duta, I.C., Sangineto, E., Sebe, N.: Video classification with densely extracted HOG/HOF/MBH features: an evaluation of the accuracy/computational efficiency trade-off. Int. J.Multimed. Info. Retr. **4**, 33–44 (2015)

32. Wang, H., Kläser, A., Schmid, C., Liu, C.L.: Dense trajectories and motion boundary descriptors for action recognition. IJCV **103**(1), 60–79 (2013)

33. Wang, H., Oneata, D., Verbeek, J., Schmid, C.: A robust and efficient video representation for action recognition. IJCV **119**, 219–238 (2015)
34. Wang, H., Schmid, C.: Action recognition with improved trajectories. In: ICCV (2013)
35. Wang, H., Schmid, C.: LEAR-INRIA submission for the THUMOS workshop. In: ICCV Workshop (2013)
36. Wang, H., Ullah, M.M., Klaser, A., Laptev, I., Schmid, C.: Evaluation of local spatio-temporal features for action recognition. In: BMVC (2009)
37. Wang, L., Xiong, Y., Wang, Z., Qiao, Y.: Towards good practices for very deep two-stream convnets. arXiv preprint arxiv:1507.02159 (2015)
38. Yue-Hei Ng, J., Hausknecht, M., Vijayanarasimhan, S., Vinyals, O., Monga, R., Toderici, G.: Beyond short snippets: deep networks for video classification. In: CVPR (2015)
39. Zach, C., Pock, T., Bischof, H.: A duality based approach for realtime TV-L 1 optical flow. In: Pattern Recognition (2007)
40. Zhu, J., Wang, B., Yang, X., Zhang, W., Tu, Z.: Action recognition with actons. In: ICCV (2013)

Structure-Aware Image Resizing for Chinese Characters

Chengdong Liu, Zhouhui Lian[(⊠)], Yingmin Tang, and Jianguo Xiao

Institute of Computer Science and Technology, Peking University,
No. 128 Zhongguancun North Street, Haidian District, Beijing 100080, China
{liuchengdong,lianzhouhui,tangyingmin,xiaojianguo}@pku.edu.cn

Abstract. This paper presents a structure-aware resizing method for Chinese character images. Compared to other image resizing approaches, the proposed method is able to preserve important features such as the width, orientation and trajectory of each stroke for a given Chinese character. The key idea of our method is to first automatically decompose the character image into strokes, and then separately resize those strokes naturally using a modified linear blend skinning approach and as-rigid-as-possible shape interpolation under the guidance of structure information. Experimental results not only verify the superiority of our method compared to the state of the art but also demonstrate its effectiveness in several real applications.

Keywords: Image resizing · Chinese characters · Structure features

1 Introduction

As the number of display devices with different resolutions increases rapidly, how to effectively resize a given image into visually pleasing ones with new ratio aspects has become an important research topic in areas of Image Processing, Computer Graphics, etc. So far, large numbers of algorithms for image resizing have been reported. However, most of these methods can only handle natural images while some others such as character images are rarely under consideration. Character images are quite different against natural images, since they are typically binary, texture-free and with rich structure information. When those images are screwed or stretched only vertically or horizontally using standard resizing approaches, important character features especially structure properties may be seriously distorted. For Chinese characters, important character features are the so-called structure features including shape, width, orientation and trajectory of each stroke. Standard resizing approaches are generally applied to stretch or screw an image of character. Nevertheless, it bears intolerant distortions especially in stroke width and stroke shape, which is demonstrated in Fig. 1.

Effective resizing methods for Chinese characters are critical in many real applications. For instance, there are some researchers trying to synthesis all

© Springer International Publishing AG 2017
L. Amsaleg et al. (Eds.): MMM 2017, Part I, LNCS 10132, pp. 379–390, 2017.
DOI: 10.1007/978-3-319-51811-4_31

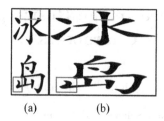

(a)   (b)

Fig. 1. Characters with the meaning of *Iceland* in Chinese (a) are stretched three times horizontally with standard method (b), with which stroke shape (marked in green) and stroke width (marked in red) are not well preserved. (Color figure online)

Chinese characters based on hundreds of human-generated characters by automatically extracting components from input characters and optimally assembling resized components [1,2]. However, when resizing component images, due to the limitation of existing approaches, many important features of those components are distorted and thus corresponding synthesized characters are unsatisfactory (see Fig. 2). Artistic designing of Chinese characters is also in demand of proper resizing method since standard image resizing approaches inevitably result in distortions when screwing or stretching these images to different aspect ratios.

Motivated by the work [3] that achieves real-time deformation, this paper proposes a novel method called structure-aware image resizing, which is designed to resize Chinese character images in a completely different manner compared to other existing approaches. The key idea is to first automatically decompose the character image into strokes, and then separately resize those strokes naturally using a modified Linear Blend Skinning (LBS) approach and As-Rigid-As-Possible (ARAP) shape interpolation under the guidance of structure information. Experimental results not only verify the superiority of our method compared to the state of the art but also demonstrate its effectiveness in some applications.

2   Related Work

Image resizing is one of the most important research topics of Computer Vision and Computer Graphics. It attracts so much attention that large numbers of methods have been proposed. Standard resizing approaches with the main idea of interpolation [4] resize an image uniformly to a target size, ignoring its content and structure. This results in distortions when resizing an image to different ratio aspects, as shown in Fig. 1(b).

Besides the standard image resizing approaches, there exist many other resizing methods to process natural pictures. Cropping [5,6] is another popular approach for non-uniformly resizing, but obviously we cannot discard any part of a character image in order to maintain its completeness.

Image retargeting [7] processes the foreground and the background separately. The foreground and background components are labeled based on the

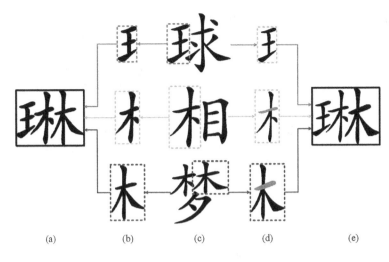

Fig. 2. Resizing the components of the characters in the middle to synthesize a new character via our method (a) and standard method (e). (d) demonstrates that components resized by standard method are quite different in stroke width, and strokes with yellow color should be parallel. (Color figure online)

salience map of an image at first. Then the size of the foreground remains unchanged while the background is resized with standard method. Finally separated components are combined to obtain the result. However, it has similar problem with Cropping as the foreground and background components are of the same importance.

Seam Carving [8] and its modified versions [9,10] perform pretty well for natural images by removing or adding a seam of image iteratively according to the salience map. Nonetheless, they fail in character images containing rich structure and little texture information.

Image warping [11] is considered to be one of the best image resizing methods that preserve structure properties. This method was then further improved to become shape-preserving [12] and line-structure-preserving [13]. However, all of these existing methods are not well-suited to deal with the resizing task for Chinese character images mainly due to the rich and complicated structure information they contain.

3   Approach

By combining several computer graphic algorithms, we process each stroke separately and assemble them to form the final result. As shown in Fig. 3, given an image of Chinese character with the objective resizing scale, we obtain the resized image with the following four steps:

- **Stroke Extraction and Handle Selection:** Decompose the character image into strokes and then obtain each stroke's pure skeleton, on which key points are selected as handles to control deformations;
- **Triangulation and Constrained Weight Calculation:** Apply Delaunay triangulation for each stroke separately and then the constrained Bounded Biharmonic Weights (BBW) on meshes are calculated;
- **Structure-Guided Deformation:** Deform each stroke naturally according to the target scale using a modified Linear Blend Skinning (LBS) approach and As-Rigid-As-Possible (ARAP) shape interpolation under the guidance of structure information to minimize distortions and preserve structure properties;
- **Stroke Assembling:** Assemble the deformed strokes properly to generate the final character image.

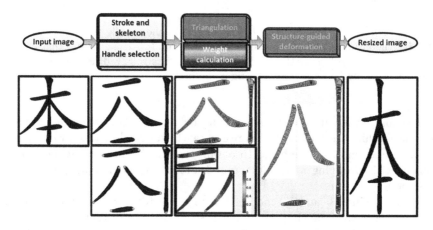

Fig. 3. Overview of the proposed method. The Chinese character *'ben'* is resized to twice in height. Images from the second column to the fifth column correspond to the four steps mentioned above. Red lines in column 2 are the skeletons of the decomposed character and the yellow points represent the handles. (Color figure online)

3.1   Stroke Extraction and Handle Selection

As we know, Chinese characters are all composed of a limited number of simple strokes. With the help of Coherent Point Drift (CPD) [14,15], a given character image can be automatically decomposed into stroke level and then the pure skeleton (i.e., trajectory) of each stroke is obtained. Here, strokes are classified to 33 categories and the pure skeletons of different kinds of strokes have different *key points*, such as starting points, ending points and turning points. *Key points* of each stroke are selected as point handles in other steps to control deformations, as shown in Fig. 4(a). A pair of two adjacent handles locate at the endpoints of the skeleton of each stroke segment, which is the basic element of Chinese characters. Here, we denote the set of handles as H for a stroke.

3.2   Triangulation and Constrained Weight Calculation

Let P be the contour points of a stroke, we apply Delaunay Triangulation [16] for P to get a mesh consisting of triangles T. Due to the limitation of our automatic skeleton extraction algorithm, selected handles sometimes may not be precisely located in desired positions. Therefore, we design a rule to automatically modify their locations based on triangulation results and then update the handle set H.

Afterwards, we compute Bounded Biharmonic Weights (BBW) [3] constrained according to the special properties of Chinese characters for all contour points. $\omega_i : P \rightarrow \mathbb{R}$, which represents the weights associated with the ith handle H_i, can be obtained by solving the following minimization problem.

$$\operatorname*{argmin}_{\omega_i, i=1,2,\ldots,m} \sum_{i=1}^{m} \frac{1}{2} \int_H \|\Delta \omega_i\|^2 \, dV \tag{1}$$

subject to:

$$\omega_i|_{H_j} = \delta_{ij}$$

$$\sum_{i=1}^{m} \omega_i(p) = 1 \qquad\qquad \forall p \in P$$

$$0 \le \omega_i(p) \le 1, i = 1, \ldots, m, \qquad \forall p \in P$$

$$\omega_i(p) = 1 \qquad \text{if } p \in t \text{ and } t \text{ encircles } H_i$$

where m denotes the number of handles in H, $t \in T$ and δ_{ij} is Kronecker's delta. We add the last constraint to the original BBW scheme [3] (namely, Eq. (1) subject to the first three constraints) to prevent each stroke segment from being influenced by handle points of other segments (see Fig. 4).

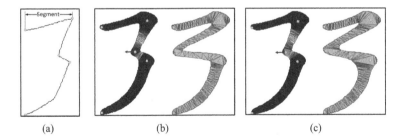

(a)   (b)   (c)

Fig. 4. Select handles based on the skeleton of a stroke (a). Constrained BBW (b) outperforms the original one (c) since with the proposed constrain energies do not propagate into unrelated segments.

Then, we use linear finite elements to discretize the constrained variational problem (1) and estimate the weights for all handle points via the quadratic programming solver provided in Matlab [17].

3.3 Structure-Guided Deformation

After the calculation of constrained BBW, we utilize LBS [18] to perform the deformation because all vertices can be linearly arranged with the affine transformations of the handles and the constrained weights. We move the handles according to the resizing scale and some heuristic rules specifically designed for Chinese characters to preserve important features and structure information. The heuristic rules are applied so that we can deform the strokes just by moving handles based on the desired resizing scale. Strokes are divided to stroke segments that can be classified into 7 categories: *'heng'*, *'shu'*, *'pie'*, *'na'*, *'dian'*, *'ti'* and *'gou'*. As shown in Fig. 5, the seven categories of stroke segments are clustered into three groups based on the features to be preserved.

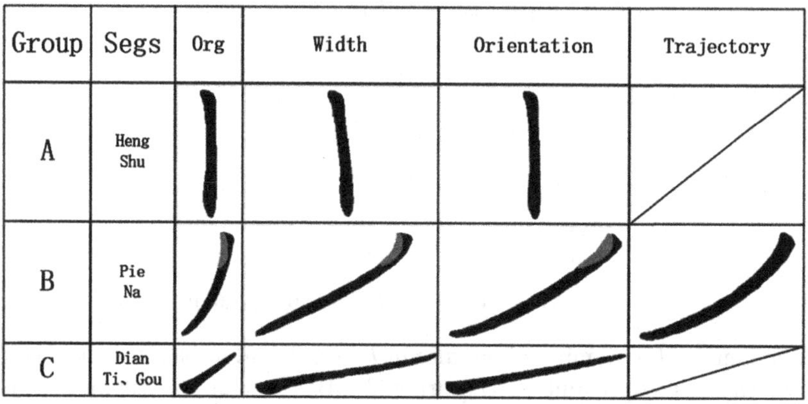

Group	Segs	Org	Width	Orientation	Trajectory
A	Heng Shu				
B	Pie Na				
C	Dian Ti، Gou				

Fig. 5. Groups of stroke segments and important features of each group to be preserved. Here, we resize stroke segments in the third column to three times horizontally. The last three columns show the resizing results when corresponding features are preserved. E.g., images in the fifth column are results when the stroke width and orientation are preserved. (Color figure online)

One important feature needs to be preserved is orientation, which consists of two aspects: the slope and flatness of the stroke segment. Handles of a segment in Group A should be slightly modified to prevent the slope from changing too much, otherwise the segment *'heng'* will be quite steep, for example, when stretched vertically. For segments in other groups, we rotate the handles to eliminate the distortions around them (see the fourth column in Fig. 5). For segments in Group B or Group C with handles h_1 and h_2, we should calculate the rotation angles for both of them:

$$Angle = \arccos(\frac{x_2 - x_1}{Dis(h_1, h_2)}) - \arccos(\frac{x_2' - x_1'}{Dis(h_1', h_2')}), \qquad (2)$$

where (x_1, y_1) and (x_2, y_2) are the original coordinates of h_1 and h_2, while (x_1', y_1') and (x_2', y_2') are the new coordinates of them after the translation of handle points, respectively. Differences between the third column and fourth column for Group B in Fig. 5 are labeled in red, from which we can see that our method preserves the orientation of a segment to keep the concavity of the contour in labeled areas. Note that handles at turning points should not be rotated since they are shared by two conjoint segments and the rotation could result in more serious distortion. With the help of LBS, the deformed stroke is obtained based on the constrained weights, affine transformation of each handle.

After above-mentioned processing, the skeleton trajectory of a deformed stroke segment in Group B will be different against the desired trajectory (see Fig. 6). To solve this problem, we adopt the As-Rigid-As-Possible (ARAP) shape interpolation [19] to naturally modify the deformed stroke. To be specific, we first evenly divide the stroke into a number of (e.g., 11) sub-segments (shown in Fig. 6(d)), and translate the mass centers of those sub-segments to the desired positions. Then, we rotate the direction of each sub-segment around its center to make it fitting well with the desired trajectory (see Fig. 6(d)). At last, similar as [20], the ARAP method is applied to get the final shape (see Fig. 6(e)). We define a mapping $F : T \rightarrow SEG$, where T represents the triangle set and SEG denotes the sub-segment set of the stroke. $F(i) = j$ indicates that the triangle T_i belongs to SEG_j, a 2D mesh consisting of N_{f_j} triangles $\{T_{j1}, T_{j2}, \ldots, T_{jN_{f_j}}\}$ and N_{v_j} vertices $\{V_{j1}, V_{j2}, \ldots, V_{jN_{v_j}}\}$. Then we resize the extracted skeleton of stroke using the standard method and select N_c points from the resized skeleton to get the new centers of sub-segments C', where N_c is the number of elements in the original sub-segments C. Next, the center of each sub-segment SEG_j is translated to C'_j to get the new sub-segment SEG'_j. Simultaneously, we estimate the orientations of SEG_j and SEG'_j according to C and C'. The rotation matrix R for each sub-segment can be obtained based on the difference of the orientations. After rotating the triangles in SEG_j, we update SEG'_j with N_{f_j} triangles $\{T'_{j1}, T'_{j2}, \ldots, T'_{jN_{f_j}}\}$ and $3 \times N_{v_j}$ vertices $\{V'_{j1}(1), V'_{j1}(2), V'_{j1}(3), V'_{j2}(1), V'_{j2}(2), V'_{j2}(3), \ldots, V'_{jN_{v_j}}(1), V'_{jN_{v_j}}(2), V'_{jN_{v_j}}(3)\}$. We denote T as the set of source triangles in the source sub-segments, and T' as the target triangles set. Let the vertices of T_i be $\{V_{i1}, V_{i2}, V_{i3}\}$ and those of T'_i be $\{V'_{i1}, V'_{i2}, V'_{i3}\}$. The affine transformation is defined as

$$V'_f = A_i V_f + L_i$$
$$= \begin{bmatrix} a_{i1} & a_{i2} \\ a_{i3} & a_{i4} \end{bmatrix} \begin{bmatrix} x_f \\ y_f \end{bmatrix} + \begin{bmatrix} lx_i \\ ly_i \end{bmatrix}, \tag{3}$$

where $V'_f = [x'_f, y'_f]$, $f \in \{i_1, i_2, i_3\}$. If T_i belongs to SEG_j, we have $A_i = R_j$. Let the final triangles be $\tilde{T}$ and the final vertices of $\tilde{T}_i$ be $\{\tilde{V}_{i1}, \tilde{V}_{i2}, \tilde{V}_{i3}\}$. The affine transformation from T_i to $\tilde{T}_i$ can be calculated by

$$\tilde{V}_f = \tilde{A}_i V_f + \tilde{L}_i, \tag{4}$$

where $\tilde{V}_f = [\tilde{x}_f, \tilde{y}_f]^T$ and $f \in \{i_1, i_2, i_3\}$.

Final locations of vertices are determined by minimizing the quadratic error between matrix A_i and $\tilde{A}_i$. The minimization problem has a unique solution if we consider a vertex as a constant and solve it with a sparse QR solver [20]. Figure 6 demonstrates the resizing work flow of the stroke segment 'pie', which is horizontally stretched to three times.

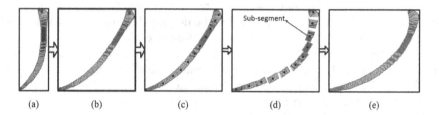

(a)	(b)	(c)	(d)	(e)

Fig. 6. Workflow of structure-guided deformation: (a) original stroke, (b) immediate result with LBS and rotation, (c) dividing the stroke into sub-segments, (d) translating and rotating triangles, (e) final result after ARAP interpolation.

3.4   Stroke Assembling

Connectivity of strokes is also an important feature to be preserved, since we need to assemble those deformed strokes to get the final resized character image. Two strokes are connected when they share a common area which is close enough (2.5 times the average stroke width of the character) to an endpoint handle. For each stroke pair, we determine whether they are connected and select the shared contour points of them. Relative positions of the shared pixels in the deformed strokes are calculated, according to which we move the handle points to make them connected (Fig. 7). Note that orientation of segments in Group A must be maintained when assembling strokes.

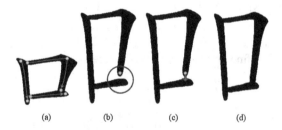

(a)	(b)	(c)	(d)

Fig. 7. Stretching the character image (a) twice vertically to obtain the final result after assembling the strokes (d). (b) is the immediate result before we move the handles to preserve the connectivity.

4   Experiments

We conduct a number of experiments to verify that the proposed method is able to preserve the important features of Chinese characters in different resizing scales and different font libraries (i.e., Kai Ti, Fang Song, Hei Ti and Pei An Ti). The algorithm is implemented in matlab on a PC with a 3.40 Hz Intel Duo CPU and 8 GB RAM. The average time consumed by our method to resize a 250 × 250 character image is about 4 s and the cost can be decreased to less than 2 s if only we sample uniformly from the contour points with few distortions induced.

Figure 8 shows the resizing results of our method and other existing approaches for Chinese character images. As we can see, methods based on seam carving [8] and warping [11] both result in intolerable distortions. The standard image resizing method obtains the most satisfactory results among existing methods we compare. However, it fails to maintain the structure features of each stroke, which is critical in many applications. By taking structure information into account, our method is able to preserve all important features for Chinese characters. Figure 9 shows that our method obtains satisfactory results with different resizing scales. Important features are better preserved compared to those with standard method when stretched.

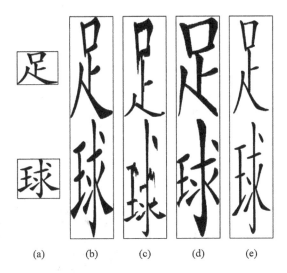

    (a)   (b)   (c)   (d)   (e)

Fig. 8. Resizing the characters with the meaning of football to three times in height with different methods. (a) original images, (b) standard method, (c) seam carving, (d) warping and (e) our method.

As shown in Fig. 10, our resizing method is well suited for Chinese character in different font libraries. For instance, the orientation has been preserved so that segments 'heng' won't be too steep even if it is stretched too much vertically. What's more, the width of all stroke segments keeps the same with each other and the detailed information near the key points of each stroke barely changes. Thus, we can use the proposed method to automatically resize all Chinese characters in a given font library to different aspect ratios and easily generate a set of new font libraries that are visually pleasing.

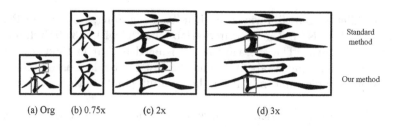

Fig. 9. Resizing the Chinese character $'ai'$, shown in (a), to different scales (b)–(d).

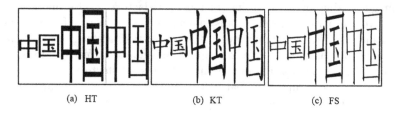

Fig. 10. Results of characters with the meaning of China in three font libraries (original images in the left) resized 3 times in height by standard method (middle) and our method (right).

Another application of our method is to generate a Chinese font library based on a small number of input characters. Figure 11 demonstrates the automatically generated characters by assembling best-suited components selected from other input characters. Since our method preserves important features for those components when resizing them to desired scales, synthesized characters based on our method appear much better than those using the standard resizing approach.

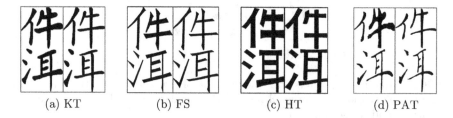

Fig. 11. Synthesized characters of different font libraries based on standard method (left) and our method (right).

We also apply our method to resize English alphabet images. We are familiar with English alphabets resized with standard method, as show in Fig. 12. However the width of alphabets is quite different from the original ones. Our method can preserve the stroke width and writing direction for the trajectory of each alphabet, and keep the coherence of the shape of each stroke.

Fig. 12. Resizing the English alphabets (top) twice in width with standard method (middle) and our method (bottom).

5 Discussion and Conclusion

In this paper, we presented a structure-aware image resizing method for Chinese character images. Mainly due to the consideration of structure information and the utilization of several advanced shape processing techniques, the proposed method is able to preserve important features such as the width, orientation, shape and trajectory of each stroke when resizing a given Chinese character into arbitrary aspect ratios.

Experiments carried out on a large number of Chinese characters in several different font libraries demonstrated the superiority of our method compared to other state-of-the-art approaches. We also applied our approach to resize English alphabets and obtain satisfactory results. We believe that, by introducing some properly-designed rules, the proposed method can be generalized to handle image resizing tasks for characters of other types of language systems.

Acknowledgements. This work was supported by National Natural Science Foundation of China (Grant Nos.: 61472015, 61672056 and 61672043), National Hi-Tech Research and Development Program (863 Program) of China (Grant No.: 2014AA015102), Beijing Natural Science Foundation (Grant No.: 4152022) and National Language Committee of China (Grant No.: ZDI135-9).

References

1. Xu, S., Lau, F.C., Cheung, W.K., Pan, Y.: Automatic generation of artistic Chinese calligraphy. IEEE Intell. Syst. **20**(3), 32–39 (2005)
2. Zong, A., Zhu, Y.: StrokeBank: automating personalized Chinese handwriting generation. In: AAAI, pp. 3024–3030 (2014)
3. Jacobson, A., Baran, I., Popovic, J., Sorkine, O.: Bounded biharmonic weights for real-time deformation. ACM Trans. Graph. **30**(4), 78 (2011)
4. Lehmann, T.M., Gonner, C., Spitzer, K.: Survey: interpolation methods in medical image processing. IEEE Trans. Med. Imaging **18**(11), 1049–1075 (1999)
5. Chen, L.Q., Xie, X., Fan, X., Ma, W.Y., Zhang, H.J., Zhou, H.Q.: A visual attention model for adapting images on small displays. Multimedia Syst. **9**(4), 353–364 (2003)

6. Santella, A., Agrawala, M., DeCarlo, D., Salesin, D., Cohen, M.: Gaze-based interaction for semi-automatic photo cropping. In: Proceedings of the SIGCHI Conference on Human Factors in Computing Systems, pp. 771–780. ACM (2006)
7. Setlur, V., Takagi, S., Raskar, R., Gleicher, M., Gooch, B.: Automatic image retargeting. In: Proceedings of the 4th International Conference on Mobile and Ubiquitous Multimedia, pp. 59–68. ACM (2005)
8. Avidan, S., Shamir, A.: Seam carving for content-aware image resizing. ACM Trans. Graph. (TOG) **26**, 10 (2007)
9. Achanta, R., Süsstrunk, S.: Saliency detection for content-aware image resizing. In: 2009 16th IEEE International Conference on Image Processing (ICIP), pp. 1005–1008. IEEE (2009)
10. Dong, W., Zhou, N., Paul, J.C., Zhang, X.: Optimized image resizing using seam carving and scaling. ACM Trans. Graph. (TOG) **28**, 125 (2009)
11. Wang, Y.S., Tai, C.L., Sorkine, O., Lee, T.Y.: Optimized scale-and-stretch for image resizing. ACM Trans. Graph. (TOG) **27**, 118 (2008)
12. Zhang, G.X., Cheng, M.M., Hu, S.M., Martin, R.R.: A shape-preserving approach to image resizing. Comput. Graph. Forum **28**, 1897–1906 (2009). Wiley Online Library
13. Chang, C.H., Chuang, Y.Y.: A line-structure-preserving approach to image resizing. In: 2012 IEEE Conference on Computer Vision and Pattern Recognition (CVPR), pp. 1075–1082. IEEE (2012)
14. Sun, H., Lian, Z., Tang, Y., Xiao, J.: Non-rigid point set registration for Chinese characters using structure-guided coherent point drift. In: 2014 IEEE International Conference on Image Processing (ICIP), pp. 4752–4756. IEEE (2014)
15. Lian, Z., Xiao, J.: Automatic shape morphing for Chinese characters. In: SIGGRAPH Asia 2012 Technical Briefs, p. 2. ACM (2012)
16. Shewchuk, J.R.: Triangle: engineering a 2D quality mesh generator and Delaunay triangulator. In: Lin, M.C., Manocha, D. (eds.) WACG 1996. LNCS, vol. 1148, pp. 203–222. Springer, Heidelberg (1996). doi:10.1007/BFb0014497
17. Jacobson, A., Tosun, E., Sorkine, O., Zorin, D.: Mixed finite elements for variational surface modeling. Comput. Graph. Forum **29**, 1565–1574 (2010). Wiley Online Library
18. Kavan, L., Collins, S., Žára, J., O'Sullivan, C.: Geometric skinning with approximate dual quaternion blending. ACM Trans. Graph. (TOG) **27**(4), 105 (2008)
19. Alexa, M., Cohen-Or, D., Levin, D.: As-rigid-as-possible shape interpolation. In: Proceedings of the 27th Annual Conference on Computer Graphics and Interactive Techniques, pp. 157–164. ACM Press/Addison-Wesley Publishing Co. (2000)
20. Lian, Z., Godil, A., Xiao, J.: Feature-preserved 3D canonical form. Int. J. Comput. Vis. **102**(1–3), 221–238 (2013)

Supervised Class Graph Preserving Hashing for Image Retrieval and Classification

Lu Feng, Xin-Shun Xu$^{(\boxtimes)}$, Shanqing Guo$^{(\boxtimes)}$, and Xiao-Lin Wang

School of Computer Science and Technology, Shandong University, Jinan, China
lasiafeng@gmail.com, {xuxinshun,guoshanqing,xlwang}@sdu.edu.cn

Abstract. With the explosive growth of data, hashing-based techniques have attracted significant attention due to their efficient retrieval and storage reduction ability. However, most hashing methods do not have the ability of predicting the labels directly. In this paper, we propose a novel supervised hashing approach, namely Class Graph Preserving Hashing (CGPH), which can well incorporate label information into hashing codes and classify the samples with binary codes directly. Specifically, CGPH learns hashing functions by ensuring label consistency and preserving class graph similarity among hashing codes simultaneously. Then, it learns effective binary codes through orthogonal transformation by minimizing the quantization error between hashing function and binary codes. In addition, an iterative method is proposed for the optimization problem in CGPH. Extensive experiments on two large scale real-world image data sets show that CGPH outperforms or is comparable to state-of-the-art hashing methods in both image retrieval and classification tasks.

Keywords: Hashing · Image retrieval · Image classification · Similarity search

1 Introduction

With the development of the Internet and electronic devices, data (e.g., texts, images and videos) is growing rapidly. The explosive growth of such data brings great challenges to people, e.g., how to search and store the data. For the similarity search task, traditional methods can work well by comparing original features when the data size is small. However, they become infeasible on large scale data because of their big computational cost of similarity calculation. Thus, in these years, a lot of efforts have been devoted to this problem. Among those proposed methods, hashing based methods have shown their great power to handle such large scale applications [22,29,30]. Hashing based similarity search methods transform original visual features into a low dimensional binary space, while at the same time preserve the certain data properties (e.g., local structure, semantic similarity) as much as possible. By computing the pairwise Hamming distance of binary codes, approximate nearest neighbor search can be efficiently performed

© Springer International Publishing AG 2017
L. Amsaleg et al. (Eds.): MMM 2017, Part I, LNCS 10132, pp. 391–403, 2017.
DOI: 10.1007/978-3-319-51811-4_32

on massive data corpus. They have efficient constant query time and can also reduce storage by storing compact codes of data points. Thus, in these years, hashing based methods have attracted more attention; many methods have been proposed. Moreover, hashing has become a popular approach in some large scale vision problems such as feature descriptor learning [21], information retrieval [14,31,32] and object recognition [23].

Generally, existing hashing methods can be divided into two categories [27]: data-independent methods [2,8] and data-dependent methods [5,10]. For data-independent methods, hash functions are designed without using any training data. Representative data-independent methods include locality-sensitive hashing (LSH) [2], shift-invariant kernels hashing (SIKH) [18] and many other extensions [11,12]. LSH randomly generates linear hash function to approximate cosine similarity. Nonetheless, the LSH family normally requires longer code length to guarantee search performance, which causes higher storage cost and limits the scalability of the whole algorithm.

For data-dependent methods, their hashing functions are learned from training data. The data-dependent methods can be further divided into two categories [24]: unsupervised and semi-supervised/supervised methods. Unsupervised methods try to preserve the metric (e.g., Euclidean) structure between data points only using their feature information. Representative unsupervised methods include spectral hashing (SH) [28], principal component analysis based hashing (PCAH) [25], iterative quantization (ITQ) [5], anchor graph hashing (AGH) [16], isotropic hashing (Iso-Hash) [9] and an affinity-preserving k-means hashing (KMH) [6], etc. On the contrary, semi-supervised/supervised hashing methods try to leverage some semantic (label) information for hash function learning. For example, Canonical Correlation Analysis with Iterative Quantization (CCA-ITQ) [5] treats original samples and labels as two views and uses CCA to learn hashing projections. Minimal loss hashing (MLH) [17] introduces a hinge-like loss function depending on semantic similarity information and learns binary codes based on structured prediction. Supervised hashing with kernels (KSH) [15] maps data to hashing codes whose Hamming distances are minimized on similar pairs and simultaneously maximized on dissimilar pairs. Binary reconstructive embeddings (BRE) [10] learns hashing functions by minimizing the reconstruction error between original semantic similarities and the Hamming distances of the corresponding binary embeddings. Shen et al. proposed supervised discrete hashing (SDH) [20] in which the binary codes can be used for classification problem. Li et al. proposed a method for face image retrieval and attributes prediction [13].

From the above, we can find that many hashing methods have been proposed. However, most of them are proposed for retrieval task instead of classification. This means that we cannot predict the labels with the obtained binary codes. Although Shen et al. [20] proposed SDH which can do retrieval and classification simultaneously, it does not consider the relationship between samples and cannot make sure that similar hashing codes predict the same class. Thus, the test accuracy of binary codes is still not compared to that of real-value features sometimes. Moreover, the discrete optimization problem is hard to tackle.

Thus, if a hashing method can efficiently tackle retrieval and classification tasks simultaneously, it would be much useful and practical. Motivated by this, in this paper, we propose a novel efficient supervised hashing method, i.e., Class Graph Preserving Hashing (CGPH), which can well embed semantic information into hash functions and directly predict the labels of query data using the learned hashing codes. Specifically, in CGPH, we incorporate the supervised label information for jointly learning effective hash functions and correlations between tags and hashing codes to preserve the consistency between labels and hashing codes. In this way, the hashing codes can be mapped into label space and the label consistency will be preserved. In addition, to preserve the class similarity between samples, we also construct the class mapping based on manifold regularization by semantic neighbor graph. Experimental results on two real-world image data sets with semantic labels show that CGPH can outperform or is comparable to other state-of-the-art methods for image retrieval and classification tasks.

The rest of this paper is organized as follows. The proposed GCPH method is presented in Sect. 2. Section 3 gives the experimental settings and results. The conclusions are given in Sect. 4.

2   Class Graph Preserving Hashing

2.1   Formulation

Suppose we have n training samples $X = \{x_1, x_2, ..., x_n\} \in R^{d \times n}$, where d is the dimensionality of image feature. Without loss of generality, we suppose that all samples are zero-centered. Denote the groundtruth label matrix as $Y = \{y_1, y_2, ..., y_n\} \in \{0, 1\}^{c \times n}$, where c is the total classes of labels, $y_{ij} = 1$ if x_j belongs to class i, otherwise $y_{ij} = 0$. We aim to learn a set of binary codes $B = \{b_1, b_2, ..., b_n\} \in \{-1, 1\}^{k \times n}$ to well preserve the semantic and class similarity. At the same time, we can use the binary code for classification task, i.e., getting the labels of samples. Suppose that the hashing function is $H(x) = (h_1(x), h_2(x), ..., h_c(x))$. This means that the ith bit of a binary code can be obtained by the following function.

$$b_i = sgn(H(x_i)) \tag{1}$$

In general, either linear embedding or nonlinear embedding can be used to approximate $H(x)$. In this paper, we adopt a simple but effective nonlinear embedding learning algorithm which has been employed in various hashing-based applications [10,15,20]. Denote the nonlinear embedding hashing functions as

$$H(x) = W^\top \phi(x) \tag{2}$$

where $W \in R^{m \times k}$ the mapping matrix, and $\phi(x)$ is the nonlinear embedding of x which can be given as, $\phi(x) = [\phi_1(x), ..., \phi_m(x)]^\top$. $\phi_i(x)$ is defined as:

$$\phi_i(x) = exp(- \parallel x - a_i \parallel_2^2 / 2\sigma^2) \tag{3}$$

where $\{a_i\}_{i=1}^m$ are randomly selected anchor points from X and σ is the kernel bandwidth.

2.2   Label Consistency Preservation

Images are often associated with many labels which provide valuable information for learning effective semantic hashing codes. It's a challenge how to build an appropriate bridge between label and binary codes. A direct way is to map the k-dimensional hashing codes to label space and then quantize the error between the original labels and predicted data. Thus, the learning model can be stated as follows:

$$\min_{B,C} \sum_{i=1}^{n} \mathcal{L}(y_i, Cb_i) + \lambda \Phi(C) \tag{4}$$

where $C \in R^{c \times k}$ is the mapping matrix also as the classification matrix for hashing codes. Here $\mathcal{L}(\cdot)$ is the loss function measuring the labeling approximation error in label space between the given supervised labels and prediction result by hashing codes. $\Phi(C)$ is the regularization term to avoid over-fitting. $\lambda > 0$ is a trade-off parameter.

The choice of loss function $\mathcal{L}(\cdot)$ usually depends on application domains. Three convex loss functions are used frequently such as least square loss, hinge loss and logistic loss functions. Least square loss is always adopted in quantization and classification problems and [4] shows that the ℓ_2 loss can provide comparable performance to other loss functions. Thus we choose the ℓ_2 loss as the loss function. Then the problem wrote in the compact matrix formulation is:

$$\min_{B,C} \parallel Y - CB \parallel^2 + \lambda \parallel C \parallel^2$$
$$s.t.\ B \in \{-1, 1\}^{k \times n} \tag{5}$$

It is also a multi-class linear classification model, at the same time it can handle the multi-label classification problems. By minimizing this function, the consistency between labels can be ensured.

2.3   Class Graph Preservation

For label prediction, the main purpose of C in Eq. (5) is to propagate the semantic information from the original label space to the hamming space. It is intuitive that the predicted label vectors should have ability to keep the intrinsic structure among hashing codes. In other words, if two hashing codes b_i and b_j are close in hamming space, then the predicted label vectors y_i and y_j are also close to each other. Meanwhile, if two data points x_i and x_j are semantic similar in the original feature space, the hamming distance between their hashing codes should be small. This is a bridge among original feature data, hashing codes and labels. This is referred to local invariance assumption [1], which is well studied in manifold learning theories.

Inspired by this, to preserve the intrinsic geometric structure of original data, here we first construct a semantic graph with n vertices. There is an edge between two vertices when they have at least one same label. In this paper we call this graph as class graph. Then, we can get a similarity edge matrix S, in which $s_{ij} = 1$ if there is an edge between vertices i and j, -1 otherwise. Based on this, the local invariance assumption can be formulated via a manifold regularization, that is

$$\frac{1}{2} \sum_{i,j=1}^{n} s_{ij} \parallel Cb_i - Cb_j \parallel_2^2 = tr((CB)L(CB)^{\top}) \tag{6}$$

where L is the graph Laplacian of matrix S defined as $L = D - S$, and D is a diagonal matrix whose diagonal entries are column sums of S, i.e., $D_{ii} = \sum_{i=1}^{n} s_{ij}$.

Finally, we combine Eqs. (5) and (6) together; then, we get the overall objective function as follows.

$$\min_{B,C} \parallel Y - CB \parallel^2 + \lambda \parallel C \parallel^2 + \alpha tr((CB)L(CB)^{\top})$$
$$s.t. \ B \in \{-1, 1\}^{k \times n} \tag{7}$$

where $\alpha > 0$ is a balance parameter.

2.4 Optimization

Directly minimizing the objective function in Eq. (7) is intractable because of the discrete condition, which is proved to be NP-hard. Therefore, we relax the discrete constraints of hashing codes and adopt continuous hashing functions to replace it. Then, the relaxed objective function becomes:

$$\min_{W,C} \parallel Y - CW^{\top}\phi(x) \parallel^2 + \lambda \parallel C \parallel^2 + \alpha tr((CW^{\top}\widetilde{L}WC^{\top}) \tag{8}$$

where $\widetilde{L} = \phi(x)L\phi(x)^{\top}$ which can be pre-computed.

We can find that this optimization problems can be minimized with respect to C and W by fixing the other one in an alternating manner. The detailed optimization steps are given in the following paragraphs.

Update C. If we fix W in Eq. (8), it is easy to find a solution by solving the regularized least squares problem. For instance, the gradient of C is calculated as follows:

$$-YH(x)^{\top} + CH(x)H(x)^{\top} + \lambda C + \alpha CW^{\top}\widetilde{L}W = 0 \tag{9}$$

where $H(x) = W^{\top}\phi(x)$. With the obtained gradient, then we can easily get the closed form solution of C, i.e.,

$$C = YH(x)^{\top}(H(x)H(x)^{\top} + \alpha W^{\top}\widetilde{L}W + \lambda I_k)^{-1} \tag{10}$$

Apparently, the inverse of $(H(x)H(x)^{\top} + \alpha W^{\top}\widetilde{L}W + \lambda I_k)$ exists as the matrix is symmetric positive definite.

Update W. If we fix C in Eq. (8), we can get the projection matrix W by regression. The gradient of W is given as:

$$-\phi(x)Y^\top + \phi(x)\phi(x)^\top WC^\top C + \tilde{L}WC^\top C = 0 \tag{11}$$

Then, W can be obtained by:

$$W = (\phi(x)\phi(x)^\top + \tilde{L})^{-1}\phi(x)Y^\top C(C^\top C)^{-1} \tag{12}$$

We can update C and W alternatively until we get a local optimal solution.

2.5   Binary Quantization

After we get the final projection matrix W, the hashing codes can be generated using Eq. (1). In this way, the quantization error will be $\| B - W^\top \phi(x) \|_F^2$. Inspired by the success of Iterative Quantization [5], we can further minimize the quantization error by using an orthogonal transformation. It not only preserves the optimality of relaxed solution but also provides us more flexibility to achieve better hashing codes with low quantization error. The binary quantization with orthogonal transformation is as follows:

$$\min_{B,R} \| B - (WR)^\top \phi(x) \|_F^2$$
$$s.t. \quad B \in \{-1,1\}^{k\times n}, R^\top R = I_k \tag{13}$$

The above optimization problem can be solved by updating B and R alternatively as follows.

Update B. Fixing R, B can be obtained by the following function.

$$B = sgn((WR)^\top \phi(x)) = sgn(R^\top H(x)) \tag{14}$$

Update R. Fixing B, the objective function becomes:

$$\min_{R^\top R = I_k} \| B - R^\top H(x) \|_F^2 \tag{15}$$

The above function is essentially the classis Orthogonal Procrustes problem [19], which can be solved efficiently by singular value decomposition. Then we have

$$R = UV^\top \tag{16}$$

where U and V are left and right singular values of SVD decomposition of matrix $BH(x)^\top$, respectively. We then perform the above two steps alternatively to obtain the optimal binary hashing codes. In our experiments, we find that the algorithm usually converges in about $20 - 60$ iterations. To show the proposed method clearly, we summarize it **Algorithm 1**.

Algorithm 1. Class Graph Preserving Hashing(CGPH)

Require: Training data $\{x_i, y_i\}_{i=1}^n$, number of anchor points m, code length k, maximum iteration number t, parameters λ, α and σ.

Ensure: Hashing projection matrix W, correlation matrix C and binary codes B.

1. Initialize W by eigenvalue decomposition and $R = I_k$, embed X into nonlinear space to get $\phi(x)$, calculate $\widetilde{L}$.
2. **repeat**
3.   Calculate C using Eq. (10)
4.   Compute W using Eq. (12) to form $H(x)$
5. **until** converge or reach maximum iterations.
6. **repeat**
7.   Update B using Eq. (14)
8.   Update R using Eq. (16)
9. **until** converge or reach maximum iterations.
10. **return** B, C, W, R

2.6 Complexity Analysis

The algorithm of CGPH consists of two main loops. In the first loop, we iteratively solve Eqs. (10) and (12) to obtain the optimal solution of C and W, where the time complexities are bounded by $O(nk^3 + nkm + nkc)$ and $O(nk^2 + mk^2)$. The second loop iteratively optimizes the binary hashing codes and the orthogonal transformation matrix, where the time complexities of updating B and R are bounded by $O(nk^2 + nkm + k^3)$. Thus, the total time complexity of solving GCPH is $O(nk^3 + (n+m)k^2 + nk(m+c))$, which scales linearly with n given $n \gg m > (c, k)$. For each query, the hashing time is constant, i.e., $O(dk)$.

3 Experiments

In this section, we evaluate the effectiveness of the proposed CGPH model on two data sets for image retrieval and classification tasks. We also compare it with several state-of-the-art hashing methods.

3.1 Data Sets and Baselines

The data sets we use for evaluation are **MNIST** and a multi-label data set **MIRFLICKR-25000** [7]. For all data sets, we perform normalization on feature vectors to make each vector have zero mean and equal variance. Ground truths are defined by the category information from data sets. The MNIST data set consists of 70K 784-dimensional samples associated with digits from '0' to '9'. Each image is represented by the values of pixels. The entire data set is partitioned into two subsets: a training set with 69K samples and a test set with 1 K samples. MIRFLICKR-25000 is a multi-label dataset collected from Flickr images. It has 25K image samples and a kind of annotations with 38 labels. We exclude images that are not associated with any selected labels and finally we get 24581 images; then, we extract the 512-dimensional Gist feature from each image sample. In addition, we randomly select 50 images from every class to form 1900 test samples and make the left 22681 images as the training set. We define the true neighbors of a query as the images sharing at least one labels with the query image.

In addition, we compare CGPH with five supervised hashing methods including SDH [20], CCA-ITQ [5], KSH [15], BRE [10], SSH [26] and two unsupervised methods i.e., PCA-ITQ [5] and AGH [16]. The codes of most algorithms above are kindly provided by the authors; the parameters of these algorithms are selected according to the schemes suggested by the authors. Note that BRE and KSH are too slow to run when training set is large; thus, we train them on 5k samples. For fair comparisons we also test our method on the same training set using 300 anchors, which is named as CGPH-5K.

3.2   Evaluation Criteria

To conduct fair evaluation, we follow two criteria which are commonly used in hashing retrieval: *Hamming Ranking* and *Hash Lookup*. *Hamming Ranking* ranks all the points in the database according to their Hamming distance from the query and the top k points are returned as the desired neighbors. *Hash Lookup* returns all the points within a small Hamming radius r of the query. The retrieval results are evaluated based on whether the retrieved images and query images share any common ground-truth labels. We use several metrics to measure the performance on different methods. In this paper, we calculate the MAP at top 200 which is the mean average percentage of true semantic neighbors among top 200 returned examples. A hamming radius of $r = 2$ is used to retrieve the neighbors in the case of $HashLookup$. The precision and recall of returned samples falling within in hamming radius 2 are reported.

Based on the label predicting matrix $C \in R^{c \times k}$, the label classification performance is evaluated with widely used metrics, i.e., mean accuracy for single-label and Macro-F1, instance-based Accuracy for multi-label samples. As to multi-label instance, we select a threshold for each class according to the experience for our method and SDH. To other methods, after obtaining the binary codes, we apply the effective linear SVM for classification by *LIBLINEAR* [3] solver. We also test the classification performance on continuous original feature of images.

3.3   Parameter Analysis

There are several parameters in CGPH which could affect its performance. We conduct experiments to show their influence on performance on different data sets. For bandwidth σ in RBF, we empirically set it to 0.4. The number of anchors is set to 1000 for both data sets. Figure 1 shows the results of 64 bits on both data sets with varying values of λ and α. From this figure, we can find that λ has less influence on performance than α. In addition, both precision and MAP have the same trend on MNIST when the values of parameters change.

We can also observe that CGPH obtains the best performance when $\lambda = 2$ and $\alpha = 1e - 5$ on MNIST and MIRFLICKR.

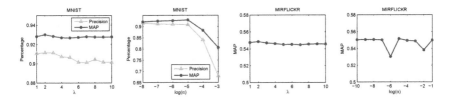

Fig. 1. Effects of λ and α in CGPH

3.4   Results and Discussions

The retrieval results of CGPH and all other compared methods on MNIST are shown in Table 1 including precision and recall within Hamming radius 2 and MAP on 64, 96 and 128 bits, respectively. From this table, we can clearly find that (1) CGPH outperforms SDM, CCA-ITQ, SSH, AGH and PCA-ITQ on all evaluation criteria when the number of training samples is 69000; (2) The results of CGPH are much better than those of KSH and BRE on all evaluation criteria when the number of training samples is 5000; (3) AGH obtains high precision values, but very small recall values. The reason is that it returns few samples within Hamming radius 2; however, most of these returned samples are true neighbors of the query sample.

The classification results on MNIST are shown in Table 2. Note that, on classification task, CGPH is compared with SDH, KSH, BRE, CCA-ITQ and SSH. From this table, we can also find that CGPH outperforms all compared methods. In addition, the results of CGPH and SDH are much better than other compared methods.

The retrieval results on MIRFLICKR with the code length varying from 8 to 128 are shown in Fig. 2. Note that, to save space, we only give the precision

Table 1. Retrieval results on MNIST (Precision, recall and MAP)

Method	#training	#anchor	Precision			Recall			MAP		
			64 bits	96 bits	128 bits	64 bits	96 bits	128 bits	64 bits	96 bits	128 bits
CGPH	5000	300	0.7977	0.7468	0.6972	0.4649	0.4451	0.4389	0.8138	0.8206	0.8246
	69000	1000	**0.9070**	**0.8893**	**0.8734**	**0.7381**	**0.7202**	**0.7078**	**0.9272**	**0.9312**	**0.9380**
SDH	69000	1000	0.9002	0.8806	0.8658	0.7197	0.6942	0.6819	0.9106	0.9173	0.9198
KSH	5000	300	0.5738	0.4997	0.4422	0.2668	0.2167	0.1844	0.1034	0.0992	0.1000
BRE	5000	-	0.1220	0.0410	0.0260	0.0036	0.0003	0.0002	0.1014	0.1214	0.1235
CCA-ITQ	69000	-	0.5024	0.2539	0.1254	0.1273	0.0662	0.0174	0.7556	0.7594	0.7592
SSH	69000	-	0.4043	0.1370	0.0030	0.0035	0.0012	0.0001	0.1954	0.1822	0.1746
AGH	69000	1000	0.8708	0.8644	0.857	0.0084	0.0059	0.0045	0.3828	0.3430	0.3180
PCA-ITQ	69000	-	0.1901	0.0850	0.0530	0.0096	0.0002	0.0001	0.4528	0.4555	0.4648

Table 2. Classification results on MNIST (accuracy)

Method	64 bits	96 bits	128 bits
CGPH	**0.943**	**0.944**	**0.947**
SDH	0.936	0.929	0.935
KSH	0.907	0.906	0.912
BRE	0.814	0.834	0.859
CCA-ITQ	0.83	0.818	0.833
SSH	0.691	0.704	0.702
Original feature	0.872		

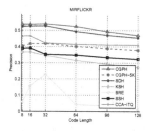 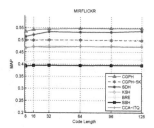

Fig. 2. Retrieval results on MIRFLICKR (Precision and MAP)

and MAP results in this Figure. From this figure, we can find that (1) CGPH outperforms all compared methods on both precision and MAP; (2) SDH outperforms other compared methods, e.g., KSH, BRE, SSH and CCA-ITQ; (3) CGPH-5k, which only uses 5000 training samples, outperforms KSH, BRE, SSH and CCA-ITQ.

In addition, we can find that, on this data set, the performance of all methods is essentially stable with the increasing of hashing bits. However, our method still achieves the best results on all code lengths.

We also test the classification performance of CGPH and all baselines on MIRFLICKR. Note that MIRFLICKR is a multi-label data set; thus, MacroF1 and instance-based accuracy are used as evaluation metrics. The detailed results are reported in Table 3. From this table, we can also observe similar results as those on MNIST, e.g., (1) CGPH obtains the best results on all cases; (2) The results of CGPH are even better than those when GIST feature is directly used for classification; (3) The results of CGPH and SDH are much better than those other compared methods.

Generally, from the results on MNIST and MIRFLICKR, we can conclude that our proposed method–CGPH performs well on both retrieval and classification tasks. Especially, it obtains better results that compared state-of-the-art hashing methods. This further confirms that CGPH can not only get the relationship between labels and binary codes but also preserve the class similarity of hashing codes.

Table 3. Multi-label classification performance on MIRFLICKR

Method	MacroF1			Accuracy		
	64 bits	96 bits	128 bits	64 bits	96 bits	128 bits
CGPH	**0.1788**	**0.18**	**0.1863**	**0.2325**	**0.239**	**0.2373**
SDH	0.1624	0.1601	0.1658	0.2089	0.2079	0.2042
KSH	0.0419	0.0451	0.0444	0.0712	0.0746	0.0741
CCA-ITQ	0.0514	0.0503	0.053	0.0815	0.0797	0.0808
BRE	0.0748	0.0795	0.085	0.1685	0.1658	0.1712
SSH	0.035	0.0366	0.0367	0.0581	0.0609	0.0616
Gist feature	0.1502			0.2207		

4 Conclusions

In this paper, we propose a novel supervised hashing method, i.e., Class Graph Preserving Hashing (CGPH), which can tackle both image retrieval and classification tasks on large scale data. In CGPH, we firstly learn the hashing functions by simultaneously ensuring the label consistency and preserving the classes similarity between hash codes. Then, we further learn the effective hash codes through the orthogonal transformation by minimizing the quantization error between hash functions and codes. In addition, an iterative solution is proposed for the optimization problem in CGPH. Extensive experiments have been conducted on two large scale real-world image data sets. The results show that CGPH outperforms or is comparable to state-of-the-art hashing methods in both image retrieval and classification tasks.

Acknowledgments. This work was partially supported by National Natural Science Foundation of China (61173068, 61573212, 91546203), Program for New Century Excellent Talents in University of the Ministry of Education, Independent Innovation Foundation of Shandong Province (2014CGZH1106), Key Research and Development Program of Shandong Province (2016GGX101044, 2015GGE27033).

References

1. Belkin, M., Niyogi, P., Sindhwani, V.: Manifold regularization: a geometric framework for learning from labeled and unlabeled examples. J. Mach. Learn. Res. **7**, 2399–2434 (2006)
2. Datar, M., Immorlica, N., Indyk, P., Mirrokni, V.S.: Locality-sensitive hashing scheme based on p-stable distributions. In: Proceedings of SCG, pp. 253–262 (2004)
3. Fan, R., Chang, K., Hsieh, C., Wang, X., Lin, C.: LIBLINEAR: a library for large linear classification. J. Mach. Learn. Res. **9**, 1871–1874 (2008)
4. Fung, G., Mangasarian, O.L.: Multicategory proximal support vector machine classifiers. Mach. Learn. **59**(1–2), 77–97 (2005)
5. Gong, Y., Lazebnik, S.: Iterative quantization: a procrustean approach to learning binary codes. In: Proceedings of CVPR, pp. 817–824 (2011)

6. He, K., Wen, F., Sun, J.: K-means hashing: an affinity-preserving quantization method for learning binary compact codes. In: Proceedings of CVPR, pp. 2938–2945 (2013)
7. Huiskes, M.J., Lew, M.S.: The MIR flickr retrieval evaluation. In: Proceedings of MIR, pp. 39–43 (2008)
8. Indyk, P., Motwani, R.: Approximate nearest neighbors: towards removing the curse of dimensionality. In: Proceedings of STOC, pp. 604–613 (1998)
9. Kong, W., Li, W.: Isotropic hashing. In: Proceedings of NIPS, pp. 1655–1663 (2012)
10. Kulis, B., Darrell, T.: Learning to hash with binary reconstructive embeddings. In: Proceedings of NIPS, pp. 1042–1050 (2009)
11. Kulis, B., Grauman, K.: Kernelized locality-sensitive hashing for scalable image search. In: Proceedings of ICCV, pp. 2130–2137 (2009)
12. Kulis, B., Jain, P., Grauman, K.: Fast similarity search for learned metrics. IEEE Trans. Pattern Anal. Mach. Intell. **31**(12), 2143–2157 (2009)
13. Li, Y., Wang, R., Liu, H., Jiang, H., Shan, S., Chen, X.: Two birds, one stone: jointly learning binary code for large-scale face image retrieval and attributes prediction. In: Proceedings of ICCV, pp. 3819–3827 (2015)
14. Liu, W., He, J., Chang, S.: Large graph construction for scalable semi-supervised learning. In: Proceedings of ICML, pp. 679–686 (2010)
15. Liu, W., Wang, J., Ji, R., Jiang, Y., Chang, S.: Supervised hashing with kernels. In: Proceedings of CVPR, pp. 2074–2081 (2012)
16. Liu, W., Wang, J., Kumar, S., Chang, S.: Hashing with graphs. In: Proceedings of ICML, pp. 1–8 (2011)
17. Norouzi, M., Fleet, D.J.: Minimal loss hashing for compact binary codes. In: Proceedings of ICML, pp. 353–360 (2011)
18. Raginsky, M., Lazebnik, S.: Locality-sensitive binary codes from shift-invariant kernels. In: Proceedings of NIPS, pp. 1509–1517 (2009)
19. Schnemann, P.: A generalized solution of the orthogonal procrustes problem. Psychometrika **31**(1), 1–10 (1966)
20. Shen, F., Shen, C., Liu, W., Shen, H.T.: Supervised discrete hashing. In: Proceedings of CVPR, pp. 37–45 (2015)
21. Strecha, C., Bronstein, A.M., Bronstein, M.M., Fua, P.: LDAHash: improved matching with smaller descriptors. IEEE Trans. Pattern Anal. Mach. Intell. **34**(1), 66–78 (2012)
22. Tang, J., Li, Z., Wang, M., Zhao, R.: Neighborhood discriminant hashing for large-scale image retrieval. IEEE Trans. Image Process. **24**(9), 2827–2840 (2015)
23. Torralba, A., Fergus, R., Freeman, W.T.: 80 million tiny images: a large data set for nonparametric object and scene recognition. IEEE Trans. Pattern Anal. Mach. Intell. **30**(11), 1958–1970 (2008)
24. Wang, J., Xu, X.-S., Guo, S., Cui, L., Wang, X.: Linear unsupervised hashing for ANN search in Euclidean space. Neurocomputing **171**, 283–292 (2016)
25. Wang, J., Kumar, S., Chang, S.: Sequential projection learning for hashing with compact codes. In: Proceedings of ICML, pp. 1127–1134 (2010)
26. Wang, J., Kumar, S., Chang, S.: Semi-supervised hashing for large-scale search. IEEE Trans. Pattern Anal. Mach. Intell. **34**(12), 2393–2406 (2012)
27. Wang, S.-S., Huang, Z., Xu, X.-S.: A multi-label least-squares hashing for scalable image search. In: Proceedings of SDM, pp. 954–962 (2015)
28. Weiss, Y., Torralba, A., Fergus, R.: Spectral hashing. In: Proceedings of NIPS, pp. 1753–1760 (2008)
29. Xu, X.-S.: Dictionary learning based hashing for cross-modal retrieval. In: Proceedings of MM, pp. 177–181 (2016)

30. Yan, T.-K., Xu, X.-S., Guo, S., Huang, Z., Wang, X.-L.: Supervised robust discrete multimodal hashing for cross-media retrieval. In: Proceedings of CIKM, pp. 1271–1280 (2016)
31. Yang, Y., Shen, F., Shen, H.T., Li, H., Li, X.: Robust discrete spectral hashing for large-scale image semantic indexing. IEEE Trans. Big Data 1(4), 162–171 (2015)
32. Yang, Y., Zha, Z.-J., Gao, Y., Zhu, X., Chua, T.-S.: Exploiting web images for robust semantic video indexing via sample-specific loss. IEEE Trans. Multimedia 16(6), 1677–1689 (2014)

Visual Robotic Object Grasping Through Combining RGB-D Data and 3D Meshes

Yiyang Zhou[1], Wenhai Wang[1], Wenjie Guan[1], Yirui Wu[2],
Heng Lai[1], Tong Lu[1(✉)], and Min Cai[3]

[1] National Key Lab for Novel Software Technology,
Nanjing University, Nanjing, China
zyy34472@gmail.com, wangwenhai362@163.com, terry_guanwenjie@163.com,
134lhforever@gmail.com, lutong@nju.edu.cn
[2] College of Computer and Information, Hohai University, Nanjing, China
wuyirui1989@163.com
[3] Riseauto Intelligent Tech, Beijing, China
caimin@riseauto.cn

Abstract. In this paper, we present a novel framework to drive automatic robotic grasp by matching camera captured RGB-D data with 3D meshes, on which prior knowledge for grasp is pre-defined for each object type. The proposed framework consists of two modules, namely, pre-defining grasping knowledge for each type of object shape on 3D meshes, and automatic robotic grasping by matching RGB-D data with pre-defined 3D meshes. In the first module, we scan 3D meshes for typical object shapes and pre-define grasping regions for each 3D shape surface, which will be considered as the prior knowledge for guiding automatic robotic grasp. In the second module, for each RGB-D image captured by a depth camera, we recognize 2D shape of the object in it by an SVM classifier, and then segment it from background using depth data. Next, we propose a new algorithm to match the segmented RGB-D shape with predefined 3D meshes to guide robotic self-location and grasp by an automatic way. Our experimental results show that the proposed framework is particularly useful to guide camera based robotic grasp.

Keywords: 3D mesh · Registration · Robotic grasping · 3D matching

1 Introduction

In the past years, robots have become much more popular in real-life applications such as industry, health care, human service and education. One reason is that new types of sensing hardware which have lower costs but higher computational performances are now becoming much cheaper nowadays. For example, RGB-D cameras (e.g., Microsoft Kinect and Intel Realsense) allow a robot to see different objects in an indoor scene with per-pixel depth information besides RGB data [1], which is particularly useful in developing environment perception systems

© Springer International Publishing AG 2017
L. Amsaleg et al. (Eds.): MMM 2017, Part I, LNCS 10132, pp. 404–415, 2017.
DOI: 10.1007/978-3-319-51811-4_33

to make robots more intelligent to walk around in unknown scenes automatically. Therefore, more researchers have been attracted in RGB-D camera based robotics explorations such as indoor environment 3D modeling [2], robot self-localization [4] and navigation [5] in the multimedia community today.

However, unlike automatic path plan research, there are few reported works on robotic grasp planning by using visual RGB-D data. This is because of the following difficulties. First, detecting proper regions on real scene objects for grasping is challenging due to the inherent limitations of vision based methods like the variations on viewpoint, size or distance of cameras and the influences from illumination conditions, which make it very difficult to decide proper points on objects for grasping. For a comparison, other systems like 3D modeling of indoor environments [2,3] need only generate 3D maps with large loops by RGB-D mapping, the accuracy of which is much lower than that of an intelligent grasp plan system. Second, objects in real scenes may have varied 3D surface shapes, which make the grasp plan task more challenging. Note that generally there are no other sensors are used for guiding grasping objects except for visual cameras here; however, a lot of new sensors can be selected in developing other systems like automatic navigation for robots.

In this paper, we present a novel method for detecting grasp regions on indoor scene objects by combing both RGB-D data and 3D mesh priors. A depth camera of Intel Realsense is used to capture source visual data composed of RGB and depth information, which are denoted by *Depth* and *Color*, respectively. Assume there are N visual object shapes of different categories for grasping, the overview of the proposed framework is shown in Fig. 1. We first categorize these objects

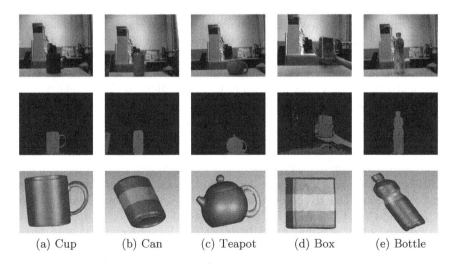

(a) Cup   (b) Can   (c) Teapot   (d) Box   (e) Bottle

Fig. 1. The first row shows the color data of five different objects captured by a Realsense camera, the second row presents their corresponding depth data, while the third row illustrates their corresponding 3D meshes with pre-defined grasp regions, which are marked the red color for guiding robotic grasp in the images of the first row. (Color figure online)

into representative shapes, and model their shape surfaces by 3D meshes as the prior knowledge of these objects. For example, we pre-define proper grasp regions on different 3D meshes as shown in the second row of Fig. 1. This is necessary due to viewpoint variations of the camera or shape variations of the object, only RGB-D data as shown in the first and the second rows of Fig. 1 are not reliable to drive accurate robot grasp. Next, after recognizing and segmenting scene objects using an SVM classifier $Classifer_{obj}$, we propose a novel algorithm to register the point clouds generated by the intrinsic parameters from $Depth$. In this step, we adopt the RANSAC strategy with the FPFH feature for matching. Finally, proper grasp regions are determined from camera captured RGB-D data by an accurate and robust way to overcome the inherent shortcomings of vision based methods for robotics. To the best of our knowledge, this is the first work for exploring grasp planning for robots by combining both RGB-D data and 3D meshes to overcome the inherent difficulties such as viewpoint and distance variations of vision based methods for accurate robotic self-location or grasp.

2   Related Work

There are a number of research using RGB-D data in developing robotic systems. These approaches can be roughly categorized into the following two classes consisting of 3D navigation and vision driven robotic hand grasping.

3D navigation systems use stereo cameras to align consecutive frames and detect loop closures for robots. Iterative Closet Point (ICP) algorithms are explored to match the best alignment between frames. Henry et al. [3] investigate how RGB-D cameras can be used for building dense 3D maps of indoor environments for robot navigation. Points in a source cloud are matched with their nearest neighboring points in a target cloud by a rigid transformation through minimizing the sum of squared spatial error between associated points. Prusak et al. [4] combine a time-of-flight camera with a CCD camera for robot self-localization. To solve the problem of no proper illumination, Prakhya et al. propose Sparse Depth Odometry (SDO) for keypoint detection, which comprises two keypoint detectors, namely, SURE and NARF, to augment RGB-D camera data for robots. Peasley and Birchfield [5] further use geometric and color information to overcome the significant limitation, namely, the sole reliance upon geometric information, of ICP algorithms used by Kinect Fusion.

Robotic object grasping is increasingly hard as the robot or the environment is not constrained. Due to the richness of stimulus, generally only visual cameras can be used to assist this task [12]. Saponaro [13] describes an approach for real-time preparation of grasping tasks based on the low-order moments of object shape on a stereo pair of images. Horaud et al. [14] present an algorithm to compute a set-point automatically from conic features interactively selected by an operator onto the object, and then use a vision based algorithm to guide grasping cylindric parts with a 6 degree of freedom robot arm. Bergamasco et al. [6] propose a technique for fitting elliptical shapes in 3D space by performing an initial 2D guess on each image followed by a multi-camera optimization. These methods show their effectiveness in developing robot systems by using visual cameras; however, knowledge priors of 3D objects are not used for guiding more accurate grasping on different types of objects.

3 Overview of the Framework

The overview of the proposed framework is shown in Fig. 2. Let $model_i$ denote the i-th (i is from 1 to N) pre-defined 3D mesh in the mesh set $Models$. For $Model_i$, the pre-defined grasp regions are represented by gp_i ($1 \leq i \leq N$). As a result, the mesh set $Models$ consists of $model_i$ and gp_i for each i in $1, 2..., N$ by

$$Models = \{< model_i, \ gp_i >\}, \ (1 \leq i \leq N) \tag{1}$$

For a depth image, its $Depth$ information is first used to segment it into object and background regions with the help of $Color$ information. For the extracted regions, we further extract key points depending on the characteristics of each object type, e.g., for a cuboid, we search for its visible vertices and edges in the image. RGB-D data are then converted to point cloud data in this step. Next, we propose a novel method for matching camera captured RGB-D data with accurately pre-defined 3D meshes for determining proper regions for grasping in the RGB-D space. By this way, image data captured by the RGB-D camera on the robot can be used to guide self-localization or grasping by a robust and accurate way.

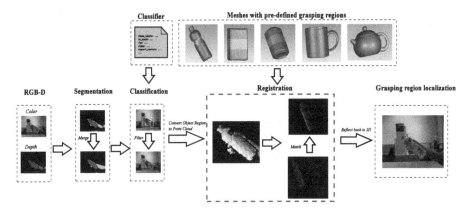

Fig. 2. The overview of the proposed framework for robotic grasping on visual objects by combining RGB-D data and 3D mesh priors. (Color figure online)

4 Pre-processing

In the pre-processing stage, the following two preparations are required, that is, (1) creating 3D meshes for each object category and define grasping region priors on these 3D meshes, and (2) training a classifier to detect and recognize visual objects from input 2D RGB images.

Object Detection from 2D Images. In this step, we train an SVM (Support Vector Machine) classifier with HOG features as in [8] for object detection. For this target, we build a dataset with the depth camera and manually label the ground truth for grasping on objects. We train altogether $N+1$ classifiers by considering indoor backgrounds as a special kind of negative class. The classifiers are denoted by $Classifer_{obj}$, the output of each classifier is $0, 1, 2, ..., N$ to give the categorization of each indoor object.

Pre-defining Grasping Priors on 3D Meshes. As well known, accurate estimation of size or shape of an object from 2D images is very challenging due to the variations on viewpoint or distance of cameras. For robotics grasping of different types of objects, a better way is to first create their 3D meshes and pre-define proper grasp regions as knowledge priors for different object types accordingly, and then use such knowledge to guide robotic self-localization and grasping by matching real time RGB-D data with the pre-defined 3D mesh. Assume there are N objects of different shapes, we build their 3D meshes separately, and annotate grasp regions gp_i on $model_i$ ($1 \leq i \leq N$) on each mesh $model_i$.

5   Visual Robotic Object Grasping

After detecting scene objects, a novel method for matching these objects with accurately pre-defined 3D meshes for determining proper regions is proposed.

5.1   Shape Analysis from RGB-D Data

We first segment each object shape from the input RGB-D image. After classifying each object into a known type, we determine grasp regions by matching real-time RGB-D data with the priors pre-defined on 3D meshes.

Object Segmentation Using Depth Information. A depth camera which is manufactured by infrared rays generally represents unreliable points as black. A black point here means that the distance from its position is either out of range or uncertain. That is, there is a recommended distance for either Kinect or Realsense cameras, the range out of which appears as black as shown in Fig. 3(b) (see the red region). Besides this, sometimes black regions are also brought due to diffuse reflections.

Considering this, we first connect non-black points in $Depth$ into groups using a region-growing method. Note that the points in $Depth$ are also clustered into groups according to whether the distance is near or far. As a result, we divide the input image into several depth connected regions DCR and black connected regions BCR with $Depth$. The segmentation results are shown in Fig. 3(c). According to the size of each region, we sort DCR by a descending order and select the *topk* regions as the main segments, which are denoted by

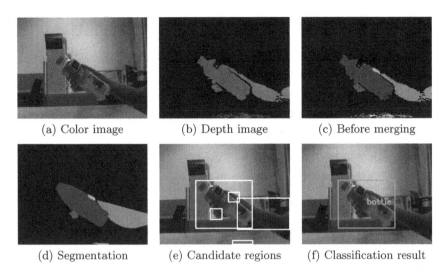

(a) Color image	(b) Depth image	(c) Before merging
(d) Segmentation	(e) Candidate regions	(f) Classification result

Fig. 3. (a) and (b) respectively show color and depth source captured by Realsense. The initial segmentations with $topk = 5$ are shown in (c). After merging small regions and black regions, we obtain the results as shown in (d). The white rectangle bounding boxes in (e) denote the $topk$ color candidate regions to calculate HOG features, which will be classified by an SVM classifier. The recognition results in (f) show that rectangle regions filtered by the classifier are marked red, while the rest white regions are removed from candidate regions after classifying. (Color figure online)

MS. We further search for convex hulls of the main segments, and define the average distance of the points in a segment as the distance $Distance$ of it.

Next, the rest small regions and black points in MS are further processed as follows. We group small regions to MS_i, $(1 \leq i \leq topk)$ as follows. If a small region sr belongs to more than one regions, we allocate sr to the nearest main region by $\arg\min_{ms}(|Distance(sr) - Distance(ms)|)$. For BCR, the points in black regions may also belong to more than one convex hull of MS. However, it is most likely that the black region belongs to MS which has the minimum distance. As a result of black points with the default value 0, the distance of black region $Distance(br)$ equals to 0. We find that black regions are due to the minimum distance Main Segmentation as well. Thus we define that br belongs to $\arg\min_{ms}(|Distance(br) - Distance(ms)|)$. By this way, we add those small regions and black regions to the same set.

Finally, we segment camera captured object from background. The proposed segmentation algorithm from the input RGB-D image is shown in Algorithm 1. Note that owing to the fact that depth data record distance for every pixel, the classic Region-Growing [9] algorithm helps us utilize depth data effectively to well segment scene objects from indoor backgrounds. After segmentation, we obtain $topk$ regions, and the example results of this step are shown in Fig. 3(d).

Algorithm 1. Segmentation from the input RGB-D image

Input:
     Depth data: *Depth*
     Number of Main Segmentations: *topk*
Output:
     Main Segmentations: MS
     $< BCR, DCR > \leftarrow$ Region-Growing($Depth$)
     Sort DCR according to the number of points
     Select *topk* segmentations as MS from DCR
     for $seg \in (DCR - MS) \cup BCR$ **do**
       for $p \in seg$ **do**
         $CS \leftarrow \{ms | ms \in MS \wedge p \ in \ ConvexHull(ms)\}$
         $SEG \leftarrow \arg \min_{ms}(distance(ms), ms \in CS)$
         $SEG \leftarrow SEG \cup \{p\}$
       end for
     end for
     return MS

Classifying Object Types. After segmenting from the input RGB-D image, we need further determine object class for each segmented region. As shown in Fig. 3(e), we calculate a white rectangle bounding box for each segmented region. These bounding boxes actually represent the ROI (Region of Interest) information, which contain our interested indoor scene objects for further verifying. For the *topk* ROIs, we determine their object classes using classic linear SVM classifier with one-versus-rest strategy. For the *i-th* segment, we first calculate its rectangle bounding box $BoundBox_i$, $(1 \leq i \leq topk)$ from the convex hull which we obtain in segmentation step. The *topk* rectangle regions are then classified by $Classifer_{obj}$. Note that for each region, it is resized to the same scale of the training samples for feature extraction (we resize it to 64×64 according to experiments). Additionally, we adopt HOG (Histogram of Oriented Gradient) [7] as the feature because HOG is based on gradient and the margin of a regular indoor object generally has significant gradient characteristic. Moreover, HOG is inexpensive and suitable for discriminating regular objects. With the extracted feature vectors, a predicted label, whose range is from 0 to N, is generated by $Classifer_{obj}$. As a result, we obtain rectangle regions labeled by our interested object shape types of $1, 2, ..., N$ as shown in Fig. 3(f) by

$$Regions = \{< r, l > | r \in Rect \wedge 0 < l \leq N\} \tag{2}$$

where $|Regions|$ represents how many regular object types need to be classified, and *rect* denotes the position and size of an ROI in the 2D image.

5.2 Matching Point Cloud with Pre-defined 3D Meshes for Guiding Robotic Grasp

In this section, we introduce how to match a recognized visual object with a pre-defined 3D mesh for driving robust robotic grasp plan. We adopt FPFH

(Fast Point Feature Histograms) [10] as the descriptor for extracting features. FPFH is actually an efficient approximation of PFH and reserves the most useful information of the latter. Essentially, FPFH has 6-DOF invariance and is robust for neighbourhood noises at different sampling densities. FPFH thus is suitable for matching point clouds here (Fig. 4).

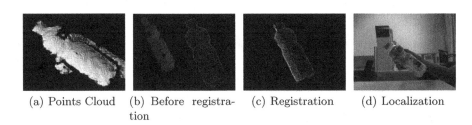

(a) Points Cloud   (b) Before registration   (c) Registration   (d) Localization

Fig. 4. We generate point cloud data from the depth information using Realsense, and then grasp regions on camera captured point cloud are guided for robots. (a) point clouds after segmentation and classification results, (b) the point cloud colored green denotes the pre-defined 3D cup mesh after downsampling, while the white shape represents the point cloud obtained by the RGB-D camera, and (c) gives the shape matching process. The grasping region is shown by yellow points in (d), and the blue rectangle show minimum bounding box of grasping region which we used to compare with ground truth. (Color figure online)

We first estimate the normals for each point both in the 3D mesh coordinate system and the world coordinate system. Specifically, for each pre-defined mesh, we estimate the normals for each point on it through its nearest neighbors. We set *radius* to search for the neighbours with an octree to speed up this process. With these normals, we then extract the FPFH features for each point on a 3D mesh. We then use camera calibration to obtain point cloud data for the segmented object from RGB-D data. Note that these point cloud data generally only contain partial 3D shape information since the back side of an object is always invisible for a RGB-D camera. Similarly, we calculate normals and extract FPFH for each point from the point cloud.

Next, we register the candidate 3D mesh with point cloud data using RANSAC [11] strategy, that is, matching FPFH descriptors to recover transform matrices, which consist of a rotation matrix R and a translation vector t. As a result, we get the rough registration and further adopt classic ICP with Euclidean distance to fine tune the registration results. The details of the method can be seen in Algorithm 2. Note that we use cosine distance to calculate the similarity between FPFH features extracted from each 3D mesh and real-time captured RGB-D data. Additionally, when the result rotate matrix R outputs $dig(1, 1, 1)$ and the translation vector equals to $(0, 0, 0)^T$, the matching process fails and a new viewpoint of the camera on the robot is required. With R and t, we gradually convert pre-defined grasp regions on a matched 3D mesh to point

---

Algorithm 2. Registration with pre-defined 3D meshes for guiding robotic grasp

---

Input:
        Point Cloud Region: $PointCloud$
        3D Object Models: $Model_{label}$
        Min Similarity: sim
        Max Acceptable Distance: mad
        Max Iterations: mi
        Acceptable Number: an
Output:
        Rotate Matrix: R, Translation Vector: t
  $R \leftarrow dig(1,1,1)$, $t \leftarrow (0,0,0)^T$
  repeat
    $mrs \leftarrow$ Random-Select($Model_{label}$)
    $pcrs \leftarrow$ Random-Select($PointCloud$)
    $similar\text{-}pairs \leftarrow \{<p1, p2>\ |Similarity(FPFH_{p1}, FPFH_{p2}) > sim \wedge p1 \in mrs \wedge p2 \in pcrs\}$
    $R, t \leftarrow$ Estimate-TransformMatrix($similar\text{-}pairs$)
    $inliers \leftarrow \{p|<p,*>\in similar\text{-}pairs\}$
    $alsoinliers \leftarrow \{p|min(distance(R*p+t,q), q \in pcrs) < mad \wedge p \in mrs - inliers\}$
  until $|alsoinliers| > an$ or Reach mi
  $R_{icp}, t_{icp} \leftarrow Classic\text{-}ICP(R * Model_{label}, PointCloud)$
  $R \leftarrow R_{icp} * R$
  $t \leftarrow R_{icp} * t + t_{icp}$
  return R, t

---

cloud captured from robot camera. The positions of a grasp point in the world coordinate can be expressed as follows:

$$position_{cloud} = \{\boldsymbol{p}_c | \boldsymbol{p}_c \leftarrow R\boldsymbol{p}_m + \boldsymbol{t}\} \tag{3}$$

where $\boldsymbol{p}_m = (x, y, z)^T$ represents the position of a grasping point on a 3D mesh.

Once the point cloud of the segmented object shape is successfully matched with a specific pre-defined 3D mesh, gp_{label} on $model_{label}$ will also be matched, which means grasp regions in the input image can be easily converted with the camera transform matrix for guiding robotic grasp.

6 Experimental Results and Discussion

To evaluate the effectiveness of the proposed framework, we collect five desktop objects as the dataset due to there are no such benchmark datasets comprising both camera captured images and scanned 3D models for guiding robotic grasp.

6.1 Dataset

We used an *Intel Realsense F200* camera on a Dell Inspiron 660 S with *Intel(R) Core(M) i3-2130 CPU@3.40 GHz*. Five regular desktop object types consist of

Cup, Can, Teapot, Box and Bottle are collected in our dataset; meanwhile, the 3D mesh of each object type is pre-scanned using a 3D scanner. We manually annotated grasp regions on each 3D mesh, which are considered as pre-defined knowledge for guiding automatic grasp in the proposed method (see the red regions of the third row in Fig. 1). The groundtruth for evaluating is composed of both object category and grasping region.

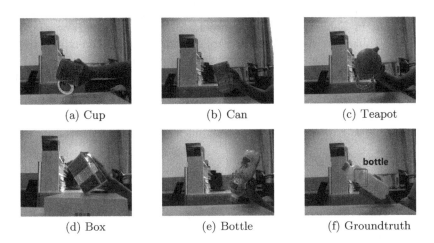

(a) Cup   (b) Can   (c) Teapot

(d) Box   (e) Bottle   (f) Groundtruth

Fig. 5. (a)-(e) show the grasping region by our method. Yellow: predicated regions for grasping, the bounding box for each region is also annotated. Note that (f) gives the groundtruth example for Bottle, the bounding box of which is marked red. (Color figure online)

6.2 Evaluation Criteria

As shown in Fig. 5(f), grasping groundtruth are manually marked by red rectangles, while the blue rectangles are the automatically calculated results using the proposed framework. To evaluate the results, we define the following criteria using Jaccard Similarity:

$$J = \frac{|P \bigcap G|}{|P \bigcup G|} \tag{4}$$

which P is the predicted region by algorithm, G is groundtruth and $|P|$ represents the area of rectangle P.

6.3 Results

In our experiment, we adopt different registration methods for comparisons. We select the classic ICP, With-Normals ICP that exploits a transformation estimated based on Point to Plane distances, and Non-linear ICP which is ICP variant that uses Levenberg-Marquardt optimization backend. We random select about 1/3 samples in our dataset to test and the left samples we utilize to train the HOG-SVM classifier for object classification. Then we take *Jaccard Similarity* as evaluation criteria to test the performance of our algorithm. In object classification step, we train a SVM classifier with train data set. Before registration, we classify detected object with SVM classifier. If the SVM classifies the sample different from the category groundtruth, (e.g. if the category is *bottle*, but the SVM output *can*) we decide *Jaccard Similarity* of sample as 0. The first columns in Table 1 gives the accuracies of object classification. The left columns in Table 1 are the average *Jaccard Similarity* with different registration algorithms. It can be found that the proposed algorithm achieves higher *Jaccard Similarity* results comparing with other ICP based methods, which shows that the proposed method is effective for automatic robotic grasp.

Table 1. Experiment of comparison different registration methods

Category	Test sample	Classification accuracy	Jaccard(Average)			
			Proposed method	ICP-classic	ICP-withnormals	ICP-nonlinear
Bottle	116	94.48%	0.657	0.309	0.307	0.316
Box	111	92.72%	0.619	0.225	0.241	0.289
Can	98	89.76%	0.597	0.445	0.372	0.329
Teacup	54	94.44%	0.695	0.042	0.132	0.177
Teapot	32	90.63%	0.658	0.165	0.237	0.215

7   Conclusion

We propose a novel framework for guiding robotic grasp plan by combing both RGB-D camera videos and pre-defined knowledge in 3D meshes. A new algorithm is proposed to match a segmented RGB-D shape with predefined 3D mesh to guide robotic self-location and grasp. Experimental results show the effectiveness of the proposed framework based on our collected dataset. To the best of our knowledge, this is the first work for exploring grasp planning for robots by combining both RGB-D data and 3D meshes to overcome the inherent difficulties such as viewpoint and distance variations of vision based methods. Our future work includes include more object types for robust robotic grasping.

Acknowledgment. The work described in this paper was supported by the Natural Science Foundation of China under Grant No. 61672273, No. 61272218 and No. 61321491, the Science Foundation for Distinguished Young Scholars of Jiangsu under Grant No. BK20160021.

References

1. Han, J., Shao, L., Xu, D., Shotton, J.: Enhanced computer vision with microsoft kinect sensor: a review. IEEE Trans. Cybern. **43**(5), 1318–1334 (2013)
2. Cheng, H., Chen, H., Liu, Y.: Topological indoor localization and navigation for autonomous mobile robot. IEEE Trans. Autom. Sci. Eng. **12**(2), 729–738 (2015)
3. Henry, P., Krainin, M., Herbst, E., Ren, X.F., Fox, D.: RGB-D mapping: using kinect-style depth cameras for dense 3D modeling of indoor enviroments. **31**(5), 647–663 (2012)
4. Prusak, A., Melnychuk, O., Roth, H., Schiller, I., Koch, R.: Pose estimation and map building with a time-of-flight-camera for robot navigation. Int. J. Intell. Syst. Technol. Appl. **5**, 355–364 (2008)
5. Peasley, B., Birchfield, S.: RGBD point cloud alignment using Lucas-Kanade data association and automatic error metric selection. IEEE Trans. Robot. **31**(6), 1–7 (2015)
6. Bergamasco, F., Cosmo, L., Albarelli, A., Torsello, A.: A robust multi-camera 3D ellipse fitting for contactless measurement. In: International Conference on 3D Imaging, Modeling, Processing, Visualization and Transmission, pp. 168–175 (2012)
7. Dalal, N., Triggs, B.: Histograms of oriented gradients for human detection, computer vision and pattern recognition (2005)
8. Chapelle, O., Haffner, P., Vapnik, N.: Support vector machines for histogram-based image classification (1995)
9. Snyder, E., Cowart, E.: An iterative approach to region growing using associative memories. IEEE Trans. Pattern Anal. Mach. Intell. **5**(3), 349–352 (1983)
10. Rusu, R.B., Blodow, N., Beetz, M.: Fast point feature histograms (FPFH) for 3D registration. In: ICRA, pp. 3212–3217 (2009)
11. Fischler, M.A., Bolles, R.C.: Random sample consensus: a paradigm for model fitting with applications to image analysis and automated cartography. Commun. ACM **24**(6), 381–395 (1981)
12. Zaharescu, A.: An object grasping literature survey in computer vision and robotics
13. Saponaro, G.: Pose estimation for grasping preparation from stereo ellipses
14. Horaud, R., Dufournaud, Y., Long, Q.: Robot stereo-based coordination for grasping cylindric parts
15. Jiang, Y., Moseson, S., Saxena, A.: Efficient grasping from RGBD images: learning using a new rectangle representation. In: 2011 IEEE International Conference on Robotics and Automation (ICRA), pp. 3304–3311. IEEE (2011)
16. Montesano, L., Lopes, M.: Active learning of visual descriptors for grasping using non-parametric smoothed beta distributions. Robot. Auton. Syst. **60**(3), 452–462 (2012)
17. Saxena, A., Driemeyer, J., Ng, A.Y.: Robotic grasping of novel objects using vision. Int. J. Robot. Res. **27**(2), 157–173 (2008)
18. Lenz, I., Lee, H., Saxena, A.: Deep learning for detecting robotic grasps. Int. J. Robot. Res. **34**(4–5), 705–724 (2015)

What Convnets Make for Image Captioning?

Yu Liu[(✉)], Yanming Guo, and Michael S. Lew

LIACS Media Lab, Leiden University, Leiden, The Netherlands
{y.liu,y.guo,m.s.lew}@liacs.leidenuniv.nl

Abstract. Nowadays, a general pipeline for the image captioning task takes advantage of image representations based on convolutional neural networks (CNNs) and sequence modeling based on recurrent neural networks (RNNs). As captioning performance closely depends on the discriminative capacity of CNNs, our work aims to investigate the effects of different Convnets (CNN models) on image captioning. We train three Convnets based on different classification tasks: single-label, multi-label and multi-attribute, and then feed visual representations from these Convnets into a Long Short-Term Memory (LSTM) to model the sequence of words. Since the three Convnets focus on different visual contents in one image, we propose aggregating them together to generate a richer visual representation. Furthermore, during testing, we use an efficient multi-scale augmentation approach based on fully convolutional networks (FCNs). Extensive experiments on the MS COCO dataset provide significant insights into the effects of Convnets. Finally, we achieve comparable results to the state-of-the-art for both caption generation and image-sentence retrieval tasks.

Keywords: Image captioning · Convolutional neural networks · Aggregation module · Long short-term memory · Multi-scale testing

1 Introduction

Image captioning is a fundamental and important task in vision-to-language research. It aims to describe an image with descriptive and meaningful sentence-level captions. The automatically generated descriptions should cover the salient content in images, including objects, actions and other relations. In early research of image captioning, it has been converted to a retrieval-based task. Those retrieval-based approaches [7,15,21] focus on mapping images to sentences based on pre-defined captions. However, they fail to generate novel sentences for unseen scenes. To address this issue, generative approaches are developed to estimate novel sentences, such as Midge [20] and Baby Talk [14].

Recently, a new paradigm for image captioning is proposed in many state-of-the-art approaches [3,12,19,26,29]. This paradigm primarily integrates a convolutional neural network (CNN) and a recurrent neural network (RNN) together. CNN is used to capture high-level image features, and RNN then generates a

---

The first two authors contributed equally to this work.

© Springer International Publishing AG 2017
L. Amsaleg et al. (Eds.): MMM 2017, Part I, LNCS 10132, pp. 416–428, 2017.
DOI: 10.1007/978-3-319-51811-4_34

sequence of words based on the image features. Typically, a rich visual representation contributes much to generating accurate image captions. However, some Convnets (CNN models) are originally trained for image classification [13,24], but not for image captioning. It thus raises an important question: *What Convnets make for image captioning?*

Our aim in this paper is to fully investigate the effects of different Convnets on image captioning. We exploit three kinds of Convnets: single-label Convenet, multi-label Convnet, and multi-attribute Convnet. (1) A single-label Convnet indicates a CNN model pre-trained on ImageNet dataset [23], such as AlexNet [13] and VGG-16 [24]. This Convnet is often used to provide one generic image representation. (2) A multi-label Convnet can predict multiple class labels given one image. It is consistent with the observation that sentence-level captions often talk about many salient objects jointly in images. Therefore, we fine-tune a multi-label Convnet on the MS COCO dataset [17] that consists of 80 object categories. Each image is annotated with multiple object labels. (3) A multi-attribute Convnet can not only reflect multiple object classes, but also describe actions and other relations about objects, for example jumping, sitting and interacting. As the attributes reflect more key words in the descriptive captions, a multi-attribute Convnet is able to narrow the gap between vision and language. We fine-tune a multi-attribute Convnet based on 300 attributes derived from MS COCO captions [17].

By observing the feature maps learned in the three Convnets, we find that their maps focus on different visual fields in images. Therefore, we propose aggregating their features together to generate a richer representation. In addition, during the test stage, we take advantage of the efficient fully convolutional networks (FCNs) [18] for multi-scale augmentation. We use two scales of FCNs that are interpreted from one pre-trained CNN. This augmentation approach can be applied to both the single Convnet and multi-Convnet aggregation. Finally, we employ the Long Short-Term Memory (LSTM) [6] to build the language model. Figure 1 shows an image captioning example from the MS COCO dataset [17]. Note that the visual feature is fed to the LSTM unit at each time step.

 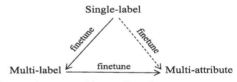

Single-label Convnet: A man is sitting on the water with a surfboard.
Multi-label Convnet: A man sitting on a boat in front of a boat.
Multi-attribute Convnet: A man and a dog on a boat.
Multi-Convnet aggregation: A man and a dog on a small boat.
Ground truth: A man and a dog on a small yellow boat.

Fig. 1. Example of image captioning using different Convnets. Each Convnet shows meaningful description. As compared to the human-written ground-truth, the multi-Convnet can generate closer result than any single Convnet.

In a nutshell, our contributions can be summarized as follows:

- We present a full comparison among the three Convnets for the image captioning task. We study the benefits of each Convnet and then integrate multiple Convnets for a richer visual representation. Our work can provide promising insights into deeply diagnosing and understanding Convnets for vision-to-language tasks.
- We employ an efficient multi-scale augmentation approach using FCNs.
- We achieve comparable results to the state-of-the-art on the MS COCO dataset, both for caption generation and image-sentence retrieval tasks.

2   Related Work

In this section, we briefly summarize related image captioning approaches based on CNN-RNN as below.

A prior work in NIC [26] employed a CNN-RNN scheme to model the image captioning problem. CNNs are used as the "encoder" to visually represent the input image with a fixed-length feature vector. Then RNNs, as the "decoder", can translate the feature vector into sentence-level captions. Similarly, other similar approaches [3,19,27] followed this CNN-RNN paradigm. Instead of only using CNN features, Jia et al. [8] added extra semantic information to each unit of the LSTM block. Jin et al. [10] integrated scene-specific contexts in order to highlight higher-level semantic information in images. In addition, Xu et al. [29] introduced a visual attention based model inspired by human visual system. The attention mechanism can automatically learn latent alignments between regions and words. Apart from the whole image captioning, there were some works focusing on region-level captioning [4,11,12]. They first localized salient regions in images and then described them with natural language.

Recent work in [31] began capturing attributes to represent visual content. Yao et al. [30] investigated the performance upper bounds based on attributes for image and video captioning. However, both of these works did not train a new CNN model based on attributes. The most similar work in [28] fine-tuned a CNN based on the task of image-attribute classification. Nevertheless, our work had several main differences from [28]:

First, we intended to add a multi-label Convnet as a bridge from a single-label to a multi-attribute Convnet (see the two solid lines in Fig. 1). Thus our multi-attribute Convnet had two-stage fine-tuning. In contrast, [28] directly fine-tuned a multi-attribute Convnet from a single-label Convnet (see the dash line in Fig. 1), and failed to study the effects of a multi-label Convnet. Second, we further evaluated the aggregation of multiple Convnets that has not been studied previously in [28]. Third, we presented an efficient multi-scale testing approach based on fully convolutional networks, as compared to using expensive region proposals in [28].

3 Proposed Approach

In this section, we will present our image captioning system in three aspects. First, we show the usage of single Convnet for capturing visual representation. Second, we find that integrating image features from three Convnets is beneficial for improving visual representation. Third, at the test stage, we use a multi-scale testing approach based on FCNs.

3.1 Convnets for Image Captioning

This part introduces the training details about the three Convnets. Notably, the multi-attribute Convnet also belongs to a multi-label classification task, but it has different training from the multi-label Convnet.

Single-Label Convnet. CNNs trained on ImageNet dataset [23] are generally used as off-the-shelf feature extractors, for instance Alexnet [13] and VGG-16 [24]. We call them as single-label Convnets, since they are originally trained for single-label classification, for example 1000 object classes in ImageNet 2012. Here we use the VGG-16 net as a single-label Convnet for our image captioning system. As the left part in Fig. 2, an image from MS COCO [17] is fed to a single-class Convnet that outputs a 1000-Dim visual feature.

Multi-label Convnet. An image caption often mentions a couple of objects in images, instead of targeting at only one salient object. We thus train a multi-label Convnet on the MS COCO dataset [17] that consists of 80 object categories. Each image in MS COCO is annotated by about 3 object labels on average. Instead of training from scratch, we fine-tune the single-label Convnet for a multi-label recognition task. We replace the original 1000-way layer with 80-way layer. The sigmoid cross-entropy function is used to compute the element-wise loss. Assume that there are K classes (e.g. 80), the total cost sums up K of sigmoid losses by

$$l_1(x) = -\sum_{k=1}^{K} y_k(x) \log p_k(x) + (1 - y_k(x)) \log(1 - p_k(x)), \tag{1}$$

where $y_k \in \{0, 1\}$ is the ground-truth label indicating the absence or presence of the category k in the input image x. $P_k(x)$ indicates the prediction probability of containing the category k. During fine-tuning, the parameters of the last fully-connected layer (i.e. the multi-class prediction layer) are initialized with gaussian filters. We initialize the learning rate of the last fully-connected layer with 0.01. The learning rates of other convolutional layers and fully-connected layers (i.e. fc6 and fc7) are initialized with 0.0001 and 0.001, respectively. The learning rate is divided by 10 after 2×10^4 iterations. The whole training will be terminated after 5×10^4 iterations. In addition, we use a weight decay of 0.0001, a momentum of 0.9, and a mini-batch size of 100. The multi-label Convnet is shown in the middle part in Fig. 2.

Multi-attribute Convnet. Apart from object categories, a descriptive caption should mention more information like actions (e.g. sit, run) and other relations

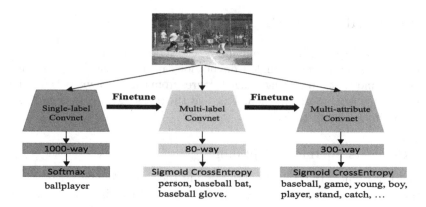

Fig. 2. Illustration of the three Convnets for visual representations. The multi-label Convnet is fine-tuned from the pre-trained single-label Convnet. The multi-attribute Convnet performs two-stage fine-tuning.

(e.g. blue, small). Hence, using a Convnet that can reflect more attributes is beneficial for narrowing the gap between visual features and language words. Based on a multi-label Convnet, we further fine-tune a multi-attribute Convnet. First, we build an attribute dictionary based on MS COCO captions dataset. In [30], they summarize three groups of atoms: entity, action and attribute. We select top-100 atoms from each group, therefore, our attribute dictionary consists of 300 words (or attributes) in total. Note that the atoms defined in [30] are renamed as attributes in our work. Then, we remake the topmost layer with a 300-way fully-connected layer, as shown in the right part in Fig. 2. Assume that G denotes the number of attributes (e.g. $G = 300$). Similarly, we compute the sigmoid cross-entropy loss by

$$l_2(x) = - \sum_{g=1}^{G} y_g(x) \log p_g(x) + (1 - y_g(x)) \log(1 - p_g(x)), \qquad (2)$$

where $y_g \in \{0, 1\}$ is the ground truth; $P_g(x)$ is the prediction probability. Since each image in MS COCO has five human-written captions, we merge five captions together to generate the ground-truth. During fine-tuning the multi-attribute model, we use the same hyper-parameters as training the multi-label CNN.

To compare the visual features from the three Convnets, we visualize their most activated feature maps learned in the fifth convolutional layer (i.e. conv5_3), as illustrated in Fig. 3. Here, we regard the feature map which has the largest average activation value as the most activated feature map. It can be seen that the three Convnets focus on different visual fields in images. This provides deep insights into diverse characteristics of the three Convnets.

| Input image | Single-label Convnet | Multi-label Convnet | Multi-attribute Convnet |

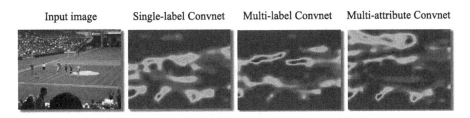

Fig. 3. Visualization of feature maps learned in the three Convnets. We select the most activated feature map in conv5_3. We can see that the three Convnets focus on different visual fields in images due to their different classification objectives.

3.2 Multi-convnet Aggregation

Since three Convnets are trained for different classification objectives and can represent different features given the input image, we propose aggregating them together to compensate the deficiency of any single Convnet feature. Although a multi-attribute Convnet may contain the same objects as in a single-label and multi-label Convnet, the aggregation feature can still improve the accurate prediction of object classes. Figure 4 illustrates the pipeline of our multi-Convnet aggregation approach.

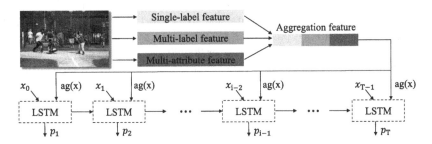

Fig. 4. The pipeline of image captioning based on multi-Convnet aggregation. The three Convnet features are concatenated together to generate an aggregation feature $ag(x)$. At each time step, both a word x_i and $ag(x)$ are fed to the LSTM unit whose output is a probability distribution for the next word.

First, the input image x is fed to three pre-trained Convnets to capture separate visual features, denoted as $sc(x), mc(x), ma(x)$. We concatenate three kinds of features to create an aggregation feature $ag(x)$ (i.e. 1380-Dim vector), where $ag(x) = [sc(x), mc(x), ma(x)]$. Then, we add this aggregation feature to the following RNN unit at each time step. We employ one-layer Long Short-Term Memory (LSTM) [6] that can alleviate the vanishing gradient problem due to its gates mechanism. Finally, at the time step t, the formulation of LSTM units with an aggregation feature can be summarized as below

$$i_t = \sigma(W_{xi}x_t + W_{vi}ag(x) + W_{hi}h_{t-1} + b_i) \tag{3}$$

$$f_t = \sigma(W_{xf}x_t + W_{vf}ag(x) + W_{hf}h_{t-1} + b_f) \tag{4}$$

$$o_t = \sigma(W_{xo}x_t + W_{vo}ag(x) + W_{ho}h_{t-1} + b_o) \tag{5}$$

$$g_t = \phi(W_{xg}x_t + W_{vg}ag(x) + W_{hg}h_{t-1} + b_g) \tag{6}$$

$$c_t = f_t \odot c_{t-1} + i_t \odot g_t \tag{7}$$

$$h_t = o_t \odot \phi(c_t) \tag{8}$$

$$p_{t+1} = Softmax(h_t) \tag{9}$$

where W and b are the weight matrices and bias terms. We refer to x_t as the input word at time step t for image x. σ and ϕ are the sigmoid and tangent activation functions. p_{t+1} is used to predict the probability distribution for the next word. Finally, the objective in LSTMs for language modeling is to minimize the following loss cost

$$-\sum_{t=0}^{T-1} \log p_t(x_{t+1}|x_t, ag(x)) + \lambda||W||_2^2 \tag{10}$$

where T is the length of the input sequence of words, and λ indicates the weight decay (Here, we follow the configuration of [3] and set λ equals 0). For notational simplicity, we just give the computation of one input image and drop the mini-batch size in the formulation. Following the hyper-parameters in [3], both the word embedding size and hidden state size are set to 1000. We use a mini-batch size of 100 image-sentence pairs. The learning rate is initialized with 0.01 and decreases to one tenth of current rate after 20,000 iterations. The whole training will be terminated after 80,000 iterations. In addition, we use a momentum of 0.9 and clip gradients of 10.

3.3 Multi-scale Testing

During the test phase, we intend to use a multi-scale augmentation approach to capture a more robust image representation, as shown in Fig. 5. We first extract a feature vector by inputting a 224×224 image to CNNs. Then, we convert one CNN model to a fully convolutional networks (FCN) [18]. FCN is quite efficient to compute regions based predictions without decreasing the ease of testing. Following [24], we set a smaller side to S and isotropically resize the other side. Here we use two scales of images, including $S = 256$ and 320. A 256×256 size can generate 2×2 feature maps at the topmost layer of FCN. We perform average pooling over these feature maps that result in another feature vector. Similarly, FCN can get one more image feature given 320×320 image size. Finally, the multi-scale feature is computed by averaging one CNN feature and two FCN features. Notably, the multi-scale testing can be used for both single Convnet and multi-Convnet aggregation. We also test more scales such as $S = 384, 512$, but no significant improvement is obtained.

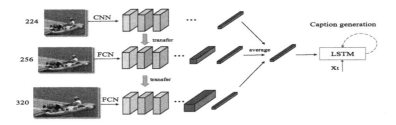

Fig. 5. The pipeline of multi-scale testing approach. Apart from the basic CNN feature, we use two extra scales based on FCNs. We compute the average over three feature vectors and feed it to LSTM units for caption generation.

4 Experiments

In this section, we evaluate our approach on the well-known MS COCO dataset [17] that consists of 82783 training images, 40504 validation images and 40775 testing images. Each image is annotated by at least five human-written captions. Following most recent works [3,12,27,28], we use 5000 images as validation set to tune hyper-parameters, and another 5000 images as test set to report results. We use the vocabulary dictionary in [3] (containing 8800 words). This dictionary is used to encode the input sequence of words. We implemented our approach based on the Caffe framework [9].

4.1 Evaluation Configuration

We evaluate our approaches on two tasks: caption generation and image-sentence retrieval. (1) For caption generation task, we evaluate our method with four metrics: BLEU [22], METEOR [2], ROUGE-L [16] and CIDEr [25]. Most of our results use a beam search of size 1 for fast evaluating. For fair comparison with the state-of-the-art, we give the results by using a beam search of size 5. (2) Image-sentence retrieval task consists of image-to-sentence retrieval and sentence-to-image retrieval. We adopt the evaluation metrics: R@K and Med r [3,12]. All metrics are computed with the public available MS COCO evaluation code [1].

We denote the three single Convnets as SL-Net, ML-Net and MA-Net. MA_ML-Net is the combination of MA-Net and ML-Net, and MA_ML_SL-Net indicates the method that aggregates the three Convnets.

4.2 Results on Caption Generation

We evaluate our approach on caption generation with 5000 test images. Table 1 shows the single-scale and multi-scale testing of three Convnets. We list the dimension of the feature vector since it is closely related with the number of LSTM parameters. It is interesting to see that, SL-Net, which utilizes the largest

Table 1. MS COCO results on caption generation by comparing three Convnets. Both single-scale and multi-scale testing are shown. Here we use a beam search of size 1.

Method	Dim	B-1	B-2	B-3	B-4	METEOR	ROUGE-L	CIDEr
Single-scale testing								
SL-Net	1000	0.651	0.474	0.333	0.229	0.214	0.483	0.703
ML-Net	80	0.664	0.487	0.345	0.241	0.213	0.487	0.717
MA-Net	300	0.686	0.516	0.374	0.266	0.228	0.506	0.796
Multi-scale testing								
SL-Net	1000	0.666	0.489	0.345	0.239	0.219	0.489	0.735
ML-Net	80	0.679	0.496	0.351	0.245	0.219	0.49	0.75
MA-Net	300	0.697	0.528	0.384	0.274	0.231	0.511	0.81

dimension feature, performs the worst among the three Convnets. This demonstrates that increasing the number of system parameters would not necessarily improve the performance.

For single-scale testing, ML-Net brings about 1% boost over the SL-Net for most evaluation metrics. This improvement is marginal compared to the MA-Net, which outperforms the SL-Net significantly over all the evaluation metrics. Notably, the boost of CIDEr shows 0.093, from 0.703 to 0.796. On the other hand, the multi-scale testing using FCN shows considerable improvement over the counterpart single-scale testing, with the same feature dimension. This verifies the efficiency and effectiveness of the multi-scale augmentation technique.

In addition to evaluating the three Convnets individually, we also explore the effect of aggregating the Convnets, as shown in Table 2. We build the multi-Convnet based on MA-Net since it is the best individual Convnet. Overall, both MA_ML-Net and MA_MC_SC-Net perform better than the individual MA-Net,

Table 2. MS COCO results on caption generation by multi-Convnet aggregation. The results are based on BLEU, METEOR (M), ROUGE-L (R) and CIDEr (C) metrics. Here we use a beam search of size 1.

Method	B-1	B-2	B-3	B-4	M	R	C
Single-scale testing							
MA-Net	0.686	0.516	0.374	0.266	0.228	0.506	0.796
MA_ML-Net	0.687	0.519	0.376	0.268	0.229	0.507	0.797
MA_ML_SL-Net	0.688	0.52	0.379	0.27	0.229	0.507	0.803
Multi-scale testing							
MA-Net	0.697	0.528	0.384	0.274	0.231	0.511	0.81
MA_ML-Net	0.703	0.537	0.393	0.282	0.234	0.516	0.846
MA_ML_SL-Net	0.704	0.54	0.398	0.287	0.236	0.519	0.848

indicating that aggregating the Convnet is beneficial for improving caption generation accuracy. This is reasonable given the fact that different Convnets focus on learning different contents, and aggregating them generally leads to a more comprehensive prediction. Furthermore, we also evaluate the multi-scale performance using FCN. Similarly, the multi-scale scheme improves the accuracy of the evaluation metric remarkably. Finally, MA_MC_SC-Net can yield a quite competitive result, such as 0.704 B-1 and 0.846 CIDEr.

Comparison with the State-of-the-Art. We compare our MA_MC_SC-Net result with current state-of-the-art methods in Table 3. It can be seen that our results delivered better results than most existing methods. Compared to [29], our method obtained the same result on Bleu-1 with the soft-attention model, slightly worse than the more sophisticated hard-attention model. But for all the other evaluation metrics, our method achieved considerably better results. Similar situation comes with [31], with which we also achieved overall competitive performance. It is worthwhile to say that, our method is not inherently conflicted with these methods (e.g. attention mechanism), and we can incorporate them together for a better achievement. Note that [28] further improved their results by extracting, clustering and selecting a large number of region proposals. Therefore, their great gains are achieved at the expense of algorithm complexity. In contrast, benefited from the high efficiency of FCNs, our multi-scale testing strategy brings negligible extra cost compared to the single-scale testing. We argue that a sophisticated region detection approach [5] is also applicable to our system, but it is out of the scope of this work. Figure 6 shows some captioning examples.

Table 3. Comparison with current state-of-the-art on MS COCO caption generation. Here we use a beam search of size 5.

Method	B-1	B-2	B-3	B-4	M	C
Karpathy et al. [12]	0.625	0.450	0.321	0.230	0.195	0.66
mRNN [19]	0.670	0.490	0.350	0.250	-	-
NIC [26]	-	-	-	0.277	0.237	0.855
LRCN [3]	0.669	0.489	0.349	0.249	-	-
gLSTM [8]	0.670	0.491	0.358	0.264	0.227	0.813
Bi-LSTM [27]	0.672	0.492	0.352	0.244	0.208	0.666
VNet-ft-LSTM [28]	0.680	0.500	0.370	0.250	0.220	0.730
Soft-Attention [29]	0.707	0.492	0.344	0.243	0.239	-
Hard-Attention [29]	**0.718**	0.504	0.357	0.250	0.230	-
Jin et al. [10]	0.697	0.519	0.381	0.282	0.235	0.838
ATT-FCN [31]	0.709	0.537	0.402	**0.304**	**0.243**	-
Ours	0.707	**0.548**	**0.410**	**0.304**	0.238	**0.895**

Ours: A man riding a wave in the ocean.
GT: A man riding a wave on a surfboard in the ocean.

Ours: A living room with a lot of furniture.
GT: Living room with furniture with garage door at one end.

Ours: A man riding a horse at a horse.
GT: A horse that threw a man off a horse.

Ours: A man holding a frisbee in a field.
GT: The man is holding the string to fly his kite.

Fig. 6. The caption generation results for some MS COCO examples by our MA_MC_SC-Net method. We show both the positive and negative examples.

4.3   Results on Image-Sentence Retrieval

We report the image-to-sentence and sentence-to-image results in Table 4. There are 5000 test images and 25,000 captions in total. Overall, our method (i.e. MA_MC_SC-Net) outperforms other state-of-the-art on both R@K and Med r.

Table 4. Image-sentence retrieval results on the MS COCO dataset. R@K: higher is better; Med r: lower is better.

Method	Image to sentence				Sentence to image			
	R@1	R@5	R@10	Med r	R@1	R@5	R@10	Med r
Karpathy et al. [12]	16.5	39.2	52.0	9.0	10.7	29.6	42.2	14.0
Bi-LSTM [27]	16.6	39.4	52.4	9.0	11.6	30.9	43.4	13.0
Ours	**16.9**	**39.8**	**53.1**	**8.0**	**12.4**	**31.5**	**44.0**	**12.0**

5   Conclusion

In this work, we studied the effects of Convnets for the image captioning task. We employed three Convnets based on single-label, multi-label, and multi-attribute classification. In addition, we integrated the three Convnets for a richer aggregation feature. During the test stage, we employed an efficient multi-scale augmentation approach. Experiments on the MS COCO dataset demonstrated that our approach achieved competitive results for both caption generation and image-sentence retrieval as compared to the state-of-the-art. In future work, we will strive to make use of the attention mechanism.

Acknowledgments. This work was supported mainly by the LIACS Media Lab at Leiden University and in part by the China Scholarship Council. We would like to thank NVIDIA for the donation of GPU cards.

References

1. Chen, X., Fang, H., Lin, T., Vedantam, R., Gupta, S., Dollár, P., Zitnick, C.L.: Microsoft COCO captions: data collection and evaluation server. CoRR abs/1504.00325 (2015)
2. Denkowski, M., Lavie, A.: Meteor universal: language specific translation evaluation for any target language. In: EACL (2014)
3. Donahue, J., Anne Hendricks, L., Guadarrama, S., Rohrbach, M., Venugopalan, S., Saenko, K., Darrell, T.: Long-term recurrent convolutional networks for visual recognition and description. In: CVPR (2015)
4. Fang, H., Gupta, S., Iandola, F., Srivastava, R.K., Deng, L., Dollar, P., Gao, J., He, X., Mitchell, M., Platt, J.C., Lawrence Zitnick, C., Zweig, G.: From captions to visual concepts and back. In: CVPR (2015)
5. Girshick, R., Donahue, J., Darrell, T., Malik, J.: Rich feature hierarchies for accurate object detection and semantic segmentation. In: CVPR (2014)
6. Hochreiter, S., Schmidhuber, J.: Long short-term memory. Neural Comput. 9(8), 1735–1780 (1997)
7. Hodosh, M., Young, P., Hockenmaier, J.: Framing image description as a ranking task: data, models and evaluation metrics. J. Artif. Intell. Res. 47(1), 853–899 (2013)
8. Jia, X., Gavves, S., Fernando, B., Tuytelaars, T.: Guiding long-short term memory for image caption generation. In: ICCV (2015)
9. Jia, Y., Shelhamer, E., Donahue, J., Karayev, S., Long, J., Girshick, R., Guadarrama, S., Darrell, T.: Caffe: convolutional architecture for fast feature embedding. In: ACM Multimedia (2014)
10. Jin, J., Fu, K., Cui, R., Sha, F., Zhang, C.: Aligning where to see and what to tell: image caption with region-based attention and scene factorization. CoRR abs/1506.06272 (2015)
11. Johnson, J., Karpathy, A., Fei-Fei, L.: Densecap: fully convolutional localization networks for dense captioning. In: CVPR (2016)
12. Karpathy, A., Fei-Fei, L.: Deep visual-semantic alignments for generating image descriptions. In: CVPR (2015)
13. Krizhevsky, A., Sutskever, I., Hinton, G.E.: Imagenet classification with deep convolutional neural networks. In: NIPS (2012)
14. Kulkarni, G., Premraj, V., Dhar, S., Li, S., Choi, Y., Berg, A.C., Berg, T.L.: Baby talk: understanding and generating image descriptions. In: CVPR (2011)
15. Kuznetsova, P., Ordonez, V., Berg, A.C., Berg, T.L., Choi, Y.: Collective generation of natural image descriptions. In: ACL (2012)
16. Lin, C.Y.: Rouge: a package for automatic evaluation of summaries. In: ACL Workshop (2004)
17. Lin, T.-Y., Maire, M., Belongie, S., Hays, J., Perona, P., Ramanan, D., Dollár, P., Zitnick, C.L.: Microsoft COCO: common objects in context. In: Fleet, D., Pajdla, T., Schiele, B., Tuytelaars, T. (eds.) ECCV 2014. LNCS, vol. 8693, pp. 740–755. Springer, Heidelberg (2014). doi:10.1007/978-3-319-10602-1_48
18. Long, J., Shelhamer, E., Darrell, T.: Fully convolutional networks for semantic segmentation. In: CVPR (2015)
19. Mao, J., Xu, W., Yang, Y., Wang, J., Huang, Z., Yuille, A.: Deep captioning with multimodal recurrent neural networks (m-RNN). In: ICLR (2015)
20. Mitchell, M., Dodge, J., Goyal, A., Yamaguchi, K., Stratos, K., Han, X., Mensch, A., Berg, A., Berg, T., Daume III, H.: Midge: generating image descriptions from computer vision detections. In: EACL (2012)

21. Ordonez, V., Kulkarni, G., Berg, T.L.: Im2Text: describing images using 1 million captioned photographs. In: NIPS (2011)
22. Papineni, K., Roukos, S., Ward, T., Zhu, W.J.: Bleu: a method for automatic evaluation of machine translation. In: ACL (2002)
23. Russakovsky, O., Deng, J., Su, H., Krause, J., Satheesh, S., Ma, S., Huang, Z., Karpathy, A., Khosla, A., Bernstein, M., Berg, A.C., Fei-Fei, L.: Imagenet large scale visual recognition challenge. IJCV 115, 211–252 (2015)
24. Simonyan, K., Zisserman, A.: Very deep convolutional networks for large-scale image recognition. In: ICLR (2015)
25. Vedantam, R., Zitnick, C.L., Parikh, D.: Cider: consensus-based image description evaluation. In: CVPR (2015)
26. Vinyals, O., Toshev, A., Bengio, S., Erhan, D.: Show and tell: a neural image caption generator. In: CVPR (2015)
27. Wang, C., Yang, H., Bartz, C., Meinel, C.: Image captioning with deep bidirectional LSTMs. In: ACM Multimedia (2016)
28. Wu, Q., Shen, C., Liu, L., Dick, A., van den Hengel, A.: What value do explicit high level concepts have in vision to language problems? In: CVPR (2016)
29. Xu, K., Ba, J., Kiros, R., Cho, K., Courville, A.C., Salakhutdinov, R., Zemel, R.S., Bengio, Y.: Show, attend and tell: neural image caption generation with visual attention. In: ICML (2015)
30. Yao, L., Ballas, N., Cho, K., Smith, J.R., Bengio, Y.: Empirical upper bounds for image and video captioning. In: ICLR (2016)
31. You, Q., Jin, H., Wang, Z., Fang, C., Luo, J.: Image captioning with semantic attention. In: CVPR (2016)

What are Good Design Gestures?
–Towards User- and Machine-friendly Interface–

Ryo Kawahata, Atsushi Shimada$^{(\boxtimes)}$, and Rin-ichiro Taniguchi

Kyushu University, Fukuoka, Japan
atsushi@limu.ait.kyushu-u.ac.jp

Abstract. This paper discusses gesture design for man–machine interfaces. Traditionally, gesture-interface studies have focused on improving performance, in terms of increasing speed and accuracy, in particular reducing false positives. Many studies neglect to consider the gestures' intrinsic machine friendliness, which can improve recognition accuracy, and user friendliness, which makes a gesture easier to use and to remember. In this paper, we investigate machine- and user-friendly gestures and analyze the results of an Internet-based questionnaire in which 351 individuals were asked to assign gestures to eight operations.

Keywords: User interface · Gesture design · User-friendly gesture · Machine-friendly gesture

1 Introduction

Man–machine interfaces play an important role in conveying intentions from a user to a machine. The keyboard and mouse have traditionally provided an interface for personal computers, because they are easy to use and the inputs are easy for the computer to interpret. Recently, gesture interfaces based on computer vision techniques have been studied [2,9,11,12] towards the realization of perceptual user interface. For example, a computer vision-based method has been studied that recognizes gestures from hand motions or shapes detected by an RGB camera [3]. A primary drawback of computer vision approaches is that the sensing area restricts the user to a narrow region where gesture recognition is available.

Recently, wearable devices such as smartwatches have become popular worldwide. A smartwatch generally has a three-axis accelerometer and three-axis gyroscope, which enable accurate measurement of motions. The limited operating region of the computer vision approach can be easily solved using wearable devices because the user carries the sensor. Therefore, wearable devices are expected to become a new mode of gesture interface. Smartwatch motion analysis can be applied to the operation of the smartwatch itself [6], or smartphone applications [8].

Many studies related to gesture interfaces focus mainly on the accuracy of gesture recognition, paying little attention to gesture design. However, good

© Springer International Publishing AG 2017
L. Amsaleg et al. (Eds.): MMM 2017, Part I, LNCS 10132, pp. 429–440, 2017.
DOI: 10.1007/978-3-319-51811-4_35

gesture design is important for enhancing usability and reducing false recognition of gestures. A user-customizable gesture interface for a TV control system was proposed in [10]. A user freely selects a gesture (e.g., hand shape, hand motion) and assigns it to an operation (e.g., volume up or volume down). Through the self-assignment of gestures, it becomes easier for a user to remember the gestures. However, candidate gestures are designed by the researchers and are not assessed based on ease of recognition by a machine (system). Some gestures that involve motions that are used frequently in daily life will be falsely recognized as one assigned to an operation. Therefore, it is necessary to consider not only ease of remembering and performing the gesture (user friendliness) but also ease of gesture recognition without false positives (machine friendliness).

Kawahata et al. surveyed research tackling the false-positive problem [4]. Several approaches have been proposed, including using a button to notify gesture start [5], using low-false-positive gestures [7], and interactively designing of low-false-positive gestures [1]. Kawahata et al. also proposed an automated method to find low-false-positive gestures from daily motions. They prepared seven primitive actions, and, based on a random forest approach, they discovered two successive actions which rarely occur in daily life. They concluded that the gestures discovered by their approach would be suitable candidate gestures for man–machine interfaces. However, they did not investigate the candidate gestures' user friendliness.

Therefore, to the best of our knowledge, there is no study which discusses gesture design from both user- and machine-friendliness perspectives. In this paper, we analyze these points through a questionnaire completed by 351 individuals. Our proposed system first extracts low-false-positive gestures based on the random forest approach proposed by Kawahata et al. [4]. Similar to this previous study, we assume that each gesture consists of two primitive actions. Then, extracted gestures are evaluated by the proposed criteria explained previously 2, to make a recommendation ranking. The ranking is adjustable by controlling the weighting of the user-friendliness scores and the machine-friendliness score. We created three ranked lists of gestures that were then presented to three groups of individuals. We investigated the effect of ranking on gesture selection, the gesture types commonly selected, and the reasons respondents gave for selecting the gestures.

2   Gesture Recommendation

We start with an explanation of the recommendation score calculations for candidate gestures (please refer to [4] for the method how to extract candidate gestures). The proposed system presents a user with a ranked list of gestures. To rank the gestures, we have to define what makes a gesture better for a user or for a machine. For a user, a smaller movement is considered 'better' (and is assigned a lower score) than a longer one. Let p_i be a motion vector of a hand, where i indicates the label of primitive action (e.g., right or left). In the proposed system, we assume that a gesture consists of two successive primitive actions,

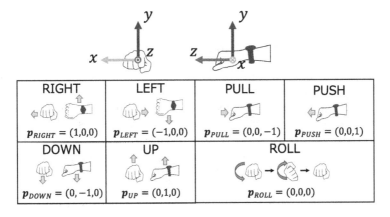

Fig. 1. Motion vector of each primitive action

so that the total movement can be represented as $||p_1|| + ||p_2||$. To evaluate the size of the movement, we introduce the score M defined as follows:

$$M = -(||p_1|| + ||p_2||) \tag{1}$$

We also introduce another score P to evaluate the difference between two positions, the start point and the end point of the gesture, defined as:

$$P = -(||p_1 + p_2||) \tag{2}$$

Figure 1 shows motion vectors of seven primitive actions. For example, when the gesture consists of primitive actions p_{right} and p_{left}, the gesture label assigned is "RIGHT_LEFT". For this example gesture, $M = -2$, indicating that the movement is large, and $P = 0$, indicating that the start and end points are the same. Note that the M becomes zero only when there is no movement.

Machines require a gesture which is easy to recognize. We utilize a recognition ratio directly as a recognition score R. Note that the recognition ratio is stored in the system when the action database was created and tested for recognition accuracy of each action. For example, when the recognition ratio is 0.9 and 0.8 for action "RIGHT" and "LEFT" respectively, the score becomes $R = 0.9 \times 0.8 = 0.72$.

To control the weighting of the user-friendliness scores and the machine-friendliness score, P, M, and R are aggregated as E using alpha blending:

$$E = \alpha R + (1 - \alpha)\frac{P + M}{2}, \tag{3}$$

where the α is a parameter to control the weighting. If we give a large value to α, machine friendliness is given a much higher weighting than user friendliness. The score E is calculated for all gestures to make a ranking.

3   Experiments

3.1   Extraction of Gesture Candidates

We measured daily motions to extract candidate gestures which rarely occur in daily life. These daily motions included hand motions made while, for example, using a computer, eating a meal, reading, writing, or walking. The daily motions were measured for ten participants on separate days. The participants were instructed to avoid deliberately using primitive actions. The total measurement time was 24 h per participant. We collected 20 samples per primitive action for training of the random forest. We followed the parameter settings written in [4].

After the training, we searched single primitive actions and pairs of successive primitive actions (gestures) from the collected daily motion sequences, and counted up the frequency of occurrence. Figure 2 shows how many times each primitive action or gesture was found. Of the six actions that occurred most frequently, five are primitive actions, and the other is the "LEFT_RIGHT" gesture. Table 1 shows examples of gestures which rarely appeared in the collected daily motions. These gestures are candidate gestures for presentation to a user.

We prepared three of gesture candidate lists by changing the weighting parameter α. Table 2 shows the three candidate lists generated by setting $\alpha = 0.1$ (user friendly), $\alpha = 0.5$ (neutral), and $\alpha = 0.9$ (machine friendly).

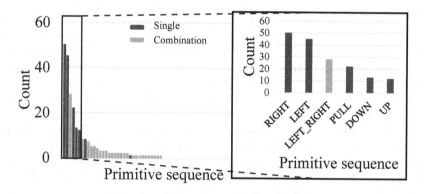

Fig. 2. Example of primitive actions

Table 1. Gesture candidates extracted from daily motions. The ROLL action is often included in the candidates.

DOWN_DOWN	RIGHT_**ROLL**	**ROLL**_RIGHT
LEFT_**ROLL**	**ROLL**_DOWN	**ROLL_ROLL**
PULL_**ROLL**	**ROLL**_LEFT	**ROLL**_UP
PULL_UP	**ROLL**_PULL	UP_PUSH
PUSH_**ROLL**	**ROLL**_PUSH	UP_**ROLL**

Table 2. Gesture candidate lists generated by $\alpha = 0.1$ (user friendly), $\alpha = 0.5$ (neutral), and $\alpha = 0.9$ (machine friendly)

Group U ($\alpha = 0.1$)		Group M ($\alpha = 0.5$)		Group S ($\alpha = 0.9$)	
Ranking	Gesture	Ranking	Gesture	Ranking	Gesture
1	DOWN_ROLL	1	DOWN_ROLL	1	DOWN_ROLL
2	ROLL_UP	2	ROLL_UP	2	DOWN_DOWN
3	DOWN_UP	3	DOWN_UP	3	DOWN_RIGHT
4	DOWN_RIGHT	4	DOWN_RIGHT	3	RIGHT_DOWN
4	RIGHT_DOWN	4	RIGHT_DOWN	5	ROLL_UP
6	RIGHT_UP	6	DOWN_DOWN	6	DOWN_UP
7	DOWN_LEFT	7	RIGHT_UP	7	RIGHT_UP
7	DOWN_PULL	8	DOWN_LEFT	8	DOWN_LEFT
7	LEFT_DOWN	8	DOWN_PULL	8	DOWN_PULL
7	PULL_DOWN	8	LEFT_DOWN	8	LEFT_DOWN
11	DOWN_PUSH	8	PULL_DOWN	8	PULL_DOWN
11	PUSH_DOWN	12	DOWN_PUSH	12	DOWN_PUSH
13	RIGHT_PUSH	12	PUSH_DOWN	12	PUSH_DOWN
14	LEFT_UP	14	RIGHT_PUSH	14	RIGHT_PUSH
14	PULL_UP	15	LEFT_UP	15	UP_UP
14	UP_PULL	15	PULL_UP	16	LEFT_UP
17	PUSH_UP	15	UP_PULL	16	PULL_UP
18	LEFT_PUSH	18	PUSH_UP	16	UP_PULL
19	DOWN_DOWN	19	UP_UP	19	PUSH_UP
20	UP_UP	20	LEFT_PUSH	20	LEFT_PUSH

3.2 User Study Settings

We investigated how each gesture was selected and assigned to a command for starting or controlling a smartphone application. We assumed a gesture interface to operate applications on a smartwatch. In the experiments, we prepared two control categories: operation of a music player, and starting an application. Each category contained four operations, for a total of eight operations to be controlled by gestures, as shown in Table 3.

We conducted an Internet-based survey to assess which gesture was likely to be assigned to each operation. In total, 351 individuals, including old and young people, responded to our survey. The attributes of the respondents are summarized in Fig. 3. We assigned respondents to one of three groups which had similar attributes of as much as possible. We asked respondents to assign a gesture to each of the eight operations. The respondents chose gestures based on the candidate list they were shown; each group was shown a different candidate list. Group U was shown the gesture list generated by $\alpha = 0.1$, Group M was

Table 3. Eight operations of two categories

category	operation	category	operation
Music player	Play music	Application start	Weather
	Stop music		Music player
	Next		Map
	Previous		Pedometer

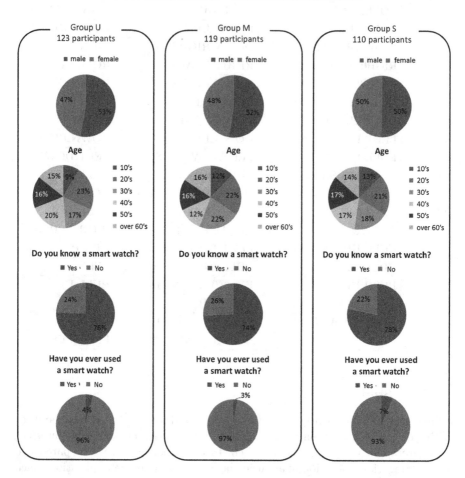

Fig. 3. Attributes of respondents

shown the gesture list generated by $\alpha = 0.5$, and Group S was shown the gesture list generated by $\alpha = 0.9$.

We asked respondents to select different gestures for each operation within the same category. For example, if the gesture "UP_DOWN" is assigned to the operation of "Play music", the same gesture cannot be assigned to any other operation in the "Music player" category. The same gesture can be assigned to an operation in the other category.

We also asked respondents to give a reason for their gesture selections. Respondents selected answers from the following list (multiple selection was allowed):

- The gesture matches the operation
- The gesture is highly ranked
- The gesture is simple
- The gesture is related to other gestures
- The gesture involves only a small movement

3.3 Results

The results of gesture assignment is shown in Table 4. The three gestures selected most commonly are summarized for each operation and respondent group. The column named "Operation(Ranking)" contains gesture labels and positions from the candidate gesture lists. The column named "Votes" shows how many respondents assigned each gesture to the operation. For example, 35 respondents in Group U assigned the gesture "DOWN_ROLL" to the operation "Play music". The "DOWN_ROLL" gesture was the top of the candidate gesture list presented to respondents in Group U. Overall, gestures tended to be selected more often if they were higher in the candidate gesture list (See the rank orders shown in parentheses).

We investigated the effectiveness of the candidate gesture list in more detail. In the ideal case, the assigned gestures are the first eight gestures on the candidate list. In the worst case, gestures from lower on the candidate list are assigned to operations. Figure 4 shows the relationship between gesture ranking and selected gestures within the rank (like an ROC curve). For example, about 70% of the eight most highly ranked gestures were assigned to operations by respondents in Group U (the red curve in the figure). The percentage was not changed even if two gestures (i.e., gestures in 9th and 10th rank) were added. The percentage is cumulative, and gradually increases as more gestures are included. There is little difference among three groups. Besides, the three curves are much better than the worst case mentioned above. These findings indicate that ranking the gestures is effective for encouraging individuals to select gestures designed by the system. Because it does not affect the gesture assignment, α can be maximized, for example by choosing a value of 0.9, to maximize machine friendliness.

It is important to understand why the results were not sensitive to changes in α. Figure 5 shows the questionnaire respondents' reasons for assigning each gesture to an operation. The answers are listed on the horizontal axis, and the frequency of the answer (i.e., how many respondents selected the answer) is indicated on the vertical axis. The most common answer was "The gesture matches the operation". Respondents tended to select gestures which were easily associated with the given operation. Some of the respondents' associations between gestures and operations are listed in Table 5. Even the most commonly assigned gestures were not chosen by all respondents. We guess that respondents had their own impressions of each operation. The second most common answer was "The

Table 4. Gestures assigned to each operation. Ranking: the rank of the gesture in the list that was presented. Votes: how many respondents assigned the gesture to the operation.

	Group U		Group M		Group S		Overall	
	Operation(Ranking)	# of votes	Operation(Ranking)	# of votes	Operation(Ranking)	# of votes	Operation(Ranking)	# of votes
Play music	DOWN_ROLL(1)	35	DOWN_ROLL(1)	39	DOWN_ROLL(1)	36	DOWN_ROLL	110
	ROLL_UP(2)	35	ROLL_UP(2)	30	ROLL_UP(5)	20	ROLL_UP	85
	RIGHT_UP(6)	11	RIGHT_UP(7)	13	PUSH_UP(19)	15	RIGHT_UP	31
Stop music	DOWN_ROLL(1)	35	DOWN_ROLL(1)	26	DOWN_DOWN(2)	40	DOWN_ROLL	80
	ROLL_UP(2)	25	ROLL_UP(2)	25	DOWN_ROLL(1)	19	DOWN_DOWN	63
	DOWN_DOWN(19)	13	DOWN_UP(3)	11	DOWN_UP(6)	5	ROLL_UP	54
Next	DOWN_RIGHT(4)	29	RIGHT_UP(7)	28	DOWN_RIGHT(3)	26	RIGHT_UP	78
	RIGHT_UP(6)	27	DOWN_RIGHT(4)	22	RIGHT_UP(7)	23	DOWN_RIGHT	77
	DOWN_UP(3)	13	DOWN_UP(3)	18	DOWN_ROLL(1)	10	DOWN_UP	34
Previous	DOWN_LEFT(7)	26	LEFT_DOWN(8)	25	LEFT_DOWN(8)	19	LEFT_DOWN	58
	LEFT_DOWN(7)	14	DOWN_LEFT(8)	17	RIGHT_DOWN(3)	15	DOWN_LEFT	57
	DOWN_RIGHT(4)	13	LEFT_UP(15)	14	DOWN_LEFT(8)	14	LEFT_UP	35
Weather	ROLL_UP(2)	25	DOWN_ROLL(1)	24	UP_UP(15)	25	ROLL_UP	67
	DOWN_ROLL(1)	24	ROLL_UP(2)	20	ROLL_UP(5)	22	DOWN_ROLL	67
	UP_UP(20)	18	UP_UP(19)	16	DOWN_ROLL(1)	19	UP_UP	59
Music player	ROLL_UP(2)	29	ROLL_UP(2)	28	DOWN_ROLL(1)	20	ROLL_UP	71
	DOWN_ROLL(1)	21	DOWN_ROLL(1)	17	ROLL_UP(5)	14	DOWN_ROLL	58
	PUSH_UP(17)	12	DOWN_UP(3)	14	DOWN_DOWN(2)	12	PUSH_UP	30
Map	DOWN_UP(3)	17	DOWN_UP(3)	13	DOWN_RIGHT(3)	15	DOWN_UP	38
	DOWN_ROLL(1)	16	ROLL_UP(2)	11	DOWN_DOWN(2)	14	DOWN_ROLL	37
	ROLL_UP(2)	13	DOWN_RIGHT(4)	10	DOWN_ROLL(1)	13	DOWN_DOWN	33
Pedometer	DOWN_DOWN(19)	14	DOWN_ROLL(1)	12	DOWN_DOWN(2)	14	DOWN_DOWN	37
	DOWN_ROLL(1)	13	UP_UP(19)	11	PUSH_UP(19)	13	DOWN_ROLL	30
	DOWN_RIGHT(4)	12	DOWN_LEFT(8)	10	ROLL_UP(5)	9	UP_UP	29

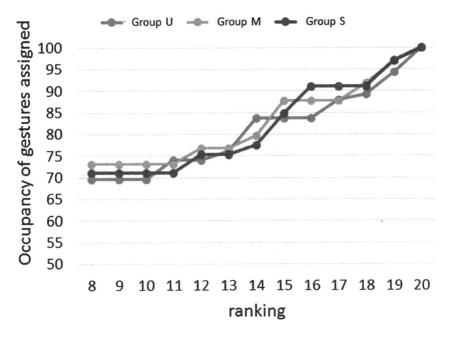

Fig. 4. Ratio of gestures assigned to operations. The ideal case is that the eight most highly ranked gestures are all assigned to operations. In such a case, the score of the vertical axis becomes 100% at the rank 8 on the horizontal axis. In the worst case, the occupancy becomes 100% when the gesture at rank 20 is assigned to an operation. (Color figure online)

Table 5. Free description about the image of operations and their associated gestures

Operation	Description	Trend of assignment
Play music	The image of right arrow	Containing "RIGHT"
Stop music	The image of remaining	Containing "DOWN"
Next	The image of skip	Containing "RIGHT" or "UP"
Weather application	The image of sky	Containing "UP"
Map application	The image of ground	Containing "DOWN"

gesture is highly ranked". Although the frequency is lower than that for "The gesture matches the operation", the rank of the proposed gestures had some effect. The other three factors were not considered so much by respondents.

Symmetrically associated operations, such as "Play music" vs. "Stop music" or "Next" vs. "Previous" tended to be assigned symmetric gestures. Figures 6 and 7 show co-occurrence matrices of gestures assigned to operations. The symmetric pair of "DOWN_LEFT" vs. "DOWN_RIGHT" was most commonly selected for the operation of "Next" vs. "Previous", as were "LEFT_UP" vs. "RIGHT_UP", "LEFT_DOWN" vs. "RIGHT_DOWN" or "LEFT_DOWN" vs. "RIGHT_UP".

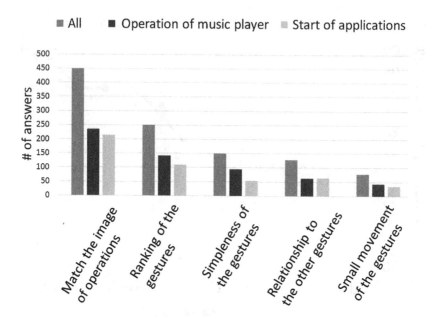

Fig. 5. Reason for gesture assignment. The vertical axis is quantity of answers. Multiple selection of answers was allowed.

Stop music

Play music	DOWN_ROLL	ROLL_UP	DOWN_UP	DOWN_RIGHT	RIGHT_DOWN	RIGHT_UP	DOWN_LEFT	DOWN_PULL	LEFT_DOWN	PULL_DOWN	DOWN_PUSH	PUSH_DOWN	RIGHT_PUSH	LEFT_UP	PULL_UP	UP_PULL	PUSH_UP	LEFT_PUSH	DOWN_DOWN	UP_UP
DOWN_ROLL	0	49	9	1	4	0	1	0	0	2	2	0	0	0	0	2	3	0	23	2
ROLL_UP	50	0	13	2	1	0	0	1	1	1	2	3	1	0	0	0	0	0	7	1
DOWN_UP	4	1	0	0	1	0	0	2	0	1	0	0	0	0	0	0	0	0	4	0
DOWN_RIGHT	3	1	0	0	2	0	4	0	0	1	0	0	0	0	0	1	0	0	1	0
RIGHT_DOWN	1	0	0	0	0	0	1	0	0	1	1	0	0	1	0	0	0	0	1	0
RIGHT_UP	3	0	1	1	11	0	2	0	4	0	0	0	0	2	0	0	0	0	4	0
DOWN_LEFT	1	0	0	1	0	0	0	0	0	0	0	0	0	0	0	0	0	0	0	0
DOWN_PULL	0	0	0	0	0	0	0	0	0	0	0	0	0	0	0	0	0	0	0	0
LEFT_DOWN	0	0	0	0	0	0	0	0	0	0	0	0	0	0	0	0	1	0	0	0
PULL_DOWN	0	0	0	0	0	0	0	0	0	0	0	0	0	0	1	0	0	0	0	0
DOWN_PUSH	1	0	0	0	0	0	0	1	0	1	0	0	0	0	0	0	0	0	0	0
PUSH_DOWN	0	0	0	0	0	0	0	0	0	1	1	0	0	0	3	1	0	0	0	0
RIGHT_PUSH	0	0	1	0	1	0	0	0	0	0	0	1	0	0	1	0	0	1	0	0
LEFT_UP	0	0	0	0	0	0	0	0	0	0	0	0	0	0	0	0	1	1	0	0
PULL_UP	2	0	0	0	0	0	0	1	0	3	1	1	0	0	0	1	0	0	0	0
UP_PULL	0	0	0	1	0	0	0	0	0	0	0	0	0	0	0	0	0	0	0	0
PUSH_UP	1	0	1	0	0	1	1	1	0	2	2	10	0	0	2	0	0	1	2	2
LEFT_PUSH	0	0	0	0	0	0	0	0	0	0	0	0	0	0	0	0	0	0	0	0
DOWN_DOWN	0	1	1	0	0	0	1	0	0	0	0	0	0	0	0	0	1	0	0	3
UP_UP	2	0	0	0	0	0	0	0	0	1	1	0	0	0	0	0	1	0	21	0

Fig. 6. Co-occurrence matrix of symmetric gestures of "Play music" vs. "Stop music".

Previous

Next	DOWN_ROLL	ROLL_UP	DOWN_UP	DOWN_RIGHT	RIGHT_DOWN	RIGHT_UP	DOWN_LEFT	DOWN_PULL	LEFT_DOWN	PULL_DOWN	DOWN_PUSH	PUSH_DOWN	RIGHT_PUSH	LEFT_UP	PULL_UP	UP_PULL	PUSH_UP	LEFT_PUSH	DOWN_DOWN	UP_UP
DOWN_ROLL	0	4	2	0	0	1	0	1	0	0	0	0	0	0	0	0	0	0	4	0
ROLL_UP	6	0	6	2	2	1	1	1	2	1	2	0	1	1	0	0	0	0	1	0
DOWN_UP	2	0	0	19	6	2	0	1	0	1	0	0	1	0	0	0	0	0	1	0
DOWN_RIGHT	3	3	1	0	15	1	47	0	1	1	0	1	0	0	1	0	0	0	1	0
RIGHT_DOWN	0	1	0	1	0	4	1	0	24	0	0	0	0	0	0	0	0	0	1	0
RIGHT_UP	0	2	2	0	9	0	7	1	24	0	0	0	1	31	0	0	0	0	0	0
DOWN_LEFT	0	1	0	4	0	0	0	0	1	0	0	0	0	1	0	0	0	0	0	0
DOWN_PULL	0	0	0	0	0	0	0	0	0	0	0	0	0	0	0	0	0	0	0	0
LEFT_DOWN	0	0	0	0	1	0	0	0	0	0	0	0	0	0	0	0	0	0	0	0
PULL_DOWN	0	0	0	0	0	0	0	0	0	0	1	0	0	0	0	0	0	0	0	0
DOWN_PUSH	0	0	0	0	0	0	0	1	0	1	0	0	0	0	0	0	0	0	0	0
PUSH_DOWN	0	0	0	0	1	0	0	0	0	0	0	0	0	0	0	0	0	0	0	0
RIGHT_PUSH	0	0	0	1	0	0	0	0	0	0	1	1	0	0	0	0	0	21	0	0
LEFT_UP	0	0	0	0	0	1	0	0	4	1	0	0	0	0	1	0	0	0	0	0
PULL_UP	0	0	0	0	0	0	0	0	0	0	0	0	0	0	0	0	0	0	0	0
UP_PULL	0	0	0	0	0	0	0	0	0	0	0	0	0	0	0	0	0	0	0	0
PUSH_UP	0	0	0	0	0	0	0	0	0	3	0	0	0	0	2	2	0	0	0	0
LEFT_PUSH	0	0	0	0	0	0	0	0	0	0	0	0	3	0	0	0	1	0	0	0
DOWN_DOWN	1	0	0	0	1	2	1	0	2	0	0	0	0	0	0	0	0	0	0	3
UP_UP	0	0	0	0	0	0	0	0	0	0	0	0	0	1	0	0	0	0	7	0

Fig. 7. Co-occurrence matrix of symmetric gestures of "Next" vs. "Previous".

4 Conclusion

In this paper, we discussed the design of user interface with gestures as an input method. Our proposed system first extracted candidate gestures based on a pattern mining approach. We defined 'gesture' as two successive primitive actions. Through long-term observation, our system suggested gestures which rarely occurred in daily life. The candidate gestures were evaluated based on their user friendliness and machine friendliness. The balance between these two perspectives was adjusted by a weighting. The candidate gestures were presented to users in a ranked list. In an Internet-based survey, respondents were asked to assign candidate gestures to eight operations. From our results, we draw the following conclusions:

- Individuals tend to select the highest ranked gestures from a list.
- It is unnecessary to give strong weighting to user-friendly gestures that involve small movements.
- Individuals tend to select gestures that are easily associated with the operations.
- Individuals tend to select symmetric gestures for symmetric operations.

In our future work, we will continue to investigate the effect of suggesting symmetric gestures. In addition, we have to evaluate how accurately the system recognizes gestures assigned by users towards practical use.

References

1. Ashbrook, D., Starner, T.: MAGIC: a motion gesture design tool. In: Proceedings of the SIGCHI Conference on Human Factors in Computing Systems, pp. 2159–2168. ACM (2010)
2. Chen, M.Y., Mummert, L., Pillai, P., Hauptmann, A., Sukthankar, R.: Controlling your TV with gestures. In: Proceedings of the International Conference on Multimedia Information Retrieval, MIR 2010, pp. 405–408. ACM, New York (2010). http://doi.acm.org/10.1145/1743384.1743453
3. Chen, Q., Georganas, N.D., Petriu, E.M.: Real-time vision-based hand gesture recognition using haar-like features. In: IEEE Instrumentation and Measurement Technology Conference Proceedings, IMTC 2007, pp. 1–6. IEEE (2007)
4. Kawahata, R., Shimada, A., Yamashita, T., Uchiyama, H., Taniguchi, R.: Design of a low-false-positive gesture for a wearable device. In: 5th International Conference on Pattern Recognition Applications and Methods, pp. 581–588 (2016)
5. Liu, J., Zhong, L., Wickramasuriya, J., Vasudevan, V.: uWave: accelerometer-based personalized gesture recognition and its applications. Pervasive Mob. Comput. 5(6), 657–675 (2009)
6. Park, T., Lee, J., Hwang, I., Yoo, C., Nachman, L., Song, J.: E-gesture: a collaborative architecture for energy-efficient gesture recognition with hand-worn sensor and mobile devices. In: Proceedings of the 9th ACM Conference on Embedded Networked Sensor Systems, pp. 260–273. ACM (2011)
7. Ruiz, J., Li, Y.: DoubleFlip: a motion gesture delimiter for mobile interaction. In: Proceedings of the SIGCHI Conference on Human Factors in Computing Systems, pp. 2717–2720. ACM (2011)
8. Ruiz, J., Li, Y., Lank, E.: User-defined motion gestures for mobile interaction. In: Proceedings of the SIGCHI Conference on Human Factors in Computing Systems, pp. 197–206. ACM (2011)
9. Shan, C.: Multimedia interaction and intelligent user interfaces. Consum. Electron. 107–128 (2010). http://www.springerlink.com/index/10.1007/978-1-84996-507-1
10. Shimada, A., Yamashita, T., Taniguchi, R.: Hand gesture based TV control system–towards both user- & machine-friendly gesture applications. In: Workshop on Frontiers of Computer Vision the 19th Japan-Korea Joint, pp. 121–126 (2013)
11. Wachs, J.P., Kölsch, M., Stern, H., Edan, Y.: Vision-based hand-gesture applications. Commun. ACM 54, 60–71 (2011). http://doi.acm.org/10.1145/1897816.1897838
12. Zabulis, X., Baltzakis, H., Argyros, A.: Vision-Based Hand Gesture Recognition for Human-computer Interaction. Lawrence Erlbaum Associates, Inc. (LEA), Mahwah (2009). http://www.ics.forth.gr/~xmpalt/publications/papers/zabulis09_book.pdf

SS1: Social Media Retrieval and Recommendation

Collaborative Dictionary Learning and Soft Assignment for Sparse Coding of Image Features

Jie Liu[1], Sheng Tang[2]([⊠]), and Yu Li[2]

[1] Beijing Advanced Innovation Center for Imaging Technology,
College of Information and Engineering, Capital Normal University,
Beijing 100048, People's Republic of China
liujie@cnu.edu.cn
[2] Key Lab of Intelligent Information Processing of Chinese Academy of Sciences
(CAS), Institute of Computing Technology, CAS,
Beijing 100190, People's Republic of China
{ts,liyu}@ict.ac.cn

Abstract. In computer vision, the bag-of-words (BoW) model has been widely applied to image related tasks, such as large scale image retrieval, image classification, and object categorization. The sparse coding (SC) method which leverages SC as a means of feature coding can guarantee both sparsity of coding vector and lower reconstruction error in the BoW model. Thus it can achieve better performance than the traditional vector quantization method. However, it suffers from the side effect introduced by the non-smooth sparsity regularizer that quite different words may be selected for similar patches to favor sparsity, resulting in the loss of correlation between the corresponding coding vectors. To address this problem, in this paper, we propose a novel soft assignment method based on index combination of top-2 large sparse codes of local descriptors to make the SC-based BoW tolerate the case of different word selection for similar patches. To further ensure similar patches select same words to generate similar coding vectors, we propose a collaborative dictionary learning method through imposing the sparse code similarity regularization factor along with the row sparsity regularization across data instances on top of group sparse coding. Experiments on the well-known public Oxford dataset demonstrate the effectiveness of our proposed methods.

Keywords: Near duplicate image retrieval · Bag of words · Dictionary learning · Sparse coding

1   Introduction

In the field of computer vision, the bag-of-words (BoW) model by treating image features as a sparse histogram over a vocabulary (or codebook), has been widely

This work was supported by National Nature Science Foundation of China (61371194, 61672361, 61572472), Beijing Natural Science Foundation (4152050, 4152012), Beijing Advanced Innovation Center for Imaging Technology (BAICIT-2016009).

L. Amsaleg et al. (Eds.): MMM 2017, Part I, LNCS 10132, pp. 443–451, 2017.
DOI: 10.1007/978-3-319-51811-4_36

applied to image related tasks, such as large scale image retrieval [1, 2], image classification and concept detection [3, 4], object categorization [5, 6]. Particularly, many previous approaches [7, 8] had paid much attention to the problem of how to leverage the BoW model as a means of image representation for image and object retrieval through matching between transformed images.

In order to increase BoW's discriminability and improve search efficiency, large vocabulary was adopted to quantize and partition the set of local feature descriptors into large set of disjoint subsets to generate large visual words (BoW with large dimension such as over 1 million size) [9]. State of the art methods indicate that large visual words can lead to better performance [6, 9] and received more and more concerns in recent years [9]. According to the types of different acceleration methods for feature coding, related work on large vocabulary can be classified into two categories [9]: Hierarchical K-Means (HKM) [10] and Approximate Nearest Neighbor (ANN) method including Approximate K-Means methods (AKM) [11], Robust AKM [12], Approximate Gaussian Mixtures (AGM) [13], etc. Mikulik [6] proposes a fine quantization with very large visual words (up to 64 million words) and demonstrated that the performances can be improved as the size of visual words increases. However, due to the heavy cost of time and memory space, large vocabulary such as 8, 001 hierarchical centroids of $8, 000 \times 128$ dimension with about 32G memory occupation in [6], produces a very heavy burden for both vocabulary construction and feature coding [9]. When confronted with applications in embedded devices with limited memory [14] such as mobile visual search, the heavy cost makes it infeasible for any large vocabulary solution.

Recently, in order to release the heavy burden for such applications, compact vocabulary was proposed to replace the large vocabulary in [9] to realize large visual words. The core idea is to exploit segmental sparse coding to generate hierarchical BoW with two learned compact dictionaries. Although sparse coding (SC) can achieve better performance than the traditional vector quantization method due to its sparsity of coding vector and lower reconstruction error in the BoW model [5, 15], it suffers from the side effect introduced by the non-smooth sparsity regularizer that quite different words may be selected for similar patches to favor sparsity, resulting in the loss of correlation between the corresponding coding vectors [5, 15].

To address this problem, in this paper, we propose a novel soft assignment method to make the SC-based BoW tolerate the case of quite different word selection for similar patches. To further ensure that similar patches select same words to generate similar coding vectors, we propose a collaborative dictionary learning (CDL) method through imposing the sparse code similarity regularization factor along with the row sparsity regularization across data instances on top of group sparse coding.

The main contributions of this paper are two-fold: First, we propose a novel soft assignment method for BoW based on index combination of top-2 large sparse codes of local descriptors which is wholly different from the traditional soft assignment proposed in [16] which uses the K-Nearest Neighbor (KNN) of the image features. Second, we propose the CDL method for more precise generation of visual words from similar image patches. In the rest of this paper, we will first introduce the related work on hierarchical BoW with segmental sparse coding, then elaborates on our proposed soft assignment and CDL method respectively followed by our experiments and conclusion.

2 Hierarchical BoW with Segmental Sparse Coding

In [9], Tang et al. presented a hierarchical BoW extraction method through segmental sparse coding to realize large visual words of 1 million size with a compact vocabulary consisting of two small dictionaries at the cost of merely 512 K bytes, about one thousandth of those of traditional large vocabulary methods such as AKM [11]. It largely reduces the memory cost while keeping high efficiency, which has particular advantages for applications in mobile devices with very limited memory such as mobile logo recognition [14]. Extracting BoW with the compact vocabulary at the mobile device can save user's network expense and reduce the heavy burden of limited network bandwidth at same time.

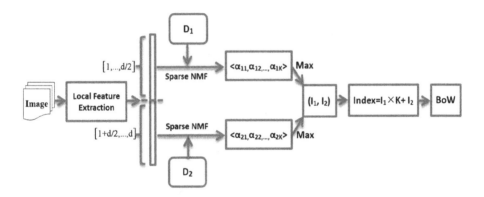

Fig. 1. Hierarchical BoW extraction with segmental sparse coding

As shown in Fig. 1, after piecewise sparse decomposition of local SIFT descriptors with two small dictionaries, they map a pair of indices (I_1 and I_2 in Fig. 1) of the dictionary's bases corresponding to the maximum elements of the two sparse codes to a large set of visual words upon the assumption that data with similar properties will share the same base corresponding to the largest sparse coefficient. Specifically, for each SIFT descriptor in a given image, they first divide it into two segments. Then, through sparse NMF decomposition of both segments with two small learned dictionaries, they can get two sparse codes α_1 and α_2 for each SIFT descriptor. Finally, they use the indices I_1 and I_2 of the maximum non-zero elements of the two sparse codes α_1 and α_2 to compute the index of the visual word (I_1, I_2) by:

$$Index = I_1 \times k + I_2 \qquad (1)$$

where $k = 1,024$ is the dictionary size. The two dictionaries ($\mathbf{D}_1$ and $\mathbf{D}_2$ in Fig. 1) are learned from 10 millions of SIFT descriptors with sparse NMF, the sizes of which are $k \times 64$, occupying only $512K$ bytes. Therefore, they can compute the BoW histogram with the index to get the large visual word of $k^2 = 1M$ size very efficiently.

3   Proposed Soft Assignment

In order to alleviate the side effect of correlation loss introduced by the non-smooth sparsity regularizer and make the above SC-based BoW [9] more robust to the problem of quite different word selection for similar patches, we propose a novel soft assignment method for BoW based on index combination of the top-2 large sparse codes of local descriptors. Additionally, descriptor-space soft assignment is efficient to compute and can gain much benefit for large visual words [17].

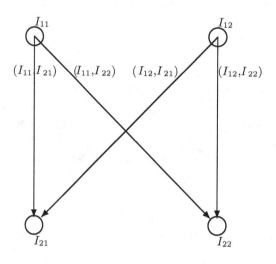

Fig. 2. Illustration of soft assignment

After careful examination of the sparse codes, we noted that if the top 2 largest elements of a sparse code are very closely similar, a slight transformation of the SIFT descriptors may cause the exchange of two corresponding dictionary bases. So we propose to use the pair combination based on the indices of the dictionary bases corresponding to the top-2 large elements of the two sparse codes. Let I_{11}, I_{12} be the index of the top-2 largest elements of sparse code learned from Dictionary D_1 as shown in Fig. 1, and I_{21}, I_{22} be those from Dictionary D_2 as shown in Fig. 1, we can assign a SIFT to 4 visual words with the 4 corresponding indices: $Index_1$: (I_{11}, I_{21}), $Index_2$: (I_{11}, I_{22}), $Index_3$: (I_{12}, I_{21}), and $Index_4$: (I_{12}, I_{22}) according to Eq. (2) as illustrated in Fig. 2.

$$\begin{cases} Index_1 = I_{11} \times k + I_{21} \\ Index_2 = I_{11} \times k + I_{22} \\ Index_3 = I_{12} \times k + I_{21} \\ Index_4 = I_{12} \times k + I_{22} \end{cases} \tag{2}$$

4 Proposed Collaborative Dictionary Learning

To further improve the precision of SIFT matching, we propose a collaborative dictionary learning (CDL) method to learn the compact dictionary for sparse coding of SIFT descriptors. Our goal is to try to get very closely similar sparse codes for a matched pair of SIFT descriptors so that we can get the same visual words for them. Thus, we can reduce the possibility of quite different words being selected for similar patches.

Group Sparse Coding (GSC) [18] can learn a dictionary which can make all the data instances in the same group share a common combination of dictionary bases. But GSC can not ensure very closely similar sparse codes for matched SIFT descriptors. Therefore, through imposing the sparse code similarity regularization factor along with the row sparsity regularization across data instances, we can achieve our goal intuitively.

The GSC tries to encode jointly all the instances in a group $\mathbf{G}$ with a dictionary $\mathbf{D}$ through solving the optimization problem [18]:

$$\min_{\mathbf{D},\mathbf{A}} \frac{1}{2} \sum_{i=1}^{|\mathbf{G}|} \left\| x_i - \sum_{j=1}^{|\mathbf{D}|} \alpha_j^i d_j \right\|^2 + \lambda \sum_{j=1}^{|\mathbf{D}|} \|\alpha_j\|_1 \tag{3}$$

$$s.t. \alpha_j^i \geq 0, \ \forall i,j.$$

where $x_i \in \mathbf{R}^l$ is the i-th data instance vector from the observed data matrix $\mathbf{X} = [x_1, \ldots, x_n] \in \mathbf{R}^{l \times n}$, and d_i is the dictionary atom (base) from dictionary $\mathbf{D} = [d_1, \ldots, d_k] \in \mathbf{R}^{l \times k}$, and l is the data dimensionality, k is the aforementioned size of the dictionary, and $\mathbf{A} = \{\alpha_j\}_{j=1}^{|\mathbf{D}|}$ is the reconstruction matrix consists of non-negative sparse codes $\alpha_j = (\alpha_j^1, \ldots, \alpha_j^{|\mathbf{G}|})$ which specifying the contribution of base d_j to each instance.

Thus, by collaboratively adding the sum of the absolute sparse code differences $|\alpha_j^{i+1} - \alpha_j^i|$ between data instances i and $i+1$ on a given base d_j into Eq. (3), our proposed CDL method can learn a dictionary to generate closely similar sparse codes for matched SIFT pairs by solving the optimization problem:

$$\min_{\mathbf{D},\mathbf{A}} \frac{1}{2} \sum_{i=1}^{|\mathbf{G}|} \left\| x_i - \sum_{j=1}^{|\mathbf{D}|} \alpha_j^i d_j \right\|^2 + \lambda \sum_{j=1}^{|\mathbf{D}|} \left[\|\alpha_j\|_1 + \sum_{i=1}^{|\mathbf{G}|-1} |\alpha_j^{i+1} - \alpha_j^i| \right] \tag{4}$$

$$s.t. \mathbf{D} \geq 0, \mathbf{A} \geq 0$$

Here, we impose the non-negativity constraints and adopt this equation to replace the standard sparse NMF in [9] for vocabulary construction. We adopt $|\mathbf{G}| = 10$ types of different transformations with the StirMark Benchmark [19] and use the RANSC matching of SIFT to automatically collect $1M$ groups of matched SIFT descriptors for collaborative learning of the two dictionaries D_1 and D_2.

5   Experiments

In order to verify the efficacy of our proposed BoW soft assignment method and collaborative dictionary learning for compact vocabulary construction, we test our method on the well-known public Oxford 5 K dataset introduced by [11], which represents images of Oxford buildings. There are 55 query images corresponding to 11 distinct buildings. All the queries are defined by a rectangle delimiting the building and are in "upright" orientation [11]. For each building, the 5062 images are annotated as relevant (good+OK), not relevant (bad), and (junk) which should not be taken into account when measuring the accuracy, because they only contain a partial view (less than 25%) of the building [11].

Table 1. MAPs for runs in the Oxford 5 K dataset

Run name	Method description	MAP (%)
Baseline [9]	Hard Assignment + Sparse NMF	50.4
Our Run1	Hard Assignment + CDL	53.7
Our Run2	Soft Assignment + Sparse NMF	55.6
Our Run3	Soft Assignment + CDL	58.2
AKM [11]	1M visual words	61.8
HKM-1 [11]	Scoring level 1	43.9
HKM-2 [11]	Scoring level 2	41.8
HKM-3 [11]	Scoring level 3	37.2
HKM-4 [11]	Scoring level 4	35.3

We have tested three runs of our method. The descriptions about each run are listed in the Table 1, and all the comparison runs use a vocabulary of $1M$ visual words without exploiting spatial information. The Mean Average Precisions (MAP) over the total 55 queries in the dataset are shown in Table 1, which shows that the MAPs of all our three runs are constantly much greater than those of all the HKM methods in [11], with the improvement of at least 9.8% (our Run1 compared with HKM-1 [11]), which proves the effectiveness of our proposed BoW method since it is essentially a kind of 2-level hierarchical quantization. Although our best MAP is still a little below that of AKM which outperforms much better than all HKM methods in precision, as aforementioned, our method has a very compact vocabulary, only one thousandth of AKM's, which is of great importance for mobile applications or embedding systems with very limited memory devices.

Furthermore, by comparing the baseline [9] v.s. Run1, and Run2 v.s. Run3, we can conclude that our proposed collaborative dictionary learning can gain MAP improvement of 3.3% and 2.6% respectively compared with standard sparse dictionary learning in [9]. Comparing the Baseline v.s. Run2, and Run1 v.s. Run3, we can conclude that our soft assignment can gain MAP improvements of 5.2% and 4.5% respectively compared with hard assignment. The above experiments verified our proposed BoW soft assignment and collaborative dictionary learning methods.

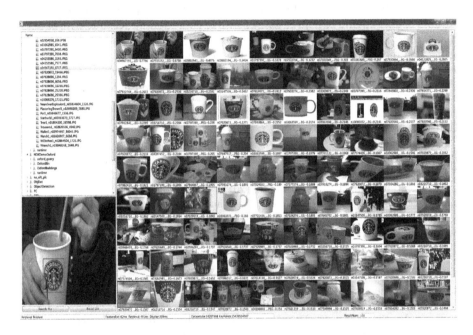

Fig. 3. Retrieval effect illustration of our system on the large scale (14.2M) ImageNet dataset

To visually show the efficacy of our method, we developed a large scale near duplicate image retrieval system on the whole ImageNet [20] with over 14.2 million images (the total number of extracted SIFT points is over 15.4 billion). As illustrated by the retrieval results in Fig. 3, queried by a small red region cropped from a given image, we can see that our method is effective and efficient on the large dataset. The feature extraction time (including GPU-SIFT extraction and BoW generation) is 62 ms, while retrieval time is 491 ms, and the memory cost (including the inverted index) is only about 63 G. During building up the inverted file index of the dataset, our system can add about 15 thousands of images per hour, a little faster than the speed (14 thousands of images per hour) of AKM implemented by FLANN [21], which indicates the high efficiency of our BoW extraction through Sparse NMF quantization.

6   Conclusion and Future Work

In this paper, in order to make the SC-based BoW tolerate the problem of quite different word selection for similar patches, we propose a novel BoW soft assignment method and a collaborative dictionary learning for more precise generation of visual words with a very compact vocabulary. Experiments on the well-known public Oxford dataset demonstrate the effectiveness of our proposed methods.

Several issues are worthy of further investigation for improving the precision: First, we will try some new promising features such as deep learning features to replace the SIFT features. Second, we plan to try some fast spatial geometric verification methods based on sparse coding to remove the false matching. Third, we want to study how to filter out unimportant image patches while only keeping distinctive ones to reduce the memory cost for large scale image retrieval.

References

1. Xie, H., Gao, K., Zhang, Y., Tang, S., Li, J., Liu, Y.: Efficient feature detection and effective post-verification for large scale near-duplicate image search. IEEE Trans. Multimedia **13**(6), 1319–1332 (2011)
2. Nie, L., Yan, S., Wang, M., Hong, R., Chua, T.-S.: Harvesting visual concepts for image search with complex queries. In: Proceedings of ACM Multimedia 2012 Conference, October 2012
3. Tang, S., Li, J.-T., Li, M., Xie, C., Liu, Y.-Z., Tao, K., Xu, S.-X.: TRECVID 2008 high-level feature extraction by MCG-ICT-CAS. In: Proceedings of TRECVID 2008 Workshop, November 2008
4. Tang, S., Zheng, Y.-T., Wang, Y., Chua, T.-S.: Sparse ensemble learning for concept detection. IEEE Trans. Multimedia **14**(1), 43–54 (2012)
5. Li, P., Lu, X., Wang, Q.: From dictionary of visual words to subspaces: locality-constrained affine subspace coding. In: 2015 IEEE Conference on Computer Vision and Pattern Recognition (CVPR), pp. 2348–2357, June 2015
6. Mikulik, A., Perdoch, M., Chum, O., Matas, J.: Learning vocabularies over a fine quantization. Int. J. Comput. Vision **103**(1), 163–175 (2013)
7. Sivic, J., Zisserman, A.: Video Google: a text retrieval approach to object matching in videos. In: Proceedings of ICCV, pp. 1470–1477 (2003)
8. Jegou, H., Douze, M., Schmid, C.: Improving bag-of-features for large scale image search. Int. J. Comput. Vis. **87**, 316–336 (2010)
9. Tang, S., Chen, H., Lv, K., Zhang, Y.D.: Large visual words for large scale image classification. In: 2015 IEEE International Conference on Image Processing (ICIP), pp. 1170–1174, September 2015
10. Nister, D., Stewenius, H.: Scalable recognition with a vocabulary tree. In: Proceedings of CVPR, pp. 2161–2168 (2006)
11. Philbin, J., Chum, O., Isard, M., Sivic, J., Zisserman, A.: Object retrieval with large vocabularies and fast spatial matching. In: Proceedings of CVPR, pp. 1–8 (2007)

12. Li, D., Yang, L., Hua, X.S., Zhang, H.J.: Large-scale robust visual codebook construction. In: ACM Multimedia 2010 (2010)
13. Avrithis, Y., Kalantidis, Y.: Approximate Gaussian mixtures for large scale vocabularies. In: Fitzgibbon, A., Lazebnik, S., Perona, P., Sato, Y., Schmid, C. (eds.) ECCV 2012. LNCS, vol. 7574, pp. 15–28. Springer, Heidelberg (2012). doi:10.1007/978-3-642-33712-3_2
14. Tang, S., Zhang, Y.D., Chen, H.: Scalable logo recognition based on compact sparse dictionary for mobile devices. In: 2015 IEEE 17th International Workshop on Multimedia Signal Processing (MMSP), pp. 1–6, October 2015
15. Yang, J., Yu, K., Gong, Y., Huang, T.: Linear spatial pyramid matching using sparse coding for image classification. In: CVPR (2009)
16. Jiang, Y.-G., Ngo, C.-W., Yang, J.: Towards optimal bag-of-features for object categorization and semantic video retrieval. In: Proceedings of ACM International Conference on Image and Video Retrieval (2007)
17. Philbin, J., Chum, O., Isard, M., Sivic, J., Zisserman, A.: Lost in quantization: improving particular object retrieval in large scale image databases. In: Proceedings of CVPR (2008)
18. Strelow, D., Bengio, S., Pereira, F., Singer, Y.: Group sparse coding. In: Neural Information Processing Systems - NIPS (2009)
19. Petitcolas, F.A.P.: Watermarking schemes evaluation. IEEE. Sig. Process. **17**(5), 117–128 (2000)
20. Deng, J., Dong, W., Socher, R., Li, L.-J., Li, K., Fei-fei, L.: Imagenet: a large-scale hierarchical image database. In: Proceedings of CVPR (2009). http://image-net.org/
21. Muja, M., Lowe, D.G.: Scalable nearest neighbor algorithms for high dimensional data. IEEE Trans. Pattern Anal. Mach. Intell. **36**, 2227–2240 (2014)

LingoSent — A Platform for Linguistic Aware Sentiment Analysis for Social Media Messages

Yuting Su[✉] and Huijing Wang

School of Electronic Information Engineering, Tianjin University,
Tianjin 300072, China
{ytsu,wanghuijing}@tju.edu.cn

Abstract. Sentiment analysis is an important natural language processing (NLP) task and applied to a wide range of scenarios. Social media
messages such as tweets often differ from formal writing, exhibiting
unorthodox capitalization, expressive lengthenings, Internet slang, etc.
While such characteristics are inherently beneficial for the task of
sentiment analysis, they also pose new challenges for existing NLP platforms. In this article, we present a new approach to improve lexicon-
based sentiment analysis by extracting and utilizing linguistic features
in a comprehensive manner. In contrast to existing solutions, we design
our sentiment analysis approach as a framework with data preprocessing, linguistic feature extraction and sentiment calculation being separate components. This allows for easy modification and extension of
each component. More importantly, we can easily configure the sentiment
calculation with respect to the extracted features to optimize sentiment
analysis for different application contexts. In a comprehensive evaluation,
we show that our system outperforms existing state-of-the-art lexicon-
based sentiment analysis solutions.

Keywords: Sentiment analysis · Opinion mining · Natural language
processing · Structured feature extraction

1 Introduction

With the emergence and popularity of social media, an increasing number of
people share their personal opinions on current affairs, products or just their
feelings and emotions through different social media platforms. Among which,
Twitter has become one of the most popular ones, with more than 645 million
registered users and about 190 million tweets per day.[1] Tweets has a maximum
length of 140 characters. Most of them contain a wide variety of non-standard
tokens including slang words, hashtags, emoticons etc. and the writing style is
often very informal. As a result, basic NLP tasks such as tokenizing, normalizing
and Part-of-Speech (POS) tagging are challenged.

---

[1] http://www.statisticbrain.com/twitter.

© Springer International Publishing AG 2017
L. Amsaleg et al. (Eds.): MMM 2017, Part I, LNCS 10132, pp. 452–464, 2017.
DOI: 10.1007/978-3-319-51811-4_37

There are two basic approaches to perform sentiment analysis. Unsupervised method relies on so-called *sentiment lexicons* which contain lists of tokens with sentiment score or polarity. While supervised method extracts features from messages and uses machine learning algorithms to train classifiers such as SVM or Naive Bayes. This method relies on large labeled datasets for training. Chikersal *et al.* showed that for a large ratio of tweets, supervised classifier cannot assign labels to them with much confidence. The "black box" nature of trained classifiers makes them usually hard to interpret and hence difficult to generalize, modify, or extend.

In this article, we therefore focus on an unsupervised, i.e., lexicon-based approach towards sentiment analysis over tweets. Although various related works exist, we argue that they under-utilize the language and writing style commonly featured on social media. To this end, we propose LingoSent, our framework for sentiment analysis that heavily utilizes linguistic features to calculate the sentiments of tweets. LingoSent consists of three main components. First, the data preprocessor can reliably tokenize and normalize tweets even in the presence of uncommon spacing and punctuation, which then also enables reliable POS tagging. Second, the linguistic feature extractor analyzes tweets with respect to a wide range of linguistic features that people use to implicitly or explicitly convey a sentiment in the message. We consider word-level, phrase-level and sentence-level features. Lastly, the sentiment calculator uses a set of sentiment lexicons together with the output of linguistic feature extraction to calculate the overall sentiment score of a tweet. We focus on a two-class output with each tweet featuring a negative or positive polarity. We evaluated LingoSent over four publicly available Twitter datasets from different resources and compared our framework with different existing solutions. Our results show that the preprocessor we customize for social media text and the linguistic features we employ achieve noticeable improvements over many other previous works.

This paper is organized as follows. Section 2 provides a review of existing works in the context of sentiment analysis for social media. Section 3 gives an overview of the architecture and main components of LingoSent. Section 4 outlines the sentiment lexicons as well as our own curated dictionaries required for data preprocessing and linguistic feature extraction steps. Section 5 introduces the data preprocessor optimized to perform low-level NLP tasks over social media messages. Section 6 details on the extraction of linguistic features from tweets. Section 7 covers the sentiment calculator component that eventually yields the output of LingoSent. Section 8 presents and discusses the results of our evaluation. Section 9 concludes this article.

2 Related Works

A large amount of works hava been conducted on Twitter sentiment analysis. Most works focus on sentence level, while some focus on term level [1] or entity level [2]. Researchers often consider sentiment analysis as a two-class or three-class classification task. The related works can also benefit large scale multimedia content analysis and retrieval [3,4].

Works using supervised method generally test different combinations of features on various classifiers. Pak and Paroubek [5] first evaluated unigrams, bigrams and trigrams on three classifiers: SVM, multinomial Naive Bayes and CRF, their results show that the best performance was achieved by multinomial Naive Bayes with bigrams. Many researches also considered POS tags as another important type of feature. Agarwal et al. [6] included the number of adjectives, adverbs, nouns, verbs and the sum over all corresponding words as features, and successfully improved the accuracy of their classifiers. Other commonly used features are Twitter specific ones: emoticons, hashtags, slangs etc. Kouloumpis et al. [7] and Saif et al. [1] took the presence or absence of positive and negative emoticons as a binary feature; Chikersal et al. [8] count the number of emoticons; Bakliwal et al. [9] and Mudinas et al. [10] replaced emoticons with particular tokens such as "Good One" or "Bad One". The result of their evaluations showed that incorporation of such explicit features help to improve the performance for sentiment analysis solutions.

As for unsupervised approach, a variety of sentiment lexicons have been proposed: AFINN [11], VADER [12], Opinion Observer [13], SentiStrength [14], SentiWordNet [15], etc. The underlying assumption is that a large amount of words have a context-independent sentiment polarity. For example, "happy", "love", "appreciate" are more likely to bear positive emotions while "sad", "hate", "miserable" express a negative mood. Most works made use of these lexicons together with some predefined rules to calculate the overall sentiment score. Among these rules, the most crucial one is aimed at negation which relates to two tasks: how to determine the scope of a negation and how to process the score of negated words. Ding et al. [16] and Taboada et al. [17] both apply sentiment analysis on reviews which, in general, are much longer and more grammatically correct than tweets. Ding et al. [16] check the adjacent two words and flip the polarity of positive and negative words, they also take "negation + neutral" pattern as a negative situation. SO-CAL proposed by Taboada et al. [17] looks backwards as long as the words/tags found are in a backward search skip list. Instead of changing the sign, they shift sentiment orientation (SO) value towards the opposite polarity by a fixed amount. In the context of tweets, Thelwall et al. [14] generated SentiStrength which reverses the sentiment scores of negated words and skipping any intervening modifiers.

3   Overview of the Proposed System—LingoSent

LingoSent is composed of three separate parts. The first component is a data preprocessing pipeline containing a tokenizer, a normalizer and a Part-of-Speech (POS) tagger. The second component is linguistic feature extractor, represents the main focus of this article. We motivate and describe the extraction of a series of word-level, phrase-level and sentence-level linguistic features. Sentiment calculator, as the last component, calculates the sentiment strength of a tweet. To this end, LingoSent uses a set of publicly available sentiment lexicons to get the basic scores of tokens and then uses the results of feature extraction to modulate these scores. The overall sentiment score of a tweet derives from the sum of the scores of all tokens. Depending on the polarity of the overall sentiment score, we finally decide whether the sentiment of a tweet is positive or negative. In the following subsections, we present and discuss each component in detail.

4   Sentiment Lexicons and Dictionaries

Like most solutions for sentiment analysis, LingoSent relies on sentiment lexicons. We currently use four lexicons: AFINN [11], Opinion Observer [13], SentiStrength [14] and VADER [12]. We favor manually generated, or at least manually curated lexicons over automatically created ones since they represent a gold-standard regarding human perception (Table 1).

Table 1. Sentiment lexicons

Lexicon	Size	Valuation	Comments
AFINN	3,382 unigrams 96 emoticons	$\{-5, -4, -3, ..., 0, ..., 3, 4, 5\}$	Emoticons, slang
Opinion observer	6,886 unigrams	$\{positive, negative\}$	Comparatives, superlatives (same values)
SentiStrength	2,547 unigrams (words + word stems)	$\{-5, -4, -3, ..., 0, ..., 3, 4, 5\}$	Emoticons, slang comparatives, superlatives (different values)
VADER	7,298 unigrams 218 emoticons	$[-4, 4]$	Emoticons, slang comparatives, superlatives (different values)

Apart from the sentiment lexicons, our system also relies on a set of dictionaries particularly for normalizing as well as the extraction of various linguistic features. Table 2 gives an overview of the used dictionaries. In this work, we limit ourselves to English tweets. To build our dictionary $\mathcal{D}^{en}$ of English words, we use the out-of-the box word files provided by Linux Ubuntu 14.04. We ignore different spellings between British and American English by merging the two corresponding word files. We use this dictionary primarily to normalize expressive lengthenings. As subsets of $\mathcal{D}^{en}$, we manually created two dictionaries of words that add or remove (sentiment) strength to or from words. Firstly, $\mathcal{D}^{en\uparrow}$ contains English intensifiers such as *very*, *extremely* or *absolutely*. In contrast, $\mathcal{D}^{en\downarrow}$ contains downtoners such as *hardly*, *slightly* or *a little bit*. The negation words dictionary $\mathcal{D}^{\overline{en}}$ contains individual words such as *not*, *neither*, *never*, contractions such as *doesn't*, *hasn't*, *won't*, as well as their common misspellings such as *doesnt*, *hasnt*, *wont*, etc. We created our Internet slang dictionary $\mathcal{D}^{slang}$ by crawling noslang.com,[2] fetching the slang words as well as the corresponding word phrases in plain English (e.g., *lol* ⇒ *laughing out loud*).

Table 2. Overview of used dictionaries

Dictionary	Description	Example
$\mathcal{D}^{en}$	English words (British/American English)	*happy, move, successfully*
$\mathcal{D}^{en\uparrow}$	Intensifier words	*very, really, extremely*
$\mathcal{D}^{en\downarrow}$	Downtoner words	*hardly, slightly, a little*
$\mathcal{D}^{\overline{en}}$	Negation words	*not, neither, never*
$\mathcal{D}^{slang}$	Internet slang words (including translations)	*lol: laugh out loud*

5   Data Preprocessing

Text processing assumes that low-level tasks such as tokenizing and normalizing performed reliably. However, off-the-shelf solutions for these tasks typically perform poorly on tweets due the informal language, spacing errors, punctuation errors, typos, etc. We therefore implemented a preprocessing pipeline optimized to handle informal writing style of social media messages.

5.1   Tokenizer

We use regular expressions to match different types of tokens and immediately label them accordingly. Here we pay special attention to those situations that require to exclude some already labeled tokens from further tokenizing steps. For example, emoticons need to be labeled before punctuation marks since emoticons often contain punctuation marks. So does token used to denote time which may contain colon, thus we can use this regular expressions: *(?:(?:([01]\?d|2[0-3]):)?([0-5]?\d):)?([0-5]?\d)\s?(am|pm|a\.m\.|p\.m\.)?* to label all kinds of formats to indicate time and then exclude them before we label punctuation marks.

5.2   Normalizer

Text normalization refers to the task of replacing nonstandard tokens with standard ones to improve the effectiveness of subsequent processing steps.

Expressive Lengthenings. Since repeated characters generally do not change the phonetic encoding of words. We index each word in $\mathcal{D}^{en}$ and $\mathcal{D}^{slang}$ based on its phonetic encoding. If the normalizer encounters a word that is not in any of

---

[2] http://www.noslang.com/dictionary/.

those dictionaries, it calculates the pairwise similarity between the input word and all candidates using the Levenshtein distance to return the best-matching candidate.

Emoticons. An emoticon is a pictorial representation of facial expressions made up of various punctuations or letters. We normalize positive emoticons and negative emoticons with *[EMOTICON+]* and *[EMOTICON-]*, respectively.

Emojis. The most commonly used emojis allow users to convey positive or negative emotions and feelings, e.g., 😞 😵 ☺. Similar to emoticons, we normalize all emojis by converting them into *[EMOJI+]*, *[EMOJI∘]* and *[EMOJI-]*.

Emotexts. Lastly, we normalize emotexts like *haha, hihi, xixi* which are popular means to convey laughter in social media texts.

5.3 Part-of-Speech (POS) Tagger

For this, we use a publicly available POS tagger which has been trained on a dataset of random English tweets [18].

6 Linguistic Feature Extraction

This section describes in detail how we extract the linguistic features to improve the accuracy of LingoSent.

6.1 Auxiliary Features

We introduce an intensifier factor IF_t for each token t, with $IF_t \geq 0$ and $IF_t = 1$ by default. With the following logarithmic function we update the factor given a set of parameters, potentially derived from linguistic features:

$$IF_t^{new} = IF_t^{old} \cdot 2^{\log_{10}\left(\sum_{p \in P} p\right)} \tag{1}$$

where the values of different parameters p can derive from, e.g., the degree of elongation of tokens due to expressive lengthenings.

6.2 Word-Level Linguistic Features

We consider three kinds of word-level features: quoted-words, all-caps and expressing lengthenings.

Quoted Words. In formal writing, users often quote individual words to make it explicit that they are sarcastic or cynical, hence flipping their original polarity of the word. We therefore distinguish between quoted phrases/sentences and quoted words and mark each token of a tweet accordingly.

All Caps. In our current implementation, we mark a token t by setting $s_t^{AllCaps} = 1$ if more than seventy-five percent of its letters are capitalized, and 0 otherwise. Then we update IF_t accordingly

$$IF_t^{new} = IF_t^{old} \cdot 2^{\log_{10}\left(1+s_t^{AllCaps}\right)} \tag{2}$$

Expressive Lengthenings. We assign each token with a scaling factor $s_t^{ExLen} \in [1, \infty\}$ to reflect the differences in length between the original and normalized token, $s_t^{ExLen} = |t_{orig}|/|t_{norm}|$. Then we update IF_t accordingly

$$IF_t^{new} = IF_t^{old} \cdot 2^{\log_{10}\left(s_t^{ExLen}\right)} \tag{3}$$

6.3   Phrase-Level Linguistic Features

For phrase-level linguistic features we mainly handle the negation words, adjective and noun pairs and degree modifiers.

Negation. We identify negation words using dictionary $\mathcal{D}^{\overline{en}}$ — ignoring words such as *"not just"* or *"not only"* and mark subsequent words as negated. We stop negating words at punctuation terminal points $(.?!)$ or non-standard tokens such as emails, URLs, emoticons, emojis and emotexts and a set of delimiter words. Then we update IF_t:

$$IF_t^{new} = IF_t^{old} \cdot (-1) \tag{4}$$

where token t is in the scope of a negation. But again, if and how negation affects the intensifier factors of the negated words is left to the user of our framework by selecting different parameters for the intensifier function.

Adjective-Noun-Pairs (ANP). Sentiment polarity of adjectives and their modified nouns maybe conflicting, thus we match adjectives with nouns using regular expressions over the string of POS tags.

Intensifiers and Downtoners. We also use regular expressions over the POS tag string together with the dictionaries of degree modifier words $\mathcal{D}^{en\uparrow}$ and $\mathcal{D}^{en\downarrow}$ to find all intensifier and downtoners and their modified words. We denote the set of degree modifiers which modify token t as $\mathcal{W}_t^{DM}$. With that, we can calculate the scaling factor s_t^{DM} with respect to intensity modifiers as:

$$s_t^{DM} = \sum_{w \in \mathcal{W}_t^{DM}} \left(1 + s_w^{AllCaps} + s_w^{ExLen}\right) \tag{5}$$

Then we update IF_t as follows:

$$IF_t^{new} = IF_t^{old} \cdot 2^{\log_{10}\left(s_t^{DM}\right)} \tag{6}$$

6.4   Sentence-Level Linguistic Features

For the third level linguist feature we focus on the terminal punctuation marks, adversative sentence, conditional sentence and subjunctive mood.

Sentence Intensifier Factor. We use the differences in the number of characters between the expressive lengthenings of terminal punctuation marks (tpm) and their standardized/normalized representation $(?, !, ?!, !?)$ to derive a scaling factor: $s_{tpm}^{ExLen} = |tpm_{orig}|/|tpm_{norm}|$, with $s_{tpm}^{ExLen} \geq 1$. With this, we update the intensifier factor SIF_t (initialized with $SIF_t = 1$):

$$\forall t, seg(t) = seg(tpm) : SIF_t^{new} = SIF_t^{old} \cdot 2^{\log_{10}\left(s_{tpm}^{ExLen}\right)} \tag{7}$$

where $seg(t)$ returns the segment number for token t.

Conjunctions (CONJ). We use regular expression to identify segments in sentences with an adversative conjunction $(but, yet, \text{etc.})$ of the form *"X but Y"*. We update the respective intensifiers to shift emphasis from the pre-conjunction to the post-conjunction clause:

$$IF_t^{new} = IF_t^{old} \cdot 2^{\log_{10}\left(sgn_t^{CONJ} \cdot k\right)} \tag{8}$$

where $sgn_t^{CONJ} = 1$ if token t is part of the pre-conjunction clause, and $sgn_t^{CONJ} = -1$ if t is part of the post-conjunction clause. Parameter k can be modified to reflect the degree of how much the emphasis is shifted from pre-conjunction clause to post-conjunction clause.

Conditionals (COND). Similar to conjunctions, we match conditional sentences with these forms: *"If X, Y"*, *"If X(,) then Y"* and *"Y, if X."*. Using different intensifier functions, we are able to specify how we want to update the intensifier factors of a token in both types of clauses.

Subjunctive Mood (SUB). We detect subjunctive mood with a wide range of words as well as word patterns according to the grammar, such as *"He thought it would be a perfect day."* We consider different verb tenses, their combination with different auxiliary verbs as well as the contractions form. For the time being, we ignore such sentences by setting the intensifier factor of all corresponding tokens to 0, i.e., $IF_t^{new} = 0$ if $sub(t) = true$, where $sub(t)$ returns *true* or *false* depending on whether token t is part of a subjective mood sentence.

7   Sentiment Calculator

As last main component of LingoSent, the sentiment calculator uses the integrated sentiment lexicons and the output of the feature extractor to assign a sentiment score to each token and eventually to the overall message.

7.1   Basic Sentiment Score

We get basic sentiment scores from lexicons set $\mathcal{L}$, let $L_t = \{\ell \in \mathcal{L} \mid t \in \ell\}$ be the subset of lexicons that contain token t. With this, we calculate the basic sentiment score s_t for token t:

$$s_t = \begin{cases} \frac{\sum_{\ell \in L_t} score(\ell,t)}{|L_t|} & if \ L_t \neq \emptyset \\ 0 & otherwise \end{cases} \tag{9}$$

where $score(l, t)$ returns the score for token t provided by sentiment lexicon ℓ. $s_t \in [-1, 1]$ since we normalize the scores from each lexicon to $[-1, 1]$.

7.2   Token-by-Token Update of Scores

First, we multiple the current sentiment score of a token with its IF_t, i.e., $s_t^{new} = s_t^{old} \cdot IF_t$. Second, we consider quoted words and negated words and flip their sentiment polarity. Therefore, if a token t is either quoted or negated we also set $s_t^{new} = (-1) \cdot s_t^{old}$.

7.3   Handling Conflicting ANPs

We establish conflicting ANPs based on the current sentiment scores of the respective tokens. In most cases, like in *"cruel humor"*, the polarity of adjective dominates the polarity of the phrase, therefore we flip the score of the noun, $s_N^{new} = (-1) \cdot s_N^{old}$. While in *"great mistake"*, the noun wins, so we flip the sentiment scores of all adjectives modifying it, $s_A^{new} = (-1) \cdot s_A^{old}$.

7.4   Handling Hashtags

Hashtag is usually concatenated phrase, thus we first split a hashtag in its most likely word phrase – e.g., *#ilovebejing* to *"i love beijing"* — using existing hashtag segmentation techniques [19]. We then simply use this new phrase as input for LingoSent to calculate its overall sentiment score.

7.5   Final Sentiment Scores

In the last step, we calculate the overall score S_T of a text message T, e.g., a tweet, by summing up the sentiment scores of all tokens: $S_T = \sum_{t \in T} s_t$.

8   Evaluation and Experiment Results

In this section, we present and discuss the evaluation of LingoSent with four real world tweets datasets from different resources: VADER-tweets, STS-GOLD, STS-TEST and SS-Tweet. If a dataset features labels outside our considered two classes *positive* and *negative*, we then map them to two classes.

8.1   Effect of Linguistic Features

We follow the straightforward approach by adding features and see how the accuracy changes. For this experiment, the baseline is the result stemming by purely the combination of all lexicons). Table 3 shows the results.

Table 3. Accuracy of LingoSent over all datasets

Accuracy (%)	VADER			STS-GOLD			STS-TEST			SS-TWEET		
	POS	NEG	ALL	POS	NEG	ALL	POS	NEG	ALL	POS	NEG	ALL
Baseline	84.57	82.04	83.79	80.22	59.49	65.93	81.87	67.23	74.65	77.61	56.48	68.85
Emoticons	85.02	82.35	84.19	80.22	59.49	65.93	81.87	66.10	74.09	78.51	56.48	69.38
Preprocessing	91.65	88.64	90.71	85.92	63.12	70.21	86.26	69.49	77.99	83.51	61.12	74.22
Word-level feature	91.72	88.64	90.76	85.62	63.20	70.26	86.26	70.06	78.28	83.43	61.12	74.18
Phrase-level feature	91.75	92.48	91.98	87.34	72.75	77.29	86.81	74.01	80.50	83.81	69.02	77.68
Sentence-level feature	**92.65**	**93.78**	**93.00**	**91.14**	**75.25**	**80.19**	**90.11**	**76.27**	**83.29**	**89.10**	**70.92**	**81.56**

The most noteworthy result is the positive effect of an optimized data preprocessing. Particularly normalizing tokens is essential to ensure that expressive lengthenings do not result in unsuccessful lookups in the sentiment lexicons. As the results show, considering more and more linguistic features does improve the results step by step, although the positive effect of a single feature can be rather small. Naturally, this limited effect is particularly pronounced for features that are less commonly used such as conflicting ANPs, adversative conjunctions, conditionals, and subjunctive mood. Since there were almost no changes, we therefore omitted their results in Table 3. All in all, the results confirm our initial expectation that the informal writing style in social media message poses both a challenge but also opens up new opportunities for the task of sentiment analysis.

8.2   Comparison with Existing Solutions

Lastly, we compared the accuracy of LingoSent with existing solutions that are publicly available: AFINN, VADER, and SentiStrength. AFINN uses the pattern object complied with all the tokens in its lexicon to match all the tokens in a tweet, if there is any matched result it then get the corresponding sentiment score in the lexicon, finally the overall sentiment score is the sum of all the scores of the matched tokens. AFINN only consider 96 emoticons and no other linguistic features. VADER increases the sentiment strength of tweets using all-caps and intensifiers by different values. And decrease the score of tweets using negations, all these addition and subtraction values are empirically derived. SentiStrength also consider capitalization, negation and degree modifiers only.

Table 4 shows the results for all solutions over all four datasets. Several observations are worth mentioning. First, the only setup that slightly outperforms LingoSent, is VADER when applied on their "own" dataset. On all other datasets, VADER yields a lower accuracy than our system. This suggests that evaluating sentiment analysis solutions should always consider multiple, but with respect to the context (here: Twitter) datasets. As general observation, most of the time, positive tweets are easier to identify than negative ones. One reason for this is that in tweets with a balanced number of sentiment-carrying tokens of opposite polarities, the scores for positive tokens are typically higher, thus dominating the final polarity. For example, in *"My best friend got sick yesterday."*, the positive score of *best* and *friend* are greater than the negative score of *sick*. Another common misclassification case is caused by none sentiment-bearing words are used in the messages, which is the intrinsic defects of lexicon-based method.

Table 4. Comparison with existing solutions

Accuracy (%)	VADER			STS-GOLD			STS-TEST			SS-TWEET		
	POS	NEG	ALL	POS	NEG	ALL	POS	NEG	ALL	POS	NEG	ALL
AFINN	81.50	80.89	81.31	76.74	55.78	62.29	79.67	64.41	72.14	74.25	49.95	64.18
VADER	**93.99**	92.40	**93.50**	79.91	55.92	63.37	81.32	64.41	72.98	76.49	48.68	64.96
SentiStrength	73.21	55.33	67.67	78.80	35.02	48.62	79.12	47.46	63.51	75.07	33.72	57.93
Our system	92.65	**93.78**	93.00	**91.14**	**75.25**	**80.19**	**90.11**	**76.27**	**83.29**	**89.10**	**70.92**	**81.56**

9 Conclusions

While sentiment analysis is an established research area and gained even more interest with the rise of social media, most existing solutions under-utilize language and linguistic features for the calculation of a sentiment score or polarity. However, particular on social media, users exhibit a language and writing style to make the sentiment, emotions, feelings and opinions more explicit. In this article, we provided a comprehensive overview of the most commonly used linguistic features, and we integrated these features into LingoSent, our current prototype of a sentiment analysis system. As the results of our evaluation show, considering linguistic features improves the accuracy of our system compared to standard approaches that often assume a more formal as well as syntactically and grammatically correct writing style.

In the context of this article, we focused on a lexicon-based, i.e., unsupervised, approach for the calculation of sentiments in tweets. This makes LingoSent independent from large datasets of labeled tweets for training. Additionally, supervised methods are typically more computationally expensive, and their accuracy strongly depends on the training data which makes their results more difficult to generalize. However, supervised methods can learn hidden pattern and features in text that cannot be uncovered using sentiment lexicons. For example, a tweet can convey a strong sentiment without containing words that are present in lexicons. In the long run, we therefore aim to extend LingoSent with a supervised component. The most interesting question here will be how to combine the unsupervised and supervised methods to get the best results. Various ways to do so are conceivable. For example, we can add the currently calculated sentiment score to the feature vector for the training process. Alternatively, we only use the classifier, if the absolute sentiment score is too low to make a meaningful decision.

Acknowledgments. This work was supported in part by the National Natural Science Foundation of China (61572356, 61472275, 61502337), the Tianjin Research Program of Application Foundation and Advanced Technology (15JCYBJC16200), the grant of China Scholarship Council (201506255073), the grant of Elite Scholar Program of Tianjin University (2014XRG-0046).

References

1. Saif, H., He, Y., Alani, H.: Alleviating data sparsity for Twitter sentiment analysis. In: Proceedings of the WWW 2012 Workshop on 'Making Sense of Microposts', vol. 838, pp. 2–9 (2012)
2. Khan, A.Z.H., Atique, M., Thakare, V.M.: Combining lexicon-based and learning-based methods for Twitter sentiment analysis. Int. J. Electron. Commun. Soft Comput. Sci. Eng. (IJECSCSE) 89 (2015)
3. Liu, A.A., Nie, W.Z., Gao, Y., et al.: Multi-modal clique-graph matching for view-based 3D model retrieval. IEEE Trans. Image Process. 25(5), 2103–2116 (2016)
4. Liu, A.A., Su, Y.T., Nie, W.Z., Kankanhalli, M.: Hierarchical clustering multi-task learning for joint human action grouping and recognition. IEEE Trans. Pattern Anal. Mach. Intell. 39(1), 102–114 (2017). doi:10.1109/TPAMI.2016.2537337
5. Pak, A., Paroubek, P.: Twitter as a corpus for sentiment analysis and opinion mining. In: Proceedings of the Seventh International Conference on Language Resources and Evaluation. European Language Resources Association (2010)
6. Agarwal, A., Xie, B., Vovsha, I., et al.: Sentiment analysis of Twitter data. In: Proceedings of the Workshop on Languages in Social Media, pp. 30–38 (2011)
7. Kouloumpis, E., Wilson, T., Moore, J.D.: Twitter sentiment analysis: the good the bad and the OMG!. In: Proceedings of the Fifth International Conference on Weblogs and Social Media, vol. 11, pp. 538–541 (2011)
8. Chikersal, P., Poria, S., Cambria, E.: SeNTU: sentiment analysis of tweets by combining a rule-based classifier with supervised learning. In: Proceedings of the International Workshop on Semantic Evaluation, SemEval, pp. 647–651 (2015)
9. Bakliwal, A., Arora, P., Madhappan, S., et al.: Mining sentiments from tweets. In: Proceedings of the 3rd Workshop in Computational Approaches to Subjectivity and Sentiment Analysis, vol. 12, pp. 11–18. Association for Computational Linguistics, Stroudsburg (2012)
10. Mudinas, A., Zhang, D., Levene, M.: Combining lexicon and learning based approaches for concept-level sentiment analysis. In: Proceedings of the First International Workshop on Issues of Sentiment Discovery and Opinion Mining, pp. 5:1–5:8. ACM, New York (2012)
11. Nielsen, Å.: A new ANEW: evaluation of a word list for sentiment analysis in microblogs (2011). CoRR, arXiv preprint: arXiv:1103.2903

12. Hutto, C.J., Vader, G.E.: VADER: a parsimonious rule-based model for sentiment analysis of social media text. In: Proceedings of the Eighth International Conference on Weblogs and Social Media. The AAAI Press (2014)
13. Liu, B., Hu, M., Cheng, J.: Opinion observer: analyzing and comparing opinions on the web. In: Proceedings of the 14th International Conference on World Wide Web, pp. 342–351. ACM, New York (2005)
14. Thelwall, M., Buckley, K., Paltoglou, G.: Sentiment strength detection for the social web. J. Am. Soc. Inf. Sci. Technol. **63**, 163–173 (2012)
15. Baccianella, S., Esuli, A., Sebastiani, F.: SentiWordNet 3.0: an enhanced lexical resource for sentiment analysis and opinion mining. In: Proceedings of the International Conference on Language Resources and Evaluation, vol. 10, pp. 2200–2204. European Language Resources Association (2010)
16. Ding, X., Liu, B., Yu, P.S.: A holistic lexicon-based approach to opinion mining. In: Proceedings of the 2008 International Conference on Web Search and Data Mining, pp. 231–240. ACM, New York (2008)
17. Taboada, M., Brooke, J., Tofiloski, M., et al.: Lexicon-based methods for sentiment analysis. Comput. Linguist. **37**, 267–307 (2011)
18. Owoputi, O., O'Connor, B., Dyer, C., et al.: Improved part-of-speech tagging for online conversational text with word clusters. In: Human Language Technologies: Conference of the North American Chapter of the Association of Computational Linguistics, pp. 380–390. Association for Computational Linguistics, Stroudsburg (2013)
19. Berardi, G., Esuli, A., Marcheggiani, D., et al.: ISTI@ TREC Microblog Track 2011: exploring the use of hashtag segmentation and text quality ranking. In: Text REtrieval and Evaluation Conference (2011)

Multi-Task Multi-modal Semantic Hashing for Web Image Retrieval with Limited Supervision

Liang Xie[1], Lei Zhu[2(✉)], and Zhiyong Cheng[3]

[1] School of Science, Wuhan University of Technology, Wuhan, China
whutxl@hotmail.com
[2] School of Information Technology and Electrical Engineering,
The University of Queensland, Brisbane, Australia
leizhu0608@gmail.com
[3] School of Computing, National University of Singapore, Singapore, Singapore
zhiyong.cheng@nus.edu.sg

Abstract. As an important element of social media, social images become more and more important to our daily life. Recently, smart hashing scheme has been emerging as a promising approach to support fast social image search. Leveraging semantic labels have shown effectiveness for hashing. However, semantic labels tend to be limited in terms of quantity and quality. In this paper, we propose Multi-Task Multi-modal Semantic Hashing (MTMSH) to index large scale social image data collection with limited supervision. MTMSH improves search accuracy via improving more semantic information from two aspects. First, latent multi-modal structure among labeled and unlabeled data, is explored by Multiple Anchor Graph Learning (MAGL) to enhance the quantity of semantic information. In addition, multi-task based Share Hash Space Learning (SHSL) is proposed to improve the semantic quality. Further, MGAL and SHSL are integrated using a joint framework, where semantic function and hash functions mutually reinforce each other. Then, an alternating algorithm, whose time complexity is linear to the size of training data, is also proposed. Experimental results on two large scale real world image datasets demonstrate the effectiveness and efficiency of MTMSH.

Keywords: Multi-modal hashing · Image retrieval · Multi-task learning · Cross-modal retrieval

1 Introduction

In recent years, with the fast development of information and network technologies, there has been a massive explosion of social multimedia on the Web. Large amounts of multimedia contents, especially images, are shared and accessed by users on Wikipedia, Flickr, Youtube and other popular social websites. Due to the nature of social images, current retrieval method should possess two basic

© Springer International Publishing AG 2017
L. Amsaleg et al. (Eds.): MMM 2017, Part I, LNCS 10132, pp. 465–477, 2017.
DOI: 10.1007/978-3-319-51811-4_38

characteristics, to meet the development tread of social multimedia. The first characteristic is multi-modality. Images on social web are usually associated with various auxiliary information, including text description, geo-tags, user information, temporal record, and etc. These heterogeneous auxiliary contents make image retrieval be a multi-modal problem. The other characteristic is the scalability, there are very large amount of images shared on the web. For example, Flickr hosts billions of images, and it has more than 3.5 million new images uploaded daily. How to efficiently process such a large number of images has became a very challenging task. In the designing of a practical retrieval method, the efficiency of training process also should be mainly considered.

Recently, substantial effort has been devoted to cope with two characteristics. For the characteristic of multi-modality, practical retrieval methods should have the ability to combine associated modalities with images [1]. Moreover, multi-modal retrieval methods also have to support queries from modality other than image, which are also named as cross-modal retrieval [2,20]. For the characteristic of scalability, hashing has gained much attention of many researchers. Hashing technique is an efficient indexing approach which can be used to solve the retrieval of large-scale web images, it converts high-dimensional data into short binary codes, which preserve the similarity of data. Then fast search can be easily implemented by XOR and bit-count operations. Machine learning based hashing methods are proven to be superior to random projection based hashing [3–5].

Multi-modal hashing [18,19] integrates the advantages of multi-modal analysis and hashing. As a result, it is able to possess both multi-modality and scalability for Web image retrieval. However, current multi-modal hashing methods do not pay much attention to bridging the 'semantic gap', which is an essential problem in multi-modal analysis. Different modalities, such as image and text, usually vary in terms of their contents and structures, even if they describe the same concept, thus they are difficult to be correlated.

Existing multi-modal hashing methods are not able to well bridge 'semantic gap', in that they cannot effectively analyze the semantic information. Unsupervised multi-modal method only use unlabeled data which lack semantic information. Supervised multi-modal hashing can bridge the 'semantic gap' by exploiting labeled training data [6]. However, they only consider using the raw semantic labels for hashing. Semantic labels in training data is usually limited in terms of quantity and quality, so performance of supervised method is significantly affected by insufficient training labels. Some semi-supervised hashing methods [7] try to supplement quantity of semantic information with graph similarity. But they still cannot improve the quality of raw labels, thus they are not able to effectively bridge 'semantic gap'.

In this paper, we propose Multi-Task Multi-modal Semantic Hashing (MTMSH) for large-scale Web image retrieval in the case of limited supervision. Since the semantic information is usually very scarce, we improve the semantic information to achieve better hashing performance. Two kinds of semantic enhancements are considered to improve semantic information. Firstly, latent structure which is hidden in both labeled and unlabeled training data, is mined by Multiple Anchor Graph Learning (MAGL). By the exploiting of unlabeled data, the quantity of semantic information labels in training data can be improved. Then, correlation between semantic labels is learned by Shared Hash Space Learning (SHSL) which uses multi-task scheme for learning. Label correlation can improve the quality of semantic information, for example, 'beach' is usually related to 'sea', if an image is labeled by 'beach', then it is likely to be labeled by 'sea'. Multi-task learning can effectively learn the label correlation for improving semantics [21,22].

MTMSH formulates SHSL and MAGL in a joint framework where hash functions and semantic labels can be simultaneously optimized. Semantic labels can be effectively enhanced by MTMSH, then enhanced semantic labels can be encodes by hash space to support more effective hashing. In the optimization of MTMSH, an efficient alternating algorithm, whose time complexity is linear to the training data, is proposed. Extensive experiments on two multi-modal image datasets collected from Web demonstrate the effectiveness and efficiency of the proposed method.

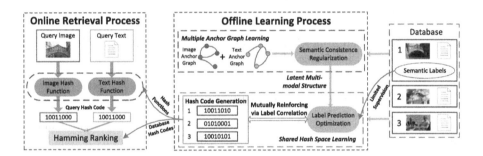

Fig. 1. The graphical illustration of MTMSH.

2   Multi-Task Multi-modal Semantic Hashing

2.1   System Overview

MTMSH involves two basic processes: offline hashing function learning and online retrieval. Figure 1 illustrates key architecture leveraging both labeled and unlabeled multi-modal data for training. The effectiveness of label prediction is enhanced via both MAGL and SHSL. In MAGL, anchor graphs of images and text are combined with proper weights. In SHSL, a unified hash space is learned using multi-task scheme, where hash space is simultaneously shared by semantic labels and different modalities. In MTMSH hash codes can interact with semantic functions, and they mutually reinforce each other. Based on the semantic enhancements and interaction, final hash functions can effectively preserve the refined semantic information.

2.2 Problem Setting

Suppose that training set consists of N multi-modal documents $x_n|_{n=1}^{N}$, each document contains M modalities, and $x_n = \{x_{n1}, \ldots, x_{nM}\}$, where x_{nm} is the feature extracted from modality m. Without loss of generality, in this work we mainly consider visual and text modality. Therefore, each document contains an image and a text, $m = 1$ denotes image and $m = 2$ denotes text. Although only two modalities are considered, our method can be extended to more modalities. Most of the training documents are not labeled, only $N_l \ll N$ documents are labeled with C semantic concepts, and their corresponding label matrix is $Y_l \in \{0, 1\}^{N_l \times C}$.

Our main aim is to learn effective hash functions which project all the modalities into a unified hamming space, and the hash function of each modality is defined as:

$$h_m = \mathrm{sgn}\,(x_m P_m - b_m) \tag{1}$$

where x_m is the feature vector of modality m. $P_m \in \mathbb{R}^{d_m \times K}$ is the hash projection matrix for modality m, K is the code length, and d_m is the feature dimension of modality m. b_m is the threshold parameter, it is computed as the mean of all $x_m P_m$ from the training data.

In order to overcome the shortage of semantic information in training data, we enhance the semantic labels in two aspects, including mining latent multi-modal structure and modeling label correlation.

2.3 Multiple Anchor Graph Learning

To solve the problem of limited supervision, we first enhance the semantic information in training data, by mining the latent multi-modal structure among both labeled and unlabeled data, to improve the semantic information. Anchor graph which is efficient and effective [8] is applied in our framework.

In the construction of anchor graphs, we randomly select N_a anchors $x_j^a|_{j=1}^{N_a}$ from the training data to approximate the data neighborhood structure. Then the truncated similarity matrix $Z \in \mathbb{R}^{N \times N_a}$ between the training data and anchors can be computed as:

$$Z_{ij} = e^{-dist(x_i - x_j^a)/\sigma} \tag{2}$$

where $dist(x_i - x_j^a)$ is the L2 distance between two points, σ is the mean of all distances. The graph matrix of each modality can be approximated by $A_m = Z_m \Lambda_m^{-1} Z_m^T$, where $\Lambda_m = \mathrm{diag}\,(Z_m^T 1)$, $Z_m \in \mathbb{R}^{N \times N_a}$ is the truncated similarity matrix of modality m.

In term of effectiveness of semantic learning, multi-modal fusion enjoys better performance than unimodal approach [9]. Thus we construct anchor graph matrices from all the modalities, then they are combined with proper weights. Different modalities have different contribution in semantic learning, thus they should have different weights. Based on this idea, we can obtain the formulation of Multiple Anchor Graph Learning (MAGL) for semantic enhancement:

$$\min \sum_{m=1}^{M} \alpha_m Tr \left(F_m^T L_m F_m \right) \tag{3}$$

where F_m is the semantic label matrix predicted by m-th modality. L_m is the Laplacian matrix, and $L_m = I - A_m$. α_m is the modal weight. $Tr(\cdot)$ denotes trace operator.

2.4   Shared Hash Space Learning

SHSL uses multi-task scheme to learn semantic correlation. In SHSL, a unified hash space is shared by semantic labels, so that similar labels can be correlated in hash space. To formulate SHSL, we particularly treat each semantic label as a task, and assume that the hash codes of training data are shared among all labels, then the semantic function is defined as:

$$F_m (Z_m) = Z_m U + HV \tag{4}$$

where $Z_m \in \mathbb{R}^{N \times N_a}$ is the multi-modal anchor matrix of modality m, $H \in \mathbb{R}^{N \times K}$ is the unified hash matrix of training data. $U \in \mathbb{R}^{N_a \times C}$ and $V \in \mathbb{R}^{K \times C}$ are the weight matrices. H is both shared by multiple modalities and semantic labels, and it is defined as;

$$H = \sum_{m=1}^{M} \alpha_m Z_m P \tag{5}$$

where $P \in \mathbb{R}^{N_a \times K}$ is the multi-modal projection matrix. α_m is the weight of modality m, it reflects modality importance in the learning process.

We aim to improve semantic functions by the shared hash space. Thus, our formulation achieves two goals: (1) minimizing classification error of labels and (2) making hash function of different modalities be consistent with the unified hash space. To this end, the objective function can be found as below:

$$\min \frac{1}{N_l} \|F_l - Y_l\|_F^2 + \lambda \|U + PV\|_F^2 + \mu \left(\|U\|_F^2 + \beta \mathcal{R}(P,V) \right)$$
$$+ \frac{\mu}{N} \sum_{m=1}^{M} \alpha_m \left(\|H_m - H\|_F^2 + \gamma \|P_m\|_F^2 \right) \tag{6}$$

where $\|\cdot\|_F^2$ is the Frobenius norm. The second term can be regarded as multi-task term which makes the unified hash space be shared by semantic labels. λ, μ, β and γ are the parameters of their corresponding regularization terms. $\mathcal{R}(P,V)$ is the regularization term for P and V, it is defined as:

$$\mathcal{R}(P,V) = \|P\|_F^2 + \|V\|_F^2 \tag{7}$$

F is the semantic label matrix, H is the unified hash matrix, H_m denotes hash matrix of modality m, they are defined as:

$$F = \sum_{m=1}^{M} \alpha_m F_m = ZU + HV$$

$$H_m = X_m P_m, \ H = ZP, \ Z = \sum_{m=1}^{M} \alpha_m Z_m \qquad (8)$$

F_l is chosen by the labeled data from F.

According to Eq. (6), we can find that not only semantic matrix F is enhanced, but also hash matrix H and H_m are optimized. H is shared by F and H_m, which means the hash space encodes both label and multi-modal correlation.

2.5   Overall Framework

By jointly considering two semantic enhancements, we have the formulation of MTMSH in the scenario of limited supervision:

$$\min \frac{1}{N_l} \|F_l - Y_l\|_F^2 + \frac{\mu}{N} \sum_{m=1}^{M} \alpha_m \left(\|H_m - H\|_F^2 + \gamma \|P_m\|_F^2 \right) + \lambda \|U + PV\|_F^2$$

$$+ \mu \left(\|U\|_F^2 + \beta \mathcal{R}(P, V) \right) + \frac{1}{N^2} \sum_{m=1}^{M} \alpha_m Tr \left(W^T Z_m^T L_m Z_m W \right) \qquad (9)$$

For simplicity, we also define:

$$B_m = Z_m^T L_m Z_m = Z_m^T \left(I - Z_m \Lambda Z_m^T \right) Z_m = Z_m^T Z_m - Z_m^T Z_m \Lambda Z_m^T Z_m \qquad (10)$$

Moreover, we can define a new weight matrix $W = U + PV$, then $F = ZW$ and $U = W - PV$. By substituting F, U, H_m and B_m into Eq. (9), the final formulation can be:

$$\min \qquad \frac{1}{N^2} \sum_{m=1}^{M} \alpha_m Tr \left(W^T B_m W \right) + \frac{1}{N_l} \|Z_l W - Y_l\|_F^2 + \lambda \|W\|_F^2$$

$$+ \mu \left(\|W - PV\|_F^2 + \beta \mathcal{R}(P, V) \right) + \frac{\mu}{N} \sum_{m=1}^{M} \alpha_m \left(\|ZP - X_m P_m\|_F^2 + \gamma \|P_m\|_F^2 \right)$$

$$s.t. \qquad \sum_{m=1}^{M} \alpha_m = 1, \alpha_m > 0 \qquad (11)$$

Z_l denotes the anchor graphs of labeled data, and it is chosen from Z.

From Eq. (11) we can find an advantage that semantic function are enhanced by two aspects, including the learning of hash functions and multi-modal anchor graphs, which effectively solve the shortage of semantics in training data. The other advantage of this framework is that hash space and label functions are interactive with each other. According to Eq. (11), semantic information can be preserved by the hash space, which will improve the performance of hashing. On the other hand, the unified hash space can model label correlation to enhance semantic information. Based on the interaction and semantic enhancements, our framework can effectively enhance the semantic labels, and it finally improves hashing performance.

2.6   Optimization

In this subsection we discuss the optimization of the objective function Eq. (11). We first compute P_m by setting the derivative of Eq. (11) w.r.t P_m to zero, then we have:

$$P_m = R_m X_m^T ZP \quad R_m = \left(X_m^T X_m + \gamma I \right)^{-1} \tag{12}$$

Substituting P_m into Eq. (11), the objective function becomes:

$$\min_{W,\alpha,P} \frac{1}{N^2} \sum_{m=1}^{M} \alpha_m Tr \left(W^T B_m W \right) + \frac{1}{N_l} \| Z_l W - Y_l \|_F^2 + \lambda \| W \|_F^2$$
$$+ \mu \left(\| W - PV \|_F^2 + \beta \mathcal{R}(P,V) \right) + \frac{\mu}{N} Tr \left(P^T DP \right) \tag{13}$$

where

$$D = \frac{1}{N} \sum_{m=1}^{M} \left(Z^T Z - Z^T X_m R_m X_m^T Z \right) \tag{14}$$

The objective function Eq. (13) is nonconvex with W, α, P and V. However, it is convex with any one of these parameters. Thus we propose an alternating algorithm for optimization. In each step, we update one parameter, and fix the rest parameters. The steps are iterated until convergence.

(1) Updating W: We fix α, P and V, by setting the derivative of Eq. (13) w.r.t W to zero, we have:

$$W = J^{-1} \left(\frac{1}{N_l} Z_l^T Y_l + \mu PV \right)$$
$$J = \frac{1}{N^2} \sum_{m=1}^{M} \alpha_m B_m + \frac{1}{N_l} Z_l^T Z_l + (\lambda + \mu) I_a \tag{15}$$

where I_a is the $N_a \times N_a$ identity matrix.

(2) Updating P: All W, α and V are fixed. We set the derivative of Eq. (13) w.r.t P to zero, and obtain the following equation:

$$DP + P \left(VV^T + \beta I_k \right) = WV^T \tag{16}$$

where I_k is the $K \times K$ identity matrix. Equation (16) is a Sylvester equation [10], we can obtain:

$$\text{vec}(P) = \left(I_k \otimes D + \left(VV^T + \beta I_k \right) \otimes I_k \right)^{-1} \text{vec}\left(WV^T \right) \tag{17}$$

(3) Updating V: We fix W, α and P, by setting the derivative of Eq. (13) w.r.t V to zero, we have:

$$V = \left(P^T P + \beta I_k\right)^{-1} P^T W \tag{18}$$

(4) Updating α: All W, P and V are fixed, by removing the irrelevant parts in (13), we can transform the objective function to:

$$\min \quad \alpha^T Q \alpha + c^T \alpha$$
$$s.t.\, \alpha^T \mathbf{1}_M = 1, \alpha > 0 \tag{19}$$

where $\alpha = [\alpha_1, \dots \alpha_M]^T$, $\mathbf{1}_M$ is an all-one column vector with length M, $Q \in \mathbb{R}^{M \times M}$ is computed by:

$$Q_{ij} = \frac{1}{N_l} Tr\left(W^T Z_{il}^T Z_{jl} W\right) + \frac{\mu}{N} Tr\left(P^T E_{ij} P\right)$$
$$E_{ij} = \sum_{m=1}^{M} \left(Z_i^T Z_j - Z_i^T X_m R_m X_m^T Z_j\right) \tag{20}$$

$c \in R^{M \times 1}$ is computed by

$$c_i = \frac{1}{N^2} Tr\left(W^T B_m W\right) - \frac{2}{N_l} Tr\left(W^T Z_{il}^T Y_l\right) \tag{21}$$

Objective function (19) is a constrained quadratic programming problem, many methods can be used to solve this problem. In this work, we use active set [11] to optimize α.

The whole optimizing algorithm for our framework is described in Algorithm 1. Since the objective function is not increased in each step, the convergence of Algorithm 1 is guaranteed. We can easily find that if the anchor size $N_a \ll N$, the optimizing algorithm can be regarded as linear to the training size N.

2.7　Out-of-Sample Extension

In the training process, we can learn semantic function, hash functions of each modality and hash codes of training data. We may have to compute the hash codes of database data which are not used for training. Database only consists of multi-modal data, and we combine image and text hash functions to compute their codes: $h = \text{sgn}\left(\sum_{m=1}^{M} \alpha_m \left(x_m P_m - b_m\right)\right)$.

3　Experiments

3.1　Datasets and Feature Preprocessing

In experiments, we use three multi-modal image datasets, including Wikipedia [12], and NUS-WIDE [13] for evaluation. Images in three datasets are all associated with text information, thus we refer each image and its associated text as

Algorithm 1. Optimizing algorithm of MTMSH

Input:
  $X_m|_{m=1}^{M}$, Y_l
Output:
  W, $P_m|_{m=1}^{M}$, P, α
1: Compute $Z_m|_{m=1}^{M}$ and $B_m|_{m=1}^{M}$ according to Eqs. (2) and (10);
2: Initialize H and V randomly;
3: Initialize α with $\alpha_m = 1/M$;
4: **while** True **do**
5:    Compute D according to Eq. (14);
6:    Update W, P and V according to Eqs. (15), (17) and (18);
7:    Compute Q and c according to Eqs. (20) and (21);
8:    Update α by solving the quadratic programming problem of Eq. (19);
9:    **if** Converge **then**
10:       Stop iteration;
11:    **end if**
12: **end while**
13: Compute $P_m|_{m=1}^{M}$ according to Eq. (12);

---

an image-text pair. The statistics of three datasets are summarized in Table 1. We use 200-D visual feature for images on Wikipedia and 100-D visual feature for images on NUS-WIDE, and the feature extraction is same to [14].

Table 1. The statistics of three multi-modal image datasets.

Datasets	Wikipedia	NUS-WIDE
Multi-modal database	2,173	201,562
Image queries	693	2,036
Text queries	693	2,036
Dimension of image feature	200	100
Dimension of text feature	30	1000
Training data	2173	10000
Labeled data	272	1,000

3.2   Implementation Details

We use two-fold cross-validation to select best values for parameters λ, μ, γ and β. They are all selected from the same candidate set $\left\{ 10^i \mid i = -7, -6, \ldots, 1, 2 \right\}$. On all three datasets, we set γ and μ to 10^{-3} and 10^{-6} respectively, and our method is not very sensitive to these two parameters. β is set to 10^{-2} on Wikipedia, 10^{-4} on MIR Flickr and NUS-WIDE. λ is set to 10^{-4} on Wikipedia, 10^{-2} on MIR Flickr and 10^{-3} on NUS-WIDE. In the anchor graph construction, we randomly select 10% of training data as anchors on Wikipedia, and select 5% of training data as anchors on MIR Flickr and NUS-WIDE.

3.3   Compared Methods and Evaluation Metrics

We compare our method to several state-of-the-art multi-modal hashing methods, including two supervised multi-modal hash methods: CVH [15] and SCMH [6], two unsupervised multi-modal methods IMH [17] and CMFH [16]. We also compare our method to the semi-supervised multi-modal hashing method MGH [7]. We adopt non-interpolated mean average precision (MAP) [17] to compare the retrieval performance, and the MAP scores are computed on the top 100 retrieved documents of each query.

In our experiments, supervised methods are only able to use labeled data for training, unsupervised and semi-supervised methods can use both labeled and unlabeled data for training. The codes of IMH, CMFH and SCMH are provided by the authors, and we implement CVH and MGH ourselves. All the parameters of compared methods are set to report the best performance.

3.4   Retrieval Results

On both two data sets, we evaluate all compared methods in two practical multi-modal retrieval tasks. The first is image query, where image example is used to search the multi-modal database. The other is text query, where text example is used to search the multi-modal database. The size of Wikipedia dataset is relatively small, so we use all database data for training. 272 training data are regarded as labeled, and the rest data are regarded as unlabeled. The size of database on NUS-WIDE is relatively large, some graph based methods such as IMH cannot use all data for training. Therefore, we randomly select 10,000 training data for all compared methods, where 1,000 data are used as labeled.

Table 2 shows the results of MAP scores in two retrieval tasks, with different code lengths, on both two datasets. We can find that MTMSH significantly outperforms other methods. An important feature of MTMSH is that its performance is consistently increased with the increasing of code length, and it also performs well at the short code length. From Table 2, we can observe the following results:

- MTMSH significantly outperforms the supervised hashing methods SCMH and CVH. In some cases, SCMH and CVH perform even worse than unsupervised CMFH and IMH. The reason is that we only use a small number of labeled data for training. This phenomenon illustrates that, in the scenario of limited supervision, directly using the scarce semantic information may obtain bad performance.
- Although MGH also uses anchor graphs to exploit latent structure of multi-modal data for hashing. It performs worse than MTMSH. The main reason is that MTMSH models the label correlation to improve semantic quality, while MGH ignores correlation between semantic labels.

Table 2. MAP scores of compared hashing methods on two datasets, the best results are marked as bold.

Task	Method	Wikipedia				NUS-WIDE			
		16	32	64	128	16	32	64	128
Image query	CVH	0.1711	0.1684	0.1671	0.1606	0.3126	0.3244	0.3244	0.3111
	IMH	0.2620	0.2330	0.2096	0.1950	0.4903	0.4801	0.4709	0.4574
	CMFH	0.2844	0.2976	0.3153	0.3131	0.3589	0.3698	0.4270	0.4534
	SCMH	0.1882	0.1917	0.1890	0.1874	0.3631	0.3635	0.3648	0.3673
	MGH	0.2691	0.2668	0.2675	0.2774	0.4693	0.4764	0.4751	0.4753
	MTMSH	**0.3340**	**0.3570**	**0.3645**	**0.3683**	**0.5702**	**0.5854**	**0.6088**	**0.6220**
Text query	CVH	0.5973	0.5894	0.5540	0.5743	0.3028	0.3214	0.3166	0.3216
	IMH	0.5618	0.5576	0.5711	0.6035	0.5075	0.4959	0.4775	0.4532
	CMFH	0.5431	0.5792	0.6099	0.6244	0.3975	0.4128	0.4364	0.4362
	SCMH	0.5993	0.6515	0.6843	0.6994	0.3724	0.3729	0.3754	0.3756
	MGH	0.5373	0.5393	0.5414	0.5632	0.5682	0.6441	0.6564	0.6534
	MTMSH	**0.7012**	**0.7196**	**0.7431**	**0.7478**	**0.6361**	**0.7051**	**0.7481**	**0.7674**

– Semantic enhancements are necessary in multi-modal retrieval with limited supervision. MTMSH exploits both label correlation and latent structure to improve semantic labels, then its hashing performance is much better than the compared methods.

Then we discuss the training efficiency of our method. Table 3 shows the training time of all methods on NUS-WIDE. CVH and SCMH cost less training time, the reason is the size of labeled training data is only 1000. Other methods use all 10000 training data, and MTMSH costs less time than other methods except CVH and SCMH. Moreover, the training time of MTMSH is not increased with the increasing of code length. Table 4 shows MAP scores and training time of different training size on NUS-WIDE. For all training sizes, we fix the number of anchors and labeled data, and hash code length is fixed to 64. We can observe that with the increasing of training size, MAP scores of two retrieval tasks are also increased. However, MAP scores are not improved significantly, which illustrates the stabilization of the hash functions learned by MTMSH

Table 3. The comparison of training time (seconds) on NUS-WIDE.

Code length	16	32	64	128
CVH	15.0	18.3	25.2	26.5
IMH	936.1	914.2	916.8	958.6
CMFH	139.2	135.9	136.8	167.4
SCMH	6.7	13.6	27.6	55.4
MGH	85.9	102.1	135.3	195.6
MTMSH	91.0	92.5	88.1	91.2

Table 4. MAP scores and training time of different training size on NUS-WIDE.

Metric	Training size		
	5k	10k	15k
MAP of image query	0.6025	0.6088	0.6207
MAP of text query	0.7392	0.7481	0.7494
Training time (seconds)	59.5	88.1	119.9

with reasonably small training set. Besides, the training time is increased linearly with the increase of training size, which confirms that the time complexity of our training process is linear to the training size.

4   Conclusions and Future Work

In this paper, we propose a novel hashing method: Multi-Task Multi-modal Semantic Hashing (MTMSH) for large scale Web image retrieval in the case that labeled data are limited. MTMSH can enhance semantic information from two aspects. It can improve quantity of semantic labels by MAGL which mines latent multi-modal structure of unlabeled and labeled data, and it can improve quality of semantic labels by SHSL which models the label correlation. Further MTMSH integrates two semantic enhancements in a joint framework, where the hash space and semantic labels are interactive with each other. Then we propose an efficient alternating algorithm whose time cost is linear to the database size to optimize MTMSH. At last, extensive experiments conducted on three Web images datasets show that our method are better than several state-of-the-art multi-modal hashing methods. And we also confirm the efficiency of our method by comparing the training time.

References

1. Meng, L., Tan, A.H., Leung, C., et al.: Online multimodal co-indexing and retrieval of weakly labeled web image collections. In: ICMR 2015, pp. 219–226 (2015)
2. Pereira, J.C., Coviello, E., Doyle, G., et al.: On the role of correlation and abstraction in cross-modal multimedia retrieval. IEEE TPAMI **36**, 521–535 (2014)
3. Weiss, Y., Torralba, A., Fergus, R.: Spectral hashing. In: NIPS 2009 (2009)
4. Liu, W., Wang, J., Ji, R., et al.: Supervised hashing with kernels. In: CVPR 2012 (2012)
5. Gong, Y., Lazebnik, S., Gordo, A., et al.: Iterative quantization: a procrustean approach to learning binary codes for large-scale image retrieval. IEEE TPAMI **35**, 2916–2929 (2013)
6. Zhang, D., Li, W.J.: Large-scale supervised multimodal hashing with semantic correlation maximization. In: AAAI 2014 (2014)
7. Cheng, J., Leng, C., Li, P., et al.: Semi-supervised multi-graph hashing for scalable similarity search. Comput. Vis. Image Underst. **124**, 12–21 (2014)

8. Liu, W., He, J., Chang, S.F.: Large graph construction for scalable semi-supervised learning. In: ICML 2010 (2010)
9. Zhu, L., Shen, J., Jin, H., et al.: Content-based visual landmark search via multi-modal hypergraph learning. IEEE Trans. Cybern. **45**, 2756–2769 (2015)
10. Bartels, R.H., Stewart, G.W.: Solution of the matrix equation AX + XB = C [F4]. Commun. ACM **15**, 820–826 (1972)
11. Murty, K.G., Yu, F.T.: Linear Complementarity, Linear and Nonlinear Programming. Heldermann, Berlin (1988)
12. Rasiwasia, N., Costa Pereira, J., Coviello, E., et al.: A new approach to cross-modal multimedia retrieval. In: ACM Multimedia (2010)
13. Chua, T.S., Tang, J., Hong, R., et al.: NUS-WIDE: a real-world web image database from National University of Singapore. In: CIVR. ACM (2009)
14. Xie, L., Zhu, L., Pan, P., et al.: Cross-modal self-taught hashing for large-scale image retrieval. Signal Process. **124**, 81–92 (2016)
15. Kumar, S., Udupa, R.: Learning hash functions for cross-view similarity search. In: IJCAI 2011 (2011)
16. Ding, G., Guo, Y., Zhou, J.: Collective matrix factorization hashing for multimodal data. In: CVPR 2014 (2014)
17. Song, J., Yang, Y., Yang, Y., et al.: Inter-media hashing for large-scale retrieval from heterogeneous data sources. In: SIGMOD. ACM (2013)
18. Lin, Z., Ding, G., Hu, M., et al.: Semantics-preserving hashing for cross-view retrieval. In: CVPR 2015 (2015)
19. Irie, G., Arai, H., Taniguchi, Y.: Alternating co-quantization for cross-modal hashing. In: ICCV 2015 (2015)
20. Nie, L., Yan, S., Wang, M., et al.: Harvesting visual concepts for image search with complex queries. In: MM 2012 (2012)
21. Song, X., Nie, L., Zhang, L., et al.: Interest inference via structure-constrained multi-source multi-task learning. In: IJCAI 2015 (2015)
22. Xie, L., Pan, P., Lu, Y., et al.: A cross-modal multi-task learning framework for image annotation. In: CIKM 2014 (2014)

Object-Based Aggregation of Deep Features for Image Retrieval

Yu Bao and Haojie Li[(✉)]

School of Software, Dalian University of Technology, Dalian, China
hjli@dlut.edu.cn

Abstract. In content-based visual image retrieval, image representation is one of the fundamental issues in improving retrieval performance. Recently Convolutional Neural Network (CNN) features have shown their great success as a universal representation. However, the deep CNN features lack invariance to geometric transformations and object compositions, which limits their robustness for scene image retrieval. Since a scene image always is composed of multiple objects which are crucial components to understand and describe the scene, in this paper we propose an object-based aggregation method over the CNN features for obtaining an invariant and compact image representation for image retrieval. The proposed method represents an image through VLAD pooling of CNN features describing the underlying objects, which make the representation robust to spatial layout of objects in the scene and invariant to general geometric transformations. We evaluate the performance of the proposed method on three public ground-truth datasets by comparing with state-of-the-art approaches and promising improvements have been achieved.

Keywords: Image retrieval · Image representation · Deep Convolutional Neural Network

1 Introduction

Content-based image retrieval, as an important problem in computer vision, has been gaining considerable attention in the last decade [5,11,14,19,24,28]. Content-based image retrieval aims to find out similar images on the premise of a given query image from image datasets. One key to advancing image retrieval tasks is to form a discriminative and compact image representation in the presence of various image transformations.

For the last decade, the traditional method of generating image representation is to aggregate local invariant features like SIFT [25] to form final representation, such as Bag of Visual Words (BoW) [22], Fisher Vectors (FV) [18], Vector of Locally Aggregated Descriptors (VLAD) [10], triangulation embedding [11], etc. They have shown robustness to image scaling, translations, occlusion and local distortions. However, these methods are based on hand-crafted local features and their performance still has much room for improvement.

© Springer International Publishing AG 2017
L. Amsaleg et al. (Eds.): MMM 2017, Part I, LNCS 10132, pp. 478–489, 2017.
DOI: 10.1007/978-3-319-51811-4_39

Apart from the hand-crafted features, it is more and more successful to extract features in a data driven manner. Recently, deep learning approaches, especially Convolutional Neural Networks (CNN) has shown powerful ability in extracting distinctive image or object features and achieved great success in general image classification and object detection tasks [6,13]. There are several image retrieval works [1,7,21,23] based on CNN features. For examples Babenko *et al.* [1] extracts output of fully connected (fc) layers of CNN on the image as the image feature and achieves impressive performance. However, Gong *et al.* [7] has empirically shown that the CNN representation is sensitive to the transformations such as scaling, rotation and translation. Because it encodes the global spatial information of the image. At the same time, an image always includes multiple objects, with different positions, poses and scales. Thus it is insufficient to describe image by extracting CNN features from the whole image for content-based image retrieval.

In order to solve the above problem, some works [7,21] proposed to extract CNN features of local image patches instead of whole image and then aggregate these features into final image representation. Although these approaches are found to be more robust to image transformations, there are also several obvious weaknesses. For example, sliding-window-based image patch extraction in [7] fails to consider the visual contents of image, which may produce redundant and noisy information. In addition, max-pooling aggregation in [21] is coarse and may loss lots of useful information.

Motivated by above approaches, we propose an object-based aggregation method over the CNN features for obtaining a more discriminative image representation for image retrieval. We treat an image as multiple objects and extract CNN features from potential objects and then aggregate them into a compact and discriminative image representation. Compared with [7,21], we use content-based approach to extract patches that are more likely to contain semantic objects and aggregate local features taking into account the relationship between objects.

The overview of the proposed method is shown in Fig. 1. For an image, we first generate object proposals which are image patches that are more likely to contain objects. Then we extract the CNN fully connected layer features of these object proposals. After that, we use VLAD model to aggregate the CNN features

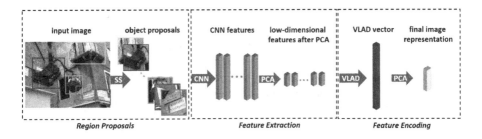

Fig. 1. The proposed pipeline on forming a global image representation.

of the object proposals extracted from an image to form a discriminative image representation that summarizes rich visual information.

We conduct extensive experiments on three well-known retrieval datasets and the results show that our proposed approach achieves superior retrieval accuracy over other state-of-the-art CNN features approaches.

The reminder of this paper is organized as follows. Section 2 briefly reviews some well-known image retrieval works based on CNN. In Sect. 3, we introduce our proposed method for image representation and retrieval. Experimental results are presented and discussed in Sect. 4. Finally, we conclude this paper in Sect. 5.

2   Related Work

Deep Convolutional Neural Networks (CNN) has made great breakthroughs which promote the development of many fields [13,15], especially in computer vision. Recently, there are several image retrieve researches based on CNN.

Many works [4,17] have proven CNN fc features own high level of visual abstraction. Babenko et al. [1] used fc features as global image representation for image retrieve and achieved a better performance than approaches based on hand-crafted features. But Gong et al. [7] experimentally showed CNN fc features are sensitive to image transformations such as scaling, rotation and translation.

To improve robustness of image representation, Gong et al. [7] proposed an approach called Multi-scale Orderless Pooling (MOP-CNN) which generates an image representation via encoding the CNN fc features from a set of patches extracted at different scales. MOP-CNN starts with resizing the image to a fixed size and extracts local image patch in sliding windows of three different sizes. Then, all patches are fed into CNN to extract their fc features. Lastly, these fc features corresponding to the same window size are pooled separately by VLAD [10], with the results further concatenated to form a global image representation. Although found to be more robust to image transformations, MOP-CNN has its weakness. The sliding-window for local patch extraction fails to consider the images contents, which may produce redundant and noisy information.

Reddy Mopuri et al. [21] proposed an approach that constructs an object-based invariant representation on top of the CNN features extracted from an image. Unlike MOP-CNN, they [21] uses Selective Search (SS) [26] to generate object proposals that are more likely to contain objects instead of sliding-window. Then the max-pooling is carried out on fc features of object proposals to form final image representation. Since the objects in the scene are crucial components to understand the underlying scene, object proposals include more visual content which is discriminative and robust. Although it is observed by [6] that each of the units of CNN features responds with a high activation to a specific type of visual input, max pooling only keeps max activations of CNN features and losses a lot of information.

Instead of generating global image representation, Sun et al. [23] directly represents images by a number of local features which are the fusion of CNN fc features and aggregated SIFT features. These features are used for retrieval without aggregation. But on account of computing and space complexity, this approach is unable to utilize numerous proposals. Therefore, it can not include rich content which can facilitate retrieval tasks.

3   Proposed Approach

In this section we will present the proposed compact image representation for image retrieval in detail. The pipeline of the proposed method is shown in Fig. 1, which is divided into three main modules: region proposals, feature extraction and feature encoding.

3.1   Object Proposals

Because the objects in the image are crucial components to understand the underlying image, we use object-based representation to describe images. Given an image, the proposed method first extracts local image patches, which are more likely to contain objects, called object proposals.

Compared to sliding-window method, we choose content-based strategy considering the structure and visual information of images. Recently there are a lot of researches [2,26] for generating a set of class independent object proposals on an image. Originally object proposals aim to provide the localization of the objects present in the image for object detection tasks and reduce the complexity of subsequent analysis compared with the sliding-window. These object proposals always have repeatability and saliency. Repeatability means that the proposals of two similar images are consistent, and saliency means that proposals contain rich discriminative information. These characters are very useful for comparison of similarity.

From many possible choices, we select Selective Search [26] in the proposed method because of its popularity and good performance in terms of detection, recall and repeatability. This approach combines the strength of both an exhaustive search and segmentation. This approach uses a bottom-up hierarchical grouping-based strategy, enabling itself naturally to generate class-independent and high-quality proposals at all scales.

3.2   Feature Extraction

The object proposals obtained are forwarded through a CNN for efficient features. It is in fact known that the features produced by CNN in the full connected layers have a high level semantic description of images and work effectively on many computer vision problems. The supremacy of the CNN features over the hand-crafted features is shown by a number of works [1,6,13]. In addition, different from local feature extraction processing, CNN has shown its fast processing speed by utilizing parallel GPUs. Although the number of proposals from an image is large, the computational efficiency also can be high due to the performance of GPUs.

CNN is usually built with a sequence of convolutional/max-pooling layers, followed by low-resolution fully-connected (fc) layers whose activations are fed to a soft-max classifier. The shallower convolutional layers pay more attention to the low-level local characteristics and max-pooling improves invariance to small-scale deformations like rotation, translation and scaling. The deeper fully-connected layers are more focused on the high-level visual abstraction.

Similar to [21] we choose the CNN architecture proposed by [13] called AlexNet, which contains 5 convolutional layers, each of which is followed by a pooling and a rectification, and 3 full connected layers. The network is implemented by the Caffe toolkit [12] and pre-trained on ImageNet [3]. Each of object proposals is reshaped to a fixed size and then passed through the network. We take the 4096-dimensional output of the seventh fully connected layer as the features of these object proposals. Finally, we conduct Principal Component Analysis (PCA) and whitening to reduce their dimensions since their dimensions are too high, which can make future computation more efficient.

3.3   Feature Encoding

At this stage, each image has a set of object level CNN features. Unlike [23] which uses the comparison of object level features which need higher computing complexity, we adopt VLAD [10] to aggregate these features into a compact and discriminative image global representation. VLAD has been shown to have great advantages in local features encoding methods [18,22,27]. Compared with BoW and simple max/average pooling methods which are coarse and may loss more information, VLAD considers the value of each dimension of the features and makes local information of images has a more detailed characterization. In addition, VLAD is orderless which helps to provide a higher robustness to the spatial layout of the objects in images.

With K centers $\{c_1, c_2, ..., c_K\}$ generated by K-means and a set of CNN features from an input image, the VLAD descriptor is constructed by assigning each proposal feature p_j to its r nearest cluster centers $rNN(p_j)$ and aggregating the residuals of the patches minus the center:

$$x = \left[\sum_{j:c_1 \in rNN(p_j)} w_{j1}(p_j - c_1), \; ..., \; \sum_{j:c_1 \in rNN(p_j)} w_{jk}(p_j - c_k) \right],$$

where w_{jk} is the Gaussian kernel similarity between p_j and c_k. For each proposal, we additionally normalize its weights to its nearest r centers to have sum one and set r = 5 in our experiments. However, the result VLAD encoding vector x is K*D dimensions, where K is the number of cluster centers and D is the dimensionality of CNN features after PCA. This is too high for many large-scale applications, so we further perform PCA on the result vectors to reduce their dimensions. The success of PCA and whitening on VLAD has been demonstrated in [8], where the two operations on the VLAD vector improve the retrieval performance remarkably. The VLAD vector after PCA is our final image representation. We compare the representation of the query image with that of the database images to retrieve the similar images. The similarity between images is computed according to distance measures such as l2 of representation vectors.

4   Experiments and Results

In this section, we report the performance of our proposed method. Experiments are carried out on a set of benchmark datasets. Our work pays more attention to have an effective image representation in our image retrieval system. For a fair comparison between the methods, we only report results on representations with relevant order of dimensions and exclude post-processing methods like spatial re-ranking and query expansion.

4.1   Dataset

We report retrieval results on following three standard benchmark datasets in the area. Sample images from these datasets are shown in Fig. 2.

Fig. 2. Sample database images. First row shows images from the *Holidays*, second from the *Oxford5K*, and third from the *UKBench* dataset.

INRIA Holidays. [9] contains 1491 images corresponding to 500 different classes captured at different places. Each class has 2–3 images describing the same object or location. One image per class is used as queries to evaluating the proposed system. Mean average precision (mAP) is the evaluation metric.

Oxford5K. [20] contains 5062 images collected from Flickr by searching for particular landmarks buildings in the oxford campus. There are 11 different landmarks, each has 5 queries for evaluating the retrieval performance through mAP.

Fig. 3. Image retrieval examples on the *Holidays* dataset. First image in each row (separated by line) is the query image and the subsequent images are the corresponding ranked similar images. The red boxes show images that are irrelevant and ranked higher than the relevant images. Note that irrelevant images ranked after the relevant images are not shown in red boxes. (Color figure online)

Table 1. Retrieval results on the *Holidays* dataset.

Method	Dimension							
	32	64	128	256	512	1024	2048	4096
VLAD [18]	48.4	52.3	55.7	-	59.8	-	62.1	55.6
Fisher vector [18]	48.6	52.0	56.5	-	61.0	-	62.65	9.5
Neural codes [1]	68.3	72.9	78.9	74.9	74.9	-	-	79.3
MOP-CNN [7]	-	-	-	-	-	-	80.2	78.9
OR-fused [23]	-	-	-	-	82.8	83.7	83.7	-
Object feature pooling [21]	74.0	80.7	85.1	87.8	88.5	86.6	85.9	85.9
Proposed	**76.6**	**81.4**	**87.5**	**91.4**	**92.0**	**89.0**	**89.7**	**89.7**

UKBench. [16] contains 10200 images corresponding to 2550 different indoor objects. Each object has images from four different viewpoints. The dataset provides a good benchmark for viewpoint changes. NS-score (averaged four times top-4 accuracy) is used to measure the retrieval accuracy.

4.2   Comparing Results

In the experiments, the object proposals are generated by the 'fast' version of Select Search [26]. Each image has an average of 2000 proposals. These proposals are described by the 4096 dimensional features obtained using AlexNet [13] pre-trained on ImageNet [3]. Through PCA-Whitening, CNN features are reduced to 512 dimensions. Then the codebook is generated by K-means with $k = 256$ centers. The resulting VLAD vectors have $256 * 512 = 131072$ dimensions. We future perform PCA on VLAD vectors and reduce them to F dimensions which can fairly compare with other methods.

Tables 1 and 2 report the mAP results for nearest neighbor retrieval of feature vectors using the Euclidean distance on Holidays [9] and Oxford5K [20] datasets. It can be shown that the proposed method achieves the best result under a variety of feature dimensions compared with other methods on two datasets. It proves that object proposals and VLAD are more effective for scene image compared with sample max-pooling [21] and sliding-window [7]. On Oxford buildings dataset, triangulation embedding [11] reports better performance, but it's hard to extend to large datasets because of its high-dimensional property. Note that it can obtain even higher results with our proposed methods with a large codebook. Finally, we show retrieval examples in Fig. 3 on Holidays dataset.

Table 3 reports the NS-Score results on UKBench [16] dataset. It is shown that our proposed method is comparable to other methods, but does not have a great improvement. The phenomenon may be explained by the type of dataset. Like examples in Fig. 2, most of the images on UKBench dataset only contain an object and the same object in different images has a big change in posture and point of view. Our method is based on objects and the CNN fc feature of object proposal is a kind of global description of an object which does not have

Table 2. Retrieval results on the *Oxford5K* dataset.

Method	Dimension								
	32	64	128	256	512	1024	2048	4096	8064
VLAD [18]	-	-	28.7	-	-	-	-	-	-
Fisher vector [18]	-	-	30.1	-	-	-	-	-	-
Neural codes [1]	39.0	42.1	43.3	43.5	43.5	-	-	54.5	-
Triangulation embedding [11]	-	-	43.3	-	-	-	-	62.4	**67.6**
Object feature pooling [21]	40.1	48.0	56.2	59.8	60.7	59.4	58.9	58.2	-
Proposed	36.1	**49.1**	**60.6**	**66.0**	**64.7**	**64.5**	**63.2**	62.7	-

Table 3. Retrieval results on the *UKBench* dataset.

Method	Dimension							
	32	64	128	256	512	1024	2048	4096
VLAD [18]	-	-	3.35	-	-	-	-	-
Fisher vector [18]	-	-	3.33	-	-	-	-	-
Neural codes [1]	3.3	3.53	3.55	3.56	3.56	-	-	3.56
Sparse-coded features [7]	**3.52**	**3.62**	3.67	-	-	-	-	-
OR-fused [23]	-	-	-	-	-	3.81	-	-
Object feature pooling [21]	3.4	3.61	**3.71**	3.77	**3.81**	3.84	3.84	**3.84**
Proposed	3.46	3.60	**3.71**	**3.78**	3.78	**3.85**	**3.87**	3.83

high robustness of the local changes. So the proposed method has advantages in multi-objects scene images, but not in single-object images.

4.3   Impacts of Parameters

In this subsection, we investigate the impacts of parameters in different datasets. We use 512-D CNN fc7 features after PCA-Whitening in all experiments and compute retrieval performance in final representation which is original VLAD vector without PCA.

Number of Centers in Codebook: It is essential to investigate the impact of the number of centers in codebook in the VLAD preprocessing stage, since it is critical to achieve a better tradeoff of performance and storage costs. We explore various numbers of centers K in VLAD for three datasets and the results are shown in Fig. 4. With the increase of K, we can see that the ability of the generated features improves but tends to be stable. Considering the storage cost and retrieval time, we adopt K = 256 as a trade-off of performance and resource consumption.

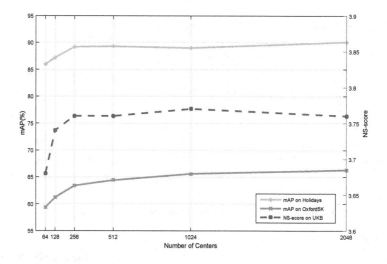

Fig. 4. Retrieval performance on various datasets for different number of centers in codebook

Number of Proposals: On step of generating object proposals, we use the 'fast' version of Selective Search [26] which provides around 2000 proposals per image as potential object locations. Those proposals are ranked by Selective Search according their priorities which indicate the possibility of containing objects. We explore considering only top N proposals for three datasets and the results are shown in Fig. 5. It can be observed that with only 100–500 proposals

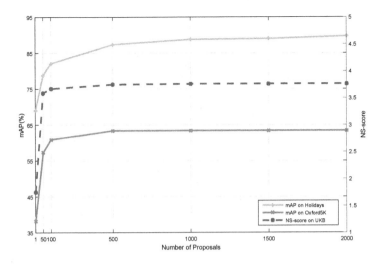

Fig. 5. Retrieval performance on various datasets for different number of object proposals per image

per image, our proposed method can get very competitive retrieval performances on all the datasets. This is mean that our method can be more efficient by only considering top priority proposals.

5   Conclusion

In this paper, we presented an object-based aggregation method over the CNN features for obtaining a more discriminative image representation for image retrieval. The representation exhibits robustness to layout of the objects in the scene and is invariant to general geometric transformations benefited from potential object proposals and effectiveness of VLAD. Meanwhile, our representation can achieve the state-of-the-art result for image retrieval task with compact codes of dimension less than 4096. In addition, the representation can be computed more efficiently by considering only few object proposals per image, which achieves competitive results as well.

In the future, we would like to optimize our approach in the following aspects. First, we will try to explore the relationship between object proposals to make the representation contain more information. Secondly, we will mine the different properties of activations of different layers in CNN and look for other complementary descriptors. Thirdly, we will investigate the appropriate method to index our compact image representations for very large scale image datasets.

Acknowledgments. This work is supported by National Natural Science Funds of China (61472059, 61428202).

References

1. Babenko, A., Slesarev, A., Chigorin, A., Lempitsky, V.: Neural codes for image retrieval. In: Fleet, D., Pajdla, T., Schiele, B., Tuytelaars, T. (eds.) ECCV 2014. LNCS, vol. 8689, pp. 584–599. Springer, Heidelberg (2014). doi:10.1007/978-3-319-10590-1_38
2. Cheng, M.M., Zhang, Z., Lin, W.Y., Torr, P.: BING: binarized normed gradients for objectness estimation at 300fps. In: Proceedings of the IEEE Conference on Computer Vision and Pattern Recognition, pp. 3286–3293 (2014)
3. Deng, J., Dong, W., Socher, R., Li, L.J., Li, K., Fei-Fei, L.: ImageNet: a large-scale hierarchical image database. In: IEEE Conference on Computer Vision and Pattern Recognition, CVPR 2009, pp. 248–255. IEEE (2009)
4. Donahue, J., Jia, Y., Vinyals, O., Hoffman, J., Zhang, N., Tzeng, E., Darrell, T.: DeCAF: a deep convolutional activation feature for generic visual recognition. In: ICML, pp. 647–655 (2014)
5. Douze, M., Ramisa, A., Schmid, C.: Combining attributes and fisher vectors for efficient image retrieval. In: 2011 IEEE Conference on Computer Vision and Pattern Recognition (CVPR), pp. 745–752. IEEE (2011)
6. Girshick, R., Donahue, J., Darrell, T., Malik, J.: Rich feature hierarchies for accurate object detection and semantic segmentation. In: Proceedings of the IEEE Conference on Computer Vision and Pattern Recognition, pp. 580–587 (2014)
7. Gong, Y., Wang, L., Guo, R., Lazebnik, S.: Multi-scale orderless pooling of deep convolutional activation features. In: Fleet, D., Pajdla, T., Schiele, B., Tuytelaars, T. (eds.) ECCV 2014. LNCS, vol. 8695, pp. 392–407. Springer, Heidelberg (2014). doi:10.1007/978-3-319-10584-0_26
8. Jégou, H., Chum, O.: Negative evidences and co-occurences in image retrieval: the benefit of PCA and whitening. In: Fitzgibbon, A., Lazebnik, S., Perona, P., Sato, Y., Schmid, C. (eds.) ECCV 2012. LNCS, vol. 7573, pp. 774–787. Springer, Heidelberg (2012). doi:10.1007/978-3-642-33709-3_55
9. Jégou, H., Douze, M., Schmid, C.: Hamming embedding and weak geometric consistency for large scale image search. In: Forsyth, D., Torr, P., Zisserman, A. (eds.) ECCV 2008. LNCS, vol. 5302, pp. 304–317. Springer, Heidelberg (2008). doi:10.1007/978-3-540-88682-2_24
10. Jégou, H., Perronnin, F., Douze, M., Sánchez, J., Perez, P., Schmid, C.: Aggregating local image descriptors into compact codes. IEEE Trans. Pattern Anal. Mach. Intell. **34**(9), 1704–1716 (2012)
11. Jégou, H., Zisserman, A.: Triangulation embedding and democratic aggregation for image search. In: Proceedings of the IEEE Conference on Computer Vision and Pattern Recognition, pp. 3310–3317 (2014)
12. Jia, Y., Shelhamer, E., Donahue, J., Karayev, S., Long, J., Girshick, R., Guadarrama, S., Darrell, T.: Caffe: convolutional architecture for fast feature embedding. In: Proceedings of the 22nd ACM International Conference on Multimedia, pp. 675–678. ACM (2014)
13. Krizhevsky, A., Sutskever, I., Hinton, G.E.: ImageNet classification with deep convolutional neural networks. In: Advances in Neural Information Processing Systems, pp. 1097–1105 (2012)
14. Nie, L., Wang, M., Zha, Z., Li, G., Chua, T.S.: Multimedia answering: enriching text QA with media information. In: Proceedings of the 34th International ACM SIGIR Conference on Research and Development in Information Retrieval, pp. 695–704. ACM (2011)

15. Nie, L., Wang, M., Zhang, L., Yan, S., Zhang, B., Chua, T.S.: Disease inference from health-related questions via sparse deep learning. IEEE Trans. Knowl. Data Eng. **27**(8), 2107–2119 (2015)
16. Nister, D., Stewenius, H.: Scalable recognition with a vocabulary tree. In: 2006 IEEE Computer Society Conference on Computer Vision and Pattern Recognition (CVPR 2006), vol. 2, pp. 2161–2168. IEEE (2006)
17. Oquab, M., Bottou, L., Laptev, I., Sivic, J.: Learning and transferring mid-level image representations using convolutional neural networks. In: Proceedings of the IEEE Conference on Computer Vision and Pattern Recognition, pp. 1717–1724 (2014)
18. Perronnin, F., Dance, C.: Fisher kernels on visual vocabularies for image categorization. In: 2007 IEEE Conference on Computer Vision and Pattern Recognition, pp. 1–8. IEEE (2007)
19. Perronnin, F., Liu, Y., Sánchez, J., Poirier, H.: Large-scale image retrieval with compressed fisher vectors. In: 2010 IEEE Conference on Computer Vision and Pattern Recognition (CVPR), pp. 3384–3391. IEEE (2010)
20. Philbin, J., Chum, O., Isard, M., Sivic, J., Zisserman, A.: Object retrieval with large vocabularies and fast spatial matching. In: 2007 IEEE Conference on Computer Vision and Pattern Recognition, pp. 1–8. IEEE (2007)
21. Reddy Mopuri, K., Venkatesh Babu, R.: Object level deep feature pooling for compact image representation. In: Proceedings of the IEEE Conference on Computer Vision and Pattern Recognition Workshops, pp. 62–70 (2015)
22. Sivic, J., Zisserman, A.: Video Google: a text retrieval approach to object matching in videos. In: Proceedings of the Ninth IEEE International Conference on Computer Vision, pp. 1470–1477. IEEE (2003)
23. Sun, S., Zhou, W., Tian, Q., Li, H.: Scalable object retrieval with compact image representation from generic object regions. ACM Trans. Multimed. Comput. Commun. Appl. (TOMM) **12**(2), 29 (2016)
24. Tang, J., Hong, R., Yan, S., Chua, T.S., Qi, G.J., Jain, R.: Image annotation by kNN-sparse graph-based label propagation over noisily tagged web images. ACM Trans. Intell. Syst. Technol. (TIST) **2**(2), 14 (2011)
25. Tang, S., Zheng, Y.T., Wang, Y., Chua, T.S.: Sparse ensemble learning for concept detection. IEEE Trans. Multimed. **14**(1), 43–54 (2012)
26. Uijlings, J.R., van de Sande, K.E., Gevers, T., Smeulders, A.W.: Selective search for object recognition. Int. J. Comput. Vis. **104**(2), 154–171 (2013)
27. Wang, J., Yang, J., Yu, K., Lv, F., Huang, T., Gong, Y.: Locality-constrained linear coding for image classification. In: 2010 IEEE Conference on Computer Vision and Pattern Recognition (CVPR), pp. 3360–3367. IEEE (2010)
28. Yang, Y., Shen, F., Shen, H.T., Li, H., Li, X.: Robust discrete spectral hashing for large-scale image semantic indexing. IEEE Trans. Big Data **1**(4), 162–171 (2015)

Uyghur Language Text Detection in Complex Background Images Using Enhanced MSERs

Shun Liu[1], Hongtao Xie[1(✉)], Chuan Zhou[1,2], and Zhendong Mao[1]

[1] National Engineering Laboratory for Information Security Technologies,
Institute of Information Engineering, Chinese Academy of Sciences,
Beijing 100093, China
xiehongtao@iie.ac.cn
[2] University of Chinese Academy of Sciences, Beijing 100049, China

Abstract. Text detection in complex background images is an important prerequisite for many image content analysis tasks. Actually, nearly all the widely-used methods of text detection focus on English and Chinese while some minority languages, such as Uyghur language, are paid less attention by researchers. In this paper, we propose a system which detects Uyghur language text in complex background images. First, component candidates are detected by the channel-enhanced Maximally Stable Extremal Regions (MSERs) algorithm. Then, most non-text regions are removed by a two-layer filtering mechanism. Next, the remaining component regions are connected into short chains, and the short chains are expanded by an expansion algorithm to connect the missed MSERs. Finally, the chains are identified by a Random Forest classifier. Experimental comparisons on the proposed dataset prove that our algorithm is effective for detecting Uyghur language text in complex background images. The F-measure is 84.8%, much better than the state-of-the-art performance of 75.5%.

Keywords: Uyghur language · Text detection · The channel-enhanced maximally stable extremal regions

1 Introduction

Text in images usually contains valuable information. With the development of computer vision and image retrieval [1], many researchers pay attention to the field of recognizing the text information in images [2–7]. However, how to extract the texts quickly and accurately from images is still a difficult problem, because of the complex background and variations of fonts and colors. It is important and essential to detect the text in images so that we can obtain the useful information we need [8]. Though many papers published in some journals or conferences are around the theme of text detection, most of their points concentrate on English or Chinese. Some scholars and groups turn into the research of text detection and recognition of the minority languages in recent years, still they account for a low proportion.

This paper focuses on Uyghur text detection in the complex background images. For 8 to 11 millions of people using Uyghur language in total all over the world [9], the

© Springer International Publishing AG 2017
L. Amsaleg et al. (Eds.): MMM 2017, Part I, LNCS 10132, pp. 490–500, 2017.
DOI: 10.1007/978-3-319-51811-4_40

applications of Uyghur text detection and recognition in images are very important. Uyghur belongs to the Karluk branch of the Turkic language family, which also includes languages such as Uzbek. In addition to the influence from the other Turkic languages, Uyghur has also been impacted strongly by Persian and Arabic historically [9]. The characteristics of Uyghur language make it different from other languages. Some of them are as follows:

- The shape of Uyghur language character is not square, and the characters in one word always conglutinate together.
- Most characters (20 out of 32) have one dot, two dots, three dots, or zigzags associated with the character and can be above, below, or inside the character.
- No upper or lower cases exist in Uyghur language.
- Uyghur language sentences always have one "baseline" in the middle of text, as shown in Fig. 1.

 baseline

Fig. 1. The baseline of Uyghur language text.

In text detection, the Maximally Stable Extremal Regions (MSERs) methods [3, 4, 10] are widely used, because of their computational efficiency and stability. The methods perform well but have problems on blurry images or characters with low contrast [5]. In this paper, a channel-enhanced MSERs algorithm is introduced to detect the Uyghur text regions in complex background images. Almost all the text regions can be drawn out by the algorithm. However, the apparent pitfall of the MSERs algorithm is that most of the detected MSERs are in fact repeating with each other and the noises (the regions are not what we want). The key point after the algorithm is to remove the repetitions and noises.

In this paper, we propose a robust and accurate channel-enhanced MSERs-based text detection method. First, the channel-enhanced MSERs algorithm is used to get all the word candidates. Then, a two-layer filtering mechanism is proposed to remove the non-text regions and identify the text regions. In the first layer, some heuristic rules are used to prune the repetitions and some simple non-text regions. In the second layer, one word classifier trained by Histogram of Oriented Gradient (HOG) [11] features is proposed to identify the text regions among the rest MSERs. Next, a text connection algorithm is proposed to connect the adjacent words and a text chain expansion algorithm is used to expand the existing chains to complete ones. Finally, a Random Forest [12] classifier with a free threshold trained by a set of gradient features is used to eliminate the non-text chains. The chains passed the classifier are the final results.

By integrating the above ideas, we build an accurate and robust complex background Uyghur text detection system. The system is evaluated on a dataset named IMAGE570 and has achieved the result of 84.8% in F-measure, which is much higher than the state-of-the-art performance of 75.5%.

The rest of this paper is organized as follows. Section 2 describes the structure and details of the system. Section 3 introduces the dataset and evaluation protocol. Experiments are presented in Sect. 4 and Sect. 5 gives the conclusion.

2 Methodology

By incorporating the advantages of MSER-based methods and the characteristics of Uyghur language, we propose a novel channel-enhanced MSER-based Uyghur text detection method. In this section, the details of the proposed framework are elaborated.

2.1 System Overview

The system includes four stages: (1) Component Extraction, (2) Component Analysis, (3) Chain Linking and (4) Chain Analysis. These stages can be further categorized into two procedures: *grouping* and *pruning*, as shown in Fig. 2. In the grouping procedure, pixels are first formed to connected components and then these connected components are aggregated to form chains. In the pruning procedure, non-text components and chains are successively identified and eliminated by the filtering algorithms.

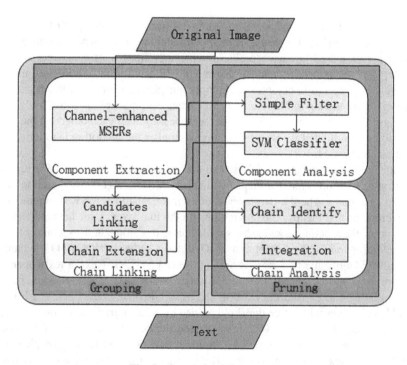

Fig. 2. Structure of the system.

2.2 Component Extraction

To extract components from an image, MSERs algorithm is adopted for its effectiveness and efficiency. According to Mates et al. [6], an extremal region is a connected component of an image whose pixels have either higher or lower intensity than its outer boundary pixels. The intensity contrast is measured by increasing intensity values, and controls the region's area. The MSERs algorithm is generally used in grayscale image. The intensity contrast between some objects and the background becomes weak after the transformation from color image to grayscale image, so some objects are discarded. Besides, many objects contrast with the background more obvious in one of R, G and B channel than in grayscale image. A method called the channel-enhanced MSERs algorithm is proposed in this paper. To keep the complete information of the color images and get as many text regions as possible, the MSERs algorithm is employed in R, G and B channel respectively, and all the bounding boxes are drawn out on the color image. The regions with same bounding boxes would be deleted by comparing their center's coordinate and corners' coordinates, and the remaining rectangles are recorded. Almost all the object regions can be obtained through the algorithm. One example of using channel-enhanced MSERs is shown in Fig. 3.

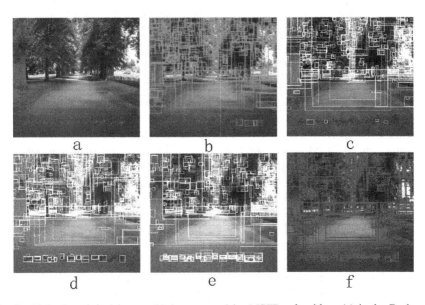

Fig. 3. (a) is the original image. (b) is processed by MSERs algorithm. (c) is the B channel processed by MSERs. (d) is the G channel processed by MSERs. (e) is the R channel processed by MSERs and (f) is processed by the channel-enhanced MSERs. (Color figure online)

After processed by the channel-enhanced MSERs algorithm, there are many overlapped regions and non-text regions, which is the common drawback of connected component methods. The motivation of our next step is to move these repeating regions and non-text regions, to get the text regions.

2.3   Component Analysis

Since many regions are left after the Component Extraction stage, we identify and eliminate the MSERs that are unlikely to be text regions in the stage of the Component Analysis. Towards this goal, we devise a two-layer filtering mechanism.

The first layer is a filter consists of a set of heuristic rules to take out the repetitions and some simple noises. For the repetitions, if two regions overlap together and the intersection area is over 80% of the union area, the one whose MSER variation is larger is pruned. If they have the same variation, the one whose area is smaller is pruned. For one text region, in terms of the characteristics that the Uyghur word always containing some but not many little components, the number of contained regions is not big. A threshold number is set. If the number of one region containing little regions is more than the threshold, the region is regarded as noise and deleted. In the second layer of the filtering mechanism, a classifier is employed to identify the left components and reject the non-text components that are hard to be removed in the first layer. The classifier is trained using the HOG features of the components. Due to the results of many experiments, the MSERs area all resized into 24×32 (*height* $\times$ *width*) pixels. For Uyghur components detection, 8×8 cells blocks of 4×4 pixel cells and 9 bins work best. A Support Vector Machine (SVM) with polynomial kernel is chosen as the strong classifier throughout the study, which is trained using the positive text regions and the negative random regions cut from the images in the training set of dataset.

2.4   Chain Linking

After the Component Analysis stage, the text regions and a small quantity of non-text regions are left. The appropriate component candidates should be connected to chains by the following steps. Firstly, the word candidates are linked into pairs. Whether two candidates can be linked into a pair is determined by their horizontal position, their height value and the distance between them. The two candidates must close enough, which implies having similar height and horizontal position. After checking the pre-requisites, the applicable regions are connected together.

However, the short chains are sometimes only parts of the text in image. A small number of text regions are discarded by the first layer of the filtering mechanism in Component Extraction stage as noises, since they contain too many little components. Secondly, the missed MSERs containing words would be aggregated into the short chains by the following expansion algorithm. For the purpose of connecting the missed MSERs into text chains, every short text chain searches forward and backward in the sequence. When one MSER, that is not the existing text candidates, is found in the left or right of a text chain, some simple requirements would be checked, including horizontal position, proximal height value and the distance between it and the chain. If the region meets these criteria, it would be judged by the SVM classifier used in the Component Analysis. If it is decided as a text region by the classifier, the MSER would be linked into the chain.

Finally, after the expansion of short chains, the expanded chains would be connected into complete ones. For two chains, if they are close or even crossed together and are in same horizontal position, they are connected together. When exact the chains linking, the rectangles around chains are modified. At last, the remaining chains are the chain candidates. Figure 4 shows the procedure.

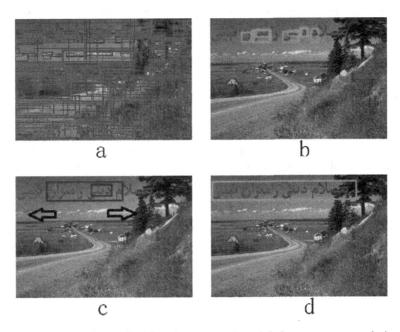

Fig. 4. (a) All the MSERs. (b) The character regions left in component analysis stage. (c) Character regions connect into short chains. (d) The two short chains are expanded, and then connect together, getting the chain candidates.

2.5   Chain Analysis

The candidate chains formed at the previous stage might contain some false positives, which are combinations of text regions and background clutters or the non-text regions left from Component Analysis stage. In this stage, a Random Forest classifier is trained to identify the chains formed at last stage. A collection of chain-level features, capturing the differences of textural properties between text chains and non-text chains, are used to train the classifier. The features extraction steps are discussed in the following. First, all the chains are resized into 26×122 pixels. For the color image, the gradient value of each pixel of each color channel is calculated separately, and the one with the largest norm is specified as the color image pixel's gradient. After the calculation, a two-dimension array with the gradient value is established. Then, the gradient value is normalized by the L_2-norm,

$$v \rightarrow v/\sqrt{\|v\|_2^2 + \varepsilon^2} \tag{1}$$

where v is the un-normalized gradient vector, and ε is a small number (set to be 0.1). Next, Pooling is carried out. Successive 2-by-2 patches are extracted, and the max value and min value in these patches are chosen to form the feature vector.

The probability of one chain is the fraction of votes for positive text from the trees. The chains with probabilities lower than a threshold θ are eliminated. If one chain is repeated with others, the one with largest area is left and others are removed. If one chain contains any short chains, these contained short chains are deleted. After all the processing above, the rest chains are the results.

3   Dataset and Evaluation

In this section, a new image dataset with horizontal texts is organized. Most of texts in the images are written in Uyghur language, and the few rest texts are written in Chinese or English. The dataset is named as IMAGE570 because it contains 570 complex background images in total. These images are downloaded from the Internet. IMAGE570 is divided into two parts: 370 images of them are used for training and the rest 200 images are used for evaluation. This dataset is challenging because of both the diversity of the texts and the complexity of the background in the images. There may be one or more Uyghur text lines on one image. The backgrounds may contain vegetation (e.g. trees and grasses), repeated patterns (e.g. windows and bricks) and colorful sundries (e.g. household items and flowers), which may be not so distinguishable from text. Some examples of IMAGE570 are showed in Fig. 5.

Fig. 5. Examples of IMAGE570. (Color figure online)

The evaluation method is same as the one used in [6]. Precision, recall and F-measure are used to evaluate the system.

4 Experiments

In this section, experiments are conducted to evaluate the proposed system. First, we compare the overall performance with the state-of-the-art methods on our IMAGE570 dataset in Sect. 4.1. Then, some experiments are performed in Sect. 4.2 to test the accuracy of the SVM classifier used in Component Analysis stage.

4.1 Performance on IMAGE 570

The performance is appraised by the proposed evaluation protocol. We compare the result of our algorithm on IMAGE570 with the method proposed by Yin [3], which is the state-of-the-art multi-language detection system. But the system in this paper is only trained by Uyghur language. The precision and recall in the Table 1 are only about Uyghur language detection.

From Table 1, we can see that the precision, recall and f-measure of our method all have higher values and the F-measure can be increased by 12.3%, compared to the Yin's method. The improvements by our method mainly gain from some facts in the following:

- The powerful channel-enhanced MSERs detector is able to detect most text regions. The expansion algorithm in Chain Linking stage expands some incomplete chains to complete ones. These two factors result in a high recall.
- The high capability of the two-layer filtering mechanism in Component Analysis identifies the text components from non-text components, leading to a large improvement on precision.

Table 1. Performance (%) on IMAGE570.

	Precision	Recall	F-measure
Our method	85.0	84.7	84.8
Yin [3]	76.0	75.0	75.5

Figure 6 shows some successful detection results on a number of challenging cases in IMAGE570. It indicates that our system is effective to large variation in texts including different font size, low contrast and strong noise background effects. Figure 7 shows two failure cases in our experiments. For the first one, some text regions are expurgated as noises by the classifier in Component Analysis, because of the influence of water shimmer and the existing chain cannot be extended by the extension algorithm in Chain Linking stage. The second one is pruned by the two-layer filter in Chain Analysis, because of the change of bounding boxes.

Fig. 6. The successful detection cases.

Fig. 7. The failure detection cases.

4.2   Tests on SVM Classifier

A SVM classifier is used to prune MSERs in the Component Analysis stage. To test the relationship between the classifier accuracy with different kernels and the training samples quantity, 2000 images are cut from the test set of IMAGE570 to form SVM test set. The SVM test set is divided into two parts: word region set and non-word region set. The word region set consists of 1000 images in total. Different non-word regions have different sizes and backgrounds, in total 1000 images. The SVM classifier is trained with LINEAR kernel, Polynomial (POLY) kernel and Radial Basis Function (RBF) kernel separately. The SVM training images include two kinds as the followed described: half of them are word region images and half of them are non-word region images. The number of SVM training images changes from 5000 to 12000 accompanied by the change of the accuracy of SVM classifier.

The specific data comparisons are showed in Fig. 8. And the highest accuracy of Poly SVM is 97.95%. The accuracy of LINEAR SVM decreases with the increase of training number. When the quantity is 5000, the Linear SVM gets the highest accuracy 92.05%. The accuracy of RBF SVM grows but is lower than that of POLY SVM all the time. When the quantity is 12000, the RBF SVM gets the highest accuracy 94.20%. Both of them are lower than the highest accuracy of Poly SVM.

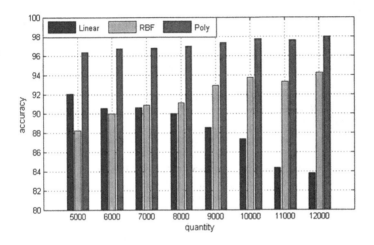

Fig. 8. Accuracy (%) of Linear SVM, RBF SVM and Poly SVM.

5 Conclusion

A robust system for Uyghur text detection in complex images is presented in the paper. The main contribution of the system lies in effectively using MSER algorithm in three channels to detect the character candidates and having strong robustness and highly discriminative capability to distinguish Uyghur language text from a large amount of non-text components. The state-of-the-art performance on the benchmark dataset convincingly verifies the efficiency of the proposed method. In future, we would build an end-to-end system [13] by modifying the method proposed in this paper and use our method in multimedia implication systems [14–16].

Acknowledgement. This work is supported by the National Nature Science Foundation of China (61303171, 61502477, 61502479), the "Strategic Priority Research Program" of the Chinese Academy of Sciences (XDA06031000).

References

1. Xie, H., Zhang, Y., Tan, J., Guo, L., Li, J.: Contextual query expansion for image retrieval. IEEE Trans. Multimedia **16**(4), 1104–1114 (2014)
2. Nie, L., Yan, S., Wang, M., Hong, R., Chua, T.S.: Harvesting visual concepts for image search with complex queries. In: ACM MM (2012)
3. Yin, X.C., Yin, X., et al.: Robust text detection in natural scene images. IEEE Trans. Pattern Anal. Mach. Intell. **36**(5), 970–983 (2013)
4. Huang, W., Qiao, Yu., Tang, X.: Robust scene text detection with convolution neural network induced MSER trees. In: Fleet, D., Pajdla, T., Schiele, B., Tuytelaars, T. (eds.) ECCV 2014. LNCS, vol. 8692, pp. 497–511. Springer, Heidelberg (2014). doi:10.1007/978-3-319-10593-2_33
5. Neumann, L., Matas, J.: Real-time scene text localization and recognition. In: IEEE CVPR, vol. 157, no. 10, pp. 3538–3545 (2012)
6. Yao, C., et al.: Detecting texts of arbitrary orientations in natural images. In: IEEE CVPR, pp. 1083–1090 (2012)
7. Zhang, Z., Shen, W., Yao, C., Bai, X.: Symmetry-based text line detection in natural scenes. In: IEEE CVPR, Boston, MA, June 2015
8. Xie, H., Gao, K., Zhang, Y., Tang, S., Li, J., Liu, Y.: Efficient feature detection and effective post-verification for large scale near-duplicate image search. IEEE Trans. Multimedia **13**(6), 1319–1332 (2011)
9. http://en.wikipedia.org/wiki/Uyghur_language
10. Matas, J., et al.: Robust wide-baseline stereo from maximally stable extremal regions. IVC **2** (10), 761–767 (2002)
11. Dalal, N., Triggs, B.: Histograms of oriented gradients for human detection. In: IEEE CVPR, vol. 1, pp. 886–893 (2005)
12. Leo, B.: Random forests. Mach. Learn. **45**(6), 422–432 (2001)
13. Wang, T., Wu, D.J., Coates, A., et al.: End-to-end text recognition with convolutional neural networks. In: 2012 21st International Conference on Pattern Recognition (ICPR), pp. 3304–3308. IEEE (2012)
14. Chen, Z., Feng, B., Xie, H., Zheng, R., Xu, B.: Video to article hyperlinking by multiple tag property exploration. In: Gurrin, C., Hopfgartner, F., Hurst, W., Johansen, H., Lee, H., O'Connor, N. (eds.) MMM 2014. LNCS, vol. 8325, pp. 62–73. Springer, Heidelberg (2014). doi:10.1007/978-3-319-04114-8_6
15. Chen, Z., Chen, Y., Gao, X., et al.: Unobtrusive sensing incremental social contexts using fuzzy class incremental learning. In: International Conference on Data Mining (2015)
16. Gao, X., Chen, Z., Tang, S., Zhang, Y., Li, J.: Adaptive weighted imbalance learning with application to abnormal activity recognition. Neurocomputing **173**, 1927–1935 (2016)

SS2: Modeling Multimedia Behaviors

CELoF: WiFi Dwell Time Estimation in Free Environment

Chen Yan, Peng Wang[✉], Haitian Pang, Lifeng Sun, and Shiqiang Yang

National Laboratory for Information Science and Technology,
Department of Computer Science and Technology,
Tsinghua University, Beijing, China
{yan-c15,pht14}@mails.tsinghua.edu.cn,
{pwang,sunlf,yangshq}@tsinghua.edu.cn

Abstract. WiFi wireless access has been the basic living need for smart phone users in the era of mobile multimedia. A large number of WiFi hotspots have also developed into an important infrastructure of multimedia accessing in smart city. Learning the dynamic features of free-environment WiFi connections is of great help to both the customization of WiFi connection service and the strategy of mobile multimedia. While mobility prediction attracts much interest in human behavior research which is more focused on fixed environments such as university, home and office, etc., this paper investigates more challenging public regions like shopping malls. A WiFi dwell time estimation method is proposed from a crowdsourcing view, to tackle the lack of contextual information for a single individual in such free environments. This is achieved by a context-embedded longitudinal factorization (CELoF) method based on multi-way tensor factorization and experiments on real dataset demonstrate the efficacy of the proposed solution.

Keywords: WiFi dwell time estimation · Context modeling · Tensor factorization · Crowdsourcing

1 Introduction

The ubiquitous employment of mobile digital devices and population's habit of continuous multimedia content accessing have made WiFi one of the main accessing methods to wireless networks. As an important part of information infrastructure for smart cities, the public WiFi access has harnessed cities with sensing facilities which can reflect user mobility patterns and group features. Research in these fields has both potentials and challenges for city planing, public services, commercial recommendation, advertisement, etc.

Within human behavior research, mobility prediction has attracted much interest due to its enormous value in various applications. Most prior approaches

This work was part-funded by 973 Program under Grant No. 2011CB302206, National Natural Science Foundation of China under Grant Nos. 61272231, 61472204, 61502264, Beijing Key Laboratory of Networked Multimedia.

L. Amsaleg et al. (Eds.): MMM 2017, Part I, LNCS 10132, pp. 503–514, 2017.
DOI: 10.1007/978-3-319-51811-4_41

to mobility prediction are mainly focused on transitions among several frequently visited places [3,5,6,9,10,18] such as "home", "office", "dorm", "library", etc. These approaches cannot adapt to prediction tasks in more general public regions (such as shopping malls and airports), which involve a large number of customers and passers-by with relatively irregular movement patterns. More severe, such environments result in shorter traces for each of these guests from which it is more challenging to build an accurate mobility model for each individual based on existing methods due to deficiency of per-user historical data.

In this paper, we attempt to investigate such more challenging scenarios, i.e., WiFi dwell time prediction in free environments, which is a less exploited research question. The use of "free environment" has two meanings: First, it involves arbitrary locations which lead to more random human traces and less historical recordings for individuals. This challenges the accurately modeling of user mobility. Second, there are no other contextual information available in free environment, such as sensor readings recorded by smart phones or wearable devices. Only group WiFi connection records are feasible to estimate WiFi dwell time.

In order to tackle the above challenges of WiFi dwell time estimation in free environments, we propose a method from a crowdsourcing view, to alleviate the lack of contextual information for single individuals. This is carried out by factorize and then re-estimate the modeled multi-way tensor. The contribution of this work can be summarized as:

- The estimation of WiFi dwell time is investigated for free environments such as in public regions. This is more practical in building real-world applications where there lack of heterogenous sensor readings and only WiFi connection records are available.
- Context-embedded longitudinal factorization (CELoF) method is proposed to tackle the constrains of limited contextual information. In CELoF, the estimation of WiFi dwell time is crowdsourced to make full use of contextual information from group people.
- A set of experiments on a dataset collected in real-world settings to validate the effectiveness of the above.

The rest of the paper is organized as follows. In Sect. 2, we review the related work. Then we analyze the challenge and propose our algorithm in Sect. 3. The detailed description of algorithm is given in Sect. 4. In Sect. 5, we present a set of experiments including a description of the dataset we used and a discussion of results. We finish with conclusions and future work in Sect. 6.

2   Related Work

Current research on the analysis of human mobility traces is mainly dependent on the localization of Global Position System (GPS), Bluetooth sniffering, WiFi access record in local regions, etc. The relevant research to the work reported in this paper can be summarized as follows:

Markov Model. As a kind of state space models, different variants of Markov models have been proposed and applied in mobility data analysis. In [19], authors employed k-order Markov model to predict human mobility in WiFi network environment. [14] also proposed a method using Markov Chain to analyze the mobility data generated from phone call records and made predictions on human locations. In [16], Hidden Markov Model was applied to predict the transition between different locations where users have already stayed for a while. Similarly, [1] applied a mixed Markov-Chain model in human mobility prediction.

Semi-Markov Method. Because Markov methods cannot fully analyze the dynamics of temoral features such as when to arrive, how long a user has stayed, etc., [13] proposed to use semi-Markov method to model the temporal features. Similar method is also validated in [24] to describe the mobility and transitions between different locations. Similar as [8,23] proposed a hierarchical hidden Markov methods to better understand the mobility behaviors and this is demonstrated in GPS trace dataset.

Spatio-Temporal Model. Given the fact that the temporal and spatial patterns are processed separately in most methods of human behavior modeling, [9] integrated the temporal and spatial information with a smoothing method to tackle the high uncertainty of temporal feature. Similarly, [4] proposed a semantic place prediction framework with which human mobility behaviors are incorporated from both aspects of time and space. The key idea of this method is to extract high dimensional features in order to fully reflect the semantic contents of places.

It is not hard for us to notice the shortcomings of these related work. The precisely modeling of human mobility is based on the assumption that there are enough historical records to characterize the mobility patterns of individuals. This assumption probably holds in their reported work carried out in experimental environments in which the consecutive recording of mobility traces are possible for experiment participants, or in specific regions such as universities where users' mobility dynamics concentrates in several dominant places like lab, dorm, restaurant, etc. However, the application of these methods to more arbitrary domains like public areas is still questionable, due to various factors like larger number of users, sparse historical data, randomness of human movements, etc.

Another stream of human behavior measurement is carried out by utilizing sensing capabilities embedded in smartphones or wearable devices such as digital camera, light sensor, accelerometer, electromagnetic compass, etc., which have enabled high-resolution human behavior insights. Lifelogging [20] is the term used to describe such process of automatically, and ambiently, digitally recording human behaviors, using a variety of sensor types. The richness of sensory contextual information makes it practical in characterizing and measuring the occurrence of human behavior [20] and in so doing to further gain accurate prediction of behavior patterns [17,22]. In [15], the sensing facilities are applied in predicting how quickly a person will move through a space such as in a cafe. A support vector machine (SVM) was employed in [15] to classify multi-sensory

features such as acceleration, compass direction, etc. and then apply feature
similarities for prediction. Though effectiveness has been demonstrated in [15]
through live experiments with real users at a university cafe, the disadvantage
is obvious in that the prediction depends on a variety of sensor readings which
is impossible in free environments. Though [15] has the same goal as our work
reported in this paper, what differentiates this work from our own is that we
exploit the dwell time prediction in free WiFi environment such as public areas
without the needs of extra sensor readings, making it more applicable in realistic
sittings.

3   Motivation and Proposed Solution

As an important task in human behavior analysis, the prediction of stay lengths
of users in a specific location could open numerous application feasibilities like
optimizing the utilization of wireless internet resources in multimedia streaming, location choosing for business, commercial recommendation, etc. For example, the streaming strategy can better allocate necessary bandwidth to a video
caching device if how long the watcher will stay is recognized as *a priori* contextual knowledge. Similarly, owners in a shopping mall might target merchandise
promotion more appropriately if a customer's dwell time in his/her shop can be
precisely estimated in advance. Due to its great value, we focus on the investigation of dwell time prediction in this paper.

The challenge of prediction for free environment can be depicted from Fig. 1,
in which the cumulative distribution function (CDF) of the trajectory lengths
is shown for all users in the dataset collected in a shopping mall. The average
length of the trajectories equals to 9.1 and this significantly differs from some
previous work (e.g., the dataset in [19] contains WiFi traces collected in campus
and is widely used in many previous work), where users tend to have much
longer visiting history for pattern learning (e.g., the median number is 494 in
[19]). The prevailing predictors of human mobility proposed so far try to estimate
for a given user by inspecting the historical traces of this specific individual. As
a result, these methods tend to make inaccurate predictions if the past traces
are limited for characterizing the mobility patterns of this user.

While neighborhood-based methods show effectiveness in capturing the relationships between users, hence to aid the predictions based on similar neighbors,
these methods also fail to work when there are fewer observed trajectories to
correctly measure neighborhood. When most users only connect to a limited
number of WiFi hotspots, user trajectories are very sparse, making it difficult to
predict using traditional crowdsourcing methods. This is especially severe when
users connect to new WiFi hotspots they have never visited before. According
to [25], the factorization of matrix and tensor can be applied to tackle the data
sparsity problems. The intuition behind is that the latent factor model can effectively capture the global context information and is capable in generalizing in
predicting the missing values through multiplying the decomposed matrices.

To tackle the above described problem, we proposed an factorization
method to decompose and make predicts within a high-order tensor framework.

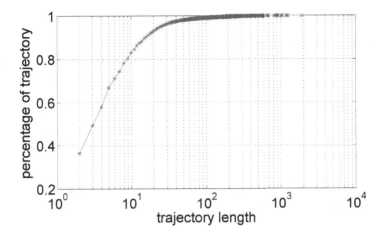

Fig. 1. CDF of trajectory length.

The contextual representation of crowd connections to WiFi access points (APs) is inherently the construction of a multi-dimensional array in which the dynamic evolution of user-AP interactions can be organized as a tensor. For example, the phenomenon of group users connecting to a set of WiFi APs can be modeled as a $N \times M$ matrix, where N and M are numbers of users and WiFi APs respectively. Taking into account of the time stamps of WiFi connections, the problem is indeed represented as a matrix time series as shown in Fig. 2. In this time series, the relationships between users and APs are modeled as an ordered, finite collection of matrices that all share the same dimensionality. That is, by discretizing the longitudinal evolution into L slotted time intervals, each temporal slice in such tensor reflects the state of user-AP connections. Since the contexts are implicitly modeled in the longitudinal interaction of user and AP-specific features which can be generated by tensor factorization, we name this method context-embedded longitudinal factorization (CELoF), which is now described.

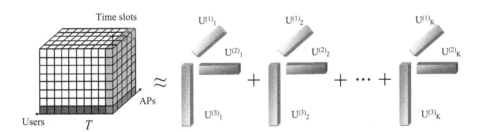

Fig. 2. The paradigm of tensor representation of WiFi connections and its factorization.

4   Context-Embedded Longitudinal Factorization

In order to flexibly leverage the underpinning contexts for dwell time prediction, we proposed a context-embedded longitudinal factorization method to model the temporal evolution of WiFi connections. This is motivated by [7], in which an SVD-based matrix factorization algorithm is proposed for predicting activity attendance. Different with [7], we extend the factorization to the decomposition of a multi-way weighted non-negative tensor [21] which integrates the connection contexts into a single framework.

The task of CELoF is to find the latent features to represent the components of contextual tensor $T \in \mathbb{R}^{N \times M \times L}$ as shown in Fig. 2. The 3-way tensor can then be approximated by Tucker Decomposition (TD) [2] as $T \approx G \times_1 U^{(1)} \times_2 U^{(2)} \times_3 U^{(3)}$, where $G \in \mathbb{R}^{R \times S \times T}$, $U^{(1)} \in \mathbb{R}^{N \times R}$, $U^{(2)} \in \mathbb{R}^{M \times S}$ and $U^{(3)} \in \mathbb{R}^{L \times T}$. $\times_i (i = 1, 2, 3)$ denotes tensor-matrix multiplication operators with the subscript i specifying which dimension of the tensor is multiplied with the given matrix. As a particular case of the general Tucker Decomposition, the Canonical Decomposition (CD) [2] is derived from TD model by constraining that each factor matrix has the same number of columns, i.e., the length of latent features has the fixed value of K. The CD model simplifies the approximation of tensor T as a sum of 3-fold outer-products with rank-K decomposition:

$$\hat{T} = \sum_{f=1}^{K} U^{(1)}_{\cdot f} \otimes U^{(2)}_{\cdot f} \otimes U^{(3)}_{\cdot f} \tag{1}$$

When the valence is 2, T is simply a 2-mode tensor (matrix) and Eq. (1) degenerates as

$$\hat{T} = \sum_{f=1}^{K} U^{(1)}_{\cdot f} \otimes U^{(2)}_{\cdot f} = U^{(1)} U^{(2)T} \tag{2}$$

The CD approximation factorization defined above can be solved by optimizing the cost function defined to quantify the quality of the approximation. Different forms of cost function and corresponding optimization can be applicable to this problem. Euclidian distance is employed in this paper to define the cost function, which can be embedded with the weighted form

$$F = \frac{1}{2} \| T - \hat{T} \|^2_W = \frac{1}{2} \sum_{ijk} w_{ijk} (T_{ijk} - \sum_{f=1}^{K} U^{(1)}_{if} \otimes U^{(2)}_{jf} \otimes U^{(3)}_{kf})^2$$

$$\text{s.t. } U^{(1)}, U^{(2)}, U^{(3)} \geq 0 \tag{3}$$

where $W = (w_{ijk})_{N \times M \times L}$ denotes the weight tensor and $\| \cdot \|^2_F$ denotes the Frobenius norm, i.e., the sum of squares of all entries in tensor. The nonnegative constraints guarantee each component described by $U^{(1)}$, $U^{(2)}$ and $U^{(3)}$ are additively combined. In the work reported in this paper, we employ the multiplicative method [12, 21] to iteratively solve the optimization problem defined in Eq. (3):

$$U_{if}^{(1)} \leftarrow U_{if}^{(1)} \sum_{jk} (W \circ T)_{ijk} U_{jf}^{(2)} U_{kf}^{(3)} / \sum_{jk} (W \circ \hat{T})_{ijk} U_{jf}^{(2)} U_{kf}^{(3)}$$

where $\circ$ denotes element-wise multiplication. The updating of $U^{(2)}$ and $U^{(3)}$ can be achieved in the similar manner. After convergence, each elements in tensor T can be estimated through Eq. (1). For example, if we assign T_{ijk} as the dwell time of user i connecting to AP j in time interval k, tensor T is actually the dwell time tensor and the unknown elements can be predicted through the above CELoF method. The performance of dwell time prediction using CELoF is evaluated in Sect. 5.

5 Experiments and Evaluation

5.1 Experimental Setup

Dataset Summary. We collected the mobility trace from NextWiFi[1], a recently launched commercial wireless service provider. To understand the relationship between physical contexts and mobility patterns, we analyzed the data collected at 30 WiFi APs deployed in a four-floor shopping mall (public region) [26]. The NextWiFi dataset contains 76,933 unique human trajectories over the period of one month. To be more detailed, these 30 APs are deployed in four types of places which are restaurant, clothing store, office and public infrastructure, so we can get insights into the impact of different places on users' stay length prediction.

We select a subset of the NextWiFi dataset which contains 2101 active users, all of 30 WiFi APs over the whole period. We use the time interval of one hour in our experiment, hence a tensor of dimension 2101 * 30 * 168 is built for one week whose element T_{ijk} is represented as stay time length of user i at AP j in time interval k. Similar as in [15], we also evaluated the performance of our algorithm on different dwell time-classes. We first divide dwell time into three categories: ≤ 1 min (short stay), 1–10 min (normal stay) and ≥ 10 min (long stay). A finer granularity of division is also tested, including: ≤ 1 min (very short stay), 1–5 min (short stay), 5–10 min (normal stay), 10–20 min (long stay) and ≥ 20 min (very long stay). This is to validate the robustness on various discretized time scales.

Baselines. As shown in Sect. 2, Markov method is a typical solution for human mobility prediction. The effectiveness of Markov predictors has been demonstrated in [19] and low-order Markov predictors performed equally well compared to high-order ones. [3] also demonstrated that the Markov predictor performs well on WiFi mobility data in terms of accuracy. As one baseline method, the Markov model is trained with the form $M_u(l, l') = P(l[t+1] = l'|l[t] = l)$, where $P(l[t+1] = l'|l[t] = l)$ stands for the first-order transition probability between two AP locations l and l' at consecutive time slots of t and $t+1$ for user u. More details of predictions using Markov can be found in [3,19].

---

[1] http://www.nextwifi.cn/wifind/.

In order to evaluate the discriminative performance of the proposed method, we also employed two widely used classifiers: support vector machine (SVM) [15] and boosted decision tree (BDT) [11]. Since the solution is proposed and targeted for free environment where only limited features can be sensed, we extract five most discriminative features for these classifiers according to our experimental comparison, namely user ID, WiFi hotspot ID, connecting time, average stay length of a specific user and average stay length of all users at a specific WiFi hotspot. Since the extracted features can reveal the dwelling similarities which characterize the user's behavior, by employing these features, the classifiers can recognizes such similarities and categorize the user's dwell time.

5.2   Results and Evaluation

In our experiment, we randomly selected 80% data for training and the remaining 20% are left for testing. As we can see from Eq. (3), the only parameters of the proposed algorithm are the weight tensor W and the length of latent feature K. In the implementation, we assign larger value of $w_{ijk} = 1$ for elements in T which are already known, while smaller value of $w_{ijk} = 0.05$ for elements to be predicted. Though these can be chosen empirically, $w_{ijk} = 0.05$ and $K = 25$ are determined through 5-fold cross validation on training set.

Table 1. WiFi hotspot dwell time prediction performances.

Methods	3-category			5-category		
	Precision	Recall	F1	Precision	Recall	F1
CELoF	**0.57**	**0.59**	**0.58**	**0.48**	**0.52**	**0.50**
2d-SVD	0.54	0.55	0.55	0.46	0.51	0.48
Markov	0.53	0.52	0.53	0.31	0.29	0.30
SVM	0.55	0.58	0.56	0.44	0.50	0.46
BDT	0.55	0.57	0.56	0.42	0.43	0.43

The performance of dwell time prediction is shown in Table 1, in which CELoF is compared to baselines with metrics of precision, recall and F1-score. As we can see for the table, CELoF outperforms the baselines on all metrics in both 3 and 5-category classifications. In 3-category classification, machine learning methods work slightly worse but are significantly outperformed by CELoF in the 5-category task. This demonstrates the robustness of CELoF in performing predictions on different time scales. In both circumstances, Markov models performers the worst showing its deficiency in free environments due to the lack of per-user mobility traces as we previously discussed. The performance of 2d-SVD is also compared in Table 1 using degenerated 2d CELoF as formalized in Eq. (2). In 2d-SVD, the temporal information is ignored by the absence of time dimension, leading to traditional matrix factorization. The less satisfactory performance of 2d-SVD also shows the effects of modeling longitudinal dynamics in tensor by CELoF.

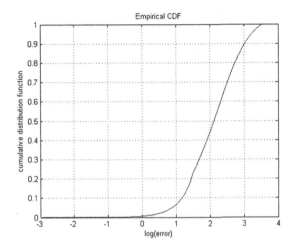

Fig. 3. CDF of prediction error.

Since the dwell durations are employed as tensor entries in CELoF, the dwell time prediction error (time length offset) can be further analyzed. The cumulative distribution function of error is shown in Fig. 3 in which the x-axis represents logarithmic scale of prediction errors (in minutes). As shown in Fig. 3, about half of the prediction errors can be controlled in the range of less than 2 min in the experiment.

The temporal pattern of prediction errors by CELoF for different types of WiFi hotspots is shown in Fig. 4. According to Fig. 4, prediction errors is much larger at midnight compared to daytime. This makes sense because the volume of training records at midnight is much smaller, hence it is more difficult to automatically self-learn the contextual patterns from such limited WiFi connections. In Fig. 4, the mispredictions for office APs are much smaller compared to the other AP types. One interpretation is that people connecting to office APs usually have regular mobility pattern which makes it easier to predict.

5.3   Discussions

The difficulty of dwell time prediction for free WiFi environment is shown by the reported results. As reflected in Table 1, the best accuracy achieved among various methods is only around 0.5 for 5-category prediction. Though difficult, we found that the proposed CELoF method can alleviate the burdens and achieve better results by taking advantage of crowd connection contexts when individuals' trajectories are limited. According to the derivation of factorization

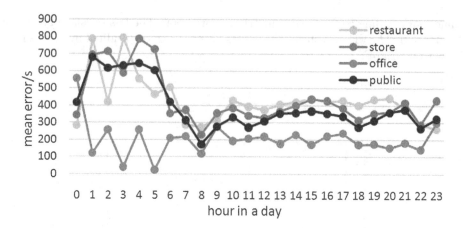

Fig. 4. Temporal pattern of prediction errors for different types of WiFi hotspots.

in Sect. 4, CELoF can be easily extended to an arbitrary-order tensor framework in order to incorporate richer context information to further improve the performance.

6   Conclusions and Future Work

While dwell time prediction is becoming increasingly important due to the widely deployment of WiFi hotspots in public areas, we propose a crowdsourced prediction algorithm to alleviate the challenges in free environments which are featured with limited but irregular trajectories of individuals. Based on a multi-way tensor factorization, the proposed context-embedded longitudinal factorization (CELoF) method implicitly models the temporal evolution of user-AP connection interactions and further applies the embedded contextual correlations to improve prediction accuracies. CELoF has been shown to provide state-of-the-art results in such realistic scenarios according to experiment. Future work can be enrichment of the CELoF with more contexts to further improve the prediction performance.

References

1. Asahara, A., Maruyama, K., Sato, A., Seto, K.: Pedestrian-movement prediction based on mixed Markov-chain model. In: Proceedings of the 19th ACM SIGSPATIAL International Conference on Advances in Geographic Information Systems (GIS 2011), pp. 25–33 (2011)
2. Bader, B.W., Kolda, T.G.: Efficient MATLAB computations with sparse and factored tensors. SIAM J. Sci. Comput. **30**(1), 205–231 (2007)
3. Baumann, P., Kleiminger, W., Santini, S.: How long are you staying?: predicting residence time from human mobility traces. In: Proceedings of 19th Annual International Conference on Mobile Computing and Networking. ACM (2013)

4. Huang, C.M., Ying, J.J.C., Tseng, V.S.: Mining users' behaviors and environments for semantic place prediction. In: Nokia Mobile Data Challenge 2012 Workshop (2012)
5. Chon, Y., Shin, H., Talipov, E., Cha, H.: Evaluating mobility models for temporal prediction with high-granularity mobility data. In: PERCOM. IEEE (2012)
6. Do, T.M.T., Gatica-Perez, D.: Contextual conditional models for smartphone-based human mobility prediction. In: Proceedings of 2012 ACM Conference on Ubiquitous Computing. ACM (2012)
7. Rong, D., Zhiwen, Y., Mei, T., Wang, Z., Wang, Z., Guo, B.: Predicting activity attendance in event-based social networks: content, context and social influence. In: Proceedings of 2014 ACM International Joint Conference on Pervasive and Ubiquitous Computing (UbiComp 2014), pp. 425–434 (2014)
8. Eagle, N., Pentland, A.S.: Reality mining: sensing complex social systems. Pers. Ubiquit. Comput. **10**(4), 255–268 (2005)
9. Gao, H., Tang, J., Liu, H.: Mobile location prediction in spatio-temporal context. In: Nokia Mobile Data Challenge Workshop (2012)
10. Ghosh, J., Beal, M. J., Ngo, H. Q., Qiao, C.: On proling mobility and predicting locations of wireless users. In: Proceedings of 2nd International Workshop on Multi-hop Ad Hoc Networks: From Theory to Reality. ACM (2006)
11. Krumm, J., Rouhana, D.: Placer: semantic place labels from diary data. In: 2013 ACM International Joint Conference on Pervasive and Ubiquitous Computing, UBICOMP 2013, Zurich, Switzerland, pp. 163–172, 8–12 September 2013
12. Lee, D.D., Sebastian Seung, H.: Learning the parts of objects by non-negative matrix factorization. Nature **401**(1999), 788–791 (1999)
13. Lee, J.-K., Hou, J.C.: Modeling steady-state and transient behaviors of user mobility: formulation, analysis, and application. In: Proceedings of 7th ACM International Symposium on Mobile Ad Hoc Networking and Computing. ACM (2006)
14. Lu, X., Wetter, E., Bharti, N., Tatem, A.J., Bengtsson, L.: Approaching the limit of predictability in human mobility. Sci. Rep. **3**, 2923 (2013)
15. Manweiler, J., Santhapuri, N., Choudhury, R. R., Nelakuditi, S.: Predicting length of stay at WiFi hotspots. In: INFOCOM, 2013 Proceedings IEEE, pp. 3102–3110 (2013)
16. Mathew, W., Raposo, R., Martins, B.: Predicting future locations with hidden Markov models. In: Proceedings of 2012 ACM Conference on Ubiquitous Computing. ACM (2012)
17. Rubin, J., Eldardiry, H., Abreu, R., Ahern, S., Du, H., Pattekar, A., Bobrow, D. G.: Towards a mobile and wearable system for predicting panic attacks. In Proceedings of 2015 ACM International Joint Conference on Pervasive and Ubiquitous Computing (UbiComp 2015), pp. 529–533 (2015)
18. Scellato, S., Musolesi, M., Mascolo, C., Latora, V., Campbell, A.T.: NextPlace: a spatio-temporal prediction framework for pervasive systems. In: Lyons, K., Hightower, J., Huang, E.M. (eds.) Pervasive 2011. LNCS, vol. 6696, pp. 152–169. Springer, Heidelberg (2011). doi:10.1007/978-3-642-21726-5_10
19. Song, L., Kotz, D., Jain, R., He, X.: Evaluating location predictors with extensive Wi-Fi mobility data. In: IEEE INFOCOM. IEEE (2004)
20. Wang, P., Smeaton, A.F.: Using visual lifelogs to automatically characterize everyday activities. Inf. Sci. **230**, 147–161 (2013)
21. Wang, P., Smeaton, A.F., Gurrin, C.: Factorizing time-aware multi-way tensors for enhancing semantic wearable sensing. In: He, X., Luo, S., Tao, D., Xu, C., Yang, J., Hasan, M.A. (eds.) MMM 2015. LNCS, vol. 8935, pp. 571–582. Springer, Heidelberg (2015). doi:10.1007/978-3-319-14445-0_49

22. Yu, M.-C., Tong, Y., Wang, S.-C., Lin, C.-J., Chang, E.Y.: Big data small footprint: the design of a low-power classifier for detecting transportation modes. Proc. VLDB Endow. **7**(13), 1429–1440 (2014)
23. Yu, S.-Z., Kobayashi, H.: A hidden semi-Markov model with missing data and multiple observation sequences for mobility tracking. Sig. Process. **83**(2), 235–250 (2003)
24. Yuan, Q., Cardei, I., Wu, J.: An efficient prediction-based routing in disruption-tolerant networks. IEEE Trans. Parallel Distrib. Syst. **23**(1), 19–31 (2012)
25. Zheng, Y., Capra, L., Wolfson, O., Yang, H.: Urban computing: concepts, methodologies, and applications. ACM Trans. Intell. Syst. Technol. **5**(3), 38:1–38:55 (2014)
26. Pang, H., Wang, P., Gao, L., Tang, M., Huang, J., Sun, L.: Crowdsourced mobility prediction based on spatio-temporal contexts. In: 2016 IEEE International Conference on Communications, ICC 2016 (2016)

Demographic Attribute Inference from Social Multimedia Behaviors: A Cross-OSN Approach

Liancheng Xiang[1,2], Jitao Sang[1,2], and Changsheng Xu[1,2(✉)]

[1] National Lab of Pattern Recognition, Institute of Automation, CAS,
Beijing 100190, China
{liancheng.xiang,jtsang,csxu}@nlpr.ia.ac.cn
[2] University of Chinese Academy of Sciences, Beijing 100049, China

Abstract. This study focuses on exploiting the dynamic social multimedia behaviors to infer the stable demographic attributes. Existing demographic attribute inference studies are devoted to developing advanced features/models or exploiting external information and knowledge. The conflicts between dynamicity of behaviors and the steadiness of demographic attributes are largely ignored. To address this issue, we introduce a cross-OSN approach to discover the shared stable patterns from users' social multimedia behaviors on multiple Online Social Networks (OSNs). The basic assumption for the proposed approach is that, the same user's cross-OSN behaviors are the reflection of his/her demographic attributes in different scenarios. Based on this, a coupled projection matrix extraction method is proposed for solution, where the cross-OSN behaviors are collectively projected onto the same space for demographic attribute inference. Experimental evaluation is conducted on a self-collected Google+ and Twitter dataset consisting of four types of demographic attributes as gender, age, relationship and occupation. The experimental results demonstrate the effectiveness of cross-OSN based demographic attribute inference.

Keywords: Cross-OSN · Stable demographic attribute inference · Dynamic behavior

1 Introduction

Along with the explosive prevalence of social media, more and more people are now being engaged in various Online Social Networks (OSNs) to create and share huge volume of multimedia information. Towards the goal of efficient digital information management and customized social media services, user modeling from the social multimedia behaviors has become more urgent and important than ever before. User models distribute in different aspects, ranging from demographic attributes (e.g., age, gender, marriage status, occupation), personal interests (e.g., politics, technology, music, sports), to social networking status, mobility patterns, consuming patterns, emotional orientation, etc. Among them,

© Springer International Publishing AG 2017
L. Amsaleg et al. (Eds.): MMM 2017, Part I, LNCS 10132, pp. 515–526, 2017.
DOI: 10.1007/978-3-319-51811-4_42

demographic attributes recording the basic and intrinsic user information constitute the most fundamental dimensions to build generic user models, and thus are widely applied in practical information services.

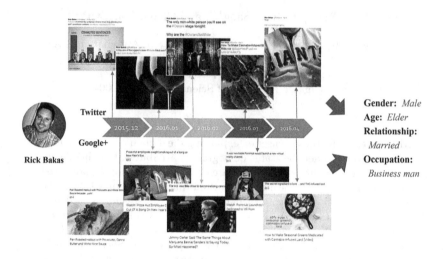

Fig. 1. An illustrative example to infer demographic attributes of a common user "Rick Bakas" from social multimedia behaviors.

Recent years have witnessed extensive studies on inferring user demographic attributes from their social multimedia behaviors [2,4,7,9,10,14]. Most of these studies either developed advanced features and models or exploited external information and knowledge. For example, Rao et al. [10] exploited sociolinguistic features and n-gram models to infer users' demographic attributes including gender, age and regional origin from their Twitter behaviors. Fang et al. [4] identified the inter-relation between different demographic attributes and proposed a multi-task learning scheme for relational attribute inference on Google+. However, so far as we know, a critical problem has been ignored and remained unexplored till now: *the contradiction between the dynamicity of observed social multimedia behaviors and the relative stable demographic attributes.* As illustrated in the left of Fig. 1, users' social multimedia behaviors are significantly dynamic with changing focuses from time to time. On one hand, the above existing demographic attribute inference studies generally take user dynamic behaviors at different time periods as a whole, which inevitably leads to information loss in user modeling and fails to capture the underlying correlation between the dynamic behaviors and the stable demographic attributes. On the other hand, studies in personal interest modeling have tackled the dynamicity problem by separating user behaviors into different time sessions to estimate the evolving interests over time [12]. In the context of demographic attribute inference, since the demographic attributes such as gender, age, marriage status and occupation are static or remain unchanged during a long period of time, the methodologies from dynamic interest modeling also cannot be directly used.

The solution goes to two lines to address the contradiction between dynamic behaviors and stable demographic attribute: one is to discover the stable patterns from user behaviors during an enough long time, and the other is to identify the shared patterns from user behaviors under different circumstances. This exploratory study is focusing on solution along the second line. Today's social media users are now using a multitude of OSN services, e.g., following real-time hot events on Twitter, communicating with his/her friends on Facebook/Google+, and subscribing and watching videos on YouTube, which has called attention from bunches of researchers. Abel et al. [1] introduced a cold-start recommendation solution by aggregating user proles in Flickr, Twitter and Delicious. In [11], the real-time and socialized characteristics of the Twitter tweets was exploited to facilitate video applications in YouTube. Deng et al. [3] incorporated user information from Google+ to facilitate personalized YouTube video recommendation. Yan et al. [13] proposed a united YouTube video recommendation framework via cross-network collaboration in which users auxiliary information on Twitter are exploited to address the typical problems in single network-based recommendation solutions. This cross-OSN scenario also provides a natural testbed to explore the shared user behavior patterns towards demographic attribute inference. It is reasonable to assume that it is the unique and stable demographic attributes that explain and lead to the disparate and dynamic social multimedia behaviors on various OSNs (illustrated in Fig. 1).

We propose a cross-OSN demographic attribute inference approach to realize the above assumption. In particular, we consider Google+ and Twitter, the popular Social Networking Site (SNS) and microblogging site, as the test OSNs in our study. In either of them, users are allowed to post texts, pictures, and videos. The training stage of the proposed approach consists of two steps. Firstly, users' social multimedia behaviors are processed to construct user feature representation for each OSN. Secondly, with the ground-truth demographic attribute as supervision, we utilize a coupled projection matrix extraction method to identify the shared patterns among the same individual's OSN respective representations and obtain the correlation between the demographic attribute space and the behavior feature spaces. At the test stage, given the observed user behaviors on different OSNs, we first extract user features and then infer his/her demographic attributes by mapping onto the derived coupled projection matrixes. Experiments on a real-world dataset demonstrate the superior performance of cross-OSN demographic attribute inference over the single-OSN based solutions.

The main contributions of this paper can be summarized as follows:

- We propose and tackle with the problem of stable demographic attribute inference from dynamic social multimedia behaviors.
- A cross-OSN approach is presented to identify the shared behavior patterns towards demographic attribute inference.

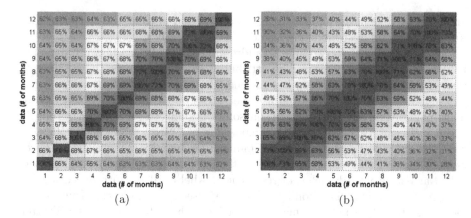

Fig. 2. The user behavior similarities between different time pairs in Google+ and Twitter from 2014.12 to 2015.12.

2   Data Collection and Analysis

In this section, we first introduce how we collect our cross-OSN user dataset, and then conduct some quantitative analysis on investigating the dynamicity of users' social multimedia behaviors.

2.1   Data Collection

To construct a dataset with user account linkage between different OSNs, we start from Google+ website where users are willing to share their user accounts on other OSNs, and collect 1,478 users with accounts on both Google+ and Twitter. These 1,478 users are recorded as *common users* in the rest of this paper. For each of the common users, we further download his/her recent 2,000 social posts (including both the texts and attached images) and the user profile from Google+ and Twitter, respectively. As a result, the collected cross-OSN dataset consists of 1,622,247 social activities in Google+ and 2,572,546 multimedia tweets in Twitter for the 1,478 common users.

2.2   User Behavior Analysis

Due to the changing of user's interest or focus, the typical user behaviors will change with time. In this subsection, we further examine the dynamicity of the typical user behaviors in the collected dataset. We investigate this characteristic by measuring the similarities of the same users' social activities between different time periods on Google+ and Twitter, respectively. Specifically, we collect all the behavior data of the common users from December 1st, 2014 to December 1st, 2015 and calculate the average user activity similarity between each time-time pair over all users in a monthly granularity, given as follows:

$$sim(t_i, t_j) = \frac{1}{|\mathcal{U}|} \sum_{u \in \mathcal{U}} \cos(\mathbf{f}_u^{t_i}, \mathbf{f}_u^{t_j})$$

where t_i, t_j denote two different time periods (i.e., months), and $\mathbf{f}_u^{t_i}, \mathbf{f}_u^{t_j}$ are user u's derived feature representations[1] in time period t_i, t_j. $\cos(\cdot, \cdot)$ is the cosine similarity metric.

In Fig. 2, we plot the normalized similarities between all the possible time pairs with a heatmap. Both axes represent the time intervals (i.e., months) from December 1st, 2014. Red color indicates high similarity while green means dissimilarity. We can observe that in both OSNs the red-color points are mostly along the diagonal line, and the similarities get smaller when the points get farther from it. This indicates that the user behaviors are continuously changing with time going on, and just considering all user's behaviors as a whole will inevitably lead to an information loss.

3 Approach

Problem Definition: Given a collection of common users $\mathcal{U}$, we represent each user $u \in \mathcal{U}$ as a three-dimensional tuple $[\mathbf{f}_u^g, \mathbf{f}_u^t, \mathbf{a}_u]$, where $\mathbf{f}_u^g, \mathbf{f}_u^t$ are the feature representations of user u in Google+ and Twitter, respectively. $\mathbf{a}_u$ denotes the labeled user attribute set. The goal is to make use of the users' cross-OSN features $\mathbf{f}_u^g, \mathbf{f}_u^t$ and their corresponding attribute set $\mathbf{a}_u$, to learn a mapping function $\mathcal{F}(\mathbf{f}_u^g, \mathbf{f}_u^t) \rightarrow \mathbf{a}_u$ between the users' behavior feature spaces and demographic attribute space for user attribute inference.

3.1 User Feature Extraction

Textual Features: We use the stemming method and stop words elimination, and remove words with a corpus frequency less than 15 in the whole text set. To reduce the number of dimensions for feature representation, we further use an entropy-based method for discriminative word selection for each attribute category. The basic idea is to measure each word by the mutual information entropy, and choose the top 10,000 words with the highest scores. TF-IDF term weighting method is applied to weight feature appropriately.

Visual Features: Convolutional neural networks (CNNs) have become more and more popular for extracting visual features from images recently. In this work, we use the widely used VGG16 model trained on ImageNet and extract a 1000-dimensional visual feature from the top fully connected layer with Caffe [8] for each image. Since one user generally has more than one image in his/her social posts, we apply a max-pooling method on the image representations to obtain an aggregated 1000-dimensional feature vector for each user.

Finally, we concatenate the textual and visual features and obtain the final feature representation for each user.

---

[1] The feature extraction is introduced in Sect. 3.1.

3.2 Coupled Projection Matrix Extraction

In each OSN, we assume that the correlations between the user's behavior feature space and demographic attribute space are embedded in a projection matrix W, based on which the user's demographic attribute representation $\mathbf{s}_u$ can be inferred by directly projecting his/her social behavior features $\mathbf{f}_u$. This assumption can be formulated as: $\mathbf{f}_u = W\mathbf{s}_u$. The task is thus to learn this projection matrix W with observations of the training users' social behavior features and their corresponding user attribute set. This can be realized by solving an optimization problem as follows,

$$\min_{W,S} ||F - WS||_F^2 + \lambda_1 ||S - A||_F^2 + \lambda_2 ||W||_F^2 \tag{1}$$

where $F = [\mathbf{f}_1, \mathbf{f}_2, \cdots, \mathbf{f}_N]$ denotes the social behavior features which is introduced in Sect. 3.1. $A = [\mathbf{a}_1, \mathbf{a}_2, \cdots, \mathbf{a}_N]$ denotes the discrete attribute representations of all the N users in the training set, by directly expanding each user's labeled attributes as a one-hot vector, which is 0 in most dimensions and 1 in a single dimension. In this case, the ith attribute value will be represented as a vector which is 1 in the ith dimension. For example, with regard to gender, a male user would be $[1, 0]$ while a female user would be $[0, 1]$. $S = [\mathbf{s}_1, \mathbf{s}_2, \cdots, \mathbf{s}_N]$ denotes the continuous demographic attribute representations of users, which has the same size with the discrete attribute representation A. Here we update the discrete attribute representation A to a continuous form S, to better indicate the user's relative strength on different attribute types.

However, in this formulation, the contradiction between the dynamicity of observed social behaviors and relatively stable demographic attributes is not considered. To address this problem, the basic premise of our solution is to discover the shared patterns from multiple user behaviors in different OSNs. Therefore, we further modify the continuous attribute representation S in Eq. (1) as a shared factor extracted from both Google+ and Twitter, and obtain the following objective function:

$$\min_{W^g, W^t, S} \left\|F^g - W^g S\right\|_F^2 + \left\|F^t - W^t S\right\|_F^2$$
$$+\lambda_1 \left\|S - A\right\|_F^2 + \lambda_2 \left\|W^g\right\|_F^2 + \lambda_3 \left\|W^t\right\|_F^2 \tag{2}$$

where F^g, F^t are the social behavior features of all the N users in Google+ and Twitter, respectively. W^g, W^t are the coupled projection matrices in different OSNs, and $\lambda_1, \lambda_2, \lambda_3$ are three regularization parameters. In this way, the derived attribute representation S is shared across different OSNs and can reflect some stable activity patterns.

Since there are multiple variables in the objective function, we adopt an alternative algorithm to find optimal solutions for the three variables W^g, W^t and S. The key idea is to minimize the objective function w.r.t one variable while fixing the other variables, as similar to [6]. The partial derivatives of the objective function can be derived as:

$$\frac{\partial}{\partial W^g} = 2W^g(SS^T + \lambda_2 I) - 2F^g S^T \tag{3}$$

$$\frac{\partial}{\partial W^t} = 2W^t(SS^T + \lambda_3 I) - 2F^t S^T \tag{4}$$

$$\frac{\partial}{\partial S} = 2MS - 2[(W^g)^T F^g + (W^t)^T F^t + \lambda_1 A] \tag{5}$$

where $M = (W^g)^T W^g + (W^t)^T W^t + \lambda_1 I$.

Let the partial derivative equals zero, we can obtain the final updating rules as below,

$$W^g = F^g S^T (SS^T + \lambda_2 I)^{-1} \tag{6}$$

$$W^t = F^t S^T (SS^T + \lambda_3 I)^{-1} \tag{7}$$

$$S = M^{-1}[(W^g)^T F^g + (W^t)^T F^t + \lambda_1 A] \tag{8}$$

Based on this, we update W^g, W^t and S iteratively until convergence or maximum iteration.

3.3   User Attribute Inference

With the derived coupled projection matrices W^g and W^t, given a new user with his/her social behavior features $\mathbf{f}^g$ and $\mathbf{f}^t$, we can estimate the user's unique demographic attributes as:

$$\mathbf{s}^* = \min_s \left\| \mathbf{f}^g - W^g \mathbf{s} \right\|_F^2 + \left\| \mathbf{f}^t - W^t \mathbf{s} \right\|_F^2 \tag{9}$$

Besides, when the projection matrices are obtained, we can also roughly infer the demographic attributes for the typical users with observed social behaviors in one single OSN, by solving the objective function as follows:

$$\mathbf{s}^* = \min_{\mathbf{s}} \left\| \mathbf{f} - W\mathbf{s} \right\|_F^2 \tag{10}$$

where $\mathbf{f}$ and W correspond to the user's social behavior features and the projection matrix in single Google+ or Twitter.

With the estimated user attribute representation $\mathbf{s}$ for each user, we rank the attribute values in each attribute type and choose the biggest one as the final inference result for the attribute type.

4   Experiments

4.1   Experimental Setting

We consider four types of user demographic attributes, i.e., gender, age, relationship, and occupation. The attribute values are defined according to a comprehensive study on Google+ and Twitter data and a survey of previous work on user attribute inference [4,5,10], shown in Table 1. Gender, age and relationship are binary attributes, while occupation is classified into 15 groups, such as student, IT person and entertainer. Since there is no available groundtruth of user attribute values, we build the evaluation dataset by manually labeling the attributes for each user. To ease the annotation task, some user platforms such as Facebook[2], Wikipedia[3] are utilized as referenced sources for accurately

---

[2] http://www.facebook.com/.
[3] http://www.wikipedia.org/.

Table 1. User attribute definition.

Attribute name	Attribute values
Gender	1-Male; 2-Female
Age	1-Young ($\leq$30); 2-Elder ($>$30)
Relationship	1-Unmarried; 2-Married
Occupation	1-Student; 2-IT person, Software Engineer, Geek; 3-Entertainer, Actor, Comedian, Musician, Model, TV show host; 4-Writer, Journalist Editor, Blogger, TV news host, Critics Lawyer; 5-Politician; 6-Ath-lete; 7-Business man, Economist, Entrepreneur; 8-Scientist; 9-Photographer; 10-Doctor; 11-Chef, eater, cook; 12-Engineer, Specialist, Designer; 13-Teacher; 14-Artist, Religious people, Critic; 15-Other

annotating the user attributes. Each user is annotated by three active social network users, and their attribute values are determined as ground-truth only when at least two annotators agree on it.

In this paper, we introduce a cross-OSN demographic attribute inference approach which can infer stable demographic attributes using dynamic social multimedia behaviors. Each of the four attributes is inferred independently. To demonstrate the effectiveness of the proposed coupled projection matrix extraction (CPME) method, we compare it with the popular support vector machine (SVM) method and projection matrix extraction (PME) method using features extracted from single Google+ or Twitter. And a cross-OSN approach, the SVM method using features concatenating the two OSNs' features, acts as a comparison as well. All are listed as follows:

- **Support Vector Machine Using Google+ Feature or Twitter Feature (SVM_Google+/SVM_Twitter):** The method uses features extracted from Google+ or Twitter to train an attribute classifier for each attribute type.
- **Support Vector Machine Using Both OSNs Feature (SVM_Both OSNs):** This is a contrastive cross-OSN attribute inference method which uses a joint feature of Google+ feature and Twitter feature.
- **Projection Matrix Extraction Using Google+ Feature or Twitter Feature (PME_Google+/PME_Twitter):** This refers to the single-OSN attribute inference method described in Eq. (1). It represents the typical method which does not consider the dynamicity problem.

In all experiments, we use 10-fold cross-validation to compare different methods and adopt *accuracy* as the final evaluation metrics. As for the parameters of the proposed model, we find the optimum parameter values according to a separate validation set, i.e., $\lambda_1 = 0.1$, $\lambda_2 = \lambda_3 = 0.3$.

4.2   Experimental Results and Analysis

Table 2 shows the *accuracy* of different methods for inferring user demographic attributes, from which we can make the following observations. (1) Different OSNs contribute differently to the inference of these four types of attributes. For both the SVM and PME methods, age attribute and occupation attribute can be more accurately inferred from Google+, while gender attribute and relationship attribute benefit more from Twitter. A proper selection of typical OSNs is thus important in inferring the user demographic attributes. (2) The SVM method leverages the rich cross-OSN user data do not shows obvious advantage comparing with SVM method using features extracted from single Google+ or Twitter. The accuracy of gender attribute even drops when using both OSNs feature. (3) Although the PME method is not that effective compared with the SVM method in exploiting the same single-OSN information, the proposed CPME method can further improve the performance not only over all the single-OSN methods but also the cross-OSN SVM method. This in turn shows the advantage of the proposed method in addressing the contradiction problem between the dynamic user behaviors and stable demographic attributes from a cross-OSN view. (4) As a binary classification problem, the accuracy of relationship attribute seems relatively low no matter which method is used. It is a rather difficult problem to infer user relationship attribute because of the comparatively weak correlations between users behavior and relationship attribute. Even so, the proposed CPME method has greatly enhanced the relationship attribute inference accuracy 5.95% than that of the SVM method using cross-OSN user data which achieves highest accuracy in baselines.

Table 2. Accuracy of different methods for user demographic attribute inference.

	Gender	Age	Relationship	Occupation
SVM_Google+	0.74737	0.69333	0.58301	0.39104
SVM_Twitter	0.76003	0.68000	0.58492	0.37258
SVM_Both OSNs	0.75748	0.71667	0.58758	0.41238
PME_Google+	0.74245	0.66301	0.56092	0.38161
PME_Twitter	0.75521	0.65710	0.57236	0.36681
CPME	**0.79325**	**0.75085**	**0.62250**	**0.48070**

In Fig. 3, we present some demographic attribute inference by the proposed CPME model. Some users use different images as profile photo in different OSNs, while some users use the same images. We can see that the model accurately infers most of the user attributes, such as attribute "male", "young", "unmarried", "IT person" and so on.

Fig. 3. Predicted user demographic attributes of the testing sample users. Failure instances are highlighted with red. (Color figure online)

4.3 Discussion

In the above-mentioned comparison experiment, more user data in both OSNs are used in the cross-OSN method when testing, which may not be a fair comparison indeed for single-OSN methods. Moreover, it is also very difficult to collect the cross-OSN user data for every user. In a real-world application, we usually can only get access to the user data in one single OSN for attribute inference. As illustrated in Sect. 3.3, with the derived coupled projection matrices, the proposed CPME method can also handle this kind of users with single-OSN behavior data. Therefore, we further test the accuracy of the proposed CPME method on three different user settings, i.e., given test users' data only on Google+, only on Twitter, and on both of them. The result is shown in Table 3, we can see that: (1) Even only with user data in one single OSN for inference, the performance is still higher than the single-OSN methods in Table 2, which may be due to the reason that the potential and stable relevance between different OSN feature spaces are captured in the proposed coupled projection matrix extraction

Table 3. Performance comparison of the proposed CPME method on three different user settings when inferring.

	Gender	Age	Relationship	Occupation
Google+	0.77227	0.738176	0.61200	0.41988
Twitter	0.78495	0.74324	0.59620	0.45270
Both OSNs	**0.79325**	**0.75085**	**0.62250**	**0.48070**

method. (2) By providing more user data in both OSNs, the performance can be further improved. This indicates that more user data available for inference contributes to more accurate attribute prediction.

5   Conclusion

In this paper, we introduce a cross-OSN approach for demographic attribute inference from social multimedia behaviors. The proposed approach is validated to effectively discover the shared stable behavior patterns and achieve superior performance. We also explored the potential of employing single-OSN behavior data for robust demographic attribute inference by utilizing the implicit correlations between users' cross-OSN behaviors. In the future, we will be working along three lines: (1) The cross-OSN approach in dynamic user attribute inference can be compared and examined. (2) The stable patterns from long-time user behaviors for demographic attribute inference can be identified. (3) The proposed method shows its superiority in user demographic attribute inference compared to the SVM method, even if the SVM method is a standard classification method while the proposed method with square loss function is not a strict classification method. Hence, we will try to use more distinguishing features and other suitable loss function.

Acknowledgements. This work is supported by National Natural Science Foundation of China (Nos. 61432019, 61225009, 61303176, 61272256, 61373122, 61332016).

References

1. Abel, F., Araújo, S., Gao, Q., Houben, G.-J.: Analyzing cross-system user modeling on the social web. In: Auer, S., Díaz, O., Papadopoulos, G.A. (eds.) ICWE 2011. LNCS, vol. 6757, pp. 28–43. Springer, Heidelberg (2011). doi:10.1007/978-3-642-22233-7_3
2. Chen, X., Wang, Y., Agichtein, E., Wang, F.: A comparative study of demographic attribute inference in twitter. In: Ninth International AAAI Conference on Web and Social Media (2015)
3. Deng, Z. Sang, J. Xu, C.: Personalized video recommendation based on cross-platform user modeling. In: 2013 IEEE International Conference on Multimedia and Expo (ICME), pp. 1–6. IEEE (2013)

4. Fang, Q., Sang, J., Xu, C., Hossain, M.S.: Relational user attribute inference in social media. Multimedia IEEE Trans. **17**(7), 1031–1044 (2015)
5. Filatova, E., Prager, J.: Occupation inference through detection and classification of biographical activities. Data Knowl. Eng. **76**, 39–57 (2012)
6. Gao, H., Tang, J., Hu, X. Liu, H.: Content-aware point of interest recommendation on location-based social networks. In: AAAI, pp. 1721–1727 (2015)
7. Huang, Y., Yu, L., Wang, X., Cui, B.: A multi-source integration framework for user occupation inference in social media systems. World Wide Web **18**(5), 1247–1267 (2015)
8. Jia, Y., Shelhamer, E., Donahue, J., Karayev, S., Long, J., Girshick, R., Guadarrama, S. Darrell, T.: Caffe: convolutional architecture for fast feature embedding. In: Proceedings of ACM International Conference on Multimedia, pp. 675–678. ACM (2014)
9. Pennacchiotti, M. Popescu, A.-M.: Democrats, republicans and starbucks afficionados: user classification in Twitter. In: Proceedings of 17th ACM SIGKDD International Conference on Knowledge Discovery and Data Mining, pp. 430–438. ACM (2011)
10. Rao, D., Yarowsky, D., Shreevats, A., Gupta, M.: Classifying latent user attributes in Twitter. In: Proceedings of 2nd International Workshop on Search and Mining User-Generated Contents, pp. 37–44. ACM (2010)
11. Roy, S.D., Mei, T., Zeng, W., Li, S.: Socialtransfer: cross-domain transfer learning from social streams for media applications. In: Proceedings of 20th ACM International Conference on Multimedia, pp. 649–658. ACM (2012)
12. Sang, J., Lu, D., Xu, C.: A probabilistic framework for temporal user modeling on microblogs. In: Proceedings of 24th ACM International on Conference on Information and Knowledge Management, pp. 961–970. ACM (2015)
13. Yan, M., Sang, J., Xu, C.: Unified YouTube video recommendation via cross-network collaboration. In: Proceedings of 5th ACM on International Conference on Multimedia Retrieval, pp. 19–26. ACM (2015)
14. Zheleva, E. Getoor, L.: To join or not to join: the illusion of privacy in social networks with mixed public and private user profiles. In: Proceedings of 18th International Conference on World Wide Web, pp. 531–540. ACM (2009)

Understanding Performance of Edge Prefetching

Zhengyuan Pang[1(✉)], Lifeng Sun[1], Zhi Wang[2], Yuan Xie[3], and Shiqiang Yang[1]

[1] Department of Computer Science and Technology,
Tsinghua University, Beijing, China
pangzy12@mails.tsinghua.edu.cn
[2] Graduate School at Shenzhen, Tsinghua University, Beijing, China
[3] Department of Computer Science, Indiana University Bloomington,
Bloomington, USA

Abstract. When using online services, the time that users wait for the requested content to be downloaded from online servers to local devices can significantly influence user experience. To reduce user waiting time, the content which are likely to be requested in the future can be pre-downloaded to the local cache on edge proxies (i.e. *edge prefetching*).

This paper addresses the performance issues of prefetching at edge proxies (*e.g.* Wi-Fi Access Points (APs), cellular base stations). We introduce an AP-based prefetching framework and study the impact of several factors on the benefit and the cost of this framework based on trace-driven simulation experiments. Useful insights which can be used to guide the design of prediction algorithms and edge prefetching systems are gained from our experimental results. First, increasing prediction window size of the prediction algorithms used by mobile applications can significantly reduce user waiting time. Second, the cache size is important to reducing user waiting time before a certain threshold. Third, the ratio of correct predictions to all actual requests (i.e. *recall*) can reduce user waiting time linearly while the ratio of correct predictions to all predictions (i.e. *precision*) will influence the traffic cost, so a trade-off should be made when designing a prediction algorithm.

Keywords: Edge prefetching · Wi-Fi Access Point · Performance · Verification

1 Introduction

Internet traffic from mobile devices has grown rapidly in the past few years. According to the Cisco Visual Networking Index, global mobile data traffic increased by 69% in 2014 [8]. Besides, up to 55% of the total mobile data traffic is mobile video traffic [8]. For online services (*e.g.* browsing web pages or watching online videos) on mobile devices, waiting time is one of the most important metrics of user experience [3,14]. For example, experiments at Amazon showed that every 100 ms increase in the load time of Amazon.com would decrease the revenue by 1%. Google also found that an extra 500 ms in search page generation

© Springer International Publishing AG 2017
L. Amsaleg et al. (Eds.): MMM 2017, Part I, LNCS 10132, pp. 527–539, 2017.
DOI: 10.1007/978-3-319-51811-4_43

time dropped its number of requests by 20% [10]. In order to reduce user waiting time, prefetching is widely adopted to avoid real-time data downloading from network.

Prefetching can be implemented on mobile clients or proxy servers. In client-initiated prefetching approaches, such as [7], one of the most concerned challenges is how to balance the benefit and the cost. First of all, because of the limited resources on mobile devices, e.g. battery life, storage capability and so on, unnecessary consumption introduced by prefetching will harm user experience. Moreover, the concurrent prefetching requests of multiple mobile clients will compete for network resources.

Due to these limitations, some research works propose proxy-based prefetching schemes [5,13]. The basic idea of proxy-based prefetching is pre-downloading the data which are likely to be requested by users in the future to the cache storage on a proxy (e.g. a wireless router or a home gateway). When an actual request arrives, it is served by the proxy so that the time of downloading relevant data from external network is saved. Prefetching on proxies can reduce resource competition by scheduling prefetching tasks from different devices. Meanwhile, extra battery or memory consumption of mobile devices can be avoided since the prefetching tasks are executed on proxies.

There are some existing works aiming to study the performance of proxy-based prefetching systems. In [9], the authors give a measurement study of the cost, benefit and performance of a variety of prefetching algorithms. However, this work only considers prefetching web content without verification of the effectiveness of these algorithms on other types of content, such as online videos. The authors of [11] propose and evaluate an integrated prefetching and caching approach for adaptive video streaming over HTTP, but their evaluation does not include the impact of the prediction algorithm on improving user experience by prefetching.

In this paper, we focus on the performance issues of a Wi-Fi AP-based prefetching framework. The basic scenario can be described as follows: the prefetching engine is located on a smart Wi-Fi AP equipped with a Linux-based OS and a large storage [2] so that it can be used as a lightweight edge server; prefetching tasks, such as priority controlling, resource provisioning, content caching, etc., can be executed on the AP. To avoid interfering with normal user requests, we introduce an opportunistic prefetching strategy, which means the prefetching proceeds only during the *idle* periods of the network.

We focus on three fundamental aspects of this AP based prefetching framework:

- What is the benefit of opportunistically prefetching on Wi-Fi APs?
- What cost will this framework incur?
- What factors will influence the benefit and the cost of this framework?

The main contributions of our work are as follows:

▷ We introduce an AP-based prefetching framework as the example of edge prefetching, including the architecture and the Earliest Deadline First (EDF) prefetching policy.

▷ We study the benefit and the cost of this framework using trace-driven simulation experiments.

▷ We gain several useful insights from the experimental results. First, prediction window size and cache size has significant impact on prefetching performance. Second, prediction recall will influence the prefetching performance and prediction precision will affect the traffic cost. Third, AP-based prefetching can improve user experience where the network bandwidth is limited.

The rest of the paper is organized as follows. In Sect. 2, we describe the architecture and key elements of the framework. In Sect. 3, we present the numerical results and useful insights obtained from trace driven simulations. Finally, conclusions are drawn in Sect. 4.

2   Framework Overview

In this section we introduce the architecture and key elements of the AP-based prefetching framework.

2.1   Architecture

Figure 1 illustrates the elements of the AP-based prefetching framework. Other than generating predictions on the AP, we introduce another scheme in which each client generates the predictions of future requests by its own. This is because that there may be plenty of applications (abbr. *app*) installed on a mobile device, such as browsers, video players, news readers, etc., and each app may correspond to a specific prediction model. More importantly, each app can collect and analyze the preference of the user independently. Thus generating predictions by each app can gain higher accuracies. There are already some prediction methods that have proven to be effective, *e.g.* [12]. Based on this premise, the prefetching process is as follows:

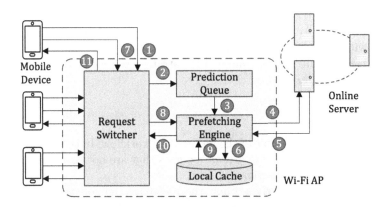

Fig. 1. Wi-Fi AP-based prefetching.

- *Mobile devices* are connected to the AP through wireless network such as 802.11 g. Predictions, denoted by $\mathcal{P} = \{p_1, p_2, \ldots, p_k\}$, are generated by the apps installed on the mobile devices. The actual requests sent by mobile clients are denoted by $\mathcal{R} = \{r_1, r_2, \ldots, r_n\}$. A typical prediction p_i of a request r_i includes the predicted arrival time (denoted by d_i) of r_i, the target URL (denoted by u_i) and the predictor information (such as the device ID).
- The *request switcher*, the *prediction queue* and the *prefetching engine* are located on the AP. Predictions are sent to the request switcher (①), then they are put in the prediction queue (②). The predictions in the queue are scheduled based on the prefetching policy. When the network is idle, the chosen predictions are processed by the prefetching engine (③). The prefetching engine makes the prefetching decision and manages the prefetching process. The requested data of the processed predictions are pre-downloaded from online servers (④, ⑤) and stored in the *local cache* (⑥).
- When an actual request arrives (⑦), the request is forwarded to the prefetching engine (⑧). If the requested data are in the local cache, the request is served by the cached data directly (⑨, ⑩, ⑪), otherwise the data will be downloaded from online servers normally (④, ⑤). Note that if a predicted request arrives, yet its requested data are not completely prefetched, then only the remaining data need to be downloaded.

2.2 Prefetching Schedule

The prefetching policy (as shown in Algorithm 1) is based on the EDF scheduling policy and the arrival time of each request is treated as scheduled deadline here. In Algorithm 1, $\mathcal{F}$ denotes the ongoing prefetching task and $\mathcal{T}$ denotes the system time. The prefetching engine keeps monitoring the network status, and once the network is *idle* and there is no ongoing prefetching task, it checks the prediction queue and picks the one with the earliest deadline. If a normal request arrivals, the ongoing prefetching task is suspended until the network is idle again. Note that in order to use all available bandwidth to download the requested data of the prediction as soon as possible, there is only *one* prefetching task active at a time.

The cache storage on the AP is managed by the prefetching engine. Entries in the cache with predicted arrival time ahead of current $\mathcal{T}$ are marked as *expired*. When the cache is full, it uses a simple LRU strategy to remove the least recently used entry from the *expired* entries.

2.3 Workflow

Figure 2 illustrates an example of the prefetching workflow. In the framework, we mark correct predictions as *hit*, arrival requests that are not correctly predicted as *miss*, and incorrect predictions as *false*.

From Fig. 2 we can clearly see how the prefetching engine works. Moreover, we can find some details which are helpful for us to analyze the performance of the framework:

Algorithm 1. The EDF Prefetching Policy

```
 1: procedure EDF–PREFETCHING(R, P, F, T)
 2:     if R = φ then
 3:         if F then
 4:             F.proceed()
 5:         else
 6:             P.sort(dᵢ)
 7:             for pᵢ ∈ P do
 8:                 if dᵢ < T then
 9:                     P.pop(pᵢ)
10:                 else
11:                     F ← pᵢ
12:                     F.start()
13:                     P.pop(pᵢ)
14:                     break
15:                 end if
16:             end for
17:         end if
18:     else
19:         F.suspend()
20:     end if
21: end procedure
```

- If a request is correctly predicted (i.e. *hit*) and there is enough idle time of the network before its arrival, the downloading time of the requested data can be entirely concealed from the user. The user only needs to wait for the data to be transferred from the AP to the mobile device. Since the network bandwidth between the mobile device and the AP is usually one order of magnitude higher than the external network, this waiting time is negligible.
- Not only completely (*e.g.* $1, 3, 7$) and partially (*e.g.* $4, 5$) prefetched requests can be accelerated, but some *miss* requests as well (*e.g.* 2). The reason is that request 2 and 5 will compete for bandwidth when there is no prefetching (t3 ∼ t4 in Fig. 2a). However, with prefetching, request 5 will end earlier, so that request 2 can take advantage of the exclusive bandwidth since t3, thus ends earlier as well. Meanwhile, some *miss* requests can not be accelerated, such as 8.
- Incorrect predictions (i.e. *false*) may cause unnecessary traffic overhead (*e.g.* $9, 10$) and a waste of network idle time, but they will not impact the waiting time of other requests or predictions directly.
- The prefetching will change the network status. Specifically, prefetched requests can release network bandwidth because the requested data need not to be downloaded from the content servers through external network. For example, the time period between t1 and t2 becomes idle as request 1 is prefetched (Fig. 2b).

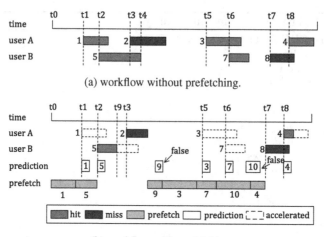

(a) workflow without prefetching.

(b) workflow with prefetching.

Fig. 2. An example of prefetching workflow. Solid squares denote the user requests, hollow squares denote the predictions and dashed squares denote the accelerated parts of the requests. Resource competition will occur in overlapping regions of requests.

2.4 Use Case

There are lots of scenarios in which the AP-based prefetching framework can be deployed, such as a house or a cafe. Residents of a house or customers in a cafe generally share an Internet entry, *e.g.* a wireless router. The prefetching framework can be deployed on this entry so that every mobile client connected to it can benefit from the prefetching scheme.

3   Numerical Results

In this section, we measure the benefit and the cost of the AP-based prefetching framework using trace-driven simulations.

3.1   Experimental Setup

Basic Assumptions.   In our experiments, we make some reasonable assumptions:

- Because the AP is located in the same wireless LAN (Local Area Network) as mobile devices, the transmission time of data from the AP to mobile devices is negligible compared to the downloading time from the external network. So we regard the former as zero in our experiments, in other words, we only consider the downloading time below.
- In order to simulate resource competition of requests, we assume that concurrent active download processes share the bandwidth equally.

- Since the prediction algorithms on mobile devices are not the major concern in this paper, we assume that all clients deploy similar prediction methods in the same kind of apps. These similar prediction methods have the same performance features (i.e. prediction window size, prediction accuracy) when generating predictions.

Metrics. The performance metrics used in our experiments are as follows:

Waiting Time (WT): average time interval between the arrival of user requests at the AP and the acquisition of the requested content. According to the first assumption, this waiting time is mainly the downloading time from the external network.

Waiting Time Reduction (WTR): the reduced waiting time with prefetching compared with that without prefetching, measured in percentage. The waiting time reduction is calculated as follows:

$$WTR = \frac{WT_{noprefetch} - WT_{prefetch}}{WT_{noprefetch}} \tag{1}$$

Because we focus on the effect of reducing user waiting time in the framework, so we use WTR to evaluate the prefetching performance in the following subsections.

Traffic Increase (TI): the ratio of increased downloading traffic on AP with prefetching to the traffic without prefetching, measured in percentage and calculated as follows:

$$TI = \frac{Traffic_{noprefetch} - Traffic_{prefetch}}{Traffic_{noprefetch}} \tag{2}$$

Recall: the ratio of correct predictions to all arrived requests.

Precision: the ratio of correct predictions to all predictions.

Recall and precision both indicate the prediction accuracy, they are calculated as follows:

$$Recall = \frac{hit}{hit + miss} \tag{3}$$

$$Precision = \frac{hit}{hit + false} \tag{4}$$

$\mathcal{W}$: prediction window size measured in seconds which denotes the time interval between the generation of a prediction and the actual arrival of the corresponding request.

Simulation Setup. We test the prefetching performance of two types of content in our experiments: web pages and online videos. We first study distributions of data size of these two types of content using real-world datasets and then use the distributions to generate empirical data in our experiments.

The mobile web page dataset is downloaded from *HTTP Archive* [1], which are publicly available. We download the archives of year 2015, which are from

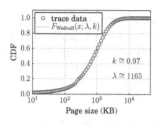
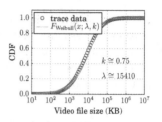
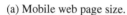

(a) Mobile web page size.   (b) YouTube video file size.

Fig. 3. CDF of file size. (a) Distribution of web page size. (b) Distribution of YouTube video file size.

Jan. 1, 2015 to Oct. 15, 2015, containing a total of 94048 records. Each record includes a lot of information, such as the URL, the total size, the page rank, etc. Here we just use the total size of each page in our experiments.

We model the distribution of video file size using a YouTube video file dataset which is from [4] and publicly available online now. The dataset contains some fundamental characteristics (*e.g.* video duration, bitrate, etc.) of 1.6 million YouTube videos. We calculate the file size of each video by multiplying the video duration with bitrate.

As illustrated in Fig. 3, the size of both types of data follows the Weibull distribution. The *shape* parameter (k) and the *scale* parameter (λ) are identified as $k_{page} \cong 0.97$, $\lambda_{page} \cong 1165$ for web page data and $k_{video} \cong 0.75$, $\lambda_{video} \cong 15410$ for YouTube video data. Furthermore, in order to simplify the web browsing model, we assume that all objects of a web page are assembled into a single file and thus can be downloaded by one request.

In our experiments, we set the number of requests received by the AP following a Poisson distribution, which means that the time intervals between two consecutive requests follow an exponential distribution [6]. The time of the process we simulate is 1 h and in this process users keep sending requests to the AP, each of which requests a web page or a video file. The default parameters are set to $\mathcal{W} = 200\,\text{s}$, bandwidth $= 10\,\text{Mbps}$, cache size $= 200\,\text{MB}$.

3.2   Results

In this subsection we present the experimental results and summarize some useful insights by analyzing these results.

Impact of Prediction Window Size. We first evaluate the impact of prediction window size ($\mathcal{W}$) on the average user waiting time (WT). We measure the variation of WT with different values of $\mathcal{W}$ for both types of requests. In this experiment, we set recall $= 1.0$, precision $= 1.0$, and other parameters are set to be default values. For each value of $\mathcal{W}$, the simulation is repeated 20 times. The results are shown in Fig. 4.

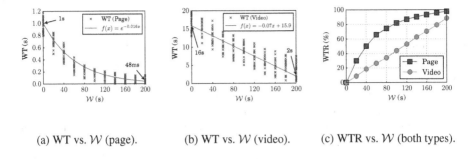

(a) WT vs. $\mathcal{W}$ (page).   (b) WT vs. $\mathcal{W}$ (video).   (c) WTR vs. $\mathcal{W}$ (both types).

Fig. 4. The impact of prediction window size. In (a) and (b), each marker (*cross*) denotes the WT in one simulation. In (c), each marker (*square* or *circle*) denotes the average WTR of 20 simulations of the same $\mathcal{W}$.

Figure 4 depicts that $\mathcal{W}$ has different impact on the two types of requests. WT of web page requests decreases exponentially as $\mathcal{W}$ increases while WT of video requests decreases linearly as $\mathcal{W}$ increases. These different patterns can be observed more explicitly in Fig. 4c when we look at the change of WTR. Under the settings used in this experiment, when $\mathcal{W}$ is larger than 200 s, WT of page requests and video requests are reduced to about 48 ms and 2 s respectively. This is a considerable improvement to user experience compared to the WT without prefetching (i.e. 1 s and 16 s). Thus, increasing prediction window size of the prediction algorithms used by mobile applications can significantly improve prefetching performance.

Impact of Prediction Accuracy. Intuitively, prediction accuracy (*recall* and *precision*) is important to prefetching performance. To verify this, we evaluate

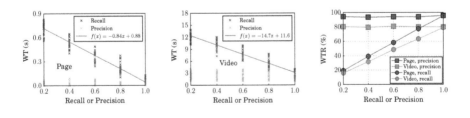

(a) WT vs. prediction accuracy (page).   (b) WT vs. prediction accuracy (video).   (c) WTR vs. prediction accuracy (both types).

Fig. 5. The impact of prediction accuracy. In (a) and (b), each marker (*cross*) denotes the WT of one simulation corresponding to a prediction recall value (blue) or a prediction precision value (green). In (c), each marker denotes the average WTR of 20 simulations of the same recall value (*circle*) or precision value (*square*). (Color figure online)

the impact of prediction accuracy on user waiting time. First, we set precision $=1.0$ and treat the recall as a variable. Then we set recall $=1.0$ and treat the precision as a variable. For each value of the variable, the simulation is repeated 20 times. Other parameters are set to be default values and the results are shown in Fig. 5.

From this figure we can see that WT of both types of requests decreases linearly as prediction recall increases, while prediction precision has little impact on prefetching performance. The explanation to this is that *miss* requests which are not in the prediction list will not be prefetched, therefore most of them can not be accelerated by prefetching (a few exceptions are explained in the example in Sect. 2.3). The lower the recall is, the more unaccelerated requests there are, which will result in larger WT. In contrast, the precision reflects the ratio of *false* predictions, which will not actually be sent by terminal devices, so the prediction precision only increases the unnecessary traffic, but do not influence the WT of *hit* requests. This implies that prediction recall is more important for a prediction algorithm to obtain better prefetching performance.

Impact of Cache Size. In this subsection, we evaluate the impact of cache size on prefetching performance. We set recall $=0.7$, precision $=0.6$ and other parameters are set to be default values. For each value of the cache size, the simulation is repeated 20 times, and the average WTR is adopted.

The results shown in Fig. 6 indicate that the prefetching performance is poor when the cache size is too small. This is reasonable because that the prefetched data can not be stored in the cache when there is not enough space and previous data have not expired. However, when the cache size reaches a threshold, previous data will expire before the arrival of new data. In such conditions, enlarging cache size can not improve the prefetching performance further. Under the settings of this experiment, the thresholds are 4 MB and 80 MB for web page data and video data respectively. These results indicate that the threshold which is important to prefetching performance depends on the distribution of data size and access pattern of requests. When designing an AP-based prefetching system, the cache size should be taken into serious consideration according to the content type.

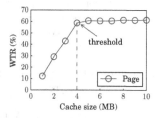

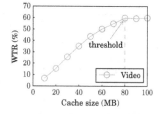

(a) WTR vs. cache size (page).   (b) WTR vs. cache size (video).

Fig. 6. Impact of cache size.

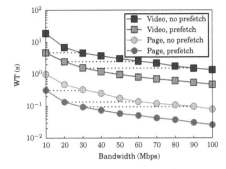
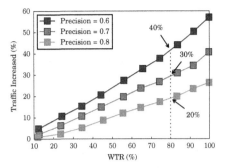

Fig. 7. WT vs. bandwidth.   **Fig. 8.** Traffic increased vs. WTR.

Impact of Network Bandwidth. In this subsection we study how network bandwidth influences user waiting time with or without prefetching. From the experimental results shown in Fig. 7, we find that WT decreases almost linearly as the bandwidth increases for both types of requests. Another interesting observation is that there is some correlation between WT with prefetching and that without prefetching. For example, WT of web page requests with prefetching under bandwidth of 10, 20, 30 Mbps are close to WT of web page requests without prefetching under bandwidth of 30, 60, 90 Mbps, and the same correlation can be found from the results of video data as well (as depicted by dotted lines in Fig. 7). These correlations imply that using AP-based prefetching can obtain the approximate acceleration effect as improving the bandwidth by 3 times. We call this a *bandwidth expansion* effect and the magnitude that the bandwidth expands is termed as *bandwidth expansion factor*. In our experiments, the bandwidth expansion factor is 3. This conclusion proves that AP-based prefetching has great potential in practical scenarios, *e.g.* places where the Internet bandwidth is limited.

Traffic Increased (TI). There are mainly two parts of cost introduced by the prefetching framework. One is the storage cost on the AP, which is exactly the cache size we discuss in the previous subsection. The other one is the additional downloading traffic introduced by false predictions. According to Eqs. (3) and (4), when the precision is constant, the predictor can improve the prediction recall by increasing the number of predictions. This method can improve the overall prefetching performance but it will introduce more network traffic cost because of the increased false predictions. Figure 8 shows the correlation between traffic increase and the performance improvement (WTR).

From this figure we can see that TI increases linearly as WTR increases and higher precisions will yield the same prefetching performance with lower traffic increase. More specifically, if the service provider wants to reduce WT by 80%, the extra traffic will be 20%, 30% and 40% when the precision are 0.8, 0.7 and

0.6 respectively (as annotated in Fig. 8). From previous subsection we find that precision has slight impact on prefetching performance, but it can be seen here that better precision can save the traffic cost of the framework. From this point, the service provider can make trade-offs according to the benefit and the cost of the whole system.

4   Conclusion

In this paper we address the performance issues of edge prefetching. We introduce an AP-based prefetching framework as the example. The impact of multiple factors on the performance and the cost of this framework is studied through trace-driven simulations. Several insights which can be used to guide the design of edge prefetching systems and prediction algorithms used on mobile applications are obtained from the experimental results. For example, this AP-based prefetching system can be deployed at places where the Internet bandwidth is limited to improve user experience. Besides, the prediction algorithms should be designed under the consideration that the prediction window size and the prediction recall will influence the acceleration effect of prefetching while the prediction precision will affect the traffic cost of the prefetching system. Moreover, the cache size of the prefetching system should be determined according to the content type which the system intends to cache.

Acknowledgments. This work is supported in part by NSFC under Grant Nos. 61472204 and 61402247, SZSTI under Grant No. JCYJ20140417115840259, and supported by Beijing key lab of networked multimedia.

References

1. HTTP Archive Mobile, October 2015. http://mobile.httparchive.org/
2. Xiaomi MiWiFi, April 2015. http://www.mi.com/miwifi/
3. Chi, H.Y., Chen, C.C., Cheng, W.H., Chen, M.S.: Ubishop: commercial item recommendation using visual part-based object representation. Multimedia Tools Appl. 1–23 (2015)
4. Deneke, T., Haile, H., Lafond, S., Lilius, J.: Video transcoding time prediction for proactive load balancing. In: 2014 IEEE International Conference on Multimedia and Expo (ICME), pp. 1–6. IEEE (2014)
5. Guan, X., Choi, B.Y.: Push or pull? Toward optimal content delivery using cloud storage. J. Netw. Comput. Appl. **40**, 234–243 (2014)
6. Gündüz, Ş., Özsu, M.T.: A Poisson model for user accesses to web pages. In: Yazıcı, A., Şener, C. (eds.) ISCIS 2003. LNCS, vol. 2869, pp. 332–339. Springer, Heidelberg (2003). doi:10.1007/978-3-540-39737-3_42
7. Higgins, B.D., Flinn, J., Giuli, T.J., Noble, B., Peplin, C., Watson, D.: Informed mobile prefetching. In: Proceedings of 10th International Conference on Mobile Systems, Applications, and Services, pp. 155–168. ACM (2012)
8. Index, C.V.N.: Global mobile data traffic forecast update, 2014–2019. White paper, February 2015

9. Jiang, Y., Wu, M.Y., Shu, W.: Web prefetching: costs, benefits and performance. In: Proceedings of 7th International Workshop on Web Content Caching and Distribution (WCW 2002), Boulder, Colorado. Citeseer (2002)
10. Kohavi, R., Longbotham, R.: Online experiments: lessons learned. Computer **40**(9), 103–105 (2007)
11. Liang, K., Hao, J., Zimmermann, R., Yau, D.K.: Integrated prefetching and caching for adaptive video streaming over HTTP: an online approach. In: Proceedings of 6th ACM Multimedia Systems Conference, pp. 142–152. ACM (2015)
12. Lymberopoulos, D., Riva, O., Strauss, K., Mittal, A., Ntoulas, A.: Pocketweb: instant web browsing for mobile devices. In: ACM SIGARCH Computer Architecture News, vol. 40, pp. 1–12. ACM (2012)
13. Teng, W.G., Chang, C.Y., Chen, M.S.: Integrating web caching and web prefetching in client-side proxies. IEEE Trans. Parallel Distrib. Syst. **16**(5), 444–455 (2005)
14. Tsai, T.H., Cheng, W.H., You, C.W., Hu, M.C., Tsui, A.W., Chi, H.Y.: Learning and recognition of on-premise signs from weakly labeled street view images. IEEE Trans. Image Process. **23**(3), 1047–1059 (2014)

User Identification by Observing Interactions with GUIs

Zaher Hinbarji[✉], Rami Albatal, and Cathal Gurrin

Insight Centre for Data Analytics, Dublin City University, Dublin, Ireland
zaher.hinbarji@dcu.ie
https://www.insight-centre.org

Abstract. Given our increasing reliance on computing devices, the security of such devices becomes ever more important. In this work, we are interested in exploiting user behaviour as a means of reducing the potential for masquerade attacks, which occur when an intruder manages to breach the system and act as an authorised user. This could be possible by using stolen passwords or by taking advantage of unlocked, unattended devices. Once the attacker has passed the authentication step, they may have full access to that machine including any private data and software. Continuous identification can be used as an effective way to prevent such attacks, where the identity of the user is checked continuously throughout the session. In addition to security purposes, a reliable dynamic identification system would be of interest for user profiling and recommendation. In this paper, we present a method for user identification which relies on modeling the behaviours of a user when interacting with the graphical user interface of a computing device. A publicly-available logging tool has been developed specifically to passively capture human-computer interactions. Two experiments have been conducted to evaluate the model, and the results show the effectiveness and reliability of the method for the dynamic user identification.

Keywords: User behavior modeling · GUI usage analysis · Identification · Multimodal interaction modeling · Masquerade detection

1 Introduction

Computer security is an increasingly important topic and there are many types of possible attacks on the integrity of a computing system. One type is the masquerade attack, in which an intruder tries to imitate a legitimate user's behaviour in order to pass the security system and get access to privileged computer resources. This can be seen as a problematic special case of the intrusion detection domain. It could simply happen when the intruder takes advantage of unlocked or unattended computers or when he/she gets access to the legitimate user's password somehow and uses it to bypass the initial layer of authentication. Thus, it's important to have a dynamic authentication mechanism in place which keeps checking the user's identity throughout the session not only at the

© Springer International Publishing AG 2017
L. Amsaleg et al. (Eds.): MMM 2017, Part I, LNCS 10132, pp. 540–549, 2017.
DOI: 10.1007/978-3-319-51811-4_44

beginning of the session as a static authentication system does. In addition to the security purposes, dynamic identification systems would have an important additional value to most personalisation and recommendation systems that need to identify the current user before providing their services [6].

Instead of modeling *what* the user usually does on the computer, our main focus here is to model *how* the user interacts with the computing resource. Compared to command-line interface (CLI) based systems, those based on graphical user interface (GUI) give the user more freedom and provide alternative methods to complete tasks [8]. Even very simple operations such as copy-paste, writing short words and switching between application windows can be done in significantly different ways among users in GUI based systems. For instance, to copy a piece of text, a user could first use the mouse to select the text, press the copy command and then move the mouse to the destination and paste the text eventually. Whereas, another user could perform all the operations through keyboard shortcuts. Others could use both keyboard and mouse together to achieve this simple task. As we see, there are many choices available to the users to achieve their tasks in such systems and therefore there is potential to model this behaviour and use it as an identifier of users. This also helps us to see clearly the benefit of a comprehensive identification model that takes advantage of the different forms of interaction data coming from all the available sources to increase identification accuracy and to reduce identification time.

To detect the real identity of the current user of a computer, our approach depends on the manner the user interacts with the GUI of the system. It captures the statistical characteristics of the usage patterns and compares them to the previously built profiles. The comparison is done using a support vector machine classifier [5]. The advantage of our approach is that it doesn't depend on the observed usage patterns of mouse or keyboard separately, but rather it utilises all the available modalities to identify the user. This is done continuously during an entire computing session. Specifically, the purpose of our proposed identification system is to answer the question: to which user from our database does this observed interaction behaviour belong? Our research shows that computer usage patterns have enough discriminatory power to reveal the user identity. Although we are mainly interested here in identification, our features are engineered to be extensible and applicable to other domains. This is to conform to our future research plans of automatically building enhanced user profiles based on the interaction data, which has potential to extract new insights about a user's preferences, skills or even stress level or mood.

The contribution of this paper is a new model for dynamic user identification that utilises multimodal GUI-based features. In addition, two experiments have been conducted based on two databases collected by a logging tool that has been developed specifically for this purpose and released for public use. This paper is organised as follows. In the next section we present some of the most relevant work in the area. In Sect. 3 we present the three types of GUI features that we are extracting for user identification, and then our experiments are introduced and presented in Sect. 4.

2   Related Work

The user study on the use of interactive computing systems [4] from 1974 could be considered as the first user behaviour tracking experiment, even though the data was not used for identification but rather for investigating how users were using mainframe computer.

The work of Anderson in 1980 formed the first step toward automatic behaviour modelling for intrusion detection [16], in which he automatically collected log files, resource and command usage information to profile users behaviour and make sure they are not engaged in harmful actions [2]. Most of the previous work in this area was based on the set of commands used by the users in command-line based systems. A typical target of this approach is UNIX operating system. The familiarity with the command line interfaces varies a lot among users, including the awareness and usage of their features. According to studies, users tend to be consistent with their choice of commands during their everyday tasks. The patterns of commands sequences generated by users are studied in [14] and proven to be useful for abnormal user behaviour detection. The login host, the login time, the command set, and the command execution time were used to profile and identify users with a low error rate [19]. In [18], 15,000 commands from about 70 users were captured. Statistical methods were applied to capture the intrusion such as Hybrid multistep Markov and Bayes one-step Markov. This was extended by [15] with Naive Bayes classifier, a 56% improvement in masquerade detection was achieved at a corresponding false-alarm rate of 1.3%. Other researchers have also focused on the area of command line profiling such as in [17,20].

Nowadays, most modern operating systems and software are deploying graphical user interfaces instead of the legacy command line interface. This shift has not been followed by an equivalent increase of research work in the area of behaviour profiling based on GUI interaction. Still, some work has been done here. In [9], time between windows switching, time between new windows, number of windows open at once and number of words in window title were employed as features. The same dataset of the previous work has been utilised in [13] using symbolic learning. The mean and the standard deviation of 8 raw features of mouse interaction were used in [8]. These raw features were the number of right/left mouse clicks, mouse distance, mouse speed and mouse angle (four directions). This work was based on 3 users, and adding more features based on the running processes and keyboard usage statistics did not improve the results significantly.

Other approaches have focused only on the mouse usage patterns. For instance, in [11] the mathematical properties of the user's mouse curves have been modeled for identification. Ahmed and Traore's work in [1] utilised features related to movement speed, movement direction, traveled distance to produce user signature. In addition, many techniques can be found in literature for analysing the user's keyboard typing patterns. Bergadano et al. [3], used the duration of typing two (di-graph) and three (tri-graph) consecutive characters of a text to construct the user profile. Gunetti and Picardi [10] extended the approach of [3] to work on free text. Curtin et al. [7] used only common characters and special keys. The disadvantage of such single-input-device identification systems is the poor performance when the user is not using that input device long enough to generate the needed interaction data, which may affect both accuracy and identification time. On the other hand, a comprehensive identification system that takes advantage of all kinds of multimodal interaction data (mouse, keyboard and any other available sources) has a better potential to detect the user identity in shorter session length and regardless of what input device the user is using. These are the advantages of our approach.

Another domain that could be close to our work is program profiling, which typically employs system calls to model the normal behaviour of the program, and then tries to detect suspicious behaviours based on the difference between the current captured behaviour and the previous modelled one. We think program profiling is an easier problem, since programs are designed to do specific tasks with limited freedom, unlike human whose behaviour is unpredictable.

In this work, we present our approach for user identification and profiling based on a multitude of multimodal GUI interaction data. Our approach provides richer user behavioural model, and our experiments show that it achieves accurate and reliable user identification.

3   An Approach to Behaviour Modeling

In order to test our approach in a real life scenario, our model uses low-level computer events collected from users during their normal computer work. To make this possible and to overcome the lack of a comprehensive data set, we designed a logging tool, LoggerMan, that works passively in the background and captures all types of interaction events [12]. It is released publicly for researchers and can be downloaded via LoggerMan.org website. Our tool allows the user to either store the logged data locally on their computer or to sync it automatically to our server. Table 1 shows the different channels of low-level computer events that our logging tool can capture: mouse, keyboard, application, window and clipboard. However we discarded the data coming from window and clipboard modules as we believe that such data is more related to *what* the user is doing (e.g., window's title and clipboard's content) not *how* the user is using the machine. We will describe later in Sect. 4 the two datasets we have collected for this research.

Our goal is to recognise the user currently interacting with the machine based on interaction behaviour characteristics. To achieve this, we calculate statistical features of the computer events we intercept. We then concatenate these features to form input vectors (called the behavioural signature vectors) that are introduced to a classification algorithm to classify the interactions and output the detected identity as a result. As we introduced earlier, our features are selected to be meaningful and readable by human as possible. That's mainly to be utilised later in our profiling research as well, which could give us insights about the user's interaction skills, preferences or even behaviour change. In order

Table 1. Different sources of interaction events

Module	Desciption	Format
Mouse	Mouse actions	x,y,type,timestamp
Keyboard	Typed keys	key_code,timestamp
Apps	Used apps	app_name,timestamp
Windows	Opened windows	app,win_title,timestamp
Clipboard	Copy-paste actions	timestamp

to study the time needed to identify a user, we introduce the concept of session. We define the session as *consecutive computer events captured during a specific time range*. We will measure the model's accuracy based on different session lengths, measured in seconds. In the following sub sections, we will describe the 22 features that our model uses.

3.1 Keyboard Features

The reason for logging keyboard features is to have an overview of how a user uses the keyboard in general. We are trying here to detect the user's statistical typing patterns. Handedness (right-handed/left-handed) and the way of typing (one hand or both hands are used) are both important attributes to be considered here. To handle this, our model uses 4 different features related to the average typing speed of 2 consecutive keys depending on their location on the keyboard. In other words, how fast can the user type two keys both located to the right side of the keyboard? How that can be compared if they both located to the left or each one in a different side? The complete set of keyboard features our model utilises (ten features in total) are as follows:

- percentage of delete keys usage
- percentage of control keys usage
- minimum typing speed
- average typing speed
- maximum typing speed
- average typing speed of numbers keys
- typing speed of 2 consecutive right-side keys
- typing speed of 2 consecutive left-side keys
- typing speed of 2 consecutive right-to-left keys
- typing speed of 2 consecutive left-to-right keys

3.2 Mouse Features

The purpose of the mouse features is to capture the patterns a user follows during his/her interaction with the computer using a pointing device and they represent the mouse usage characteristics of the user. The following are the ten mouse features our model uses:

- Average and standard deviation of the clicking delay (the time the user takes between pressing down the mouse's button and releasing it).
- Normalised histogram of the count of different mouse actions performed (move, click, scroll and drag-drop) by the user.
- Average and standard deviation of the mouse actions distances in pixel (does the user tend to perform long or short actions?).
- Average and standard deviation of the mouse actions speeds (how fast does the user move the mouse?).

3.3   Application Features

The applications the user normally uses can differ from day to day and from task to task. Thus, we can't rely on them for identification purposes. However, the duration the user spends on a single application can be considered as a behavioural attribute of the user profile. It can be seen as a measurement of how multitasking the user is. In our approach, we take the average and the standard deviation of applications usage time in the targeted session as features.

4   Evaluation

As described earlier, the objective of our approach is to detect the identity of the user based on statistical features of the observed computer events. To do that, we use the behavioural signature vector. To show the discrimination power of each component (mouse, keyboard, application) separately, we will first present our result by considering only the sub features related to each single component alone and then we will follow that with the results of combining all the features of all components in single signature vector. As a result, our final classifier takes 22 feature long vectors as input and outputs a single value representing the recognised user. A support vector machine with radial basis kernel function (RBF) was used as a classifier. To see how the classification accuracy changes according to session's length (the duration of interaction), we employed different values of sessions duration: 2, 5, 10, 15, 20, 25 and 30 min. Two datasets have been collected to evaluate our approach:

- The first dataset consists of the interaction data of seven subjects (professional computer users) collected during their work hours.
- The second dataset contains the interaction data of only two subjects (from the same seven above) collected while they were using their personal computers (different machines) out of work hours.

Both of the datasets are three consecutive weeks long (for each user) and gathered during normal computer use. For the best of our knowledge these are the first datasets that provide mouse, keyboard and app events together. Table 2 shows the number of computer events collected by each user for each dataset. The description and format for each event are presented in Table 1. Two experiments have been conducted based on the above datasets.

Table 2. Number of computer events collected for each user

	User	Mouse	Keyboard	App
Dataset 1	1	275456	165354	3537
	2	136495	87180	5420
	3	61972	90661	1052
	4	92675	144620	1417
	5	55843	48068	1076
	6	467710	69013	12684
	7	170178	50454	2515
Dataset 2	1	90518	13152	700
	2	35901	11975	412

4.1   Experiment 1: Identifying the Individual

The first dataset has been used in this experiment i.e. the interaction data of the seven users while using their work computers as described earlier. The evaluation is performed using 5-fold cross validation. Figure 1 compares the classification accuracy of mouse, keyboard and application features each separately and also shows the results after combining all the components together in one input vector. As we can see, the model's accuracy is around 82% after only 2 min of interaction and it gets higher as the session's length increases, up to almost 90% after 30 min. That's because a long session means more behavioural patterns observed and as a result more accurate statistical features. However, the longer the session is, the more time the intruder has to manipulate the system before being identified as an intruder. Also we can see how the accuracy increases when the final features vector (all the features) is used, especially when compared to the app-only features.

4.2   Experiment 2: Environment Agnostic Analysis

The goal of this experiment is to explore whether the model is able to identify the user even if the data was collected on a different machine than the one used for training the model. This helps to ensure that the model has learned the usage patterns of the user independently from the environment used for data collection, i.e., that the process is GUI environment agnostic. We utilise the second dataset for this experiment, since it contained data from two users (in dataset one) who have been logging data on different computing devices for dataset two. After training the model on the first dataset which contains the interaction data of all the seven users, we then used the second dataset to test the accuracy of the model. Thus, all the seven users are in competition in this experiment although the test set contains data of two of them only. The mouse, keyboard and application features all have been fused into one input vector here (as was the case with the 'all' line in Fig. 1). Figure 2 illustrates the identification

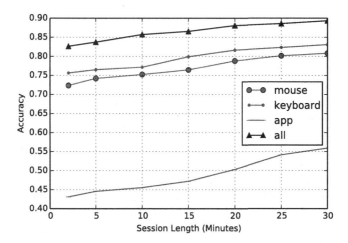

Fig. 1. Identification accuracy comparison according to experiment 1

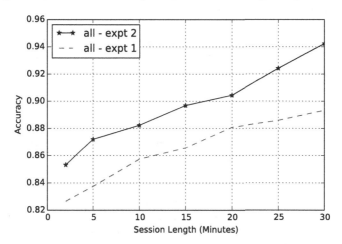

Fig. 2. Identification accuracy of experiment 2 by using all features combined

accuracy of the second experiment. The results show that the model's accuracy has not dropped even though the test data was collected from a totally different machine than the one the model has been trained on, as well as the user using the computer in a different environment. The marginally better result of this experiment's curve is because the model has seen more training samples while training on the whole first dataset, unlike the previous experiment in which the 5-fold cross validation was used.

5   Conclusions

In this paper we have proposed an approach to model the user interaction with the computer in GUI based system. A logging tool has been developed to build the required dataset and released publicly. Two experiments have been conducted to evaluate our approach. Unlike other identification and authentication systems that rely on the data coming from single input device (mouse or keyboard), our model exploits all the available multimodal interaction data coming from mouse, keyboard and the app usage to recognise user identity. This can be done dynamically throughout the session which helps to prevent a masquerade attack. This approach can also be used to support recommendation and personalisation systems as most of them need an identification phase as a start.

Future research should extend the dataset to a larger number of users to check the model scalability and consistency over time. In addition, an online machine learning model could be important here to continuously update the learned usage profiles of the users to keep them consistent with any future change to the user's style or skills of computer usage.

Acknowledgments. This publication has emanated from research conducted with the financial support of Science Foundation Ireland (SFI) under grant number SFI/12/RC/2289.

References

1. Ahmed, A.A.E., Traore, I.: A new biometric technology based on mouse dynamics. IEEE Trans. Dependable Secur. Comput. **4**(3), 165–179 (2007)
2. Anderson, J.P.: Computer Security Threat Monitoring and Surveillance. James P. Anderson Co. (2002)
3. Bergadano, F., Gunetti, D., Picardi, C.: User authentication through keystroke dynamics. ACM Trans. Inf. Syst. Secur. **5**(4), 367–397 (2002)
4. Boies, S.J.: User behaviour on an interactive computer system. IBM Syst. J. **13**, 2–18 (1974)
5. Boser, B.E., Guyon, I.M., Vapnik, V.N.: A training algorithm for optimal margin classifiers. In: Proceedings of the Fifth Annual Workshop on Computational Learning Theory, COLT 1992, pp. 144–152. ACM, New York (1992)
6. Carmagnola, F., Cena, F.: User identification for cross-system personalisation. Inf. Sci. **179**(1–2), 16–32 (2009)
7. Curtin, M., Villani, M., Ngo, G., Simone, J., Fort, H.S., Cha, S.: Keystroke biometric recognition on long-text input: a feasibility study. In: International Workshop on Scientific Computing and Computational Statistics (2006)
8. Garg, A., Rahalkar, R., Upadhyaya, S., Kwiaty, K.: Profiling users in GUI based systems for masquerade detection. In: 2006 IEEE Information Assurance Workshop, pp. 48–54 (2006)
9. Goldring, T.: User profiling for intrusion detection in windows NT. Comput. Sci. Stat. **35** (2003)
10. Gunetti, D., Picardi, C.: Keystroke analysis of free text. ACM Trans. Inf. Syst. Secur. **8**(3), 312–347 (2005)

11. Hinbarji, Z., Albatal, R., Gurrin, C.: Dynamic user authentication based on mouse movements curves. In: He, X., Luo, S., Tao, D., Xu, C., Yang, J., Hasan, M.A. (eds.) MMM 2015. LNCS, vol. 8936, pp. 111–122. Springer, Heidelberg (2015). doi:10.1007/978-3-319-14442-9_10

12. Hinbarji, Z., Albatal, R., O'Connor, N.E., Gurrin, C.: LoggerMan, a comprehensive logging and visualization tool to capture computer usage. In: Tian, Q., Sebe, N., Qi, G.-J., Huet, B., Hong, R., Liu, X. (eds.) MMM 2016. LNCS, vol. 9517, pp. 342–347. Springer, Heidelberg (2016). doi:10.1007/978-3-319-27674-8_31

13. Kaufman, K.A., Cervone, G., Michalski, R.S.: An application of symbolic learning to intrusion detection: preliminary results from the LUS methodology. Reports of the Machine Learning and Inference Laboratory, MLI 03-2, George Mason University, Fairfax, VA (2003)

14. Lane, T., Brodley., C.: An application of machine learning to anomaly detection. In: Proceedings of the 20th National Information Systems Security Conference, pp. 366–377 (1997)

15. Maxion, R., Townsend, T.: Masquerade detection using truncated command lines. In: Proceedings of the International Conference on Dependable Systems and Networks, DSN 2002, pp. 219–228 (2002)

16. Pannell, G., Ashman, H.: Anomaly detection over user profiles for intrusion detection (2010)

17. Ryan, J., Jang Lin, M., Miikkulainen, R.: Intrusion detection with neural networks. In: Advances in Neural Information Processing Systems, pp. 943–949. MIT Press (1998)

18. Schonlau, M., Dumouchel, W., Ju, W.-H., Karr, A.F., Theusan, M., Vardi, Y.: Computer intrusion: detecting masquerades. Stat. Sci. **16**(1), 58–74 (2001)

19. Dao, V., Vemuri, R., Templeton, S.: Profiling users in the UNIX OS environment. In: International ICSC Conference on Intelligent Systems and Applications (2000)

20. Yeung, Y.D., Ding, Y.: Host-based intrusion detection using dynamic and static behavioral models. Pattern Recogn. **36**, 229–243 (2003)

Utilizing Locality-Sensitive Hash Learning for Cross-Media Retrieval

Jia Yuhua[1]([✉]), Bai Liang[1], Wang Peng[2], Guo Jinlin[1], Xie Yuxiang[1],
and Yu Tianyuan[1]

[1] Science and Technology on Information Systems Engineering Laboratory,
National University of Defense Technology, Changsha 410073, China
jiayuhua11@outlook.com, gjlin99@gmail.com, yxxie@nudt.edu.cn,
xabpz@163.com, yutianyuan92@163.com
[2] National Laboratory for Information Science and Technology,
Department of Computer Science and Technology, Tsinghua University,
Beijing 100084, China
pwang@tsinghua.edu.cn

Abstract. Cross-media retrieval is an imperative approach to handle
the explosive growth of multimodal data on the web. However, existed
approaches to cross-media retrieval are computationally expensive due
to the curse of dimensionality. To efficiently retrieve in multimodal data,
it is essential to reduce the proportion of irrelevant documents. In this
paper, we propose a cross-media retrieval approach (FCMR) based on
locality-sensitive hashing (LSH) and neural networks. Multimodal information is projected by LSH algorithm to cluster similar objects into the
same hash bucket and dissimilar objects into different ones, using hash
functions learned through neural networks. Once given a textual or visual
query, it can be efficiently mapped to a hash bucket in which objects
stored can be near neighbors of this query. Experimental results show
that, in the set of the queries' near neighbors obtained by the proposed
method, the proportions of relevant documents can be much boosted, and
it indicates that the retrieval based on near neighbors can be effectively
conducted. Further evaluations on two public datasets demonstrate the
effectiveness of the proposed retrieval method compared to the baselines.

Keywords: Cross-media retrieval · Neural networks · Locality-sensitive
hashing · Multimodal indexing

1 Introduction

In the era of multimedia big data, a huge amount of multimodal information has
been generated continuously which brings huge cross-media retrieval requirements such as using textual description to search for visual media like images or
videos, and *vice versa*. For example, an entity on Wikipedia is usually described
with texts accompanied with images, for which retrieval usually involves in

© Springer International Publishing AG 2017
L. Amsaleg et al. (Eds.): MMM 2017, Part I, LNCS 10132, pp. 550–561, 2017.
DOI: 10.1007/978-3-319-51811-4_45

cross-media indexing and learning. Compared to single-media retrieval, cross-media retrieval focuses on mining the correlations between semantics obtained from heterogenous modalities.

Most of previous method for cross-media retrieval are computationally expensive due to the curse of dimensionality. More seriously, such approaches suffer when querying from the whole document collection which usually contains massive documents irrelevant to the query. The locality-sensitive hashing (LSH) [8] is a well performed ANN algorithm to improve the search speed significantly. Recently, LSH has been widely adopted in image or text retrieval and achieves satisfactory results [13]. The key idea of LSH is to hash objects using several locality-sensitive hash functions to ensure that for each function the probability of collision is much higher for objects that are relevant than for those that are irrelevant [16]. Then, near neighbors can be determined by hashing the query into hash buckets and elements stored in these buckets would be near neighbors, i.e., irrelevant elements to the query can be reduced with a high probability. LSH algorithm is utilized in this paper to guarantee linear time of the proposed approach for cross-media retrieval.

In this work, we propose a fast cross-media retrieval approach (FCMR) based on locality-sensitive hashing and neural networks. The main contribution of FCMR lies in that it can reduce the proportion of irrelevant objects remarkably thus improve the query efficiency and accuracy. FCMR involves two stages, including LSH and hash function learning. In LSH stage, multimodal data like texts and images are hashed into buckets by LSH algorithm. The task of this stage is to hash relevant objects into the same bucket. In the hash function learning stage, the hash functions for data projection are learned with neural networks. Finally, when performing a query task, the query is mapped to hash buckets with locality-sensitive hash functions or hash functions learned by neural networks and objects stored in these buckets are identified as near neighbors of the query. In the returned neighbor list, the proportion of relevant objects is much higher and accurate retrieval can be performed effectively.

2   Related Work

Cross-media retrieval has been a hot research topic recently and many approaches have been proposed to deal with this challenging task. Most of previous work can be categorized into two types by modelling cross-modal correlations.

Topic-Based Models. Approaches of this kind model the similarity across different modalities through topic proportion analysis. Typical research includes LDA [1], Corr-LDA [2], MDRF [3], CMLDA [15] etc. LDA-based methods assume the different modality data has same topic proportions or same topic numbers, or has one-to-one topic correspondences. Similarly, correspondence LDA [2] aims to find the topic-level relationships between images and text annotations. In [3], a combination of directed and undirected probabilistic graphical model (MDRF) was proposed for retrieving images with short text descriptions, using a Markov random field over LDA. More recently, the work [15] proposes a multimedia social event summarization framework to automatically generate visualized summaries from the microblog stream of multiple media types.

Subspaces-Based Models. This category of approaches tries to seek subspaces to maximize the correlations between two different sets of multidimensional variables. Typical approaches based on this model include CCA [4], GMA [5], T-V CCA [6] and Bi-CMSRM [7] etc. These approaches usually involves two stages: firstly, features for each modality are extracted separately, and secondly, CCA is utilized to build a lower-dimensional common representation space. CCA is a method of data analysis used to discover a subspace of multiple data spaces. Sharma et al. present a generic framework for multi-modal feature extraction techniques, called Generalized Multiview Analysis (GMA) [5]. The work [6] proposes a three-view CCA model by introducing a semantic view to produce a better separation for multi-modal data of different classes in the learned latent subspace. Bi-CMSRM [7] introduces an intermediate latent space and optimizes bi-directional ranking examples such that the trained model is applicable for the reverse direction of cross-media retrieval.

A common weakness of the related work is that the document collection lacks of better organization before the query is performed, resulting in more unrelated documents included and checked. Instead, the proposed method in this paper can remove numerous unrelated documents before performing the query in order to obtain a set of near neighbors on which the retrieval can be better performed.

3   Fast Cross-Media Retrieval Model

To simplify the notation and algorithm description, we take images and texts as two exemplar modalities to describe the proposed fast cross-media retrieval, and the method can be easily extended to other modalities. In this section, the mapping of visual objects into hash buckets is first presented which is followed by the learning of hash functions in order to project textual objects into these hash buckets.

3.1   Model Overview and Notations

The FCMR with one hash table derived by LSH is shown in Fig. 1. As shown in Fig. 1, there are two stages in FCMR. In LSH stage, we project the image samples into hash buckets by LSH algorithm. In hash functions learning stage, the hash functions are learned to project text counterparts into their corresponding hash buckets. To be more specific, the task of FCMR is first to project the image samples into hash buckets of m hash tables $G = [g_1, g_2, ..., g_m] \in R^{k \times m}$, where g_j denotes the hash table and k is the length of hash code. Then all the hash functions $Ht = \left[Ht^{(1)}, Ht^{(2)}, ..., Ht^{(m)}\right]$ are learned to project the text samples into these m hash tables by neural networks, where $Ht^{(j)}$ denotes hash functions of the j-th hash table.

Here, we assume all vectors are column vectors. $T = [t_1, t_2, ..., t_{nt}] \in R^{dt \times nt}$ stands for the matrix representation of text data samples, where dt is the dimension of the text feature space (e.g., vocabulary size of bag-of-words (BoW)), nt

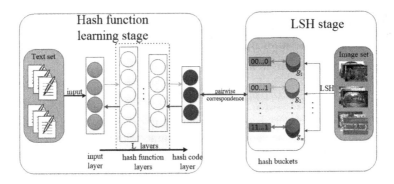

Fig. 1. The FCMR, illustrated with one hash table

is the number of total samples in T. Similarly, $P = [p_1, p_2, ..., p_{np}] \in R^{dp \times np}$ stands for image samples, where dp denote the dimension of the image feature space (e.g., vocabulary size of bag-of-visual-words (BoVW)), np is the number of the image samples in P. In the paper, each t_i is associated with p_i, which means $nt = np = n$. We use n instead of nt and np in the following sections.

3.2 Structure Description of FCMR

Given m hash tables derived by LSH algorithm, we need design m NNs to learn corresponding hash functions projecting text samples to the m hash tables. Based on bridging of the learned hashing functions, cross-media retrieval can be achieved. For example, if a query is described in the form of text, we can project it into hash buckets of m hash tables with the learned Ht and perform exact retrieval from image samples contained in these buckets. On the other hand, text retrieval given an image query can be carried out in the same manner. Two structural components in the proposed FCMR including LSH and neural network are now described.

Locality-Sensitive Hashing. Locality-sensitive hashing (LSH) [8] is a randomized algorithm for searching approximately nearest neighbors in high dimension spaces. It ensures that for each locality-sensitive hash function the probability of collision is much higher for objects that are close to each other than for those that are far apart. In FCMR, LSH is employed to conduct the data correlation structures according to which near neighbors can be efficiently determined by hashing the query into hash buckets.

According to [9], the definition of LSH function can extended and verified as the form of

$$h_r(p_i) = \begin{cases} 1, \text{ if } r^T p_i \geq 0, \\ 0, \text{ else} \end{cases} \tag{1}$$

where the hyper plane vector r follows a zero-mean multi-Gaussian $N(0, 1)$ distribution. After applying the sign function, a image sample p_i can be projected to its hash code. The final length of k hash codes for p_i are obtained by repeating Eq. (1) k times, which can be formalized as

$$g(p_i) = \langle h_1(p_i), h_2(p_i)..., h_k(p_i) \rangle, (0 < i \leq n) \qquad (2)$$

while Eq. (1) denotes the hash code of image p_i in one hash table, its hash codes in all m hash tables can be generated in the same manner.

Using LSH, image samples $p_i(1 \leq i \leq n)$ are projected into hash buckets of m hash tables to ensure related samples are mapped to the same buckets with high probability. Recall that in our scenario, each t_i should be associated with p_i as introduced in Sect. 3.1 because t_i and p_i are two different modalities describing the same semantics. Therefore, each related t_i should be associated with a hash bucket in each of the m hash tables. In this way, we can obtain training examples for neural networks to learn the hash functions $Ht^{(j)}$ that project the text samples into their corresponding hash buckets in j-th hash table. We denote $(t_i, Y_i^{(j)})(i \in 1, 2,...,n, j \in 1, 2,...,m)$ as a training example, where $Y_i^{(j)}$ is the hash code of the j-th hash bucket in the hash table containing p_i.

Neutral Network Structure of FCMR. Inspired by [10], we design the network structure of FCMR as shown in Fig. 1. The hash functions are learned in terms of m NNs with the same structure. Each NN$^{(j)}(j \in 1, 2,...,m)$ consists of L layers: one layer with dt neurons for the input data, one layer with k neurons for the output (hash codes) predictions, and the rest $L - 2$ layers for the hash functions. For each $t_i \in T$, we forward t_i layer-by-layer through NN$^{(j)}$ to generate the representation of each layer, i.e., $t_i^{(1)}, ..., t_i^{(L)}$. The $(l + 1)$-th layer takes $t_i^{(l)}$ as the input and a projection function is used to transform it to $t_i^{(l+1)}$:

$$t_i^{(l+1)} = f^{(l+1)}(W^{(l+1)}t_i^{(l)}) \qquad (3)$$

where $t_i^{(l)} \in R^k$ and $t_i^{(l+1)} \in R^k$ are the feature representation in the l-th and $(l + 1)$-th layer, respectively. $W^{(l+1)}$ is the projection matrix and $f^{(l+1)}(\cdot)$ is the activation function, which is usually the *sigmoid* or *tanh* function for $l = 1$ to $L - 1$, and is the *softmax* function for $l = L$. Hash function $Ht^{(j)}$ takes t_i as input, forwards t_i to the hash code layer (the L-th layer), and outputs the k-dimensional binary hash codes:

$$Ht^{(j)}(t_i) = sign(t_i^{L-1}) \qquad (4)$$

where $t_i^{(L-1)} \in R^k$ is a k-dimensional real-value vector, and we can then use $sign$ function to convert t_i^{L-1} to a binary hash code.

Nevertheless, the *sign* function is not differentiable, and thus is hard to optimize directly. Similar with most of other hashing approaches [10,11], we remove the *sign* function at the hash function learning stage and add it back at the testing stage. Considering that $Ht^{(j)}(t_i)$ should be same to $Y_i^{(j)}$ for training sample $(t_i, Y_i^{(j)})$, i.e., $t_i^{(L-1)}$ and $Y_i^{(j)}$ should be as close as possible, we define a loss function based on the least square of errors (SE):

$$SE(t_i^{(L-1)}, Y_i^{(j)}) = \frac{1}{2} \left\| t_i^{(L-1)} - Y_i^{(j)} \right\|_F^2 \tag{5}$$

where $t_i^{(L-1)}$ is the estimated hashing value by the neural network before we apply sign function.

3.3   Learning Algorithm of FCMR

The learning of FCMR consists of locality-sensitive hashing and network training. We can get training examples $(t_i, Y_i^{(j)})(i \in 1, 2, ..., nt, j \in 1, 2, ..., m)$ from the locality-sensitive hashing stage to train $NN^{(j)}$ learning hash function mapping t_i to $Y_i^{(j)}$. Network training consists of pre-training and fine-tuning. Pre-training provides a good parameters initialization and preventing the learned neural network from being trapped in a bad local optimum. In FCMR, we choose the stacked autoencoder (SAE) to pre-train each layer of the $NN^{(j)}$ sequentially [12]. Then, we fine-tune the parameters by optimizing a loss function according to the constraint as described in Sect. 3.2. The loss of the output layer will be back propagated to its former layers to make the hash function learned by neural networks effective. Finally, we use the sum squared of errors (SSE) for all text samples in T and minimize the overall loss function as follows:

$$SSE(t_i^{(L-1)}, Y_i^{(j)}) = \frac{1}{2} \sum_{i=1}^{n} \left\| t_i^{(L-1)} - Y^{(j)}_i \right\|_F^2 \tag{6}$$

The learned hash functions $Ht^{(j)}$ is expected to map the text samples into their corresponding hash buckets in the j-th hash table. The training of FCMR is conducted by the classical back propagation (BP) algorithm. After $NN^{(j)}$ was trained, we finally obtain the hash functions $Ht^{(j)}$ using Eq. (4) in the testing stage. The overall procedures of FCMR is given in Algorithm 1.

4   Experiments

FCMR is proposed to increase the proportion of relevant documents in document collection which we finally retrieve in. To show its effectiveness, we compare the proportion of related documents in the whole document collection with that of in the neighbor-list generated by FCMR. We report the experiments in both scenarios of image-query-text retrieval and text-query-image retrieval.

Algorithm 1

Input: data set T, P, k, m
Output: The hash functions Ht and LSH hash functions
1: **for** j=1 to m **do**
2:     $(t, Y^{(j)}) \leftarrow \mathrm{LSH}(P)$
3:     $\mathrm{NN}^{(j)} \leftarrow \mathrm{SAE}(T)$
4:     **repeat**
5:         Pick a random training example $(t_i, Y_i^{(j)})$
6:         Make a gradient step for SSE:

$$SE(t_i^{(L-1)}, Y_i^{(j)}) = \tfrac{1}{2} \left\| t_i^{(L-1)} - Y_i^{(j)} \right\|_F^2$$

7:         Update $\mathrm{NN}^{(j)}$
8:     **until** stopping criteria is met
9: **end for**

4.1   Experiment Setup

Datasets. We use two real-world data sets Wikipedia Feature Articles[1]. and NUS-WIDE[2], each of which contain images and texts. The details of the data sets are listed in Table 1. The Wikipedia Featured Articles contains 2,866 images with 5,000 associated tags from Wikipedia. Each image with the corresponding tags has one of the 10 concepts as the ground-truth. For the text modality, the corresponding tags of each image are represented by a 5,000-d BoW. For the image modality, 1,000-d BoVW are extracted for each image. NUS-WIDE consists of 269,648 Flickr documents. Each document contains one text-image pair and is labeled by several of 81 semantic categories. For the text modality, the corresponding tags of each image are represented by a 1,000-d BoW. For the image modality, 500-d BoVW are extracted for each image.

Table 1. The details of the data sets used in the experiments

Data set	Wiki	NUS-WIDE
BoVW vocabulary size	1,000	500
BoW vocabulary size	5,000	1,000
Average of words/image	117.5	7.73
Training set size	1,000	100,000
Validation set size*	866/866	5,000/20,000
Testing set size*	1,000/1,000	5,000/149,648

* Partitions are ordered by query/database set respectively, and the query set are randomly sampled from the database set.

---

[1] http://www.svcl.ucsd.edu/projects/crossmodal/
[2] http://lms.comp.nus.edu.sg/research/NUS-WIDE.htm.

Evaluation Metrics. To evaluate the proposed retrieval method, the following evaluation metrics are adopted:

Recall: The accuracy of retrieval results are measured by recall in order to check whether the nearest neighbors of each query sample are in the neighbor-list or not.

Selectivity [14]: Selectivity is the averaged fraction of the whole data set that is returned in the neighbor-list. Multiplying the selectivity by n gives the expected number of elements returned as potential related documents.

Proportion: The proportion is the percentage of documents related to the query in the neighbor-list.

MAP: As a widely adopted metrics in evaluating the retrieval performance, mean average precision (MAP) criterion is used in our experiment.

Parameter Settings. For the proposed FCMR method, we empirically setup the NNs with the number of layers $L = 3$, which has already achieves satisfactory results with a simple structure. The parameters of these NNs are initialized by SAE. The activation function for the output layer is set to the *softmax* function and other layers is set to the *sigmoid* function. We training the value of parameters k (the length of hash code) and m (the number of hash table) of FCMR on a random sampled validation set whose size is listed in Table 1. For Wikipedia Featured Articles dataset, k and m are trained in the range of $\{1, 2 ..., 7\}$ and $\{1, 2 ..., 15\}$. For NUS-WIDE dataset, k and m are trained in the range of $\{2, 3 ..., 9\}$ and $\{1, 2 ..., 20\}$. Finally those parameters values with the best performances are chosen as the final setting.

4.2   Results and Discussions

From Fig. 2, we can see the evaluation of the different length of hash code (k) for a varying number of hash tables (m) from both query directions on Wikipedia Featured Articles and NUS-WIDE. We can see that the performance of FCMR from both query directions has similar trend but FCMR performs better in the direction of Image-query-Text. Figure 2 shows that using a short length of hash code is sufficient to achieve high recall but results in worse selectivity. Higher numbers of hash tables leads to better trade-offs between search quality and selectivity. Comparing Fig. 2(a,b) to Fig. 2(c,d), we find FCMR performs better in Wikipedia Featured Articles. Recall that in the NUS-WIDE dataset, one image is associated with only about 8 annotated words on average which degrades the performance with short texts.

Queries are divide into different categories to compare the performances of FCMR for each category. For considering the trade-off of recall and selectivity, we fixed $k = 3$ and $m = 14$ for Wikipedia Featured Articles and fixed $k = 4$ and $m = 20$ for NUS-WIDE. The overall performances in terms of recall, selectivity and proportion of FCMR are shown in Table 2. As we can see from Table 2, the proportion of relevant documents in the neighbor list generated by FCMR is approximately 2 times than that in the whole document collection,

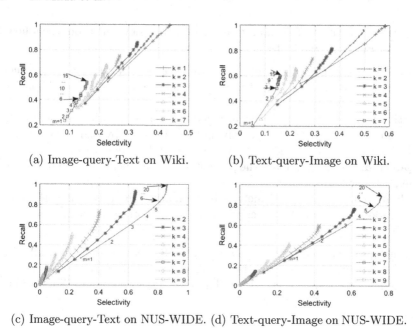

(a) Image-query-Text on Wiki.   (b) Text-query-Image on Wiki.

(c) Image-query-Text on NUS-WIDE. (d) Text-query-Image on NUS-WIDE.

Fig. 2. Evaluation of the different length of hash code (k) for a varying number of hash tables (m).

Figs. 3 and 4 report similar results in terms of different categories. Table 2 also demonstrates that FCMR performs better for long texts (Wikipedia) in the direction of Image-query-Text. Figure 3 depicts the average proportion of documents related to query for 10 categories on Wikipedia Featured Articles. It can be seen that FCMR behaves very well in all categories of Wikipedia Featured Articles. In Fig. 4, the proportion of relevant documents for 60 typical concepts is demonstrated in both query directions on NUS-WIDE dataset. From Fig. 4, we can see that the proposed method performs well over most categories except for "book", "garden", "glacier" and "leaf". It can be easily found that there are few samples in the four categories and most of texts belong to the four categories contain very few words. This phenomenon demonstrate that the qualities of multimodal data might limit the performance of FCMR.

Finally, we fixed $k = 3$ and $m = 14$ for Wikipedia Featured Articles and fixed $k = 4$ and $m = 20$ for NUS-WIDE and combined FCMR with two state-of-the-art approaches (CCA [4], Bi-CMSRM [7]) for cross-media retrieval, i.e., we use CCA and Bi-CMSRM to retrieve in the neighbor list we obtained by FCMR. We evaluate the cross-media retrieval performance and report results in terms of MAP@all in Tables 3 and 4. As is shown in Tables 3 and 4, the retrieval performance of using FCMR outperforms the baseline approaches significantly. When combined with FCMR, both CCA and Bi-CMSRM can be much improved. This improvement relies on the neighbor-list we obtained by FCMR which contains more relevant documents.

Table 2. The overall performance of FCMR

Dataset	Task	Recall	Selectivity	Proportion/original proportion
Wiki	Image-query-Text	0.8621	0.3263	2.5451
(k = 3, m = 14)	Text-query-Image	0.8506	0.3867	2.1317
NUS-WIDE	Image-query-Text	0.7331	0.4044	1.7893
(k = 4, m = 20)	Text-query-Image	0.7331	0.4250	1.6155

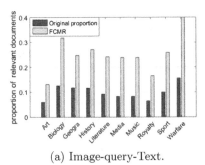

(a) Image-query-Text.

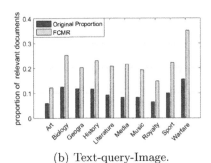

(b) Text-query-Image.

Fig. 3. The performance in terms of proportion over 10 concepts on Wiki.

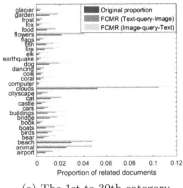

(a) The 1st to 30th category.

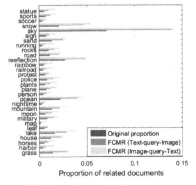

(b) The 31st to 60th category.

Fig. 4. The performance in terms of proportion over 60 concepts on NUS-WIDE.

Table 3. The performance comparison in terms of MAP@all scores on the Wiki dataset. The results shown in boldface are the best results.

Method	Text-query-Image	Image-query-Text
CCA	0.1433	0.1451
FCMR-CCA	**0.1972**	**0.2012**
Bi-CMSRM	0.2123	0.2528
FCMR-Bi-CMSRM	**0.2849**	**0.3051**

Table 4. The performance comparison in terms of MAP@all scores on the NUS-WIDE dataset. The results shown in boldface are the best results.

Method	Text-query-Image	Image-query-Text
CCA	0.0851	0.0883
FCMR-CCA	**0.1067**	**0.1106**
Bi-CMSRM	0.1453	0.2380
FCMR-Bi-CMSRM	**0.1862**	**0.2961**

5   Conclusions

In this paper, we propose a cross-media retrieval approach based on locality-sensitive hashing and neural networks named FCMR, which increase the proportion of returned documents related to the query from which final retrieval can be performed more accurately. Experimental results on the two public data sets demonstrate the effectiveness of FCMR performance in terms of various evaluation metrics. Finally, we combined FCMR with two state-of-the-art approaches and demonstrated FCMR can improve their effectiveness by introducing the advantages of the proposed method.

Acknowledgements. Thanks to the Natural Science Foundation of China under Grant No. 61571453, No. 61502264, and No. 61405252, Natural Science Foundation of Hunan Province, China under Grant No. 14JJ3010, Research Funding of National University of Defense Technology under grant No. ZK16-03-37.

References

1. Blei, D.M., Ng, A.Y., Jordan, M.I.: Latent dirichlet allocation. J. Mach. Learn. Res. **3**, 993–1022 (2003)
2. Blei, D.M., Jordan, M.I.: Modeling annotated data. In: Proceedings of the 26th Annual International ACM SIGIR Conference on Research and Development in Informaion Retrieval, pp. 127–134. ACM (2003)
3. Jia, Y., Salzmann, M., Darrell, T.: Learning cross-modality similarity for multinomial data. In: 2011 IEEE International Conference on Computer Vision (ICCV), pp. 2407–2414. IEEE (2011)
4. Hardoon, D.R., Szedmak, S., Shawe-Taylor, J.: Canonical correlation analysis: an overview with application to learning methods. Neural comput. **16**(12), 2639–2664 (2004)
5. Sharma, A., Kumar, A., Daume, H., Jacobs, D.W.: Generalized multiview analysis: a discriminative latent space. In: IEEE Conference on Computer Vision and Pattern Recognition, pp. 2160–2167 (2012)
6. Gong, Y., Ke, Q., Isard, M., Lazebnik, S.: A multi-view embedding space for modeling internet images, tags, and their semantics. Int. J. Comput. Vis. **106**, 210–233 (2013)
7. Wu, F., Lu, X., Zhang, Z., et al.: Cross-media semantic representation via bi-directional learning to rank. In: Proceedings of the 21st ACM International Conference on Multimedia, pp. 877–886. ACM (2013)

8. Kulis, B., Grauman, K.: Kernelized locality-sensitive hashing for scalable image search. In: 2009 IEEE 12th International Conference on Computer Vision, pp. 2130–2137. IEEE (2009)
9. Weems, M.A.: Kernelized locality-sensitive hashing for fast image landmark association (2011)
10. Zhuang, Y., Yu, Z., Wang, W., et al.: Cross-media hashing with neural networks. In: Proceedings of the ACM International Conference on Multimedia, pp. 901–904. ACM (2014)
11. Kumar, S., Udupa, R.: Learning hash functions for cross-view similarity search. In: IJCAI Proceedings-International Joint Conference on Artificial Intelligence, vol. 22, no. 1, p. 1360 (2011)
12. Bengio, Y.: Learning deep architectures for AI. Found. Trends Mach. Learn. **2**(1), 1–127 (2009)
13. Li, J.-Y., Li, J.-H.: Fast image search with locality-sensitive hashing and homogeneous kernels map. Sci. World J. **2015**, 350676 (2015)
14. Paulev, L., Jgou, H., Amsaleg, L.: Locality sensitive hashing: a comparison of hash function types and querying mechanisms. Pattern Recogn. Lett. **31**(11), 1348–1358 (2010)
15. Bian, J., Yang, Y., Zhang, H., et al.: Multimedia summarization for social events in microblog stream. IEEE Trans. Multimedia **17**(2), 216–228 (2015)
16. Lv, Q., Josephson, W., Wang, Z., et al.: Multi-probe LSH: efficient indexing for high. In: Proceedings of the International Conference on Very Large Data Bases, pp. 950–961 (2007)

SS3: Multimedia Computing for Intelligent Life

A Sensor-Based Official Basketball Referee Signals Recognition System Using Deep Belief Networks

Chung-Wei Yeh, Tse-Yu Pan, and Min-Chun Hu$^{(\boxtimes)}$

Department of Computer Science and Information Engineering,
National Cheng Kung University, Tainan, Taiwan
{roy,felix,trimy}@mislab.csie.ncku.edu.tw

Abstract. In a basketball game, basketball referees who have the responsibility to enforce the rules and maintain the order of the basketball game has only a brief moment to determine if an infraction has occurred, later they communicate with the scoring table using hand signals. In this paper, we propose a novel system which can not only recognize the basketball referees' signals but also communicate with the scoring table in real-time. Deep belief network and time-domain feature are utilized to analyze two heterogeneous signals, surface electromyography (sEMG) and three-axis accelerometer (ACC) to recognize dynamic gestures. Our recognition method is evaluated by a dataset of 9 various official hand signals performed by 11 subjects. Our recognition model achieves acceptable accuracy rate, which is 97.9% and 90.5% for 5-fold Cross Validation (5-foldCV) and Leave-One-Participant-Out Cross Validation (LOPOCV) experiments, respectively. The accuracy of LOPOCV experiment can be further improved to 94.3% by applying user calibration.

1 Introduction

In a basketball game, when a violation or a foul event occurs, referees have to make judgments according to the rules and communicate with the scorer using gestures as shown in Fig. 1, so that the scoring table can record correct and immediate information about the game. However, manual communication between referees and the scorer sometimes causes misunderstandings and hence delays the progress of the game. With the advances of sensor and computer technology, human-computer interaction (HCI) system becomes more and more popular in our daily life and it occurs to us that the HCI technology can be used to facilitate the interaction between referees and the scorer. In this work, we propose a real-time dynamic hand gesture recognition system which uses electromyography (EMG) and accelerometer (ACC) signals to analyze several commonly used basketball referees' signals. Our system will automatically recognize the gestures and record the corresponding events; therefore, the only thing the scorer has to do is to check whether the recorded results shown on the screen are accurate or not, which makes the recording process more efficient.

© Springer International Publishing AG 2017
L. Amsaleg et al. (Eds.): MMM 2017, Part I, LNCS 10132, pp. 565–575, 2017.
DOI: 10.1007/978-3-319-51811-4_46

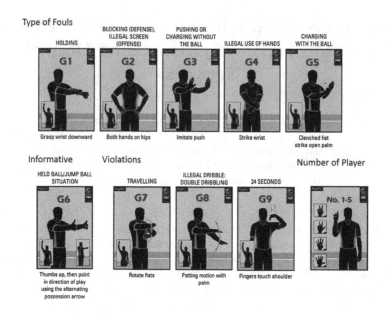

Fig. 1. The basketball referees' official signals. (From https://www.fiba.com/)

The researches related to gesture recognition can be divided into vision-based and sensor-based approaches. Vision-based approaches purely use a RGB camera to recognize hand gestures, so that these methods do not work well if there are complex colors, textures or motions in the background environment [3,8]. On the other hand, sensor-based approaches apply signals of electromyograms (EMGs), electroencephalograms (EEGs), and electrooculograms (ECGs) to analyze human gestures. For example, surface EMG (sEMG) sensor is placed on the surface skin of a human arm to acquire rich information about the muscle movements, which can be utilized to sense and decode human muscular activity like subtle finger configurations, hand shapes, and wrist movements [7,10]. Moreover, ACC, which measures both dynamic accelerations like vibrations and static accelerations like gravity, has been successfully equipped in many devices for simple and supplementary control applications [11]. Because of the complex background around the basketball court, sensor-based approaches are more reliable than vision-based approaches to recognize the referees' gestures. In 2012, Thalmic Labs introduced a wearable device named Myo, which is equipped with eight sEMG sensors and one Inertial Measurement Unit (IMU) sensor to measure the trajectory of hand movements and simply recognize 5 types of hand gestures. Our idea is to apply Myo armband to facilitate the game recording process.

The applications of sensor-based gesture recognition can be found in several areas like sign language recognition [12], human computer interaction [1,2,9], etc. It is challenging to recognize gestures by sEMG signals because of two reasons: (1) sEMG signals might vary significantly from person to person even though they performed the same gesture. (2) The sensors should be placed at precise locations on the muscle, otherwise the acquired signals might be very different even the same person performs the same gesture. Fortunately, since the acceleration signals implicitly represent the patterns of gesture trajectories, we are able to construct a robust recognition system by fusing the three-axis ACC signals and multichannel sEMG signals.

The contributions of this work are summarized as follows:

(1) We propose a system framework to robustly analyze and recognize gestures made by basketball referees. (2) Time-domain features have been widely used in gesture analysis due to its simplicity of calculation [1,9,12]. In this work, we further use deep belief networks (DBN) to learn more representative features for gesture recognition, and combine both the DBN-based features and the time-domain features to achieve more accurate recognition results. (3) We investigate several different classification models, including Hidden Markov Model (HMM), Support Vector Machine (SVM) and Artificial Neural Network (ANN), and found that SVM has the best classification ability in our application. (4) We collect a large Myo-based hand gesture dataset, which contains 4620 sequential signals of 42 kinds of gestures. To the best of our knowledge, it is the first organized Myo-dataset and we release it to advance the progress of HCI technology.

The rest of the manuscript is organized as follows. Section 2 illustrates our system framework and elaborates the signal processing and feature extraction modules in detail. Experimental results and discussions are presented in Sect. 3. Conclusions and future work are given in Sect. 4.

2   System Framework

Figure 2 illustrates our system framework. Myo transmits signals through Bluetooth 4.0 protocol, which can transmit data to a receiver far away (more than 100 m). Therefore, its transmission range can cover the whole basketball court, and our system is able to acquire signals during the whole game. In this section, we will show how to use signals measured by Myo to recognize hand gestures.

2.1   Signal Acquisition and Preprocessing

The eight-channel sEMG and three-axis ACC signals are captured with a fixed sampling rate of 200 Hz and 50 Hz, respectively. Each recorded dynamic gesture is about 2.5 s. Since signals captured from Myo may be affected by external noise or artifacts, there are three steps to be applied to eliminate noises. First, the mean of both sEMG and ACC raw signals are subtracted from the data to eliminate DC offset of the signals. Later, in order to get the signal envelope of both sEMG and ACC signals, full-wave rectification is applied to all data points (i.e., each signal value is set to the absolute value). These adjusted signals are finally fed into a low-pass filter to remove noises.

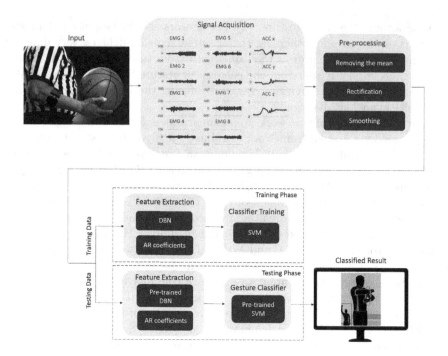

Fig. 2. The proposed framework.

2.2   Feature Extraction

DBN is a probabilistic generative model, or alternatively a type of deep neural network, composed of multiple layers of Restricted Boltzmann machines (RBMs), where the output from a lower-level RBM is the input to a higher-level RBM. RBMs are a typical neural network, with connections to each other across layers, but no two units of the same layer are linked. In this work, we apply DBN to train representative features for hand gesture signals. An efficient greedy layer-wise training algorithm is used to pre-train each layer of the DBN. For a RBM having binary-valued hidden and visible units, the energy function is defined as

$$E(v, h; \theta) = -\sum_{i=1}^{I}\sum_{j=1}^{J} w_{ij} v_i h_j - \sum_{i=1}^{I} a_i v_i - \sum_{j=1}^{J} b_j h_j, \qquad (1)$$

where θ is the model parameters, w_{ij} is the connection weight between visible unit v_i and hidden unit h_i, a_i and b_j are the bias term, I and J are the numbers of visible and hidden units, respectively. The joint probability distributions over visible and hidden units are defined in terms of the energy function

$$p(v, h; \theta) = \frac{exp(-E, h; \theta)}{Z}, \qquad (2)$$

where $Z = Z(\theta) = \sum_{h,v} exp(-E(v,h;\theta))$ is a partition function. The marginal probability of a visible unit of booleans is the sum over all possible hidden layer configurations

$$p(v;\theta) = \sum_h \frac{exp(-E,h;\theta)}{Z}. \qquad (3)$$

Since the RBM has the shape of a bipartite graph, there is no connections between the same layer units. The visible units and the hidden units are mutually independent. That is, the conditional probability is given by

$$p(h_j = 1|v;\theta) = \delta(\sum_{i=1}^{I} w_{ij}v_i + b_j), \qquad (4)$$

$$p(v_i = 1|h;\theta) = \delta(\sum_{j=1}^{J} w_{ij}h_j + a_i), \qquad (5)$$

where $\delta(x) = 1/(1+exp(-x))$. The weights update required to perform gradient descent in the log-likelihood can be obtained as

$$\Delta W_{ij} = E_{data}(v_i h_j) - E_{model}(v_i h_j), \qquad (6)$$

where $E_{data}(v_i h_j)$ is the expectation observed in the training set and $E_{model}(v_i h_j)$ is an expectation with respect to the distribution defined by the model. The expectation $E_{model}(v_i h_j)$ cannot be computed in less than exponential time using Gibbs sampling. The Contrastive Divergence (CD) algorithm due to Hinton is quite efficient and greatly reduces the variance of the estimates used for learning.

A DBN consists of multi-layer RBMs from bottom to up. When training a DBN, we train the RBMs layer by layer, and then fine-tune all the parameter of the DBN together. There are various kinds of features for the classification of the EMG have been considered in previous works which utilized time-domain, frequency domain, and time-frequency-domain. According to [5], autoregressive model (AR) coefficients of the EMG signals with a typical order of 4–6 yield good performance for myoelectric control. In our work, we combine both the DBN-based features and the time-domain features to achieve more accurate recognition results. After preprocessing, basketball referees' signals fragments with 300 dimensions of each sEMG channel and 75 dimensions of each ACC channel were sent into DBNs to extracted features. There were 11 DBNs corresponding to each sEMG or ACC channel. Each DBNs was constructed by stacking 4 RBMs. For those 4-layer DBNs corresponding to sEMG channel, it has 300 visible units, 500 units in hidden layer 1, 250 units in hidden layer 2, 125 units in hidden layer 3, and 3 units in 4. For DBNs corresponding to ACC channel has 75 visible units, 300 units in hidden layer 1, 150 units in hidden layer 2, 75 units in hidden layer 3, and 3 units in 4. After training DBNs, the 33 normalized features extracted by 11 DBNs and 44 normalized 4-order AR model coefficients constitute a feature vector.

2.3 Classification

We investigate several different classification models, including HMM, SVM and ANN, and found that SVM has the best classification ability in our application. SVM is a supervised learning model which is widely used for pattern recognition. The goal of SVM is to find the optimal hyperplane which has the largest margin between the two classes. If the decision boundary is highly non-linear, SVM will apply kernel functions which project the original data to a high-dimensional feature space where the data is linear-separable. The kernel functions enable SVM to classify in a high-dimensional feature space without ever computing the coordinates of the data in that space, but rather by simply computing the inner products between the images of all pairs of data in the feature space. In this work, the Radial Basis Function kernel (RBF) is selected, which is defined as

$$K(x, x') = exp(-\frac{\|x - x'\|^2}{2\sigma^2})\tag{7}$$

where $\|x - x'\|^2$ may be recognized as the squared Euclidean distance between the two feature vectors. σ is a free parameter.

3 Experimental Results

We implemented the proposed method on Matlab R2014a and used LibSVM and DEEBNET [6]. All experiments were run on 64-bit Windows 7 with the Intel Core i7-3770 CPU and 16GB RAM. Eleven able-bodied subjects (seven males and four females) were asked to participate in recording process by performing a set of 9 basketball referees' signals. Each signal was recorded ten times by each subjects. In total, there are 990 gestures signals in our dataset.

3.1 DBN Structure

In this section, we examined the influence of parameters in the DBN model and tried to find suitable type of each RBM in the DBN and the proper number of RBMs for the basketball referee signals dataset.

– **Type of RBM:** In Keyvanrad's work [6], they constructed a DBN model which is composed of 4 RBMs to extract features for both MNIST and ISO-LET datasets and achieved good recognition performance. Therefore, we also constructed our DBN model with 4 RBMs and examined proper types for each RBM based on our dataset. For each RBM in the DBN, the nodes can be assigned to either Binary nodes or Gaussian nodes. Table 1 shows the accuracy of using different combinations of RBM types, and we found that using the combination of Gaussian-Binary-Binary-Gaussian RBMs to construct the DBN can achieve better performance.

Table 1. Recognition accuracy for DBN models which built by different combinations of RBM type.

Basketball referee signal dataset	
DBN structure (1st-2nd-3rd-4th RBM)	Accuracy (%)
Gaussian - Binary - Binary - Binary	78.9
Binary - Binary - Binary - Gaussian	85.5
Gaussian - Binary- Binary - Gaussian	88.4
Gaussian - Gaussian - Gaussian - Gaussian	87.3

- **Number of RBMs:** Based on the previous experiment, we found that using Binary RBMs to construct the middle layers (i.e. layers between the first and the last layer) of the DBN might result in better recognition rate. Hence, we used different numbers of Binary RBMs to construct middle layers of DBN models and tried to examine the best number of RBMs. The experimental results in Table 2 show that using 4 RBMs (containing 2 Binary RBMs for the middle layers) is most suitable for our dataset.

Table 2. Recognition accuracy for the DBN models using different numbers of Binary RBMs to construct the middle layers.

DBN structure	Accuracy (%)
Gaussian - 1 Binary RBM - Gaussian	86.8
Gaussian - 2 Binary RBMs - Gaussian	88.4
Gaussian - 3 Binary RBMs - Gaussian	78.6

3.2   Recognition Result

After we built a DBN model which was suitable for our dataset, we evaluated the performance of our gesture recognition system for 5-fold Cross Validation and Leave-One-Participant-Out Cross Validation by determining its accuracy in discriminating the 9 different basketball referee signal classes.

- **5-fold Cross Validation (5-foldCV):** All basketball referees' signals are divided into 5 sections. Testing is done by cross-validation individually for each section. The training set for each validation fold consists of all sections but one section. The remaining section is used as the testing set. The average recognition accuracy achieves 97.9% among five sections, as shown in Table 3.
- **Leave-One-Participant-Out Cross Validation (LOPOCV):** The training set is composed of the signal recordings of all subjects but one in the cross-validation, leaving the recordings of the remaining subject as testing set. The average recognition accuracy achieves 90.5% for LOPOCV in Table 4.

Table 3. 5-foldCV recognition accuracy.

	G1	G2	G3	G4	G5	G6	G7	G8	G9	Section accuracy (%)
Section 1	95.5	100.0	100.0	90.0	95.5	100.0	95.5	86.4	100.0	95.5
Section 2	100.0	100.0	100.0	100.0	100.0	100.0	95.5	100.0	95.5	98.5
Section 3	100.0	100.0	100.0	95.5	100.0	100.0	100.0	100.0	100.0	97.0
Section 4	100.0	100.0	100.0	81.8	100.0	95.5	100.0	95.5	95.5	97.5
Section 5	100.0	100.0	95.5	100.0	100.0	95.5	100.0	95.5	100.0	98.0
Gesture accuracy (%)	99.1	100.0	99.1	93.6	99.1	98.2	98.2	95.5	98.2	**97.9**

Table 4. LOPOCV recognition accuracy.

	G1	G2	G3	G4	G5	G6	G7	G8	G9	Section accuracy (%)
Subject 1	80.0	100.0	100.0	70.0	100.0	80.0	100.0	100.0	100.0	92.2
Subject 2	100.0	100.0	100.0	10.0	100.0	100.0	100.0	100.0	100.0	90.0
Subject 3	90.0	100.0	100.0	100.0	100.0	70.0	100.0	80.0	100.0	93.3
Subject 4	90.0	100.0	100.0	90.0	40.0	100.0	100.0	100.0	90.0	90.0
Subject 5	100.0	100.0	90.0	30.0	100.0	100.0	100.0	90.0	100.0	90.0
Subject 6	100.0	100.0	100.0	100.0	80.0	90.0	80.0	90.0	100.0	93.3
Subject 7	100.0	100.0	100.0	50.0	100.0	100.0	100.0	100.0	100.0	94.4
Subject 8	80.0	100.0	60.0	100.0	100.0	90.0	100.0	100.0	100.0	92.2
Subject 9	100.0	100.0	100.0	60.0	90.0	50.0	100.0	40.0	100.0	82.2
Subject 10	100.0	100.0	100.0	90.0	100.0	100.0	100.0	100.0	100.0	98.9
Subject 11	100.0	100.0	70.0	70.0	70.0	80.0	90.0	90.0	40.0	78.9
Gesture accuracy (%)	94.5	100.0	92.7	70.0	89.1	87.3	97.3	90.0	93.6	**90.5**

However, considering the basketball games usually referees by some professional people with experience or license, it is necessary to customize the gesture recognition system for those referees in order to get a better recognition result. Therefore, we customize our system for referees by calibrating by themselves. When a referee calibrates the system, he or she must perform the nine referee signals once a time. Later, the calibration gesture fragments are merged into the training set. In LOPOCV with calibration experiment, the training set is composed of the signal recordings of all subjects except one, but we add one signal recordings of each gesture of the remaining subject as calibration. The testing set still consists of signal recordings of the remaining subject. In Table 5, we achieve an average recognition accuracy of 94.3%. In order to show the performance by our proposed method, we implement several other proposed methods in recent years, and two of them use hidden markov model (HMM) with gaussian mixture models (GMMs) [2,9]. We also try some time-domain features and use SVM as a classifier. The results are shown in Table 5.

Table 5. Recognition accuracy by comparing among different features.

	AR+DBN +SVM (Calibrated)	DBN +SVM	MAV+AR +SVM	M. Georgi et al. [2]	A.-A. Sarnadani et al. [9]
G1	95.5	86.4	47.3	60.9	50.0
G2	100.0	96.4	80.9	73.6	72.7
G3	97.3	90.9	82.7	79.1	77.3
G4	89.1	56.4	57.3	39.1	46.4
G5	92.7	87.3	63.6	59.1	61.8
G6	87.3	74.5	61.8	77.3	80.0
G7	97.3	97.3	70.0	99.1	99.1
G8	94.5	80.9	72.7	86.4	86.4
G9	95.5	82.7	68.2	70.9	76.4
Accuracy (%)	**94.3**	83.6	67.2	71.7	72.2

MAV:Mean Absolute Value

The proposed method is capable of distinguishing gestures based on their sEMG and ACC signals, although sEMG signals varies from person to person. However we observed that there are some similar gestures, which means there is a large confusions between these gestures (e.g., ILLEGAL USE OF HANDS and JUMP BALL). In these gestures, the motions of right hand are fist-based gesture, which cause the highly similarity between gestures and more tend to misclassification. However, the number of misclassification decrease significantly through a one-time calibration procedure. Due to the limitation of hardware, the user only wear one Myo armband on their right forearm, which may cause getting limited information and lead to wrong classification result.

We also construct a classifier of the signals of number of players through one to five (as shown in Table 6). However, due to the high similarity of myoelectric among these gestures, the average recognition accuracy is 54.3%. Even through a user calibration, the average recognition accuracy merely achieves 75.5%, which is not enough to be used in a real-time system.

Table 6. Confusion matrix of the number of player.

		Predicted number (Person-independent)				
		Number 1	Number2	Number3	Number4	Number 5
Ground truth	Number 1	**77**	9	11	7	6
	Number 2	8	**56**	11	24	11
	Number 3	14	9	**42**	19	26
	Number 4	0	16	9	**58**	27
	Number 5	1	0	26	23	**60**
		Predicted number (Calibrated)				
		Number 1	Number2	Number3	Number4	Number 5
Ground truth	Number 1	**101**	2	5	0	2
	Number 2	5	**75**	15	14	1
	Number 3	10	5	**64**	11	20
	Number 4	0	9	9	**75**	17
	Number 5	2	3	13	14	**78**

4   Conclusions and Future Work

In this paper, a sensor-based basketball referees' signals classification method using deep belief networks is proposed. We devise a method to utilize DBNs to extract features, with 4-order AR model coefficients composed into the feature vector, and then SVM is used for classification. We evaluated the proposed recognition method for 5-fold Cross Validation (5-foldCV) and Leave-One-Participant-Out Cross Validation (LOPOCV), and our method achieves the average accuracies of 97.9% and 90.5%, respectively, which outperforms the existing works using HMM or simply SVM. Furthermore, due to customizing a recognition system for referees themselves to get a better recognition results, we also evaluate our method on person-independent with calibration experiment whose accuracy achieves 94.3%, and it represents that we can dramatically enhance our recognition accuracy by user calibration. In the future, we will focus on enlarging the quantity of referee signals and improving the accuracy of number of players.

Acknowledgments. This research was supported by the Ministry of Science and Technology (contracts MOST-105-2221-E-006-066-MY3 and MOST-103-2221-E-006-157-MY2), Taiwan.

References

1. Ahsan, M.R., Ibrahimy, M.I., Khalifa, O.O.: Electromyography (EMG) signal based hand gesture recognition using artificial neural network (ANN). In: The 4th IEEE International Conference on Mechatronics (ICOM), pp. 1–6 (2011)

2. Georgi, M., Amma, C., Schultz, T.: Recognizing hand and finger gestures with IMU based motion and EMG based muscle activity sensing. In: International Conference on Bio-inspired Systems and Signal Processing, pp. 99–108 (2015)

3. Han, J., Shao, L., Dong, X., Shotton, J.: Enhanced computer vision with microsoft kinect sensor: a review. IEEE Trans. Cybern. **43**(5), 1318–1334 (2013)

4. Hinton, G.E., Osindero, S., Teh, Y.-W.: A fast learning algorithm for deep belief nets. Neural Comput. **18**(7), 1527–1554 (2006)

5. Hu, X., Nenov, V.: Multivariate AR modeling of electromyography for the classification of upper arm movements. Clin. Neurophysiol. **115**(6), 1276–1287 (2004)

6. Keyvanrad, M.A., Homayounpour, M.M.: A brief survey on deep belief networks and introducing a new object oriented MATLAB toolbox (DeeBNet V2.1). arXiv preprint, arXiv:1408.3264 (2014)

7. Kim, J., Mastnik, S., Andre, E.: EMG-based hand gesture recognition for realtime biosignal interfacing. In: The 13th ACM International Conference on Intelligent User Interfaces, pp. 30–39 (2008)

8. Pan, T.-Y., Lo, L.-Y., Yeh, C.-W., Li, J.-W., Liu, H.-T., Hu, M.-C.: Real-time sign language recognition in complex background scene based on a hierarchical clustering classification method. In: The 2nd IEEE International Conference on Multimedia Big Data (BigMM), pp. 64–67 (2016)

9. Samadani, A.-A., Kulic, D.: Hand gesture recognition based on surface electromyography. In: The 36th International Conference of the IEEE Engineering in Medicine and Biology Society, pp. 4196–4199 (2014)

10. Saponas, T.S., Tan, D.S., Morris, D., Balakrishnan, R.: Demonstrating the feasibility of using forearm electromyography for muscle-computer interfaces. In: The ACM SIGCHI Conference on Human Factors in Computing Systems, pp. 515–524 (2008)

11. Wang, J.-S., Chuang, F.-C.: An accelerometer-based digital pen with a trajectory recognition algorithm for handwritten digit and gesture recognition. IEEE Trans. Ind. Electron. **59**(7), 2998–3007 (2012)

12. Zhang, X., et al.: A framework for hand gesture recognition based on accelerometer and EMG sensors. IEEE Trans. Syst. Man Cybern.-Part A: Syst. Hum. **41**(6), 1064–1076 (2011)

Compact CNN Based Video Representation for Efficient Video Copy Detection

Ling Wang, Yu Bao, Haojie Li[(✉)], Xin Fan, and Zhongxuan Luo

Dalian University of Technology, Dalian, China
hjli@dlut.edu.cn

Abstract. Many content-based video copy detection (CCD) systems have been proposed to identify the copies of a copyrighted video. Due to storage cost and retrieval response requirements, most CCD systems represent video contents using sparsely sampled features, which tends to lose information to some extend and thus results in unsatisfactory performance. In this paper, we propose a compact video representation based on convolutional neural network (CNN) and sparse coding (SC) for video copy detection. We first extract CNN features from the densely sampled video frames and then encode them into a fixed length vector via the SC method. The proposed representation presents two advantages. First, it is compact while is regardless of the sampling frame rate. Second, it is discriminative for video copy detection by encoding the densely sampled frames' CNN features. We evaluate the performance of proposed representation on video copy detection over a real complex video dataset and marginal performance improvement has been achieved as compared to state-of-the-art CCD systems.

Keywords: Video copy detection · Convolutional neural network · Sparse coding · Video level representation · Dense sampling

1 Introduction

A copy is a duplicate segment of video derived from another video, by means of various transformations. The task of content-based video copy detection is to determine if a given video (query) has its copy in a set of testing videos. Analyzing and comparing the features between the querying and testing videos are the usual methods to cover this task. Copy detection has a wide range of potential applications such as copyright control, business intelligence, etc., thus has attracted lots of research efforts over the last decade [1, 20].

The duplicate segments may be as long as the origin video or even shorter than 1 s with some distortions. These distortions include simulated camcording, picture in picture, insertions of pattern, compression, etc. [16]. Variety distortions bring great challenges to the copy detection problem. To address these challenges, TREC Video Retrieval Evaluation (TRECVID) released a content-based copy detection benchmark with a large collection of synthetic queries. It launched

© Springer International Publishing AG 2017
L. Amsaleg et al. (Eds.): MMM 2017, Part I, LNCS 10132, pp. 576–587, 2017.
DOI: 10.1007/978-3-319-51811-4_47

the CCD competition task from 2008 to 2011 [16]. Many solutions have been proposed to tackle this problem. Among them, the most popular way is to use local features like SIFT [11] to match and find similar frame pairs, followed by temporal alignment [3]. In these methods, inverted files and Bag-of-words (BoW) representation are widely adopted for fast frame matching. Actually some of them have achieved near-perfect performance [16].

As a simulated dataset, the TRECVID can not accurately reflect real copy videos. There are more complicated visual transformations and more complex temporal structures in real copy videos. The performance of current state-of-the-art approaches is far away from satisfactory for real video copy detection, which remains CCD still an open issue to the multimedia research community as Jiang et al. evaluated [8].

In order to avoid unaffordable computational and storage burdens, most of the current copy detection works extract video features on sparsely sampled frames, e.g. sampling one or two frames per second [3,8]. We argue that sparse sampling could miss much useful information as most frames are dropped. Thus, if we can encode more information of a video segment, the detection performance will be improved. Meanwhile, deep learning approaches, especially convolutional neural network (CNN) has recently shown powerful ability in extracting distinctive image or object features. It achieved great success in general image classification, object detection tasks and semantic analysis [10,12,24]. Jiang's initial attempts [9] of using CNN features for copy detection also demonstrated its promising advantages over existing traditional methods.

Motivated by the above observations, we propose a novel video level representation which encodes the CNN features of densely sampled frames of a short time video into a compact descriptor, for real video copy detection. We first extract the CNN full-connect (fc) layer features for the sampled frames of a short video and then reduce the features dimension by using PCA. After that, sparse coding method is adopted to sparsely assign each frame feature into a set of M codes. So we can get an M-dimensional sparse vector for each frame. After these steps, a max-pooling operation is performed on each component of these vectors. A compact video level representation is finally derived.

The contributions of this paper are two-fold. (1) The proposed novel video level representation is compact which needs less storage and enables fast retrieval of similar video clips. Its dimension is independent to the sampling frame rate. (2) By encoding the densely sampled frames CNN features, the proposed copy detection method significantly outperforms state-of-the-art traditional approaches; it also achieves competitive precision to lately CNN based method [9] but improve the recall rate about 10%, which is more practical to some applications such as web video monitoring and tracking.

The reminder of this paper is structured as follows. Section 2 briefly reviews some well-known video copy detection works and approaches. In Sect. 3, we introduce our method for video feature representation and copy detection. Experimental results are presented and discussed in Sect. 4. Finally, we conclude this paper in Sect. 5.

2 Related Work

Video copy detection has attracted a lot of research interests in recent years. The approaches to this task mainly include two categories: local feature based methods and global feature based methods. Local feature based methods are widely leveraged by most works. For instance, LBP is used as visual feature in [15], LBP-CS [5] and SIFT are used by Douze et al. [3]. In [3], the local features are clustered into visual words and the image is represented with a BoW model. Then inverted file is adopted to index images for fast retrieval. In [6], hamming Embedding (HE) is employed to further divide the clusters into sub-spaces to improve the accuracy of local feature matching. Meanwhile, weak geometry consistency (WGC) [6] is used to help eliminating the wrong matching between features. Zhao et al. [25] adopt BoW, HE and WGC to effectively annotate web videos via near-duplicate video detection. All these above works extract local features on sparsely sampled frames for the purpose of low memory usage and efficiently retrieval. However, since one single video frame can produce about one thousand local features, it is still a heavy burden to storage and retrieve the large number of frames' features of a very large video dataset.

As a solution, employing global features could significantly decrease the number of features. Wu et al. [22] extract color histogram as frame features to detect similar videos. In [4], a Fisher Vector (FV) [13] alike representation is proposed to aggregate local features to a global feature. Meanwhile, CNN has shown their absolute advantages in image representation and high speed processing. In [9], Jiang and Wang sample frames at fixed time interval and extract CNN based features for each sampled frame. There are two ways to implement CNN features in his work. One of them is the standard CNN which uses Caffe [7] toolkit with AlexNet [10]. A 4,096-dimensional fc feature is extracted to present a frame. The other way extracts local image patch features using a supervised CNN structure called Siamese convolutional neural network (SCNN). Then image patches are described by features which are ranged from 64-dimensional to 512-dimensional. All these features are organized into a fast retrieval structure to do the match processing. Finally the matching results are aligned to the original videos by temporal network [17] according to their temporal consistency. Comparing to the traditional Hough Voting Alignment, Jiang [8] proves that temporal network is more suitable for temporal alignment.

As can be summarized from the above works, most approaches extract features on sparsely sampled frames rather than densely sampled frames. Much useful information is given up to make a heavy concession for considering the memory usage and efficiency retrieval. We will show in the experiments that this will cause the degrade of performance to some extend due to information loss. To overcome this disadvantage, our target is to find a representation which could aggregate more information and keep the final representation compact and discriminative.

3   Compact Video Representation

We will present the proposed compact video representation for short time video clips in detail. The outline of our video copy detection system is shown in Fig. 1, which is divided to three main steps: frame feature extraction, video feature encoding and video segment matching. Video feature encoding includes compression and aggregation. In this paper we pay more attention to the first two steps which are marked by blue arrows in Fig. 1. The matching step (i.e., fast retrieval and temporal alignment) will be introduced briefly.

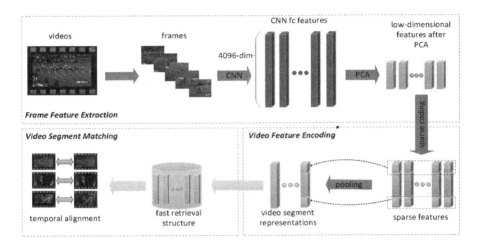

Fig. 1. The outline of our video copy detection system which includes three parts: frame feature extraction, video feature encoding and matching.

3.1   Frame Feature Extraction

Traditional works sparsely sample frames from videos for keeping balance between accuracy and efficiency. On the contrary, our starting point is gathering more information into final representations. So the first step of our method is to sample frames from query and database videos densely.

The next step is to describe sampled frames via frame-based features. Different from local feature extraction processing, CNN has shown its fast processing speed by utilizing parallel GPUs. Although the number of frames becomes more than ten times than the sampling methods in [3,8], the processing time may be equal to or less than the previous methods due to the performance of GPUs. We utilize the Caffe toolkit [7] to implement deep learning algorithm on the densely sampled frames. According to [23], we use the deeper network architecture, i.e. VGG-16layers, which is the winner of VGG ILSVRC 2014 classification task [14]. This network contains 16 weight layers: 13 convolutional layers and 3 fully-connected layers. And other five max-pooling layers are inserted after some convolutional layers.

Finally, we conduct PCA-whitening on the CNN fc features since the dimension of 4,096-D fc features are too high. Up to now, a frame is represented by a low-dimensional feature vector x with length k. We will explore the appropriate value of k in the experiments then.

3.2   Video Feature Compression and Aggregation

We aim to use compact representation to describe the densely sampled features. Directly calculating the distance between frame features is very sensitive to the noise in visual feature and the result is easily influenced by even one dimension of noise feature [18]. Since sparse coding, which models data vectors by the sparse linear combinations of the basis dictionary, could reserve the main components of vectors and make it possible to compactly represent the vectors. There are several works use SC and gains great performance [19]. We compress the frame features using sparse coding in our method. Sparse coding [2] can be regarded as

$$\min_{D,s(i)} \sum_i \left\| Ds^{(i)} - x^{(i)} \right\|_2^2 + \lambda \left\| s^{(i)} \right\|_1 \qquad \text{s.t. } \left\| D^{(m)} \right\|_2^2 = 1, \forall m = 1, 2, \ldots, M \qquad (1)$$

where D indicates the overcomplete dictionary, i.e. $M > k$, k is the dimension of feature x. In this work, we set M be four times of k unless otherwise noted. s is the target sparse representation.

We investigated some solutions of sparse decomposition problem and found that the Orthogonal Matching Pursuit (OMP) [21] is the most suitable method for our processing:

$$\min_{D,s(i)} \sum_i \left\| Ds^{(i)} - x^{(i)} \right\|_2^2 + \lambda \left\| s^{(i)} \right\|_1 \qquad \text{s.t. } \left\| D^{(m)} \right\|_2^2 = 1, \forall m = 1, 2, \ldots, M \qquad \text{and } \left\| s^{(i)} \right\|_0 \le T, \forall i \qquad (2)$$

In this case, s could have at most T non-zeros items. We employ K-SVD to train the dictionary. According to the greediness of OMP, the number of non-zeros in s could be close or equal to T. This property is helpful to us while other sparse decomposition methods may make our final features become excessively sparse, which is insufficient to distinguish different video segments.

At this stage, each frame is essentially a sparse feature. The extracted sparse features are then pooled and aggregated into a compact representation for the specified length of video segment. Different from event detection and video classification, video copy detection task needs fine-grained time interval representation for accurately aligning the time line between copies and original video. As shown in Fig. 2, we take 1 s interval in our method.

Video pooling could be divided to three categories: max-pooling, mean-pooling and sum-pooling. Since sparse coding describe a feature by linear combination of its principal basis, it is better to conserve the maximal item among a short time video. Thus, keeping the component-wise maximum by max-pooling

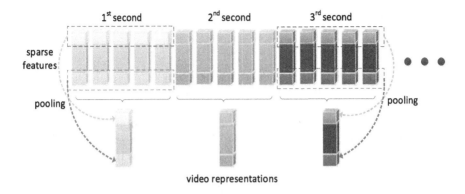

Fig. 2. Illustration of video pooling

is our choice. Because there are some negative items in the sparse feature by the OMP algorithm, we make a small adjustment for max-pooling:

$$v^m = s_i^m, \forall m \in M$$
$$\text{s.t.} \ \max_i abs\,(s_i^m)\,, i \in 1, 2, \ldots, n \qquad (3)$$

where v is the video representation with the dimension M. s_i is the sparse vector for the i-th sampled frame of the one-swcond interval clip. n is the total number of sampled frames among the one-second interval clip. By this way, we can reserve the most important part of each component whether or not it is positive.

After pooling and aggregating, we finally get a series of video level representations, each of which represents a one-second interval time video segment.

3.3   Video Segment Matching

The final step of video copy detection is to compare the query video with database videos and identify the most similar segment pairs between them. The matching method contains fast retrieval and temporal alignment.

We leverage the normal KD-tree to store our features and do fast retrieval. Although we don't specially investigate its efficiency, there is huge potential of our sparse video features. First, the sparse feature needs less storage and could make it possible to reside all features in memory. Second, a large number of floating point arithmetic is no more needed due to the large amount of matching between zero and zero.

It's necessary to link our video level representations to a longer video segment because each feature only describes a short-time interval clips. The longer video segment is then associated with a starting timestamp and a ending timestamp of a testing video. Following [8], we employ the temporal network to align our matched video segments and adopt the following formula to measure the similarity score between two features:

$$score = e^{-dis^2} \qquad (4)$$

where dis is the Euclidean distance between these two features.

4   Experiments

4.1   VCDB Dataset

In our experiments, we utilize the latest released copy detection dataset, namely VCDB dataset [8] to evaluate the proposed method. The VCDB includes core-dataset and distraction-dataset which are all collected from two video-sharing websites: YouTube and MetaCafe. The core-dataset contains over than 500 videos with 9,236 partial copies and the distraction-dataset includes 100,000 videos. We mainly evaluate our video level representations in the core-dataset. Same to the baseline method in [8], all the segments of the 9,236 pairs are considered as a query. If both the two segments among a detected video pairs had intersection time with a ground-truth pair, they would be considered as a correct pairs in spite of the length of overlapped time window. Because a video segment with one single copy frame can be fully demonstrated as a copy pair. We use the precision and recall to measure our features' performance:

$$precision = \frac{|\text{correctly retrieved segments}|}{|\text{all retrieved segments}|} \tag{5}$$

$$recall = \frac{|\text{correctly retrieved segments}|}{|\text{ground-truth copy segments}|} \tag{6}$$

4.2   Experimental Results and Comparisons

We show the results of the proposed method and also compare it with some state-of-the-art systems from several aspects. Our method achieves the best results with the following settings: (1) all frames are used to generate our final representation without sampling; (2) the dimension of features at the frame feature extraction step is set to be 512-D, which results in a 2,048-D video segment representation; (3) the number of non-zero components in the sparse representation is controlled to be at most 32. These settings are both utilized on the fc6 and fc7 layers of VGG-16layer network. We adjust the threshold of the segment pairs' matching score to draw the precision-recall curves.

The comparing of our method, the baseline system [8], standard CNN and SCNN [9] are shown in Fig. 3. As we can see, our methods achieve remarkable performance both on fc6 and fc7, while features extracted based on fc6 works better than on fc7. This may suggest that fc6 is more suitable than fc7 for video copy detection task.

The green curve represents the baseline method which is proposed in [8] and utilizes local features, i.e., SIFT and temporal network to detect copy pairs. This approach is widely used by previous work and has shown near-perfect performance in TRECVID benchmark, however, it is far away from satisfactory in real complexity copy detection.

The red curve shows the result of standard CNN method [9] that extracts features using CNN with the AlexNet and directly uses the 4,096-D features on fc6 for retrieving. It also proposed a fusion method by combining SCNN and

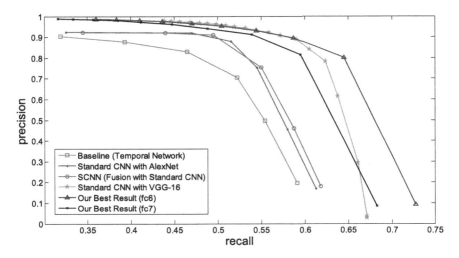

Fig. 3. Precision-recall curves for different methods on the core-dataset of VCDB. (Color figure online)

standard CNN to extract frame features in [9]. Our method performs better than these two CNN based methods, owing to the densely sampled frames feature with more information and the deeper network with better descriptive power.

To demonstrate the effectiveness of dense sampling strategy and compact video feature representation, we also implemented standard CNN method on fc6 using the VGG-16layers network and show its results with the orange curve in Fig. 3. Comparing to it, our method (blue curve) significantly increases the recall rate while maintains a good precision rate, which is more practical to some applications such as web video monitoring and tracking where low miss is more important.

4.3   Impacts of Parameters

Sparsity. We investigated several sparse decomposition algorithms and found that the sparsity of our features obviously affects the performance. We employ the OMP algorithm in our experiments due to the controllable sparsity, and adjust the number of non-zero components in the sparse frame features to produce different sparsity of final video representations. In Fig. 4, the parameter T indicates the maximum number of non-zero components in a sparse frame feature. We conduct experiments on different T while fix the rest parameters to see the performance changes with the sparsity of our representation. As can be seen from Fig. 4, fc6 performs better than fc7 and T exhibits little influence on fc7. For fc6, the larger value of T results in higher performance. However, the performance improves slowly when the value of T increases 32 from 16. Therefore, we set T to 32 in all our experiments.

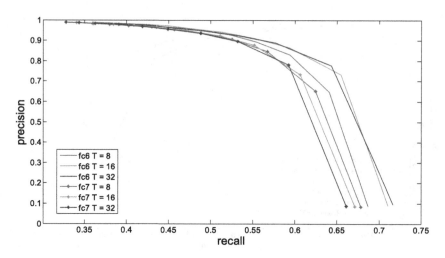

Fig. 4. Results comparison for different sparsity of features

Dimension. The dimension of features after PCA-whitening could influence the dimension of our final representations. It's essential to investigate the impact of the dimensions on the final detection results. In Fig. 5, only the dimension of PCA is changed. From the figure, we can see that the performance is increasing with the rise of dimension. However, the storage cost and retrieval time will both growing exponentially. In our method, we adopt 512-D as a trade-off of performance and resource consumption, which leads to a 2,048-D final video representation.

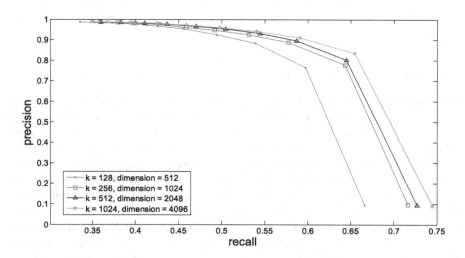

Fig. 5. Results comparison for different dimension of features

4.4   Analysis of the Impact of Sampling Rate

We are the first to utilize densely sampled frames to generate video representations for video copy detection, it is necessary to investigate whether more frames bring better discrimination. To do so, we test the proposed method on fc6 with different sampling rate in the pooling step. The best F1-measure (i.e., the harmonic mean of precision and recall) is used to evaluate the performance.

Table 1. Detection results with different sampling rate of frames

1-frame per sec	2-frames per sec	1/5 frames	1/3 frames	1/2 frames	All frames
0.6695	0.6879	0.6994	0.6995	0.7026	**0.7038**

Table 1 shows that the detection performance increases consistently with the sampling rate, and when the sampling rate increases to 1/5, the performance becomes relatively steady. This is reasonable because the adjacent frames are extremely similar and contain much redundant information. However, 1/5 of 1 s video means 5 to 6 frames, which is denser than any existing methods and could bring much more storage cost and much longer retrieval time to these methods. By fusing the sparse coding and max-pooling strategies, our method elaborately encode the densely sampled frame features into a compact yet discriminative representation.

5   Conclusions

We have proposed a novel compact video representation to detect video copies from large video collections. More and discriminative information is embedding to our final representation through sparse coded CNN features extracted from densely sampled video frames. Experimental results on the VCDB show that this presentation is advantageous over recent state-of-the-art CCD approaches. We significantly improve the recall rate about 10% with higher precision rate. In the future, we will investigate the effective indexing strategy for the proposed representation to fast and accurately retrieve very large scale video dataset.

Acknowledgements. This work is supported by National Natural Science Funds of China (61472059, 61428202).

References

1. Chou, C.L., Chen, H.T., Lee, S.Y.: Pattern-based near-duplicate video retrieval and localization on web-scale videos. IEEE Trans. Multimedia **17**(3), 382–395 (2015)
2. Coates, A., Ng, A.Y.: The importance of encoding versus training with sparse coding and vector quantization. In: Proceedings of the 28th International Conference on Machine Learning (ICML-2011), pp. 921–928 (2011)

3. Douze, M., Jégou, H., Schmid, C.: An image-based approach to video copy detection with spatio-temporal post-filtering. IEEE Trans. Multimedia **12**(4), 257–266 (2010)
4. Douze, M., Jégou, H., Schmid, C., Pérez, P.: Compact video description for copy detection with precise temporal alignment. In: Daniilidis, K., Maragos, P., Paragios, N. (eds.) ECCV 2010. LNCS, vol. 6311, pp. 522–535. Springer, Heidelberg (2010). doi:10.1007/978-3-642-15549-9_38
5. Heikkilä, M., Pietikäinen, M., Schmid, C.: Description of interest regions with local binary patterns. Pattern Recogn. **42**(3), 425–436 (2009)
6. Jegou, H., Douze, M., Schmid, C.: Hamming embedding and weak geometric consistency for large scale image search. In: Forsyth, D., Torr, P., Zisserman, A. (eds.) ECCV 2008. LNCS, vol. 5302, pp. 304–317. Springer, Heidelberg (2008). doi:10.1007/978-3-540-88682-2_24
7. Jia, Y., Shelhamer, E., Donahue, J., Karayev, S., Long, J., Girshick, R., Guadarrama, S., Dar-rell, T.: Caffe: convolutional architecture for fast feature embedding. In: Proceedings of the 22nd ACM International Conference on Multimedia, pp. 675–678. ACM (2014)
8. Jiang, Y.-G., Jiang, Y., Wang, J.: VCDB: a large-scale database for partial copy detection in videos. In: Fleet, D., Pajdla, T., Schiele, B., Tuytelaars, T. (eds.) ECCV 2014. LNCS, vol. 8692, pp. 357–371. Springer, Heidelberg (2014). doi:10.1007/978-3-319-10593-2_24
9. Jiang, Y.G., Wang, J.: Partial copy detection in videos: a benchmark and an evaluation of popular methods. IEEE Trans. Big Data **2**(1), 32–42 (2016)
10. Krizhevsky, A., Sutskever, I., Hinton, G.E.: Imagenet classification with deep convolutional neural networks. In: Advances in Neural Information Processing Systems, pp. 1097–1105 (2012)
11. Lowe, D.G.: Distinctive image features from scale-invariant keypoints. Int. J. Comput. Vis. **60**(2), 91–110 (2004)
12. Ren, S., He, K., Girshick, R., Sun, J.: Faster R-CNN: towards real-time object detection with region pro-posal networks. In: Advances in Neural Information Processing Systems, pp. 91–99 (2015)
13. Sánchez, J., Perronnin, F., Mensink, T., Verbeek, J.: Image classification with the fisher vector: theory and practice. Int. J. Comput. Vis. **105**(3), 222–245 (2013)
14. Simonyan, K., Zisserman, A.: Very deep convolutional networks for large-scale image recognition. arXiv preprint arXiv:1409.1556 (2014)
15. Song, J., Yang, Y., Huang, Z., Shen, H.T., Hong, R.: Multiple feature hashing for real-time large scale near-duplicate video retrieval. In: Proceedings of the 19th ACM International Conference on Multi-media, pp. 423–432. ACM (2011)
16. U.S. National Institute of Standards and Technology: Trec video retrieval evaluation. http://www-nlpir.nist.gov/projects/tv2011/#ccd
17. Tan, H.K., Ngo, C.W., Hong, R., Chua, T.S.: Scalable detection of partial near-duplicate videos by visual-temporal consistency. In: Proceedings of the 17th ACM International Conference on Multi-media, pp. 145–154. ACM (2009)
18. Tang, J., Hong, R., Yan, S., Chua, T.S., Qi, G.J., Jain, R.: Image annotation by k nn-sparse graph-based label propagation over noisily tagged web images. ACM Trans. Intell. Syst. Technol. (TIST) **2**(2), 14 (2011)
19. Tang, S., Zheng, Y.T., Wang, Y., Chua, T.S.: Sparse ensemble learning for concept detection. IEEE Trans. Multimedia **14**(1), 43–54 (2012)
20. Thomas, R.M., Sumesh, M.: A simple and robust colour based video copy detection on summarized videos. Procedia Comput. Sci. **46**, 1668–1675 (2015)

21. Tropp, J.A., Gilbert, A.C.: Signal recovery from random measurements via orthogonal matching pursuit. IEEE Trans. Inf. Theor. **53**(12), 4655–4666 (2007)
22. Wu, X., Hauptmann, A.G., Ngo, C.W.: Practical elimination of near-duplicates from web video search. In: Proceedings of the 15th ACM International Conference on Multi-media, pp. 218–227. ACM (2007)
23. Xu, Z., Yang, Y., Hauptmann, A.G.: A discriminative CNN video representation for event detection. In: Proceedings of the IEEE Conference on Computer Vision and Pattern Recognition, pp. 1798–1807 (2015)
24. Yang, Y., Zhang, H., Zhang, M., Shen, F., Li, X.: Visual coding in a semantic hierarchy. In: Proceedings of the 23rd ACM International Conference on Multimedia, pp. 59–68. ACM (2015)
25. Zhao, W.L., Wu, X., Ngo, C.W.: On the annotation of web videos by efficient near-duplicate search. IEEE Trans. Multimedia **12**(5), 448–461 (2010)

Cross-Modal Recipe Retrieval:
How to Cook this Dish?

Jingjing Chen, Lei Pang, and Chong-Wah Ngo$^{(\boxtimes)}$

Department of Computer Science, City University of Hong Kong,
Kowloon Tong, Hong Kong
{jingjchen9-c,leipang3-c}@my.cityu.edu.hk, cscwngo@cityu.edu.hk

Abstract. In social media users like to share food pictures. One intelligent feature, potentially attractive to amateur chefs, is the recommendation of recipe along with food. Having this feature, unfortunately, is still technically challenging. First, the current technology in food recognition can only scale up to few hundreds of categories, which are yet to be practical for recognizing ten of thousands of food categories. Second, even one food category can have variants of recipes that differ in ingredient composition. Finding the best-match recipe requires knowledge of ingredients, which is a fine-grained recognition problem. In this paper, we consider the problem from the viewpoint of cross-modality analysis. Given a large number of image and recipe pairs acquired from the Internet, a joint space is learnt to locally capture the ingredient correspondence from images and recipes. As learning happens at the region level for image and ingredient level for recipe, the model has ability to generalize recognition to unseen food categories. Furthermore, the embedded multi-modal ingredient feature sheds light on the retrieval of best-match recipes. On an in-house dataset, our model can double the retrieval performance of DeViSE, a popular cross-modality model but not considering region information during learning.

Keywords: Recipe retrieval · Cross-modal retrieval · Multi-modality embedding

1 Introduction

Food recognition is generally regarded as a hard problem, due to diverse appearances of food as a result of non-rigid deformation and composition of ingredients. Recently, the problem has started to capture more attention [1–4] partly due to the success of deep learning technologies. The accuracy of food recognition can be as high as 80% on the benchmark datasets such as Food101 [2], FoodCam-256 [5] and VIREO Food-172 [6]. The success gives light to the development of techniques for auto dietary food tracking [1,7,8] and nutrition estimation [9], which has long been recognized as a challenge not only in multimedia [8,10] but also health and nutritional science [11].

© Springer International Publishing AG 2017
L. Amsaleg et al. (Eds.): MMM 2017, Part I, LNCS 10132, pp. 588–600, 2017.
DOI: 10.1007/978-3-319-51811-4_48

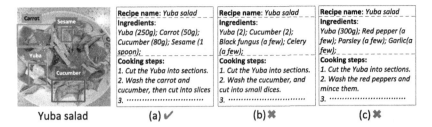

Yuba salad   (a) ✔   (b) ✖   (c) ✖

Fig. 1. Although recipe (a), (b) and (c) are all about "Yuba salad", only recipe (a) uses the exactly same ingredients as the dish picture. Retrieving best-match recipe requires fine-grained analysis of ingredient composition.

Nevertheless, the existing efforts are mostly devoted to recognizing a predefined set of food categories, ranging from 100 to 256 categories [2,3,5,6]. Extending to large-scale recognition, for example tens of thousands food categories, remains an area yet to be researched. In this paper, we pose food recognition as a problem of recipe retrieval. Specifically, given a food picture, of whether the category has been seen in the training model, the aim is to retrieve a recipe for the food. The advantages of having recipe, rather than the name of food category, as output are numerous. Sharing food pictures in social media has been a trend. The ability to recommend recipes along will benefit users who want to cook a particular dish, and the feature is yet to be available. In addition, recipe provides rich information, such as cooking methods, ingredients and their quantities, which can facilitate the estimation of food balance and nutrition facts. The challenge of recipe retrieval, nevertheless, comes from the fact that there could be many recipes named under the same categories, each of which differs in the composition of ingredients. Figure 1 shows an example, where recommending the right recipe for "Yuba Salad" indeed requires also fine-grained recognition of ingredient composition.

This paper explores the recent advances in cross-modality learning for addressing the aforementioned problems. Specifically, given food pictures and their associated recipes, our aim is to learn a model that captures their correspondence by learning a joint embedding space for visual-and-text translation. We exploit and revise a deep model, stacked attention network (SAN) [12], originally proposed for visual question-answering for our purpose. The model learns the correspondence through assigning heavier weights to the attended regions relevance to ingredients extracted from recipes. For the task of recipe retrieval, fortunately the learning does not require much effort in labeling training examples. There are already millions of food-recipe pairs, uploaded by professional and amateur chefs, on various cooking websites, which can be freely leveraged for training. We demonstrate that using these online resources, a fairly decent model can be trained for recipe retrieval with minimal labeling effort. As input to SAN includes ingredients, the model has higher generalization ability in recognizing food categories unseen during training, as long as all or most ingredients are known. Furthermore, as ingredient composition is considered in SAN, the chance of retrieving the best-match recipes is also enhanced. To this end, the contribution of this paper lies in addressing of food recognition as a recipe retrieval problem. Under this umbrella, the problem is turned into cross-modality feature learning, which can integrally model three inter-related problems: scalable food recognition, fine-grained ingredient recognition and best-match recipe retrieval.

2 Related Work

Analysis of recipes has been studied from different perspectives, including retrieval [6,13,14], classification [15,16] and recommendation [17]. Most of the approaches employ text-based analysis based upon information extracted from recipes. Examples include extraction of ingredients as features for cuisine classification [15] and taste estimation [16]. More sophisticated approaches model recipes as cooking graphs [13,18] such that graph-based matching can be employed for similarity ranking of recipes. The graph, either manually or semi-automatically constructed from a recipe, represents the workflow for cooking and cutting procedures of ingredients. In [13], multi-modality information was explored, by late fusion of cooking graphs and low-level features extracted from food pictures, for example-based recipe retrieval. Few works have also studied cross-modality retrieval [6,14,17]. In [17], recognition of raw ingredients was studied for cooking recipe recommendation. Compared to prepared food where ingredients are mixed or even occlude each other, raw ingredients are easier to recognize. In [14], classifier-based approach was adopted for visual-to-text retrieval. Specifically, the category of food picture is first recognized, followed by retrieval of recipes under a category. As classifiers were trained from UPMC Food-101 dataset [2], retrieval is only limited to 101 food categories. The issues in scalability and finding best-match recipes are not addressed. The recent work in [6] explored ingredient recognition for recipe retrieval. Using ingredient network as external knowledge, the approach is able to retrieve recipes even for unseen food categories. Different from [6], this paper aims to learn a joint space that can inherently capture the visual-text commonality for retrieval.

Cross-modality analysis has been actively researched for multimedia retrieval [19–21]. Frequently employed algorithms include canonical correlation analysis (CCA) [22] and partial least squares (PLS) [23], which find a pair of linear transformation to maximize the correlation between data from two modalities. CCA, in particular, has been extended to three-view CCA [24], semantic correlation matching (SCM) [19], deep CCA [25] and end-to-end deep CCA [26] for cross-modality analysis. Among variants of model, deep visual semantic embedding (DeViSE) [20] is generally used and usually exhibits satisfactory performance. These models, nevertheless, consider image-level features, such as $fc7$ extracted from deep convolutional network (DCNN), and usually ignore regional features critical for fine-grained recognition. One of the exceptions is deep fragment embedding (DFE) proposed in [21], which aligns image objects and sentence fragments while learning the visual-text joint feature. However, the model is not applicable here for requiring of R-CNN [27] for object region detection. In food domain, there is yet to have any algorithm for robust segmentation of ingredients, which can be fed into DFE for learning.

3   Stacked Attention Network (SAN)

Figure 2 illustrates the SAN model, with visual and text features respectively extracted from image and recipe as input. The model learns a joint space that boosts the similarity between images and their corresponding recipes. Different from [12], where the output layer is for classification, we modify SAN so as to maximize the similarity for image-recipe pairs. As SAN considers spatial information, attention map can be visualized by back projection of embedded feature into image.

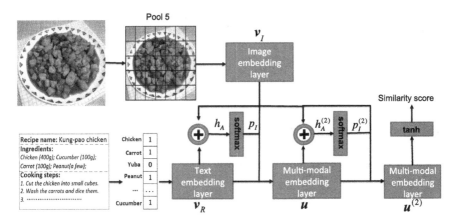

Fig. 2. SAN model inspired from [12] for joint visual-text space learning and attention localization.

3.1   Image Embedding Feature

The input visual feature is the last pooling layer of DCNN – *Pool5* – that retains the spatial information of the original image. The dimension of *Pool5* feature is $512 \times 14 \times 14$, corresponding to 14×14 or 196 spatial grids of an image. Each grid is represented as a vector of 512 dimensions. Denote f_I as the *Pool5* feature and is composed of regions f_i, $i \in [0; 195]$. Each region f_i is transformed to a new vector or embedding feature as following:

$$v_I = tanh(W_I f_I + b_I) \tag{1}$$

where $v_I \in \mathbb{R}^{d \times m}$ is the transformed feature matrix, with d as the dimension of new vector and $m = 196$ is the number of grids or regions. The embedding feature of f_i is indexed by i-th column of v_I, denoted as v_i. The transformation is performed region-wise, $W_I \in \mathbb{R}^{d \times 512}$ is the transformation matrix and $b_I \in \mathbb{R}^d$ is the bias term.

3.2   Recipe Embedding Feature

A recipe is represented as a binary vector of ingredients, denoted as $r \in \mathbb{R}^t$. The dimension of vector is t corresponding to the size of ingredient vocabulary. Each entry in r indicates the presence (1) or absence (0) of a particular ingredient in a recipe. As Pool5 feature, the vector is embedded into a new space as following

$$v_R = tanh(W_R r + b_r) \tag{2}$$

where $W_R \in \mathbb{R}^{d \times t}$ is the embedding matrix and $b_r \in \mathbb{R}^d$ is the bias vector. Note that, for joint learning, the embedding features of recipe ($v_R \in \mathbb{R}^d$) and Pool5 region (i-th column of v_I) have the same dimension.

3.3   Joint Embedding Feature

The attention layer is to learn the joint feature by trying to locate the visual food regions that correspond to ingredients. There are two transformation matrices, $W_{I,A} \in \mathbb{R}^{k \times d}$ for image I and $W_{R,A} \in \mathbb{R}^{k \times d}$ for recipe R, mimicking the attention localization, formulated as following:

$$h_A = tanh(W_{I,A} v_I \oplus (W_{R,A} v_R + b_A)) \tag{3}$$

$$p_I = softmax(W_P h_A + b_P) \tag{4}$$

where $h_A \in \mathbb{R}^{k \times m}$, $p_I \in \mathbb{R}^m$, $W_P \in \mathbb{R}^{1 \times k}$. Note that p_I aims to capture the attention, or more precisely relevance, of image regions to a recipe. The significance of a region f_i is indicated by the value in the corresponding element $p_i \in p_I$.

The joint visual-text feature is basically generated by adding the embedding features v_I and v_R. To incorporate attention value, regions v_i are linearly weighted and summed (Eq. 5) before the addition operation with v_R (Eq. 6), as following:

$$\tilde{v}_I = \sum_{i=1}^{m} p_i v_i \tag{5}$$

$$u = \tilde{v}_I + v_R \tag{6}$$

where $\tilde{v}_I \in \mathbb{R}^d$, and $u \in \mathbb{R}^d$ represents the joint embedding feature.

As suggested in [12], progressive learning by stacking multiple attention layers can boost the performance, but will heavily increase the training cost. We consider two-layer SAN, by feeding the output of first attention layer, $u^{(1)}$, into the second layer to generate new joint embedding feature $u^{(2)}$ as following

$$h_A^{(2)} = tanh(W_{I,A}^{(2)} v_I \oplus (W_{R,A}^{(2)} u + b_A^{(2)})) \tag{7}$$

$$p_I^{(2)} = softmax(W_P^{(2)} h_A^{(2)} + b_P^{(2)}) \tag{8}$$

$$\tilde{v}_I^{(2)} = \sum_i p_i^{(2)} v_i \tag{9}$$

$$u^{(2)} = \tilde{v}_I^{(2)} + u \tag{10}$$

As $p_I^{(2)}$ indicates the region relevancy, the attention map can be visualized by back projecting the attention value p_i to its corresponding region f_i, followed by upsampling to the original image size with bicubic interpolation.

3.4 Objective Function

To this end, the similarity between food image and recipe is generated as following:

$$S<\boldsymbol{v}_I,\boldsymbol{v}_R>= tanh(W_{u,s}\boldsymbol{u}^{(2)} + b_s) \tag{11}$$

where $W_{u,s} \in \mathbb{R}^d$ and $b_s \in \mathbb{R}$ is bias. $S<\boldsymbol{v}_I,\boldsymbol{v}_R>$ outputs a score indicating the association between the embedding features of image and recipe. The learning is based on the following rank-based loss function with a large margin form as the objective function:

$$\mathcal{L}(W, D_{trn}) = \sum_{(\boldsymbol{v}_I,\boldsymbol{v}_R^+,\boldsymbol{v}_R^-)\in D_{trn}} max(0, \triangle + S<\boldsymbol{v}_I,\boldsymbol{v}_R^-> -S<\boldsymbol{v}_I,\boldsymbol{v}_R^+>) \tag{12}$$

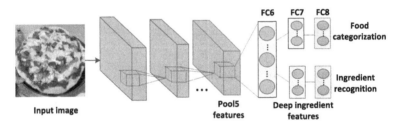

Fig. 3. Multi-task VGG model in [6] offering *pool5* and deep ingredient features for cross-modal joint space learning.

The training set, D_{trn}, consists of triples in the form of $(\boldsymbol{v}_I, \boldsymbol{v}_R^+, \boldsymbol{v}_R^-)$, where $\boldsymbol{v}_R^+$ $(\boldsymbol{v}_R^-)$ is true (false) recipe for food v_I. The matrix W represents the network parameters, and $\triangle \in (0,1)$ controls the margin in training and is cross-validated.

4 Experiments

4.1 Settings and Evaluation

Here we detail the parameter setting of SAN. The dimension of embedding feature is set to $d = 500$ for both *Pool5* regional and recipe feature, while the dimension for h_A is $k = 1,024$ for Eqs. 3 and 7. Through cross-validation, the hyper parameter $\triangle$ for the loss function is set as 0.2. SAN is trained using stochastic gradient descent with momentum set as 0.9 and the initial learning rate as 1. The size of mini-batch is 50 and the training stops after 10 epochs. To prevent overfitting, dropout [28] is used. The *pool5* feature can be extracted from any DCNN models. We employed the multi-task VGG released by [6], which reported the best performances on two large food datasets, VIREO Food-172 [6] and UEC Food-100 [3]. The model, as shown in Fig. 3, has two pathways, one for classifying 172 food categories while another for labeling 353 ingredients. For a fair comparison, all the compared approaches in the experiment are using multi-task VGG features, either *pool5* or deep ingredient feature (*fc7*), as shown in Fig. 3.

As the task is to find the best possible recipe given a food picture, the following two measures are employed for performance evaluation:

- Mean reciprocal rank (MRR): MRR measures the reciprocal of rank position where the ground truth recipe is returned, averaged over all the queries. This measure assesses the ability of the system to return the correct recipe at the top of the ranking. The value of MRR is within the range of $[0, 1]$. A higher score indicates a better performance.
- Recall at Top-K (R@K): R@K computes the fraction of times that a correct recipe is found within the top-K retrieved candidates. R@K provides an intuitive sense of how quickly the best recipe can be located by investigating a subset of the retrieved items. As MRR, a higher score also indicates a better performance.

4.2   Dataset

The dataset is composed of 61,139 image-recipe pairs crawled from the "Go Cooking"[1] websites. Each pair consists of a recipe and a picture of resolution 448×448. The dataset covers different kinds of food, like Chinese dishes, snacks, dessert, cookies and Chinese-style western food. Each recipe includes the list of ingredients and cooking procedure. As the recipes were uploaded by amateurs, the naming of ingredients is not always consistent. For example, "carrot" is sometimes called as "carotte". We manually rectified the inconsistency and compiled a list of 5,990 ingredients, both visible and non-visible (e.g., "honey"), from these recipes. The list, represented as a binary vector indicating the presence or absence of particular ingredients in a recipe, serves as input to the SAN model. Note that in some cases the cooking and cutting methods are directly embedded into the name of ingredient, for example, "tofu" and "tofu piece", "egg" and "steamed egg".

The dataset is split into three sets: 54,139 pairs for training, 2,000 pairs for cross-validation, and 5,000 pairs for testing. Furthermore, we selected 1,000 images from the testing set as queries to search against the 5,000 recipes. The queries are sampled in such a way that there are around 45% of them (446 queries) belonging to food categories unknown to SAN and multi-task VGG models. In addition, around 85% of the queries have more than one relevant recipe. We recruited a homemaker, who has cooking experience, to manually pick the relevant recipes for each of the 1,000 queries. The homemaker was

---

[1] https://www.xiachufang.com.

instructed to label relevant recipes based on title similarity in recipes, titles that are named differently because of geography regions or sharing almost the same cooking procedure with similar key ingredients. For example, the dish "sauteed tofu in hot and spicy sauce" is sometimes called as "mapo tofu" in the restaurant menu. In the extreme case, some queries have more than 60 relevant recipes. On average each query has 9 number of relevant recipes. Note that the testing queries are designed in these ways so as to verify the two major claims in this paper, i.e., the degree in which the learnt model can generalize to unseen food categories (Sect. 4.4) and the capability in finding the best-matched recipe (Sect. 4.5).

4.3 Performance Comparison

We compared SAN to both shallow and deep models for cross-modal retrieval as following. The inputs to these models are the deep ingredient feature ($fc7$) of multi-task VGG model and the ingredient vector of 5,990 dimensions. The $Pool5$ feature is not used due to high dimensionality ($14 \times 14 \times 512$). As reported in [29], simply concatenating the features from 14×14 grids performs worse than $fc7$ in visual recognition.

- Canonical Correlation Analysis (CCA) [22]: CCA is a classic way of learning latent subspace between two views or features by maximizing the correlation between them. Two linear mapping functions are learnt for projections of features into subspace.
- Partial Least Squares (PLS) [23]: Similar to CCA, PLS learns two linear mapping functions between two views. Instead of using cosine similarity as in CCA, PLS uses dot product as the function for measuring correlation.
- DeViSE [20]: DeViSE is a deep model with two pathways which respectively learn the embedded features of recipe-image pairs to maximize their similarities. Note that, instead of directly using word2vec as in [20], the embedded feature of ingredients is learnt from the training set of our dataset. This is simply because word2vec is learnt from documents such as news corpus [30] and lacks specificity in capturing information peculiar to ingredients. Different from SAN, DeViSE is not designed for attention region localization.
- DeViSE++: We purposely included a variant of DeViSE, which takes the hand-cropped regions of food as input to the deep model. The cropping highlights the target food region and basically removes the background or irrelevant part of food pictures. The aim of using DeViSE++ is to gate the potential improvement over DeViSE when only food region is considered, and more importantly, to justify the merit of SAN in identifying appropriate attention region in comparison to hand-cropped region.

Table 1 lists the results of different approaches. Deep models basically outperform shallow models in terms of recall at the depth of 20 and beyond. In contrast to PLS, which does not perform score normalization, CCA manages to outperform DeViSE in terms of MRR and R@K for $K < 20$. Among all these approaches, the proposed model SAN consistently exhibits the best performance

Table 1. MRR and R@K for recipe retrieval. The best performance is highlighted in bold font.

Method	MRR	R@1	R@5	R@10	R@20	R@40	R@60	R@80	R@100
CCA	0.055	0.023	0.079	0.123	0.182	0.262	0.329	0.371	0.413
PLS	0.032	0.009	0.039	0.073	0.129	0.219	0.284	0.338	0.398
DeViSE	0.049	0.016	0.060	0.108	0.182	0.300	0.391	0.456	0.524
DeViSE++	0.05	0.016	0.059	0.105	0.174	0.307	0.404	0.471	0.531
SAN	**0.115**	**0.048**	**0.161**	**0.249**	**0.364**	**0.508**	**0.601**	**0.671**	**0.730**

across all the measures. Compared to DeViSE, SAN achieves a relative improvement of 130% in MRR and doubles its performance at R@20, which is fairly impressive.

Despite the encouraging performance by SAN, the value of R@1 is only around 0.05. Figure 4 shows some successful and near-miss examples. The first two pictures show query images where all visible ingredients are clearly seen. SAN manages to retrieve the ground-truth recipe at top-1 rank in such cases. In the third example, SAN ranks "grilled salmon" higher than "fried salmon" as the current model does not consider cooking attributes. In addition, SAN overlooks the beef and peanuts which are mixed and partially occluded by salmon, while confused by the ingredients of similar appearance, i.e., caviar and red pepper, bean sprout and basil. The last query image shows an example of how non-visible ingredients, flour in this example, affect the ranking. The flour is used to make the dish into round shape, and this knowledge does not seem to be learnt by SAN.

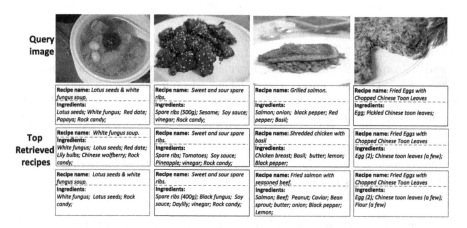

Fig. 4. Examples of top-3 retrieved recipes (ranked from top to bottom). Ground-truth recipe is marked in green. The ingredients in different colours have different meanings: green – true positive, purple – true positive but non-visible in dish, red – false positive. (Color figure online)

Fig. 5. (a) Examples contrasting the manually cropped region (green bounding box), (b) the learnt attention region (masked in white) by SAN. (Color figure online)

Another result worth noticing is that there is no performance difference between DeViSE and DeViSE++. While DeViSE is not designed for attention localization, the model seems to have the ability to exclude irrelevant background regions from recognition. To provide further insights, Fig. 5 shows some examples visualizing the attention regions highlighted by SAN and in contrast to hand-crafted regions. In the first example, the region attended by SAN is about the same as the region manually cropped. In this case, DeViSE+ and SAN use to have similar performance. The next two examples highlight the superiority of SAN in excluding soup and foil as attention regions, which cannot be not easily done by simple region cropping. SAN significantly outperforms DeViSE in such examples. Finally, the last example shows a typical case that SAN only highlights part of dishes as attention. While there is no direct explanation of why certain food regions are ignored by SAN for joint space learning, it seems that SAN has the ability to exclude regions that are vague and hard to be recognized even by human.

4.4 Finding the Best Matches Recipes

Recalled that around 85% of query images have more than one relevant recipe. This section examines the ability of SAN in identifying the best (or ground-truth) recipe from the testing set composed of 5,000 recipes. To provide insights, we select the queries that retrieval at least one relevant recipe (excluding ground-truth recipe) within the top-5 position for analysis. We divide the selected queries into 7 groups based on the number of relevant recipes. Table 2 lists the performance. As can be seen from the table, the difficulty of finding best-match is proportional to the number of relevant recipes. Compared of DeViSE, SAN generally shows better performance for R@1. As the number of recipes increases, they tie in performance. Nevertheless, while looking deeper into the list, SAN consistently outperforms DeViSE in terms of R@5 and R@10. Two main reasons that ground truth recipe are not ranked higher are due to occluded ingredients and use of different non-visible ingredients. Two such examples include the last two pictures in Fig. 4.

Table 2. Performance comparison between SAN and DeViSE in retrieving best-match recipes.

Recipe #	Query #	R@1		R@5		R@10	
		SAN	DeViSE	SAN	DeViSE	SAN	DeViSE
2–3	33	0.21	0.15	0.67	0.48	0.82	0.76
4–7	66	0.18	0.17	0.56	0.53	0.70	0.67
8–11	54	0.17	0.15	0.54	0.30	0.60	0.50
11–15	38	0.13	0.08	0.47	0.39	0.63	0.55
16–30	48	0.06	0.06	0.46	0.39	0.62	0.52
31–61	25	0.08	0.08	0.28	0.26	0.44	0.44

4.5   Generalization to Unknown Categories

Table 3 further shows the performance of SAN to unseen categories. As expected, the performance is not as good as that for the food categories known to SAN and multi-task VGG. When the ingredients of unknown food categories are previously seen and can be correctly identified, SAN performs satisfactorily. In contrast, when some ingredients, especially key ingredients, are unknown, the model will likely fail to retrieval relevant recipes.

Table 3. Generalization of SAN to unseen food categories.

	Query #	MRR	R@1	R@5	R@10	R@20	R@40	R@60	R@80
Known category	554	0.125	0.054	0.175	0.263	0.394	0.535	0.623	0.698
Unknown category	446	0.103	0.04	0.143	0.231	0.327	0.475	0.572	0.637

5   Conclusion

We have presented a deep model for learning the commonality between image and text at the fine-grained ingredient level. The power of model comes from the ability to infer attended regions relevant to ingredients extracted from recipes. This peculiarity enables retrieval of best-match recipes even for unseen food category. The experimental results basically verify our claims that the model can deal with unknown food categories to the extent that at least key ingredients are seen during the training. In addition, SAN exhibits consistently better performance than DeViSE, showing the advantage of fine-grained ingredient analysis at the regional level for best-match recipe retrieval.

The current model can be extended to explicitly model cooking attributes, which could address some limitations identified in the experiments. In addition, as the attention layers couple both visual and text features, the embedding features cannot be offline indexed and have to be generated on-the-fly when the query image is given. This poses limitation on retrieval speed for online application, which is an issue needs to be further researched.

Acknowledgments. This work was supported in part by the National Natural Science Foundation of China (No. 61272290), and the National Hi-Tech Research and Development Program (863 Program) of China under Grant 2014AA015102.

References

1. Meyers, A., Johnston, N., Rathod, V., Korattikara, A., Gorban, A., Silberman, N., Guadarrama, S., Papandreou, G., Huang, J., Murphy, K.P.: Im2calories: towards an automated mobile vision food diary. In: ICCV, pp. 1233–1241 (2015)
2. Bossard, L., Guillaumin, M., Van Gool, L.: Food-101-mining discriminative components with random forests. In: ECCV, pp. 446–461 (2014)
3. Matsuda, Y., Hoashi, H., Yanai, K.: Recognition of multiple-food images by detecting candidate regions. In: ICME (2012)
4. Beijbom, O., Joshi, N., Morris, D., Saponas, S., Khullar, S.: Menu-match: restaurant-specific food logging from images. In: WACV, pp. 844–851 (2015)
5. Kawano, Y., Yanai, K.: Foodcam-256: a large-scale real-time mobile food recognitionsystem employing high-dimensional features and compression of classifier weights. In: ACM MM, pp. 761–762 (2014)
6. Chen, J., Ngo, C.-W.: Deep-based ingredient recognition for cooking recipe retrieval. In: ACM MM (2016)
7. Kitamura, K., Yamasaki, T., Aizawa, K.: Food log by analyzing food images. In: ACM MM, pp. 999–1000 (2008)
8. Aizawa, K., Ogawa, M.: Foodlog: multimedia tool for healthcare applications. IEEE Multimedia **22**(2), 4–8 (2015)
9. Zhang, W., Qian, Y., Siddiquie, B., Divakaran, A., Sawhney, H.: Snap-n-eat: food recognition and nutrition estimation on a smartphone. J. Diab. Sci. Technol. **9**(3), 525–533 (2015)
10. Ruihan, X., Herranz, L., Jiang, S., Wang, S., Song, X., Jain, R.: Geolocalized modeling for dish recognition. TMM **17**(8), 1187–1199 (2015)
11. Probst, Y., Nguyen, D.T., Rollo, M., Li, W.: mhealth diet and nutrition guidance. mHealth (2015)
12. Yang, Z., He, X., Gao, J., Deng, L., Smola, A.: Stacked attention networks for image question answering. arXiv preprint arXiv:1511.02274 (2015)
13. Xie, H., Yu, L., Li, Q.: A hybrid semantic item model for recipe search by example. In: IEEE International Symposium on Multimedia (ISM), pp. 254–259 (2010)
14. Wang, X., Kumar, D., Thome, N., Cord, M., Precioso, F.: Recipe recognition with large multimodal food dataset. In: ICMEW, pp. 1–6 (2015)
15. Su, H., Lin, T.-W., Li, C.-T., Shan, M.-K., Chang, J.: Automatic recipe cuisine classification by ingredients. In: Proceedings of the 2014 ACM International Joint Conference on Pervasive, Ubiquitous Computing, pp. 565–570. Adjunct Publication (2014)

16. Matsunaga, H., Doman, K., Hirayama, T., Ide, I., Deguchi, D., Murase, H.: Tastes and textures estimation of foods based on the analysis of its ingredients list and image. In: Murino, V., Puppo, E., Sona, D., Cristani, M., Sansone, C. (eds.) ICIAP 2015. LNCS, vol. 9281, pp. 326–333. Springer, Heidelberg (2015). doi:10.1007/978-3-319-23222-5_40
17. Maruyama, T., Kawano, Y., Yanai, K.: Real-time mobile recipe recommendation system using food ingredient recognition. In: Proceedings of the ACM International Workshop on Interactive Multimedia on Mobile and Portable Devices, pp. 27–34 (2012)
18. Yamakata, Y., Imahori, S., Maeta, H., Mori, S.: A method for extracting major workflow composed of ingredients, tools and actions from cooking procedural text. In: 8th Workshop on Multimediafor Cooking and Eating Activities (2016)
19. Rasiwasia, N., Pereira, J.C., Coviello, E., Doyle, G., Lanckriet, G.R.G., Levy, R., Vasconcelos, N.: A new approach to cross-modal multimedia retrieval. In: ACM MM, pp. 251–260 (2010)
20. Frome, A., Corrado, G.S., Shlens, J., Bengio, S., Dean, J., Mikolov, T. et al.: Devise: a deep visual-semantic embedding model. In: NIPS, pp. 2121–2129 (2013)
21. Karpathy, A., Joulin, A., Li, F.F.: Deep fragment embeddings for bidirectional image sentence mapping. In: NIPS, pp. 1889–1897 (2014)
22. Hardoon, D.R., Szedmak, S., Shawe-Taylor, J.: Canonical correlation analysis: an overview with application to learning methods. Neural Comput. 16(12), 2639–2664 (2004)
23. Rosipal, R., Krämer, N.: Overview and recent advances in partial least squares. In: Saunders, C., Grobelnik, M., Gunn, S., Shawe-Taylor, J. (eds.) SLSFS 2005. LNCS, vol. 3940, pp. 34–51. Springer, Heidelberg (2006). doi:10.1007/11752790_2
24. Gong, Y., Ke, Q., Isard, M., Lazebnik, S.: A multi-view embedding space for modeling internet images, tags, and their semantics. IJCV 106(2), 210–233 (2014)
25. Andrew, G., Arora, R., Bilmes, J.A., Livescu, K.: Deep canonical correlation analysis. In: ICML, pp. 1247–1255 (2013)
26. Yan, F., Mikolajczyk, K.: Deep correlation for matching images and text. In: CVPR, pp. 3441–3450 (2015)
27. Girshick, R., Donahue, J., Darrell, T., Malik, J.: Rich feature hierarchies for accurate object detection and semantic segmentation. In: CVPR, pp. 580–587 (2014)
28. Srivastava, N., Hinton, G.E., Krizhevsky, A., Sutskever, I., Salakhutdinov, R.: Dropout: a simple way to prevent neural networks from overfitting. J. Mach. Learn. Res. 15(1), 1929–1958 (2014)
29. Donahue, J., Jia, Y., Vinyals, O., Hoffman, J., Zhang, N., Tzeng, E., Darrell, T.: Decaf: a deep convolutional activation feature for generic visual recognition. In: ICML, pp. 647–655 (2014)
30. Mikolov, T., Dean, J.: Distributed representations of words and phrases and their compositionality. In: NIPS (2013)

Deep Learning Based Intelligent Basketball Arena with Energy Image

Wu Liu[1], Jiangyu Liu[2], Xiaoyan Gu[3(✉)], Kun Liu[1], Xiaowei Dai[2],
and Huadong Ma[1]

[1] Beijing Key Laboratory of Intelligent Telecommunications
Software and Multimedia, Beijing University of Posts and Telecommunications,
Beijing 100876, China
{liuwu,liu_kun,mhd}@bupt.edu.cn
[2] Zepp Labs, Inc., Beijing 100080, China
{jiangyu,xiaowei}@zepplabs.com
[3] Information of Information Engineering,
Chinese Academic of Science, Beijing 100093, China
guxiaoyan@iie.ac.cn

Abstract. With the development of computer vision and artificial intelligence technologies, the "Intelligent Arena" is becoming one of the new-emerging applications and research topics. Different from conventional sports video highlight detection, the intelligent playground can supply real-time and automatic sport video broadcast, highlight video generation, and sport technological analysis. In this paper, we have proposed a deep learning based intelligent basketball arena system to automatically broadcast the basketball match. First of all, with multiple cameras around the playground, the proposed system can automatically select the best camera to supply real-time high-quality broadcast. Furthermore, with basketball energy image and deep conventional neural network, we can accurately capture the scoring clips as the highlight video clips to supply the wonderful actions replay and online sharing. Finally, evaluations on a built real-world basketball match dataset demonstrate that the proposed system can obtain 94.59% accuracy with only 45 ms processing time (i.e., 10 ms live camera selection, 30 ms hotspot area detection, and 5 ms BEI+CNN) for each frame. As the outstanding performance, the proposed system has already been integrated into the commercial intelligent basketball arena applications.

Keywords: Intelligent Arena · Multiple cameras · Basketball energy image · Deep conventional neural network

1 Introduction

With the development of computer vision, mobile network, and artificial intelligence technology, the "Intelligent Arena" becomes one of the new-emerging applications and research topics. The intelligent arena technologies investigate

© Springer International Publishing AG 2017
L. Amsaleg et al. (Eds.): MMM 2017, Part I, LNCS 10132, pp. 601–613, 2017.
DOI: 10.1007/978-3-319-51811-4_49

to integrate internet, software, and hardware technology with sport arena to supply real-time, automatic, and intelligent sport video broadcast, highlight video generation, and professional sport technological analysis. In practice, with multi-cameras around the playground, the video highlight detection and analysis technologies can replace program directors to select the live camera, detect the hotspots, and generate the highlight video clips. With intelligent arena, the ordinary sport fans can broadcast their match without high cost. Moreover, they can also real-time share their hotspot sport videos on the social network, which can attract more people to appreciate their outstanding Performance and improve the fun of sports.

However, the intelligent arena also faces many great challenges. Firstly, with multi-camera around the playground, the system must real-time select the live camera and generate the highlight clips. More important, as shown in Fig. 1, different from professional sports venue, the common playgrounds are often outdoor and built with small interval, which means that one camera may capture more than one match at the same time. Therefore, the cameras may record multiple players, auditions, and cheers from different matches. It is hard to discriminate the main playground from the others. Moreover, as the audiences just stand around the playground and can walk around the playground at anytime, effectively differentiating the players from the audiences is another challenge. Finally, the match video analysis is seriously disturbed by the occlusion, lighting, background, and audiences in the playground.

In industry, many existing companies have tried to supply the intelligent arena applications. For example, as shown in Fig. 1, the "Intelligent Arena" mobile application[1] tries to supply real-time match broadcast for every ordinary player, which mainly depends on the multi professional cameras around the playground to realize automatic follow capture of each player. Differently, the "Beikantai" application supplied by "ZEPP" sports[2] tries to broadcast sport match with the mobile cameras of audiences. Moreover, the "Beikantai" can generate the highlight reels with the label of users. However, the existing applications mainly try to supply the match broadcast servers, few of them can automatically generate the highlight videos and analysis the technical details. Besides, "Catapult Sports" and "WSC Sports Technologies" mainly serve the professional sport match and teams.

In the academia, most existed works focus on how to select highlight video clips, which can be classified into four categories. (1) *Audio based methods* [16] — the cheer voices of audiences have been always important clues for video highlight detection. However, in the amateurish match, there is few audiences in most time. In addition, in the crowd playground, it is hard to differentiate the cheers from which match. (2) *Motion based methods* [10] — this kind of methods utilize the global motion feature such as frame difference and optical flow to estimate the highlight level. Nevertheless, these motion features are too coarse to catch the most highlight clips, such as shoot and scoring shots. (3) *Object*

---

[1] "Intelligent Arena," http://www.huiti.com.

[2] "Beikantai," http://www.kantai.tv.

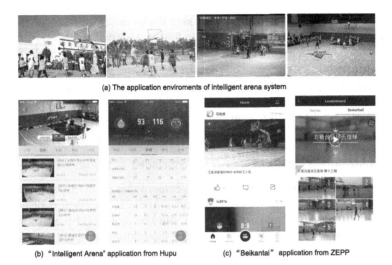

(a) The application enviroments of intelligent arena system

(b) "Intelligent Arena" application from Hupu   (c) "Beikantai" application from ZEPP

Fig. 1. The common match environments of intelligent arena system and existed intelligent arena applications.

and people detection based methods [3,13–15] — these methods try to detect and track the balls or athletes to analysis the match and generate the highlight videos. However, it is very hard to detect the balls and athletes in the outdoor environments. (4) *Search based methods* [2,17,19] — recently, some methods try to search the similar well edited videos from Internet to assist highlight video generation. However, these methods are not only too complicated for intelligent arena, but also not suitable for amateurish match. Unlike the traditional sport video highlight generation, the basketball video highlight generation in intelligent arena system has unique targets: (1) with multi-cameras installed around the playground, the method must automatically select the camera that best records the match in real-time to supply high-quality broadcast; (2) the proposed system must automatically detect the highlight clips such as shooting, scoring, and quick attack with the selected camera; and (3) the system can summarize a long basketball match video into variable-length short highlight clips, which can be quickly downloaded and shared by users.

In this paper, we investigate the deep learning based intelligent basketball arena system with energy image, which can automatically broadcast the basketball match, generate the highlight videos, and detect the scoring in real-time. First of all, the pyramid histogram of gradients (HOG) features are employed as motion features to automatically select the camera that best records the match to implement the real-time match broadcast. Next, to generate the highlight videos, we utilize the aggregate channel features (ACF) to detect the backboard and basket areas as the hotspot area. With the hotspot area, we can decrease most of the disturbance in the background. More important, in the hotspot area, the proposed system can generate the basketball energy image (BEI) which records the spatial-temporal track of basketballs to accurately detect the shoot and scoring moment with deep convolutional neural network (CNN). Finally, the highlight videos are generated from the scoring clips. To comprehensively evaluate the proposed system, we collect 20 h real-world basketball match videos and compare with the state-of-the-art methods and human. The results show that with similar recall, the proposed method can achieve better precision than human. Moreover, 12 volunteers are invited to evaluate the attractiveness of the generated highlight videos, which demonstrate the effectiveness and efficient of the propose method.

The contribution of this paper can be concluded as follows:

- We have designed an innovative multi-camera based intelligent basketball arena system that towards real-time and automatic sports broadcast, highlight video generation, and score statistics in the real-world playground.
- We proposed a basketball energy images generation method to record the basketball's spatial-temporal track, which is integrated with deep learning method to detect the scoring clip.
- As the high accuracy (96.55%) and speed (45 ms/frame), the proposed method will be integrated into "Beikantai" to supply intelligent basketball arena.

2   System Implementation

2.1   System Overview

The overview of our proposed intelligent basketball arena system is shown in Fig. 2. (1) We set up multi-cameras around the basketball arena to capture the entire match process, which can capture all the details of the match. (2) From the multi-cameras, the system must select one as the broadcast camera. For the balance of accuracy and speed, we extract the pyramid HoG from the differential image as motion features. Then the support vector machine (SVM) model is trained to predict the highlight score. The camera with the highest highlight score is selected as the broadcast camera. With the game progresses, the selected cameras will be smoothly switched to keep catching the most highlight scenes for the automatic match broadcast. (3) Moreover, for each selected camera, the ACF detector is employed to detect the basketball backboard, hoop and basket as the hotspot area, which can decrease the disturbance of the unrelated motion in the background. (4) With the continuous hotspot areas, we can generate the BEIs which records the spatial-temporal tracks of basketball. (5) With BEIs, we train the CNN model to detect the scoring moment. The proposed system can instantly generate the highlight video clips with scoring moments. (6) Finally, the basketball match can be real-time broadcasted and shard on the Internet. We will describe each component in detail.

2.2   Live Camera Selection

With multi-cameras around the arena, the intelligent arena system needs to automatically control them, and determine which view to be

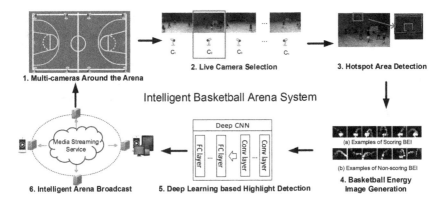

Fig. 2. The framework of the proposed intelligent basketball arena system.

broadcasted [1,5]. As the accuracy and speed requirements, we choose to extract the light-weight and robust features. After receiving videos from all the cameras, the system extracts the video frames, and resizes them into a fixed scale (smaller than the original size) for decreasing the noise and accelerating the calculation. Then the graying and Gaussian smooth operations are employed to reduce the influence from the record environment. After that, for the sequential frames from one camera, the system applies the image subtraction technique to extract the differential images. Next, the differential image is divided into 3×3, 6×6, and 12×12 equal-sized blocks. Then the system calculates the motion ratio in each individual block and entire frame for representing the motion intensity in the basketball match. The motion ratio is calculated as the proportion of the motion pixels (differential value larger than 25) to the number of entire pixels. The 189 dimensional motion intensity features represent the intensity of the motion.

In general, the more intense the motion is, it is more possible to capture a more wonderful content in this camera. However, in some special cases, such as fast break and backdoor cut, it may have a small motion ratio. To avoid missing these special motions, we need to analyze the moving direction and speed in specific moving directions to select the best camera. Besides, the moving direction and speed can also eliminate the influence from the background audiences. In our system, we extract the Pyramid HOG feature from the differential image to describe the moving direction information. In the implementation, we extract the three level Gaussian pyramid of the differential image, and extract 8,586 dimensions HOG features. In summary, we totally extract 8,775 dimensions motion features, which contain 189 dimensions motion intensity features and 8,586 dimensions motion orientation features.

Finally, we employ the LibSVM to learn the camera selection model. Our system does not depend on one fixed learning method, any efficient supervised learning method can be easily integrated into the system. In the training, we select 104,065 video frames as the training data, which contains 52,137 positive (highlight) frames and 51,928 negative (non-highlight) frames, then apply the cross-check method to obtain the optimal parameters. In practical applications, for each frame we will use the learned model to calculate its highlight score.

If in ten frames, one camera has more than seven frames with the highest score, the model will automatically select this camera as the match broadcasting camera. If there are multiple such cameras, the system will automatically select the optimal one which has the most positive frames. In our servers, it totally needs less than 10 ms to select the camera for broadcast.

2.3   Basketball Highlight Video Generation

Hotspot Area Detection. After selecting the camera that best records the match, we will detect the scoring clips as the candidate highlight clips. However, as the fast and diverse body interactive actions among multi-players and complicated occlusion, lighting, and background situation, it is hard to detect the player who is shooting or scoring in the playground. Instead, we will detect the basketball backboard, hoop, and basket as the hotspot area because its importance for scoring and non-interfering from other motion. In the implementation, the backboard can be manually marked in the system setup stage. However, as the cameras may move during the match, we design a hotspot area detection method with ACF detector [4]. We employ the ACF detector as its good balance between the efficiency and object detection. Given the video frame I, the ACF compute 10 channels (e.g., HoG, LUV color channels) from the image, and sum every block of pixels in each channel. Then the channels are smoothed and final features are single pixel lookups in the aggregated channels. Next Boosting is used to train decision trees with these features to distinguish object from background. In our task, we label the areas of basketball backboard, hoop, and basket as the positive example, and then use the ACF implementation in Piotr's Computer Vision Matlab Toolbox [4] to train the ACF detector. Finally, the AdaBoost [6] is trained to detect the basketball backboard, hoop, and basket areas. In our servers, it totally needs less than 30 ms to detect the hotspot area for one frame.

Basketball Energy Image Generation. After obtained the hotspot area, it is much easier to detect the shooting action through the motion ratio, because basketball is the maximum likelihood to cause obvious motion in the hotspot area. However, it is still very hard to distinguish the scoring from the shooting, especially only according to one frame. After obtaining the backboard area, we found that a scoring ball would finish its track passing through the basketball hoop. Differently, a non-scoring ball would end its track at the edge of the backboard area. Hence recoding the balls motion track can identify whether this shooting is scoring. Moreover, the extracted motion track must not only record the track that the ball passes the backboard, but also contain the time information to judge whether it passes through the hoop from up. For this target, we propose a spatial-temporal track record method named BEI. After detecting the hotspot area, the BEI image is computed as the Algorithm 1:

Algorithm 1. *BEI generation process*

Require: the sequence of video frames's hotspot area H.
Ensure: basketball energy image BEI.
 1: $BEI \leftarrow$ an empty image whose size equals H
 2: $s \leftarrow 0$, $n \leftarrow 0$, $\tau_1 \leftarrow 0.012$, $\tau_2 \leftarrow 3$
 3: **while** each $H_{i>0}$ in H and $n < \tau_2$ **do**
 4:     calculate the difference $D_i = H_i - H_{i-1}$
 5:     calculate the motion ratio M_i in D_i
 6:     **if** $M_i > \tau_1$ **then**
 7:         $s \leftarrow s + 1$
 8:         for each pixel $P_{D_i}(x,y) > 0$, set $P_{BEI}(x,y) = s$
 9:     **else**
10:         $n \leftarrow n + 1$
11:     **end if**
12: **end while**
13: **for** each $P_{BEI}(x,y)$ in BEI **do**
14:     $P_{BEI}(x,y) \leftarrow P_{BEI}(x,y) \times 255/s$
15: **end for**
16: **return** BEI

---

(1) For two consecutive frames, we calculate the frame difference image in the hotspot area. After smooth, we can obtain the motion ratio. If the r is larger than a threshold (e.g., 0.012 in our experiments), the frame is regarded as a candidate shooting frame.

(2) For the consecutive candidate shooting video frames, we fuse all the difference images of hotspot in one image. In this image, we can obtain the track of the basketball.

(3) However, only with the motion track, it is hard to distinguish whether the basketball flies through the hoop or just across it. So temporal information is also needed as reference. Therefore, we normalize the motion track with temporal information. When generating the BEI, we set the pixel value $P_{BEI}(x,y)$ by

$$P_{BEI}(x,y) = t_i/T * 255, \tag{1}$$

where T represents the total length of the shooting video clip, t_i is the motion generated time.

As shown in Fig. 3, in the BEI, except the zero pixels, the darkest part in the track represents the entering area, and the brightest part is the leaving area. As a result, BEI records the balls' spatial-temporal track information, which indicates the ball's entering place, passing area, and leaving locations. As examples shown in Fig. 3, the balls' spatial-temporal information have obvious difference between the scoring and non-scoring shooting. After obtaining the BEI, we can directly train a CNN model to effectively identify the scoring. In our implementation, we utilize the AlexNet network with four convolutional layers and two full connection layers [7]. The parameters are trained on 862 real-world basketball match clips with stochastic gradient descent implemented in Caffe.

(1) The examples of scoring BEI

(2) The examples of non-scoring BEI

Fig. 3. The examples of scoring and non-scoring BEI.

Finally, with the scoring possibility, we can rank the video clips and select the top N clips to generate the highlight clips. The N can be determined by the length requirement of the highlight videos. With the highlight videos, users can timely share their wonderful performance on the social media to attract more attention.

3   Evaluations

To comprehensive evaluate the performance of the intelligent basketball arena, we build a comprehensive dataset which contains 20 h real-world match videos. The videos are all real basketball matches recorded by different cameras around the arenas in Beijing and Guangzhou. The videos contain 6 different arenas include indoor, outdoor, multi-arenas, and so on. Some examples of the video can be found in Fig. 1.

3.1   Scoring Detection Evaluation

To evaluate the proposed BEI based scoring detection method, we invited ten volunteers to label 476 scoring clips and 571 shooting but non-scoring clips. In our experiments, the number of train and test videos is 862 and 185 respectively. We have compared our method with four different methods as follow:

(1) **Single Frame based CNN Method (CNN)** [9]. Instead of BEI, this method directly trains the CNN model with single frame. For each shooting clip, it detects all the frames and judges whether scoring through the late fusion.
(2) **Optical Flow based Method (OF)** [11]. Instead of BEI, this method extracts the optical flow from the shooting clips in the hotspot area as features, and trains the SVM model to detect the scoring.
(3) **3-D CNN based Method (C3D)** [18]. Instead of BEI, this method directly uses the clips as the input of C3D, which trains a 3-D CNN to predict the scoring.

(4) **BEI + Pyramid HOG + SVM (BEI+SVM).** This method also uses the proposed BEI. Differently, instead of CNN, it extracts the motion and Pyramid HOG features described in Sect. 2.2 and trains a SVM model to predict the scoring.

(5) **The proposed method (BEI+CNN).** The proposed method in this paper, which employs the BEI as input, and trains a deep CNN model to predict the scoring.

Table 1. Comparison of different scoring detection methods on real-world basketball match videos.

Methods	Accuracy	Precision	Recall	F1	Time(ms/frame)
CNN [9]	83.24%	95.45%	69.32%	80.31%	40
OF [11]	64.35%	68.06%	60.78%	64.21%	8
C3D [18]	90.27%	98.66%	81.31%	89.15%	12
BEI+SVM	92.86%	92.23%	93.60%	92.91%	5
BEI+CNN	**94.59%**	**96.55%**	**92.31%**	**94.38%**	**5**

We employ the accuracy, precision, recall, and F1 value to evaluate the overall performance of the five different methods. The results are shown in Table 1. From the results, we can find that as only using spatial or motion information respectively, the CNN and OF methods cannot well detect the scoring clips. Nonetheless, we still find that the spatial information is more useful than the motion feature as the scoring ball must pass through the hoop and basket. Furthermore, although the C3D method can fuse the spatial and motion information in the 3D deep neural network, the performance of C3D is worse than the proposed BEI based method. The reason is that sufficiently training the 3D deep neural network needs massive data to mine the spatial-temporal pattern in scoring clip. Differently, the BEI method can well solve the data limitation problem. That is, the BEI fuses the spatial-temporal trajectory of basketball in one image to help deep neural network quickly capture the discriminative information in scoring clips. Therefore, the proposed BEI+CNN method achieves the best performance, which demonstrates our conclusions. Finally, the BEI+CNN is better than BEI+SVM because the CNN can automatically learn commendable features from the training dataset. Admittedly, the gap between them is small, which further demonstrates the effectiveness of the proposed BEI.

For the time complexity, we test the performance of the proposed method on a server with Xeon E5-2660v3 2.6 GHz CPU, four NVIDIA TESLA K80 GPU cards, 256G DDR4 memory, and 1T SSD disk. From the average process time for each frame, we can find that the proposed method uses the least time — only 5 ms per frame. Added with the camera selection time (i.e., 10 ms) and hotspot area detection time (i.e., 30 ms), it totally needs 45 ms to generate the highlight videos, which means that the proposed method can process the sport video in real-time.

3.2 Subjective Evaluation of Highlight Video Generation

As the highlight video generation is a very subjective task, we randomly invited 12 users to judge the performance of three different highlight video generation methods: (1) interestingness-driven summary method (IDS) [8]; (2) the pairwise deep ranking based method (PDR) [20]; and (3) the proposed scoring based method (BEI). The subjects include three female and nine male company staff and college students, with ages ranging from 22 to 36. They are all basketball fans and often watch the professional basketball match broadcast like National Basketball Association (NBA). In addition, eight of them usually play the basketball in the public arena. From the interview, we find that all the subjects think the proposed intelligent basketball arena is very cool when they first see it. They all want to use the real-time broadcast function to live their match and share their highlight moment on the social media. After that, the subjects are asked to accomplish the following tasks.

- Task 1. The subjects are asked to watch 185 test video clips, which contain 84 scoring and 101 non-scoring clips, to judge whether is scoring. All test videos clips only contain the shooting segment extracted from the 20 h match videos. Therefore, all clips are less than 15 s length. To simulate the real-world scene, each subject only has one chance to watch the videos, who cannot slow or repeat them. The sequences of the videos are given randomly. Then we can get a human baseline for basketball scoring detection.
- Task 2. All subjects are asked to watch eight basketball matches. Each video is nearly 10 min length. Then they will watch three highlight videos generated from PRD, IDS, and BEI respectively. The review process is random order and blindly. The generated videos have same length. Then, a questionnaire was given to each subject to evaluate the (1) coverage: which summaries better covers the progress of the video; (2) presentation: which summaries better distills and presents the essence of the video, and (3) attractiveness: which is more attractive. 1~5 indicate the worst to the best level [12].

For Task 1, from the results shown in Table 2, we can find that if only watching the basketball match one time, the humans cannot accurately find all the scoring shots. Especially, the precision of the proposed BEI based method is higher than HUMAN. From the false examples, we find that the human cannot well differentiate the swish (scoring) and air ball (non-scoring). The proposed method can distinguish the two situations with the shake of hook and basket. Nonetheless, the proposed method miss more scoring shots than human. Therefore, we need to further decrease the miss cases.

For Task 2, the quantitative evaluation of user satisfaction scores with these generated highlight videos is listed in Table 3. This indicates the advantages of BEI over the other two highlight video generation applications. First of all, we find that the users would like to watch the scoring clips than others, which demonstrate our motivation. Moreover, the proposed method not only accurately detects the scoring, but also selects the camera which captures the best scene of the shooting. Therefore, the subjects give 4.4/5.0 for its attractiveness.

Table 2. Comparison of scoring detection performance with human and the proposed method on real-world basketball match videos.

Methods	Accuracy	Precision	Recall	F1	Time(ms/frame)
HUMAN	97.25%	95.92%	98.24%	97.07%	37
BEI+CNN	94.59%	96.55%	92.31%	94.38%	5

Table 3. A summary of the subjective evaluation of highlight video generation: (a) IDS, (b) PDR, and (c) BEI. 1~5 indicate the worst to the best level.

ID	Question	IDS [8]	PDR [20]	BEI (proposed)
1	Coverage	3.3	3.6	3.9
2	Presentation	3.1	3.6	4.1
3	Attractiveness	3.2	3.8	4.4

Differently, as the other two methods are seriously disturbed by the occlusion, lighting, background, and audiences in the playground, they cannot well generate the highlight videos. Besides, most of the subjects give a positive response when they are asked whether they will use this application and recommend it to their friends. Moreover, the subjects provide comments, such as "detect the steal and block shot as highlight clips," "add sharing search results function," and so on.

4   Conclusion

In this paper, we propose a deep learning based intelligent basketball arena system with energy image, which can automatically broadcast the basketball match, generate the highlight videos, and detect the scoring in the real-world amateur basketball match. To real-time broadcast the match with multiple cameras around the playground, the system extracts Pyramid HOG as motion features to choose the camera which captures the most wonderful scenes. Then a ACF detector based method is implemented to detect the backboard, hoop, and basket as the hotspot area. In the hotspot area, the proposed system can generate the BEI, which records the spatial-temporal track of basketballs to accurately detect the shoot and scoring moment with deep CNN. Finally, the highlight videos are generated from the scoring clips. The comprehensively evaluations demonstrate that the proposed method can significantly outperform the state-of-the-art methods. Moreover, the highlight videos generated by the proposed method obtains the highest scores in the subjective evaluation. In particular, the proposed system will be integrated into the commercial products, which can be further evaluated by the massive sports fans. In the future, we will try to detect the steal and block shots as highlight clips, and add more technical-tactics analysis in our system.

Acknowledgement. This work is partially supported by the National High Technology Research and Development Program of China (2014AA015101), the National Natural Science Foundation of China (No. 61602049), the Funds for Creative Research Groups of China (No. 61421061), the Beijing Training Project for the Leading Talents in S&T (ljrc 201502), and the Fundamental Research Funds for the Central University (No. 2016RC43).

References

1. Chen, C., Wang, O., Heinzle, S., Carr, P., Smolic, A., Gross, M.: Computational sports broadcasting: automated director assistance for live sports. In: IEEE ICME, pp. 1–6 (2013)
2. Chu, L., Jiang, S., Wang, S., Zhang, Y., Huang, Q.: Robust spatial consistency graph model for partial duplicate image retrieval. IEEE Trans. Multimedia **15**(8), 1982–1996 (2013)
3. Chu, L., Wang, S., Liu, S., Huang, Q., Pei, J.: Alid: scalable dominant cluster detection. Proc. VLDB Endowment **8**(8), 826–837 (2015)
4. Dollár, P.: Piotr's Computer Vision Matlab Toolbox (PMT). https://github.com/pdollar/toolbox
5. Foote, E., Carr, P., Lucey, P., Sheikh, Y., Matthews, I.: One-man-band: a touch screen interface for producing live multi-camera sports broadcasts. In: ACM Multimedia, pp. 163–172 (2013)
6. Friedman, J., Hastie, T., Tibshirani, R., et al.: Additive logistic regression: a statistical view of boosting. Ann. Stat. **28**(2), 337–407 (2000)
7. Gan, C., Wang, N., Yang, Y., Yeung, D., Hauptmann, A.G.: Devnet: a deep event network for multimedia event detection and evidence recounting. In: IEEE CVPR, pp. 2568–2577 (2015)
8. Gygli, M., Grabner, H., Riemenschneider, H., Gool, L.: Creating summaries from user videos. In: Fleet, D., Pajdla, T., Schiele, B., Tuytelaars, T. (eds.) ECCV 2014. LNCS, vol. 8695, pp. 505–520. Springer, Heidelberg (2014). doi:10.1007/978-3-319-10584-0_33
9. Krizhevsky, A., Sutskever, I., Hinton, G.E.: Imagenet classification with deep convolutional neural networks. In: NIPS, pp. 1106–1114 (2012)
10. Liu, A., Xu, N., Su, Y., Lin, H., Hao, T., Yang, Z.: Single/multi-view human action recognition via regularized multi-task learning. Neurocomputing **151**, 544–553 (2015)
11. Liu, C.: Beyond pixels: exploring new representations and applications for motion analysis. Ph.D. thesis, Cambridge, MA, USA (2009)
12. Liu, W., Mei, T., Zhang, Y.: Instant mobile video search with layered audio-video indexing and progressive transmission. IEEE Trans. Multimedia **16**(8), 2242–2255 (2014)
13. Liu, W., Mei, T., Zhang, Y., Che, C., Luo, J.: Multi-task deep visual-semantic embedding for video thumbnail selection. In: IEEE CVPR, pp. 3707–3715 (2015)
14. Lucey, P., Bialkowski, A., Carr, P., Morgan, S., Matthews, I.A., Sheikh, Y.: Representing and discovering adversarial team behaviors using player roles. In: IEEE CVPR, pp. 2706–2713 (2013)
15. Maksai, A., Wang, X., Fua, P.: What Players do with the ball: a physically constrained interaction modeling. In: IEEE CVPR (2016)
16. Oldfield, R., Shirley, B., Cullen, N.: Demo paper: audio object extraction for live sports broadcast. In: IEEE ICME Workshop, pp. 1–2 (2013)

17. Sun, M., Farhadi, A., Seitz, S.: Ranking domain-specific highlights by analyzing edited videos. In: Fleet, D., Pajdla, T., Schiele, B., Tuytelaars, T. (eds.) ECCV 2014. LNCS, vol. 8689, pp. 787–802. Springer, Heidelberg (2014). doi:10.1007/978-3-319-10590-1_51

18. Tran, D., Bourdev, L., Fergus, R., Torresani, L., Paluri, M.: Learning spatiotemporal features with 3D convolutional networks. In: IEEE ICCV, pp. 4489 4497 (2015)

19. Yang, H., Wang, B., Lin, S., Wipf, D., Guo, M., Guo, B.: Unsupervised extraction of video highlights via robust recurrent auto-encoders. In: IEEE ICCV, pp. 4633–4641 (2015)

20. Yao, T., Mei, T., Rui, Y.: Highlight detection with pairwise deep ranking for first-person video summarization. In: IEEE CVPR, pp. 982–990 (2016)

Efficient Multi-scale Plane Extraction
Based RGBD Video Segmentation

Hong Liu$^{(\boxtimes)}$, Jun Wang, Xiangdong Wang, and Yueliang Qian

Key Laboratory of Intelligent Information Processing and Beijing Key
Laboratory of Mobile Computing and Pervasive Device, Institute of Computing
Technology, Chinese Academy of Sciences, Beijing 100190, China
{hliu,wangjun,xdwang,ylqian}@ict.ac.cn

Abstract. To improve the robustness and efficiency of RGBD video segmentation, we propose a novel video segmentation method combining multi-scale plane extraction and hierarchical graph-based video segmentation. Firstly, to reduce depth data noise, we extract plane structures of 3D RGBD point clouds in three levels including voxel, pixel and neighborhood with geometry and color features. To solve uneven distribution of depth data and object occlusion problem, we further propose multi-scale voxel based plane fusion algorithm and use amodal completion strategy to improve plane extraction performance. Then hierarchical graph-based RGBD video segmentation is used to segment the rest of the non-plane pixels. Finally, we fuse above plane extraction and video segmentation results to get final RGBD video scene segmentation results. The qualitative and quantitative results of plane extraction and RGBD scene video segmentation show the effectiveness of proposed methods.

Keywords: Plane extraction · RGBD · Video segmentation · Amodal completion

1 Introduction

Segmentation is a basic process for object detection and scene understanding. Besides image segmentation, video segmentation has quick progress recently, which aims to group pixels with similar appearance and spatiotemporal continuity in video volume [1–6]. Super-voxel based video segmentation methods are widely adopted. Xu and Corso [7] surveyed super-voxel video segmentation into five categories including segmentation by weighted aggregation [8], graph-based [9], hierarchical graph-based [11], mean shift [1], and Nystrom normalized cuts [12]. These methods mostly focus on RGB video, which lack depth information of scene.

RGBD videos came from Kinect sensor can provide both RGB and depth image sequences of front scene, which can conveniently obtain 3D RGBD point clouds. Steven et al. [13] extended the hierarchical graph-based approach [9] from RGB to RGBD video. This method firstly segment consecutive frames using the depth and motion information, then an over-segmentation of the frames is done while respecting depth boundaries. Histograms of the resulting regions are used to build a hierarchical segmentation of the spatiotemporal volume, which can yield a particular segmentation depending on the

© Springer International Publishing AG 2017
L. Amsaleg et al. (Eds.): MMM 2017, Part I, LNCS 10132, pp. 614–625, 2017.
DOI: 10.1007/978-3-319-51811-4_50

desired segmentation level output. The final step performs a bipartite graph matching with overlapping frames to enforce the consistency of region identities over time. But this method relies on initial segment of depth data, which will result over segmentation in clutter scenes, especially for big plane structures with large depth distance.

RGBD videos are usually captured in indoor scene. In man-made indoor environment, there exist many planes which contain much structural information. Extracting these planes could be very helpful in RGBD scene segmentation. To improve the robustness and effectiveness, we proposed a novel RGBD video segmentation framework based on multi-scale voxel based plane extraction. We first extract plane regions in the 3D RGBD point cloud and segment the rest non-plane pixels using hierarchical graph-based approach [13]. Finally we fusion the above segmentation results to get the final video segmentation results.

There are many plane extraction methods. In order to make use of 3D data, Holz et al. [17] computed local surface normal of point clouds using integral images. And then the points were clustered, segmented, and classified in both normal space and spherical coordinates. Dube and Zell [18] used Randomized Hough Transformation to extract planes from depth images. This algorithm could real-time run on a mobile platform. However it can only detect planes, and cannot segments out the planes. Wang et al. [14] proposed a two-step plane segmentation algorithm which combines the speed of voxel-wise cluster and the accuracy of pixel-wise process. But this method only used single voxel size, which is not robust to uneven depth noise from depth sensor. They further proposed a multi-scale plane extraction method [15]. This method used two sizes of voxel, small size for near region and large size for far region of 3D point cloud, which used different voxel size for different regions. But due to the noise of depth data, there are several outlier depth points in segmented plane areas. Another problem is the occlusion of objects. In occlusion situation some plane regions will be divided into several plane regions, such as when a large wall region is occluded by a table. And the above methods only use depth information to segmentation plane area, which don't make full use of color information to improve the robustness of segmentation.

This paper proposes a robust multi-scale voxel based plane extraction method on 3D RGBD point clouds. To solve the outlier points problem of segmentation, initial planes are extracted in three levels including voxel, pixel and neighborhood with geometry and color features to improve the Wang's method [14]. To solve uneven distribution of depth data problem, we propose multi-scale voxel based plane segmentation algorithm on the whole 3D point cloud and fusion the results using depth and color features. To solve the occlusion problem, we further use amodal completion strategies based on RGBD on the plane segmentation results to combine the divided plane regions that belong to one plane to improve plane extraction performance.

We summarize our contributions as followings:

Firstly, we propose a robust multi-scale voxel based plane extraction algorithm on 3D RGBD point clouds to reduce the influence of depth noise and object occlusion for plane segmentation. Secondly, we propose a RGBD video segmentation framework, which uses hierarchical graph-based video segmentation method to segment the rest of non-plane pixels and fuse the results of plane extraction to obtain the final segmentation results. The qualitative and quantitative results of plane extraction and scene segmentation show the effectiveness of proposed methods.

2 Our Framework

The framework of our whole system is shown in Fig. 1. At first, the input RGBD data is preprocessed. The position of the RGB camera and the depth camera are not the same, so they should be aligned to the same camera coordinate. The second step is multi-scale voxel plane segmentation. Firstly, the 3D RGBD point cloud is divided into 3D voxels with two different scales. Then three-level plane extraction method is used to segment plane regions on above two different voxel grids respectively. Then multi-scale voxel fusion algorithm is used to merge above plane segmentation results using color and depth information. Finally, amodal completion is used to solve the problem of occlusion problem. The third step is to segment the rest non-plane regions using hierarchical graph-based video segmentation method using RGBD data. The final step is to fuse the results of plane segmentation and non-plane segmentation results to get the final RGBD video segmentation results.

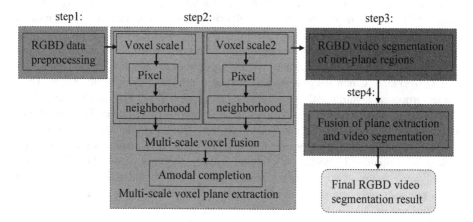

Fig. 1. The framework of our system

3 Our Main Work

3.1 Multi-scale Voxel Based Plane Extraction (MPE)

Plane Extraction Based on Three Levels

Wang et al. [14] proposed a two-stage plane segmentation algorithm on 3D point clouds including voxel level and pixel level. But Wang's method only used depth data, which may bring many isolated noise pixels in each detected plane for the noise or missing of depth data. We propose a three-level plane segmentation method to improve Wang's method as step 2 in Fig. 1 shows. Our first two levels are similar to Wang's two stages. To achieve sparse sampling and speed up plane extraction, 3D point cloud are firstly divided into uniform voxels and least square method is used to fit a plane of each voxel. Then region growing algorithm is used to cluster the voxels of plane. At pixel level, accurate judgment is done for each pixel in un-clustered voxels based on the distance from the point to the nearest plane.

In the third level, we project the results of above two stages back into 2D image and fuse those non-plane pixels to certain plane based on the differences of color feature within 8 neighborhood pixels. And we use adaptive threshold strategy according to the distance of objects.

$$P = \sum_{i=1}^{8} a_i \tag{1}$$

$$a_i = \begin{cases} 1 & if \ |R_{a_i} - R_c| + |G_{a_i} - G_c| + |B_{a_i} - B_c| < th1 \\ 0 & otherwise \end{cases} \tag{2}$$

Where, $R_c, G_c, B_c, R_{a_i}, G_{a_i}, B_{a_i}$ represent three RGB channels of current pixel and the no i th neighborhood pixel. $th1$ is the threshold of pixel difference. If P is larger than threshold $th2$, this pixel will be added into the neighborhood plane structure.

This three-level strategy makes use of color and scene geometry features, which reduces mangy isolated noise pixels within extracted planes compared with Wang's results.

Multi-scale Voxel Based Plane Combination Algorithm

The quality of the 3D point cloud data varies according to the distance from object to camera. The data near the sensor will have less noise and higher resolution and the data far away have more noise and lower resolution. We further use a multi-scale voxel strategy to solve this problem. Different from the method [15], they used different voxel size on the different 3D point cloud part. We built a dense voxel grid with short size and a sparse voxel grid with large size on the whole 3D point cloud. Then above three-levels plane extraction is used for each voxel grid. Finally, we use multi features, including difference of color histogram, surface normal direction, space covering rate and spatial distribution, to merge overlapping plane areas from above two results of voxel grid.

The details of fusion algorithm are shown in following algorithm. We use small voxel scale C_1 and large voxel scale C_2 to divide 3D point cloud and use the three-level plane extraction method based on RGBD to extract plane regions respectively. We norm the plane extraction results from small scale as r_1 and large scale as r_2. For the results from small scale contains more small plane areas, so we make r_1 as the priority plane extraction results and make r_2 as the auxiliary plane results. For each plane region $P2_j$ in r_2, we build a queue Q_j to merge. For each plane region $P1_i$ in r_1, we test the similarity difference with each plane region in r_2 according to formulation (3). To combine these two results, we proposed a similarity difference function including color histogram difference, the difference of surface normal direction, space covering differences, spatial distribution, etc., as formulation (3) shows.

$$SD = \omega_1 * Dif_{cSAD} + \omega_2 * Dif_{cos} + \omega_3 Dif_{cr} + \omega_4 Dif_{cent} \tag{3}$$

Where, ω_1, ω_2, ω_3, ω_4 are weights for different features. Dif_{cSAD} is color histogram statistic difference, Dif_{cos} is cosine similarity difference, Dif_{cr} is the plane cover differences in 3D point cloud and Dif_{cent} is the difference of planar centroid position.

If $P1_i$ and $P2_j$ have smallest similarity difference, we will add $P1_i$ into the merge queue Q_j of $P2_j$. When we get the total queue to merge, we further calculate the similarity difference between each two planes in the queue and merge the planes with small SD threshold.

Because the noise of depth data are different according to the distance between the object and the camera, so fixed parameters and threshold are difficult to get robust plane segmentation results for the whole 3D scene. This paper used the adaptive sliding adjustment to get the weight parameters as following:

$$\omega_i = \frac{curD - minD}{maxD - minD} \omega_i \ or \ \left(1 - \frac{curD - minD}{maxD - minD}\right)\omega_i \tag{4}$$

Where ω_i is the weight of different factors in formula (3), $curD$ is the distance between the center of current plane region and the camera, $minD$ and $maxD$ are the minimum distance and maximum distance between the scene objects and the camera. The details of fusion algorithm are shown in following Algorithm 1.

Algorithm 1: Multi-scale Plane Combination

for $\forall P2_j \in r_2$
  build list Q_j
for $\forall P1_i \in r_1$
  for $\forall P2_j \in r_2$
    get SD$(P1_i, P2_j)$
    if SD$(P1_i, P2_j)$ >maxSD
      maxSD = SD$(P1_i, P2_j)$, index = j
  if maxSD >th1
    add $P1_i$ to Q_{index}
for $\forall Q_j$
  for $\forall P1 \in Q_j$ & $P2 \in Q_j$
    get SD$(P1, P2)$
    if SD$(P1, P2) >$ th4
      merge$(P1, P2)$

Amodal Completion (AC)

In complex indoor scenes, some plane structures are usually occluded by objects. Figure 2 gives some plane extraction results in two frames from one RGBD indoor scene (scene_1) compared with Wang's method [14]. As the first column of Fig. 2(b) and (c) show the wall is segmented into several different plane structures for the blocking of table legs. We adopt Amodal Completion (AC) [10] to further merge plane segmentation results using the differences of color consistency, normal direction, location distance as shown in following formula.

$$SD = \omega_1 * Dif_{cSAD} + \omega_2 * Dif_{cos} + \omega_3 Dif_{dst} \qquad (5)$$

Figure 2(d) shows our plane extraction results using multi-scale voxel and Amodal Completion strategy, which can combine the wall region as a whole plane and improve the final plane extraction performance. And the results of Fig. 2(b) also shows the small voxel scale will bring much more small planes, which maybe over segmentation such as wall parts. Figure 2(c) shows the large voxel scale may bring loss plane detection, such as the plane regions of box in the right corner and the table regions with some small object, which will influence the plane feature in certain voxels. And our

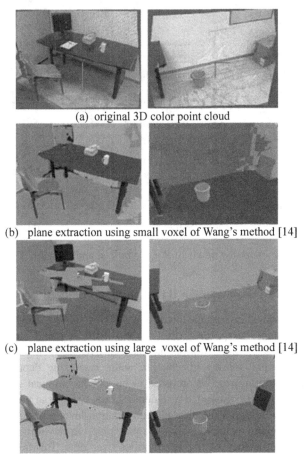

(a) original 3D color point cloud

(b) plane extraction using small voxel of Wang's method [14]

(c) plane extraction using large voxel of Wang's method [14]

(d) plane extraction by our multi-scale voxel based plane extraction (MPE)

Fig. 2. Some plane extraction results of Wang's method and our method in scene_1 (Color figure online)

multi-scale voxel fusion can make use of the result from difference voxel scales and final improve the performance of plane extraction results as Fig. 2(d) shows.

Figure 3 also shows plane extraction results in other two indoor scenes, which shows the effectiveness of your multi-scale voxel segmentation and amodal completion.

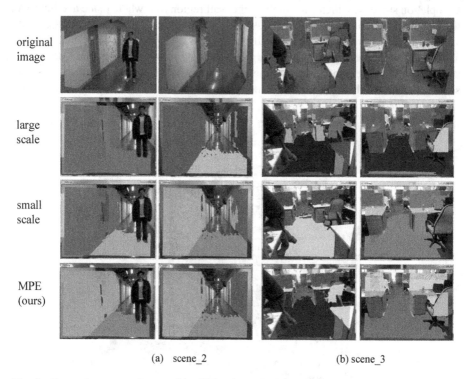

original
image

large
scale

small
scale

MPE
(ours)

(a) scene_2   (b) scene_3

Fig. 3. Some plane extraction results of Wang's method and our method (Color figure online)

3.2 RGBD Video Segmentation Based on Plane Extraction

Steven [13] presented a scalable algorithm for segmenting RGBD videos using a hierarchical graph-based approach. this method firstly segment 8 consecutive frames using the depth and motion information, then an over-segmentation of the frames is done using color and motion information while respecting depth boundaries. Histograms of the resulting regions are used to build a hierarchical segmentation of the spatiotemporal volume represented as a dendrogram, which can then yield a particular segmentation depending on the desired segmentation level output. The final step performs a bipartite graph matching with the 8 previous frames with 4 frames overlapping to enforce the consistency of region identities over time. But this method relies on initial segment of depth data, which will over segmentation in clutter scenes, especially for big plane structures with large depth distance.

In our RGBD video segmentation system, we first extract plane structures in scene and then adopt Steven's method to divide the non-plane pixels. After we get the video segmentation of non-plane points, we combine the results of plane extraction and non-plane video segmentation to get the final video segmentation results. We select adjacent overlapping regions as candidate for merging. Here difference of color histogram, center distance and overlap proportion of these regions are used to determine whether to merge.

4   Experimental Results

To test our proposed method, this paper use Microsoft Kinect to capture RGBD videos from several indoor scenes. The resolution of RGB and Depth image is 640 * 480 pixels. In our experiment, we developed a wearable device and fix the RGBD camera on the belt of the backpack in front of chest [19]. Using this device, we capture multiple RGBD video sequences in front of the user while walking in office, table tennis room and corridor scenes.

We designed two experiments to evaluate the effectiveness of our proposed method with the state of the art methods. (1) Multi-scale voxel plane extraction (MPE) in plane level compared with Wang's method [14]. (2) RGBD video segmentation based on plane extraction and video segmentation compared with Steven's method [13]. The detailed results are as followings.

4.1   Quantitative Analysis of Plane Extraction

In order to analyze the accuracy of proposed multi-scale plane extraction based on visual and geometrical features, we design a quantitative comparison test. We select one RGBD video (scene_1) from office scene with 270 RGB and depth images. We use LabelMe tool of MIT [16] to manually label one frame pre 10 on RGB video as Fig. 4 shows. Each frame contains 4 to 10 plane areas and there are total 136 labeled planar regions for test video. If one plane is divided into several regions as occlusion, we will label them as one plane with same label number. We compare our plane extraction

Fig. 4. Some labeled samples using LabelMe tool (Color figure online)

method with Wang's method using three commonly used metrics including recall rate R, accuracy P and F1 value.

We match the detected planes with labeled planes. If the ratio of overlapping area is greater than 70%, this detected plane is correct. The quantitative results of plane extraction from our proposed multi-scale method and Wang's method are shown as Table 1 and Fig. 5. Wang's method with small voxel scale (W_small, 10 cm) will get more small planar blocks. And it's easy to split a large plane into different planar structures with the noisy depth data and lead to false positives. So the accuracy P is the lowest as 61.0%. Wang's method with large voxel scale (W_large, 25 cm) has low recall rate R as 63.9% for mistaking a part of planes as non planar areas, especially for small planar structures, such as display, box. And our multi-scale voxel plane extraction method (MPE) fuses results of small scale (10 cm) and large scale (25 cm) based on the visual and geometric features. Our multi-scale plane fusing strategy can improve the F1 value from 0.726 (W_small) and 0.699 (W_large) to 0.895, which achieves about 17−20% improvement compared with Wang's method.

Table 1. Results of plane extraction

	P (%)	R (%)	F1
W_Small [14]	61.0	89.7	0.726
W_Large [14]	77.0	63.9	0.699
MPE (ours)	85.3	94.1	0.895

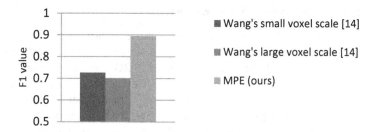

Fig. 5. Quantitative comparison of plane extraction from Wang's method and our MPE (Color figure online)

The experimental results show that the proposed method can combine the advantages of both small and large scales to improve the recall rate, and amodal completion strategy can reduce over segmentation and false positives causes by noise and occlusion. The qualitative comparison results of plane extraction are shown in Figs. 2 and 3.

4.2   RGBD Video Segmentation and Analysis Based on Plane Extraction

To evaluate the performance of proposed RGBD video segmentation based on plane extraction, we compare our method with Steven's method [13].

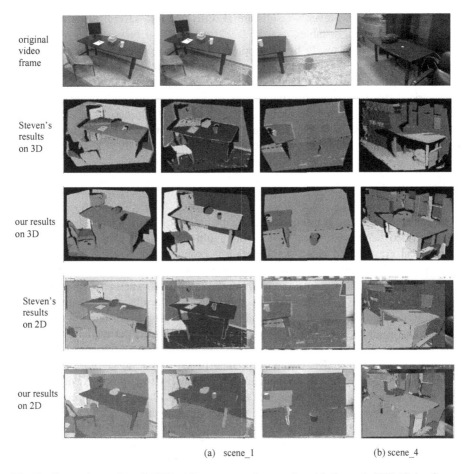

original
video
frame

Steven's
results
on 3D

our results
on 3D

Steven's
results
on 2D

our results
on 2D

(a) scene_1   (b) scene_4

Fig. 6. Comparison of our RGBD video segmentation results with Steven's [13] (Color figure online)

In this experiment, we select two RGBD video sequences from different office scenes. Figure 6 shows the results of RGBD video segmentation and different area labeled with different color. The first three columns of Fig. 6 show the results from scene_1 and the last column from scene_4. The first line is the original video frames. The second and the third lines are the video segmentation results on 3D point cloud and the last two lines show the video segmentation results from 3D back to 2D image. The second and the forth lines are came from Steven's method and the third and the fifth lines are results of our methods.

From the results, we can see the desktop and ground regions were divided into several areas and the segmented areas have some isolated noise pixels from Steven's method. This maybe because Steven's video segmentation method relies on the initial segmentation results of depth data and they didn't consider the influence of uneven distribution of depth data, so bring over segmentation and not robust to depth noise.

And the wall areas in Steven's results are also divided into several plane blocks for object occlusion, such as tables.

Figure 6 shows our results by the proposed RGBD video segmentation method can segment the whole desktop and the wall as a whole plane region, which is more robust and efficient than Steven's method. Our proposed approach both fuses visual and geometrical structure characteristic of multi-scale voxel plane extraction and use adaptive threshold strategy according to the distance of objects. This strategy can effectively solve the inconsistency problem of depth data distribution and amodal completion strategy can improve the occlusion problem.

5 Conclusions

This paper focuses on the robust video segmentation of RGBD sequences. A novel RGBD video segmentation method fusing multi-scale voxel plane extraction and hierarchical graph-based video segmentation is proposed. Firstly, a three-level plane extraction method is used to segment plane regions on voxel, pixel and neighborhood levels to reduce isolated points and improve the robustness. Then a multi-scale voxel fusion algorithm is proposed to merge plane segmentation results of different voxel size with color and geometric features. To further solve the problem of occlusion, amodal completion is adopted to improve the performance of plane extraction. Secondly, hierarchical graph-based video segmentation method is used to segment the rest non-plane regions on RGBD data and the final fusion is used to get the final RGBD video segmentation results. Experimental results show that the proposed method can effectively improve the robustness and effectiveness of plane extraction and RGBD video segmentation compared with the state of the art method.

Acknowledgments. This work is supported in part by Beijing Natural Science Foundation (4142051) and National Key Technology R&D Program of China (2014BAK15B02).

References

1. Paris, S., Durand, F.: A topological approach to hierarchical segmentation using mean shift. In: CVPR (2007)
2. Shi, J., Malik, J.: Normalized cuts and image segmentation. TPAMI **22**(8), 888–905 (2000)
3. Sharon, E., Galun, M., Sharon, D., Basri, R., Brandt, A.: Hierarchy and adaptivity in segmenting visual scenes. Nature **442**(7104), 810–813 (2006)
4. Felzenszwalb, P.F., Huttenlocher, D.P.: Efficient graph-based image segmentation. IJCV **2** (59), 167–181 (2004)
5. Stuckler, J., Behnke, S.: Efficient dense rigid-body motion segmentation and estimation in RGB-D video. IJCV **113**(3), 233–245 (2015)
6. Song, J.K., Gao, L.L., Pusca M.M., et al.: Joint graph learning and video segmentation via multiple cues and topology calibration. In: ACM MM (2016)
7. Xu, C., Corso, J.J.: Evaluation of super-voxel methods for early video processing. In: CVPR (2012)

8. Corso, J.J., Sharon, E., et al.: Efficient multilevel brain tumor segmentation with integrated Bayesian model classification. IEEE Trans. Med. Imaging **27**(5), 629–640 (2008)
9. Felzenszwalb, P.F., Huttenlocher, D.P.: Efficient graph-based image segmentation. IJCV **59**(2), 167–181 (2004)
10. Gupta, S., Arbeláez, P., Malik, J.: Perceptual organization and recognition of indoor scenes from RGB-D images. In: CVPR (2013)
11. Grundmann, M., Kwatra, V., et al.: Efficient hierarchical graph-based video segmentation. In: CVPR (2010)
12. Fowlkes, C., Belongie, S., et al.: Spectral grouping using the Nystrom method. TPAMI **26**(2), 214–225 (2004)
13. Steven, H., Stan B., et al.: Efficient hierarchical graph-based segmentation of RGBD videos. In: CVPR (2014)
14. Wang, Z., Liu, H., Qian, Y., Xu, T.: Real-time plane segmentation and obstacle detection of 3D point clouds for indoor scenes. In: Fusiello, A., Murino, V., Cucchiara, R. (eds.) ECCV 2012. LNCS, vol. 7584, pp. 22–31. Springer, Heidelberg (2012). doi:10.1007/978-3-642-33868-7_3
15. Wang, Z., Liu, H., Wang, X.D., Qian, Y.L.: Segment and label indoor scene based on RGB-D for the visually impaired. In: MMM (2014)
16. labelme.csail.mit.edu
17. Holz, D., Holzer, S., Rusu, R.B., Behnke, S.: Real-time plane segmentation using RGB-D cameras. In: Röfer, T., Mayer, N.,Michael, Savage, J., Saranlı, U. (eds.) RoboCup 2011. LNCS (LNAI), vol. 7416, pp. 306–317. Springer, Heidelberg (2012). doi:10.1007/978-3-642-32060-6_26
18. Dube, D., Zell, A.: Real-time plane extraction from depth images with the randomized Hough transform. In: ICCV Workshops (2011)
19. Liu, H., Wang, J., Qian, Y. L., Wang, X.D.: iSee: obstacle detection and feedback system for the blind. In: UbiComp (2015)

Human Pose Tracking Using Online Latent Structured Support Vector Machine

Kai-Lung Hua[1], Irawati Nurmala Sari[1], and Mei-Chen Yeh[2]($\boxtimes$)

[1] Department of Computer Science and Information Engineering,
National Taiwan University of Science and Technology, Taipei City, Taiwan
[2] Department of Computer Science and Information Engineering,
National Taiwan Normal University, Taipei City, Taiwan
myeh@csie.ntnu.edu.tw

Abstract. Tracking human poses in a video is a challenging problem and has numerous applications. The task is particularly difficult in realistic scenes because of several intrinsic and extrinsic factors, including complicated and fast movements, occlusions and lighting changes. We propose an online learning approach for tracking human poses using latent structured Support Vector Machine (SVM). The first frame in a video is used for training, in which body parts are initialized by users and tracking models are learned using latent structured SVM. The models are updated for each subsequent frame in the video sequence. To solve the occlusion problem, we formulate a Prize-Collecting Steiner tree (PCST) problem and use a branch-and-cut algorithm to refine the detection of body parts. Experiments using several challenging videos demonstrate that the proposed method outperforms two state-of-the-art methods.

Keywords: Human pose tracking · Latent structured SVM · Online learning · Body parts

1 Introduction

Human pose tracking is important in a variety of domains, including human-computer interaction, human activity recognition, video surveillance, gaming, and medical imaging. The task involves estimating the positions of the main body components in a video sequence. Human pose tracking in realistic scenes is challenging, considering the person under tracking may vary in viewpoints and the movement of body parts can be fast and complicated. For example, limbs can move speedily, creating a wide range of motions.

Recent approaches use online learning techniques to deal with these problems. In [8], Lim et al. proposed an online learning approach to detect a human body by foreground/background segmentation and track the human pose from videos captured by moving cameras. The algorithm is based on a sequential bayesian filtering process. This method is robust to background clutter, similar colors between background and human body parts, pose and illumination changes. However, the method can not handle cases when occlusions occur in the tracking process.

© Springer International Publishing AG 2017
L. Amsaleg et al. (Eds.): MMM 2017, Part I, LNCS 10132, pp. 626–637, 2017.
DOI: 10.1007/978-3-319-51811-4_51

Huang et al. addressed the occlusion problem by building multiple reference models for key poses of the tracked person [5]. Though the authors overcame the situation when the occlusions were closely touching the subject, they encountered new problems when a human body was fully covered for few consecutive frames.

The N-BMD method proposed in [10] generates multiple candidates of human poses in an image by using a part-based model. Then, the best human pose that does not overlap with other human poses is identified. In particular, the authors use dynamic programming to obtain the best human pose in an image. This method also involves a simple greedy algorithm for instantiating multiple candidates of human poses and an iterative process including the following steps to obtain the best human pose in an image: searching over the large space of human pose samples, instantiating a part-based model for obtaining human pose candidates, computing the single best human pose by dynamic programming, removing all human poses that overlap with the best human pose. The method has two disadvantages. First, part-based models are not sufficiently robust to cluttered background when the models are applied for tracking. Moreover, the method can neither deal with the cases when two poses partially overlap; therefore, it is not effective for tracking human poses for multiple people.

In this paper, we propose a robust human pose tracking method using online latent structured SVM. An exemplar outcome of the proposed method is shown in Fig. 1. We consider a body part contains a few latent parts, and use both information to obtain a more accurate localization of body parts. Using both latent and body parts handles in general the occurrence of significant pose changes and complicated movements. Furthermore, we employ an adaptive search radius during tracking to address the wide range of body part motion. To address the occlusion problem, we consider a body part as a connected subgraph with the maximal detection score. This is achieved by applying branch-and-cut that solves the PCST problem, which can be used accordingly for addressing the maximum-weight connected subgraph (MWCS) problem. The proposed method can be used to re-identify human poses, even though the previous frames contain fully occluded body parts. Only the first image is used as the training image, human poses in the subsequent frames are automatically tracked. Moreover, the detected human parts are consequently used to enhance our models.

The use of latent parts is particularly well-suited for human pose tracking because of the following reason. Latent parts provide additional information for obtaining more accurate location of body part. Even though only one or a few (not all) latent parts are detected, they are helpful for the detection of body parts. In other words, our model does not rely on a single region to trigger a positive body part classification. Instead, it scores a region using the whole region (body part) and its corresponding four sub-regions (latent parts), and associates with each sub-region a latent variable that indicates whether the sub-region represents the target body part. Because each latent part model is trained independently, the detection of latent part can be easier in comparison with the whole body part when a significant pose change occurs in the video sequence.

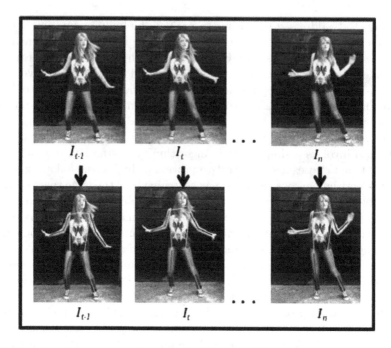

Fig. 1. Results of the proposed approach. Given a video sequence capturing one or more moving people and a target person, our approach tracks the skeletons of the target.

In the remainder of this paper, we present the technical details of the human pose tracking approach powered by structured SVM. The approach consists of both a training and a testing procedure, and a mechanism of updating the model parameters on the fly. We finally validate the effectiveness of this approach by conducting experiments on several challenging video clips, and conclude the paper with a short discussion.

2   Related Work

Several human pose tracking techniques have been developed in the literature [5,8,11,12]. Although those approaches have been shown to be effective for some videos, certain pre-processing steps are essential and required before training the tracking models.

A few works successfully detect human body parts, including [4,13,16]. However, these methods are limited to processing only static images. The latent structured SVM is used in [6] for 3D human pose tracking. Our approach differs in that it focuses on 2D human pose tracking and can detect and track human body parts in images.

In terms of the design of tracking model, our approach is most similar to the part-based visual tracking method [17] that can be used to create models for general objects. We opt for humans and improve [17] for human pose tracking.

In summary, the proposed method has several merits described as follows; other approaches, in general, do not have them all. The model is adaptive, online, capable of detecting body parts more robustly (because of the use of latent parts), and handles occlusions well (by formulating a PCST problem for that case).

3   Approach

As in existing tracking applications we use the first frame as a training image providing ground truth data of body parts, and track the human poses from the second frame to the end of the input video. The latent structured SVM is used to train a model that distinguishes positive and negative samples. During tracking, the target body part location is estimated by searching for the maximum classification score in the vicinity around the estimate from the previous frame, where the classification score consists of the scores of body parts and latent parts. To address the occlusion problem, we further refine body part detection by using branch-and-cut. We describe the technical details in the subsections.

3.1   Training

Given an input video $V = \{I_1, I_2, I_3, ...I_n\}$ with n frames, the training process involves only the first frame I_1, along with the ground truth data of body parts B_{GT}^i. A body part is represented by a bounding box $B^i = (C^i, R^i, W, H)$, where C^i, R^i are the x and y coordinates of the upper-left corner and W, H are the width and height of the bounding box. Each body part B_k^i is divided into four latent parts $b_l^{i,j}$, as shown in Fig. 2. Each latent part is again described by a bounding box $b^{i,j} = (c^{i,j}, r^{i,j}, w, h)$.

After we have body and latent part samples, we extract the Haar-like and histogram of oriented gradients (HOG) features from them. More specifically, the following local features are computed: $\phi_1(I_1, b^{i,j})$ representing the appearance of the i-th body part and j-th latent part, $\phi_2(I_1, B_k^i)$ represents the appearance of the i-th body part, and $\phi_3(B_k^i, b_l^{i,j})$ reflects the compatibility between the body and latent part.

Next, we use latent structured SVM for learning the tracking models. Latent SVM has been used previously [3,18,19] for offline learning alone. We extend latent SVM for an online learning setting, consisting of a four-stage training process: model of latent part $(u^{i,j})$, model of body part (v^i), model of correlation between body and latent parts $(v^{i,j})$, and model of whole body (w^i).

Let the label cost $\Delta(B_{GT}^i, B_k^i)$ denote the dis-similarity between the ground truth B_{GT}^i and body part samples B_k^i, and it is computed as:

$$\Delta(B_{GT}^i, B_k^i) = 1 - \frac{(B_{GT}^i \cap B_k^i)}{(B_{GT}^i \cup B_k^i)}. \tag{1}$$

Similarly, the label cost $\Delta(b_{GT}^{i,j}, b_l^{i,j})$ measures the dis-similarity between the ground truth $b_{GT}^{i,j}$ and latent part samples $b_l^{i,j}$ and is computed in the way similar to Eq. 1. These label costs are used to compute the bounding box overlap ratio introduced in [2]. A smaller label cost value implies two similar bounding boxes.

Finally, we train four models as follows:

1. Latent part model $(u^{i,j})$:

$$\operatorname*{argmin}_{u^{i,j}} \frac{\lambda}{2} \left\| u^{i,j} \right\|^2 + \frac{1}{M} \sum_{l=1}^{M} \left[(\Delta b_{GT}^{i,j}, b_l^{i,j}) + \phi_1(I_1, b_l^{i,j}) - \phi_1(I_1, b_{GT}^{i,j}) \right] \quad (2)$$

2. Body part model (v^i):

$$\operatorname*{argmin}_{v^i} \frac{\lambda}{2} \left\| v^i \right\|^2 + \frac{1}{N} \sum_{k=1}^{N} \left[(\Delta B_{GT}^i, B_k^i) + \phi_2(I_1, B_k^i) - \phi_2(I_1, B_{GT}^i) \right] \quad (3)$$

3. Correlation model between body and latent part $(v^{i,j})$.

$$\operatorname*{argmin}_{v^{i,j}} \frac{\lambda}{2} \left\| v^{i,j} \right\|^2 + \frac{1}{N} \sum_{k=1}^{N} \left[(\Delta B_{GT}^i, B_k^i) + \phi_3(B_k^i, b_l^{i,j}) - \phi_3(B_{GT}^i, b_{GT}^{i,j}) \right] \quad (4)$$

4. Model of final body part (w^i).

$$\operatorname*{argmin}_{w^i} \frac{\lambda}{2} \left\| w^i \right\|^2 + \frac{1}{N} \sum_{k=1}^{N} \left[(\Delta B_{GT}^i, B_k^i) + \max_{B_k^i} f(I_1, B_k^i, b_l^{i,j}) - \max_{B_{GT}^i} f(I_1, B_{GT}^i, b_{GT}^{i,j}) \right] \quad (5)$$

where,

$$f(I_1, B_k^i, b_l^{i,j}) = \sum_{j=1}^{J} \left[\phi_1(I_1, b_l^{i,j}) \right] + \left[\phi_2(I_1, B_k^i) \right] + \sum_{j=1}^{J} \left[\phi_3(B_k^i, b_l^{i,j}) \right] \quad (6)$$

$$f(I_1, B_{GT}^i, b_{GT}^{i,j}) = \sum_{j=1}^{J} \left[\phi_1(I_1, b_{GT}^{i,j}) \right] + \left[\phi_2(I_1, B_{GT}^i) \right] + \sum_{j=1}^{J} \left[\phi_3(B_{GT}^i, b_{GT}^{i,j}) \right]. \quad (7)$$

3.2 Testing

The testing process involves applying the four models on the second frame I_2 to the end frame I_n. We start with the body part locations previously tracked and search for the bounding boxes with the best score using a sliding window approach inside a predefined search radius. For example, the best latent part $b_t^{i,j} = b_{t,l}^{i,j} argmax \left[u_t^{i,j}, \phi_1(I_t, b_{t,l}^{i,j}) \right]$. Similarly, we search for the best score of body part and correlation by using the corresponding features, and compute the final body part as follows.

$$\mathrm{BP}_t^i = \operatorname*{argmax}_{B_{t,k}^i} \left[\sum_{j=1}^{J} \operatorname*{argmax}_{b_{t,l}^{i,j}} \left\langle \left[u_t^{i,j}, \phi_1(I_t, b_{t,l}^{i,j}) \right] + \left[v_t^{i,j}, \phi_3(B_{t,k}^i, b_{t,l}^{i,j}) \right] \right\rangle + \left[v_t^i, (\phi_2(I_t, B_{t,k}^i)) \right] \right]$$
$$(8)$$

Once we determine the final body part in the current frame, the new scores are fed into the model as inputs. The four models are updated accordingly.

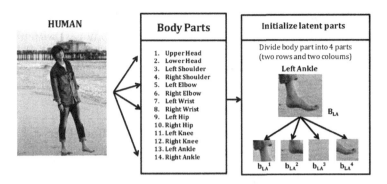

Fig. 2. Representation of a human body. Left: input image. Center: body parts. Right: latent parts

3.3 Features

This work uses the Haar-like [14] and HOG features [1] to describe the appearance of body and latent parts. A simple rectangular Haar-like feature is computed as the difference of the sum of pixels of areas inside the rectangle, which can be at any position and scale within the image. In particular, we use 2 types of Haar-like features, including 2- and 3-rectangle features. For HOG feature, each image is partitioned into 5×5 pixel blocks, from which the HOG feature is extracted.

3.4 Handling the Occlusions

Interacting with other objects in a video often interferes human pose tracking if the human body under tracking is occluded by objects. Therefore, we use a post-processing step to refine the result of body part detection.

We formulate a prize-collecting Steiner tree (PCST) problem for localizing the best-scoring region representing concise body parts. We over-segment each body part into several sub-regions by using superpixel segmentation [7]. Next, a graph is constructed in which each sub-region corresponds to a vertex and two vertices are connected if two sub-regions are adjacent. The vertex score is obtained by feeding the sub-region into our structured SVM model.

To obtain the best-scoring region, a branch-and-cut algorithm [9] is applied to efficiently solve the PCST problem. This method localizes the best-scoring region by identifying all possible subsets of connected vertices, and the summed profit score of the vertices are maximal. After we obtain best-scoring regions, they are used to update the structured SVM model. We consider the best-scoring regions in the current image as positive samples, which contains several sub-regions, and other sub-regions are used as negative samples.

Next, we compute the similarity score between the m-last tracked body parts and body part in the current image. The Bhattacharyya coefficient [15] is used to compute the score. Then, the average similarity score is used to determine if a body part is absent (due to occlusions) in the current frame. If it is occluded, we increase the radius of sampler and track again this body part within a larger search area. If the body part is still missing, we stop tracking this body part in the current frame, and reset the radius of sampler for processing the next frame.

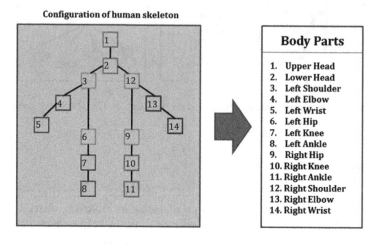

Fig. 3. Configuration of a human body skeleton.

Finally, we have collected all the bounding boxes of body parts. The precise body part (including joint position) location is defined as the center point of the body part box. Our configuration of a human skeleton contains 14 body parts (i.e., 14 bounding boxes, including knees, elbows, and shoulders) and 13 edges connecting them, demonstrated in Fig. 3. The skeleton is automatically built by connecting the centers of the detected body parts.

4   Experimental Results

4.1   Setup

We compare the proposed approach with two state-of-the-art methods–N-BMD [10] and APE [16]. All of the methods can be used to generate human skeletons from images. Five challenging videos—*girl, pitching, boy, two people,* and *siska*—were used in the experiments, each containing a few combinations of the following factors: complicated movements, a rotating body, illumination changes, similar color between body parts and background, occlusion and cluttered background. The test videos were collected as follows. The video *pitching* was used in [10]; *girl, boy, two people* and *siska* were selected and downloaded from YouTube[1] in order to evaluate the performance of multiple people and serious occlusion conditions. Snapshots of each test video are shown in Fig. 5.

---

[1] https://www.youtube.com.

We used two metrics—probability of correct pose (PCP) and hausdorff distance (HD)—to evaluate the tracking performance. The PCP criteria used in [10] was calculated based on how localized body parts overlapped ground-truth data. We used strict PCP, in which a body part is considered correct if both of its segment endpoints lie within 50% of the length of the annotated endpoints (ground truth). HD measured how a skeleton was correctly localized. The HD criteria measures the distance of two skeletons represented by sets of 2D points. A smaller HD value indicates a better tracking performance.

In the experiments we used a search radius $r = 30$ in the tracking process. The search radius of body part tracker and latent part tracker are same for all videos.

4.2   Results

The tracking performances of APE [16], N-BMD [10] and the proposed approach are summarized in Tables 1 and 2, showing the body part and skeleton localization accuracies respectively. The proposed method performed the best among the methods under comparison. In particular, APE [16] tracked a human pose by detection, and did not consider the temporal continuity in adjacent frames. N-BMD [10] is a part-based model that often fails when cluttered background and complicated movements occur, which frequently appear in *girl* and *pitching*. The proposed method using the latent structured SVM technique is more effective because more information were used to obtain accurate location of body parts, providing additional robustness to cluttered background and complicated movements. In terms of the part representation, APE [16] used HOG feature alone and N-BMD [10] used raw pixel values, both are not robust to illumination changes (e.g., a challenge in *boy*). The proposed method used two types of features–Haar-like and HOG features. This combination can handle the illumination problem well.

The test video *two people* captures two people with similar poses. Since APE [16] estimates human pose in the static images and does not consider the temporal continuity in adjacent frames, the identified people and the corresponding skeleton are not always targeting the same person. As for N-BMD [10], though this approach considers temporal context from neighboring frames, it does not work well when there are multiple human poses in a frame. The proposed method, however, selects and trains the best pose from the first frame of video, and uses subsequently the detection results for each of the following frames. Therefore, it can successfully centralize the tracking process only for the best pose.

For test video *Siska*, some body parts are covered by an object occasionally. Since both N-BMD [10] and APE [16] methods always return their best solution for all the body parts. When a body part is missing, these two methods suffer the performance degradation. On the other hand, the proposed method addresses this challenging issue and detects the potential body part. It examines its validity via comparing the m-last tracked body part obtained from the previous images.

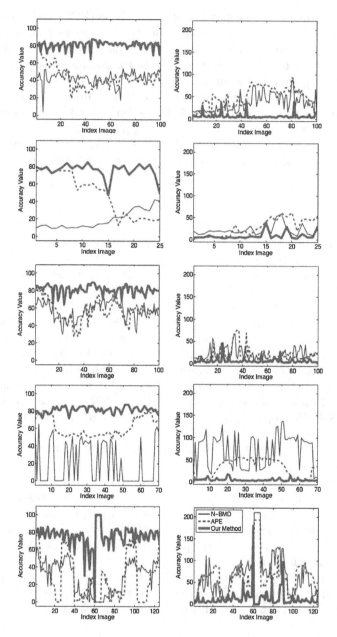

Fig. 4. The figures show the performance comparison using PCP (left) and HD (right) of three methods—N-BMD [10] (blue line), APE [16] (green dashed line), and our method (red thick line). The test videos used in the experiment are (from top to bottom): *girl*, *pitching*, *boy*, *two people*, and *siska*. A higher PCP and a lower HD value indicate a higher tracking accuracy. The proposed method performs the best among the approaches under comparison. (Color figure online)

Fig. 5. Qualitative tracking results of N-BMD [10] (first column), APE [16] (second column), and our method (third column).

Figure 4 displays the frame-level accuracy for both measures of the test videos. The proposed method consistently has a higher PCP and a lower HD than the other approaches. Qualitative tracking results of these methods are displayed in Fig. 5 and the supplementary video. The proposed method can successfully track the human body parts in challenging video sequences containing heavy pose variation, shape deformation, partial occlusion and *etc.*

The improvement margin can be explained threefold. First, the modeling of latent parts, in addition to body parts, delivered more accurate detection results. Using both latent and body parts handles well the occurrence of significant pose changes and complicated movements. Second, we employed an adaptive search radius during tracking to overcome the wide range of body part motion. Third,

Table 1. Body part tracking accuracy in PCP. A larger PCP value indicates a better performance.

No.	Dataset	N-BMD [10]	APE [16]	Our method
1	Girl	42.62	43.16	**80.76**
2	Pitching	19.09	47.99	**75.73**
3	Boy	57.21	57.15	**81.32**
4	Two people	20.53	61.87	**81.69**
5	Siska	29.06	27.91	**75.48**

Table 2. Skeleton tracking accuracy in HD. A smaller HD value indicates a better performance.

No.	Dataset	N-BMD [10]	APE [16]	Our method
1	Girl	35.83	50.21	**6.75**
2	Pitching	21.81	28.32	**9.34**
3	Boy	21.54	20.89	**6.5**
4	Two people	71.23	33.27	**5.52**
5	Siska	69.51	60.82	**13.58**

we address the occlusion problem by formulating a PCST problem for that case. Thus, our method can successfully re-identified human poses, even though the previous frames contained fully occluded body parts.

5 Conclusion

We propose an online learning approach for tracking human poses via latent structured SVM, involving the use of part models describing a person by local body parts, latent parts and their correlations. The models are adaptively adjusted when more frames are processed during tracking. To address the occlusion problem, we formulate body part detection as a PCST problem and solve it by branch-and-cut. We test our tracking system on five challenging videos, each with a combination of difficult conditions, and show that the proposed method outperforms two state-of-the-art methods.

References

1. Dalal, N., Triggs, B.: Histograms of oriented gradients for human detection. In: ICCV, pp. 886–893 (2005)
2. Everingham, M., Van Gool, L., Williams, C.K.I., Winn, J., Zisserman, A.: The PASCAL Visual Object Classes (VOC) challenge. IJCV **88**(2), 303–338 (2010)
3. Felzenszwalb, P.F., Girshick, R.B., McAllester, D., Ramanan, D.: Object detection with discriminatively trained part-based models. TPAMI **32**(9), 1627–1645 (2010)

4. Huang, C.H., Boyer, E., Ilic, S.: Robust human body shape and pose tracking. In: 3DV, pp. 287–294 (2013)
5. Huang, C., Boyer, E., Navab, N., Ilic, S.: Human shape and pose tracking using keyframes. In: CVPR, pp. 3446–3453 (2014)
6. Ionescu, C., Li, F., Sminchisescu, C.: Latent structured models for human pose estimation. In: ICCV, pp. 2220–2227 (2011)
7. Li, Z., Wu, X.M., Chang, S.F.: Segmentation using superpixels: a bipartite graph partitioning approach. In: CVPR, pp. 789–796 (2012)
8. Lim, T., Hong, S., Han, B., Hee Han, J.: Joint segmentation and pose tracking of human in natural videos. In: ICCV (2013)
9. Ljubic, I., Weiskircher, R., Pferschy, U., Klau, G.W., Mutzel, P., Fischetti, M.: An algorithmic framework for the exact solution of the prize-collecting steiner tree problem. Math. Program. **105**(2–3), 427–449 (2006)
10. Park, D., Ramanan, D.: N-best maximal decoders for part models. In: ICCV, pp. 2627–2634 (2011)
11. Ramakrishna, V., Kanade, T., Sheikh, Y.: Tracking human pose by tracking symmetric parts. In: CVPR, pp. 3728–3735 (2013)
12. Tian, J., Li, L., Liu, W.: Multi-scale human pose tracking in 2D monocular images. J. Comput. Commun. **2**(2), 78–84 (2014)
13. Tian, Y., Zitnick, C.L., Narasimhan, S.G.: Exploring the spatial hierarchy of mixture models for human pose estimation. In: Fitzgibbon, A., Lazebnik, S., Perona, P., Sato, Y., Schmid, C. (eds.) ECCV 2012. LNCS, vol. 7576, pp. 256–269. Springer, Heidelberg (2012). doi:10.1007/978-3-642-33715-4_19
14. Viola, P., Jones, M.: Rapid object detection using a boosted cascade of simple features. In: CVPR, pp. 511–518 (2001)
15. Wang, X., Ning, C., Shi, A., Lv, G.: An improved similarity measure in particle filters for robust object tracking. In: CISP, pp. 46–50 (2013)
16. Yang, Y., Ramanan, D.: Articulated pose estimation with flexible mixtures-of-parts. In: CVPR (2011)
17. Yao, R., Shi, Q., Shen, C., Zhang, Y., van den Hengel, A.: Part-based visual tracking with online latent structural learning. In: CVPR (2013)
18. Yu, C.N.J., Joachims, T.: Learning structural svms with latent variables. In: ICML, pp. 1169–1176 (2009)
19. Zhu, L., Chen, Y., Yuille, A.L., Freeman, W.T.: Latent hierarchical structural learning for object detection. In: CVPR, pp. 1062–1069 (2010)

Micro-Expression Recognition by Aggregating Local Spatio-Temporal Patterns

Shiyu Zhang[1,3], Bailan Feng[2], Zhineng Chen[3(✉)],
and Xiangsheng Huang[3]

[1] Beijing Institute of Technology, Beijing, China
signal926@163.com
[2] Shannon Cognitive Computing Laboratory, 2012 Labs,
Huawei Technologies, Co., Ltd, Shenzhen, China
fengbailan@huawei.com
[3] Institute of Automation, Chinese Academy of Sciences, Beijing, China
{zhineng.chen,xiangsheng.huang}@ia.ac.cn

Abstract. Micro-expression is an extremely quick facial expression that reveals people's hidden emotions, which has become one of the most important clues for lies as well as many other applications. Current methods mostly focus on the micro-expression recognition based on the simplified environment. This paper aims at developing a discriminative feature descriptor that are less sensitive to variants in pose, illumination, etc., and thus better implement the recognition task. Our novelty lies in the use of local statistical features from interest regions in which AUs (Action Units) indicate micro-expressions and the combination of these features for the recognition. To this end, we first use a face alignment algorithm to locate the face landmarks in each video frame. The positioned face is then divided to several specific regions (facial cubes) based on the location of the feature points. In the following, the movement tendency and intensity in each region are extracted using optical flow orientation histogram and Local Binary Patterns from Three Orthogonal Planes (LBP-TOP) feature respectively. The two kinds of features are concatenated region-by-region to generate the proposed local statistical descriptor. We evaluate the local descriptor using state-of-the-art classifiers in the experiments. It is observed that the proposed local statistical descriptor, which is located by the facial spatial distribution, can capture more detailed and representative information than the global features, and the fusion of different local features can inspire more characteristics of micro-expressions than the single feature, leading to better experimental results.

Keywords: Micro-expression · Face alignment · Optical flow · LBP-TOP · Feature fusion

1 Introduction

Micro-expression is defined as a brief facial movement that reveals an emotion that a person tries to conceal [1]. In contrast to macro-expressions, a micro-expression is featured by its short duration, which lasts for 1/25 s to 1/5 s [2]. These fleeting facial expressions usually have low intensity: it might be so brief for the facial muscles to

© Springer International Publishing AG 2017
L. Amsaleg et al. (Eds.): MMM 2017, Part I, LNCS 10132, pp. 638–648, 2017.
DOI: 10.1007/978-3-319-51811-4_52

become fully stretched with suppression. Because of the short duration and low intensity, it is usually imperceptible or neglected by the naked eyes [3]. To better analyze micro-expressions and to reveal people's hidden emotions, an automatic micro-expression recognition system is in great need.

Micro-expressions are not easy to elicit and code, which makes it difficult to build large datasets. Nowadays, there are already many facial expression datasets [15] but micro-expression datasets are rare. To our knowledge, there are only two spontaneous micro-expression databases so far, SMIC [5] and CASME [12]/CASME II [13], while there are several faked or posed "micro-expression" dataset before. These spontaneous datasets are all in small size and in uneven distribution (e.g. disgust videos are twice as surprise), which is limited by the elicitation of micro-expressions and time-consuming coding, so these datasets can only be enlarged bit by bit [12].

On account of the limitation of datasets and characteristic of micro-expressions, researches on automatic micro-expression recognition move slowly in recent years, and there are only several works on automatic recognition of micro-expressions. They mainly focus on investigating feature descriptors that are capable of capturing subtle facial variations, which are crucial to the success of micro-expression recognition. Among all the popular feature descriptors, Local Binary Patterns from Three Orthogonal Planes (LBP-TOP) has been proved to be an effective descriptor. It captures dynamic textures illustrating the micro-expression movements along time, and has been chosen as the baseline by many researchers [4–6]. On the other hand, optical flow and optical strain features capture the relative amount of muscular movements on faces within a time interval, which are also advantageous to subtle expression recognition [7, 14]. As a result, the two kinds of features have been widely used in existing studies (Figs. 1 and 2).

The two kinds of features describe micro-expressions to some extent. However, we argue that they still could be improved for the following reasons. On one hand, some researchers extract features from the global face area [6] while some others extract features in equal-sized blocks divided directly in rows and columns [12, 13], which both ignored the relative position of facial parts and mapping connections between these regions. Besides, these rough strategies only apply to full front view of faces without consideration for variations in poses. Moreover, there are redundant information within

Fig. 1. An example of 'surprise' frames sequence from SMIC.

Fig. 2. An example of 'disgust' frames sequence from CASME II.

the regions regardless of emotion movements. This redundancy causes an increase in computational complexity, and also intuitively results in a less discriminative set of features. On the other hand, the optical flow based features are sensitive to noise and illumination change. While micro-expressions related to only certain facial regions, extracting optical flow without considering these spatial clues trends to be noisy, especially the face is not in a frontal view or with certain variants.

According to these issues, we believe that it is a more practical way towards accurate micro-expression recognition by paying more attention to specified face regions rather than the whole images or equal-sized blocks regardless of local face movements. To this end, we propose a novel local feature based descriptor for micro-expression recognition in this paper. The descriptor focuses on spatio-temporal patterns in specific face units, which is located by precisely positioning of facial landmarks. It thus tolerates to variations in face poses. The feature descriptor combines optical flow orientation histogram and classical LBP-TOP in a novel way to detect micro-expression automatically. To suppress noises in low intensity movements, the orientation histogram descriptor of sparse flow is employed, which is obvious a more robust statistical motion estimation. (the left half in Fig. 3). We also compute the global features as the baseline, i.e., extract features from the whole face area (the right half in Fig. 3). Experiments on a public micro-expression dataset show that with the proposed local feature descriptor, we achieve a 7% performance gain than global features in classification accuracy.

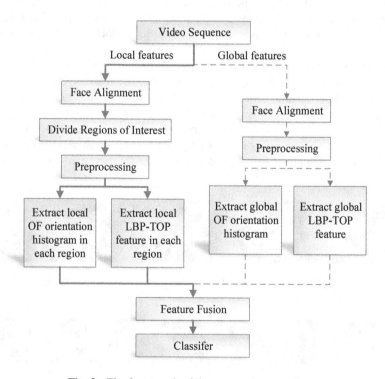

Fig. 3. The framework of the proposed approach.

Main contributions of this paper are threefold: First, we propose an AU-based region division strategy for micro-expression recognition, which requires accurate face alignment to locate the face landmarks and crop specific regions (facial cubes) automatically. The strategy can facilitate the process of micro-expression recognition to be applied with more camera angles. Second, optical flow orientation histogram is extracted to characterize statistical movement tendency and intensity in each interest region, which is more robust than classical optical flow methods. Third, the two statistical feature descriptors, optical flow orientation histogram and LBP-TOP, are aggregated from both the facial regions and time dimension, which are able to exploit more detailed representation of micro-expression than a single feature.

2   Methods

In this section, we present the feature descriptor extraction pipeline of the proposed approach. First, we calculate the face regions of interest located by face landmarks. Then, the optical flow statistical feature and LBP-TOP feature are introduced. Last, we fuse the two features according to the facial regions in order, generating the proposed descriptor.

2.1   Region of Interest

Generally, micro-expressions on face are not always in a fixed position, so a fully automatic and robust face alignment method is the first step to locate facial micro-expressions accurately.

In this paper, we used a fast and robust face alignment method based on the concept of Project-Out Cascaded Regression [8]. It uses generative models of facial shape and appearance fitted via cascaded regression in a subspace orthogonal to the learned appearance variation. The method can locate face key points, track the head pose and facial expression under different angles with high stability in real time speed.

Following the facial action coding system (FACS) [9] that decomposes facial expressions in terms of 46 component movements, we selected the most representative 9 regions automatically located by the facial landmarks above. The corner points of the specific region are calculated automatically with feature points as the center of the region and eye width as the unit length of the regions. To extract subtle facial motion in each region of interest, 3D facial cubes are created for each region.

The regions of interest are divided as shown in Fig. 4. The related FACS Action Units (AUs) are listed in Table 1.

2.2   Optical Flow Statistics (OF Statistics)

Optical flow is a well-known motion estimation technique that is based on the brightness conservation principle [10]. However, traditional optical flow methods are highly sensitive to noises and brightness changes, because it is assumed that all temporal intensity changes are due to motion only. This assumption is not always true for

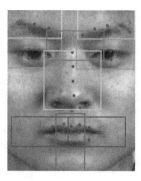

Fig. 4. Facial regions divided by the location of facial landmarks.

Table 1. The nine facial regions and corresponding FACS AUs.

Number of region	Region of interest	FACS AU
1, 2	Eyebrows	AU4, AU5
3, 4	Eyes	AU6, AU7
5, 6	Mouth corners	AU12, AU15, AU24
7	Chin	AU17, AU25
8	Between eyes	AU1, AU2
9	Nose	AU9, AU10

the micro-expression, as it often appears with face movements. So we extend the orientation histogram of optical flow to sequence statistical descriptor which extracts both the spatial and dynamic motion information of micro-expressions.

The computation of differential optical flow is by measuring the spatial and temporal changes of intensity to find a matching pixel in the next frames. The optical flow gradient equation is often expressed as:

$$I_t + vI_x + uI_y = 0 \tag{1}$$

$$I_x = \frac{\partial I}{\partial x}, I_y = \frac{\partial I}{\partial y}, I_t = \frac{\partial I}{\partial t} \tag{2}$$

The method we chose to solve the original optical flow constrain equation was given by Lucas-Kanade. We resize all the image sequence into size of 183 * 149 pixels and apply a median filtering pre-processing to reduce random noises. Flows of a sample pre-processed frame are shown in Fig. 5.

After obtaining the positions and velocities of flows, we characterized the flow with orientation histogram weighted by the magnitude of velocity. The histogram provides an intuitive and descriptor for motion orientations quantized to n principal directions. The angles can be calculated as:

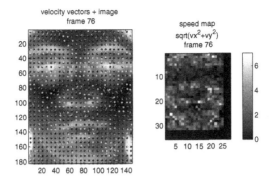

Fig. 5. Optical flows estimated from a pre-processed frame.

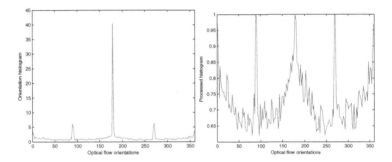

Fig. 6. Global optical flow orientation histograms.

$$a = \arctan\left(\frac{v_y}{v_x}\right) \tag{3}$$

The distribution of global optical flow orientations is shown in the left histogram in Fig. 6. We can find that the motion orientations occur most commonly in the range around 180° which also contains the tiniest movements. To better analyze the orientation distribution, we applied the sigmoid function to the left histogram and achieved the right histogram within the range of zero to one, which can reflect a more intuitive difference between frames. As for the local optical flow statistics, we estimate the weighted orientation histogram in each cube and connect them together as a 1 * 9n-dimensional descriptor.

2.3 Local Binary Patterns from Three Orthogonal Planes (LBP-TOP)

As an extension of the basic LBP, Local Binary Pattern on Three Orthogonal Planes (LBP-TOP) was proposed by Guoying and Matti [11] for dynamic texture analysis in the spatio-temporal domain. The LBP-TOP code is extracted from the XY, XT and YT planes, which are denoted as XY − LBP, XT − LBP and YT − LBP, for all pixels,

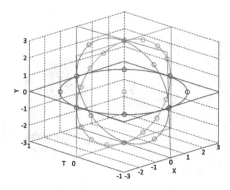

Fig. 7. Different radii and number of neighbor points on three planes.

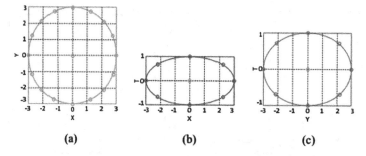

Fig. 8. Detailed sampling for Fig. 5. (a) XY plane, $R_X = 3$, $R_Y = 3$, $P_{XY} = 16$, (b) XT plane, $R_X = 3$, $R_T = 1$, $P_{XY} = 8$, (c) YT plane, $R_Y = 3$, $R_T = 1$, $P_{XY} = 8$.

and statistics of three different planes are obtained, and then concatenated into a single histogram. The XT and YT planes contain information of how the gray values of a row or a column of pixels change along the time dimension.

The radii in axis X, Y and T, and the number of neighbor points in the XY, XT and YT planes can also be different. They can be marked as: $R_X, R_Y, R_T, P_{XY}, P_{XT}, P_{YT}$, as shown in Figs. 7 and 8. Then we can select different parameters on the three orthogonal planes in terms of different uses.

The corresponding feature is written as $LBP - TOP_{R_X,R_Y,R_T,P_{XY},P_{XT},P_{YT}}$. Suppose the coordinates of the center pixel $g_{t_c,c}$ is (x_c, y_c, t_c). Then we can use the following formulas to compute the pixel coordinates of each orthogonal plane.

The coordinates of $g_{XY,p}$ are given by:

$$(x_c - R_X \sin(2\pi p/P_{XY}), y_c + R_Y \cos(2\pi p/P_{XY}), t_c)$$
$$p = 0, 1, \ldots, P_{XY} - 1. \tag{4}$$

The coordinates of $g_{XT,P}$ are given by:

$$(x_c - R_X \sin(2\pi p/P_{XT}), y_c, t_c - R_T \cos(2\pi p/P_{XT}))$$
$$p = 0, 1, \ldots, P_{XT} - 1. \tag{5}$$

The coordinates of $g_{YT,P}$ are given by:

$$(x_c, y_c + R_Y \cos(2\pi p/P_{YT}), t_c - R_T \sin(2\pi p/P_{YT}))$$
$$p = 0, 1, \ldots, P_{YT} - 1. \tag{6}$$

We extract LBP-TOP features from the 9 regions respectively and concatenate the local features to form a 1 * 1593 dimensional (1 * 177 dimensional for each region cube) row vector.

2.4 Feature Aggregation

We normalize the two kinds of features to the same scale, respectively, and then concatenate the features region-by-region. By adopting this early fusion strategy, subtle spatio-temporal variants captured by the two kinds of features are appropriately integrated, resulting in a discriminative descriptor aggregating local spatio-temporal patterns from both appearance and motion aspects. Effectiveness of the descriptor will be demonstrated in the following experiments.

3 Experiments

3.1 Dataset

We used the most recent and publicly available spontaneous micro-expression dataset CASMEII [12] for experiments. CASMEII includes 26 candidates (mean age of 22.03 years) and 247 video clips classified to five classes: happiness, disgust, surprise, repression and others, the number of each class is shown in Table 2 [13]. The dataset has a face resolution of around 280×340 pixels, which is the largest resolution so far. Compared with SMIC, the dataset has been improved in face resolution and fixed illumination. The samples are tagged with the onset and offset frames, AUs and emotions, which provides an objective description of local facial movements.

Table 2. The number and criteria of each class.

Emotion	Criteria	N
Happiness	Either AU6 or AU12	33
Disgust	One of AU9, AU10 or AU4 + AU7	60
Surprise	AU1 + 2, AU25 or AU2	25
Regression	AU15 or AU17 alone or in combination	27
Others	Other emotion-related facial movements	102

In the experiments below, we also selected use CASME II for training and evaluate our method on the five classes of micro-expressions: happiness, surprise, disgust, repression and others.

3.2   Comparison with Different Classifiers

The classifier is an essential building block of micro-expression recognition. To evaluate performance of different classifiers for local features and determine the best one, we compared our method with the same local LBP-TOP feature extraction method and three different classifiers: k-Nearest Neighbor (KNN), SVM with a RBF kernel and Random Forest (RF) with 200 trees.

Leave-one-subject-out (LOSO) and leave-one-video-out (LOVO) cross-validation were used to get a more comprehensive view of the performance of each method. We empirically select $R_X = 1$, $R_Y = 1$, $R_T = 4$, $P_X = P_Y = P_T = 8$ for parameters of LBP-TOP. The performance is shown in Table 3.

Table 3.   Accuracies (in %) with different classifiers using the same feature

Method	Local LBP-TOP + KNN	Local LBP-TOP + SVM	Local LBP-TOP + RF
LOVO	54.51	42.35	**56.08**
LOSO	34.83	43.32	43.92

As seen in Table 2, KNN achieves a good result under LOVO cross validation but the poorest result under LOSO cross validation. It is a convenient method for classification only when the training set is complete enough. SVM classifies most of the test samples as "others", which means that it is susceptible to the unequal distribution of examples. RF achieves a balance between the classification accuracy and stability for the local LBP-TOP feature, because RF does not expect linear features or even features that interact linearly. In conclusion, we chose RF as the classifier for the following experiments.

3.3   Comparison with Different Features

Considering samples unequally distributed in the five micro-expression classes, we separate the five classes of clips into training set and test set with the same ratio of 1:1 respectively, so that we can guarantee each class of micro-expressions are trained sufficiently.

We test LBP-TOP and OF statistics using the global and local features respectively with the same classifier RF. The results are shown in Table 4.

Table 4.   Accuracies (in %) of different features using the same classifier random forest

Single feature	Global LBP-TOP	Local LBP-TOP	Global OF statistics	Local OF statistics
Accuracy	53.17	**57.14**	47.29	38.76

Table 5. Accuracies (in %) of merging different features using the same classifier random forest

Multiple feature	Global LBP-TOP + Global OF statistics	Local LBP-TOP + Global OF statistics	Local LBP-TOP + Local OF statistics
Accuracy	55.56	59.72	**62.50**

Compared with the global LBP-TOP features, the improvement in accuracy is apparent by using local features. It is also observed that global OF statistics outperform local OF statistics with an increase of 8.5%. The probable reason is that noises in local region have a greater impact on the observation for subtle facial movements than the global statistics.

We then evaluate LBP-TOP and OF statistics jointly with different fusion combinations, also with the same classifier RF. The results are given in Table 5.

As can be seen above, the fusion of local LBP-TOP and local OF statistics features manages to obtain an increase of 5.36% over the single local LBP-TOP feature, and 23.74% over the single local OF statistics feature. It shows that the combination of different local features may inspire more characteristics of micro-expressions than the single feature, which may be robust to complex environments. Moreover, the combination of local features also performs better than both the combination of global features (i.e., global LBP-TOP plus with global OF statistics) and the mixed combinations of local and global features (local LBP-TOP plus with global OF statistics), showing that the aggregation of local spatio-temporal patterns indeed more accurately captures face movements related to micro-expressions.

4 Conclusions

We propose a novel approach to recognize micro-expressions in real environment by fusing local statistical features, i.e., local LBP-TOP and local OF statistics. The proposed method is able to achieve 62.5% accuracy for a five-class classification on CASMEII dataset. The results obtained above are still preliminary. But it basically reveals that local statistical features within interest regions based on the facial spatial distribution can capture more detailed and representative information than the traditional ones. It is also more suitable to the real environment with different views.

For future works, since different facial regions and different types of features may contribute differently in the recognition, weights of both facial regions and features can be separately evaluated in the modeling process, aims at further improve the recognition performance. Besides, large variations in illumination and poses should be taken into consideration to avoid being mixed up with micro facial movements, so a robust feature extraction method which can reveal the nature characteristic of micro-expressions is still in great demand.

Acknowledgment. The research is supported by the National Natural Science Foundation of China (#61303175, #61573356).

References

1. Ekman, P.: Lie catching and micro expressions. In: The Philosophy of Deception (2009)
2. Matsumoto, D., Hwang, H.S.: Evidence for training the ability to read micro expressions of emotion. Motiv. Emot. **35**(2), 181–191 (2011)
3. Ekman, P., Friesen, W.V.: Nonverbal leakage and clues to deception. Psychiatry Interpers. Biol. Process. **32**(1), 88–106 (1969)
4. Guo, Y., Tian, Y., Gao, X., et al.: Micro-expression recognition based on local binary patterns from three orthogonal planes and nearest neighbor method. In: IEEE International Joint Conference on Neural Networks, pp. 3473–3479 (2014)
5. Li, X., Pfister, T., Huang, X., et al.: A spontaneous micro-expression database: inducement, collection and baseline. In: IEEE International Conference and Workshops on Automatic Face and Gesture Recognition, pp. 1–6 (2013)
6. Wang, Y., See, J., Phan, R.C., et al.: Efficient spatio-temporal local binary patterns for spontaneous facial micro-expression recognition. PloS ONE **10**(5), e0124674 (2015)
7. Liong, S.-T., See, J., Phan, R.C.-W., Ngo, A.C., Oh, Y.-H., Wong, K.: Subtle expression recognition using optical strain weighted features. In: Jawahar, C.V., Shan, S. (eds.) ACCV 2014. LNCS, vol. 9009, pp. 644–657. Springer, Heidelberg (2015). doi:10.1007/978-3-319-16631-5_47
8. Tzimiropoulos, G.: Project-out cascaded regression with an application to face alignment. In: IEEE Conference on Computer Vision and Pattern Recognition, pp. 3659–3667 (2015)
9. Ekman, P., Friesen, W.V.: Facial action coding system (FACS): a technique for the measurement of facial actions. Q. J. Exp. Psychol. (1978)
10. Black, M.J., Anandan, P.: The robust estimation of multiple motions: parametric and piecewise-smooth flow fields. Comput. Vis. Image Underst. **63**(1), 75–104 (1996)
11. Guoying, Z., Matti, P.: Dynamic texture recognition using local binary patterns with an application to facial expressions. IEEE Trans. Pattern Anal. Mach. Intell. **29**(6), 915–928 (2007)
12. Yan, W.J., Wu, Q., Liu, Y.J., et al.: CASME database: a dataset of spontaneous micro-expressions collected from neutralized faces. In: IEEE Conference on Automatic Face and Gesture Recognition, Shanghai, pp. 1–7 (2013)
13. Yan, W.J., Li, X., Wang, S.J., et al.: CASME II: an improved spontaneous micro-expression database and the baseline evaluation. PLoS ONE **9**(1), e86041 (2014)
14. Brizzi, J., Goldgof, D.B., Sarkar, S.: Optical flow based expression suppression in video. pp. 1817–1821 (2014)
15. Valstar, M.: Automatic facial expression analysis. In: Mandal, M.K., Awasthi, A. (eds.) Understanding Facial Expressions in Communication, pp. 293–307. Springer, Berlin (2015)

egoPortray: Visual Exploration of Mobile Communication Signature from Egocentric Network Perspective

Qing Wang[1]([✉]), Jiansu Pu[2]([✉]), Yuanfang Guo[3], Zheng Hu[1], and Hui Tian[1]

[1] State Key Laboratory of Networking and Switching Technology,
School of Information and Communication Engineering,
Beijing University of Posts and Telecommunications, Beijing 100876, China
{wangqingval,huzheng,tianhui}@bupt.edu.cn
[2] CompleX Lab, Web Sciences Center, Big Data Research Center,
University of Electronic Science and Technology of China, Chengdu 611731, China
jiansu.pu@uestc.edu.cn
[3] State Key Laboratory of Information Security, Institute of Information
Engineering, Chinese Academy of Sciences, Beijing 100093, China
eeandyguo@connect.ust.hk

Abstract. The coming big data era calls for new methodologies to process and analyze the huge volumes of data. Visual analytics is becoming increasingly crucial in data analysis, presentation, and exploring. Communication data is significant in studying human interactions and social relationships. In this paper, we propose a visual analytics system named egoPortray to interactively analyze the communication data based on directed weighted ego network model. Ego network (EN) is composed of a centered individual, its direct contacts (alters), and the interactions among them. Based on the EN model, egoPortray presents an overall statistical view to grasp the entire EN features distributions and correlations, and a glyph-based group view to illustrate the key EN features for comparing different egos. The proposed system and the idea of ego network can be generalized and applied in other fields where network structure exits.

Keywords: Communication network · Ego network · Visual analytics · Communication signature

1 Introduction

The booming of information and communication technologies nurtures the big data era [1]. Among all these large volumes of data, communication data records the behaviors of how people communicate with each other and how they organize their social networks. The accumulation of such digital records provides a new approach for studying the social networks, human dynamics, and other interesting topics [2,3]. For example, the Call Detail Records (CDRs) can be

© Springer International Publishing AG 2017
L. Amsaleg et al. (Eds.): MMM 2017, Part I, LNCS 10132, pp. 649–661, 2017.
DOI: 10.1007/978-3-319-51811-4_53

used to study the human communication behaviors and human mobility [4–6]. Besides, the digital communication records are perfect for analysing Ego Networks (ENs), which examine the social relationships between a target individual (ego) and its direct contacts (alters) [7]. The key idea of ego network is paying more attention to individuals rather than the overall networks. In-depth insights can be obtained from studying the properties of ego communication networks (ECNs) [8].

With the coming of the forth paradigm [9], new methodologies are in urgent need. Visual analytics is an innovate approach, and it is becoming increasingly popular in data science [10]. Visual representations and interactive techniques take advantage of the human eye's broad bandwidth and pathway into the mind to allow users to see, explore, and understand huge amounts of information at once [11]. Sophisticated visual analytics system can highlight useful data thus convey large amounts of information in a more efficient way. This enables decision making with less cognitive efforts, thereby help the analysts adapt to the big data era.

In this paper, we propose egoPortray to study the ECNs based on ego network model. Specifically, we extract the ECNs from the communication data, and portray the ECNs with six network metrics. In order to visually explore the ego networks, we further design two views for interactive investigations: the first view is the macroscopic statistical view, which use the interactive scatter design to capture the holistic correlations and distributions of different ECN features for the entire data. The second view is the microscopic group view, which use glyph-based design to compare different ECNs from different groups. In summary, we build ECNs based on the communication data and further design a visual analytics system for interactively exploring from macroscopic and microscopic scales.

The rest of this paper is organized as follows. Section 2 presents the related research and makes comparisons. Section 3 describes the data and the methods applied in this paper. Section 4 presents the system overview and the detailed design of the proposed visual analytics system. The whole paper will be concluded in Sect. 5.

2   Related Work

The widespread of mobile communication accumulates the relevant data so that we are able to study the social networks at large scale [12,13]. Onnela et al. [14] uncovered the existence of the weak tie effect and further demonstrated its significance to the network's structural integrity by analyzing the weighted mobile communication networks. Eagle et al. [13] found it possible to infer 95% of friendships accurately based only on the mobile communication data. Miritello et al. [15] uncovered the time constraints and communication capacity by studying the individual's communication strategies. Saramäki et al. [16] showed that individuals have robust and distinctive social signatures that can persist over-time. Wang et al. [17] studied the communication network from ego perspective, and found that ECN size played a crucial role in affecting its structure properties. As illustrated above, much attentions have been paid in uncovering the overall features of the mobile communication networks while only limited studies on ECNs have been reported.

Ego network has also been a heated topic in the information visualization community recently. Shi *et al.* [18] proposed a new 1.5D visualization design to reduce the visual complexity of dynamic networks. Liu *et al.* [19] raised a constrained graph layout algorithm on dynamic networks to prune, compress, and filter the networks in order to reveal the salient part of the network. Wu *et al.* [20] presented a visual analytics system named egoSlider for exploring and comparing dynamic citation networks from 3 levels. Cao *et al.* [21] proposed TargetVue, which applied glyph-based design in detecting the anomalous users of online communication system via unsupervised learning. Liu *et al.* [22] introduced egoComp, the storyflow-like links design, to compare two ego networks. Among all these diverse literatures, most studies mainly focus on exploring the spatial and temporal features of communication behaviors. However, the directions of communications are also important in understanding social relationships [23], and the studies on directed ego network topological features are still insufficient.

In this paper, we propose a two-level visual analytics system egoPortray based on weighted directed EN model, which provides a macroscopic statistical view to display various ego network features and a microscopic multi-feature view to visually compare grouped users. Different from the 1.5D egocentric dynamic network visualization [18], egoPortray does not visualize the communication behaviors directly, but shows the EN properties. In order to display large networks, EgoNetCloud [19] proposed algorithms to compress the networks while egoPortray shows the statistical features of the ego networks. egoSlider explored the citation networks from 3 scales by applying node-link, time-line, and glyph-based designs, but such designs did not support visualizing large networks (with million nodes). TargetVue [21] utilized the glyph-based design to illustrate the top anomalous users whilst showed limited overall ECN information. egoComp [22] applied storyflow-like links into node-link graph to compare the alters from 2 ego networks and egoPortray supports comparing a group of users. Different from the above researches, egoPortray proposed the directed weighed ego network model and visualized ego network properties instead of displaying the communication behaviors directly. Visualizing ego network properties also enables egoPortray to present very large networks and compare more egos at the same time.

3   Data and Methods

The call detail records are collected by mobile operators for billing and network traffic monitoring. The basic information of such data contains the anonymous IDs of callers and callees, time stamps, call durations, and so on. In this study, the data set is provided by one of the largest mobile operators in China. It covers 7 million people of a Chinese provincial capital city for half a year spanning from Jan. to Jun. 2014. According to the operator the users choose, all the users can be divided into two categories, namely, the *local* users (customers of the mobile operator who provide this data set) and the *alien* users (customers from the other operators). The reason for such distinction is that the communication behaviors of *alien* users are not recorded completely by this dataset. As a result, we only focus on the *local* users whose entire calling behaviors are included within the dataset. The basic statistics of the mobile communication data are summarized in Table 1.

Table 1. Basic statistics of the mobile communication networks.

Time	N_t (*total* users)	N_l (*local* users)	L_t (*total* links)
Jan.	6520121	751643	32521180
Feb.	6234877	742504	27600221
Mar.	6481767	783751	32720452
Apr.	6526250	777486	32383231
May	6561107	787614	34119390
Jun.	6531076	787156	33461297

Mobile communication is important in maintaining social relationships nowadays [16,24]. Different from the reciprocity nature of communication in off-line life, mobile communication is intrinsically directed. The directions of communication are significant in understanding the relationships among people and the information diffusion process, especially for digital social networks [23]. Therefore, the mobile communication network can be modeled as a directed graph $G(V, E)$ with the number of nodes and links being $|V| = N$ and $|E| = L$, respectively. Link weight is defined as w_{ij} for a directed link l_{ij}, which is the number of calls that user i has made to user j. It is the link strength between two users. The directed weighted ego network model is built for all the egos within the communication graph, and it is composed of all the direct contacts of an centered ego as well as all the directed weighted links between them. The ego network model and the metrics applied are demonstrated in Fig. 1, the definitions of the metrics will be given in the following paragraphs.

The directions of communication divided the alters into two sets for ego i's ECN: the in-contact set C_i^{in} and the out-contact set C_i^{out}. The sizes of C_i^{in} and C_i^{out} are in-degree k_i^{in} and out-degree k_i^{out}, respectively. k_i^{out} represents the ECN size ego i maintains while k_i^{in} reflects the influence of ego i in the network. In this paper, we mainly focus on k^{out}, because it represents the number of alters an ego intends to spend cognitive resources to maintain. We further define the node weight of an ego as $W_i = \sum_{j \in C_i^{out}} w_{ij}$ to indicate the total amount of cognitive resources an ego spend on maintaining his/her social relationships. In fact, the call durations are also important in communication behaviors and the link weight in duration perspective can be defined as $Wd_i = \sum_{j \in C_i^{out}} wd_{ij}$, where wd_{ij} is the call duration from i to j. To further investigate the properties

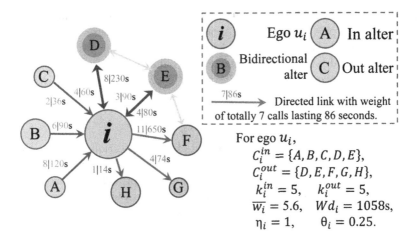

Fig. 1. The ego network structure and the network metrics.

of the ECNs, another three metrics are also introduced, namely, average node weight $\overline{w}$, attractiveness balance η, and tie balance θ.

For ego i, the average node weight $\overline{w_i}$ is defined as:

$$\overline{w_i} = \frac{1}{k_i^{out}} \sum_{j \in C_i^{out}} w_{ij},$$ (1)

where w_{ij} is the weight of link l_{ij}, and k_i^{out} is the size of ECN. This metric indicates the average emotional closeness between an ego and the alters [16,24].

Considering the communication directions, we need to pay attention to the structural balance between in-contacts and out-contacts. Large number of in-contacts indicates the attractiveness of this ego to the network while large number of out-contacts indicates the attractiveness of the network to this ego. Thus we introduce the attractiveness balance (AB) to measure such relationships. It is defined in a straight forward way:

$$\eta_i = \frac{k_i^{in}}{k_i^{out}}.$$ (2)

The attractiveness balance $\eta = 1$ means that the number of contacts a user calls is equal to the number of contacts who call him/her, suggesting the balance of the attractiveness. Large η implies strong attractiveness of an ego whilst small η refers to a weaker attractiveness.

Apart from the attractiveness balance, communication direction also distinguishes bidirectional alters (who appear in both C_i^{in} and C_i^{out}) from the unidirectional ones (who only appear in either C_i^{in} or C_i^{out}). Usually, the reciprocal relationships are stronger than the unidirectional relationships, thus they can be viewed as strong and weak ties [25]. Without lost of generosity, strong ties in this paper suggest reciprocal intentions of forming the relationships thus have larger chance to provide mutual support than the weak ties. In order to measure the balance between strong and weak ties within the ECN, we introduce another structural balance metric named tie balance (TB), which is defined as the Jaccard distance [26] between C_i^{in} and C_i^{out}. Mathematically, it reads:

$$\theta_i = \frac{|C_i^{in} \cap C_i^{out}|}{|C_i^{in} \cup C_i^{out}|}. \tag{3}$$

$\theta = 1$ means all of ego i's direct contacts have bidirectional links with ego i, while $\theta = 0$ means ego i has no reciprocal contacts. The above two kinds of ECNs are all extremely imbalance. Strong relationships can provide support while weak relationships can provide diverse information, and people tend to organize their ECNs with a balanced proportion of strong and weak relationships [27].

4   Visualization and Experiments

In this section, the visual analytics system egoPortray will be presented and demonstrated with experiments. The system overview, user interface, and the proposed two views will be illustrated and discussed consecutively.

4.1   System Overview

By analyzing the communication data interactively from macroscopic to microscopic perspective, egoPortray can be used to discover the overall data distributions and correlations as well as compare different ECNs. With these designs and functions, egoPortray can be applied in analyzing user behaviors such as anomalous user detection. Thus the most significant requirements of this system are: (1) extracting the ECN models from the raw communication data; (2) calculating the specified metrics for the ECNs; (3) conducting some statistical analysis and application algorithms like anomalous ranking; (4) visualizing the overall distributions of ECNs; (5) comparing different ECNs.

Figure 2 illustrates the system architecture and the data processing pipeline of egoPortray. It mainly consists of four modules: the data storage module, the processing module, the analysis module, and the visual representation module.

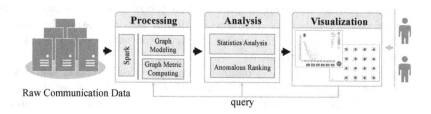

Fig. 2. The system overview and data processing pipeline.

Among them, the data storage module stores all the raw communication data (as described in Data section). These data are subsequently sent to the processing module which is built on Apache Spark [28] to get the ECNs by cleaning and processing. ECNs are stored as instances which contains the information of interactions between ego and alters. With such data, the analysis module can conduct the basic statistical analysis and specific computing tasks, *e.g.* similarity ranking, anomalous ranking, and filtering. After all these procedures, the results are visualized in the visualization module, where an user interface is designed to present the two views.

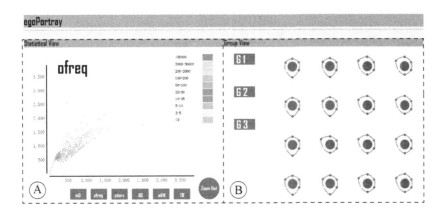

Fig. 3. The user interface of the egoPortray.

The user interface of egoPortray is illustrated in Fig. 3. In this figure, area A shows the statistical view which presents the distributions of the ECNs' features and their correlations with the ECN size. Analysts can zoom in and zoom out to interactively explore the correlation patterns in this view. Area B presents the group view, in which a few egos are taken out as groups for comparison. The design of each view will be presented and discussed in the following two sub-sections.

4.2 Statistical View

Due to the large quantity of users (more than millions), it is almost impossible to present all the users on the screen directly. Granted that it is possible, such large volume of information will overrun and distract the analysts. Statistical distributions are more efficient and practical than directly visualizing such communication data for grasping the overall information. According to Wang's research [17], the size of ECN plays a crucial role in affecting other ECN properties, thus it is better to show the correlation information rather than merely distributions. From this point, we combine the traditional distribution diagram with the correlation diagram in the statistical view.

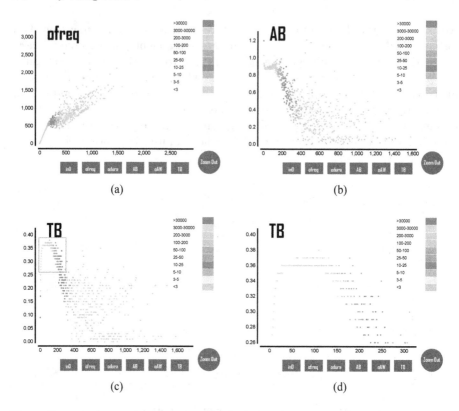

Fig. 4. Statistical view of egoPortray. (a) Distribution of W and the correlation between k^{out} and W; (b) Distribution of η and the correlation between k^{out} and η; (c) Distribution of θ and the correlation between k^{out} and θ; (d) Zoom-in of (c) of $k^{out} \in (0, 300)$ and $\theta \in (0.26, 0.40)$.

The overall statistical view is shown in Fig. 4(a), and the buttons below are used for selecting different ECN properties: "inD" for k^{in}, "ofreq" for W, "odura" for Wd, "AB" for η, "oAW" for $\overline{w}$, and "TB" for θ. X-axis and y-axis corresponds to the ECN size (k^{out}) and the selected ECN property, respectively. Each point in the main view stands for a number of egos with the same ECN size, the x-coordinate is the ECN size and the y-coordinate is the average value of the egos' selected property. The color encodes the number of egos within one point, and the legend is placed on the top right corner. This view shows how the users are distributed according to the ECN size and the selected property, and the correlations between ECN size and the selected property.

4.3   Group View

With the help of the statistical view, we can figure out some specific user groups we are interested in (one or several points in the main view). In order to explore the egos within each group and compare different ECNs at the same time, we develop the multi-feature group view. Glyph-based design is intrinsically suitable for visualizing such multidimensional data [29]. The advantages of the glyph design are flexibility, elasticity, and easy for comparing. In this view, each ECN is visualized as a multi-feature glyph and the glyphs are densely packed for comparison in a matrix.

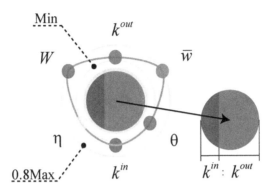

Fig. 5. Glyph design of the group view. (Color figure online)

The basic design of the feature glyph can be found in Fig. 5, where the six small rounds encode six normalized metrics of the ECN, and they are evenly allocated around the centered round. The six properties are: the ECN size (blue round at 0°), the average node strength (green round at 60°), the tie balance (red round at 120°), the in-degree (purple round at 180°), the attractiveness balance (yellow round at 240°), and the node weight on frequency perspective (brown round at 300°). The background rings represent the minimum and 80% of the maximum value for all the metrics. The large round in the center is split into red and blue parts, and the split position indicates the ratio of k^{in} to k^{out} (*i.e.* η). All the small rounds will be connected by a smooth curve to form the main part of the glyph, thus different ECNs will be mapped to different glyphs. This design emphasize the attractiveness balance of ECN.

Based on the glyph design, the group view is presented in Fig. 6. As illustrated, different egos have different glyphs. There are totally 24 egos illustrated and they are placed according to the groups they belong. Different groups are separated by dashed lines, they are labeled as "G1", "G2", and "G3" (each for two columns). This view is useful in comparing different egos in the same and different groups.

4.4 Visual Results

In egoPortray, statistical view helps the analysts to explore the correlations between different ECN properties and the ECN size. In Fig. 4(a), *ofreq* cannot keep the same increasing speed with the increase of ECN size after some k^{out}.

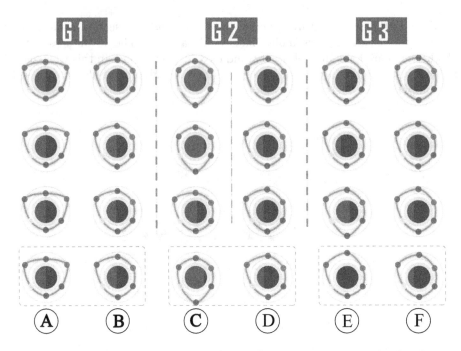

Fig. 6. Group view of egoPortray, users from different groups are visualized as feature glyphs.

In Fig. 4(b), the correlation shows different patterns for different ECN size intervals, but it is not easy to find the exact turning point visually. The similar patterns can be found in Fig. 4(c), which states the correlation between TB and ECN size varies with the increase of k^{out}. Thus in Fig. 4(d), we zoom in to examine the exact turning point of the scatter diagram. With the help of the above four steps, we can roughly divide the egos in three groups according to the correlations between the ECN properties and the ECN size. The groups are "G1": $0 < k^{out} \leq 50$ (different patterns for W, η and θ), "G2": $50 < k^{out} \leq 250$ (same pattern for η and θ but different from W), and "G3": $250 < k^{out}$ (same pattern for η and θ but different from W). Such results are agree with the results calculated by K-means algorithm on multi-ECN-property clustering [30].

To further investigate the ECN properties within each group, egoPortray proposes the group view. In Fig. 6, egos are listed in columns according to the groups they belong, and the egos of the last row (within the dashed box) are taken as examples for demonstrating, which are marked by A, B, C, D, E, F. Among them, A and B belong to group 1, C and D belong to group 2, E and F belong to group 3. A has large $\overline{w}$ and W (brown and green rounds are far away from the center), which means A frequently calls a small number of users. Different from A, B calls a small number of alters not so frequently (green round is close to the center). Both of them have balanced number of incoming alters and outgoing alters (the centered large round). In group 2, C has a large number of incoming alters (large proportion of blue area in the centered round), D has more outgoing alters (large proportion of red area in the centered round). In group 3, both egos have extremely large ECNs (the centered round is almost red), but F has a larger θ and larger $\overline{w}$ compared with E. By comparing the egos from different groups, we can see in group 1: With balanced ECN, A and B make lots of calls to the limited alters, this is because they only have small number of alters, thus have enough time and cognitive resources to keep all the social relationships strong enough. When it comes to group 2, users have diversified ECNs, some of them have very large number of incoming alters while others have more outgoing alters, and most of them make lots of calls (brown rounds on the outer background ring). As in group 3, they have similar features, like large number of calls made, larger ECN size (small blue round on the top), and small average number of calls to the alters (low green round). This means the average emotional closeness between ego and alters becomes weak in this group, *i.e.*, egos decrease their average social strength with alters when they have large ECNs. This agrees with the results in [17].

5   Conclusion

In summary, this paper brings about egoPortray, a visual analytics system for analyzing the communication data from ego network perspective. By proposing the directed weighted ego network model, this paper presents a macroscopic statistical view to illustrate the overall distributions and correlations for ECN properties, and develop a microscopic glyph-based group view for ECN comparison. Visual results show that egoPortray can present the data distributions and correlations, and further compare different ECNs from different groups. They also illustrate the potentials of this system in studying communication behaviors. Our design can be generalized to different scenarios where network model applies, and scale to different network sizes. As future works, the temporal information of the ego communication signatures can also be taken into EN models to better explore the communication behaviors. Another potential research direction is to present the geo-information of the ego and alters at the same time to improve mobility predictions.

Acknowledgments. The authors acknowledge Zhi-Yao Teng for useful discussions. This work was partially supported by the National Natural Science Foundation of China (Grant Nos. 61302077, 61421061, 61433014, 11222543 and 61502083) and the Key Project of the National Research Program of China (Grant No. 2012BAH41F03). Qing Wang acknowledges the joint cooperation project of BUPT and China Telecom Beijing Institute.

References

1. Manyika, J., Chui, M., Brown, b., Bughin, J., Dobbs, R., Roxburgh, C., Byers, A.H.: Big data: the next frontier for innovation, competition, and productivity. http://www.mckinsey.com/business-functions/business-technology/our-insights/big-data-the-next-frontier-for-innovation

2. Barabási, A.L.: The origin of bursts and heavy tails in human dynamics. Nature **435**, 207–211 (2005)
3. Borgatti, S.P., Mehra, A.M., Brass, D.J., Labianca, J.: Network analysis in the social sciences. Science **323**, 892–895 (2009)
4. Song, C., Qu, Z., Blumm, N., Barabási, A.-L.: Limits of predictability in human mobility. Science **27**, 1018–1021 (2010)
5. Miritello, G., Moro, E., Lara, R.: Dynamical strength of social ties in information spreading. Phys. Rev. E. **83**, 045102 (2011)
6. Toole, J.L., Herrera-Yaqüe, C., Schneider, C.M., González, M.C.: Coupling human mobility and social ties. J. R. Soc. Interface **12**, 20141128 (2015)
7. Roberts, S.G.B., Dunbar, R.I.M.: Communication in social networks: effects of kinship, network size, and emotional closeness. Pers. Relatsh. **18**, 439–452 (2011)
8. Fisher, D.: Using egocentric networks to understand communication. IEEE Internet Comput. **9**, 20–28 (2005)
9. Hey, T.: The fourth paradigm – data-intensive scientific discovery. In: Kurbanoğlu, S., Al, U., Erdoğan, P.L., Tonta, Y., Uçak, N. (eds.) Communications in Computer and Information Science, Berlin (2012)
10. Keim, D., Andrienko, G., Fekete, J.D., Görg, C., Kohlhammer, J., Melançon, G.: Visual analytics: definition, process, and challenges. In: Kerren, A., Stasko J.T., Fekete, J.D., North, C. (eds.) Information Visualization – Human-Centered Issues and Perspectives, Berlin (2008)
11. Mazza, R.: Introduction to Information Visualization. Springer, London (2009)
12. Onnela, J.P., Saramäki, J., Hyvönen, J., Szabó, G., Menezes, M.A., Kaski, K., Barabási, A.L., Kertész, J.: Analysis of a large-scale weighted network of one-to-one human communication. New J. Phys. **9**, 179–206 (2007)
13. Eagle, N., Pentland, A.S., Lazer, D.: Inferring friendship network structure by using mobile phone data. Proc. Natl. Acad. Sci. U.S.A. **106**, 15274–15278 (2009)
14. Onnela, J.P., Saramäki, J., Hyvönen, J., Szabó, G., Kaski, K., Kertész, J., Barabási, A.L.: Structure and tie strengths in mobile communication networks. Proc. Natl. Acad. Sci. U.S.A. **104**, 7332–7336 (2007)
15. Miritello, G., Moro, E., Lara, R., Martínez-López, R., Belchamber, J., Roberts, S.G.B., Dunbar, R.I.M.: Time as a limited resource: communication strategy in mobile phone networks. Soc. Networks **35**, 89–95 (2013)
16. Saramäki, J., Leicht, E.A., López, E., Roberts, S.G.B., Reed-Tsochas, F., Dunbar, R.I.M.: Persistence of social signatures in human communication. Proc. Natl. Acad. Sci. U.S.A. **111**, 942–947 (2014)
17. Wang, Q., Gao, J., Zhou, T., Hu, Z., Tian, H.: Critical size of ego communication networks. EPL **114**, 58004 (2016)
18. Shi, L., Wang, C., Wen, Z., Qu, H., Liao, Q.: 1.5D egocentric dynamic network visualization. IEEE Trans. Vis. Comput. Graphics. **21**, 624–637 (2015)
19. Liu, Q., Hu, Y., Shi, L., Mu, X., Zhang, Y., Tang, J.: EgoNetCloud: event based egocentric dynamic network visualization. In: IEEE Conference on Visual Analytics Science and Technology (VAST 2015), pp. 65–72. IEEE Press, Chicago (2015)
20. Wu, Y., Pitipornvivat, N., Zhao, J., Yang, S., Huang, G., Qu, H.: EgoSlider: visual analysis of egocentric network evolution. IEEE Trans. Vis. Comput. Graphics. **22**, 260–269 (2016)
21. Cao, N., Shi, C., Lin, S., Lu, J., Lin, Y., Lin, C.: TargetVue: visual analysis of anomalous user behaviors in online communication systems. IEEE Trans. Vis. Comput. Graph. **22**, 280–289 (2016)

22. Liu, D., Guo, F., Deng, B., Wu, Y., Qu, H.: EgoComp: a node-link based technique for visual comparison of ego-network. http://vacommunity.org/egas2015/papers/IEEEEGAS2015-DongyuLiu.pdf
23. Brzozowski, M.J., Romero, D.M.: Who should i follow? recommending people in directed social networks. In: 5th International AAAI Conference on Weblogs and Social Media (ICWSM), pp. 458–461. AAAI Press, Barcelona (2011)
24. Zhou, W.X., Sornette, D., Hill, R.A., Dunbar, R.I.M.: Discrete hierarchical organization of social group sizes. Proc. R. Soc. B. **272**, 439–444 (2005)
25. Zhu, Y., Zhang, X., Sun, G., Tang, M., Zhou, T., Zhang, Z.: Influence of reciprocal links in social networks. PLoS One **9**, e103007 (2014)
26. Levandowsky, M., Winter, D.: Distance between sets. Nature **234**, 34–35 (1971)
27. Brown, J.J., Reingen, P.H.: Social ties and word-of-mouth referral behavior. J. Consum. Res. **14**, 350–362 (1987)
28. Spark. http://spark.apache.org/
29. Ward, M.O.: A taxonomy of glyph placement strategies for multidimensional data visualization. Inf. Vis. **1**, 194–210 (2002)
30. Jain, A.K.: Data clustering: 50 years beyond k-means. J. Pattern Recogn. **31**, 651–666 (2010)

i-Stylist: Finding the Right Dress Through Your Social Networks

Jordi Sanchez-Riera[1]([✉]), Jun-Ming Lin[1,2], Kai-Lung Hua[2],
Wen-Huang Cheng[1], and Arvin Wen Tsui[3]

[1] Research Center for Information Technology Innovation,
Academia Sinica, Taipei, Taiwan
{jsan3386,whcheng}@citi.sinica.edu.tw
[2] Department of CSIE, National Taiwan University of Science and Technology,
Taipei, Taiwan
{b10115015,hua}@mail.ntust.edu.tw
[3] Industrial Technology Research Institute, Hsinchu, Taiwan
arvin@itri.org.tw
http://mclab.citi.sinica.edu.tw/

Abstract. Searching the Web has become an everyday task for most
people. However, the presence of too much information can cause information overload. For example, when shopping online, a user can easily be
overwhelmed by too many choices. To this end, we propose a personalized
clothing recommendation system, namely i-Stylist, through the analysis
of personal images in social networks. To access the available personal
images of a user, the i-Stylist system extracts a number of characteristics
from each clothing item such as CNN feature vectors and metadata such
as color, material and pattern of the fabric. Then, these clothing items
are organized as a fully connected graph to later infer the personalized
probability distribution of how the user will like each clothing item in a
shopping website. The user is able to modify the graph structure, e.g.
adding and deleting vertices by giving feedback about the retrieved clothing items. The i-Stylist system is compared against two other baselines
and demonstrated to have better performance.

Keywords: Personalized recommendation · Clothing items · Dress
style · Social network · Convolutional neural networks (CNN)

1 Introduction

User centered applications are becoming more and more popular due to the
multiple possibilities that can offer in personalized services [3,7,23]. For example,
in online shops, a personalized service query can save a lot of browsing time
by retrieving in first positions those items that are more relevant to the user
(usually online shops contain thousands of items). Let's assume we want to buy
a cloth item and we start browsing in the website category "clothes". At the
same time, "clothes" can be divided into multiple sub-categories (i.e. "shoes",

© Springer International Publishing AG 2017
L. Amsaleg et al. (Eds.): MMM 2017, Part I, LNCS 10132, pp. 662–673, 2017.
DOI: 10.1007/978-3-319-51811-4_54

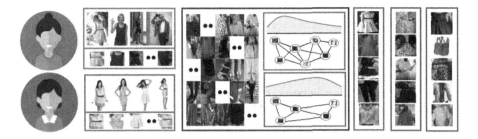

Fig. 1. Two different users provide personal images to the i-Stylist system. With those images, visual and textual information is extracted to generate a personalized model and calculate the probability distribution of website items for each user. Website items are sorted and shown to the users for a system feedback.

"pants", "jackets", etc.), and for each of these sub-categories, the items can come in different colors, styles, fabrics and patterns. In a non-personalized search, the items displayed by the website are sorted by an unknown criteria, which can make the search for a desired item long and tedious as the user might have to browse among many items. In a personalized search taking into account the individual's user preferences, the items displayed by the website will be sorted, which presumably the desired item will appear in the first positions of the user search.

Most recommender systems rely on information either from a similar users or from the popularity[1] of items to learn the user's individual preferences and predict the desired items. However, these assumptions are not always valid due to privacy issues (e.g., not allowed to access information about other users) or are not generalizable (the most popular item is not necessarily the item that a certain user likes). Therefore, we propose a clothing recommendation system, namely i-Stylist, which is based on the user's own images to predict the most likable items, cf. Fig. 1. From the personal images of a user in his/her social networks, the i-Stylist system will generate a probability distribution for all the items in a certain online shopping website. The probability distribution is derived from the user's personalized graph model that contains visual information (deep learning features) and semantic information (clothing properties such as fabric, pattern, etc.). In addition, each user graph model can be modified dynamically when the user selects and discards items from the retrieved search list.

The rest of the paper is organized as follows: Sect. 2 surveys the related recommender systems and clothing retrieval algorithms. Section 3 describes the datasets used amd Sect. 4 presents the proposed algorithm. Section 5 conducts the experiments and Sect. 6 draws the conclusions.

---

[1] Usually a popular item is defined by the number of likes, links, etc., that an image or user can have.

2 Related Work

When a website wants to recommend an item to a user, one of the typical strategies is to recommend the items that are most popular among other users [4]. Assuming that similar users will have similar preferences is also of great interest to stablish a relationship between the new user and the past users. Looking at the interactions between users through social media can be useful to explore the relationship strengths and define an item priority list as shown in [2]. However, it is also possible to infer the likability of an item using solely information of the own user. For example, [6] proposed a system based on selfies to predict user personality and then infers the priority list of items. Similarly, based on the popularity of user images, [25] proposed to predict the popularity of a person for an item recommendation. However, working with personal images or data extracted from social networks presents some inconveniences, e.g., unable to access to the user information (restricted due to privacy settings).

Another possibility to have information about the users is to analyze their behaviors when browsing websites. This information is much more easy to collect and rapidly can be made an inventory of, for example, which sites are visited, how many people liked some photos, which items are purchased (in case of shopping website) among other metadata. This is the case of [17] which uses the browsing history of a user to generate a preference score of a shopping website item. Then, the computed score is mixed with the price and popularity of the item and finally based on these parameters to infer the ranking scores for the website items. This approach makes some assumptions that are generally true, but not always valid. For example, preferences of other users might tell most popular items, but they might not be an item that the targeted user likes or is willing to buy.

Even if the above discussed ideas are applicable in any recommender system, there is an extensive use when it comes to clothes and fashion domain. Hence, the popularity concept can be applied to detect different clothing styles or suggest some clothing items for a user (also the purpose of our work). The latter is the aim of [10], where topic modeling techniques are used to classify the image clothes with the purpose of recommending lower body clothes that are a good match for a given upper body images. In this case though, the images don't correspond with an actual user, being the outfits recommended independent of the personal user preferences. Similarly, [22] makes clothing recommendations using a Markov random field (MRF) model, but the personal user information is still not considered.

In addition to the above approaches, the most common paradigm for predicting likability is the use of a regression model. For example, [8] used a regression model to recommend a set of clothes. [9] exploited support vector regression (SVR) to predict user preferences learnt from different mid-level features. Also, [14] made use of mid-level clothing attributes to learn a latent SVM. Again, none of the methods presented takes into consideration the specificities of the user.

Therefore, we propose an algorithm that is based on information provided by the user, rather than information based on popularity of items or the item itself, to model a personalized recommender system that is able to retrieve preferred clothing items from a clothing shopping website. Based on the properties extracted from the user images, the proposed system will configure a graph model to assign a probability to each clothing item belonging to the website.

3   Datasets

For the experiments we will use three datasets: **Fabric & Pattern**, **People Images** and **Street2shop** datasets. **Fabric & Pattern** dataset is collected to be able to extract material and pattern metadata of a given cloth. **People Images** dataset is collected to have a collection of user images. And **Street2shop** dataset is a publicly available dataset that contains clothing images from different shopping websites. The details of these datasets are given below (Fig. 2).

(a) Fabric & Pattern   (b) People Images   (c) Street2shop

Fig. 2. Sample images of the different datasets used.

Fabric and Pattern. In this dataset, we collect data by Google Images using keywords like the name of category fabric and category pattern followed by adjectives: "cloth", "clothing", "clothes" and "look". In total, we collected 200 images for each combination of category keyword and adjective. Afterwards, the images that are either too small or not related with the keywords are removed. In the category fabric we use the keywords: cotton, denim, fur, lace, leather, silk, tweed and wool. For the category pattern we use the keywords: animal print, zebra, leopard, argyle, checkered, dotted, floral, herringbone, houndstooth, paisley, pinstripes, plaid, print, striped and tartan. The categories and procedures to obtain the data are similar to the ones in [1].

People Images. This dataset consists of personal model images extracted from the Royalty Free images website[2]. From the website we crawl 100 image profiles and only those images with upper body detected and only one face detected are kept (around 20 images per profile). Then, we segment each image with

---

[2] http://peopleimages.com.

the clothing parser [26] to obtain the location and label of the different clothing garments the user is wearing.

Street2shop. This dataset consists of 20,357 real world photos of outfit posts and 404,683 shop photos from 25 different online clothing retailers. One characteristic of this dataset that makes it different from the others is the ground truth provided. The location of a clothing item in the photo as well as the exact match for that item in the online clothing retailer photos is given. The images we use belonging to the following categories: bags, dresses, footwear, leggings, outerwear, pants, skirts and tops. More details of the dataset can be found in [11].

4   The Proposed Method

4.1   Calculate the Item Probability Distribution

Given a set of images $I^U = \{I_1^U, I_2^U, .., I_P^U\}$ that belong to a user U, we want to find a probability distribution $P(I^S|U)$ of a set of images $I^S = \{I_1^S, I_2^S, .., I_Q^S\}$ from a cloth shopping website S, that indicates the likability of each image with respect the user U. Each user image contains the user itself in a stand-up position wearing some clothes, which are parsed using [26] to extract the different cloth items $I_i^U = \{c_1, c_2, .., c_N\}$. Each cloth item c_i in I^U and each image in I^S are computed as a 5-tuple feature vector consisting of: a deep learning feature DL 4096, a color histogram $Color$ 100, the category CAT, the material MAT, and the pattern PAT labels which each cloth item belongs to, cf. Sect. 4.3. Then the user cloth items c_i are organized as a fully connected graph $\mathcal{G}_U = (\mathcal{V}, \mathcal{E})$ where the vertices correspond to each one of the 5-tuple feature vector of a cloth. This process is illustrated in Fig. 3.

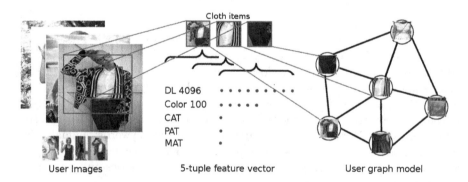

Fig. 3. The user images are parsed and segmented into different cloth items. For each cloth item a ClothCNN and color histogram features are extracted. Moreover, the labels: category, material and pattern are computed. This 5-tuple information will be a vertex of a fully connected graph representing the user model. Each cloth item will be a vertex of the graph.

Therefore, the probability distribution $P(I^S|\mathcal{G}_U)$ of each shopping image I^S for each user U is defined as:

$$P(I^S|\mathcal{G}_U) = \sum_{j=1}^{|V|} (w_{ij}^{DL4096} f(I_i^S, v_j) + w_{ij}^{Color100} f(I_i^S, v_j) \tag{1}$$
$$+ w_{ij}^{CAT} f(I_i^S, v_j) + w_{ij}^{PAT} f(I_i^S, v_j) + w_{ij}^{MAT} f(I_i^S, v_j))$$

where $w_{ij}^{DL4096} = w_{ij}^{Color100} = e^{-\frac{1}{2\sigma^2} f(I_i^S, v_j)}$ measures the similarity of cloth shopping website images I_i^S with the graph vertex v_j, $\sigma^2 = \frac{1}{|V|^2} \sum_{i,j \in V} f(I_i^S, v_j)$ and $w_{ij}^{CAT} = w_{ij}^{PAT} = w_{ij}^{MAT} = TF - IDF(I_i^S, \mathcal{G}_u)$. In abuse of notation we define $f(I_i^S, v_j)$ to be $f^{DL4096}(I_i^S, v_j)$, $f^{Color100}(I_i^S, v_j)$, $f^{CAT}(I_i^S, v_j)$, $f^{PAT}(I_i^S, v_j)$ and $f^{MAT}(I_i^S, v_j)$, since each component of the 5-tuple feature is a vector. Then $f(I_i^S, v_j)$ is the cosine distance of each component for the 5-tuple feature between cloth item I_i^S and the graph vertex v_j.

4.2   Update the Item Probability Distribution

In most of the cases the number of images provided by the user I^U is small $P < 10$. Therefore, a mechanism is necessary to improve and refine the recommender system by updating the probability distribution of each image in I^S. The steps to update the probability distribution are enumerated below:

1. Introduce user images I^U and generate user graph $\mathcal{G}_U$.
2. Calculate cloth shopping items probability distribution $P(I^S|\mathcal{G}_U)$.
3. Output the list of images recommended to the user I_{RU}^S.
4. User selects images they like from the I_{RU}^S image set.
5. Update the user graph $\mathcal{G}_U$ and repeat the process.

Once the probability distribution for each item is computed according to Eq. 1, the recommender system can output and show the retrieved list of items. This list is generated taking the first ten images with maximum probability distribution and in addition ten extra cloth items are selected randomly from all categories of the Street2shop dataset. The reason to include random items from all dataset is to guarantee the discovery of new potential clothing items that are not similar to the clothes given by the user, which is presumably small.

When the user selects a new cloth, a new vertex v_n is computed and added to the graph model if $\forall v_i \in \mathcal{G}_U f(I_i^S, v_k) > \delta$, where δ is the parameter to measure the distance between two vertices.

4.3   Compute Cloth Image Feature Vectors

As stated previously, a 5-tuple vector is computed for each cloth item extracted from user images I^U and from the cloth website images I^S. The features of 5-tuple vector are designed to better fit the addressed problem. From one side, state-of-the-art low level features from a finetuned CNN will provide a global interpretation of the image. On the other side, common characteristics such as color, category, pattern and material of cloth are included to reinforce the description of the cloth present in the image.

Deep Learning Features (DL 4096). Deep learning based features has demonstrated to be reliable and outperform many manual tuned ones [21]. However, training a CNN network requires huge amount of images and time. Many CNN architectures provide already pre-trained weights on ImageNet dataset. We choose an architecture defined in [20] to make an initial test. The features, **Overfeat**, are extracted at the layer (fc19) and are $\mathbb{R}^{4096}$ dimensional vectors. However, to obtain features more adjusted to our **Street2shop** dataset, we fine-tune the architecture defined in [13]. The features are also in $\mathbb{R}^{4096}$ dimensional space, corresponding to the last layer of the network. We will refer to these features as **ClothCNN** features. The latter features are used in the experiments.

Color Histograms (Color 100). Color is a very important characteristic of any cloth item and can influence the user decision whether he/she likes or not a certain cloth. Therefore, each cloth item will be represented by a color histogram in $\mathbb{R}^{100}$ dimensional space computed by a standard bag of words approach. First a set of keypoints are obtained for each image and then, for each keypoint the hue SIFT descriptor [24] is computed using the code provided by [19]. K-means is used to cluster all descriptors from all images into 100 clusters. Finally, a nearest neighbor algorithm is used to obtain the final histogram representing the color of a cloth item.

Cloth Items Metadata (CAT, MAT, PAT). Other important factor in a cloth item is the label of *category*, *fabric* and *pattern* of a cloth. See Sect. 3 for a description of the labels. In the case of cloth label *category*, these labels are given as the ground truth parameter in the **Street2shop** dataset. For the other datasets, the parsing algorithm [26] will determine the label *category*. Since categories from the parsing algorithm and from the **Street2shop** dataset have not direct correspondence, we use the latter categories as our labels. The other two labels, *fabric* and *pattern*, are determined to train a linear SVM for each one of the categories. The images used to train the SVM are from the **Fabric & Patterns** dataset. Using the 90% of data for training and 10% for testing, the average recognition rate (ARR) obtained is 68% and 62% for *fabric* and *pattern* respectively. Once the labels are computed for each image, it turns to convert each label into a vector which will allow us to define a distance between two different label tags. This is achieved by using the algorithm in [16].

5 Experiments

For the experiments we conduct two kind of evaluations. First, we investigate among different image feature representations to obtain an optimal cloth dataset embedding. Second, given the embedding for the set of images I^S and the user set of images I^U, we compare the results of our proposed method **Graph (i-Stylist)** with two other baseline methods: **Random** and **SVM**.

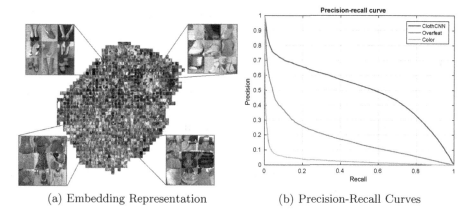

(a) Embedding Representation   (b) Precision-Recall Curves

Fig. 4. This figure shows (a) the embedding of the cloth website shopping images according to the **ClothCNN** network and (b) the precision-recall curves of three different embeddings: **Overfeat**, **ClothCNN** and color histograms **Color 100**.

5.1 Shopping Cloth Items Embedding

To evaluate the embedding of deep learning features we use two common parameters in retrieval systems: precision and recall. Precision (also called positive predictive value) is the fraction of retrieved instances that are relevant, while recall (also known as sensitivity) is the fraction of relevant instances that are retrieved. To compute these parameters we split the **Street2shop** dataset in training and testing sets. For each image on the test set we iterate over $k = \{1, .., N\}$, where N is the size of the test set. If I_k has a certain category C, N_T^C is the total number of items belonging to category C, D^C is the total number of items correctly predicted in category C and C_T is the total number of items of category C, then we compute precision as $P = D^C/N_T^C$ and recall as $R = D^C/C_T$.

In Fig. 4 we show the embedding of the shopping website images using the **ClothCNN** architecture and the precision-recall curves for **Overfeat**, **ClothCNN** and color histograms **Color 100** feature vectors. The details about these three feature vectors are given in Sect. 4.3. To display the images of the dataset in a 2D plot, we transform the dimensionality of the **ClothCNN** features from $\mathbb{R}^{4096}$ to $\mathbb{R}^2$ by means of the t-SNE [15] method.

5.2 Baselines

We briefly introduce the two baselines below used for comparisons.

Random. In this baseline the probability distribution of cloth i in set I^S will be given by $P(I_i^S|I^U) = rand()$ where $rand()$ is a random variable that has values between zero and one. The results are sorted in an ascending manner.

SVM. In this baseline we follow an approach similar to [8]. For each cloth item c_k in set I^U, we will define a vector $v_{c_k} = (nc_{DL}, nc_{col}, l_{CAT}, l_{MAT}, l_{PAT})$

where l_{CAT}, l_{MAT}, and l_{PAT} are the labels for **CAT**, **MAT**, and **PAT** respectively. $nc_{DL} = \min_k f^{DL4096}(c_k, I_i^S)$ and $nc_{col} = \min_k f^{Color100}(c_k, I_i^S)$. Each user will have associated a SVM model which is trained with $v_{c_k}^{pos} = \{\min_k f^{DL4096}(c_k, I^S)\}_{1..N}$ positive and $v_{c_k}^{neg} = \{\max_k f^{DL4096}(c_k, I^S)\}_{1..N}$ negative vectors, where $N = 20$ is the total number of vectors collected for training. Then a linear SVM is trained with those vectors. Everytime each user selects new cloth items from the algorithm output, the process of collecting positive/negative vectors is repeated adding to the set of user items I^U the new items and SVM is re-trained. Finally, the probability distribution for each cloth in I^S is defined as $P(I_i^S|I^U) = SVM(v_{c_k}^{pos}, v_{c_k}^{neg})$, where the scores of SVM are transformed to probabilities based on logistic regression proposed by [18]. The results are sorted in an ascending manner.

5.3   Results from People Images Dataset

Given the images of a user I^U in the **People Images** dataset, these images are parsed to extract the different cloth items which will be introduced as input to the recommender system. Then, the recommender system will show a list of suggested images from the cloth shopping website dataset and the user will select the images he/she likes. This process is iterated and can be repeated as many times as the user wishes. For our experiments we set the maximum number of iterations to be 20. In order to make the comparison we will run three different recommender systems, our proposed **Graph** system and the two baselines **SVM** and **Random**, which will output three lists l_{Graph}, l_{SVM} and l_{Random} of suggested images. The length of each one of the lists is set to 10.

Since a quantitative and objective evaluation is not possible due to the nature of the problem, to evaluate the performance of all three systems we follow a methodology proposed in [12]. This methodology also has been used in [2,5].

The key idea is to have several independent[3] reviewers that can express their opinions on several aspects related to the recommender system. In concrete, to evaluate and compare our proposed method we will ask these reviewers to give a mark between 1 (bad) and 5 (good) to the following questions:

Q1. The recommender system helps me to discover new cloth items?
Q2. The cloth items recommended to me are similar to my input cloth images?
Q3. The recommender can be trusted?

These questions are designed to obtain a global and detailed evaluation of how our proposed algorithm and baselines performance. While questions **Q1** and **Q3** refer to global performance of the recommender system, question **Q2** refers to each cloth shopping website item presented in the lists l_{Graph}, l_{SVM} and l_{Random}.

Figure 5 shows the mean marks given by the reviewers for iterations 1, 10 and 20 for the three retrieved lists for all the users. Some observations can be made from the statistical results:

---

[3] Different from the authors of this paper.

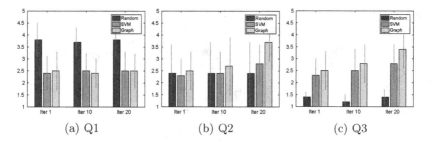

(a) Q1   (b) Q2   (c) Q3

Fig. 5. Mean mark results among all users and reviewers for questions **Q1**, **Q2** and **Q3** and methods **Random**, **SVM** and **Graph** (i-Stylist).

In question **Q1**, we can observe that the reviewers consider the **Random** method the best to discover new clothes. This is somehow expected, since select random clothes from the entire Street2shop dataset offers more variability than the retrieved clothes similar to the ones given by the user. The other two methods **SVM**, **Graph** have very similar performance when it comes to discover new clothes.

In question **Q2**, the results between different iterations are more relevant. We can observe that when increasing the number of iterations the **SVM** and **Graph** methods perform better. Specially, **Graph** method outperforms all others because the model has strong priority on clothes given by the user, whether the **SVM** method is still too general and fails to put similar clothes in the first positions of the suggested list. Probably, much more iterations would be necessary to achieve similar performance to **Graph** method.

In question **Q3**, all reviewers agree that for method **Random** the suggestions can not be trusted. However, results for **SVM** and **Graph** are quite similar. One of the reasons is that meanwhile **Graph** retrieves cloths more similar to the ones introduced by the user, the method **SVM** retrieves cloths more similar between them. Therefore, the global coherence for the two methods make the reviewer to see as similar methods.

To give a better insight of the results obtained by the three methods in iterations 1, 10, and 20, in Fig. 6, we show some images introduced by the user, and the first three items in result lists for each method/iteration. We can see that the images retrieved from the method **Graph** are more similar (color, style, shape) to those introduced by the user. For the **SVM** method, in iteration 10 we observe a coherence (all are shirts), but for the other two iterations results are more similar to the **Random** method. As for the latter, we can observe that the cloth items suggested not only differ in color and style, but also in category of clothes. No shoes or bags images are introduced by the user, but the method still retrieves shoes and bags. However, this is totally expected due to be a random selection of the whole cloth shopping dataset.

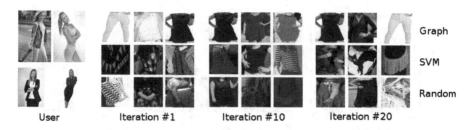

Fig. 6. Results images for the methods **Graph**, **SVM** and **Random** at iterations 1, 10, and 20 given some user images I^U. Note we don't show all images introduced by user but only display the first three images of each retrieved list.

6   Conclusions

We presented a personalized clothing recommendation system, i-Stylist. Given a few images from a user, the i-Stylist system can suggest some items from a cloth shopping website. We demonstrated that the recommender system is able to generate reasonable results and can improve as the user gives some feedback by selecting some images from the retrieved lists. The graph model proposed is used as an intermediate step to generate a personalized probability distribution for each image in the **Street2shop** dataset. The features used combine state-of-the-art fine-tuned CNN features as well as metadata obtained from SVM models. The comparisons performed suggest that a more user oriented model **Graph** can be better than a general linear regression model **SVM** where user preferences are faded among similar clothes.

Acknowledgement. This work is funded by ITRI Grant #105-W100-21A1, under the project "Big Data Technologies and Applications (2/4)" of the Industrial Technology Research Institute of Taiwan, R.O.C.

References

1. Bossard, L., Dantone, M., Leistner, C., Wengert, C., Quack, T., Gool, L.: Apparel classification with style. In: Lee, K.M., Matsushita, Y., Rehg, J.M., Hu, Z. (eds.) ACCV 2012. LNCS, vol. 7727, pp. 321–335. Springer, Heidelberg (2013). doi:10.1007/978-3-642-37447-0_25
2. Carmel, D., Zwerdling, N., Guy, I., Ofek-Koifman, S., Har'el, N., Ronen, I., Uziel, E., Yogev, S., Chernov, S.: Personalized social search based on the user's social network. In: ACM IKM (2009)
3. Chi, H.Y., Chen, C.C., Cheng, W.H., Chen, M.S.: Ubishop: commercial item recommendation using visual part based object representation. Multimed. Tools Appl. **75**, 16093–16115 (2015)
4. Chi, H.-Y., Cheng, W.-H., Chen, M.-S., Tsui, A.W.: MOSRO: enabling mobile sensing for real-scene objects with grid based structured output learning. In: Gurrin, C., Hopfgartner, F., Hurst, W., Johansen, H., Lee, H., O'Connor, N. (eds.) MMM 2014. LNCS, vol. 8325, pp. 207–218. Springer, Heidelberg (2014). doi:10.1007/978-3-319-04114-8_18

5. Dooms, S., De Pessemier, T., Martens, L.: A user-centric evaluation of recommender algorithms for an event recommendation system. In: RecSys (2011)
6. Guntuku, S.C., Qiu, L., Roy, S., Lin, W., Jakhetiya, V.: Do others perceive you as you want them to?: modeling personality based on selfies. In: ASM (2015)
7. Hidayati, S.C., Hua, K., Cheng, W., Sun, S.: What are the fashion trends in new york? In: ACM Multimedia Conference (2014)
8. Hu, Y., Yi, X., Davis, L.S.: Collaborative fashion recommendation: A functional tensor factorization approach. In: ACM Multimedia (2015)
9. Huang, C., Wei, C., Wang, Y.F.: Active learning based clothing image recommendation with implicit user preferences. In: ICME Workshops (2013)
10. Iwata, T., Watanabe, S., Sawada, H.: Fashion coordinates recommender system using photographs from fashion magazines. In: IJCAI (2011)
11. Kiapour, M.H., Han, X., Lazebnik, S., Berg, A.C., Berg, T.L.: Where to buy it: matching street clothing photos in online shops. In: ICCV (2015)
12. Knijnenburg, B.P., Willemsen, M.C.: Evaluating recommender systems with user experiments. In: Ricci, F., Rokach, L., Shapira, B. (eds.) Recommender Systems Handbook, pp. 309–352. Springer, Boston (2015)
13. Krizhevsky, A., Sutskever, I., Hinton, G.E.: Imagenet classification with deep convolutional neural networks. In: NIPS (2012)
14. Liu, S., Feng, J., Song, Z., Zhang, T., Lu, H., Xu, C., Yan, S.: Hi, magic closet, tell me what to wear! In: Proceedings of ACM Multimedia (2012)
15. van der Maaten, L., Hinton, G.E.: Visualizing high-dimensional data using t-SNE. JMLR **9**, 2579–2605 (2008)
16. Mikolov, T., Chen, K., Corrado, G., Dean, J.: Efficient estimation of word representations in vector space. CoRR abs/1301.3781 (2013)
17. Nguyen, H.T., Almenningen, T., Havig, M., Schistad, H., Kofod-Petersen, A., Langseth, H., Ramampiaro, H.: Learning to rank for personalised fashion recommender systems via implicit feedback. In: Prasath, R., O'Reilly, P., Kathirvalavakumar, T. (eds.) MIKE 2014. LNCS (LNAI), vol. 8891, pp. 51–61. Springer, Heidelberg (2014). doi:10.1007/978-3-319-13817-6_6
18. Platt, J.C.: Probabilistic outputs for support vector machines and comparisons to regularized likelihood methods. In: Advances in Large Margin Classifiers (1999)
19. van de Sande, K.E.A., Gevers, T., Snoek, C.G.M.: Empowering visual categorization with the GPU. IEEE Trans. Multimed. **13**(1), 60–70 (2011)
20. Sermanet, P., Eigen, D., Zhang, X., Mathieu, M., Fergus, R., LeCun, Y.: Overfeat: Integrated recognition, localization and detection using convolutional networks. In: ICLR (2014)
21. Sharif Razavian, A., Azizpour, H., Sullivan, J., Carlsson, S.: CNN features off-the-shelf: an astounding baseline for recognition. In: CVPR Workshops (2014)
22. Simo-Serra, E., Fidler, S., Moreno-Noguer, F., Urtasun, R.: Neuroaesthetics in fashion: modeling the perception of fashionability. In: CVPR (2015)
23. Tsai, T.H., Cheng, W.H., You, C.W., Hu, M.C., Tsui, A.W., Chi, H.Y.: Learning and recognition of on-premise signs (opss) from weakly labeled street view images. IEEE Trans. TIP **23**(3), 1047–1059 (2014)
24. Weijer, J.V.D., Gevers, T., Bagdanov, A.: Boosting color saliency in image feature detection. IEEE Trans. PAMI **28**, 150–156 (2005)
25. Yamaguchi, K., Berg, T.L., Ortiz, L.E.: Chic or social: visual popularity analysis in online fashion networks. In: ACM ICM (2014)
26. Yamaguchi, K., Kiapour, M.H., Ortiz, L.E., Berg, T.L.: Retrieving similar styles to parse clothing. IEEE Trans. PAMI **37**(5), 1028–1040 (2015)

SS4: Multimedia and Multimodal Interaction for Health and Basic Care Applications

Boredom Recognition Based on Users' Spontaneous Behaviors in Multiparty Human-Robot Interactions

Yasuhiro Shibasaki[1], Kotaro Funakoshi[2(✉)], and Koichi Shinoda[1]

[1] Tokyo Institute of Technology, Meguro, Japan
shibasak@ks.cs.titech.ac.jp, shinoda@cs.titech.ac.jp
[2] Honda Research Institute Japan Co., Ltd., Wako, Japan
funakoshi@jp.honda-ri.com

Abstract. To recognize boredom in users interacting with machines is valuable to improve user experiences in human-machine long term interactions, especially for intelligent tutoring systems, health-care systems, and social assistants. This paper proposes a two-staged framework and feature design for boredom recognition in multiparty human-robot interactions. At the first stage the proposed framework detects boredom-indicating user behaviors based on skeletal data obtained by motion capture, and then it recognizes boredom in combination with detection results and two types of multiparty information, i.e., gaze direction to other participants and incoming-and-outgoing of participants. We experimentally confirmed the effectiveness of both the proposed framework and the multiparty information. In comparison with a simple baseline method, the proposed framework gained 35% points in the F1 score.

Keywords: Gesture · Posture · Gaze · Spoken dialogue

1 Introduction

With recent technology advancement, new kinds of computer applications are emerging such as intelligent tutoring systems and social assistants [9]. These systems are expected to engage in long term interactions or dialogues with users to assist them for a variety of purposes including care giving. Sustaining long term human-machine interactions requires rapport building through affective state recognition of users [3,7,8]. In this paper, we focus on boredom in users who are interacting with such systems or robots. Although there are related but kind of opposite notions of boredom: curiosity and interest, we go for boredom because dialogues with machines including robots tend to be tedious due to the limitations of immature technologies. Automatic assessment of boredom would be highly valuable in such situations [15] and in fact we can implement various measures which are triggered by recognition of boredom in users, such as joking, encouraging, inserting small talk, etc.

© Springer International Publishing AG 2017
L. Amsaleg et al. (Eds.): MMM 2017, Part I, LNCS 10132, pp. 677–689, 2017.
DOI: 10.1007/978-3-319-51811-4_55

It is well known that non-verbal information is important or even rather dominant in communication [13]. There has been several approaches to boredom recognition with non-verbal communication cues such as head positions [10], posture [5,14]. There also has been a number of researchers trying to detect user's curiosity in customer service applications, interest detection in one-to-one interactions, and in meetings [11,16].

Following the previous works, this paper also relies on non-verbal information. Fundamentally visual information is used. Here, the most major type of information is skeletal data obtained by motion capture, which captures users' spontaneous physical behaviors, i.e., gestures and postures, that possibly indicate users' boredom. Our proposed two-stage boredom recognition framework detects these boredom indicating behaviors, or boredom behaviors for short, at the first stage, and then recognizes boredom in combination of the detection results and other multiparty-based information. The use of this multiparty-based information in boredom recognition is the major contribution of this paper, as all the previously proposed methods focus on the target user only as far as the authors know, while multiparty settings in human-machine interaction are increasing in popularity and importance [12,17]. We experimentally confirm the effectiveness of both the proposed boredom recognition framework and the multiparty information. In comparison with a simple baseline method, the proposed framework gained 35% points in the F1 score.

Another contribution of this paper is the use of separate boredom behavior detectors for body parts (i.e., trunk, legs, arms, and head). We demonstrate that using multiple detectors for different body parts only can achieve a performance nearly equal to that of using a whole body-based boredom behavior detector. This feature is advantageous to build a robust boredom recognizer against occlusions and changes in interaction settings (e.g., standing and sitting settings). It is also favorable to retrain only weak detectors with additional data.

In what follows, first Sect. 2 describes a human-robot dialogue corpus that we used in our work. Then Sect. 3 explains our annotation scheme of boredom states and boredom behaviors, and the results of annotation to the dialogue corpus. Section 4 gives the details of the proposed boredom recognition framework and feature design. Section 5 shows the results of two offline evaluation experiments and an error analysis. Finally Sect. 6 concludes this paper.

2   Multiparty Dialogue Corpus

We use the same multimodal corpus as that described in [17], which now includes 30 sessions of human-robot interactions between a robot and multiple human participants.

The situation of the data collection is illustrated in the left of Fig. 1. The 30 sessions were collected in a Wizard-of-Oz (WOZ) manner, and were recorded using a Microsoft Kinect sensor (V1) and a microphone. In each session, up to 3 participants could interact with a robot for 25 min. Each initially stayed in one of the two waiting spaces and was individually directed to join or leave at arbitrary

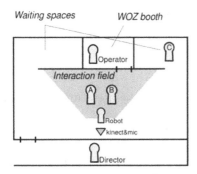 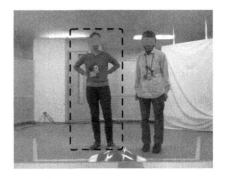

Fig. 1. (Left) Data collection setting. (Right) Illustration of boredom behaviors. The left-sided person (marked by a dotted square) spontaneously exhibits typical boredom-related behaviors with her arms and legs.

points in time by an experimental director via an earphone and a walkie-talkie using a dedicated channel to the participant. A group of 3 people who knew each other well (i.e., friends or families) participated in a session. Thus 30 different groups of 90 people (45 males and 45 females) joined the data collection. The ages of them varied from 18 to 65.

The robot was controlled by a wizard, i.e., a human operator, who was hidden from the participants. As they passed by, the robot called out to them, introduced itself, then invited them to play a game. If they agreed, the rules were explained and, if understood, the game begun. The game was a '20-questions' scenario where the robot thinks of an object and the participants could ask yes/no questions. The game was successful if one of the participants guessed the correct object in 20 turns. Multiple games could be played in a session.

In advance of each data collection session, the robot was explained to participants as an English conversation trainer robot under development. Thus, the robot spoke only English via TTS, while the participants could speak either English, Japanese or a mixture of English and Japanese. The wizard understood both English and Japanese. This setting was a potential source of rather frequent boredom in participants because quite a few participants had trouble understanding the robot's speech in English.

Using the ELAN annotation tool, professional annotators annotated the gaze, speech recipients, participation states (passing, observing, or participating), dialogue acts, and transcribed the speech of each participant and the robot.

3 Boredom Annotation

We additionally annotated the corpus described in Sect. 2 on boredom from two points of view, i.e., boredom states and boredom behaviors. Boredom states represent whether participants are bored or not based on third person observations. Boredom behaviors are spontaneous behaviors (gestures and postures) of users that could be a signal of boredom. The picture in Fig. 1 shows examples of boredom behaviors. In this section, the annotation protocols and results of the two types of annotation are explained in sequence.

3.1 Boredom States

We developed an original annotation schema based on the literature [4, 10] and our trials and errors.

First the recorded videos were split into contiguous non-overlapping intervals of 5 s. Then for each session, by observing the audio (speech) and video (body movement) interval by interval in reverse order[1], two annotators independently annotated per participant all the intervals whether the participant was bored or not according to the annotator's intuitive judgement based on the common-sense notion of boredom[2]. When an annotator could not judge whether the participant was bored or not, the annotator chose 'cannot say'. Hereafter, we call each pair of an interval and a participant 'sample', which is the data unit of processing and evaluation.

The results of two annotators on a target person in an interval were merged to fix the final boredom state of each sample, i.e., BORED, MAYBE, or NOT BORED. Only a sample which was judged as 'bored' by both of the two annotators, was assigned the label of 'BORED', vice versa for 'NOT BORED'. The remaining samples were assigned the label of 'MAYBE'.

As a result, we obtained 893 BORED samples, 2,582 MAYBE samples, and 7,265 NOT BORED samples, in total 10,740 samples. The Cohen's Kappa value (calculated per participant and averaged) was 0.441, that could be interpreted as a moderate agreement between annotators.

3.2 Boredom Behaviors

Next we annotated the samples defined in Sect. 3.1 on the existence of boredom-related behaviors. We define boredom behaviors as either spontaneous gestures, postures or miscellaneous body movements that can indicate one's boredom. Table 1 shows the list of boredom behaviors observed in the corpus. As shown in the list, each of those behaviors can be ascribed to one or several body parts of trunk, legs, arms, and head. As for examples of non boredom behaviors, we can list such as showing excitement by lifting one's hands, touching another participant, etc.

Note that boredom behaviors do not necessarily mean one's boredom. Therefore, the binary annotation of boredom behaviors (any of them is observable or

---

[1] This is intended to make an annotator's judgement on an interval independent of the neighboring intervals as much as possible in a reasonable cost of annotation. In forward annotation, the influence of the judgement of an interval on the judgement of the next interval could be larger. The playback of each interval was of course forwarded.

[2] Although it is known that there are several types of boredom [6], we employ a naive notion of boredom for the sake of simplicity.

Table 1. List of observed boredom behaviors (T: Trunk, L: Legs, A: Arms, H: Head).

Spontaneous behavior	T	L	A	H	Spontaneous behavior	T	L	A	H
Swinging body	✓				Fixing hair or clothes			✓	
Stretching body	✓				Touching face or ears			✓	
Twisting body	✓				Putting chin in hand			✓	
Tilting body backward	✓				Fumbling with fingers			✓	
Switching the pivot foot		✓			Folding arms			✓	
Putting weight on one foot		✓			Grasping hips			✓	
Crossing legs		✓			Rubbing hands			✓	
Stamping		✓			Looking aside from participants				✓
Wandering		✓			Looking at a non-participant				✓
Crouching to touch knees		✓	✓		Closing eyes				✓
Squatting to clasp knees		✓	✓		Leaning head to one side				✓
Touching the floor		✓	✓		Yawning				✓

nothing is observable) were done independently of the boredom state labels. We did not ask the annotators to label the kinds of boredom behaviors that they recognized, because annotating those kinds which were not exclusive to each other was expensive.

As well as the case of boredom states, each sample was annotated by two annotators independently of each other, and the final decision on the sample was made by merging the judgements of the two. Only when two annotators coincided on the existence of any boredom behavior in a sample, the sample was regarded as a positive example of boredom behaviors. The Cohen's Kappa value (calculated per participant and averaged) was 0.662, that could be interpreted as a substantial agreement.

3.3 Boredom Behavior Part and Duration

As mentioned above each boredom behavior could be ascribed to some body parts. Thus we classified each positive sample of boredom behaviors into 5 categories of BB-Trunk, BB-Legs, BB-Arms, BB-Head, and BB-W.B. BB-W.B. means that any type of boredom behavior is recognizable over the whole body. That is, if one classifies a sample into any of BB-Trunk, BB-Legs, BB-Arms, or

Table 2. Per-category numbers of available samples.

Category	strong	weak	none	Total
Boredom state	366	952	2,448	3,766
BB-Trunk	46	134	3,586	
BB-Legs	265	1,203	2,298	
BB-Arms	150	942	2,674	
BB-Head	161	782	2,823	
BB-W.B	2,772	238	756	

BB-Head, it is also classified as BB-W.B. Again the types of boredom behaviors are not exclusive and thus any sample can be classified into multiple categories aside from BB-W.B.

We also classified the positive samples in terms of behavior duration. Our samples are based on five-second intervals. While some samples contain substantial boredom behaviors, some samples contain very brief behaviors. Those samples containing only short behaviors could be noises when we build behavior detectors assuming five-second interval inputs. The classification was done based on whether a behavior was observable longer than 2.5 s or not.

In combination of these two bases of classification, the positive samples of boredom behaviors were classified into 10 classes ({BB-*} × {strong, weak}). In addition to these classifications, the quality of skeletal data obtained by Kinect in each sample was visually inspected, and only the qualified samples were held out for latter experiments. As a result, we obtained 3,766 qualified samples in total. These classifications and inspections were done by the first author.

The distributional details of these qualified samples over the parts and duration classifications are shown in Table 2[3]. One may think that the number of strong BB-W.B. samples (2,722) seems to be too many in comparison to other strong samples. This is because even if a sample is classified as weak with regards to a body-part (either BB-Trunk, BB-Legs, BB-Arms, or BB-Head), it could contain boredom behaviors longer than 2.5 s in total as a combination of multiple boredom behaviors.

4   Boredom Recognition

4.1   Recognition Framework

We propose a motion-based two-staged boredom recognition framework as shown in Fig. 2. The framework receives as input skeletal information and face directions

---

[3] The meanings of strong/weak are different between the boredom state and the other boredom behavior categories (BB-*). For the boredom state, 'strong', 'weak', and 'none' respectively mean the samples with boredom state labels 'BORED', 'MAYBE', and 'NOT BORED'. For the other behavior categories, 'strong' means the samples that contain relevant behaviors more than 2.5 s. 'weak' means those less than 2.5 s, and 'none' means those with no relevant behaviors.

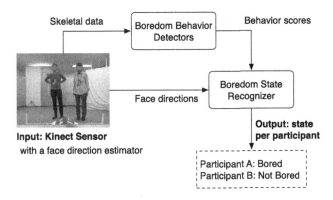

Fig. 2. Proposed framework of boredom recognition.

Table 3. Features for boredom behavior detection.

Part	Feature (3-dimensional directional unit vector)
Trunk	From the robot to the target user's hip center
Legs	From the user's right hip to the right knee
	From the user's right knee to the right foot
	From the user's left hip to the left knee
	From the user's left knee to the left foot
Arms	From the user's right shoulder to the right elbow
	From the user's right elbow to the right hand
	From the user's left shoulder to the left elbow
	From the user's left elbow to the left hand
Head	From the user's neck (shoulder center) to the head

of every participant in a sample. Then boredom behavior detectors dedicated to different body parts calculate scores for each participant based on skeletal data. Finally a boredom state recognizer outputs classification results based on boredom behavior scores, face directions, and the information of incomings and outgoings of participants.

4.2 Boredom Behavior Detection

Boredom behavior detection requires pattern recognition on sequential motion data. HMM (Hidden Markov Model) is a popular framework for gesture recognition. Although we simply use forward-chaining linear topology HMMs, which are referred to as 'conventional HMMs' in [2], we propose to build separate detectors for different body parts. This is expected to increase robustness in detection. In this paper, we build five detectors, each of which corresponds to one of the five behavior categories defined in Sect. 3.3.

Table 3 shows the features used by the five detectors, i.e., BBD-Trunk, BBD-Legs, BBD-Arms, BBD-Head, and BBD-W.B. Detector BBD-W.B. uses all the features in Table 3, while each of the others uses only the relevant features to it.

These features are 3-dimensional directional unit vectors computed from the skeletal data of one participant. Therefore, BBD-Trunk and BBD-Head handle 3-dimensional input, BBD-Legs and BBD-Arms handle 12-dimensional input, and BBD-W.B. handle 30-dimensional input.

4.3   Boredom State Recognition

The output of boredom recognition in our proposed framework comes from the output of the boredom state recognizer, which uses the scores from the boredom behavior detectors and two other kinds of multiparty information.

The boredom state recognizer is implemented as a classifier. In this paper, we will compare the two classification algorithms of Support Vector Machines (SVM) and Random Forests (RF).

The recognizer uses 18 dimensional features based on three types of information as shown in Table 4. The first type of information is the boredom behavior detection scores, or likelihoods, that are given by the five HMMs.

The second type of information is the incoming and outgoing information of participants. Incoming or outgoing information of a person is a significant event in interactions. Usually it evokes an off-game dialogue between the person and the robot to make the person join the game, and often refreshes a constipated situation. (especially when another person joins the game between a robot and a person). Therefore it is likely that this kind of information affects the boredom state of a participant. We take in this kind of information as the elapsed time from the last incoming or outgoing person.

The third is gaze score, that is, whether a participant is looking at the robot or another participant. We define the gaze score as the cosine similarity between two directional vectors, one vector represents the face direction of a person in question, and the other represents a target entity in a relative position from the person. If a person in question is looking at the target straight on, the gaze score

Table 4. Features for boredom state recognition.

Category	Feature description
Behavior detection	Scores from BBD-{Trunk, Legs, Arms, Heads, W.B.} (5 dim.)
Human in-and-out	Time from the last incoming or outgoing of a human (1 dim.)
Gaze score	Max. & means of 0th, 1st, 2nd-order values to the robot (6 dim.)
	Max. & means of the same values to the other user (6 dim.)

value becomes 1. If the person is looking away from the target orthogonally, it becomes 0. Here we use two gaze scores: (1) the gaze score to the robot, and (2) the gaze score to another person who is also participating in the game. From each gaze score, we extract six feature values per sample. First we compute the 1st and 2nd-order differentials of gaze score in a sample. Then the maximums and means of the raw (0th-order), 1st-order, 2nd-order differential values in a sample are obtained.

5   Evaluation

We experimentally evaluated the framework and features proposed in Sect. 4 using the data samples described in Sect. 3.

5.1   Baseline Method

As a baseline method, we adopt a single RF-based classifier. The classifier basically used all the same information available from the input sensor, however, in more naive ways. Multiparty information is not incorporated.

The features used were the maximums and means of the 1st-order and 2nd-order differentials of six joint positions, that is, head, left elbow, right elbow, left knee, right knee, and hip center, resulting in 72 dimensions. In addition, the maximums and means of the 0th-order, 1st-order and 2nd-order differentials of individual face direction (pitch, yaw, and roll) were used, too. The final feature vectors for the baseline method had 90 dimensions.

5.2   Experimental Settings

We used the 'strong' samples in Table 2 as positive examples and the rest as negative examples. The sampling rate of skeletal data was 30 fps. Face direction in the data samples was estimated by the face tracker implementation in the Kinect SDK library. The human in-and-out feature was computed based on the manual annotation of participation states described in Sect. 2.

We used the Kaldi speech recognition toolkit to implement the HMM-based boredom behavior detectors. The 12-state left-to-right HMM structure was used for all the detectors. Emission probabilities were modeled by 64-mixture Gaussian distributions.

As for the implementations of SVM and RF used in the boredom state recognizer, the Weka 3.6 data mining software was used. The kernel of SVM was RBF. The values of parameter C and γ were 6 and 0.1, respectively. We set the number of trees of RF to 18 according to preliminary experiments. The other parameters were set to the default values of Weka.

All the experiments were performed in 10-fold cross validation. Both the baseline and the proposed method used the same parameter values.

5.3   Results

Boredom Behavior Detection. First we confirmed that the HMM-based detectors worked properly. The left table in Table 5[4] shows the unit performances of the five boredom behavior detectors in the metrics of precision, recall and F1. F1 is the harmonic average of precision and recall. The detectors presented reasonable performances of about .80 in the F1 score.

Table 5. (Left) Results of boredom behavior detection. (Right) Results of boredom state recognition.

Detector	Prec.	Rec.	F1	#targets	#residuals
BBD-Trunk	.70	.66	.68	46	3,720
BBD-Legs	.84	.75	.79	265	3,501
BBD-Arms	.75	.82	.78	150	3,616
BBD-Head	.77	.83	.80	161	3.605
BBD-W.B.	.77	.79	.79	2,772	994

Recognizer	Prec.	Rec.	F1
Baseline	.52	.36	.43
Proposed-SVM	.79	.58	.67
Proposed-RF	.91	.69	.78
Oracle-BBD	.95	.85	.89

Boredom State Recognition. The right table in Table 5 shows the results of the final boredom recognition. The proposed method achieved considerably better results than the naive baseline. In this experiment, RF worked better than SVM with the RBF kernel.

Oracle-BBD shows the results of the boredom state recognizer that used not the scores from the boredom behavior detectors but the boredom behavior annotation (1 if any relevant boredom behavior exists, otherwise 0). This shows that improving the boredom behavior detectors will increase the performance up to 11% points in the F1 score. The gain from the Proposed-RF to Oracle-BBD is noticeable in recall. This indicates that enlarging the data size will probably improve the performance. On the other hand, gaining the remaining 11% points (i.e., 100 − 89) demands improvements in boredom state recognition itself, e.g., incorporating other types of features. This is to be discussed in Sect. 5.4.

When we removed the BBD-W.B. score feature, the performance in the F1 score degraded less than 1 point. This suggests that we do not need to build a boredom detector that monitors all the participants' bodies. BBD-W.B. requires BB-W.B. annotation, but it is cumbersome for annotators to annotate BB-W.B. more than other specific ones. Moreover, using multiple detectors for different body parts only is advantageous to build a robust boredom recognizer as pointed out in Sect. 1.

---

[4] '#targets' means the numbers of samples in column 'strong' shown in Table 2. '#residuals' means the sums of the numbers of the 'weak' and 'none' samples.

5.4   Error Analysis

We looked into the samples that were recognized improperly.

Boredom behavior detection errors were found with scenes in which a participant exhibits the boredom behavior of 'rubbing hands'. The samples containing such a scene were not detected as a boredom behavior. Because the precision of skeleton recognition with a Kinect sensor was not precise enough at the ends of limbs, we did not include hands and feet in the features for boredom behavior detection. Such a sample requires these features to be recognized. Furthermore, we may need a fine hand/finger motion sensor that is comparable to a data glove.

Other detection errors were found with scenes in which participants are looking aside. The samples containing such a scene were not detected as a boredom behavior. So far we use face direction information only. A more precise gaze detection may be necessary.

In one scene, a person showed obvious inactivity as the lack of response to the other participant's encouragement to actively join the game. Human annotators recognize this scene as an indication of boredom, but the current proposed features do not. To deal with such a situation, we have to incorporate more interaction-oriented features. At least the use of speech signals is essential as we used only visual information.

Some participants, by habit, show spontaneous behaviors that are defined as boredom behaviors in general while they are not bored at all. In one scene, a person who was concentrating on a quiz showed typical boredom behaviors such as 'swinging body' and 'putting weight on one foot'. In another scene, two persons converse cheerfully but one of them was showing boredom behaviors such as 'Grasping hips'. Thus our recognizer wrongly recognized boredom in both cases. To deal with such cases we will have to adapt to different personalities.

6   Conclusion

This paper proposed a two-staged boredom recognition framework and a feature design based on users' physical motions, and, by offline experiments, demonstrated the effectiveness of them in a multiparty human-robot interaction setting.

On recognizing boredom in users, we attended to their spontaneous physical behaviors that could indicate boredom. Therefore, we annotated an existing simulated multiparty human-robot dialogue game corpus not only on boredom states (bored/cannot say/not bored) but also on the existence of boredom-indicating behaviors, or boredom behaviors in short, such as folding arms, wandering, etc. We also designed a manual annotation scheme to label data by third persons in consideration of subjectivity and vagueness of 'boredom'. Our proposed annotation scheme showed a moderate agreement in the annotation of boredom states and a substantial agreement in the annotation of boredom behaviors.

At the first stage of the recognition framework, boredom behavior detectors that are dedicated to different body parts output scores about existence of boredom behaviors based on users' sequential skeletal data in a five-second interval. We adopted 12-state left-to-right 64-mixture Gaussian HMM to implement detectors. In a unit test, the implemented detectors showed substantial performances of .68–.79 F1 scores.

At the second stage, a classifier tells whether users in a given interval are bored or not using the scores from the first stage and multiparty-oriented information of face directions and incoming and outgoing participants. We adopted Random Forests to implement a classifier, and obtained .78 in the F1 score, gaining 35% points from the baseline. However, we selectively used only samples with enough quality of skeletal data in the evaluation experiments. This point must be cautiously considered in interpreting the results.

As for the results of error analysis, limitations in the implemented boredom behavior detectors were identified, that is, inabilities of recognizing fine motions of hands, fingers, and eyes. One major issue identified in final boredom recognition was incorporating more interaction-oriented features and contextual features. At least the use of speech signals must be explored in future work. Another issue was handling participants' individual habits and adapting to their personalities, or traits [1], as some of them frequently showed boredom behaviors while they were not bored.

In future work, we will work on these issues, and evaluate the proposed boredom recognition in an online interaction between a fully automated system and actual users.

References

1. Bixler, R., D'Mello, S.: Detecting boredom and engagement during writing with keystroke analysis, task appraisals, and stable traits. In: 2013 International Conference on Intelligent User Interfaces, pp. 225–234 (2013)
2. Brand, M., Oliver, N., Pentland, A.: Coupled hidden markov models for complex action recognition. In: IEEE Computer Society Conference on Computer Vision and Pattern Recognition, pp. 994–999 (1997)
3. Breazeal, C., Aryananda, L.: Recognition of affective communicative intent in robot-directed speech. Auton. Robots **12**, 83–104 (2002)
4. Castellano, G., Pereira, A., Leite, I., Paiva, A., McOwan, P.W.: Detecting user engagement with a robot companion using task and social interaction-based features. In: 2009 International Conference on Multimodal Interfaces, pp. 119–126 (2009)
5. D'Mello, S., Chipman, P., Graesser, A.: Posture as a predictor of learner's affective engagement. In: 29th Annual Conference of the Cognitive Science Society, pp. 905–910 (2007)
6. Goetz, T., Frenzel, A.C., Hall, N.C., Nett, U.E., Pekrun, R., Lipnevich, A.A.: Types of boredom: an experience sampling approach. Motiv. Emot. **38**(3), 401–419 (2014)
7. Gratch, J., Wang, N., Okhmatovskaia, A., Lamothe, F., Morales, M., Werf, R.J., Morency, L.-P.: Can virtual humans be more engaging than real ones? In: Jacko, J.A. (ed.) HCI 2007. LNCS, vol. 4552, pp. 286–297. Springer, Heidelberg (2007)
8. Hudlicka, E.: To feel or not to feel: the role of affect in human-computer interaction. Int. J. Hum. Comput. Stud. **59**(1), 1–32 (2003)

9. Jacko, J.A.: Human Computer Interaction Handbook: Fundamentals, Evolving Technologies, and Emerging Applications. CRC Press, Boca Raton (2012)
10. Jacobs, A.M., Fransen, B., McCurry, J.M., Heckel, F.W., Wagner, A.R., Trafton, J.G.: A preliminary system for recognizing boredom. In: 4th ACM/IEEE International Conference on Human-Robot Interaction, pp. 299–300 (2009)
11. Kennedy, L.S., Ellis, D.P.: Pitch-based emphasis detection for characterization of meeting recordings. In: 2003 IEEE Workshop on Automatic Speech Recognition and Understanding, pp. 243–248 (2003)
12. Liu, Y., Chee, Y.S.: Intelligent pedagogical agents with multiparty interaction support. In: IEEE/WIC/ACM International Conference on Intelligent Agent Technology, pp. 134–140 (2004)
13. Mehrabian, A.: Silent messages. Wadsworth, Belmont (1971)
14. Mota, S., Picard, R.W.: Automated posture analysis for detecting learner's interest level. In: Computer Vision and Pattern Recognition Workshop, pp. 49–49 (2003)
15. Pantic, M., Caridakis, G., André, E., Kim, J., Karpouzis, K., Kollias, S.: Multimodal emotion recognition from low-level cues. In: Petta, P., Pelachaud, C., Cowie, R. (eds.) Emotion-Oriented Systems, pp. 115–132. Springer, Heidelberg (2011)
16. Qvarfordt, P., Beymer, D., Zhai, S.: RealTourist – a study of augmenting human-human and human-computer dialogue with eye-gaze overlay. In: Costabile, M.F., Paternò, F. (eds.) INTERACT 2005. LNCS, vol. 3585, pp. 767–780. Springer, Heidelberg (2005). doi:10.1007/11555261_61
17. Sugiyama, T., Funakoshi, K., Nakano, M., Komatani, K.: Estimating response obligation in multi-party human-robot dialogues. In: 2015 IEEE-RAS 15th International Conference on Humanoid Robots, pp. 166–172 (2015)

Classification of sMRI for AD Diagnosis with Convolutional Neuronal Networks: A Pilot 2-D+ε Study on ADNI

Karim Aderghal[1,3], Manuel Boissenin[1], Jenny Benois-Pineau[1(✉)], Gwenaëlle Catheline[2], and Karim Afdel[3]

[1] ENSEIRB, LaBRI, Laboratoire Bordelais de Recherche en Informatique, Université de Bordeaux, 351, cours de la Libération, 33405 Talence cedex, France
Manuel.Boissenin@gmail.com, jenny.benois@labri.fr
[2] CNRS UMR 5287 - INCIA Institut de Neurosciences cognitives et intégratives d'Aquitaine Université Victor Segalen, 146 rue Léo Saignat, 33076 Bordeaux cedex, France
[3] LabSIV, Université Ibn Zhor, BP 32/S, Agadir 80000, Maroc

Abstract. In interactive health care systems, Convolutional Neural Networks (CNN) are starting to have their applications, e.g. the classification of structural Magnetic Resonance Imaging (sMRI) scans for Alzheimer's disease Computer-Aided Diagnosis (CAD). In this paper we focus on the hippocampus morphology which is known to be affected in relation with the progress of the illness. We use a subset of the ADNI (Alzheimer's Disease Neuroimaging Initiative) database to classify images belonging to Alzheimer's disease (AD), mild cognitive impairment (MCI) and normal control (NC) subjects. As the number of images in such studies is rather limited regarding the needs of CNN, we propose a data augmentation strategy adapted to the specificity of sMRI scans. We also propose a 2-D+ε approach, where only a very limited amount of consecutive slices are used for training and classification. The tests conducted on only one - saggital - projection show that this approach provides good classification accuracies: AD/NC 82.8% MCI/NC 66% AD/MCI 62.5% that are promising for integration of this 2-D+ε strategy in more complex multi-projection and multi-modal schemes.

Keywords: Alzheimer's disease · AD · Convolutional neural network · CNN · Deep learning · Structural magnetic resonance imaging · sMRI · Hippocampus · Computer-aided diagnosis · CAD

1 Introduction

With the world population aging, the care of neurodegenerative diseases has become crucial and computer-aided interactive systems can play a key role in the medical practice. Alzheimer's disease (AD) is the most prevalent neurodegenerative brain disease affecting the elders, its prevalence is estimated to be around 5% after 65 years old and a staggering 30% for the more than 85 years

© Springer International Publishing AG 2017
L. Amsaleg et al. (Eds.): MMM 2017, Part I, LNCS 10132, pp. 690–701, 2017.
DOI: 10.1007/978-3-319-51811-4_56

old in developed countries [1]. From now to 2050 it is estimated that 0.64 Billion people in the world will be diagnosed with AD. Aside of being a major social and economical issue, see insert below, its effects are devastating not only for the diseased but also for the families that have the heavy duty of taking care of the patient.

France: about 900 000 diseased, costs are estimated at more than €14.3 Billions (€5.3 Billion of medical and paramedical cost + €9 Billion of social-medical cost) furthermore, the opportunity costs are estimated at €14 Billions [4].

Europe: more than 7 Million people are suffering directly from the Alzheimer's disease (AD), costs are evaluated to €71 Billion of direct spendings and a prudent opportunity cost of 89 Billion euros which represents more than 22000€ per sufferer per year [3].

USA: 5.4 Million people, $150 Billion dollars per year, 18 Billion hours of unpaid care, a contribution to the nation valued at over $220 Billion, for a more extended picture you may refer to [2].

When speaking about computer-aided diagnosis systems, different modalities for establishment of the diagnosis can be used: such as different cognitive, neurological and psychological tests as the mini mental state (MMS) evaluation, the Modified Mini Mental evaluation (3MS), *etc.*

To reinforce the diagnostic, various technical and biological testing can also be performed; different biomarkers[1] can be identified in the cerebrospinal fluid (CSF) after a lumbar puncture, and, different image modalities, such as MRI, tractilograpy (also known as Diffusion Tensor Imaging (DTI)) and Positron Emission Tomography (PET), are used to complement and reinforce the diagnosis of an AD.

In this article we focus on using CNN on the hippocampal region, which shape is known to change significantly at the inception and with the progress of an AD. The modality of choice is sMRI scans for which relatively large image databases are available since they are one of the best clinical candidate for routine analysis. Before integrating multimodal systems, assessing the performances of CNN on a single modality allows us to better apprehend its discriminating power. We use a dataset from ADNI (see Sect. 3) of labeled scans of patients with AD, MCI subjects and normal control subjects (NC). The paper is organized as follows: in Sect. 2 we briefly review related works, in Sect. 3 we present the Open dataset ADNI, in Sect. 3.2 we shortly explain the extraction of the hippocampal region, in Sect. 4 we present the characteristics of our network and evaluate various domain-dependent data augmentation strategies. Results and discussion are given in Sect. 5. Section 6 contains conclusions and improvement perspectives.

---

[1] The main ones being the tau, beta amyloid protein and phophorylated tau proteins.

2 Related Works

Despite its recent tremendous successes, CNNs are only starting to be used for CAD[2] in classification of brain images, and the literature does not provide many attempts to use CNN for diagnosis and prognosis of AD. Thus we have found only two previous works [9,11] closely related to our approach.

In [9] MRI of the whole brain have been used, 150 features are first learned using a sparse auto encoder on $5 \times 5 \times 5$ image patches. These features are then used for the first and only convolutional layer of their CNN. Features were not further trained during the last back-propagation training stage of the final network, the convolutional layers are followed by a max-pooling layer, a fully connected layer of 800 units and the output units. Results are promising: three-way classification: 89.47%, AD vs. NC: 95.39%, AD vs. MCI: 86.84%, NC vs. MCI: 92.11% on an ADNI dataset of 755 patients in each one of the three classes, for a total of 2,265 scans. In their experiments 3-D patches provided better classification results than 2-D patches. The main distinctions from our work is that we focus on a specific part of the brain while they considered the whole brain, we use more than one convolutional layer and we did not pre-train features.

Another study using 3-D CNN [11] confirms that the usage of CNN is a good choice for classifying MRI scans as belonging to NC/MCI/AD individuals. The 3-D CNN was used on the whole brain and initialised with convolutional auto-encoders, training was done on the CADementia database and the resulting CNN was tested on 210 scans of the ADNI database. Comparisons with other techniques using various image modalities confirm that both the choice of using sMRI and CNN is relevant.

Nonetheless, several other previous attempts to use deep neural nets have already been made using multiple modalities. For instance, in [12] multiple indicators from multiple modalities: MRI, PET scans and CSF biomarkers were fused to evaluate the state of the patient. The inputs consist of 93-region-of-interest-based-volumetric features extracted from PET and MRI scans plus 3 bio-markers from the CSF for a total amount of 189 features, principal component analysis (PCA) on these features, with selection of the discriminative one, was performed then, a deep belief network (DBN), consisting of three hidden layers with hidden units of 100-50-20, was trained and dropout was used in the multi-task learning (MTL) to fine tune the network. The last layer was then used as a new feature representation on which SVM is applied to classify between AD vs. non-AD. Multiple learning schemes have been used and they evaluated their impact by using them or not and measuring the differences in the final classification accuracy. This method, dubbed impact evaluation, showed that dropout and MTL have the major impacts on their performance. The proposed method achieved 91.4%, 77.4%, 70.1%, and 57.4% accuracies for AD vs. HC, MCI vs. HC, AD vs. MCI, and cMCI vs. sMCI classifications, respectively.

---

[2] Computer Aided Diagnosis.

A similar study, combining multiple modalities and using artificial networks as an element of its learning algorithms is [13].

These studies [13,15–17], focusing on prognosis, classify stable MCI (sMCI) vs. progressive MCI (pMCI), also called MCI converters (cMCI) *i.e.* MCI patient that will be diagnosed with AD in a relatively short period of time. In [16] an accuracy of 65% for sMCI vs. pMCI was obtained, 70% in [15] 74% in [17] and 83% accuracy for sMCI vs. cMCI in [13].

Despite many attempts made to use visual patched features for classification [14], which have shown for a while state of the art results, CNN approaches are consistently superseding them, setting new standards for state of the art recognition. We will speak of filters instead of features in the remainder of this paper. This is often explained by the fact that filters relevant to the application are learnt and not designed and as such are more specific to the application domain on which they have been trained.

In the above cited studies various brain regions were used: the whole brain, the hippocampal region or the cingulate posterior cortex together with the hippocampus, *etc.* In our work we focus on the hippocampal region.

Indeed, some parts of the brain present more modifications than others at the onset of AD. It was shown, using morphometric techniques on the hippocampus, in [6] that studying the hippocampus can give good prediction of the evolution from MCI to AD. The accuracy of these predictions appears to be comparable to classification methods operating at the whole brain level highlighting the relevance of this region. Pennanen *et al.* [8] showed that using stepwise discriminant function analysis the hippocampal volume can be used to classify AD vs. NC with an accuracy of 90.7% and AD vs. MCI with 82.3%. However, for MCI vs. NC, the volume of the entorhinal cortex provided better accuracy with 65.9% against 59.7% with the hippocampal volume. This 2003 study comprised 59 NC, 65 MCI and 48 AD subject from Finland. There could be a lateralization of the illness, also, as higher resolution scans of the hippocampus show not all of its parts are affected equally. While the first element, if deemed useful, could be taken into account during the training of the network, CNN would naturally take into account the second element.

3   Dataset and Its Preprocessing

3.1   Presentation of the Used Dataset

To be able to have a classifier that generalises well enough it has to be trained on a large enough dataset. This is specially true for CNN classifiers. Thanks to the ADNI collaborative effort it is now possible to work on larger datasets although they might not be large enough for our chosen artificial network architecture.

We used the screening dataset of the ADNI[3] dataset which consists of scans collected from different acquisition sites using various brands of 1.5 T scanners.

---

[3] http://www.adni.loni.usc.edu/.

For our preliminary tests we do not consider images for patients that have been followed up and whose diagnosis has seldom been revised[4]. Our dataset description is given in Table 1. MMSE values are the results of the Mini Mental State Examination. Although the dataset is not completely standard it is very similar to the one used in [17]. As such it provides a first reasonable indication of the results that could be expected on that dataset which will be used on future work for more rigorous comparisons.

Table 1. Dataset composition of the 815 scans

Category	♯ of scans	Age [range]/μ (σ)	Gender (M/F)	MMSE [range]/μ (σ)
AD	188	[55 91]/75.4(± 7.52)	99/89	[18 27]/23.3 (± 2.03)
MCI	399	[55 89]/74.9(± 7.3)	256/143	[23 30]/27 (± 1.78)
NC	228	[60 90]/76(± 5.02)	118/110	[25 30]/29.1 (± 1.0)

3.2   Extraction of the Hippocampal Region from sMRI Scans

The extraction of the hippocampal region consists of a sequence of stages as in [14] which are summed up in Fig. 1: first MRI scans from the ADNI databasis were selected, they were then aligned with the Montreal Neurological Institute (MNI) template representing an "average brain". Their voxel intensity was

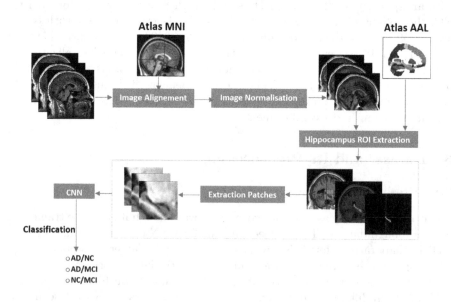

Fig. 1. Data extraction flow chart

---

[4] Cf the diagnostic summary of the ADNI1 dataset.

normalised in order for similar structures to have similar intensities. Then the Region of Interest (ROI) was extracted using the Automated Anatomical Labeling (AAL) atlas. These regions represent the 3-D bounding boxes of the hippocampus in each brain scan. From the extracted regions data augmentation was performed to increase the size of the dataset. This stage is explained in more details in the next section.

4 CNN Architecture and Data Augmentation Strategies

4.1 Architecture

The caffe [10] framework was used to implement the architecture of our network. Our CNN architecture is described in Fig. 2. It consists of two convolutional and max-pooling layers. Rectified linear units (ReLU) were used as activation functions. The two output units are fully connected to the output of the second convolutional layer consisting of $7 \times 7 \times 64$ units. The loss function is a softmax. The first convolutional layer is constituted of 32 filters and the second one of 64 filters. The limited depth of the architecture is conditioned by a small size (28×28) of hippocampal ROIs, extracted from the brain scans. The input is thus a $28 \times 28 \times 3$ layer.

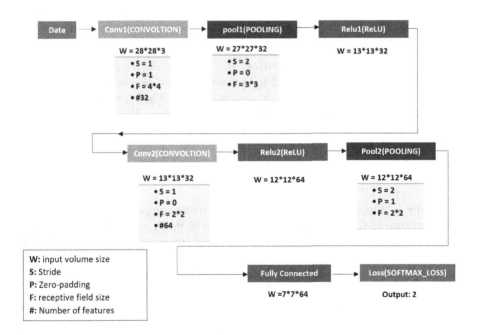

Fig. 2. Architecture of our CNN

Here is an explanation of the name of the technique (2-D+ϵ); instead of using one sagittal slice of the hippocampus and to take into account inter-subject morphological differences we used three adjacent slices. Let us recall that these inter-subject differences are kept intact during the alignment which is a rigid-body affine transformation (translation, rotation and scaling) and that for a given sagittal slice, from one subject to another, slightly different parts of the hippocampus can be captured. Thus, from an implementation perspective, the input layer of the network constituted of $28 \times 28 \times 3$ units receives data from three 28×28 saggital central slices of the hippocampal region. Additionally, during data augmentation we considered translations of these 3 layers. The translation orthogonal to the sagittal plane led us to consider the 2 slices adjacent to the 3 above-mentioned central slices. As a result the network is trained considering the 5 sagittal slices at the center of the hippocampal region. Since we are not training our network directly on the 3-D region that encompasses the hippocampus, which would be of size $28 \times 28 \times 25$, and to distinguish the considered input from this more general input we called the technique 2-D+ϵ.

4.2   Data Augmentation

The selected dataset has then been split as follows: 60% for the training, 20% for validation, 20% to test the trained classifier. As the CNN classifier requires a large amount of data for training, we propose the following data augmentation scheme that are coherent with issues related to the scan acquisition process.

Data Augmentation (DA) Strategies. The first way to augment the data consists in blurring it, as proposed for general-purpose images in [10]. This imitates possible contrast variations in original scans. The blurring was fulfilled with 3×3, 5×5, 7×7 Gaussian filters with (weak) spread parameter: $\sigma = 0.7$, 0.7, 0.6 respectively.

Our second augmentation technique is the translation of the hippocampal ROI by ± 1 pixel in each dimension providing thus 7 times more data than the original one. This considers possible variations dues to alignment imprecisions of scans on the MNI template.

Finally, a "flipping" technique was used. Since the hippocampus is a symmetrical structure of the brain, for each scan twice as much information is obtained by flipping the left hippocampus to match the right one. This is summed up in Table 2.

Note, that the three DA techniques were applied only to the training and validation data. In our work we follow a scenario where no preprocessing is operated on a new brain scan submitted for classification; as such no DA is performed on the test dataset. Table 3 presents early scores obtained on different metrics of the network on the original data with a ten-fold DA (additional blurred images, translated images and flipping). The over-sampling of the MCI category is noticeable for the sensitivity of AD/MCI and the specificity of MCI/NC. This suggests to balance the dataset.

Table 2. Data augmentation: G is the Gaussian blur, T is the translation, F is the flip

	O	G(×3)	T(×6)	(O+T+F)×2
AD	188	564	1128	3760
MCI	399	1197	2394	7980
NC	228	684	1368	4560
Total	815			16300

O: number of original scans
G: number of Gaussian filtered scan
T: translated scans
(O+G+T)×2: total number of data input

Table 3. Binary classifications with augmented data (10x): flip, translation, blur

	AD/NC	AD/MCI	MCI/NC
Accuracy	83.7%	66.5%	64.9%
Sensitivity	79.16%	*36.76%*	76.9%
Specificity	87.2%	79.01%	*45.2%*
Scans	AD (188), MCI (399), NC (228)		

Balancing the Data. Balancing the data consists of taking a similar number of samples for each category. We first have studied three balancing strategies on AD (188 scans) vs. NC (228 scans) classification task, as these two classes are more easily separable. The three techniques of data balancing that we used are: (i) the reduction of the number of original scans from the over-sampled category. Here 188 scans are taken for both AD and NC categories; (ii) the random duplication of the original data from the under-sampled category. Here 228 scans were used for both AD and NC categories; (iii) the random reduction of the augmented data from the category having more data, thus keeping the data variety of the dataset: 188*2*28 = 10528 scans. The results are given in Table 4.

Table 4. Data balancing, (1) simple data reduction, (2) data augmentation by duplication of original scans, (3) randomized reduction of the augmented data

	1	2	3
Accuracy	82.8%	71.3%	79.3%
Sensitivity	79.68%	70.31%	82.81%
Specificity	85.93%	75%	75.86%

From this table we can see, that the first data balancing technique by a random data reduction is the most effective. Hence we have applied it to the more challenging classification problems of AD vs. MCI and MCI vs. NC. Results are given in Table 5. Despite a drop in accuracy, linked to the loss of training data, and showing the necessity of a data augmentation strategy, the three metrics are more balanced compared to results without data balancing.

Table 5. AD vs. MCI and MCI vs. NC with and without a roughly equilibrated number of scans (reduction balancing) with blurred images

	AD vs. MCI		MCI vs. NC	
	Unbalanced	Balanced	Unbalanced	Balanced
Accuracy	66.5%	63.22%	64.9%	58.23%
Sensitivity	36.7%	60%	76.9%	63.33%
Specificity	79.1%	67.14%	45.2%	52.5%
Scans	AD(188), MCI(399)	(188), MCI(199)	MCI(399), NC(228)	MCI(199), (228)

Blurring the Dataset. To augment the dataset, a filter is used to blur scans. These newly generated scans are added to the training and validation datasets. Comparing the accuracy of the resulting trained network, all other parameters being kept equal, showed an increase in accuracy for AD vs. NC which seemed to validate the approach (Table 6).

However, if we consider MCI vs. NC the results are opposite; for some reason, maybe related to small enough differences to be discarded by blurring, the blurring seems to affect negatively accuracy and specificity for this binary classification (Table 6). A similar effect appear for specificity in the result section when comparing the second and third part of Table 7.

Table 6. AD/NC and MCI/NC with and without additional *blurred* images with reduction data balancing

	AD/NC		MCI/NC	
	0 blurring	+blurring	0 blurring	+blurring
Accuracy	80.7%	83.7%	69.1%	*58.23%*
Sensitivity	74.28%	79.1%	52%	52.5%
Specificity	86.45%	87.2%	84.21%	*63.33%*
Scans	AD (188), NC (228)		MCI (199), NC (228)	

5   Results and Discussion

Results of the Method. The following table presents the test results of the CNN. We used a 56-fold Data Augmentation (DA) (×7 from translations, ×4 from blurring and ×2 from symmetry) and reduction data balancing that balances sensitivity and specificity at the expense of a slight reduction in accuracy. Tests were performed on non augmented data first except for the "flipping" technique. Then translation was added and finally blurring. In each binary classification task the number of original samples for training the two-class network was the same: 188 for AD vs. NC, and 228 for MCI vs. NC and 188 for AD vs. MCI.

Table 7. Impact of the data augmentation on results

Without data augmentation but for the flip (x2)			
	AD vs. NC	MCI vs. NC	AD vs. MCI
Accuracy	82.2%	61.8%	64.7%
Sensitivity	88.3%	53.0%	70.0%
Specificity	76.6%	68.3%	60.0%
Data augmentation using translation and flipping (x14)			
	AD vs. NC	MCI vs. NC	AD vs. MCI
Accuracy	76.6%	60.8%	60.6%
Sensitivity	73.4%	60.6%	57.8%
Specificity	81.2%	62.5%	64.0%
Trained with blurred, translated and flipped images (x56)			
	AD vs. NC	MCI vs. NC	AD vs. MCI
Accuracy	82.8%	66.0%	62.5%
Sensitivity	79.6%	73.7%	60.0%
Specificity	85.9%	58.7%	64.0%

Contrary to what we expected, accuracies are reduced by the translation DA scheme, see part 1 and 2 of Table 7. Because the stride is 1 for the receptive fields of the first convolution layer, translations, except for borders, have mostly no effects on the data over which filters are trained, additional tests with stride 2, that will provide a blindness effect compensated by the translations DA strategy are needed to have a finer interpretation of what is happening. The fact that border effects have such an impact on the formation of filters might rise some question about the validity of inter subject variability hypothesis, however from a creativity point of view it might also led to some innovation. Nonetheless all metrics, except specificity for MCI/NC, are boosted back by augmenting the dataset with blurred images (second and third part of Table 7). It remains to determine which blurring parameters provide best results.

Comparison. To put our results into perspective the baseline categorisation, obtained by measuring a single feature which is the volume of the hippocampus, can be used. It was shown [16] that an accuracy of 80% can be obtained to categorize AD/NC. And a 78% accuracy can be obtained for pMCI/NC with this same feature. This can be compared to the 2 first columns of Table 7. In our study we used only five grey-value patches that are coarse segmentations of the hippocampus on a saggital projection.

In [9] accuracies obtained are: 89.47% for 3-way classification, 95.39% for AD/NC 86.84% for AD/MCI and 92.11% for MCI/NC. Nevertheless the whole brain scan has been used to obtain these results. In [11] results in accuracy are respectively 89.1%, 97.6%, 95% and 90.8%, the classifier has also been trained

on whole brain scans but this time on the CAD Dementia dataset and tested on
210 scans of the ADNI dataset.

6   Conclusions

From this pilot study a few pathways are emerging. Since CNN discriminative
power notoriously depends on the size of training data, we have artificially augmented the dataset by translating and blurring it. While the first DA technique
by translation was not conclusive the DA technique using blurring has shown
a strong potential of improvement for our type of classifier. Additionally, we
measured the influence of a few data balancing techniques and showed that balancing the data by suppressing excess data from one category was providing best
results.

Despite using only a subset of the hippocampal region, through the "2D+ϵ"
approach, encouraging accuracy results were obtained: AD/NC 82.2% MCI/NC
66.0% AD/MCI 62.5% that validate the approach and confirm independent and
similar works on the topic [9,11]. While figures for MCI binary tasks classification
remain to be improved, this should naturally occur with the different following
improvement perspectives.

Considering the input of the networks the direct perspective of this research
is to further this study with coronal and transversal slices. Combining these data
could lead to a new type of feature/kernel. The more classical avenues of 3-D
convolutional network and the usage of unsupervised pretrained filters could also
be followed.

From the point of view of the network architecture, a first improvement would
be to add a fully connected layer of n units (n to be determined) just before the
softmax layer.

Finally, more investigations on the translations DA strategy has to be carried
out. Tweaking the stride parameters of the network architecture could be key.

Acknowledgments. We thank Dr. Pierrick Coupé for discussion and insights. This
research is supported by the PHC Toubkal joint research program between France and
the Kingdom of Morocco (TBK/15/23). Data used in the preparation of this article
were obtained from the Alzheimer's Disease Database Initiative (ADNI) database. A
complete listing of ADNI investigators can be found at: www.loni.ucla.edu/ADNI/
Collaboration/ADNI_Autorship_list.pdfdatabase

References

1. Ramaroson, H., Helmer, C., Barberger-Gateau, P., Letenneur, L., Jean-François, D.:
   Prévalence de la démence et de la maladie d'Alzheimer chez les personnes de 75 ans
   et plus: données réactualisées de la cohorte Paquid. Rev. Neurol. (Paris) **159**, 405–
   411 (2003)
2. Gaugler, J., James, B., Johnson, T., Scholz, K., Weuve, J.: Alzheimer's disease
   facts and figures (2016). Alzheimer association

3. Paul-Ariel, K., Katalin, E., László, G., Kristian, K., Alan, J., Anders, G., Linus, J., David, M., Hannu, V., Anders, W.: Impact socio-économique de la maladie d'Alzheimer et des maladies apparentées en Europe, Gérontologie et société (n° 128–129), pp. 297–318 (2009)
4. Bérard, A., Fontaine, R., Aquino, J.-P., Plisson, M.: Combien coûte la maladie d'Alzheimer? Fondation Médéric Alzheimer, September 2015
5. Arbabshirani, M.R., Plis, S., Sui, J., Calhoun, V.D.: Single subject prediction of brain disorders in neuroimaging: promises and pitfalls. Neuroimage, pii: S1053–8119(16)00210-X. doi:10.1016/j.neuroimage.2016.02.079. [Epub ahead of print], 21 March 2016
6. Costafreda, S.G., Dinov, I.D., Tu, Z., et al.: Automated hippocampal shape analysis predicts the onset of dementia in mild cognitive impairment. NeuroImage $56(1)$, 212–219 (2011). doi:10.1016/j.neuroimage.2011.01.050
7. Tang, X., Holland, D., Dale, A.M., Younes, L., Miller, M.I.: Shape abnormalities of subcortical and ventricular structures in mild cognitive impairment and Alzheimer's disease: detecting, quantifying, and predicting. Hum. Brain Mapp. $35(8)$, 3701–3725 (2014). doi:10.1002/hbm.22431
8. Pennanen, C., Kivipelto, M., Tuomainen, S., Hartikainen, P., Hänninen, T., Laakso, M.P., Vainio, P.: Hippocampus and entorhinal cortex in mild cognitive impairment and early AD. Neurobiol. Aging $25(3)$, 303–310 (2004)
9. Payan, A., Predicting, M.G.: Alzheimer's disease: a neuroimaging study with 3D convolutional neural networks (2015). arXiv preprint arXiv:1502.02506
10. Yangqing, J., Shelhamer, E., Donahue, J., Karayev, S., Long, J., Girshick, R., Guadarrama, S., Darrell, T.: Caffe: convolutional architecture for fast feature embedding. In: Proceedings of the 22nd ACM International Conference on Multimedia, pp. 675–678. ACM (2014)
11. Hosseini-Asl, E., Keynto, R., El-Baz, A.: Alzheimer's disease diagnostics by adaptation of 3-D convolutional network. In: IEEE ICIP Conference (2016)
12. Li, F., Tran, L., Thung, K.H., Ji, S., Shen, D., Li, J.: A robust deep model for improved classification of AD/MCI patients. IEEE J. Biomed. Health Inform. $19(5)$, 1610–1616 (2015). doi:10.1109/JBHI.2015.2429556. Epub 4 May 2015. PubMed PMID:25955998; PubMed Central PMCID: PMC4573581
13. Heung-Il, S., Seong-Whan, L., Dinggang, S.: Latent feature representation with stacked auto-encoder for AD/MCI diagnosis. Brain Struct. Funct. $220(2)$, 841–859 (2015). doi:10.1007/s00429-013-0687-3. Epub 22 Dec 2013. PubMed PMID: 24363140; PubMed Central PMCID: PMC4065852
14. Ahmed, O.B., Benois-Pineau, J., Allard, M., Amar, C.B., Catheline, G.: Classification of Alzheimer's disease subjects from MRI using hippocampal visual features. Multimedia Tools Appl. $74(4)$, 1249–1266 (2015)
15. Tong, T., Wolz, R., Gao, O., Guerrero, R., Hajnal, J.V., Rueckert, D.: Multiple instance learning for classification of dementia in brain MRI. Med. Image Anal. $18(5)$, 808–818 (2014). ISSN 1361-8415. http://dx.doi.org/10.1016/j.media.2014.04.006
16. Wolz, R., Julkunen, V., Koikkalainen, J., Niskanen, E., Zhang, D.P., Rueckert, D., Alzheimer's Disease Neuroimaging Initiative: Multi-method analysis of MRI images in early diagnostics of Alzheimer's disease. PloS one $6(10)$, e25446 (2011)
17. Coupé, P., Eskildsen, S.F., Manjón, J.V., Fonov, V.S., Pruessner, J.C., Allard, M., ADNI: Scoring by nonlocal image patch estimator for early detection of Alzheimer's disease. NeuroImage: Clin. $1(1)$, 141–152 (2012)

Deep Learning for Shot Classification in Gynecologic Surgery Videos

Stefan Petscharnig$^{(\boxtimes)}$ and Klaus Schöffmann

Alpen-Adria-Universität Klagenfurt,
9020 Klagenfurt, Austria
{stefan.petscharnig,ks}@itec.aau.at

Abstract. In the last decade, advances in endoscopic surgery resulted in vast amounts of video data which is used for documentation, analysis, and education purposes. In order to find video scenes relevant for aforementioned purposes, physicians manually search and annotate hours of endoscopic surgery videos. This process is tedious and time-consuming, thus motivating the (semi-)automatic annotation of such surgery videos. In this work, we want to investigate whether the single-frame model for semantic surgery shot classification is feasible and useful in practice. We approach this problem by further training of AlexNet, an already pre-trained CNN architecture. Thus, we are able to transfer knowledge gathered from the Imagenet database to the medical use case of shot classification in endoscopic surgery videos. We annotate hours of endoscopic surgery videos for training and testing data. Our results imply that the CNN-based single-frame classification approach is able to provide useful suggestions to medical experts while annotating video scenes. Hence, the annotation process is consequently improved. Future work shall consider the evaluation of more sophisticated classification methods incorporating the temporal video dimension, which is expected to improve on the baseline evaluation done in this work.

Keywords: Multimedia content analysis · Convolutional neural networks · Deep learning · Medical shot classification

1 Introduction

Advances in the field of endoscopic surgery not only enable physicians to perform minimally invasive surgeries, but they are also capable of revisiting video material recorded from every surgery performed. These recorded videos can be used for documentation, analysis, and education. In order to provide an efficient video retrieval system supporting the aforementioned tasks, we need to be able to classify the nature of a shot, which may show a surgery action (e.g., coagulation of the ovary) or diagnostic investigation (e.g., shot of the uterus).

As many surgery actions can be bound to the use of certain surgery devices and the organs are discriminative, we consider frame-based shot classification with a convolutional neural network (CNN) sufficient for certain use cases, such

© Springer International Publishing AG 2017
L. Amsaleg et al. (Eds.): MMM 2017, Part I, LNCS 10132, pp. 702–713, 2017.
DOI: 10.1007/978-3-319-51811-4_57

as suggestion of annotations for an expert reviewer in order to simplify the annotation process. For other use cases such as the automatic detection of the start of the surgery, we conjecture that temporal information has to be included in order to provide results which are of practical use. With these considerations in mind, this work addresses the following research goal:

Evaluate the predictive performance of shot classification for the use case of gynecologic surgery videos using a single-frame CNN model.

In order to cope with this task, we have manually annotated 18 h of specific video material from gynecological surgeries. We use this material as ground truth for training and testing a deep CNN of an existing architecture with pre-trained initial weights (AlexNet [4]). We evaluate the fine tuned model's performance for classification of individual frames in gynecologic surgery videos with means of precision and recall for 14 different semantic content classes. Eventually, we deal with a case study of classifying suturing shots, in order to show the practical applicability of our results. This work is novel as to the best of our knowledge there is no work that performs a semantic shot classification within gynecologic surgery videos. Our results indicate that CNN models are of practical worth in the very specific domain of gynecologic laparoscopic surgery videos. Furthermore, we think that the insights gathered from the generated model and the frame based classification evaluation can be used as a base for the application of methods which also incorporate the temporal dimension of the videos, such as histogram-based event detection or Recurrent Neural Networks (RNNs).

The remainder of this paper is structured as follows: Sect. 2 shows other CNN approaches in medical use cases. Section 3 deals with issues relevant for CNN training and validation, this is data and fine tuning methodology of the used CNN, whereas Sect. 4 has a focus on the presentation of the results. Eventually, Sect. 5 concludes the paper by summarizing the results and outlining future work.

2 Related Work

For the use case of interstitial lung diseases, [5] provides a simple CNN model containing a single convolutional layer. They yield per-class precision and recall between 0.8 and 0.9 for classification into five classes (normal, emphysema, ground glass, fibrosis, and micronodules) outperforming the SIFT feature as well as Restricted Boltzmann Machines. The authors of [2] propose a deep CNN model containing five convolutional layers for the classification of CT images into seven classes of interstitial lung diseases (healthy, ground glass opacity, micronodules, consolidation, reticulation, and honeycombing). Their results imply that, for this use case, their CNN approach outperforms other CNNs as well as state of the art methods using handcrafted features. In [12], a multi-stage deep learning framework is presented. With this framework, the problem of body-part recognition is solved. In total, they achieve best performance regarding recall, precision and f-score compared against logistic regression, SVMs, and CNNs. The importance of CNNs in medical applications is also apparent from their use within other applications such as nucleus segmentation [11], polyp detection in colonoscopy videos [6], microcalcification detection in digital breast tomosynthesis [8], mitosis detection in breast cancer histology [1], and short-term breast cancer risk prediction [7]. Our work is delimited to the aforementioned research as in contrast to the classification of a state (e.g., healthy or consolidation, type of tissue), we aim at classifying both, diagnostic classes and surgical events, i.e. surgical actions. Furthermore there haven't been any efforts made regarding the classification of images extracted from laparoscopic surgery videos.

The authors of [9,10] deal with fine tuning and transfer learning effects of CNNs. These pieces of work are based on the use cases of lymph node detection, interstitial lung disease classification, polyp detection and image quality assessment in colonoscopy, pulmonary embolism detection in computed tomography images, and intima-media boundary segmentation in ultrasonographic images. Their results imply that CNNs are suitable for computer aided diagnosis problems, and transfer learning from large-scale annotated natural image datasets is beneficial for performance. We base the selection and further training of an already existing CNN architecture [4] for our use case on these results.

3   Learning Content Classes

As a base for our ground truth, we use 24 gynecological surgery videos (approximately 26 h of raw video data) showing myoma resection and endometriosis treatment. For our dataset, we chose classes corresponding with frequent events in gynecologic surgery, e.g. suturing, or diagnosis of extent of a disease. In collaboration with surgeons from LKH Villach, we identified 14 semantic content classes either categorizing an action in the surgery (*suction & irrigation, suture, dissection (blunt), cutting, cutting (cold), sling, coagulation, injection*) or a certain diagnostic aspect, in particular a clearly visible organ (*uterus, ovary, oviduct, liver, colon*), as well as *blood* accumulations. These aforementioned two types of classes (surgery actions and diagnostic shots) are mixed corresponding to a real world scenario, as usually both types occur in a single surgery video. Hence, we think it is not convenient to separate those classes. As the identified content classes describe basic surgical actions and anatomical structures in gynecology, inter-expert disagreement is negligible and non-surgeons are able to identify those classes correctly with instructions from surgeons. Hence, the annotation process itself was executed by ourselves on the shot-level. We manually annotated 842 video shots consisting of 798228 different video frames which were chosen according to the given instructions.

Table 1 characterizes the different classes of the annotated dataset in dimensions of number of shots, number of frames, average duration, standard deviation, and semantic description. Please note the high standard deviation of the video durations of the surgical action classes. This indicates that the duration of the single actions heavily depends on the circumstances of the surgery and the individual patient's anatomy and general assumptions on a basis of video sequence length can not be made.

Table 1. An overview on the annotated dataset: class id, class name, number of shots, number of frames, average duration in seconds, standard deviation of duration in seconds, and class description.

ID	Class	Shots	Frames	avg [s]	sd [s]	Description
1	Suction & Irrigation	190	94260	19.8	28.8	Application of the suction and irrigation tube
2	Suture	25	102152	163.4	154.6	Process of suturing
3	Dissection (blunt)	53	69689	52.5	75.5	Blunt dissection of tissue (e.g. by tearing it apart)
4	Cutting	64	111756	69.8	89.2	Thermally dissect tissue (e.g. with mono-polar electrodes)
5	Cutting (cold)	103	208210	80.8	74.2	Dissect tissue with a sharp instrument (e.g. scissors)
6	Sling	8	6634	33.1	24.3	Dissection of large parts of tissue with an electrical sling
7	Coagulation	176	85932	19.5	17.3	Application of coagulation in order to close a wound
8	Blood	52	12909	9.9	12.2	Noticeable amount of blood visible
9	Uterus	46	24358	21.1	39.3	Clearly visible uterus
10	Ovary	57	31607	22.1	26.3	Clearly visible ovary
11	Oviduct	15	8379	22.3	25.1	Clearly visible oviduct
12	Liver	31	25483	32.8	60.5	Clearly visible liver
13	Colon	12	9832	32.7	43.0	Clearly visible colon
14	Injection	10	7027	28.1	10.2	Injection with a needle

We provide example frames for three of the annotated classes: *cutting*, *suture*, and *sling*. Please note the similarity of the chosen frames between different classes (c.f. Figs. 1 and 2) as well as the dissimilarity within a class (c.f. Figs. 3 and 4). The classes and their semantics as well as number of shots, average shot duration, standard deviation of the shot duration, and the number of extracted frames are presented in Table 1. In total, the manually annotated ground truth contains 842 shots labeled as the best matching class out of the dataset's classes. All shots together consist of 798228 frames. We derive the best matching class implicitly by camera positioning and current action, e.g., the action in the center of the image or the organ which is inspected by a surgeon is the action or object of interest.

With the surgical task classes, there is the issue that a shot contains frames which could be classified as a diagnostic class as well. For example, suturing the ovary may contain images with the ovary without a surgical needle, or the suture

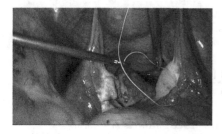

Fig. 1. Suture

Fig. 2. Sling

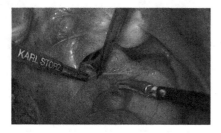

Fig. 3. Cutting

Fig. 4. Cutting

is not clearly visible. On the one hand, this frame does not look like it belongs to a suturing shot, but on the other hand it indeed does belong to the suturing shot as the image was recorded in its context. For the annotation of our dataset, we chose to stick to the latter case and annotate such frames as the surgical task by defining begin and end of the surgical action. Each frame from beginning until the end of a shot is labeled with the corresponding shot label for the class it belongs to. Due to this circumstance, the dataset also may contain blurry frames or frames in which instruments may cover huge parts of the camera. We argue that these frames are nonetheless part of the corresponding shot and thus correctly labeled.

For **training the CNN**, we used the caffe framework [3] and the CNN architecture of AlexNet [4]. The network architecture contains eight trainable layers: five convolutional and three fully connected layers. We modified the number of neurons on the output layer in order to match our use case, the classification into 14 content classes. Furthermore, the architecture features MAX-pooling, local response normalization (LRN), and dropout. For a detailed description of the CNN architecture, please refer to [4]. We split the dataset in half by choosing half the shots of each class belonging to the training set, the other half to the test set. We do not split 80-20 in favor of a more diverse test set. Please note that the class distribution of the dataset is highly imbalanced: the three classes *dissection (blunt)*, *cutting (cold)*, and *cutting* combined constitute approximately 50% of the annotated frames (see Table 1. We use oversampling in order to balance the training set by duplicating shots of the under-represented classes until the number of training images per class is approximately balanced. This results in a training set size of 861016 images generated out of 429011 unique frames. The neural network expects image patches of 227×227 pixels. At training time, we re-size single frames on-the-fly to a height of 256 pixels while preserving the aspect ratio. We then perform a central crop to a width of 256 pixels. The result of this operation is a 256×256 image. In each epoch, a random 227×227 patch out of such a resulting image is extracted and fed to the network as training input. In order to cope with this huge amount of data, we extended the Caffe framework with a pre-fetching data layer performing the extraction of video frames and the above described data preparation and augmentation on-the-fly. We do not arbitrarily shuffle the input images, because then video decoding becomes a serious performance bottleneck. As a trade-off, we shuffle the shots at the beginning of each epoch and switch to extracting frames from the next shot at least every 100 frames.

The training was performed on a machine featuring an Intel(R) Core(TM) i7-5960X CPU 3.00 GHz processor, 64 GB of DDR-4 RAM, a Samsung SSD 850 pro and a NVIDIA GeForce GTX TITAN X graphics card. This system took approximately an hour of training time per epoch. We fine-tuned the network with a batch size of 600 images for 100 epochs as the loss function has stabilized at this point. We set the base learning rate of the fully connected layers to 0.1 as the low-level image features are pre-trained already. All other layers used a learning rate of 0.00001.

4   Evaluation

We evaluate the performance of the network by means of precision and recall on a per-class basis. Our test set contains 369217 images extracted from gynecological surgery videos and manually annotated as discussed above. The distribution of the images over the classes within the test set is approximately equal to the distribution in the whole dataset. In particular the classes *suction & irrigation, suture, dissection (blunt), cutting, cutting (cold),* and *coagulation* are represented by approximately 11%, 9%, 10%, 15%, 26%, and 10% respectively.

The $n \times n$ confusion matrix depicted in Table 2 summarizes the single-frame prediction results for our $n = 14$ classes. Columns in the table denote the predicted class referred to by its class ID, while rows indicate the true class. Cell colors illustrate prediction percentages relative to the number of examples for a class. For example, the darkest shade (which is used in row 6 and column 6) indicates that the network predicted for over 80% of the test images of class 6 (indicated by the row) their true class 6 (column).

For the calculation of precision, recall, and f-value of class i, we determine TP_i (true positive classification of class i) NP_i (number of false positive predictions for class i), and FN_i (number of false negative predictions of class i).

To evaluate precision and recall in a class-based manner has the advantage that the imbalance of the classes in the test set is taken into account. Over all

Table 2. Confusion matrix for the fine tuned model. Rows denote true class id, columns predicted class id. Colors indicate prediction accuracy relative to the number of test examples per class. For the semantics of the class IDs please refer to Table 1.

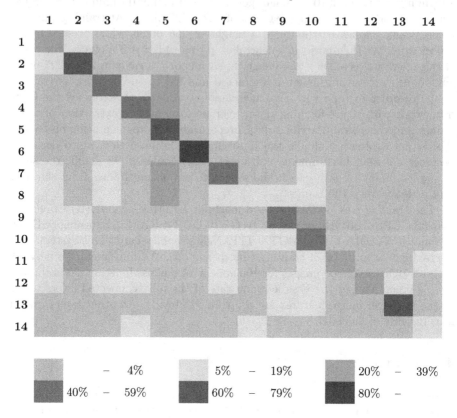

classes, we achieve an average precision of 0.422, an average recall of 0.430, and an average f-value of 0.41. In the following we discuss the per-class results and try to explain the mis-classifications. We also calculate the probability that the true class is among the top three predictions. We refer to this probability as Recall@3, whose class-based average is 0.699. For the per class results for precision, recall, recall@3, and f-value, please refer to Table 3. In total, we correctly classified 179289 out of 368400 images resulting in an overall prediction accuracy of 48.67%. For 288200 images the true label was in the top three predictions, resulting in an overall top three prediction accuracy of 78.23%. These accuracy values are higher than the average per-class recall, as the dataset bias is not taken into account. We do not consider specificity, as we consider the use of this measure for the performance evaluation of multi-class problems as problematic. We discuss the results of selected individual classes in the following.

Suction & irrigation (1). For the application of the suction an irrigation tool, we have a recall of 0.270, which is clearly below the mean recall of 0.430. The

Table 3. Performance overview for the fine tuned model. For the semantics of the class IDs please refer to Table 1.

Class ID	1	2	3	4	5	6	7
Precision	0.441	0.611	0.407	0.732	0.576	0.695	0.295
Recall	0.270	0.672	0.411	0.419	0.661	0.858	0.403
Recall@3	0.720	0.885	0.764	0.639	0.913	0.932	0.877
f-value	0.335	0.64	0.409	0.533	0.616	0.768	0.34
Class ID	8	9	10	11	12	13	14
Precision	0.155	0.447	0.314	0.174	0.634	0.277	0.156
Recall	0.168	0.421	0.509	0.250	0.231	0.625	0.122
Recall@3	0.566	0.664	0.752	0.410	0.382	0.909	0.352
f-value	0.161	0.434	0.389	0.205	0.339	0.384	0.137

high number of false positive classifications of the (true) classes suturing and coagulation as suction & irrigation is explainable due to the fact, that suction and irrigation is done within the same scene in order to clean tissue. The precision of this class is with 0.411 about average. For a given image of suction & irrigation, the network frequently classified it as coagulation, cutting (cold), ovary, suture, blood, and oviduct. We observe that the network recognized the most-frequent contexts and regions of suction and irrigation which in fact, are the ones mentioned above.

Suture (2). For frames extracted from suturing shots, the network achieves a precision of 0.611 and a recall of 0.672 resulting in one of the best performances among the classes. We conclude that this is due to the fact that the surgical needle is exclusively used for the procedure of suturing and the suture itself is clearly visible in the image in many examples. The frequent mis-classification of suturing as coagulation my be explained by two circumstances: (1) the tool for coagulation is used to position tissue and suture (which, of course, is a wrong prediction), as well as (2) the fact that it might become necessary to perform coagulation during suturing (which is essentially a correct prediction). The classification of *suture* as *ovary* is comprehensible for suturing shots of the ovary (contrary to other diagnostic classes which only feature parts of sutured tissue or multiple candidates for diagnostic classes). Furthermore, an example for a confusion of suturing and sling is given in Figs. 1 and 2. This confusion originates in the fact that the suture may look similar to the sling. Another frequent mis-classification is the class *suction & irrigation*. While suturing, tissue may be positioned and then neither suture nor surgical needle are visible in the center of the image. Furthermore, the tool used for positioning contributes to this classification, as it looks similar to the tool used for suction & irrigation.

Dissection (blunt), cutting, and cutting (cold) (3-5). For these three dissection classes combined, the achieved results are above average. The most frequent classifications for examples of these classes are either the other two

Table 4. Detailed classification results for suturing shots **Shot1** and **Shot2**. Cell values denote the (rounded) percentage of frames classified as the class which is determined by the column.

Class ID	1	2	3	4	5	6	7
S1	0.053	0.756	0.000	0.000	0.034	0.097	0.024
S2	0.000	0.415	0.001	0.000	0.347	0.195	0.002
Class ID	8	9	10	11	12	13	14
S1	0.001	0.005	0.021	0.000	0.000	0.008	0.000
S2	0.006	0.000	0.031	0.001	0.001	0.000	0.000

classes of the coagulation class. These results may originate from tool usage, as the used tools in this four classes are similar.

Sling (6). With a recall of 0.858, precision of 0.695, and f-value of 0.768, this class yields the best performance. We infer that this originates in the appearance of the electrical sling used for the thermal separation which is only visible in these shots.

Coagulation (7). For the class of coagulation, our net achieved a recall similar to the average recall and a precision slightly less than the average of the net. In most cases where the net mis-classified coagulation examples, the predicted classes were classes with a high probability that either coagulation is part of the scene, or the same tool is used.

Diagnostic shots (8-13). In total, the diagnostic shot classes *Blood, Uterus, Ovary, Oviduct, Liver, and Colon* performed with mean recall 0.67, average precision 0.334, and average f-value 0.319 slightly worse than the average performance. At first glance this result is surprising, as these classes do not need temporal information and seem to be identifiable from a single frame more easily than surgery actions. Deeper analysis of the results shows that these classes frequently were identified as surgical actions *cutting (cold), suction & irrigation, suturing,* and *coagulation.* This may originate in the fact, that the network also identified these organs within a surgery action shot, as for example suturing the ovary is very frequent among suturing scenes.

Injection (14). With a recall of 0.122, precision 0.156, and f-value of 0.137, this class yields the worst results. In most test cases the injection was either classified as suture or oviduct. We assume that these result can be improved with further training examples.

Suture (2)- shot classification. We use the trained model for a case study of a naive majority-based approach to shot classification. We classify each frame of the shot. These frames then vote for their predicted classes. The class with the majority of votes determines the shot's class. We select two different shots of suturing the ovary, which were neither part of the training nor part of the test set. The first shot (**S1**, 5000 frames) is a suturing shot in which surgical

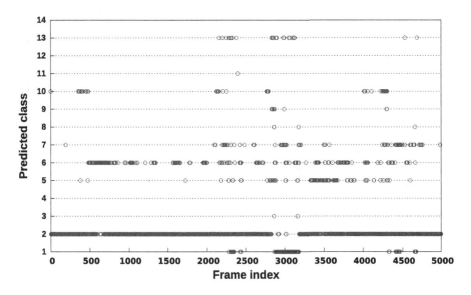

Fig. 5. Frame-wise classification result overview for suturing shot 1 (S1)

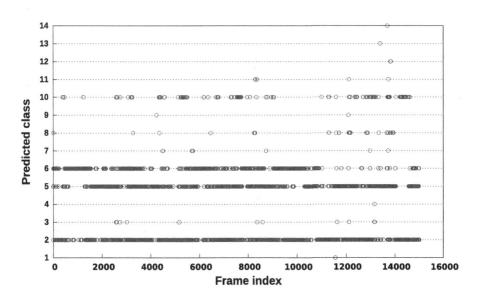

Fig. 6. Frame-wise classification result overview for suturing shot 2 (S2)

needle and suture are in the center of the image almost all the time. The second shot (**S2**, 20000 frames) is chosen to be more challenging, as suture and surgical needle are poorly visible over large parts of the shot. For these two example shots S1 and S2, we achieve a recall of 0.756 and 0.415 respectively. For class-based classification results, please refer to Table 4. Figures 5 and 6 show the detailed classification results for each frame of the shots S1 and S2. For parts of both shots, the net classifies the suturing scene as *ovary*(10) which is the best 'false' classification possible as the ovary is sutured in these specific shots. For shot S1, there are two clusters of false classification for the classes *sling* (6) and *suction & irrigation* (1) (c.f. Fig. 5). In total, such clusters do not influence the majority decision, which clearly favors the classification of *suture* (0.756). For shot S2, *suture* (0.415), *sling* (0.347), and *coagulation* (0.195) are the most frequent classification suggestions. As shown in Fig. 6, the classification oscillates between these three classes during the whole shot. However, there is still a clear majority for *suture*. Such oscillations may become problematic, when the number of classifications for two or more classes is approximately equal. We declare the detailed investigation of such cases to future work.

5   Conclusion

In this work, we perform a detailed investigation of gynecological surgery shot classification performance of a deep CNN. In particular, we trained AlexNet [4], an existing CNN architecture with pre-trained initial weights for this very special context of gynecologic surgery videos. We achieve an average precision of 0.42, average recall of 0.43, average f-value of 0.41, and an accuracy of 48.67%. The average recall for the top 3 predictions is 69.9%, the accuracy in the top 3 prediction is 78.23%. We want to emphasize the novelty of this work, as this is the very first semantic content classification in the field of gynecologic surgery, a special area of laparoscopic surgery. Out results show that reasonable content classification is possible in this domain. Furthermore, our results are of practical relevance for use cases like suggestion of annotation in order to speed-up the manual annotation process. Future work concerns the investigation whether data augmentation to low density classes instead of simple duplication improves the results significantly and is thus a more suitable technique for imbalanced data sets in the gynecologic domain. Another topic to future work is a deeper analysis of the predefined network, i.e. the variation of filter number and removal of layers as well as filter visualization. Moreover, approaches for event detection based on histograms of class confidences, high level CNN features (general versus domain-specific model), as well as recurrent neural network models. We also aim at the recognition of surgery actions from action-specific motion sequences.

Acknowledgments. This work was supported by Universität Klagenfurt and Lakeside Labs GmbH, Klagenfurt, Austria and funding from the European Regional Development Fund and the Carinthian Economic Promotion Fund (KWF) under grant KWF 20214 u. 3520/ 26336/38165.

References

1. Albarqouni, S., Baur, C., Achilles, F., Belagiannis, V., Demirci, S., Navab, N.: AggnNet: deep learning from crowds for mitosis detection in breast cancer histology images. IEEE Trans. Med. Imaging **35**(5), 1313–1321 (2016)
2. Anthimopoulos, M., Christodoulidis, S., Ebner, L., Christe, A., Mougiakakou, S.: Lung pattern classification for interstitial lung diseases using a deep convolutional neural network. IEEE Trans. Med. Imaging **35**(5), 1207–1216 (2016)
3. Jia, Y., Shelhamer, E., Donahue, J., Karayev, S., Long, J., Girshick, R., Guadarrama, S., Darrell, T.: Caffe: convolutional architecture for fast feature embedding. In: Proceedings of the 22nd ACM International Conference on Multimedia, MM 2014, pp. 675–678, New York, NY, USA. ACM (2014)
4. Krizhevsky, A., Sutskever, I., Hinton, G.E.: Imagenet classification with deep convolutional neural networks. In: Bartlett, P., Pereira, F., Burges, C., Bottou, L., Weinberger, K. (eds.) Advances in Neural Information Processing Systems 25, pp. 1106–1114 (2012)
5. Li, Q., Cai, W., Wang, X., Zhou, Y., Feng, D.D., Chen, M.: Medical image classification with convolutional neural network. In: 13th International Conference on Control Automation Robotics and Vision (ICARCV), pp. 844–848. IEEE (2014)
6. Park, S.Y., Sargent, D.: Colonoscopic polyp detection using convolutional neural networks. In: SPIE Medical Imaging, p. 978528. International Society for Optics and Photonics (2016)
7. Qiu, Y., Wang, Y., Yan, S., Tan, M., Cheng, S., Liu, H., Zheng, B.: An initial investigation on developing a new method to predict short-term breast cancer risk based on deep learning technology. In: SPIE Medical Imaging, p. 978521. International Society for Optics and Photonics (2016)
8. Samala, R.K., Chan, H.P., Hadjiiski, L.M., Cha, K., Helvie, M.A.: Deep-learning convolution neural network for computer-aided detection of microcalcifications in digital breast tomosynthesis. In: SPIE Medical Imaging, p. 97850Y. International Society for Optics and Photonics (2016)
9. Shin, H.C., Roth, H.R., Gao, M., Lu, L., Xu, Z., Nogues, I., Yao, J., Mollura, D., Summers, R.M.: Deep convolutional neural networks for computer-aided detection: CNN architectures, dataset characteristics and transfer learning. IEEE Trans. Med. Imaging **35**(5), 1285–1298 (2016)
10. Tajbakhsh, N., Shin, J.Y., Gurudu, S.R., Hurst, R.T., Kendall, C.B., Gotway, M.B., Liang, J.: Convolutional neural networks for medical image analysis: full training or fine tuning? IEEE Trans. Med. Imaging **35**(5), 1299–1312 (2016)
11. Xing, F., Xie, Y., Yang, L.: An automatic learning-based framework for robust nucleus segmentation. IEEE Trans. Med. Imaging **35**(2), 550–566 (2016)
12. Yan, Z., Zhan, Y., Peng, Z., Liao, S., Shinagawa, Y., Zhang, S., Metaxas, D.N., Zhou, X.S.: Multi-instance deep learning: discover discriminative local anatomies for bodypart recognition. IEEE Trans. Med. Imaging **35**(5), 1332–1343 (2016)

Description Logics and Rules for Multimodal Situational Awareness in Healthcare

Georgios Meditskos$^{(\boxtimes)}$, Stefanos Vrochidis, and Ioannis Kompatsiaris

Information Technologies Institute,
Centre for Research and Technology - Hellas, Thessaloniki, Greece
{gmeditsk,stefanos,ikom}@iti.gr

Abstract. We present a framework for semantic situation understanding and interpretation of multimodal data using Description Logics (DL) and rules. More precisely, we use DL models to formally describe contextualised dependencies among verbal and non-verbal descriptors in multimodal natural language interfaces, while context aggregation, fusion and interpretation is supported by SPARQL rules. Both background knowledge and multimodal data, e.g. language analysis results, facial expressions and gestures recognized from multimedia streams, are captured in terms of OWL 2 ontology axioms, the de facto standard formalism of DL models on the Web, fostering reusability, adaptability and interoperability of the framework. The framework has been applied in the eminent field of healthcare, providing the models for the semantic enrichment and fusion of verbal and non-verbal descriptors in dialogue-based systems.

Keywords: Multimodal data · Ontologies · Rules · Situation awareness

1 Introduction

A key requirement in multimodal domains is the ability to integrate the different pieces of information (modalities), so as to derive high-level interpretations. More precisely, in such environments, information is typically collected from multiple sources and complementary modalities, such as from multimedia streams (e.g. using video analysis and speech recognition), lifestyle and environmental sensors [15]. Though each modality is informative on specific aspects of interest, the individual pieces of information themselves are not capable of delineating complex situations. Combined pieces of information on the other hand can plausibly describe the semantics of situations, facilitating intelligent situation awareness.

In parallel, the demand for context-aware user task support has proliferated in the recent years across a multitude of application domains, ranging from healthcare and smart spaces to transportation and energy control. A key challenge in such applications is to abstract and fuse the captured *context* in order to elicit an adequate understanding of user actions [5]. In healthcare, for example, wearable and ambient sensors, coupled with profile information and clinical knowledge can be used to improve the quality of life of care recipients and provide useful insights to clinical experts for personalized interventions and care solutions [23].

© Springer International Publishing AG 2017
L. Amsaleg et al. (Eds.): MMM 2017, Part I, LNCS 10132, pp. 714–725, 2017.
DOI: 10.1007/978-3-319-51811-4_58

Given the inherent requirement in multimodal environments to aggregate low-level information and integrate domain knowledge, it comes as no surprise that Semantic Web technologies have been acknowledged as affording a number of highly desirable features. More precisely, the OWL 2 ontology language [10] has been extensively used to capture context elements (e.g. profiles, events, activities, locations, postures and emotions) and their pertinent relations, mapping observations and domain knowledge to class and property assertions in the Description Logics (DL) [3] theory, fostering integration of information at various levels of abstraction and completeness [14]. The generated models encapsulate formal and expressive semantics, harvesting several benefits brought by ontologies, e.g. modelling of complex logical relations, sharing information from heterogeneous sources, sound and complete reasoning engines.

This paper describes an ontology-based framework for context awareness and conversation understanding in multimodal natural language interfaces. The framework allows the semantic enrichment of verbal and non-verbal information coming from multiple devices and acquisition methods, e.g. from multimedia analysis, following a knowledge-driven methodology for observation aggregation, linking and situation interpretation. The contributions of our work can be summarized in the following:

- We alleviate the lack of inherent temporal reasoning support in DL and OWL 2 by adopting an a-temporal approach for subsumption reasoning and multimodal data fusion based on time-windows.
- We propose an iterative combination of DL reasoning and rules to enhance the reasoning capabilities of the framework.
- We use SPARQL queries (CONSTRUCT graph patterns) as the underlying rule language of the framework, overcoming the lack of a standard rule language that runs directly on top of RDF and OWL ontologies.
- Due to the dynamic and open nature of ontologies, the framework is modality-agnostic, in the sense that it is not tight to specific domains and data sources but it can be extended, adapted and used in a variety of situations.

We illustrate the capabilities of the framework through its integration into a dialogue-based agent for conversational assistance in healthcare. More specifically, elderly use the dialogue system (usually at home) to acquire information and suggestions related to basic care and healthcare (e.g. symptoms, treatments, etc.). A key challenge in this domain is the effective fuse of verbal and non-verbal communication modalities, e.g. deictic gestures and spoken utterances, in order to disambiguate and interpret user input during the interaction with the agent.

The rest of the paper is structured as follows: Sect. 2 begins with a basic background on the DL theory and OWL ontologies. It continues with a discussion on ontology-based context-aware solutions, explaining basic concepts and challenges. Section 3 describes the proposed framework, providing details on the representation and interpretation layers, as well as on the hybrid reasoning scheme and the role of SPARQL. Section 4 explicates through an example use case from the ongoing simulated evaluation of the framework in the healthcare domain and Sect. 5 concludes the paper and outlines next steps.

2   Background and Related Work

2.1   Description Logics

Description Logics (DL [3]) is a family of knowledge representation formalisms characterised by logically grounded semantics and well-defined reasoning tasks. The main building blocks are *concepts* (or classes), representing sets of objects, *roles* (or properties), representing relationships between objects, and *individuals* (or instances) representing specific objects. Starting from atomic concepts, arbitrary complex concepts can be described through a rich set of constructors that define the conditions of concept membership. DL provides, among others, constructs for concept inclusion ($C \sqsubseteq D$), equality ($C \equiv D$) and assertion ($C(a)$), as well as role inclusion ($R \sqsubseteq S$) and assertion ($R(a, b)$).

The semantics of a DL language is formally defined through an interpretation I that consists of a nonempty set Δ^I (the domain of interpretation) and an interpretation function $\cdot^I$, which assigns to every atomic concept A a set $A^I \subseteq \Delta^I$ and to every atomic role R a binary relation $R^I \subseteq \Delta^I \times \Delta^I$. Table 1 shows the syntax and semantics of some of the most common DL constructors. For example, the class of all deictic gesture observations that point to the head can be defined as $\texttt{PointsToHeadGesture} \equiv \texttt{DeicticGesture} \sqcap \exists\texttt{hasBodyPart}.\{\texttt{head}\}$.

Besides formal semantics, DL comes with a set of powerful reasoning services, for which efficient, sound and complete reasoning algorithms are available. For example, through *subsumption*, one can derive implicit taxonomic relations among concepts. *Satisfiability* and *consistency checking* are useful to determine whether a knowledge base is meaningful at all. *Instance realization* returns all concepts from the knowledge base that a given individual is an instance of.

2.2   OWL 2 Ontologies

An ontology is a set of precise descriptive statements about some part of the world (usually referred to as the domain of interest). Precise descriptions satisfy several purposes: most notably, they prevent misunderstandings in human communication and they ensure that software behaves in a uniform, predictable way and works well with other software[1].

Table 1. Examples of concept and role constructors in DL.

Name	Syntax	Semantics
Intersection	$C \sqcap D$	$C^I \cap D^I$
Union	$C \sqcup D$	$C^I \cup D^I$
Universal quantification	$\forall R.C$	$\{a \in \Delta^I \mid \forall b.(a, b) \in R^I \rightarrow b \in C^I\}$
Existential quantification	$\exists R.C$	$\{a \in \Delta^I \mid \exists b.(a, b) \in R^I \wedge b \in C^I\}$

---

[1] https://www.w3.org/TR/owl2-overview/.

The Web Ontology language (OWL/OWL 2) [10] is a knowledge represen-tation language widely used within the Semantic Web community for creating ontologies. The design and semantics of OWL 2 have been strongly influenced by DL[2]. Some basic notions are: (a) *axioms*, the basic statements that an OWL ontology expresses, (b) *entities*, elements used to refer to real-world objects, and (c) *expressions*, combinations of entities to form complex descriptions.

In principle, every OWL 2 ontology is essentially a collection of such basic "pieces of knowledge". Statements that are made in an ontology are called axioms, and the ontology asserts that its axioms are true. However, despite the rich primitives, there are certain limitations that amount to the DL style model theory used to formalise semantics, and particularly the tree model property [16] conditioning DL decidability. For example OWL 2 can model only domains where objects are connected in a tree-like manner. In order to leverage OWL's limited relational expressiveness, research has been devoted to the integration of OWL with rules (e.g. SWRL [12], SPIN [13]). User-defined rules on top of the ontology allow expressing richer semantic relations that lie beyond OWL's expressive capabilities and couple ontological and rule knowledge [9].

2.3 Ontology-Based Context Awareness and Fusion

Congruous with the open nature of context-awareness, where information at var-ious levels of abstraction and completeness has to be integrated, ontologies have attracted growing interest as means for modelling and reasoning over contex-tual information in various domains [2,14]. For example, BeAware! [4] provides a framework for context awareness in road traffic management; [26] proposes an ontology-based framework for context-aware activity recognition in smart homes. A survey on context awareness from an IoT perspective is presented in [17], whereas challenges and opportunities in applying Semantic Web technolo-gies in context-aware pervasive applications are discussed in [27].

A common characteristic in all cases above is the use of ontologies for domain modelling. Ontology languages, such as OWL 2, share a common understanding of the structure and semantics of information, enabling knowledge reuse and inferencing. Capitalizing on the expressivity of the models, several approaches define one or more interpretation layers in order to elicit an understanding of the situation. For example, in the domain of natural language interfaces and dialogue-based systems [24], ontologies provide the vocabulary and semantics for content disambiguation [6,7], such as WordNet[3] and BabelNet[4]. Ontologies have been also used in NLP information extraction contexts for coreference resolu-tion in textual input [19,22]. In the domain of multimodal fusion, ontologies are used to fuse multi-level contextual information [8]. For example, [18] presents a framework for coupling audio-visual cues with multimedia ontologies. Rele-vant approaches are also described in [1] for various multimedia analysis tasks.

---

[2] In this paper, OWL 2 is used to refer to OWL 2 DL ontologies interpreted using the Direct Semantics [20].

[3] http://wordnet-rdf.princeton.edu/.

[4] http://babelnet.org/rdf/page/.

SMARTKOM [25] partially uses ontologies to fuse information in multimodal dialogue systems, combining speech, gesture and facial expressions.

Similar to the aforementioned approaches, we use OWL 2 ontologies for modelling context types and their relationships in terms of DL concept class constructors. However, we argue that the constructors provided by DL, and hence by OWL 2, are sometimes inadequate to facilitate effective multimodal fusion. Certain modelling and reasoning limitations, such as the tree-model property mentioned above or the lack of temporal reasoning, render OWL 2 insufficient to address practical fusion requirements, such as the assertion of property fillers for unconnected instances, as we demonstrate in Sect. 4. Our framework leverages OWL 2 limited expressivity through an intelligent, multi-tier hybrid scheme of DL reasoning that follows a context-aware fusion and interpretation solution along with the use of SPARQL CONSTRUCT graph patterns [11] as the underlying rule language of the framework.

3   Semantic Fusion Framework

The aim of the Semantic Fusion Framework (SFF) is to aggregate context types and couple them with background knowledge. SFF does not impose any restriction on the modalities that can be fused, provided that the underlying ontologies support their representation. As such, SFF consists of two core tiers:

- **Representation tier:** Provides the knowledge structures needed to capture the semantics and structure of the various modalities, as well as the semantics of the domain model that drives the fusion task.
- **Interpretation tier:** Implements the fusion logic, capitalizing on OWL 2 DL reasoning and custom interpretation rules that combine the available input and generate additional inferences.

The conceptual architecture of SFF is depicted in Fig. 1. In the following sections, we further elaborate on the specifics of each tier.

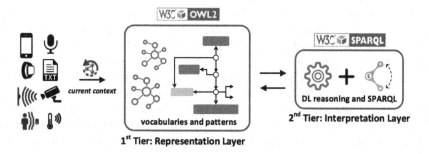

Fig. 1. Conceptual architecture of the Semantic Fusion Framework (SFF)

3.1   Domain and Context Descriptors

As mentioned above, SFF is modality-agnostic since it is not tight to specific context types. In that sense, contextual information may be collected from a variety of sources, such as ambient and wearable sensors (e.g. temperature and proximity observations), multimedia analysis, such as text analysis (named entities and concepts), video analysis (e.g. location, gestures), etc. All this information needs to be mapped on domain entities to enable the derivation of contextual descriptors that best satisfy and interpret the context.

We use the term "observation" to abstractly refer to the root of the context type hierarchy. Figure 2 depicts a lightweight vocabulary for modelling context types. The ontology extends the `leo:Event` concept of LODE [21] to benefit from existing vocabularies to describe events and observations. Property assertions about the temporal extension of the observations and the agent (actor) are allowed, reusing core properties of LODE. Figure 2 also depicts the relationship between the upper-level domain and context models. More precisely, the `Context` class is provided that allows one or more `contains` property assertions referring to observations. In terms of DL semantics, the `Context` class is defined as:

$$\text{Context} \equiv \exists \text{contains.Observation} \tag{1}$$

classifying instances with `contains` property assertions in `Context`. As we demonstrate in Sect. 4, the adaptation of the framework in different domains involves the extension of the `Context` concept, specifying the observation types that designate complex situations of interest that need to be recognized. Intuitively, instances of the `Context` concept define set of observations, designating the *current context* that needs to be classified and interpreted.

3.2   DL Reasoning and SPARQL

The interpretation tier defines the way atomic observations can lead to the derivation of high-level interpretations. For this task, we group observations into a single `Context` instance, creating the *current context*, which is then fed into the DL reasoner for subsumption reasoning and context classification. In principle, the current context is built taking into account the temporal extension of observations, along with background information pertinent to the domain. We present an example current context definition in Sect. 4.

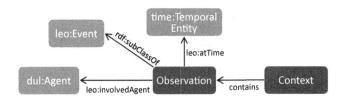

Fig. 2. Upper-level domain and context structures in SFF

However, apart from context classification, an important reasoning requirement in multimodal fusion is the propagation of property fillers among incoming
observations, e.g. the injection of the body part where a deictic gesture points to,
which is derived after fusion with spoken utterances. Due to the tree-model property, DL reasoning is not able to update property fillers for unconnected instances
(observations). SFF uses SPARQL CONSTRUCT graph patterns to enrich the reasoning capabilities of the framework, implementing certain fusion requirements,
according to the entities and relations involved.

The hybrid reasoning algorithm is depicted in Fig. 3. Assuming that G is the
RDF/OWL graph with context observations, Q is the set with all SPARQL
CONSTRUCT graph patterns, R_{DL} is the OWL 2 DL reasoning module and
R_{SPARQL} is the SPARQL query engine, the algorithm in Fig. 3 enriches G with
additional interpretations. More specifically, the algorithm implements an iterative combination of DL reasoning and SPARQL query execution. Initially, the
DL reasoning module is used over G for subsumption reasoning and realization
(line 2). The derivations are added back to G that is now used as the underlying
graph for the SPARQL reasoning module. When all SPARQL queries have been
executed (lines 3 to 5), a reasoning iteration has been completed. The algorithm
terminates when no SPARQL inferences are derived after an iteration.

4   Use Case: Reference Resolution

We describe the simulated evaluation of SFF that involves the conversation of
users with a dialogue-based agent at a home or a nursing environment in order
to acquire treatment suggestions about problems they have. We describe the
ontologies and rules needed to disambiguate referring expressions, taking into
account non-verbal modalities, e.g. deictic gestures. In the simulated example,
the user touches his head and says "It hurts here!". By fusing pointing gestures,
the agent can conclude that the user has a headache and it can provide relevant
treatment suggestions.

4.1   Domain Ontologies

It is assumed that SFF acquires contextual information about body gestures
(e.g. deictic gestures to the head) and verbal events (e.g. entities and concepts
extracted through language analysis) through respective multimedia analysis

```
1: repeat
2:      G ← G ∪ R_DL(G)
3:      for all q ∈ Q do
4:          G ← G ∪ R_SPARQL (q, G)
5:      end for
6: until R_SPARQL (q, G) = ∅
```

Fig. 3. Skeleton of the hybrid context interpretation algorithm.

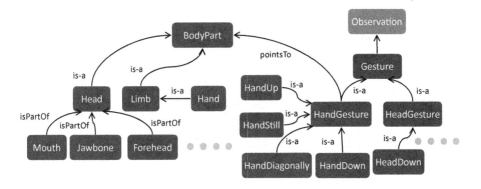

Fig. 4. Excerpt of the gesture and body part ontologies.

modules (in this case, from video and audio data). Figure 4 depicts the specialization of the **Observation** hierarchy (Fig. 2) for modelling gestures. The emphasis is placed on the **HandGesture** concept that allows **pointsTo** property assertions about the body part where the hand points to. Additional complex concepts are defined (not visualised in Fig. 4) by composing existing contexts, e.g. **HeadReference** ≡ **HandGesture** ⊓ ∃**pointsTo.Head**. As far as language analysis is concerned, the current deployment capitalizes on the results of a frame-based formalisation of natural language utterances using DOLCE-DnS Ultralite patterns[5]. Figure 5 depicts the relevant ontology. For example, a gesture event pointing to the head can be represented as:

```
:g1 a :HandGesture;
  :pointsTo [rdf:type :Head];
  :atTime [...].
```

which is further classified as **HeadReference**, based on the axiom defined above. Likewise, the verbal event corresponding to the example can be represented as:

```
:fs1 a :InformSpeechAct, :PerceptionBodyFrameSituation;
  dul:isSettingFor :h1.
:h1 a [:Hurt rdfs:SubClassOf dul:Event];
  dul:hasParticipant :d1.
:d1 a :Here;
  dul:isClassifiedBy [:bp1 a :BodyPart], [:sd1 a :SpatialDeictic].
```

As illustrated, the example utterance is an **InformSpeechAct** about a physical experience (i.e. a **PerceptionBodyFrameSituation**), where the affected body part, i.e. the object classified as **BodyPart**, is not named explicitly but instead implied by a deictic referring expression.

---

[5] http://ontologydesignpatterns.org/.

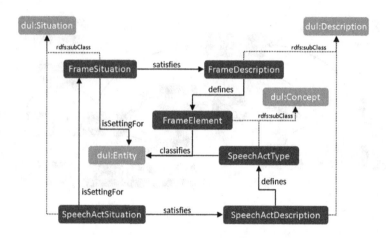

Fig. 5. The upper level ontology for representing verbal analysis results.

4.2 Context Models and Fusion

As already discussed, SFF needs to build the current context. In our example, whenever a `FrameSituation` is sent to the SFF framework, SPARQL queries retrieve neighbouring events that overlap a fixed time interval around it, e.g. $[-2s, +2s]$. The overlapped observations form the current context, which is fed into the ontology reasoner to interpret it. The current context is defined as

$$\texttt{CurrentContext} \equiv \texttt{Context} \sqcap \exists \texttt{contains}.\texttt{FrameSituation} \qquad (2)$$

where `Context` is given by (1). In order to model the situation when the user feels pain, `CurrentContext` is further specialised as:

$$\texttt{PainContext} \equiv \texttt{CurrentContext} \sqcap \exists \texttt{contains}.(\texttt{FrameSituation} \\ \sqcap \exists \texttt{isSettingFor}.\texttt{Hurt}) \qquad (3)$$

According to (3), if the current context contains a `FrameSituation` that is associated with a `Hurt` conceptualisation, it is classified in the `PainContext` class. Assuming that `fs1` is part of the current context, SFF interprets it as a `PainContext` situation, since `sf1` satisfies the complex class description in (3).

In addition, provided that the `Hurt` instantiation of the `FrameSituation` is also associated with a body part, the current context can be further classified in the `Headache` class, defined as:

$$\texttt{Headache} \equiv \texttt{PainContext} \sqcap \exists \texttt{contains}. \\ (\texttt{FrameSituation} \sqcap \exists \texttt{isSettingFor}. \\ (\texttt{Hurt} \sqcap \exists \texttt{isAssociatedWith}. \\ (\texttt{Head} \sqcap \exists \texttt{isClassifiedBy}.\texttt{BodyPart}))) \qquad (4)$$

As such, if the user explicitly mentions the body part, then the `FrameSituation` can be directly classified by the underlying ontology reasoner as a `Headache`. In our example, however, the user does not explicitly refer to the head, but instead points to it, while using the deictic referring expression "here". As a result, the inferred `PainContext` is associated with a non-body part entity, which moreover is classified as `SpatialDeictic`. In this case, SFF needs to take into account the fact that there is an underspecified body part in the `FrameSituation` that requires additional contextual information, and in particular non-verbal one, in order to resolve the ambiguity and provide an appropriate feedback. The logic to derive such inferences is beyond the expressivity provided by OWL 2. In this case, SFF uses a fusion rule to resolve this ambiguity. The following SPARQL rule implements the fusion of language analysis results with hand gestures to body parts, so as to fill the missing body part fillers.

```
CONSTRUCT {
    ?p isAssociatedWith ?bodypart.
}
WHERE {
    ?c a PainContext;
       contains [isSettingFor ?p].
    ?p a Hurt;
       isAssociatedWith ?bp.
    ?bp a SpatialDeictic.
    ?c contains [a HandGesture; pointTo ?bodypart].
}
```

Having updated the context with the inferred body part, the DL reasoner can now classify the current context in the `Headache` class, based on (4). As such, through the combination of the DL and SPARQL modules, SFF interprets the current situation as a headache, propagating it to subsequent modules to retrieve suggestions and provide feedback to the end user.

5   Conclusions

In this work, we presented SFF, an ontology-driven framework that couples DL reasoning and rules for multimodal fusion. The focus has been given on the interpretation of conversational contexts in dialogue-based systems, fusing non-verbal (i.e. gestures) and verbal features extracted from multimedia data for situation awareness. Ontologies are used to formally capture context types and background knowledge, while fusion and interpretation is reduced on the efficient combination of DL reasoning and SPARQL query execution.

We also described the simulated evaluation of SFF for reference resolution. We are currently collecting data for evaluating the framework using real-world conversations. In parallel, we are working towards further enrichment of the fusion and interpretation capabilities of the framework, so as to support for additional use cases, e.g. tasking into account emotions and facial expressions. It is also important to mention that the identification of the current conversational context does not take into account uncertainty. Our plan is to investigate lightweight probabilistic and non-monotonic reasoning schemes to enhance the interpretation capabilities of SFF.

Acknowledgments. This work has been partially supported by the H2020-645012 project "KRISTINA: A Knowledge-Based Information Agent with Social Competence and Human Interaction Capabilities".

References

1. Atrey, P.K., Hossain, M.A., El Saddik, A., Kankanhalli, M.S.: Multimodal fusion for multimedia analysis: a survey. Multimedia Syst. **16**(6), 345–379 (2010)
2. Attard, J., Scerri, S., Rivera, I., Handschuh, S.: Ontology-based situation recognition for context-aware systems. In: Proceedings of the 9th International Conference on Semantic Systems, I-SEMANTICS 2013, pp. 113–120. ACM (2013)
3. Baader, F., Calvanese, D., McGuinness, D.L., Nardi, D., Patel-Schneider, P.F. (eds.): The Description Logic Handbook: Theory, Implementation, and Applications. Cambridge University Press, Cambridge (2003)
4. Baumgartner, N., Gottesheim, W., Mitsch, S., Retschitzegger, W., Schwinger, W.: Beaware! - Situation awareness, the ontology-driven way. Data Knowl. Eng. **69**(11), 1181–1193 (2010)
5. Bettini, C., Brdiczka, O., Henricksen, K., Indulska, J., Nicklas, D., Ranganathan, A., Riboni, D.: A survey of context modelling and reasoning techniques. Pervasive Mob. Comput. **6**(2), 161–180 (2010)
6. Damljanović, D., Agatonović, M., Cunningham, H., Bontcheva, K.: Improving habitability of natural language interfaces for querying ontologies with feedback and clarification dialogues. Web Seman. Sci. Serv. Agents World Wide Web **19**, 1–21 (2013)
7. Denaux, R., Dimitrova, V., Cohn, A.G.: Interacting with ontologies and linked data through controlled natural languages and dialogues. In: Do-Form: Enabling Domain Experts to Use Formalised Reasoning-AISB Convention 2013, pp. 18–20. Society for the Study of Artificial Intelligence (2013)
8. Dourlens, S., Ramdane-Cherif, A., Monacelli, E.: Multi levels semantic architecture for multimodal interaction. Appl. Intell. **38**(4), 586–599 (2013)
9. Eiter, T., Ianni, G., Krennwallner, T., Polleres, A.: Rules and ontologies for the semantic web. In: Baroglio, C., Bonatti, P.A., Małuszyński, J., Marchiori, M., Polleres, A., Schaffert, S. (eds.) Reasoning Web. LNCS, vol. 5224, pp. 1–53. Springer, Heidelberg (2008). doi:10.1007/978-3-540-85658-0_1
10. Grau, B.C., Horrocks, I., Motik, B., Parsia, B., Patel-Schneider, P., Sattler, U.: OWL 2: the next step for OWL. Web Seman. Sci. Serv. Agents World Wide Web **6**(4), 309–322 (2008)
11. Harris, S., Seaborne, A.: SPARQL 1.1 query language, W3C recommendation, 21 March 2013. http://www.w3.org/TR/sparql11-query/
12. Horrocks, I., Patel-Schneider, P.F., Boley, H., Tabet, S., Grosof, B., Dean, M.: SWRL: a semantic web rule language combining OWL and RuleML. National Research Council of Canada and Stanford University, Technical report, May 2004
13. Knublauch, H., Hendler, J.A., Idehen, K.: SPIN - overview and motivation. W3C member submission, World Wide Web Consortium, February 2011
14. Kokar, M.M., Matheus, C.J., Baclawski, K.: Ontology-based situation awareness. Inf. Fusion **10**(1), 83–98 (2009). Special Issue on High-level Information Fusion and Situation Awareness

15. Lahat, D., Adali, T., Jutten, C.: Multimodal data fusion: an overview of methods, challenges, and prospects. Proc. IEEE **103**(9), 1449–1477 (2015)
16. Motik, B., Cuenca Grau, B., Sattler, U.: Structured objects in OWL: representation and reasoning. In: Proceedings of the 17th International Conference on World Wide Web (WWW 2008), pp. 555–564. ACM, New York (2008)
17. Perera, C., Zaslavsky, A., Christen, P., Georgakopoulos, D.: Context aware computing for the internet of things: a survey. IEEE Commun. Surv. Tutorials **16**(1), 414–454 (2014)
18. Perperis, T., Giannakopoulos, T., Makris, A., Kosmopoulos, D.I., Tsekeridou, S., Perantonis, S.J., Theodoridis, S.: Multimodal and ontology-based fusion approaches of audio and visual processing for violence detection in movies. Expert Syst. Appl. **38**(11), 14102–14116 (2011)
19. Prokofyev, R., Tonon, A., Luggen, M., Vouilloz, L., Difallah, D.E., Cudré-Mauroux, P.: SANAPHOR: ontology-based coreference resolution. In: Arenas, M., et al. (eds.) ISWC 2015. LNCS, vol. 9366, pp. 458–473. Springer, Heidelberg (2015). doi:10. 1007/978-3-319-25007-6_27
20. Schneider, M., Rudolph, S., Sutcliffe, G.: Modeling in OWL 2 without restrictions. In: Proceedings of the 10th International Workshop on OWL: Experiences and Directions, Co-located with 10th Extended Semantic Web Conference (2013)
21. Shaw, R., Troncy, R., Hardman, L.: Lode: linking open descriptions of events. In: 4th Asian Conference on the Semantic Web, Shanghai, China, pp. 153–167 (2009)
22. Sleeman, J., Finin, T.: Type prediction for efficient coreference resolution in heterogeneous semantic graphs. In: 2013 IEEE Seventh International Conference on Semantic Computing (ICSC), pp. 78–85. IEEE (2013)
23. Solanas, A., Patsakis, C., Conti, M., Vlachos, I.S., Ramos, V., Falcone, F., Postolache, O., Pérez-Martínez, P.A., Di Pietro, R., Perrea, D.N., et al.: Smart health: a context-aware health paradigm within smart cities. IEEE Commun. Mag. **52**(8), 74–81 (2014)
24. Sonntag, D.: Ontologies and Adaptivity in Dialogue for Question Answering, vol. 4. IOS Press, Heidelberg (2010)
25. Wahlster, W.: Dialogue systems go multimodal: the SmartKom experience. In: Wahlster, W. (ed.) SmartKom: Foundations of Multimodal Dialogue Systems, pp. 3–27. Springer, Heidelberg (2006)
26. Wongpatikaseree, K., Ikeda, M., Buranarach, M., Supnithi, T., Lim, A.O., Tan, Y.: Activity recognition using context-aware infrastructure ontology in smart home domain. In: Knowledge, Information and Creativity Support, pp. 50–57 (2012)
27. Ye, J., Dasiopoulou, S., Stevenson, G., Meditskos, G., Kontopoulos, E., Kompatsiaris, I., Dobson, S.: Semantic web technologies in pervasive computing: a survey and research roadmap. Pervasive Mob. Comput. **23**, 1–25 (2015)

Speech Synchronized Tongue Animation by Combining Physiology Modeling and X-ray Image Fitting

Jun Yu[✉]

Department of Automation,
University of Science and Technology of China, Hefei 230026, China
harryjun@ustc.edu.cn

Abstract. This paper proposes a speech synchronized tongue animation system from text or speech. Firstly, an anatomically accurate physiological tongue model is built, and then produces tremendous tongue deformation samples according to the randomly input muscle activation samples. Secondly, these input and output samples are used to train a neural network for establishing the relationship between the muscle activation and tongue contour deformation. Thirdly, the neural network is used to estimate the non-rigid tongue movement parameters, namely tongue muscle activations, from a collected X-ray tongue movement image database of Mandarin Chinese phonemes after removing the rigid tongue movement, and then the estimation results are used for constructing the tongue physeme (the sequences of the tongue muscle activations and the rigid movement) database corresponding to the Mandarin Chinese phoneme database. Finally, the physemes corresponding to the phonemes extracted from input text or speech are blended to drive the physiological tongue model for producing the speech synchronized tongue animation according to the durations of phonemes. Simulation results demonstrate that the synthesized tongue animations are visually realistic and approximate the tongue medical data well.

Keywords: Tongue animation · Physical modeling · Visual speech synthesis

1 Introduction

Audio-visual speech synthesis generates speech synchronized animation of articulators, including tongue, jaw, lips, etc. Compared with facial animation [1, 2], the modeling and animation of the tongue have not been well researched. The first reason is the tongue movements are not always visible during speech, which make its importance being overlooked. The second reason is those technologies suitable for capturing tongue shape, such as Magnetic Resonance Imaging (MRI), Electromagnetic Articulography (EMA) and ultrasound, cannot record the dense and continuous 3D tongue shapes easily. Without sufficient tongue samples, realistic tongue animation can hardly be synthesized. However, building a precise tongue model and synthesizing realistic tongue animation are important. It can be used to study articulatory kinematics, pronunciation patterns, and speech motor control [3] in linguistics, and supply the visualization help of diagnostic for dysphonia patients [4] in medical science.

© Springer International Publishing AG 2017
L. Amsaleg et al. (Eds.): MMM 2017, Part I, LNCS 10132, pp. 726–737, 2017.
DOI: 10.1007/978-3-319-51811-4_59

The model for simulating tongue animation can be classified into three types. (1) A parametric tongue model controls its movements by several parameters which represent shape attributes of the tongue [5]. For speech animation, the parameters have to be manually adjusted, in order to make the model deformation to fit the tongue shape of a viseme. (2) Statistical models extract the degrees of freedom of tongue deformations from collected tongue data, and use these degrees of freedom to control tongue deformations [6, 7]. (3) Physiological models simulate the tongue's anatomical structure (the tongue geometry and muscle architecture) and biomechanical properties (the elastic property of tongue soft tissue and the contraction properties of muscle) so that the deformation of tongue models is driven by physiological control parameters, such as muscle activations [8]. The control parameters associated with a particular pronunciation can be estimated from the collected tongue data of this pronunciation.

For parametric model, the accuracy of synthesized tongue animation depends on whether the model can be deformed through adjusting control parameters to approximate real tongue shapes. However, it is very hard to design suitable control parameters for presenting complex tongue deformations [9, 10]. For statistical models, the diversity of generated tongue animation is determined by the space of collected tongue shapes. MRI and ultrasound are the commonly used technologies for collecting 3D tongue shapes. But ultrasound cannot capture the shape of tongue tip because of the occlusion of teeth, and it is inconvenient and inaccurate to obtain tongue shapes through MRI because it needs the subject to sustain artificially the articulation for a long time for capturing tomographic images [11]. As far as we know, there is not such a database that contains the static or dynamic 3D tongue shapes of all phonemes (the smallest linguistic unit) of a certain language. However, 2D acquisition techniques, such as X-ray, are capable of recording the dynamic movement data of tongue contour; and thus can be used to build the database of 2D dynamic tongue movements.

Under this background, we propose an audio-visual speech synthesis system, which can synthesize 3D tongue animation by using the 2D dynamic tongue data with the fact that the movements perpendicular to the midsaggital plane are relatively much smaller and negligible during speech. Its framework is as follows.

Firstly, we build an accurate physiological tongue model based on anatomy and biomechanics. By given muscle activations, this model can produce corresponding tongue deformations through finite-element method (FEM) simulation. In order to make use of the 2D tongue data (X-ray images), we produce tremendous muscle activation samples $(A_1, \cdots, A_n)$ to driven the physiological model for simulating tongue deformations, and then we record the 2D contour deformations $(P_1, \cdots, P_n)$ on the midsaggital plane of the tongue model. Based on the input/output training samples (A_i, P_i), a Radial Basis Function (RBF) network is trained to establish the relationship between the muscle activation and tongue contour deformation directly.

Secondly, by detecting the tongue contours in the 2D tongue X-ray images of Mandarin Chinese phonemes, we obtain the tongue rigid displacement D and local movement of tongue contour on each frame. With this local movement, the corresponding muscle activation is estimated by fitting the RBF network. For the whole image sequences of each phoneme, we obtain the muscle activation sequences and rigid displacement sequences, namely physeme $M\{A_1, A_2 \cdots, D_1, D_2 \cdots\}$, which represents the primitives used to synthesize the visemes of physiology based visual speech synthesis. Then a phoneme-to-physeme database of Mandarin Chinese is built.

Finally, when phonemes and durations are is extracted from the input text or speech, corresponding physemes are selected from the database, and blended to drive the physiological tongue model for synthesizing speech synchronized animation.

Our work has following advantages. Firstly, the input of animation is the physemes, which have physiological meanings, as well as the physiological model and simulation framework truly embody the anatomy and biomechanics of the tongue. Secondly, the neural network is used to transform the 2D dynamic tongue data into 3D physemes, which overcomes the problem of the lack of a 3D tongue database.

2   Physiological Tongue Model

2.1   Marking Muscle Geometry and Fiber Structure by Anatomy

The tongue includes connective tissue and muscle. The connective tissues contains blood vessels, mucosa, submucous tissue, glands and other soft tissue, and the muscles are divided into extrinsic and intrinsic ones [12]. The extrinsic muscles originate from bone and insert within the tongue, while the intrinsic muscles have both their origins and terminations within the tongue. The extrinsic muscles generally act to change the position of the tongue, while the intrinsic muscles generally act to change the shape of the tongue. Anatomical data [13] and tongue electromyography (EMG) data [14] show that there are a total of nine tongue muscles affecting the pronunciation (Fig. 1), namely anterior Genioglossus (GGA), posterior Genioglossus (GGP), Hyoglossus (HG), Styloglossus (SG), Palatoglossus (PG), Superior longitudinal (SL), Inferior longitudinal (IL), Verticalis (V), and Transversus (T).

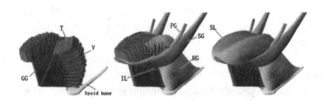

Fig. 1. Tongue muscle architecture.

Our tongue mesh model is built based on a high quality articulatory database, which is collected by Chinese Academy of Social Sciences. The database contains dynamic X-ray data, static MRI data and electromagnetic articulography data of a volunteer's articulators movements during speech. The static MRI data contains a tomographic image sequence of tongue at the rest state. Through image segmentation, image registration and 3D reconstruction technology, our tongue mesh model is built from the tomographic image sequence. Then the Delaunay tetrahedral mesh generation algorithm [15] is used to transform the surface mesh into a tetrahedral mesh (Fig. 2), which contains 3,876 nodes and 20,181 tetrahedral units.

Fig. 2. The profile of tongue mesh model.

Muscle fibers are arranged differently for different muscles, and different tongue muscles interlace complexly, so it is necessary to perform a personalized and meticulous muscle construction process. Therefore, we present an interactive muscle marking method to mark the muscle geometry and muscle fiber structure, and the method is divided into three steps.

(1) We specify one or several virtual muscle mainline for each tongue muscle, and assign one piecewise linear segment to represent each virtual muscle mainline. These virtual muscle mainline are evenly distributed in the spatial position of each muscle, and set to be consistent with the muscle's force-generating axis.

(2) We set a virtual capsule for each virtual muscle mainline, and set a loop, which can be circle, ellipse or any other geometric shapes, for each segment in the piecewise linear segment. For GGA, we set ellipses which are all perpendicular to the virtual muscle mainline as the loops, so four elliptical cylinders form the virtual muscle capsule. Those mesh nodes inside the virtual muscle capsule are marked as muscle nodes, and the space wrapped in the virtual capsule is marked as muscle geometric shape.

(3) We set a virtual fiber vector to represent the arrangement direction of fiber for each node inside the virtual capsule. The virtual fiber vector can be defined in a single direction or an azimuthally direction depending on the muscles fiber histology. Then the direction of muscle force is interpolated based on these virtual fiber vectors.

After above process, the whole tongue mesh model is separated into two parts, one is the muscular part where the fibers of different muscles are intermingled, and the other is the connective tissue part enveloping the muscles. Then we model the biomechanics of the connective tissues part and muscular part separately.

2.2 Biomechanical Modeling

According to histologic study [16], the tongue connective tissue is modeled by using a quasi-incompressible, isotropic, hyperelastic constitutive model: Mooney-rivlin constitutive model. It has been widely used to describe soft tissues deformation [17, 18], and the strain energy function is given as:

$$U = U_I(\bar{I}_1, \bar{I}_2) + U_J(J) \tag{1}$$

where U_I reflects the nonlinear, isotropic, hyperelastic properties, and U_J reflects the incompressibility. The details of the their definitions can be found in [17–19].

The active muscles will interact with surrounding soft tissues during contracting, and those inactive muscles will act as a soft tissue without active contraction mechanical behaviors. Thus the strain energy function for muscle is a sum of fiber term, and a term which is the same as that for connective tissues. It is given by:

$$U = U_I(\bar{I}_1, \bar{I}_2) + U_J(J) + U_f(\lambda_f, a) \tag{2}$$

where U_f is the strain energy function for muscle fiber, and takes into account the fiber stretch ratio in the along-fiber direction λ_f and the muscle activation a. It is based on Tang's work [20], which developed a state-of-the-art model for modeling skeletal muscle.

FEM is used for simulation, and some simulation results are compared with the collected data. Due to the lack of precise experimental data on the activity of tongue muscles during speech, it is hard to quantitatively verify the accuracy of simulation results according to muscle activations. However, the model can be evaluated by qualitatively observing the impact of specific muscle activations on the tongue shapes.

Studies [21] show that HG acts differently for front vowels and back vowels. For front vowels, a peak in HG activity occurs at about the time of lip closure, and maximum HG suppression occurs during the vocalic period. For back vowels, such as /a/ and /ɔ/, HG is activated at the vocalic period. From the X-ray image of /a/, it can be observed that the tongue moves downward with the jaw, and we can see the tongue tip curls inward slightly after removing the jaw's rigid movements. Figure 3 shows the simulation result driven by HG with the activation value at 0.6, and we can see the simulation result is consistent with the medical study [22] and X-ray image data of /a/.

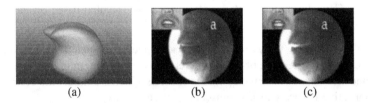

(a)   (b)   (c)

Fig. 3. (a) Deformation result of /a/. (b) X-ray image before /a/. (c) X-ray image during /a/.

We can see that the simulation results coincide well with these experimental data, and it demonstrate our tongue model can effectively embody the tongue's physiology properties, and simulate realistic tongue movement with certain muscle activations.

3   Building the Phoneme-to-Physeme Database

Firstly, tongue contours are detected in the X-ray image database of Mandarin Chinese phonemes. Secondly, a RBF network is trained to establish the relationship between the muscle activation and tongue contour deformation directly. Finally, physemes are estimated by fitting the RBF network to the detected tongue contours.

3.1   Detecting Tongue Contour Motion in X-ray Image Database

The X-ray image database contains a subject's articulatory movements when performing Mandarin Chinese pronunciation, and the image sampling frequency is 48 Hz. The pronunciations in this corpus contain 21 consonants, 39 vowels, and 144 monosyllabic words. The phonemes are represented by these consonants and vowels.

Active Appearance Model (AAM) [23] is used to track the tongue contours in X-ray images. The images, which approximately correspond to every 5th image in the image database, are used to train the AAM, and the number of bases in the trained AAM is chosen so as to keep 95% of variations. In training images, we manually mark 26 points, which correspond to those 26 points on the midsagittal plane of our tongue model. Since the X-ray image is vague and low contrast, the tracking results of the tongue contour are not very accurate. Like other studies [24] of tongue contour tracking, we manually adjust the position of those inaccurately tracked points.

In fact, the deformation of tongue contour includes the tongue local deformation and the rigid displacement of tongue. The former is driven by muscle activations, and the latter is driven by the jaw movements. AAM is also used to detect the rigid movements of jaw, which is equal to the rigid displacement of tongue. Then the local deformation of tongue is obtained by removing the rigid displacement.

3.2   RBF Network

As mentioned above, there is not a complete 3D tongue database that covers all phonemes of one language currently. However, in order to acquire tongue movement parameters associated with pronunciation, the 2D tongue database can be utilized to estimate the corresponding tongue movement parameters.

RBF network has good ability of nonlinear regression, and has been widely applied in facial animation [25]. Therefore, we use RBF network to represent the conversion relationship from muscle activation to 2D tongue contour. Firstly, we produce lots of muscle activations. For each of these nine muscles, the activation value is sampled at three levels: $\{0, 0.5, 1.0\}$, and there are $3^9=19,683$ muscle activation samples A. Then we input these samples into physiological tongue model to simulate corresponding tongue deformation results. Due to the agonist-antagonist property of tongue muscle pairs, which means some muscle pairs may have offsetting effects for moving the tongue, so that the results of some certain muscle activation combinations cannot be simulated. Finally, we obtain 16,265 successfully simulated tongue deformations, and the contour P on the midsagittal plane of these simulated deformation results. As a result, there are 16,265 muscle activation A-tongue contour P paired samples, where A is a nine dimensional vector $\{A_{GGA}, \cdots, A_{IL}\}$, and P is composed of the Y, Z coordinates of 26 points evenly distributed on the midsagittal plane of our tongue model $P = (y_1, z_1, \cdots, y_{26}, z_{26})$. A three layer RBF network is trained based on these paired samples, and represented as:

$$Prbf(A) = W \cdot \Phi(A) + B_o \tag{3}$$

The middle layer contains 250 nodes. $W(52 \times 250)$, $B_o(52 \times 1)$ are the weight matrix and bias vector of the output layer. The ith node of middle layer is given by RBF:

$$\Phi_i(A) = \exp\left(-(\|A - IW_i\| \times B_{mi})^2\right) \tag{4}$$

where IW_i is the ith row component of the weight matrix of the middle layer $IW(250 \times 9)$, and represents the mean value of the ith RBF. $\|A - IW_i\|$ represents the distance between A and IW_i. B_{mi} is the ith component of bias vector of the middle layer $B_m(250 \times 1)$, and its reciprocal represents the variance of the ith RBF.

In order to evaluation the fitting ability of our RBF network, we randomly produce 1,000 muscle activations and tongue contour paired samples, $\{A_i, P_i, i = 1, \cdots, 1,000\}$, by using our physiological tongue model. We also calculate the output of the BRF network $Prbf_1, \ldots, Prbf_{1000}$ by the muscle activations. For each $Prbf_i$ and P_i, we calculate the Root Mean Square Error (RMSE) between them as:

$$\varepsilon_i = \left(\sum_{j=1}^m \left(\sqrt{\varepsilon_{1y}^2 + \varepsilon_{1z}^2 + \cdots + \varepsilon_{26y}^2 + \varepsilon_{26z}^2}\Big/52\right)\right)\Big/m \tag{5}$$

where m is the frame number of the ith synthesized results, and $\varepsilon_{i,y}$, $\varepsilon_{i,z}$ represent the RMSE of the Y and Z coordinates of 26 points in all frames.

Figure 4 shows the statistical histogram of these 1,000 RMSE results. As can be seen from it, the most of the RMSE lie below 2 mm, while the length of our tongue model is 60 mm, so that this result shows the RBF network can approximate the transform relationship between muscle activations and tongue contour deformations well.

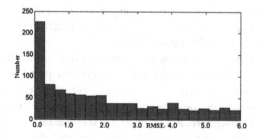

Fig. 4. The statistical histogram of 1,000 RMSE results.

3.3   Obtaining Physemes by Muscle Activation Estimation

For the X-ray sequences of a particular phoneme, a X-ray image, in which the teeth is closed and the tongue is at the rest state, is chosen as the reference frame. For X-ray image sequences of one phoneme, the local deformation of tongue contour on the midsaggital plane P is obtained after removing the detected rigid displacement D based on the reference frame [26]. Then the corresponding muscle activation $A\{A_{GGA}, A_{GGP}, \cdots, A_{IL}\}$ is estimated based on the RBF network $Prbf(A)$ as:

$$\arg \min_{A \in \Omega} \|Prbf(A) - P\| \qquad (6)$$

where Ω represents the feasible set of the muscle activations between 0 and 1. Since $Prbf(A)$ and P are come from different subjects, they are matched to each other before fitting. With the fact that different muscle activations may produce very similar tongue contour deformations, the transform relationship from muscle activations to tongue contour is many-to-one. In order to solve this problem, we take the advantages of the continuity of neighbor frames, and set the optimal solution of previous frame as the initial value of the optimization of current frame. Then we can obtain the global optimal solutions for the tongue contours in all frames associated with a particular phoneme, and the obtained solutions are the muscle activation sequences.

Then the phoneme-to-physeme database is built after obtaining the physeme $M\{A_1, \cdots, D_1, \cdots\}$ for the X-ray image sequences of all Mandarin Chinese phoneme.

Figure 6(b) shows the muscle activation estimation results for /i/. We can see that the pronunciation of /i/ is mainly derived by GGP, SG, IL, and it shows the tongue has a transformation and rotation in the YZ plane (midsaggital plane). The results are consistent with medical studies.

4   Speech Synchronized Animation

When input text or speech, the phonemes sequences and duration information are extracted, and then the physeme sequences with duration information are obtained based on the phoneme-physeme database. Moreover, the two adjacent physemes would overlap, so the sigmoid function is used to blend the overlapping portions of them. Figure 5 shows the physeme curve for Mandarin Chinese /shuei/. /shuei/ includes two phonemes /sh/ and /uei/. 50–150 ms corresponds to the segment of /sh/, 150–310 ms corresponds to the segment of /uei/. In the beginning and ending, there are two /sil/ to represent the resting state.

When input the physeme curve into our physiological tongue model, the speech synchronized animation is synthesized. The rigid displacement in physeme is used to translate and rotate the tongue model, and the muscle activations in physeme is used to simulate the tongue model deformation. As can be seen from Fig. 5, the tongue body has a slight forward movement for /sh/, and the tongue dorsum has a backward movement while the tongue tip has a backward and downward movement for /uei/.

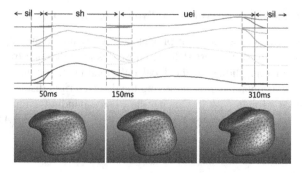

Fig. 5. The top row is the physeme curve for /shuei/. The bottom row is the key frames of /sil/, /sh/, /uei/.

5 Experiments

Experiments are conducted by a workstation with AMD Athlon (tm) II X4 640 3.01G, memory 2G, NVIDIA GT200.

5.1 Objective Evaluation

In the corpus, all consonants and vowels are used to construct the phoneme-to-physeme database, and the remaining monosyllabic words are used for objective evaluation. We select 40 monosyllabic words, and synthesize 40 tongue deformation results by using our system. The frame number of the synthesized animation is set to be equal to that of X-ray video, and the synthesis results T_S, which is the Y and Z movements of 26 points on the midsagittal plane of our tongue model, is matched to compare with the X-ray data T_X, which is the movements of 26 points on the tongue contour. The root mean square error (RMSE) of the ith point is represented as:

$$\varepsilon_i = \left(\sum_{j=1}^{40} \left(\sum_{k=1}^{F_j} \left(\sqrt{(T_{SY(jki)} - T_{XY(jki)})^2 + (T_{SZ(jki)} - T_{XZ(jki)})^2} \Big/ 2 \right) \Big/ F_j \right) \right) \Big/ 40 \tag{7}$$

where F_j is the frame number of the synthesized result of the jth monosyllabic word, $T_{SY(jki)}$ and $T_{XY(jki)}$ are the Y coordinates of the ith point in the kth frame of the synthesized result and X-ray data of the jth monosyllabic words respectively. The RMSE results of these 26 points are shown in Fig. 6. It can be observed that the fitting ability of the first half part (corresponding to tongue tip) is better than the second half part (corresponding to tongue dorsum). It indicates the ability for simulating the substantial movement of the tongue tip is better than that of the tongue back. However, the RMSE of these points is quite small in general, and the synthesized results have a good approximation to the ground truth (X-ray data).

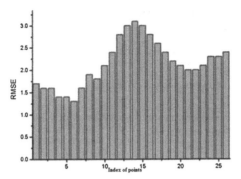

Fig. 6. The RMSE results of the 26 points.

5.2   Subjective Evaluation

In subjective evaluation, a parametric tongue model [27] is used for comparison. The parametric tongue model is formed by a Non-Uniform Rational B-Spline (NURBS) surface which consists of 70 control points and 16×7 bi-cubic patches. Through a parameterization process, these control points are controlled by several location parameters, which represent the positions of three points (Tongue Tip, Tongue Blade, Tongue Root) on this model. By ensuring the same rendering conditions (lighting, perspective angle, frame rate), we synthesize the tongue animation corresponding to 20 short phrases by using the parametric model and our physiological model respectively. For example, the synthesis results of the Mandarin Chinese phrase "yu yin dong hua", which means "speech animation", is shown in Fig. 7 (only the local movements of the tongue are demonstrated).

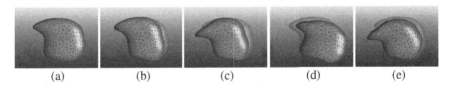

| (a) | (b) | (c) | (d) | (e) |

Fig. 7. The key frames of the synthesized animation of a Mandarin Chinese phrase. (a) /sil/. (b) /ü/. (c) /in/. (d) /ong/. (e) /ua/.

30 volunteers are recruited to participate in the subjective evaluation. They are aged 20–40 years, and have 10 women. These volunteers are asked to watch these two group of synthesized animations, and asked to score for three questions in Table 1 (5 corresponds to the most positive and 0 corresponds to the most negative). The Model 1 is our model, and the Model 2 is the comparison model. From Table 1, we can see that our synthesized tongue animation can not only achieve a high level of realism, but also approximate the real tongue movement nicely.

Table 1. Subjective evaluation results.

Question	Model	Score
Whether the shape of the model looks like a real tongue at the rest state?	1	4.0
	2	3.4
Whether the synthesized animation looks smooth and natural?	1	3.6
	2	2.9
Whether the produced tongue deformation is consistent with the tongue movements of a real person's pronunciation?	1	3.7
	2	3.1

6 Conclusion

This paper presents a speech synchronized tongue animation system by combining physiology modeling and X-ray image fitting. By building a phoneme-physeme database based on collected tongue contour information, experimental results show that realistic speech synchronized tongue animation can be synthesized. Although our system is for Mandarin Chinese, the method of it is also suitable for other languages.

In future, we will use tongue data associated with sentences to establish the context phoneme-physeme database for dealing with the co-articulation problem. The tongue animation will be embedded into a speech synchronized facial animation system.

Acknowledgement. This work is supported by the National Natural Science Foundation of China (No. 61572450, No. 61303150), the Open Project Program of the State KeyLab of CAD&CG, Zhejiang University (No. A1501), the Fundamental Research Funds for the Central Universities (WK2350000002), the Open Funding Project of State Key Laboratory of Virtual Reality Technology and Systems, Beihang University (No. BUAA-VR-16KF-12), the Open Funding Project of State Key Laboratory of Novel Software Technology, Nanjing University (No. KFKT2016B08).

References

1. Parke, F.I.: Computer generated animation of faces. In: Proceedings ACM National Conference, pp. 451–457. ACM: New York (1972)
2. Waters, K.: A muscle model for animating three dimensional facial expression. In: Stone, M. C. (ed.) Computer Graphics, vol. 21, pp. 17–24. Anaheim, CA (1987)
3. Sanguineti, V., Laboissiere, R., Payan, Y.: A control model of human tongue movements in speech. Biol. Cybern. **77**(1), 11–22 (1997)
4. Fujita, S., Dang, J., Suzuki, N., et al.: A computational tongue model and its clinical application. Oral Sci. Int. **4**(2), 97–109 (2007)
5. Modeling coarticulation in synthetic visual speech
6. Badin, P., Bailly, G., et al.: Three-dimensional linear articulatory modeling of tongue, lips and face, based on MRI and video images. J. Phonetics **30**(3), 533–553 (2002)
7. Engwall, O.: A 3D tongue model based on MRI data. In: INTERSPEECH, pp. 901–904 (2000)

8. Wilhelms-Tricarico, R.: Physiological modeling of speech production: methods for modeling soft -tissue articulators. JASA **97**(5), 3085–3098 (1995)
9. King, S.A., Parent, R.E.: A 3D parametric tongue model for animated speech. J. Vis. Comput. Anim. **12**(3), 107–115 (2001)
10. Ilie, M.D., Negrescu, C., Stanomir, D.: An efficient parametric model for real-time 3D tongue skeletal animation. In: ICC, pp. 129–132 (2012)
11. Engwall, O., Combining, M.R.I.: EMA and EPG measurements in a three-dimensional tongue model. Speech Commun. **41**(2), 303–329 (2003)
12. Miyawaki, K.: A study of the musculature of the human tongue. Annu. Bull. Res. Inst. Logopedics Phoniatrics **8**, 23–50 (1974)
13. Agur, A.M.R., et al.: Grant's Atlas of Anatomy. Lippincott Williams & Wilkins, Baltimore (2009)
14. Mac Neilage, P.F., Sholes, G.N.: An electromyographic study of the tongue during vowel production. J. Speech Lang. Hear. Res. **7**(3), 209–232 (1964)
15. Shewchuk, J.R.: Constrained Delaunay Tetrahedronlizations and provably good boundary recovery. In: IMR, pp. 193–204 (2002)
16. Takemoto, H.: Morphological analyses of the human tongue musculature for three-dimensional modeling. JSLHR **44**(1), 95–107 (2001)
17. Weiss, J.A., Maker, B.N., Govindjee, S.: Finite element implementation of incompressible, transversely isotropic hyperelasticity. CMAME **135**(1), 107–128 (1996)
18. Sifakis, E., Neverov, I., Fedkiw, R.: Automatic determination of facial muscle activations from sparse motion capture marker data. TOG ACM **24**(3), 417–425 (2005)
19. Simo, J.C., Taylor, R.L.: Quasi-incompressible finite elasticity in principal stretches. Continuum Basis Numer. Algorithms CMAME **85**(3), 273–310 (1991)
20. Tang, C.Y., et al.: A 3D skeletal muscle model coupled with active contraction of muscle fibres and hyperelastic behaviour. J. Biomech. **42**(7), 865–872 (2009)
21. Baer, T., Alfonso, P.J., Honda, K.: Electromyography of the tongue muscles during vowels in /gpvp/ environment. Ann Bull RILP **22**, 7–19 (1988)
22. Agur A M R, et al., Grant's atlas of anatomy. Lippincott Williams & Wilkins, 2009
23. Cootes, T.F., et al.: Active appearance models. TPAMI **23**(6), 681–685 (2001)
24. Laprie, Y., Berger, M.O.: Extraction of tongue contours in x-ray images with minimal user interaction. ICSLP **1**, 268–271 (1996)
25. Deng, Z., Chiang, P.Y., Fox, P. et al.: Animating blendshape faces by cross-mapping motion capture data. Interactive 3D graphics and games, pp. 43–48. ACM (2006)
26. Sock, R., Hirsch, F., Laprie, Y. et al.: An X-ray database, tools and procedures for the study of speech production. In: ISSP, pp. 41–48 (2011)
27. Yu, J., Li, A.: 3D visual pronunciation of Mandarine Chinese for language learning. In: IEEE International Conference on Image Processing, pp. 2036–2040 (2014)

Erratum to: ReMagicMirror: Action Learning Using Human Reenactment with the Mirror Metaphor

Fabian Lorenzo Dayrit[1]([⊠]), Ryosuke Kimura[3], Yuta Nakashima[1],
Ambrosio Blanco[2], Hiroshi Kawasaki[3], Katsushi Ikeuchi[2],
Tomokazu Sato[1], and Naokazu Yokoya[1]

[1] Nara Institute of Science and Technology,
Takayamacho 8916-5, Ikoma, Nara 630-0101, Japan
fabian-d@is.naist.jp
[2] Microsoft Research Asia, Building 2, No. 5 Dan Ling Street,
Haidian District, Beijing 100080, China
[3] Kagoshima University, Korimoto 1-21-24, Kagoshima 890-8580, Japan

Erratum to:
Chapter "ReMagicMirror: Action Learning Using Human
Reenactment with the Mirror Metaphor" in:
L. Amsaleg et al. (Eds.): MultiMedia Modeling, LNCS,
DOI: 10.1007/978-3-319-51811-4_25

The original version of the chapter starting on p. 303 was revised. An acknowledgement has been added. The original chapter was corrected.

The updated original online version for this chapter can be found at
DOI: 10.1007/978-3-319-51811-4_25

© Springer International Publishing AG 2017
L. Amsaleg et al. (Eds.): MMM 2017, Part I, LNCS 10132, p. E1, 2017.
DOI: 10.1007/978-3-319-51811-4_60

Author Index

Printed in the United States
By Bookmasters